Electromagnetic Field Theory Fundamentals

Guru and Hiziroğlu have produced an accessible and user-friendly text on electromagnetics that will appeal to both students and lecturers teaching this course. This lively book includes many worked examples and problems in every chapter, as well as chapter summaries and background revision material where appropriate. The book introduces undergraduate students to the basic concepts of electrostatic and magnetostatic fields, before moving on to cover Maxwell's equations, propagation, transmission, and radiation. Sections on the finite element and finite difference method, and a detailed appendix on the Smith chart, are additional enhancements.

Bhag Guru is a Professor in the ECE Department at Kettering University. He has published over 30 papers in the areas of Rotating Machinery and Electromagnetic Fields and has co-authored two books. Dr. Guru is a member of the IEEE.

Hüseyin R. Hiziroğlu is a Professor in the Department of Electrical and Computer Engineering at Kettering University, where he has been a member of faculty since 1982. He was awarded several grants by the United Nations Development Program, and has numerous publications in IEEE conferences and transactions. He is also a co-author, with Bhag Guru, of *Electric Machinery and Transformers* (1988, 1995, 2001). Dr. Hiziroglu is a senior member of IEEE.

Electromagnetic Field Theory Fundamentals

SECOND EDITION

Bhag Singh Guru and
Hüseyin R. Hiziroğlu
Kettering University

CAMBRIDGE
UNIVERSITY PRESS

CAMBRIDGE UNIVERSITY PRESS
Cambridge, New York, Melbourne, Madrid, Cape Town, Singapore, São Paulo, Delhi

Cambridge University Press
The Edinburgh Building, Cambridge CB2 8RU, UK

Published in the United States of America by Cambridge University Press, New York

www.cambridge.org
Information on this title: www.cambridge.org/9780521116022

First published 1998 by PWS Publishing Company
Second edition published 2004 by Cambridge University Press
Reprinted with corrections 2005
This digitally printed version 2009

A catalogue record for this publication is available from the British Library

ISBN 978-0-521-83016-4 hardback
ISBN 978-0-521-11602-2 paperback

Contents

Preface

Electromagnetic field theory has been and will continue to be one of the most important fundamental courses of the electrical engineering curriculum. It is one of the best-established general theories that provides explanations and solutions to intricate electrical engineering problems when other theories are no longer applicable.

This book is intended as a basic text for a two-semester sequence for undergraduate students desiring a fundamental comprehension of electromagnetic fields. The text can also be used for a one-semester course as long as the topics omitted neither result in any loss of continuity of the subject matter nor hamper the student's preparation for the courses that follows. This text may also serve as a reference for students preparing for an advanced course in electromagnetic fields.

The first edition of this book appeared in 1998 and was well accepted by both the students and the faculty. Among the numerous comments we received from the students, the one that transcends the others is that the book is written in simple, everyday language so that anyone can easily understand even the most sophisticated concepts of electromagnetic fields. We attribute such a favorable comment to the fact that the book was written from first-hand experience of class-room teaching. The development of this second edition also follows the same time-tested approach.

A thorough understanding of vector analysis is required to comprehend electromagnetic theory in a logical manner. Without much exaggeration, we can say that vector analysis is the backbone of the mathematical formulation of electromagnetic field theory. Therefore, a complete grasp of vector analysis is crucial to the comprehension of electromagnetic fields. In order to ensure that every reader begins with essentially the same level of understanding of vectors, we have devoted an entire chapter, Chapter 2, to the study of vector analysis. A great deal of emphasis is placed upon the coordinate transformations and various theorems.

In the development and application of electromagnetic field theory, the student is expected to recall from his/her memory various mathematical relationships. To help those who fail to recall some of the required mathematical formulas, we have included enough information in Appendix C on trigonometric identities, series, and integral calculus, etc.

A quick glance at the table of contents reveals that the text may be divided into two parts. The first part, which can be covered during the first semester for a two-semester sequence, introduces the students to static fields such as electrostatic fields (Chapter 3), magnetostatic fields (Chapter 5), and fields due to steady currents (Chapter 4). Because most of the applications of static fields involve both electric and magnetic fields, we decided to present such applications in one chapter (Chapter 6). We are also of the opinion that once the students grasp the basics of static fields, they can study the applications with a minimum of guidance. If time permits, the development of Maxwell's equations in both the time domain and the frequency (phasor) domain can be included in the first part of the course. This material is presented in Chapter 7, where the emphasis is laid upon the coexistence of time-varying electric and magnetic fields and the concept of average power density. Also included in this chapter are some of the applications of the time-varying fields in the area of electrical machines and transformers.

The rest of the book provides the subject matter for the second semester in a two-semester sequence, which deals with the propagation, transmission, and radiation of electromagnetic fields in a medium under various constraints. A chapter-by-chapter explanation follows.

The development of wave equation and its solution that provides an inkling of wave propagation are discussed in Chapter 8. Also explained in this chapter are the reflection and transmission of the waves at normal and oblique incidences. The wave may have perpendicular or parallel polarization. The wave incidence may involve an interface between the two conductors, the two dielectrics, a conductor and a dielectric, or a dielectric and a perfect conductor.

The transmission of energy along the transmission lines is covered in Chapter 9. Instead of postulating a distributed equivalent circuit for a transmission line, we used field theory to justify the use of such a model. The wave equations in terms of the voltages and the currents along the length of transmission line are then developed and their solutions are provided. In order to minimize reflections on the transmission line, impedance matching with the stubs is explained. Lattice diagrams are used to explain the transient behavior of transmission lines. Although the Smith chart provides a visual picture of what is happening along a transmission line, we still feel that it is basically a transmission-line calculator. We can now use pocket calculators and computers to obtain exact information on the line. For this reason, we have placed the Smith chart and its applications in Appendix A.

The propagation of guided waves within the rectangular cross-section of a waveguide is covered in Chapter 10. The conditions for the existence of transverse electric (TE) and transverse magnetic (TM) modes are emphasized. Power flow in a rectangular waveguide under various conditions is also analyzed. Also explained in this chapter are the necessary conditions for the existence of fields inside a cavity and the use of a cavity as a frequency meter.

The radiation of electromagnetic waves is the subject matter of Chapter 11. The wave equations in terms of potential functions are developed and their solutions are sought for various types of antennas. The concepts of near-zone and radiation fields are explained. Also discussed in this chapter are the directive gain and directivity of a transmitting antenna, receiving antenna and Friis equation, the radar operation and the Doppler effect.

Chapter 12 covers the computer-aided analysis of electromagnetic fields. Some of the commonly used methods discussed in this chapter are the finite-difference method, the finite-element method, and the method of moments. Computer programs based upon these methods are included in Appendix B.

Our aim was to write a detailed student-oriented book. The success of first edition attests that we have succeeded in our mission. The first edition has already been translated into two foreign languages: Chinese and Korean. We hope that this second edition will also be accepted both by the students and faculty with the same zeal and zest as the first. Our goal has been and still is to present the material in such a way that a student can comprehend it with a minimum of help from instructors. To this end, we have carefully placed numerous worked examples with full details in each chapter. These examples, clearly delineated from the textual matter, not only enhance appreciation of a concept or a physical law but also bridge the perceived gap, real or otherwise, between a formal theoretical development and its applications. We opine that examples are necessary for immediate reinforcement and further clarification of a topic. Near the end of each chapter, we have included some easy questions under the heading of "Exercises" whose answers depend upon the direct applications of the concepts covered in each section. We believe that these exercises will help to impart motivation, nurture confidence, and heighten the understanding of the material presented in each chapter. In addition, there are problems at the end of each chapter that are designed to offer a wide range of challenges to the student. The exercises and the problems are an important part of the text and form an integral part of the study of electromagnetic fields. We suggest that the student should use basic laws and intuitive reasoning to obtain answers to these exercises and problems. The practice of such problem-solving techniques instills not only confidence but also empowers a student to tackle more difficult, albeit real-life, problems. Each chapter ends with a summary as well as a set of review questions. Some of the important equations are included in the summary for an easy reference. The review questions are tailored to ensure that a student has taken to heart the basics of the material presented in that chapter. Once again, we have endeavored very hard to make this book as student friendly as possible and we welcome any suggestions in this regard.

Our experience points out that the students tend to view the theoretical development as an abstraction and place emphasis on some equations, treating them as "formulas". Soon frustrations set in as the students find that the so-called formulas are different, not only for different media but also for different configurations. The array of equations needed to compute just one field quantity intimidates them to the extent that they lose interest in the material. It then just another "difficult" course that they must pass to satisfy the requirements of a degree in electrical engineering. We believe that it is the instructor's responsibility to

- explain the aim of each development,
- justify the assumptions imperative to the development at hand,
- emphasize its limitations,
- highlight the influence of the medium,
- illustrate the impact of geometry on an equation, and
- point to some of its applications.

To attain these goals, instructors must use their own experiences in the subject and also emphasize other areas of application. They must also stress any new developments in the field while they are discussing the fundamentals. For example, while explaining the magnetic force between two current-carrying conductors, an instructor can discuss magnetically levitated vehicles. Likewise, an instructor can shed light upon the design of a microwave oven while discussing a cavity resonator.

When the subject matter is explained properly and the related equations are developed from the basic laws, the student then learns to

- appreciate the theoretical development,
- forsake intimidation,
- regain motivation and confidence, and
- grasp the power of reasoning to develop new ideas.

Whatever we have presented in this second edition, we have done so according to our own convictions and understanding of the subject matter. It is quite possible that while stating and explaining our points of view, we may have said something that may conflict with your views, for which we seek your candid opinion and constructive criticism. If your point of view helps sway our minds, we will surely include it in a revised edition of this book. For this reason, your input is highly valuable to us.

Acknowledgments

We are deeply grateful for all the help we received from Dr. A. Haq Qureshi during the development of the first edition of this book. His mastery of the subject and the yearning for its clear and accurate presentation had direct impact upon the success of the first edition. We are much beholden to reviewers of both editions for their invaluable suggestions and constructive criticisms. We are most appreciative of the following persons and their establishments for providing us various photographs that are reproduced in this book: Ellen Modock (Keithley Instruments, Cleveland, Ohio), Bernard Surtz (Andrew Corporation, Orlando Park, Illinois), Bruce Whitney (Detroit Edition, Detriot, Michigan), Homer Bartlett (Microstar Inc. Florida), and Jeremiah Chambers (Space Machine & Engineering Corp., Florida).

Above all, we could not have written this book without the unconditional support, active encouragement, and complete cooperation of our families. In appreciation of their immense sacrifices, this text is lovingly dedicated to them.

1

Electromagnetic field theory

1.1 Introduction

What is a field? Is it a scalar field or a vector field? What is the nature of a field? Is it a continuous or a rotational field? How is the magnetic field produced by a current-carrying coil? How does a capacitor store energy? How does a piece of wire (antenna) radiate or receive signals? How do electromagnetic fields propagate in space? What really happens when electromagnetic energy travels from one end of a hollow pipe (waveguide) to the other? The primary purpose of this text is to answer some of these questions pertaining to electromagnetic fields.

In this chapter we intend to show that the study of electromagnetic field theory is vital to understanding many phenomena that take place in electrical engineering. To do so we make use of some of the concepts and equations of other areas of electrical engineering. We aim to shed light on the origin of these concepts and equations using electromagnetic field theory.

Before we proceed any further, however, we mention that the development of science depends upon some quantities that cannot be defined precisely. We refer to these as fundamental quantities; they are **mass** (m), **length** (ℓ), **time** (t), **charge** (q), and **temperature** (T). For example, what is time? When did time begin? Likewise, what is temperature? What is hot or cold? We do have some intuitive feelings about these quantities but lack precise definitions. To measure and express each of these quantities, we need to define a system of units.

In the International System of Units (SI for short), we have adopted the units of kilogram (kg) for mass, meter (m) for length, second (s) for time, coulomb (C) for charge, and kelvin (K) for temperature. Units for all other quantities of interest are then defined in terms of these fundamental units. For example, the unit of current, the ampere (A), in terms of the fundamental units is coulombs per second (C/s). Therefore, the ampere is a derived unit. The newton (N), the unit of force, is also a derived unit; it can be expressed in terms of basic units as $1\text{ N} = 1\text{ kg}\cdot\text{m/s}^2$. Units for some of the quantities that we will refer to in this

Table 1.1. Derived units for some electromagnetic quantities

Symbol	Quantity	Unit	Abbreviation
Y	admittance	siemen	S
ω	angular frequency	radian/second	rad/s
C	capacitance	farad	F
ρ	charge density	coulomb/meter^3	C/m^3
G	conductance	siemen	S
σ	conductivity	siemen/meter	S/m
W	energy	joule	J
F	force	newton	N
f	frequency	hertz	Hz
Z	impedance	ohm	Ω
L	inductance	henry	H
$\mathscr{F}$	magnetomotive force	ampere-turn	A°t
μ	permeability	henry/meter	H/m
ϵ	permittivity	farad/meter	F/m
P	power	watt	W
$\mathscr{R}$	reluctance	henry^{-1}	H^{-1}

Table 1.2. Unit conversion factors

From	Multiply by	To obtain
gilbert	0.79577	ampere-turn (At)
ampere-turn/cm	2.54	ampere-turn/inch
ampere-turn/inch	39.37	ampere-turn/meter
oersted	79.577	ampere-turn/meter
line (maxwells)	1×10^{-8}	weber (Wb)
gauss (lines/cm^2)	6.4516	line/inch^2
line/inch^2	0.155×10^{-4}	Wb/m^2 (tesla)
gauss	10^{-4}	Wb/m^2
inch	2.54	centimeter (cm)
foot	30.48	centimeter
meter	100	centimeter
square inch	6.4516	square cm
ounce	28.35	gram
pound	0.4536	kilogram
pound-force	4.4482	newton
ounce-force	0.278 01	newton
newton-meter	141.62	ounce-inch
newton-meter	0.73757	pound-feet
revolution/minute	$2\pi/60$	radian/second

text are given in Tables 1.1 and 1.3. Since English units are still being used in the industry to express some field quantities, it is necessary to convert from one unit system to the other. Table 1.2 is provided for this purpose.

Table 1.3. A partial list of field quantities

Variable	Definition	Type	Unit
$\vec{\mathbf{A}}$	magnetic vector potential	vector	Wb/m
$\vec{\mathbf{B}}$	magnetic flux density	vector	Wb/m^2 (T)
$\vec{\mathbf{D}}$	electric flux density	vector	C/m^2
$\vec{\mathbf{E}}$	electric field intensity	vector	V/m
$\vec{\mathbf{F}}$	Lorentz force	vector	N
I	electric current	scalar	A
$\vec{\mathbf{J}}$	volume current density	vector	A/m^2
q	free charge	scalar	C
$\vec{\mathbf{S}}$	Poynting vector	vector	W/m^2
$\vec{\mathbf{u}}$	velocity of free charge	vector	m/s
V	electric potential	scalar	V

Table 1.4. A partial list of relationships between various field quantities

$\vec{\mathbf{D}} = \epsilon\vec{\mathbf{E}}$	permittivity (ϵ)
$\vec{\mathbf{B}} = \mu\vec{\mathbf{H}}$	permeability (μ)
$\vec{\mathbf{J}} = \sigma\vec{\mathbf{E}}$	conductivity (σ), Ohm's law
$\vec{\mathbf{F}} = q(\vec{\mathbf{E}} + \vec{\mathbf{u}} \times \vec{\mathbf{B}})$	Lorentz force equation
$\nabla \cdot \vec{\mathbf{D}} = \rho$	Gauss's law (Maxwell's equation)
$\nabla \cdot \vec{\mathbf{B}} = 0$	Gauss's law (Maxwell's equation)
$\nabla \cdot \vec{\mathbf{J}} = -\frac{\partial \rho}{\partial t}$	continuity equation
$\nabla \times \vec{\mathbf{E}} = -\frac{\partial \vec{\mathbf{B}}}{\partial t}$	Faraday's law (Maxwell's equation)
$\nabla \times \vec{\mathbf{H}} = \vec{\mathbf{J}} + \frac{\partial \vec{\mathbf{D}}}{\partial t}$	Ampère's law (Maxwell's equation)

1.2 Field concept

Prior to undertaking the study of electromagnetic fields we must define the concept of a **field**. When we define the behavior of a quantity in a given region in terms of a set of values, one for each point in that region, we refer to this behavior of the quantity as a field. The value at each point of a field can be either measured experimentally or predicted by carrying out certain mathematical operations on some other quantities.

From the study of other branches of science, we know that there are both scalar and vector fields. Some of the field variables we use in this text are given in Table 1.3. There also exist definite relationships between these field quantities, and some of these are given in Table 1.4.

The permittivity (ϵ) and the permeability (μ) are properties of the medium. When the medium is a vacuum or free space, their values are

$$\mu_0 = 4\pi \times 10^{-7} \text{ H/m}$$

$$\epsilon_0 = 8.851 \times 10^{-12} \approx 10^{-9}/36\pi \text{ F/m}$$

From the equations listed in Table 1.4, Maxwell was able to predict that electromagnetic fields propagate in a vacuum with the speed of light. That is,

$$c = (\mu_0 \epsilon_0)^{-1/2} \text{ m/s}$$

1.3 Vector analysis

Vector analysis is the language used in the study of electromagnetic fields. Without the use of vectors, the field equations would be quite unwieldy to write and onerous to remember. For example, the cross product of two vectors $\vec{\mathbf{A}}$ and $\vec{\mathbf{B}}$ can be simply written as

$$\vec{\mathbf{A}} \times \vec{\mathbf{B}} = \vec{\mathbf{C}} \tag{1.1}$$

where $\vec{\mathbf{C}}$ is another vector. When expressed in scalar form, this equation yields a set of three scalar equtions. In addition, the appearance of these scalar equations depends upon the coordinate system. In the rectangular coordinate system, the previous equation is a concise version of the following three equations:

$$A_y B_z - A_z B_y = C_x \tag{1.2a}$$

$$A_z B_x - A_x B_z = C_y \tag{1.2b}$$

$$A_x B_y - A_y B_x = C_z \tag{1.2c}$$

You can easily see that the vector equation conveys the sense of a cross product better than its three scalar counterparts. Moreover, the vector representation is independent of the coordinate system. Thus, vector analysis helps us to simplify and unify field equations.

By the time a student is required to take the first course in electromagnetic theory, he/she has had a very limited exposure to vector analysis. The student may be competent to perform such vector operations as the gradient, divergence, and curl, but may not be able to describe the significance of each operation. The knowledge of each vector operation is essential to appreciate the development of electromagnetic field theory.

Quite often, a student does not know that (a) the unit vector that transforms a scalar surface to a vector surface is always normal to the surface, (b) a thin sheet (negligible thickness) of paper has two surfaces, (c) the direction of the line integral along the boundary of a surface depends upon the direction of the unit normal to that surface, and (d) there is a difference between an open surface and a closed surface. These concepts are important, and the student must comprehend the significance of each.

There are two schools of thought on the study of vector analysis. Some authors prefer that each vector operation be introduced only when it is needed, whereas others believe that a student must gain adequate proficiency in all vector operations prior to exploring electromagnetic field theory. We prefer the latter approach and for this reason have devoted Chapter 2 to the study of vectors.

1.4 Differential and integral formulations

Quite often a student does not understand why we present the same idea in two different forms: the differential form and the integral form. It must be pointed out that the integral form is useful to explain the significance of an equation, whereas the differential form is convenient for performing mathematical operations. For example, we express the equation of continuity of current in the differential form as

$$\nabla \cdot \vec{\mathbf{J}} = -\frac{\partial \rho}{\partial t} \tag{1.3}$$

where $\vec{\mathbf{J}}$ is the volume current density and ρ is the volume charge density. This equation states that the divergence of current density at a point is equal to the rate at which the charge density is changing at that point. The usefulness of this equation lies in the fact that we can use it to calculate the rate at which the charge density is changing at a point when the current density is known at that point. However, to highlight the physical significance of this equation, we have to enclose the charge in a volume v and perform volume integration. In other words, we have to express (1.3) as

$$\int_v \nabla \cdot \vec{\mathbf{J}}\, dv = -\int_v \frac{\partial \rho}{\partial t}\, dv \tag{1.4}$$

We can now apply the divergence theorem to transform the volume integral on the left-hand side into a closed surface integral. We can also interchange the operations of integration and differentiation on the right-hand side of equation (1.4). We can now obtain

$$\oint_s \vec{\mathbf{J}} \cdot \vec{ds} = -\frac{\partial}{\partial t}\int_v \rho\, dv \tag{1.5}$$

This equation is an integral formulation of (1.3). The integral on the left-hand side represents the net outward current I through the closed surface s bounding volume v. The integral on the right-hand side yields the charge q inside the volume v. This equation, therefore, states that the *net outward current through a closed surface bounding a region is equal to the rate at which the charge inside the region is decreasing with time.* In other words,

$$I = -\frac{dq}{dt} \tag{1.6}$$

which is a well-known circuit equation when the negative sign is omitted.

The details of the preceding development are given in Chapter 4. We used this example at this time just to show that (1.3) and (1.5) are the same and that they embody the same basic idea.

1.5 Static fields

Once again we face the dilemma of how to begin the presentation of electromagnetic field theory. Some authors believe in starting with the presentation of Maxwell's equations as a basic set of postulates and then summarizing the results of many years of experimental observations of electromagnetic effects. We, however, think that the field theory should always be developed by making maximum possible use of the concepts previously discussed in earlier courses in physics. For this reason we first discuss static fields.

In the study of electrostatics, or static electric fields, we assume that (a) all charges are fixed in space, (b) all charge densities are constant in time, and (c) the charge is the source of the electric field. Our interest is to determine (a) the electric field intensity at any point, (b) the potential distribution, (c) the forces exerted by the charges on other charges, and (d) the electric energy distribution in the region. We will also explore how a capacitor stores energy. To do so, we will begin our discussion with Coulomb's law and Gauss's law and formulate such well-known equations as Poisson's equation and Laplace's equation in terms of potential functions. We will show that the electric field at any point is perpendicular to an equipotential surface and emphasize its ramifications. Some of the equations pertaining to electrostatic fields are given in Table 1.5 (see below).

Table 1.5. Electrostatic field equations

Coulomb's law:	$\vec{\mathbf{F}} = q\vec{\mathbf{E}}$		
Electric field:	$\vec{\mathbf{E}} = \dfrac{Q\vec{\mathbf{a}}_R}{4\pi\epsilon R^2}$	or	$\vec{\mathbf{E}} = \dfrac{1}{4\pi\epsilon}\displaystyle\int_v \dfrac{\rho\vec{\mathbf{a}}_R}{R^2}\,dv$
Gauss's law:	$\nabla\cdot\vec{\mathbf{D}} = \rho$	or	$\oint_s \vec{\mathbf{D}}\cdot\vec{ds} = Q$
Conservative $\vec{\mathbf{E}}$ field:	$\nabla\times\vec{\mathbf{E}} = 0$	or	$\oint_c \vec{\mathbf{E}}\cdot\vec{d\ell} = 0$
Potential function:	$\vec{\mathbf{E}} = -\nabla V$	or	$V_{ba} = -\displaystyle\int_a^b \vec{\mathbf{E}}\cdot\vec{d\ell}$
Poisson's equation:	$\nabla^2 V = -\dfrac{\rho}{\epsilon}$		
Laplace's equation:	$\nabla^2 V = 0$		
Energy density:	$w_e = \frac{1}{2}\vec{\mathbf{D}}\cdot\vec{\mathbf{E}}$		
Constitutive relationship:	$\vec{\mathbf{D}} = \epsilon\vec{\mathbf{E}}$		
Ohm's law:	$\vec{\mathbf{J}} = \sigma\vec{\mathbf{E}}$		

Table 1.6. Magnetostatic field equations

Force equation:	$\vec{\mathbf{F}} = q\vec{\mathbf{u}} \times \vec{\mathbf{B}}$	or	$d\vec{\mathbf{F}} = I\,d\vec{\ell} \times \vec{\mathbf{B}}$
Biot–Savart law:	$d\vec{\mathbf{B}} = \dfrac{\mu}{4\pi}\dfrac{I\,d\vec{\ell} \times \vec{\mathbf{a}}_r}{r^2}$		
Ampère's law:	$\nabla \times \vec{\mathbf{H}} = \vec{\mathbf{J}}$	or	$\oint_c \vec{\mathbf{H}} \cdot d\vec{\ell} = I$
Gauss's law:	$\nabla \cdot \vec{\mathbf{B}} = 0$	or	$\oint_s \vec{\mathbf{B}} \cdot d\vec{s} = 0$
Magnetic vector potential:	$\vec{\mathbf{B}} = \nabla \times \vec{\mathbf{A}}$	or	$\vec{\mathbf{A}} = \dfrac{\mu}{4\pi}\displaystyle\int_c \dfrac{I\,d\vec{\ell}}{r}$
Magnetic flux:	$\Phi = \displaystyle\int_s \vec{\mathbf{B}} \cdot d\vec{s}$	or	$\Phi = \oint_c \vec{\mathbf{A}} \cdot d\vec{\ell}$
Magnetic energy:	$w_m = \frac{1}{2}\vec{\mathbf{B}} \cdot \vec{\mathbf{H}}$		
Poisson's equation:	$\nabla^2\vec{\mathbf{A}} = -\mu\vec{\mathbf{J}}$		
Constitutive relationship:	$\vec{\mathbf{B}} = \mu\vec{\mathbf{H}}$		

We already know that a charge in motion creates a current. If the movement of the charge is restricted in such a way that the resulting current is constant in time, the field thus created is called a magnetic field. Since the current is constant in time, the magnetic field is also constant in time. The branch of science relating to constant magnetic fields is called magnetostatics, or static magnetic fields. In this case, we are interested in the determination of (a) magnetic field intensity, (b) magnetic flux density, (c) magnetic flux, and (d) the energy stored in the magnetic field. To this end we will begin our discussion with the Biot-Savart law and Ampère's law and develop all the essential equations. From time to time we will also stress the correlation between the static electric and magnetic fields. Some of the important equations that we will either state or formulate in magnetostatics are given in Table 1.6.

There are numerous practical applications of static fields. Both static electric and magnetic fields are used in the design of many devices. For example, we can use a static electric field to accelerate a particle and a static magnetic field to deflect it. This scheme can be employed in the design of an oscilloscope and/or an ink-jet printer. We have devoted Chapter 6 to address some of the applications of static fields. Once a student has mastered the fundamentals of static fields, he/she should be able to comprehend their applications without further guidance from the instructor. The instructor may decide to highlight the salient features of each application and then treat it as a reading assignment. The discussion of real-life applications of the theory makes the subject interesting.

1.6 Time-varying fields

In the study of electric circuits, you were introduced to a differential equation that yields the voltage drop $v(t)$ across an inductor L when it carries a current $i(t)$. More often than not, the relationship is stated without proof as follows:

$$v = L\frac{di}{dt} \tag{1.7}$$

Someone with a discerning mind may have wondered about the origin of this equation. It is a consequence of a lifetime of work by Michael Faraday (1791–1867) toward an understanding of a very complex phenomenon called magnetic induction.

We will begin our discussion of time-varying fields by stating *Faraday's law of induction* and then explain how it led to the development of generators (sources of three-phase energy), motors (the workhorses of the industrialized world), relays (magnetic controlling mechanisms), and transformers (devices that transfer electric energy from one coil to another entirely by induction). One of the four well-known Maxwell equations is, in fact, a statement of Faraday's law of induction. At this time it will suffice to say that Faraday's law relates the induced electromotive force (emf) $e(t)$ in a coil to the time-varying magnetic flux $\Phi(t)$ linking that coil as

$$e = -\frac{d\Phi}{dt} \tag{1.8}$$

The significance of the negative sign (*Lenz's law*) and the derivation of (1.7) from (1.8) will be discussed in detail in this text.

We will also explain why Maxwell felt it necessary to modify Ampère's law for time-varying fields. The inclusion of displacement current (current through a capacitor) enabled Maxwell to predict that fields should propagate in free space with the velocity of light. The modification of Ampère's law is considered to be one of the most significant contributions by James Clerk Maxwell (1831–1879) in the area of electromagnetic field theory.

Faraday's law of induction, the modified Ampère law, and the two Gauss laws (one for the time-varying electric field and the other for the time-varying magnetic field) form a set of four equations; these are now called *Maxwell's equations.* These equations are given in Table 1.4. Evident from these equations is the fact that time-varying electric and magnetic fields are intertwined. In simple words, a time-varying magnetic field gives rise to a time-varying electric field and vice versa.

The modification of Ampère's law can also be viewed as a consequence of the equation of continuity or conservation of charge. This equation is also given in Table 1.4.

When a particle having a charge q is moving with a velocity $\vec{\mathbf{u}}$ in a region where there exist a time-varying electric field ($\vec{\mathbf{E}}$) and a magnetic field ($\vec{\mathbf{B}}$), it experiences a force ($\vec{\mathbf{F}}$) such that

$$\vec{\mathbf{F}} = q(\vec{\mathbf{E}} + \vec{\mathbf{u}} \times \vec{\mathbf{B}}) \tag{1.9}$$

We will refer to this equation as the *Lorentz force equation.*

With the help of the four Maxwell equations, the equation of continuity, and the Lorentz force equation we can now explain all the effects of electromagnetism.

1.7 Applications of time-varying fields

Among the numerous applications of electromagnetic field theory, we will consider those pertaining to the transmission, reception, and propagation of energy. This selection of topics is due to the fact that the solution of Maxwell's equations always leads to waves. The nature of the wave depends upon the medium, the type of excitation (source), and the boundary conditions.

The propagation of a wave may either be in an unbounded region (fields exist in an infinite cross section, such as free space) or in a bounded region (fields exist in a finite cross section, such as a waveguide or a coaxial transmission line).

Although most of the fields transmitted are in the form of spherical waves, they may be considered as plane waves in a region far away from the transmitter (radiating element, such as an antenna). How far "far away" is depends upon the wavelength (distance traveled to complete one cycle) of the fields. Using plane waves as an approximation, we will derive wave equations from Maxwell's equations in terms of electric and magnetic fields. The solution of these wave equations will describe the behavior of a plane wave in an unbounded medium. We will simplify the analysis by imposing restrictions such that (a) the wave is a uniform plane wave, (b) there are no sources of currents and charges in the medium, and (c) the fields vary sinusoidally in time. We will then determine (i) the expressions for the fields, (ii) the velocity with which they travel in a region, and (iii) the energy associated with them. We will also show that the medium behaves as if it has an impedance; we refer to this as *intrinsic impedance.* The intrinsic impedance of free space is approximately 377 Ω.

Our discussion of uniform plane waves will also include the effect of interface between two media. Here we will discuss (a) how much of the energy of the incoming wave is transmitted into the second medium or reflected back into the first medium, (b) how the incoming wave and reflected wave combine to form a standing wave, and (c) the condition necessary for total reflection.

We devote Chapter 9 to the discussion of transmission of energy from one end to the other via a transmission line. We will show that when one end of the transmission line is excited by a time-varying source, the transmission of energy is in the form of a wave. The wave equations in this case will be in terms of the voltage and the current at any point along the transmission line. The solution of these wave equations will tell us that a finite time is needed for the wave to reach the other end, and for practical transmission lines, the wave attenuates exponentially with the distance. The attenuation is due to the resistance and conductance of the transmission line. This results in a loss in energy along the entire length of the transmission line. However, at power frequencies (50 or 60 Hz) there is a negligible loss in energy due to radiation because the spacing between the conductors is extremely small in comparison with the wavelength.

As the frequency increases so does the loss of signal along the length of the transmission line. At high frequencies, the energy is transmitted from one point to another via waveguides. Although any hollow conductor can be used as a waveguide, the most commonly used waveguides have rectangular or circular cross sections. We will examine the necessary conditions that must be satisfied for the fields to exist, obtain field expressions, and compute the energy at any point inside the waveguide. The analysis involves the solution of the wave equation inside the waveguide subjected to external boundary conditions. The analysis is complex; thus, we will confine our discussion to a rectangular waveguide. Although the resulting equations appear to be quite involved and difficult to remember, we must not forget that they are obtained by simply applying the boundary conditions to a general solution of the wave equation.

A transmission line can be used to transfer energy from very low frequencies (even dc) to reasonably high frequencies. The waveguide, on the other hand, has a lower limit on the frequency called the *cutoff frequency.* The cutoff frequency depends upon the dimensions of the waveguide. Signals below the cutoff frequency cannot propagate inside the waveguide. Another major difference between a transmission line and a waveguide is that the transmission line can support the *transverse electromagnetic* (TEM) mode. In practice, both coaxial and parallel wire transmission lines use the TEM mode. However, such a mode cannot exist inside the waveguide. Why this is so will be explained in Chapter 10. The waveguide can support two different modes, the *transverse electric mode* and the *transverse magnetic mode.* The conditions for the existence of these modes will also be discussed.

The last application of Maxwell's equations that we will discuss in this text deals with electromagnetic radiation produced by time-varying sources of finite dimensions. The very presence of these sources adds to the complexity of the solution of a wave equation in terms of the electric field and/or the magnetic field. However, if we develop the wave equations in terms of scalar and vector potentials, the solution of either potential function is relatively less involved. By simple algebraic manipulations, we can obtain expressions for the electric and magnetic fields. The power radiated by the sources can then be computed. We will examine the fields produced and the power radiated by straight-wire and loop antennas. We will also study how the radiation field patterns can be modified by using antenna arrays.

1.8 Numerical solutions

Each time we want to obtain an "exact" solution to a problem we are often forced to make and justify some assumptions. For instance, (a) to determine the electric field intensity within a parallel-plate capcitor we usually assume that the plates are of infinite extent so that we can apply Gauss's law, (b) to calculate the magnetic field intensity due to a long current-carrying conductor using Ampère's law we imagine that the conductor is of infinite extent, (c) to obtain the propagation characteristics and the nature of electromagnetic fields in a source-free region we visualize the fields in the form of a uniform plane wave, (d) to learn about the radiation pattern of a small linear antenna we presume that the length of the antenna is so small that the current distribution is uniform, etc. Each assumption gives rise to a special situation and the analytical solution thus obtained is precise.

In electrostatics we determined the capacitance of an isolated sphere using Gauss's law by exploiting the spherical symmetry. However, the problem becomes very complex when we try to determine the capacitance of an isolated cube. In magnetostatics, we obtained an answer for the magnetic field intensity on the axis of a circular current-carrying conductor using the Biot–Savart law. Can we follow the same technique to determine the magnetic field intensity when the current-carrying conductor has an arbitrary shape? The answer, of course, is "no" because of the nature of the integral formulation. It is also not easy to determine the radiation pattern of a current-carrying conductor of arbitrary shape using analytical methods. Likewise, a uniform plane wave cannot exist because its very existence dictates the presence of infinite energy in the medium, but the idea of a uniform plane wave is needed to get a clear picture of power flow in a region of interest.

From this discussion, it is obvious that it is not always possible to obtain an exact solution to a problem without making some simplifying assumptions. The need for a numerical solution, which is often approximate, should be clearly evident. It must be borne in mind that each numerical solution is simply an approximation of the exact differential or integral equation. How refined a numerical method we should use depends upon the accuracy of the solution required. The higher the accuracy, the more refined the numerical method must be. The accuracy of the solution further hinges on the numerical method used and the computing capability of a system.

Many methods have been used successfully to obtain solutions of those problems that cannot be easily solved using analytical techniques. The three methods we discuss in this text are the *finite-difference method*, the *finite-element method*, and the *method of moments*.

1.9 Further study

The electromagnetic field theory presented in this text is just a beginning. This information is essential not only to arouse some interest in this area but also to understand more complex developments. However, the wonderful aspect of electromagnetic field theory is that we can either predict or explain almost all electromagnetic phenomena by appropriately manipulating the four Maxwell equations, the equation of continuity, and the Lorentz force equation.

In this book we discuss only rectangular waveguides. For circular waveguides, the same wave equation transforms itself into a form called *Bessel's equation* and its solution is in terms of *Bessel functions*. The study of Bessel functions and how to express them in terms of infinite series is essential prior to discussing circular waveguides.

One of the many mapping techniques is called the *Schwarz–Christoffel transformation* and it can be used to determine the nature of the fringing field for a parallel-plate capacitor of finite dimensions. The use of this technique eliminates the assumption that each plate is of infinite extent. However, this technique is not discussed as it involves higher-level mathematics. Similarly missing from discussion at the undergraduate level is a technique known as *conformal transformation*. This technique has been applied to determine the capacitance between any two electrodes in an integrated circuit.

The general solution of an antenna is quite complex and is evident from the lifetime work of Ronald W. P. King. He has written numerous papers and published a number of books in this area. The study of scattering and radiation from various types of antennas is so captivating that it can keep a discerning intellectual overwhelmed for a long period of time.

Another captivating topic is the study of gaseous plasma invaded by electromagnetic fields. The electromagnetic fields profoundly influence the properties of a plasma because plasma contains charged particles that are nearly free. If such a study is undertaken, it will explain (a) the basic physical characteristics of plasma, and (b) the effects on waves in a plasma medium.

If working with complex mathematical equations is provisional for your proper mental exercise, then consider the study of electromagnetic field theory including the relativistic concepts. This study involves the application of the Lorentz transformation and covariant formulation of Maxwell's equations. In these equations time is treated exactly the same way as the space coordinates. Therefore, the gradient, divergence, curl, and Laplacian are all four-dimensional operators.

If we have sparked your interest in any new area of study, dive in and explore! However, to succeed in your mission, you have to first comprehend the theory presented in this text. Knowledge is not gained in an instant. Your willingness to learn fundamentals today rather than treating final equations as formulas will be amply rewarded tomorrow. You will enhance your reasoning capabilities and be able to handle more difficult problems in the future.

2

Vector analysis

2.1 Introduction

Knowledge of vector algebra and vector calculus is essential in developing the concepts of electromagnetic field theory. The widespread acceptance of vectors in electromagnetic field theory is due in part to the fact that they provide compact mathematical representations of complicated phenomena and allow for easy visualization and manipulation. The ever-increasing number of textbooks on the subject are further evidence of the popularity of vectors. As you will see in subsequent chapters, a single equation in vector form is sufficient to represent up to three scalar equations. Although a complete discussion of vectors is not within the scope of this text, some of the vector operations that will play a prominent role in our discussion of electromagnetic field theory are introduced in this chapter. We begin our discussion by defining scalar and vector quantities.

2.2 Scalar and vector quantities

Most of the quantities encountered in electromagnetic fields can easily be divided into two classes, scalars and vectors.

2.2.1 Scalar

A physical quantity that can be completely described by its magnitude is called a **scalar**. Some examples of scalar quantities are mass, time, temperature, work, and electric charge. Each of these quantities is completely describable by a single number. A temperature of 20 °C, a mass of 100 grams, and a charge of 0.5 coulomb are examples of scalars. In fact, all real numbers are scalars.

2.2.2 Vector

A physical quantity having a magnitude as well as a direction is called a **vector**. Force, velocity, torque, electric field, and acceleration are vector quantities.

A vector quantity is graphically depicted by a line segment equal to its magnitude according to a convenient scale, and the direction is indicated by means of an arrow, as shown in Figure 2.1a. We will represent a vector by placing an arrow over the letter. Thus, in Figure 2.1a, $\vec{\mathbf{R}}$ represents a vector directed from point O toward point P. Figure 2.1b shows a few parallel vectors having the same length and direction; they all represent the same vector. Two vectors $\vec{\mathbf{A}}$ and $\vec{\mathbf{B}}$ are equal; i.e., $\vec{\mathbf{A}} = \vec{\mathbf{B}}$, if they have the same magnitude (length) and direction. We can only compare vectors if they have the same physical or geometrical meaning and hence the same dimensions.

A vector of magnitude zero is called a *null vector* or a *zero vector*. This is the only vector that cannot be represented as an arrow because it has zero magnitude (length).

A vector of unit magnitude (length) is called a *unit vector*. We can always represent a vector in terms of its unit vector. For example, vector $\vec{\mathbf{A}}$ can be written as

$$\vec{\mathbf{A}} = A\vec{\mathbf{a}}_A \tag{2.1}$$

where A is the magnitude of $\vec{\mathbf{A}}$ and $\vec{\mathbf{a}}_A$ is the unit vector in the same direction of $\vec{\mathbf{A}}$ such that

$$\vec{\mathbf{a}}_A = \frac{\vec{\mathbf{A}}}{A} \tag{2.2}$$

Figure 2.1

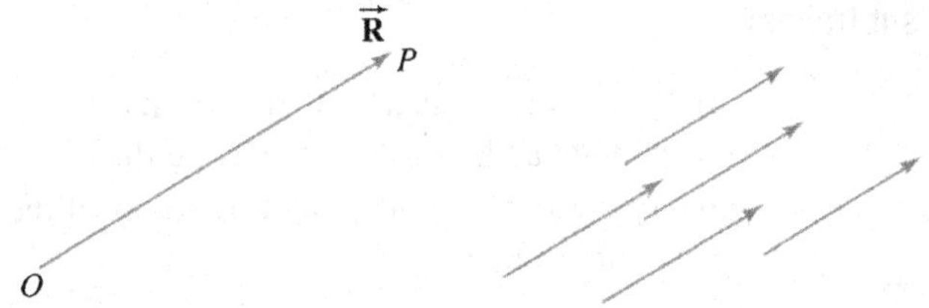

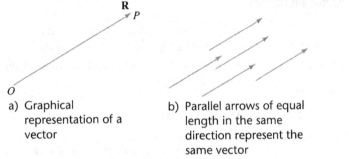

a) Graphical representation of a vector

b) Parallel arrows of equal length in the same direction represent the same vector

2.3 Vector operations

Adding, subtracting, multiplying, and/or dividing scalar quantities is second nature to most of us. For example, if we want to add two scalars having the same units, we just add their magnitudes. The process of addition in terms of vectors is not this simple, nor are subtraction and multiplication of two vectors. Note that vector division is not defined.

2.3.1 Vector addition

To add two vectors $\vec{\mathbf{A}}$ and $\vec{\mathbf{B}}$ we draw two representative vectors $\vec{\mathbf{A}}$ and $\vec{\mathbf{B}}$ in such a way that the initial point (tail) of $\vec{\mathbf{B}}$ coincides with the final

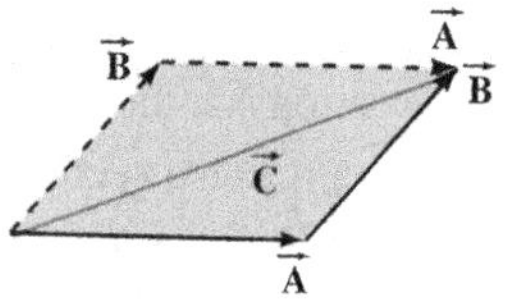

Figure 2.2 Vector addition: $\vec{C} = \vec{A} + \vec{B}$

point (tip) of $\vec{\mathbf{A}}$, as illustrated by the solid lines in Figure 2.2. The line joining the tail of $\vec{\mathbf{A}}$ to the tip of $\vec{\mathbf{B}}$ represents a vector $\vec{\mathbf{C}}$, which is the sum of the two vectors $\vec{\mathbf{A}}$ and $\vec{\mathbf{B}}$. That is,

$$\vec{\mathbf{C}} = \vec{\mathbf{A}} + \vec{\mathbf{B}} \tag{2.3a}$$

The sum of two vectors is therefore a vector. We could have drawn $\vec{\mathbf{B}}$ first and then $\vec{\mathbf{A}}$, as shown by the dotted lines in Figure 2.2. It is evident that vector addition is independent of the order in which the vectors are added. In other words, the vectors obey the **commutative law of addition**. That is,

$$\vec{\mathbf{A}} + \vec{\mathbf{B}} = \vec{\mathbf{B}} + \vec{\mathbf{A}} \tag{2.3b}$$

Figure 2.2 also provides the geometric interpretation of vector addition. If $\vec{\mathbf{A}}$ and $\vec{\mathbf{B}}$ are the two sides of a parallelogram, then $\vec{\mathbf{C}}$ is its diagonal. We can also show that vectors obey the **associative law of addition**. In other words,

$$\vec{\mathbf{A}} + (\vec{\mathbf{B}} + \vec{\mathbf{C}}) = (\vec{\mathbf{A}} + \vec{\mathbf{B}}) + \vec{\mathbf{C}} \tag{2.4}$$

2.3.2 Vector subtraction

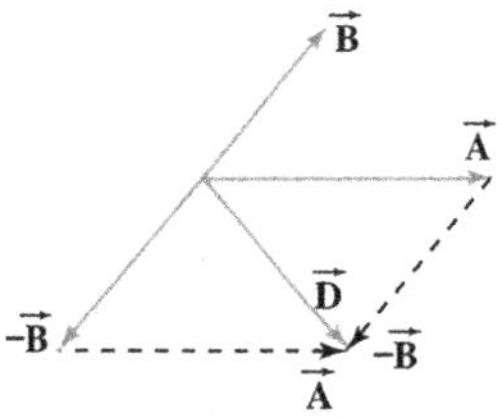

Figure 2.3 Vector subtraction: $\vec{D} = \vec{A} - \vec{B}$

If $\vec{\mathbf{B}}$ is a vector, then $-\vec{\mathbf{B}}$ (minus $\vec{\mathbf{B}}$) is also a vector with the same magnitude as $\vec{\mathbf{B}}$ but in the opposite direction. In fact $-\vec{\mathbf{B}}$ is said to be the *opposite* or *negative* of $\vec{\mathbf{B}}$. In terms of the negative of a vector, we can define vector subtraction, $\vec{\mathbf{A}} - \vec{\mathbf{B}}$, as

$$\vec{\mathbf{D}} = \vec{\mathbf{A}} + (-\vec{\mathbf{B}}) \tag{2.5}$$

Figure 2.3 shows the subtraction of $\vec{\mathbf{B}}$ from $\vec{\mathbf{A}}$.

2.3.3 Multiplication of a vector by a scalar

If we multiply a vector $\vec{\mathbf{A}}$ by a scalar k, we obtain a vector $\vec{\mathbf{B}}$ such that

$$\vec{\mathbf{B}} = k\vec{\mathbf{A}} \tag{2.6}$$

The magnitude of $\vec{\mathbf{B}}$ is simply equal to $|k|$ times the magnitude of $\vec{\mathbf{A}}$. However, $\vec{\mathbf{B}}$ is either in the same direction as $\vec{\mathbf{A}}$ if $k > 0$ or in the opposite direction from $\vec{\mathbf{A}}$ if $k < 0$. $\vec{\mathbf{B}}$ is longer than $\vec{\mathbf{A}}$ if $|k| > 1$ and shorter than $\vec{\mathbf{A}}$ if $|k| < 1$. A useful fact to remember is that $\vec{\mathbf{B}}$ is parallel to $\vec{\mathbf{A}}$, either in the same or opposite direction. Quite often, $\vec{\mathbf{B}}$ is said to be a *dependent vector*.

2.3.4 Product of two vectors

There are two useful definitions for the product of two vectors. One of them is called the *dot product*, and the other is referred to as the *cross product*.

Dot product of two vectors

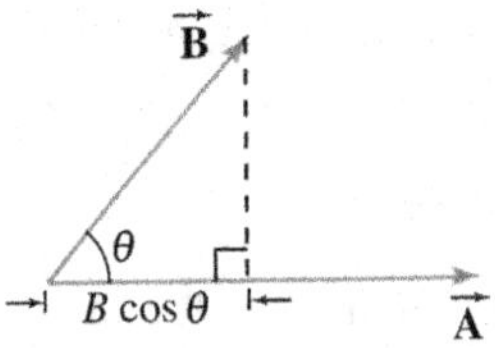

Figure 2.4 Illustration for the dot product

The **dot product** of two vectors $\vec{\mathbf{A}}$ and $\vec{\mathbf{B}}$ is written as $\vec{\mathbf{A}} \cdot \vec{\mathbf{B}}$ and is read as "$\vec{\mathbf{A}}$ dot $\vec{\mathbf{B}}$." It is defined as the product of the magnitudes of the two vectors and the cosine of the smaller angle between them, as illustrated in Figure 2.4. That is,

$$\vec{\mathbf{A}} \cdot \vec{\mathbf{B}} = AB\cos\theta \tag{2.7}$$

From equation (2.7) it is obvious that the dot product of $\vec{\mathbf{A}}$ and $\vec{\mathbf{B}}$ is a scalar. For this reason, the dot product is also known as the *scalar product*. The dot product is maximum when the two vectors are parallel. *However, if the dot product of two nonzero vectors is zero, the two vectors are orthogonal.*

Some of the basic properties of the dot product are

Commutative: $$\vec{\mathbf{A}} \cdot \vec{\mathbf{B}} = \vec{\mathbf{B}} \cdot \vec{\mathbf{A}} \tag{2.8a}$$

Distributive: $$\vec{\mathbf{A}} \cdot (\vec{\mathbf{B}} + \vec{\mathbf{C}}) = \vec{\mathbf{A}} \cdot \vec{\mathbf{B}} + \vec{\mathbf{A}} \cdot \vec{\mathbf{C}} \tag{2.8b}$$

Scaling: $$k(\vec{\mathbf{A}} \cdot \vec{\mathbf{B}}) = (k\vec{\mathbf{A}}) \cdot \vec{\mathbf{B}} = \vec{\mathbf{A}} \cdot (k\vec{\mathbf{B}}) \tag{2.8c}$$

The quantity $B\cos\theta$ in (2.7) is said to be *the component of* $\vec{\mathbf{B}}$ *along* $\vec{\mathbf{A}}$ and is commonly stated as *the scalar projection of* $\vec{\mathbf{B}}$ *on* $\vec{\mathbf{A}}$. Thus, the scalar projection of $\vec{\mathbf{B}}$ on $\vec{\mathbf{A}}$ is

$$B\cos\theta = \frac{\vec{\mathbf{A}} \cdot \vec{\mathbf{B}}}{A} = \vec{\mathbf{B}} \cdot \vec{\mathbf{a}}_A \tag{2.9}$$

By including the unit vector along $\vec{\mathbf{A}}$ in equation (2.9), we can define the **vector projection** of $\vec{\mathbf{B}}$ on $\vec{\mathbf{A}}$ as

$$B\cos\theta\, \vec{\mathbf{a}}_A = (\vec{\mathbf{B}} \cdot \vec{\mathbf{a}}_A)\vec{\mathbf{a}}_A \tag{2.10}$$

We will shortly employ (2.9) and (2.10) to determine the scalar and vector projections of a vector along three mutually perpendicular directions. Equation (2.7) can also be used to determine the angle between the two vectors $\vec{\mathbf{A}}$ and $\vec{\mathbf{B}}$ as

$$\cos\theta = \frac{\vec{\mathbf{A}} \cdot \vec{\mathbf{B}}}{AB} \tag{2.11}$$

provided $\vec{\mathbf{A}} \neq 0$ and $\vec{\mathbf{B}} \neq 0$. Using (2.7) we can also determine the magnitude of vector $\vec{\mathbf{A}}$ as

$$A = \sqrt{\vec{\mathbf{A}} \cdot \vec{\mathbf{A}}} \tag{2.12}$$

EXAMPLE 2.1 If $\vec{\mathbf{A}} \cdot \vec{\mathbf{B}} = \vec{\mathbf{A}} \cdot \vec{\mathbf{C}}$, does this imply that $\vec{\mathbf{B}}$ must always be equal to $\vec{\mathbf{C}}$?

Solution Because $\vec{\mathbf{A}} \cdot \vec{\mathbf{B}} = \vec{\mathbf{A}} \cdot \vec{\mathbf{C}}$, we can rewrite it as $\vec{\mathbf{A}} \cdot (\vec{\mathbf{B}} - \vec{\mathbf{C}}) = 0$. We can now make the following conclusions:

a) Either $\vec{\mathbf{A}}$ is perpendicular to $\vec{\mathbf{B}} - \vec{\mathbf{C}}$, or
b) $\vec{\mathbf{A}}$ is a null vector, or
c) $\vec{\mathbf{B}} - \vec{\mathbf{C}} = 0$.

Thus, only when $(\vec{\mathbf{B}} - \vec{\mathbf{C}})$ is equal to zero does $\vec{\mathbf{B}} = \vec{\mathbf{C}}$. Thus, $\vec{\mathbf{A}} \cdot \vec{\mathbf{B}} = \vec{\mathbf{A}} \cdot \vec{\mathbf{C}}$ does not always mean that $\vec{\mathbf{B}} = \vec{\mathbf{C}}$. • • •

The cross product

The **cross product** of two vectors $\vec{\mathbf{A}}$ and $\vec{\mathbf{B}}$ is written as $\vec{\mathbf{A}} \times \vec{\mathbf{B}}$ and is read as "$\vec{\mathbf{A}}$ cross $\vec{\mathbf{B}}$." The cross product is a vector that is directed normal to the plane containing $\vec{\mathbf{A}}$ and $\vec{\mathbf{B}}$ and is equal in magnitude to the product of the vectors and the sine of the smaller angle between them. That is,

$$\vec{\mathbf{A}} \times \vec{\mathbf{B}} = |AB \sin\theta|\, \vec{\mathbf{a}}_n \tag{2.13}$$

where $\vec{\mathbf{a}}_n$ is a unit vector normal to the plane of $\vec{\mathbf{A}}$ and $\vec{\mathbf{B}}$. The unit vector $\vec{\mathbf{a}}_n$ points in the direction of motion of a right-handed screw rotating from $\vec{\mathbf{A}}$ to $\vec{\mathbf{B}}$, as shown in Figure 2.5a. Another way to determine the direction of the unit vector $\vec{\mathbf{a}}_n$ is to extend the fingers of the right hand, as illustrated in Figure 2.5b. When the forefinger points in the direction of $\vec{\mathbf{A}}$ and the middle finger in the direction of $\vec{\mathbf{B}}$, then the thumb points in the direction of the unit vector $\vec{\mathbf{a}}_n$. Because the cross product yields a vector, it is also referred to as the *vector product.*

If $\vec{\mathbf{C}}$ represents the cross product of two vectors $\vec{\mathbf{A}}$ and $\vec{\mathbf{B}}$ such that

$$\vec{\mathbf{C}} = \vec{\mathbf{A}} \times \vec{\mathbf{B}} \tag{2.14}$$

then

$$C\vec{\mathbf{a}}_n = (A\vec{\mathbf{a}}_A) \times (B\vec{\mathbf{a}}_B) = (\vec{\mathbf{a}}_A \times \vec{\mathbf{a}}_B)\, AB$$

and the unit vector $\vec{\mathbf{a}}_n$ is

$$\vec{\mathbf{a}}_n = \frac{\vec{\mathbf{a}}_A \times \vec{\mathbf{a}}_B}{|\sin\theta|} \tag{2.15}$$

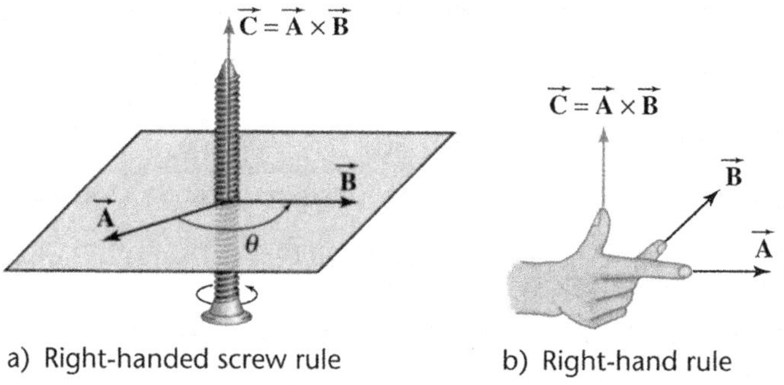

Figure 2.5 Rules to determine the direction of the cross product $\vec{\mathbf{C}} = \vec{\mathbf{A}} \times \vec{\mathbf{B}}$

By using the rules to determine the direction of the unit vector, we can show that

$$\vec{\mathbf{A}} \times \vec{\mathbf{B}} = -\vec{\mathbf{B}} \times \vec{\mathbf{A}} \tag{2.16}$$

Thus, the cross product is not commutative. Some of the other properties of the cross product are

Distributive: $$\vec{\mathbf{A}} \times (\vec{\mathbf{B}} + \vec{\mathbf{C}}) = \vec{\mathbf{A}} \times \vec{\mathbf{B}} + \vec{\mathbf{A}} \times \vec{\mathbf{C}} \tag{2.17a}$$

Scaling: $$(k\vec{\mathbf{A}}) \times \vec{\mathbf{B}} = k(\vec{\mathbf{A}} \times \vec{\mathbf{B}}) = \vec{\mathbf{A}} \times (k\vec{\mathbf{B}}) \tag{2.17b}$$

We can also show that a necessary and sufficient condition for two nonzero vectors to be parallel is that their cross product is zero.

EXAMPLE 2.2

Verify Lagrange's identity, which states that if $\vec{\mathbf{A}}$ and $\vec{\mathbf{B}}$ are two arbitrary vectors, then

$$|\vec{\mathbf{A}} \times \vec{\mathbf{B}}|^2 = A^2B^2 - (\vec{\mathbf{A}} \cdot \vec{\mathbf{B}})^2$$

Solution From the definition of the cross product of two vectors, we have

$$\begin{aligned} |\vec{\mathbf{A}} \times \vec{\mathbf{B}}|^2 &= A^2B^2 \sin^2\theta = A^2B^2(1 - \cos^2\theta) \\ &= A^2B^2 - A^2B^2\cos^2\theta \\ &= A^2B^2 - (\vec{\mathbf{A}} \cdot \vec{\mathbf{B}})^2 \end{aligned}$$

• • •

EXAMPLE 2.3

Derive the law of sines for a triangle using vectors.

Solution From Figure 2.6, we have

$$\vec{\mathbf{B}} = \vec{\mathbf{C}} - \vec{\mathbf{A}}$$

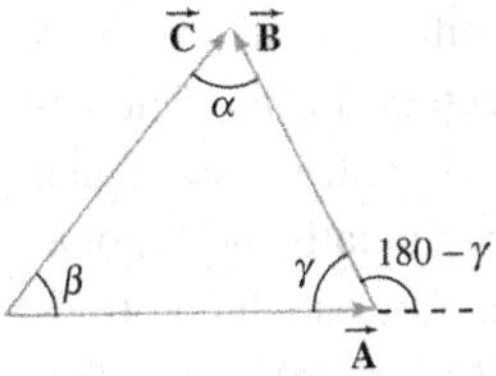

Figure 2.6

Because $\vec{\mathbf{B}} \times \vec{\mathbf{B}} = 0$, we can write

$$\vec{\mathbf{B}} \times (\vec{\mathbf{C}} - \vec{\mathbf{A}}) = 0$$

or

$$\vec{\mathbf{B}} \times \vec{\mathbf{C}} = \vec{\mathbf{B}} \times \vec{\mathbf{A}}$$

Therefore,

$$BC \sin\alpha = BA \sin(\pi - \gamma)$$

or

$$\frac{A}{\sin\alpha} = \frac{C}{\sin\gamma}$$

Similarly, we can show that

$$\frac{A}{\sin\alpha} = \frac{B}{\sin\beta}$$

Thus, we can state the law of sines for a triangle as

$$\frac{A}{\sin\alpha} = \frac{B}{\sin\beta} = \frac{C}{\sin\gamma}$$

• • •

Scalar triple product

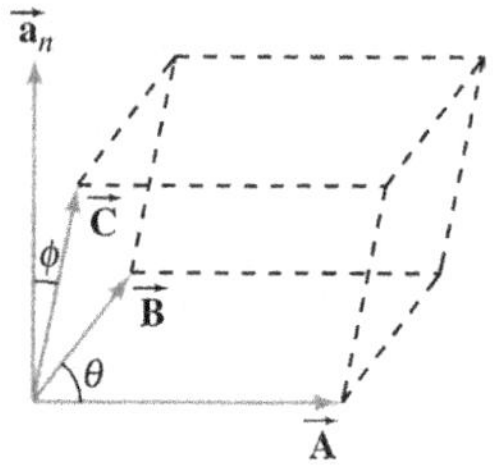

Figure 2.7 Illustration for the scalar triple product

The **scalar triple product** of three vectors $\vec{\mathbf{A}}$, $\vec{\mathbf{B}}$, and $\vec{\mathbf{C}}$ is a scalar and is computed as

$$\vec{\mathbf{C}} \cdot (\vec{\mathbf{A}} \times \vec{\mathbf{B}}) = ABC \sin\theta \cos\phi \tag{2.18a}$$

If the three vectors represent the sides of a parallelepiped, as shown in Figure 2.7, then the scalar triple product yields its volume. From (2.18a), it is evident that for three coplanar vectors, the scalar triple product is zero.

As long as the vectors appear in cyclical order, (2.18a) can also be written as

$$\vec{\mathbf{C}} \cdot (\vec{\mathbf{A}} \times \vec{\mathbf{B}}) = \vec{\mathbf{A}} \cdot (\vec{\mathbf{B}} \times \vec{\mathbf{C}}) = \vec{\mathbf{B}} \cdot (\vec{\mathbf{C}} \times \vec{\mathbf{A}}) \tag{2.18b}$$

Vector triple product

The **vector triple product** of three vectors $\vec{\mathbf{A}}$, $\vec{\mathbf{B}}$, and $\vec{\mathbf{C}}$ is a vector and is written as $\vec{\mathbf{A}} \times (\vec{\mathbf{B}} \times \vec{\mathbf{C}})$. We can show that the vector triple product is not associative. That is,

$$\vec{\mathbf{A}} \times (\vec{\mathbf{B}} \times \vec{\mathbf{C}}) \neq (\vec{\mathbf{A}} \times \vec{\mathbf{B}}) \times \vec{\mathbf{C}} \tag{2.19}$$

2.4 The coordinate systems

Up to this point we have kept our discussion quite general and used graphical representations when manipulating vectors. From a mathematical point of view it is very convenient to work with vectors when they are resolved into components along three mutually orthogonal (perpendicular) directions. In this text, we will mainly use three orthogonal coordinate systems: the *rectangular* (or Cartesian) coordinate system, the *cylindrical* (circular) coordinate system, and the *spherical* coordinate system. We shall now digress to discuss each of these coordinate systems and then resume our discussion of vectors.

2.4.1 Rectangular coordinate system

A rectangular (Cartesian) coordinate system is a system formed by three mutually orthogonal straight lines. The three straight lines are called the x, y, and z axes. The point of intersection of these axes is the origin.

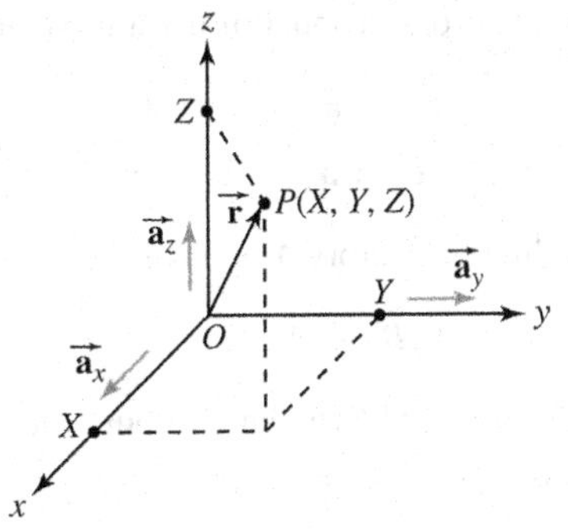

Figure 2.8 Projections of a point in a rectangular coordinate system

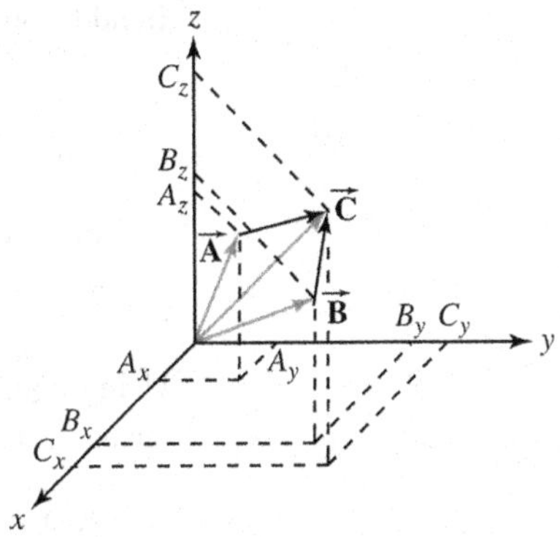

Figure 2.9 Vector addition in rectangular coordinate system

We will use the unit vectors $\vec{\mathbf{a}}_x$, $\vec{\mathbf{a}}_y$, and $\vec{\mathbf{a}}_z$ to indicate the directions of the components of a vector along the x, y, and z axes, respectively.

A point $P(X, Y, Z)$ in space can be uniquely defined by its projections on the three axes as illustrated in Figure 2.8. The *position vector* $\vec{\mathbf{r}}$, a vector directed from the origin O to point P, can be expressed in terms of its components as

$$\vec{\mathbf{r}} = X\vec{\mathbf{a}}_x + Y\vec{\mathbf{a}}_y + Z\vec{\mathbf{a}}_z \tag{2.20}$$

where X, Y, and Z are the scalar projections of $\vec{\mathbf{r}}$ on the x, y, and z axes.

If A_x, A_y, and A_z are the scalar projections of $\vec{\mathbf{A}}$, as shown in Figure 2.9, then $\vec{\mathbf{A}}$ can be written as

$$\vec{\mathbf{A}} = A_x\vec{\mathbf{a}}_x + A_y\vec{\mathbf{a}}_y + A_z\vec{\mathbf{a}}_z \tag{2.21}$$

Similarly, we can express vector $\vec{\mathbf{B}}$ as

$$\vec{\mathbf{B}} = B_x\vec{\mathbf{a}}_x + B_y\vec{\mathbf{a}}_y + B_z\vec{\mathbf{a}}_z \tag{2.22}$$

The sum of two vectors $\vec{\mathbf{A}}$ and $\vec{\mathbf{B}}$, $\vec{\mathbf{C}} = \vec{\mathbf{A}} + \vec{\mathbf{B}}$, can now be written as

$$\begin{aligned}\vec{\mathbf{C}} &= (A_x + B_x)\vec{\mathbf{a}}_x + (A_y + B_y)\vec{\mathbf{a}}_y + (A_z + B_z)\vec{\mathbf{a}}_z \\ &= C_x\vec{\mathbf{a}}_x + C_y\vec{\mathbf{a}}_y + C_z\vec{\mathbf{a}}_z\end{aligned} \tag{2.23}$$

where $C_x = A_x + B_x$, $C_y = A_y + B_y$, and $C_z = A_z + B_z$ are the components of $\vec{\mathbf{C}}$ along the $\vec{\mathbf{a}}_x$, $\vec{\mathbf{a}}_y$, and $\vec{\mathbf{a}}_z$ unit vectors.

Since the three unit vectors are mutually orthogonal, the dot product yields

$$\vec{\mathbf{a}}_x \cdot \vec{\mathbf{a}}_x = 1, \vec{\mathbf{a}}_y \cdot \vec{\mathbf{a}}_y = 1, \vec{\mathbf{a}}_z \cdot \vec{\mathbf{a}}_z = 1 \tag{2.24a}$$

and

$$\vec{\mathbf{a}}_x \cdot \vec{\mathbf{a}}_y = \vec{\mathbf{a}}_y \cdot \vec{\mathbf{a}}_z = \vec{\mathbf{a}}_z \cdot \vec{\mathbf{a}}_x = 0 \tag{2.24b}$$

In addition, the cross product of the unit vectors yields

$$\vec{\mathbf{a}}_x \times \vec{\mathbf{a}}_x = \vec{\mathbf{a}}_y \times \vec{\mathbf{a}}_y = \vec{\mathbf{a}}_z \times \vec{\mathbf{a}}_z = 0 \tag{2.24c}$$

$$\vec{\mathbf{a}}_x \times \vec{\mathbf{a}}_y = \vec{\mathbf{a}}_z, \quad \vec{\mathbf{a}}_y \times \vec{\mathbf{a}}_z = \vec{\mathbf{a}}_x, \quad \text{and} \quad \vec{\mathbf{a}}_z \times \vec{\mathbf{a}}_x = \vec{\mathbf{a}}_y \tag{2.24d}$$

The dot product of vectors $\vec{\mathbf{A}}$ and $\vec{\mathbf{B}}$ in terms of their components is

$$\vec{\mathbf{A}} \cdot \vec{\mathbf{B}} = A_x B_x + A_y B_y + A_z B_z \tag{2.25}$$

Using equation (2.25), we can compute the magnitude of vector $\vec{\mathbf{A}}$ in terms of its components as

$$A = \sqrt{\vec{\mathbf{A}} \cdot \vec{\mathbf{A}}} = \sqrt{A_x^2 + A_y^2 + A_z^2} \tag{2.26}$$

EXAMPLE 2.4

Given $\vec{\mathbf{A}} = 3\vec{\mathbf{a}}_x + 2\vec{\mathbf{a}}_y - \vec{\mathbf{a}}_z$ and $\vec{\mathbf{B}} = \vec{\mathbf{a}}_x - 3\vec{\mathbf{a}}_y + 2\vec{\mathbf{a}}_z$, find $\vec{\mathbf{C}}$ such that $\vec{\mathbf{C}} = 2\vec{\mathbf{A}} - 3\vec{\mathbf{B}}$. Find the unit vector $\vec{\mathbf{a}}_c$ and the angle it makes with the z axis.

Solution

$$\begin{aligned}\vec{\mathbf{C}} &= 2\vec{\mathbf{A}} - 3\vec{\mathbf{B}} \\ &= 2[3\vec{\mathbf{a}}_x + 2\vec{\mathbf{a}}_y - \vec{\mathbf{a}}_z] - 3[\vec{\mathbf{a}}_x - 3\vec{\mathbf{a}}_y + 2\vec{\mathbf{a}}_z] \\ &= 3\vec{\mathbf{a}}_x + 13\vec{\mathbf{a}}_y - 8\vec{\mathbf{a}}_z\end{aligned}$$

The magnitude of vector $\vec{\mathbf{C}}$, from (2.26), is

$$C = \sqrt{3^2 + 13^2 + (-8)^2} = 15.556$$

The required unit vector is

$$\vec{\mathbf{a}}_c = \frac{\vec{\mathbf{C}}}{C} = 0.193\vec{\mathbf{a}}_x + 0.836\vec{\mathbf{a}}_y - 0.514\vec{\mathbf{a}}_z$$

The angle above unit vector makes with the z axis is

$$\theta_z = \cos^{-1}\left[\frac{C_z}{C}\right] = \cos^{-1}\left[\frac{-8}{15.556}\right] = 120.95^\circ$$

• • •

EXAMPLE 2.5

Show that the following vectors are orthogonal:

$$\vec{\mathbf{A}} = 4\vec{\mathbf{a}}_x + 6\vec{\mathbf{a}}_y - 2\vec{\mathbf{a}}_z \quad \text{and} \quad \vec{\mathbf{B}} = -2\vec{\mathbf{a}}_x + 4\vec{\mathbf{a}}_y + 8\vec{\mathbf{a}}_z$$

Solution

For the two nonzero vectors to be orthogonal, their scalar product $\vec{\mathbf{A}} \cdot \vec{\mathbf{B}}$ must be zero. Computing the scalar product, we obtain

$$\vec{\mathbf{A}} \cdot \vec{\mathbf{B}} = (4)(-2) + (6)(4) + (-2)(8) = 0$$

The scalar product is zero, therefore the vectors are orthogonal. • • •

EXAMPLE 2.6

Find the vector $\vec{\mathbf{R}}$ that is directed from a point $P(x_1, y_1, z_1)$ to a point $Q(x_2, y_2, z_2)$.

Solution A vector from one point to another is referred to as a *distance vector*. Let $\vec{\mathbf{r}}_1$ and $\vec{\mathbf{r}}_2$ be the position vectors of points P and Q, as depicted in Figure 2.10, then

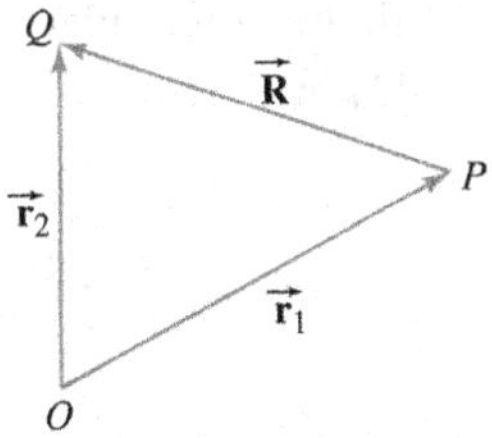

Figure 2.10 Distance vector $\vec{\mathbf{R}}$ from P to Q

$$\vec{\mathbf{r}}_1 = x_1\vec{\mathbf{a}}_x + y_1\vec{\mathbf{a}}_y + z_1\vec{\mathbf{a}}_z$$

and

$$\vec{\mathbf{r}}_2 = x_2\vec{\mathbf{a}}_x + y_2\vec{\mathbf{a}}_y + z_2\vec{\mathbf{a}}_z$$

From Figure 2.10, the distance vector $\vec{\mathbf{R}}$ from point P to Q is

$$\begin{aligned}\vec{\mathbf{R}} &= \vec{\mathbf{r}}_2 - \vec{\mathbf{r}}_1 \\ &= (x_2 - x_1)\vec{\mathbf{a}}_x + (y_2 - y_1)\vec{\mathbf{a}}_y + (z_2 - z_1)\vec{\mathbf{a}}_z\end{aligned}$$

• • •

The vector product of two vectors $\vec{\mathbf{A}}$ and $\vec{\mathbf{B}}$ can also be computed in terms of their projections onto the unit vectors. Let $\vec{\mathbf{C}} = \vec{\mathbf{A}} \times \vec{\mathbf{B}}$, then

$$\begin{aligned}\vec{\mathbf{C}} &= [A_x\vec{\mathbf{a}}_x + A_y\vec{\mathbf{a}}_y + A_z\vec{\mathbf{a}}_z] \times [B_x\vec{\mathbf{a}}_x + B_y\vec{\mathbf{a}}_y + B_z\vec{\mathbf{a}}_z] \\ &= [A_yB_z - A_zB_y]\vec{\mathbf{a}}_x + [A_zB_x - A_xB_z]\vec{\mathbf{a}}_y + [A_xB_y - A_yB_x]\vec{\mathbf{a}}_z\end{aligned}$$

The preceding equation can be conveniently expressed in terms of a determinant as

$$\vec{\mathbf{C}} = \vec{\mathbf{A}} \times \vec{\mathbf{B}} = \begin{vmatrix} \vec{\mathbf{a}}_x & \vec{\mathbf{a}}_y & \vec{\mathbf{a}}_z \\ A_x & A_y & A_z \\ B_x & B_y & B_z \end{vmatrix} \tag{2.27}$$

EXAMPLE 2.7

Calculate the volume of a parallelepiped formed by vectors $\vec{\mathbf{A}}$, $\vec{\mathbf{B}}$, and $\vec{\mathbf{C}}$ such that $\vec{\mathbf{A}} = 2\vec{\mathbf{a}}_x + \vec{\mathbf{a}}_y - 2\vec{\mathbf{a}}_z$, $\vec{\mathbf{B}} = -\vec{\mathbf{a}}_x + 3\vec{\mathbf{a}}_y + 5\vec{\mathbf{a}}_z$, and $\vec{\mathbf{C}} = 5\vec{\mathbf{a}}_x - 2\vec{\mathbf{a}}_y - 2\vec{\mathbf{a}}_z$.

Solution To compute the volume of a parallelepiped, we use the scalar triple product $\vec{\mathbf{A}} \cdot (\vec{\mathbf{B}} \times \vec{\mathbf{C}})$. With the help of (2.27), we can write the triple scalar product in the determinant form as

$$\vec{\mathbf{A}} \cdot (\vec{\mathbf{B}} \times \vec{\mathbf{C}}) = \begin{vmatrix} A_x & A_y & A_z \\ B_x & B_y & B_z \\ C_x & C_y & C_z \end{vmatrix}$$

Substituting the values, we obtain the required volume as

$$\text{volume} = \vec{\mathbf{A}} \cdot (\vec{\mathbf{B}} \times \vec{\mathbf{C}}) = \begin{vmatrix} 2 & 1 & -2 \\ -1 & 3 & 5 \\ 5 & -2 & -2 \end{vmatrix} = 57$$

• • •

2.4.2 Cylindrical coordinate system

A point $P(x, y, z)$ can also be completely represented in terms of ρ, ϕ, and z, as depicted in Figure 2.11. Note that ρ is the projection of r on the xy plane, ϕ is the angle from the positive x axis to the plane $OTPM$, and

z is the projection of r on the z axis. In this figure, r is the distance from O to P as shown. We speak of ρ, ϕ, and z as the cylindrical (circular) coordinates of point $P(\rho, \phi, z)$. From Figure 2.11, we can show that

$$x = \rho \cos \phi \qquad (2.28)$$
$$y = \rho \sin \phi$$

The coordinate surface

$$\rho = \sqrt{x^2 + y^2} = \text{constant} \qquad (2.29)$$

is a cylinder of radius ρ with the z axis as its axis, as illustrated in Figure 2.12. Thus, $0 \leq \rho \leq \infty$. The coordinate surface

$$\phi = \tan^{-1}\left(\frac{y}{x}\right) = \text{constant} \qquad (2.30)$$

is a plane hinged on the z axis, as shown in Figure 2.12. Finally, the coordinate surface

$$z = \text{constant} \qquad (2.31)$$

is a plane parallel to the xy plane. Because these surfaces intersect at right angles, they enable us to establish three mutually perpendicular coordinate axes: ρ, ϕ, and z. The corresponding unit vectors, as indicated in Figure 2.11, are $\vec{\mathbf{a}}_\rho$, $\vec{\mathbf{a}}_\phi$, and $\vec{\mathbf{a}}_z$. The angle ϕ is measured with respect to the x axis in the counterclockwise direction. Hence, ϕ varies from 0 to 2π. Note that the unit vectors $\vec{\mathbf{a}}_\rho$ and $\vec{\mathbf{a}}_\phi$ are not unidirectional; they change directions as ϕ increases or decreases. This fact must be borne in mind while integrating with respect to ϕ when the integrand has $\vec{\mathbf{a}}_\rho$ and $\vec{\mathbf{a}}_\phi$ directed components. We will reiterate this fact as the need arises.

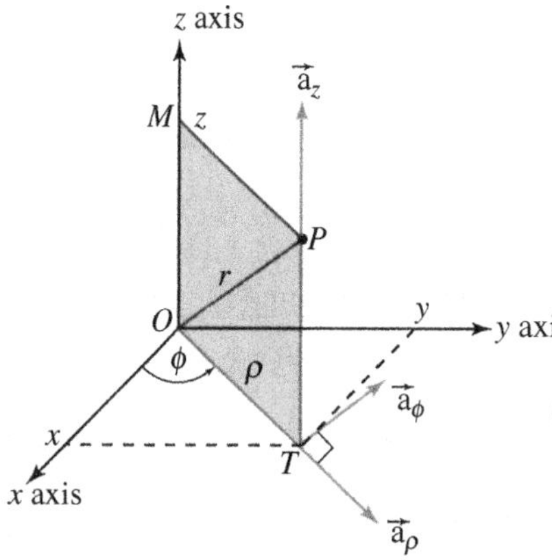

Figure 2.11 Projections of a point in a cylindrical coordinate system

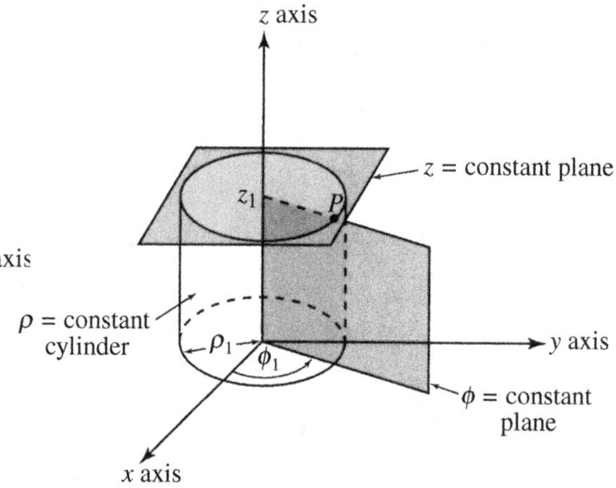

Figure 2.12 Three mutually perpendicular surfaces in the cylindrical coordinate system

If two vectors $\vec{\mathbf{A}}$ and $\vec{\mathbf{B}}$ are defined either at a common point $P(\rho, \phi, z)$ or in a $\phi =$ constant plane, we can add, subtract, and multiply these vectors as we did in the rectangular coordinate system. For example, if the two vectors at point $P(\rho, \phi, z)$ are $\vec{\mathbf{A}} = A_\rho \vec{\mathbf{a}}_\rho + A_\phi \vec{\mathbf{a}}_\phi + A_z \vec{\mathbf{a}}_z$ and $\vec{\mathbf{B}} = B_\rho \vec{\mathbf{a}}_\rho + B_\phi \vec{\mathbf{a}}_\phi + B_z \vec{\mathbf{a}}_z$, then

$$\vec{\mathbf{A}} + \vec{\mathbf{B}} = (A_\rho + B_\rho)\vec{\mathbf{a}}_\rho + (A_\phi + B_\phi)\vec{\mathbf{a}}_\phi + (A_z + B_z)\vec{\mathbf{a}}_z \tag{2.32a}$$

$$\vec{\mathbf{A}} \cdot \vec{\mathbf{B}} = A_\rho B_\rho + A_\phi B_\phi + A_z B_z \tag{2.32b}$$

and

$$\vec{\mathbf{A}} \times \vec{\mathbf{B}} = \begin{vmatrix} \vec{\mathbf{a}}_\rho & \vec{\mathbf{a}}_\phi & \vec{\mathbf{a}}_z \\ A_\rho & A_\phi & A_z \\ B_\rho & B_\phi & B_z \end{vmatrix} \tag{2.32c}$$

The dot and cross products of the unit vectors in the cylindrical coordinate system are

$$\vec{\mathbf{a}}_\rho \cdot \vec{\mathbf{a}}_\rho = 1 \qquad \vec{\mathbf{a}}_\phi \cdot \vec{\mathbf{a}}_\phi = 1 \qquad \vec{\mathbf{a}}_z \cdot \vec{\mathbf{a}}_z = 1 \tag{2.33a}$$

$$\vec{\mathbf{a}}_\rho \cdot \vec{\mathbf{a}}_\phi = 0 \qquad \vec{\mathbf{a}}_\phi \cdot \vec{\mathbf{a}}_z = 0 \qquad \vec{\mathbf{a}}_z \cdot \vec{\mathbf{a}}_\rho = 0 \tag{2.33b}$$

$$\vec{\mathbf{a}}_\rho \times \vec{\mathbf{a}}_\rho = 0 \qquad \vec{\mathbf{a}}_\phi \times \vec{\mathbf{a}}_\phi = 0 \qquad \vec{\mathbf{a}}_z \times \vec{\mathbf{a}}_z = 0 \tag{2.34a}$$

$$\vec{\mathbf{a}}_\rho \times \vec{\mathbf{a}}_\phi = \vec{\mathbf{a}}_z \qquad \vec{\mathbf{a}}_\phi \times \vec{\mathbf{a}}_z = \vec{\mathbf{a}}_\rho \qquad \vec{\mathbf{a}}_z \times \vec{\mathbf{a}}_\rho = \vec{\mathbf{a}}_\phi \tag{2.34b}$$

Transformation of unit vectors

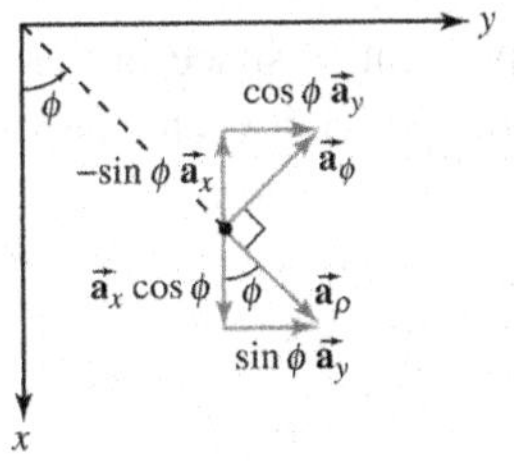

Figure 2.13 Components of $\vec{\mathbf{a}}_\rho$ and $\vec{\mathbf{a}}_\phi$ along the $\vec{\mathbf{a}}_x$ and $\vec{\mathbf{a}}_y$ directions

The projections of unit vectors $\vec{\mathbf{a}}_\rho$ and $\vec{\mathbf{a}}_\phi$ onto the unit vectors $\vec{\mathbf{a}}_x$ and $\vec{\mathbf{a}}_y$ are shown in Figure 2.13. From the projections, it is obvious that

$$\vec{\mathbf{a}}_\rho = \cos\phi\,\vec{\mathbf{a}}_x + \sin\phi\,\vec{\mathbf{a}}_y \tag{2.35a}$$

and

$$\vec{\mathbf{a}}_\phi = -\sin\phi\,\vec{\mathbf{a}}_x + \cos\phi\,\vec{\mathbf{a}}_y \tag{2.35b}$$

because $\vec{\mathbf{a}}_x \cdot \vec{\mathbf{a}}_\rho = \cos\phi$, $\vec{\mathbf{a}}_y \cdot \vec{\mathbf{a}}_\rho = \sin\phi$, $\vec{\mathbf{a}}_x \cdot \vec{\mathbf{a}}_\phi = -\sin\phi$, and $\vec{\mathbf{a}}_y \cdot \vec{\mathbf{a}}_\phi = \cos\phi$.

The unit vector transformation from the rectangular to cylindrical coordinate system can be expressed in matrix form as

$$\begin{bmatrix} \vec{\mathbf{a}}_\rho \\ \vec{\mathbf{a}}_\phi \\ \vec{\mathbf{a}}_z \end{bmatrix} = \begin{bmatrix} \cos\phi & \sin\phi & 0 \\ -\sin\phi & \cos\phi & 0 \\ 0 & 0 & 1 \end{bmatrix} \begin{bmatrix} \vec{\mathbf{a}}_x \\ \vec{\mathbf{a}}_y \\ \vec{\mathbf{a}}_z \end{bmatrix} \tag{2.36}$$

Transformation of a vector

If a vector $\vec{\mathbf{A}}$ is given in the cylindrical coordinate system, it can be expressed in the rectangular coordinate system by projecting it onto the x, y, and z axes. That is, the scalar projection of $\vec{\mathbf{A}}$ onto the

x axis is

$$A_x = \vec{\mathbf{A}} \cdot \vec{\mathbf{a}}_x = A_\rho \vec{\mathbf{a}}_\rho \cdot \vec{\mathbf{a}}_x + A_\phi \vec{\mathbf{a}}_\phi \cdot \vec{\mathbf{a}}_x + A_z \vec{\mathbf{a}}_z \cdot \vec{\mathbf{a}}_x$$
$$= A_\rho \cos\phi - A_\phi \sin\phi \qquad (2.37a)$$

Similarly, the scalar projection of $\vec{\mathbf{A}}$ onto the y axis is

$$A_y = \vec{\mathbf{A}} \cdot \vec{\mathbf{a}}_y = A_\rho \sin\phi + A_\phi \cos\phi \qquad (2.37b)$$

Finally, the scalar projection of $\vec{\mathbf{A}}$ onto the z axis is

$$A_z = \vec{\mathbf{A}} \cdot \vec{\mathbf{a}}_z = A_z \qquad (2.37c)$$

We can write (2.37) concisely in matrix form as

$$\begin{bmatrix} A_x \\ A_y \\ A_z \end{bmatrix} = \begin{bmatrix} \cos\phi & -\sin\phi & 0 \\ \sin\phi & \cos\phi & 0 \\ 0 & 0 & 1 \end{bmatrix} \begin{bmatrix} A_\rho \\ A_\phi \\ A_z \end{bmatrix} \qquad (2.38)$$

By following a similar procedure, a vector in the rectangular coordinate system can be expressed in the cylindrical coordinate system by the following transformation:

$$\begin{bmatrix} A_\rho \\ A_\phi \\ A_z \end{bmatrix} = \begin{bmatrix} \cos\phi & \sin\phi & 0 \\ -\sin\phi & \cos\phi & 0 \\ 0 & 0 & 1 \end{bmatrix} \begin{bmatrix} A_x \\ A_y \\ A_z \end{bmatrix} \qquad (2.39)$$

Note that the transformation matrix in (2.39) is the same as that in (2.36).

EXAMPLE 2.8

Write an expression for a position vector at any point in space in the rectangular coordinate system. Then transform the position vector into a vector in the cylindrical coordinate system.

Solution The position vector of any point $P(x, y, z)$ in space is

$$\vec{\mathbf{A}} = x\vec{\mathbf{a}}_x + y\vec{\mathbf{a}}_y + z\vec{\mathbf{a}}_z$$

Using the transformation matrix as given in (2.39), we obtain

$$A_\rho = x\cos\phi + y\sin\phi$$
$$A_\phi = -x\sin\phi + y\cos\phi \quad \text{and} \quad A_z = z$$

Substituting $x = p\cos\phi$ and $y = \rho\sin\phi$, we obtain

$$A_\rho = \rho, \quad A_\phi = 0, \quad \text{and} \quad A_z = z$$

Thus, the position vector $\vec{\mathbf{A}}$ in the cylindrical coordinate system is

$$\vec{\mathbf{A}} = \rho\vec{\mathbf{a}}_\rho + z\vec{\mathbf{a}}_z$$

• • •

EXAMPLE 2.9

Express the vector $\vec{\mathbf{A}} = \dfrac{k}{\rho^2}\vec{\mathbf{a}}_\rho + 5\sin 2\phi\,\vec{\mathbf{a}}_z$ in the rectangular coordinate system.

Solution Using the transformation matrix as given in (2.38) with

$$A_\rho = \frac{k}{\rho^2}, \quad A_\phi = 0, \quad \text{and} \quad A_z = 5\sin 2\phi$$

we obtain

$$A_x = \frac{k\cos\phi}{\rho^2}, \quad A_y = \frac{k\sin\phi}{\rho^2}, \quad \text{and} \quad A_z = 10\cos\phi\sin\phi$$

Substituting $\rho = \sqrt{x^2+y^2}$, $\cos\phi = \dfrac{x}{\rho}$, and $\sin\phi = \dfrac{y}{\rho}$, we obtain the desired transformation of vector $\vec{\mathbf{A}}$ as

$$\vec{\mathbf{A}} = \frac{kx}{[x^2+y^2]^{3/2}}\vec{\mathbf{a}}_x + \frac{ky}{[x^2+y^2]^{3/2}}\vec{\mathbf{a}}_y + \frac{10xy}{x^2+y^2}\vec{\mathbf{a}}_z$$

• • •

EXAMPLE 2.10

If $\vec{\mathbf{A}} = 3\vec{\mathbf{a}}_\rho + 2\vec{\mathbf{a}}_\phi + 5\vec{\mathbf{a}}_z$ and $\vec{\mathbf{B}} = -2\vec{\mathbf{a}}_\rho + 3\vec{\mathbf{a}}_\phi - \vec{\mathbf{a}}_z$ are given at points $P(3, \pi/6, 5)$ and $Q(4, \pi/3, 3)$, find $\vec{\mathbf{C}} = \vec{\mathbf{A}} + \vec{\mathbf{B}}$ at point $S(2, \pi/4, 4)$.

Solution The two vectors are not defined in the same $\phi =$ constant plane, so we cannot sum them directly in the cylindrical system. Conversion to the rectangular system is therefore necessary. For vector $\vec{\mathbf{A}}$ given at point $P(3, \pi/6, 5)$, the transformation matrix becomes

$$\begin{bmatrix} A_x \\ A_y \\ A_z \end{bmatrix} = \begin{bmatrix} \cos 30^\circ & -\sin 30^\circ & 0 \\ \sin 30^\circ & \cos 30^\circ & 0 \\ 0 & 0 & 1 \end{bmatrix} \begin{bmatrix} 3 \\ 2 \\ 5 \end{bmatrix}$$

$$\vec{\mathbf{A}} = 1.598\vec{\mathbf{a}}_x + 3.232\vec{\mathbf{a}}_y + 5\vec{\mathbf{a}}_z$$

Similarly, with $\phi = \pi/3$, the transformed vector $\vec{\mathbf{B}}$ is

$$\vec{\mathbf{B}} = -3.598\vec{\mathbf{a}}_x - 0.232\vec{\mathbf{a}}_y - \vec{\mathbf{a}}_z$$

Now we can compute $\vec{\mathbf{C}} = \vec{\mathbf{A}} + \vec{\mathbf{B}}$ in the rectangular coordinate system as

$$\vec{\mathbf{C}} = -2\vec{\mathbf{a}}_x + 3\vec{\mathbf{a}}_y + 4\vec{\mathbf{a}}_z$$

Vector $\vec{\mathbf{C}}$ can now be transformed into its components at point $S(2, \pi/4, 4)$ in the cylindrical system by making use of the transformation matrix given in (2.39). That is

$$\begin{bmatrix} C_\rho \\ C_\phi \\ C_z \end{bmatrix} = \begin{bmatrix} \cos 45^\circ & \sin 45^\circ & 0 \\ -\sin 45^\circ & \cos 45^\circ & 0 \\ 0 & 0 & 1 \end{bmatrix} \begin{bmatrix} -2 \\ 3 \\ 4 \end{bmatrix}$$

Thus, $\vec{\mathbf{C}} = 0.707\vec{\mathbf{a}}_\rho + 3.535\vec{\mathbf{a}}_\phi + 4\vec{\mathbf{a}}_z$

• • •

Note that the transformation of a vector from one coordinate system to another neither changes its magnitude nor its direction.

2.4.3 Spherical coordinate system

The last coordinate system we discuss is the spherical coordinate system. A point P in space in spherical coordinates is uniquely represented in terms of r, θ, and ϕ, as illustrated in Figure 2.14, where r is the radial distance from O to P, θ is the angle that r makes with the positive z axis, and ϕ is the angle between the positive xz and $OMPN$ planes as shown. The projection of r onto the xy plane is $OM = r\sin\theta$. From Figure 2.14 it is apparent that

$$x = r\sin\theta\cos\phi \tag{2.40a}$$

$$y = r\sin\theta\sin\phi \tag{2.40b}$$

$$z = r\cos\theta \tag{2.40c}$$

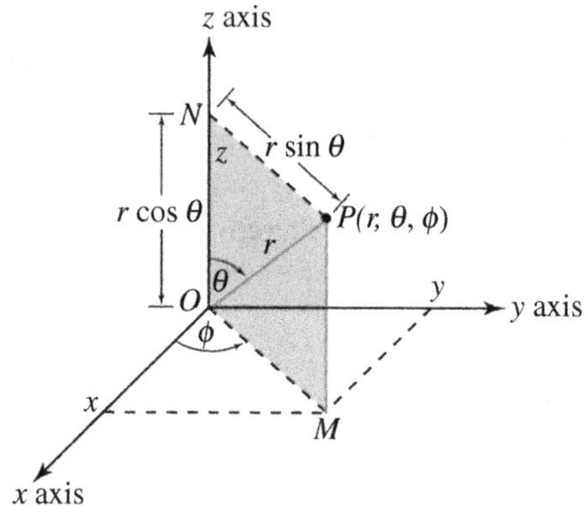

Figure 2.14 Projections of a point in a spherical coordinate system

From (2.40), we can deduce that

$$r = \sqrt{x^2 + y^2 + z^2} \tag{2.41a}$$

$$\theta = \cos^{-1}\left[\frac{z}{r}\right] \tag{2.41b}$$

$$\phi = \tan^{-1}\left[\frac{y}{x}\right] \tag{2.41c}$$

The positive direction of ϕ is that of the right-handed rotation from x to y about the z axis. Thus, ϕ varies from 0 to 2π. The positive direction of θ is from the positive z axis, where its value is zero, toward the negative z axis, where its value is π. Hence, θ varies from 0 to π. However, $0 \leq r \leq \infty$.

Through the point $P(r, \theta, \phi)$ pass the surface of a sphere of radius r, the surface of a cone of aperture θ with apex at the origin, and a plane hinged on the z axis making an angle ϕ with the xz plane, as indicated in Figure 2.15. The tangent planes to these surfaces at point P are mutually perpendicular. The unit vectors perpendicular to these intersecting planes in the increasing directions of r, θ, and ϕ are $\vec{\mathbf{a}}_r, \vec{\mathbf{a}}_\theta$,

Figure 2.15 Spherical coordinate system

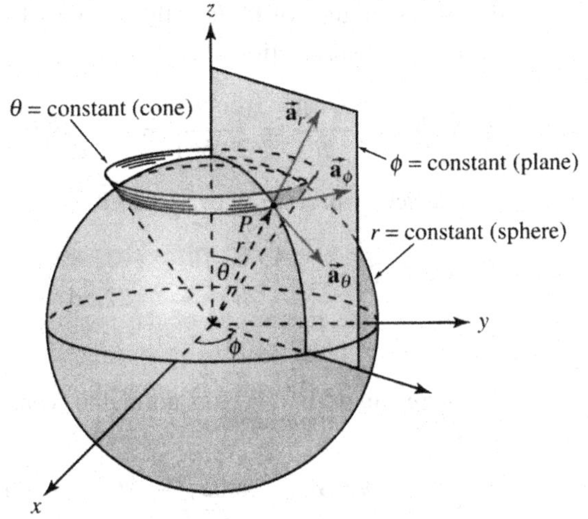

and $\vec{\mathbf{a}}_\phi$, respectively. These unit vectors are, therefore, functions of the coordinates (r, θ, ϕ). *Thus, the vector addition, subtraction, and multiplication of any two vectors in spherical coordinates can only be performed if these vectors are given at the intersection of θ = constant and ϕ = constant planes.* In other words, the vectors must be defined either at the same point or at points along the same radial line.

The scalar and vector products of the unit vectors are as follows; the student is expected to verify them.

$$\vec{\mathbf{a}}_r \cdot \vec{\mathbf{a}}_r = 1 \qquad \vec{\mathbf{a}}_\theta \cdot \vec{\mathbf{a}}_\theta = 1 \qquad \vec{\mathbf{a}}_\phi \cdot \vec{\mathbf{a}}_\phi = 1 \tag{2.42a}$$

$$\vec{\mathbf{a}}_r \cdot \vec{\mathbf{a}}_\theta = 0 \qquad \vec{\mathbf{a}}_\theta \cdot \vec{\mathbf{a}}_\phi = 0 \qquad \vec{\mathbf{a}}_\phi \cdot \vec{\mathbf{a}}_r = 0 \tag{2.42b}$$

$$\vec{\mathbf{a}}_r \times \vec{\mathbf{a}}_\theta = \vec{\mathbf{a}}_\phi \qquad \vec{\mathbf{a}}_\theta \times \vec{\mathbf{a}}_\phi = \vec{\mathbf{a}}_r \qquad \vec{\mathbf{a}}_\phi \times \vec{\mathbf{a}}_r = \vec{\mathbf{a}}_\theta \tag{2.42c}$$

EXAMPLE 2.11

Two vectors $\vec{\mathbf{A}}$ and $\vec{\mathbf{B}}$ are given at a point $P(r, \theta, \phi)$ in space as

$$\vec{\mathbf{A}} = 10\vec{\mathbf{a}}_r + 30\vec{\mathbf{a}}_\theta - 10\vec{\mathbf{a}}_\phi \quad \text{and} \quad \vec{\mathbf{B}} = -3\vec{\mathbf{a}}_r - 10\vec{\mathbf{a}}_\theta + 20\vec{\mathbf{a}}_\phi$$

Determine (a) $2\vec{\mathbf{A}} - 5\vec{\mathbf{B}}$, (b) $\vec{\mathbf{A}} \cdot \vec{\mathbf{B}}$, (c) $\vec{\mathbf{A}} \times \vec{\mathbf{B}}$, (d) the scalar component of $\vec{\mathbf{A}}$ in the direction of $\vec{\mathbf{B}}$, (e) the vector projection of $\vec{\mathbf{A}}$ in the direction of $\vec{\mathbf{B}}$, and (f) a unit vector perpendicular to both $\vec{\mathbf{A}}$ and $\vec{\mathbf{B}}$.

Solution Both vectors $\vec{\mathbf{A}}$ and $\vec{\mathbf{B}}$ are given at the same point P, so the rules of vector operations can be applied directly in the spherical coordinate system.

a) $2\vec{\mathbf{A}} - 5\vec{\mathbf{B}} = (20 + 15)\vec{\mathbf{a}}_r + (60 + 50)\vec{\mathbf{a}}_\theta + (-20 - 100)\vec{\mathbf{a}}_\phi$
$= 35\vec{\mathbf{a}}_r + 110\vec{\mathbf{a}}_\theta - 120\vec{\mathbf{a}}_\phi$

b) $\vec{\mathbf{A}} \cdot \vec{\mathbf{B}} = 10(-3) + 30(-10) + (-10)20 = -530$

c) $$\vec{\mathbf{A}} \times \vec{\mathbf{B}} = \begin{vmatrix} \vec{\mathbf{a}}_r & \vec{\mathbf{a}}_\theta & \vec{\mathbf{a}}_\phi \\ 10 & 30 & -10 \\ -3 & -10 & 20 \end{vmatrix} = 500\vec{\mathbf{a}}_r - 170\vec{\mathbf{a}}_\theta - 10\vec{\mathbf{a}}_\phi$$

d) The magnitude of $\vec{\mathbf{B}}$: $B = [(-3)^2 + (-10)^2 + (20)^2]^{1/2} = 22.561$
The scalar projection of $\vec{\mathbf{A}}$ onto $\vec{\mathbf{B}}$ is

$$\vec{\mathbf{A}} \cdot \vec{\mathbf{a}}_B = \frac{\vec{\mathbf{A}} \cdot \vec{\mathbf{B}}}{B} = \frac{-530}{22.561} = -23.492$$

e) The vector projection of $\vec{\mathbf{A}}$ onto $\vec{\mathbf{B}}$ is

$$(\vec{\mathbf{A}} \cdot \vec{\mathbf{a}}_B)\vec{\mathbf{a}}_B = \frac{(\vec{\mathbf{A}} \cdot \vec{\mathbf{a}}_B)\vec{\mathbf{B}}}{B} = \frac{-23.492}{22.561}[-3\vec{\mathbf{a}}_r - 10\vec{\mathbf{a}}_\theta + 20\vec{\mathbf{a}}_\phi]$$
$$= 3.123\vec{\mathbf{a}}_r + 10.413\vec{\mathbf{a}}_\theta - 20.825\vec{\mathbf{a}}_\phi$$

f) There are two unit vectors normal to $\vec{\mathbf{A}}$ and $\vec{\mathbf{B}}$. One of the unit vectors is

$$\vec{\mathbf{a}}_{n1} = \frac{\vec{\mathbf{A}} \times \vec{\mathbf{B}}}{|\vec{\mathbf{A}} \times \vec{\mathbf{B}}|} = \frac{500\vec{\mathbf{a}}_r - 170\vec{\mathbf{a}}_\theta - 10\vec{\mathbf{a}}_\phi}{[500^2 + 170 + 10^2]^{1/2}}$$
$$= 0.947\vec{\mathbf{a}}_r - 0.322\vec{\mathbf{a}}_\theta - 0.019\vec{\mathbf{a}}_\phi$$

The other unit vector is

$$\vec{\mathbf{a}}_{n2} = -\vec{\mathbf{a}}_{n1} = -0.947\vec{\mathbf{a}}_r + 0.322\vec{\mathbf{a}}_\theta + 0.019\vec{\mathbf{a}}_\phi$$

• • •

Transformation of unit vectors

When a set of vectors is given in spherical coordinates at different points but not along the same radial line, we have to express the vectors in rectangular coordinates in order to perform the basic vector operations. The components of the three unit vectors $\vec{\mathbf{a}}_r$, $\vec{\mathbf{a}}_\theta$, and $\vec{\mathbf{a}}_\phi$ along $\vec{\mathbf{a}}_x$, $\vec{\mathbf{a}}_y$, and $\vec{\mathbf{a}}_z$ can be obtained from Figure 2.16 (page 31). From these illustrations, we can show that

$$\begin{aligned}
&\vec{\mathbf{a}}_r \cdot \vec{\mathbf{a}}_x = \sin\theta\cos\phi, && \vec{\mathbf{a}}_r \cdot \vec{\mathbf{a}}_y = \sin\theta\sin\phi, && \vec{\mathbf{a}}_r \cdot \vec{\mathbf{a}}_z = \cos\theta \\
&\vec{\mathbf{a}}_\theta \cdot \vec{\mathbf{a}}_x = \cos\theta\cos\phi, && \vec{\mathbf{a}}_\theta \cdot \vec{\mathbf{a}}_y = \cos\theta\sin\phi, && \vec{\mathbf{a}}_\theta \cdot \vec{\mathbf{a}}_z = -\sin\theta \\
&\vec{\mathbf{a}}_\phi \cdot \vec{\mathbf{a}}_x = -\sin\phi, && \vec{\mathbf{a}}_\phi \cdot \vec{\mathbf{a}}_y = \cos\phi, && \vec{\mathbf{a}}_\phi \cdot \vec{\mathbf{a}}_z = 0
\end{aligned} \tag{2.43a}$$

These equations can be written in matrix form as

$$\begin{bmatrix} \vec{\mathbf{a}}_r \\ \vec{\mathbf{a}}_\theta \\ \vec{\mathbf{a}}_\phi \end{bmatrix} = \begin{bmatrix} \sin\theta\cos\phi & \sin\theta\sin\phi & \cos\theta \\ \cos\theta\cos\phi & \cos\theta\sin\phi & -\sin\theta \\ -\sin\phi & \cos\phi & 0 \end{bmatrix} \begin{bmatrix} \vec{\mathbf{a}}_x \\ \vec{\mathbf{a}}_y \\ \vec{\mathbf{a}}_z \end{bmatrix} \tag{2.43b}$$

Transformation of a vector

If vector $\vec{\mathbf{A}}$ is given in spherical coordinates as

$$\vec{\mathbf{A}} = A_r\vec{\mathbf{a}}_r + A_\theta\vec{\mathbf{a}}_\theta + A_\phi\vec{\mathbf{a}}_\phi$$

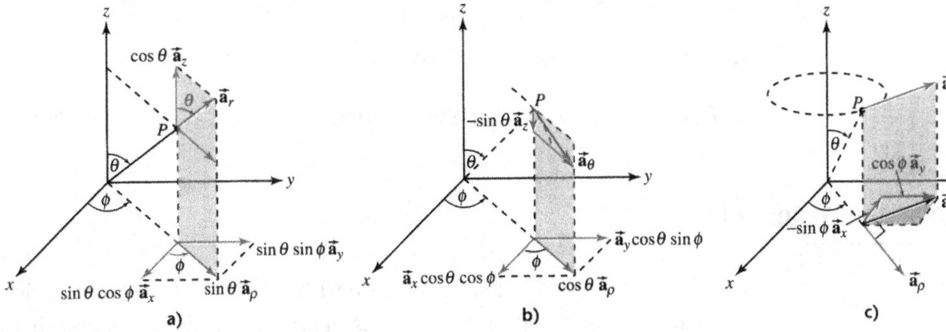

Figure 2.16 Projections of (a) $\vec{\mathbf{a}}_r$, (b) $\vec{\mathbf{a}}_\theta$, and (c) $\vec{\mathbf{a}}_\phi$ onto the unit vectors $\vec{\mathbf{a}}_x$, $\vec{\mathbf{a}}_y$, and $\vec{\mathbf{a}}_z$

we can obtain the x component of $\vec{\mathbf{A}}$ by projecting it onto the x axis as

$$\begin{aligned} A_x = \vec{\mathbf{A}} \cdot \vec{\mathbf{a}}_x &= A_r \vec{\mathbf{a}}_r \cdot \vec{\mathbf{a}}_x + A_\theta \vec{\mathbf{a}}_\theta \cdot \vec{\mathbf{a}}_x + A_\phi \vec{\mathbf{a}}_\phi \cdot \vec{\mathbf{a}}_x \\ &= A_r \sin\theta\cos\phi + A_\theta \cos\theta\cos\phi - A_\phi \sin\phi \end{aligned}$$

We can obtain the other components in a similar fashion. The results in matrix form are

$$\begin{bmatrix} A_x \\ A_y \\ A_z \end{bmatrix} = \begin{bmatrix} \sin\theta\cos\phi & \cos\theta\cos\phi & -\sin\phi \\ \sin\theta\sin\phi & \cos\theta\sin\phi & \cos\phi \\ \cos\theta & -\sin\theta & 0 \end{bmatrix} \begin{bmatrix} A_r \\ A_\theta \\ A_\phi \end{bmatrix} \tag{2.44}$$

Likewise, a vector given in the rectangular coordinate system can be expressed in terms of a vector in the spherical coordinate system by using the following matrix transformation. The student is encouraged to verify these results using the projection techniques.

$$\begin{bmatrix} A_r \\ A_\theta \\ A_\phi \end{bmatrix} = \begin{bmatrix} \sin\theta\cos\phi & \sin\theta\sin\phi & \cos\theta \\ \cos\theta\cos\phi & \cos\theta\sin\phi & -\sin\theta \\ -\sin\phi & \cos\phi & 0 \end{bmatrix} \begin{bmatrix} A_x \\ A_y \\ A_z \end{bmatrix} \tag{2.45}$$

EXAMPLE 2.12

A vector $\vec{\mathbf{F}} = 3x\vec{\mathbf{a}}_x + 0.5y^2\vec{\mathbf{a}}_y + 0.25x^2y^2\vec{\mathbf{a}}_z$ is given at a point $P(3, 4, 12)$ in the rectangular coordinate system. Express this vector in the spherical coordinate system.

Solution The vector $\vec{\mathbf{F}}$ at point $P(3, 4, 12)$ is $\vec{\mathbf{F}} = 9\vec{\mathbf{a}}_x + 8\vec{\mathbf{a}}_y + 36\vec{\mathbf{a}}_z$. Also,

$$\phi = \tan^{-1}\left[\frac{4}{3}\right] = 53.13^\circ \quad \text{and} \quad \theta = \cos^{-1}\left[\frac{12}{13}\right] = 22.62^\circ$$

Substituting the values in (2.45), we obtain

$$F_r = 37.77, \quad F_\theta = -2.95, \quad \text{and} \quad F_\phi = -2.40$$

or

$$\vec{\mathbf{F}} = 37.77\vec{\mathbf{a}}_r - 2.95\vec{\mathbf{a}}_\theta - 2.40\vec{\mathbf{a}}_\phi$$

at $P(13, 22.62°, 53.13°)$ in the spherical coordinate system. • • •

2.5 Scalar and vector fields

All the vector operations we have considered thus far are applicable to functions commonly referred to as *fields*. A field is a function that describes a physical quantity at all points in space. A physical quantity can be either a scalar or a vector; thus, a field can also be a scalar field or a vector field.

Scalar fields

A **scalar field** is specified by a single number at each point. Some well-known examples of scalar fields include temperature and pressure of a gas, the altitude above sea level, and electric potential. For example, in Chapter 3, we will show that the potential distribution within a parallel-plate capacitor (two parallel conducting plates separated by an insulating medium; see Figure 2.17a) is a linear function of the distance between the conducting plates, as illustrated in Figure 2.17b. Thus, the equipotential surfaces are planes parallel to the conductors. By the way, an equipotential surface is a surface on which there is no change in the potential.

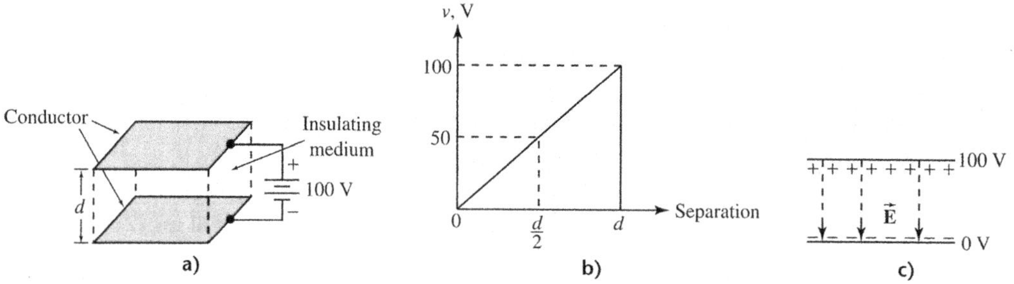

Figure 2.17 (a) Parallel-plate capacitor, (b) potential distribution, (c) electric field intensity

Vector fields

A **vector field** is specified by both a magnitude and a direction at each point in space. The velocity and acceleration of a fluid, the gravitational force, and the electric field within a coaxial cable are some examples of vector fields. We will also show in Chapter 3 that the electric field intensity within a parallel-plate capacitor is constant and is directed from the higher potential conductor toward the lower potential conductor, as shown in Figure 2.17c.

Static fields

If a field does not vary with time we refer to it as a *static field*. Static fields are also known as *time-invariant fields*. We discuss the fields produced by stationary charges (electrostatics) in Chapter 3 and the fields created by a steady motion of charges (magnetostatics) in Chapter 5.

Time-varying fields

When a field varies with time, we refer to it as a *time-varying field*. Most of this text is devoted to the study of time-varying electromagnetic (coupled electric and magnetic) fields.

Vector calculus

Before we begin our discussion of vector calculus, it is important to define the derivative of a function of one or more variables. The derivative of a scalar function $f(s)$ with respect to s is defined as

$$\frac{df}{ds} = \lim_{\Delta s \to 0} \frac{f(s + \Delta s) - f(s)}{\Delta s} \tag{2.46}$$

Let us now suppose that f is a function of two variables u and v, and each variable depends continuously on s; i.e., $f = f(u(s), v(s))$. Then the derivative of f with respect to s is defined as

$$\frac{df}{ds} = \frac{\partial f}{\partial u}\frac{du}{ds} + \frac{\partial f}{\partial v}\frac{dv}{ds} \tag{2.47}$$

where $\partial f/\partial u$ is the partial derivative of f with respect to u for a fixed value of v, and $\partial f/\partial v$ is the partial derivative of f with respect to v for a fixed value of u. From (2.46), we obtain the definition of the partial derivative of f with respect to u as

$$\frac{\partial f}{\partial u} = \lim_{\Delta u \to 0} \frac{f(u + \Delta u, v) - f(u, v)}{\Delta u} \tag{2.48}$$

We can obtain a similar expression for $\partial f/\partial v$. We will employ these equations to define the gradient and Laplacian of a scalar function.

We define the derivative of a vector field $\vec{\mathbf{F}}(s)$, a function of a scalar s, with respect to s as

$$\frac{d\vec{\mathbf{F}}}{ds} = \lim_{\Delta s \to 0} \frac{\vec{\mathbf{F}}(s + \Delta s) - \vec{\mathbf{F}}(s)}{\Delta s} \tag{2.49}$$

Now suppose $\vec{\mathbf{F}}$ is a function of position coordinates x, y, and z. Then using the definition of partial differentiation, we can write $\partial\vec{\mathbf{F}}/\partial x$ as

$$\frac{\partial \vec{\mathbf{F}}}{\partial x} = \lim_{\Delta x \to 0} \frac{\vec{\mathbf{F}}(x + \Delta x, y, z) - \vec{\mathbf{F}}(x, y, z)}{\Delta x} \tag{2.50}$$

Similar expressions for $\partial\vec{\mathbf{F}}/\partial y$ and $\partial\vec{\mathbf{F}}/\partial z$ can also be written. We will use (2.50) to define the divergence and curl of a vector.

If a scalar or a vector field of one or more variables can be differentiated, the inverse must also be true; that is, we must be able to integrate a scalar or a vector field. In fact, a physical interpretation of divergence can be obtained from the integration of a vector over the surface. To perform this integration, we must be able to define the differential surface element. Therefore, we now digress and devote the next section to defining the differential elements of length, surface, and volume in the rectangular, cylindrical, and spherical coordinate systems.

2.6 Differential elements of length, surface, and volume

In our study of electromagnetism we will often be required to perform line, surface, and volume integrations. The evaluation of these integrals in a particular coordinate system requires the knowledge of differential elements of length, surface, and volume. In the following subsections we describe how these differential elements are constructed in each coordinate system.

2.6.1 Rectangular coordinate system

A differential volume element in the rectangular coordinate system is generated by making differential changes dx, dy, and dz along the unit vectors $\vec{\mathbf{a}}_x$, $\vec{\mathbf{a}}_y$ and $\vec{\mathbf{a}}_z$, respectively, as illustrated in Figure 2.18a. The differential volume is given by the expression

$$dv = dx\,dy\,dz \tag{2.51}$$

The volume is enclosed by six differential surfaces. Each surface is defined by a unit vector normal to that surface. Thus, we can express the differential surfaces in the direction of positive unit vectors

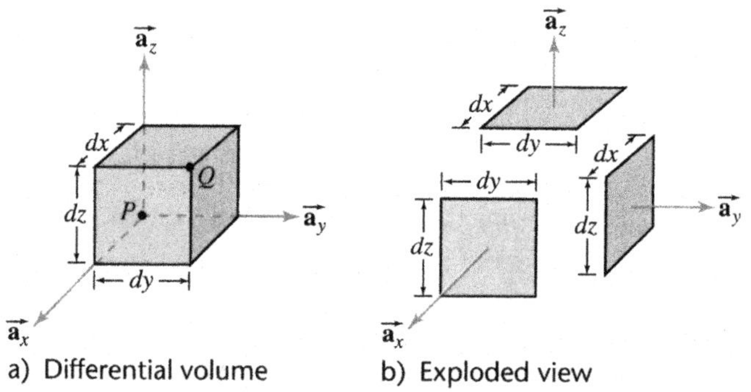

Figure 2.18 Differential elements in a rectangular coordinate system

(see Figure 2.18b) as

$$\vec{ds}_x = dy\,dz\,\vec{\mathbf{a}}_x$$
$$\vec{ds}_y = dx\,dz\,\vec{\mathbf{a}}_y \qquad (2.52)$$
$$\vec{ds}_z = dx\,dy\,\vec{\mathbf{a}}_z$$

The general differential length element from P to Q is

$$\vec{d\ell} = dx\vec{\mathbf{a}}_x + dy\vec{\mathbf{a}}_y + dz\vec{\mathbf{a}}_z \qquad (2.53)$$

2.6.2 Cylindrical coordinate system

Figure 2.19a shows the differential volume bounded by the surfaces at $\rho, \rho + d\rho, \phi, \phi + d\phi, z,$ and $z + dz$. The differential volume enclosed is

$$dv = \rho\,d\rho\,d\phi\,dz \qquad (2.54)$$

The differential surfaces in the positive direction of the unit vectors (Fig. 2.19b) are

$$\vec{ds}_\rho = \rho\,d\phi\,dz\,\vec{\mathbf{a}}_\rho$$
$$\vec{ds}_\phi = d\rho\,dz\,\vec{\mathbf{a}}_\phi \qquad (2.55)$$
$$\vec{ds}_z = \rho\,d\rho\,d\phi\,\vec{\mathbf{a}}_z$$

The differential length vector from P to Q is

$$\vec{d\ell} = d\rho\,\vec{\mathbf{a}}_\rho + \rho\,d\phi\,\vec{\mathbf{a}}_\phi + dz\,\vec{\mathbf{a}}_z \qquad (2.56)$$

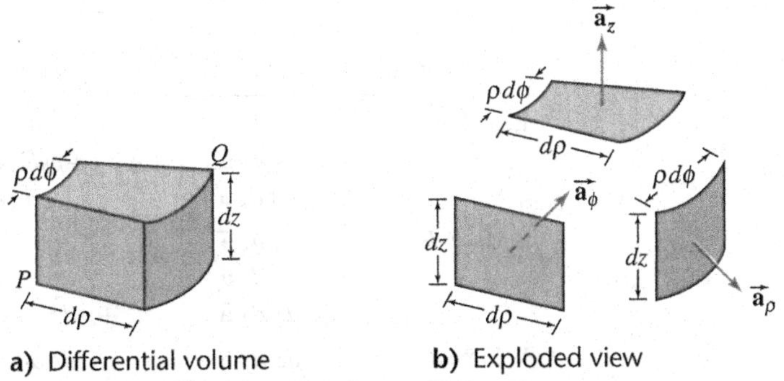

Figure 2.19 Differential elements in a cylindrical coordinate system

2.6.3 Spherical coordinate system

A differential volume element in the spherical coordinate system is obtained by incrementing $r, \theta,$ and ϕ by $dr, d\theta,$ and $d\phi$, respectively (Fig. 2.20a). The volume element is

$$dv = r^2\,dr\,\sin\theta\,d\theta\,d\phi \qquad (2.57)$$

The differential surface areas in the positive directions of the unit vectors, shown in Figure 2.20b, are

$$\begin{aligned} \overrightarrow{d\mathbf{s}}_r &= r^2 \sin\theta \, d\theta \, d\phi \, \vec{\mathbf{a}}_r \\ \overrightarrow{d\mathbf{s}}_\theta &= r \, dr \, \sin\theta \, d\phi \, \vec{\mathbf{a}}_\theta \\ \overrightarrow{d\mathbf{s}}_\phi &= r \, dr \, d\theta \, \vec{\mathbf{a}}_\phi \end{aligned} \quad (2.58)$$

The differential length vector from P to Q is

$$\vec{d\ell} = dr \, \vec{\mathbf{a}}_r + r \, d\theta \, \vec{\mathbf{a}}_\theta + r \sin\theta \, d\phi \, \vec{\mathbf{a}}_\phi \quad (2.59)$$

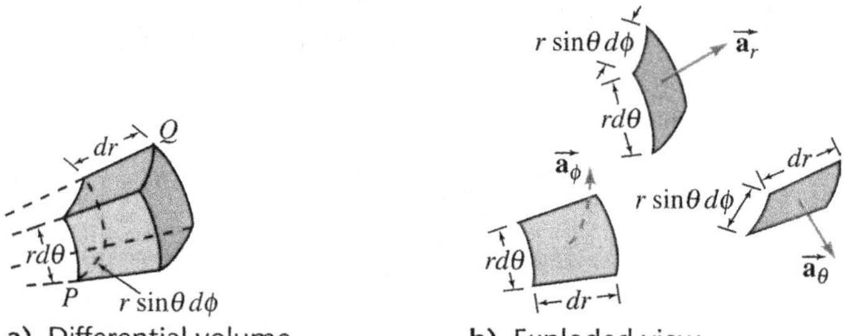

Figure 2.20 Differential elements in a spherical coordinate system

For easy reference, the differential length, surface, and volume elements for the three coordinate systems are summarized in Table 2.1.

Table 2.1. Differential elements of length, surface, and volume in the rectangular, cylindrical, and spherical coordinate systems

Differential elements	Coordinate system: Rectangular (Cartesian)	Cylindrical	Spherical
Length $\vec{d\ell}$	$dx\,\vec{\mathbf{a}}_x + dy\,\vec{\mathbf{a}}_y + dz\,\vec{\mathbf{a}}_z$	$d\rho\,\vec{\mathbf{a}}_\rho + \rho\,d\phi\,\vec{\mathbf{a}}_\phi + dz\,\vec{\mathbf{a}}_z$	$dr\,\vec{\mathbf{a}}_r + r\,d\theta\,\vec{\mathbf{a}}_\theta + r\sin\theta\,d\phi\,\vec{\mathbf{a}}_\phi$
Surface $\vec{ds}$	$dy\,dz\,\vec{\mathbf{a}}_x + dx\,dz\,\vec{\mathbf{a}}_y + dx\,dy\,\vec{\mathbf{a}}_z$	$\rho\,d\phi\,dz\,\vec{\mathbf{a}}_\rho + d\rho\,dz\,\vec{\mathbf{a}}_\phi + \rho\,d\rho\,d\phi\,\vec{\mathbf{a}}_z$	$r^2\sin\theta\,d\theta\,d\phi\,\vec{\mathbf{a}}_r + r\,dr\sin\theta\,d\phi\,\vec{\mathbf{a}}_\theta + r\,dr\,d\theta\,\vec{\mathbf{a}}_\phi$
Volume dv	$dx\,dy\,dz$	$\rho\,d\rho\,d\phi\,dz$	$r^2 dr\,\sin\theta\,d\theta\,d\phi$

2.7 Line, surface, and volume integrals

We often express the basic laws of electromagnetic fields in terms of integrals of field quantities over various curves (lines), surfaces, and volumes in a region. For example, in Chapter 3, we will define the potential function in terms of the line integral of electric field intensity.

In Chapter 4, we will define the current through a conductor as the surface integral of volume current density. A clear understanding of such spatial integrals is essential for our investigation of electromagnetic field theory. In addition, from time to time we will express our final result in integral form to shed some light on its significance. Therefore, let us make a short digression to discuss the concepts of line, surface, and volume integrals.

2.7.1 The line integral

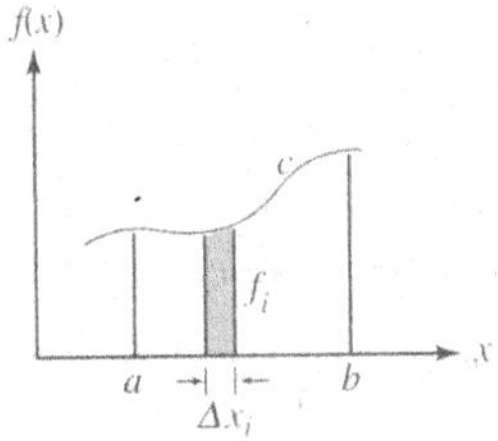

Figure 2.21 A continuous, single-valued function

Let $f(x)$ be a continuous, single-valued function of x between the limits $x = a$ and $x = b$, as shown in Figure 2.21. To define the line integral of $f(x)$, we divide the interval from a to b into n small segments, all of which approach zero in the limit. The **line integral** is then defined in terms of the limit of the sum as

$$\int_a^b f(x)\,dx = \lim_{\substack{n\to\infty \\ \Delta x_i \to 0}} \sum_{i=1}^{n} f_i\,\Delta x_i \tag{2.60}$$

where f_i is the value of $f(x)$ for the segment Δx_i such that $\Delta x_i \to 0$.

We can now extend this definition of the line integral for a general curve c in three-dimensional space, as depicted in Figure 2.22. Let us first consider a scalar field f and define its line integral from a to b along c. Again, we divide the interval between a and b into n small sections, all of which approach zero in the limit. In this case the small segments are, in fact, length vectors. The position vectors for the ith element and their lengths are shown in Figure 2.22. The line integral of f along c is then defined in the limit of the sum as

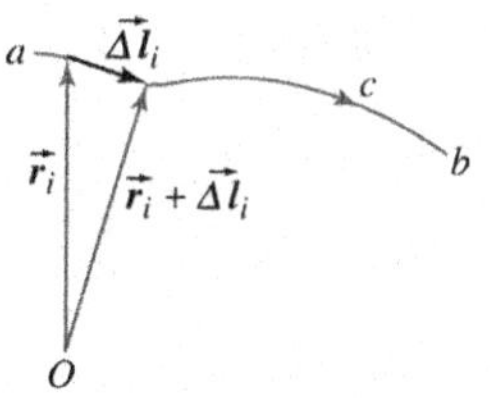

Figure 2.22 A differential length element along path *c* in three-dimensional space

$$\int_c f\,\vec{d\ell} = \lim_{\substack{n\to\infty \\ \vec{\Delta\ell}_i \to 0}} \sum_{i=1}^{n} f_i \vec{\boldsymbol{\Delta}\ell_i} \tag{2.61}$$

where f_i is the value of the scalar function f within the length segment $\vec{\boldsymbol{\Delta}\ell_i}$. It is evident that this integral is a vector.

Without repeating all the details, we can now define a scalar line integral for a vector field $\vec{\mathbf{F}}$ as

$$\int_c \vec{\mathbf{F}}\cdot\vec{d\ell} = \lim_{\substack{n\to\infty \\ \vec{\Delta\ell}_i \to 0}} \sum_{i=1}^{n} \vec{\mathbf{F}}_i \cdot \vec{\boldsymbol{\Delta}\ell_i} \tag{2.62}$$

Finally, the vector line integral of a vector field $\vec{\mathbf{F}}$ along path c is defined as

$$\int_c \vec{\mathbf{F}}\times\vec{d\ell} = \lim_{\substack{n\to\infty \\ \vec{\Delta\ell}_i \to 0}} \sum_{i=1}^{n} \vec{\mathbf{F}}_i \times \vec{\boldsymbol{\Delta}\ell_i} \tag{2.63}$$

In all of these integrals, the path of integration can be around a closed curve, in which case the points a and b coincide. Such a closed path is usually denoted by writing the integral sign as $\oint$.

EXAMPLE 2.13

If $\vec{\mathbf{A}} = (4x + 9y)\vec{\mathbf{a}}_x - 14yz\vec{\mathbf{a}}_y + 8x^2z\vec{\mathbf{a}}_z$, evaluate $\int_c \vec{\mathbf{A}} \cdot \vec{d\ell}$ from $P(0, 0, 0)$ to $Q(1, 1, 1)$ along the following paths:

a) $x = t$, $y = t^2$, and $z = t^3$

b) The straight lines from (0, 0, 0) to (1, 0, 0) then to (1, 1, 0) and finally to (1, 1, 1)

c) The straight line joining $P(0, 0, 0)$ to $Q(1, 1, 1)$

Solution

a) $\vec{\mathbf{A}} \cdot \vec{d\ell} = (4x+9y)\,dx - 14yzdy + 8x^2z\,dz$. Because $x = t$, $y = t^2$, and $z = t^3$, $dx = dt$, $dy = 2t\,dt$, and $dz = 3t^2\,dt$. By direct substitution, we obtain

$$\int_c \vec{\mathbf{A}} \cdot \vec{d\ell} = \int_{t=0}^{1} [4t + 9t^2 - 28t^6 + 24t^7]\,dt = 4$$

b) Along the paths specified there are three regular curves, as shown in Figure 2.23. Thus, we have to evaluate the integral along each path separately. Path c_1: $y = 0$, $dy = 0$, $z = 0$, $dz = 0$, and $0 \le x \le 1$.

$$\int_{c_1} \vec{\mathbf{A}} \cdot \vec{d\ell} = \int_0^1 4x\,dx = 2$$

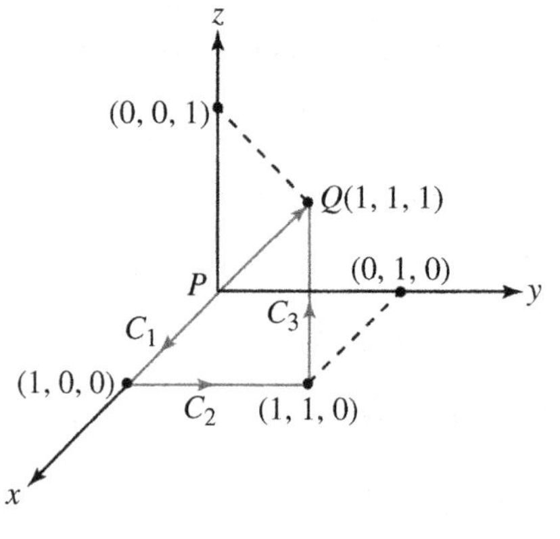

Figure 2.23 Illustration of paths of integration for Example 2.13

Path c_2: $x = 1$, $dx = 0$, $z = 0$, $dz = 0$, and $0 \le y \le 1$.

$$\int_{c_2} \vec{\mathbf{A}} \cdot \vec{d\ell} = 0$$

Path c_3: $x = 1$, $dx = 0$, $y = 1$, $dy = 0$, and $0 \le z \le 1$.

$$\int_{c_3} \vec{\mathbf{A}} \cdot \vec{d\ell} = \int_0^1 8z\,dz = 4$$

Thus, the line integral from P to Q along the three paths is

$$\int_c \vec{\mathbf{A}} \cdot \vec{d\ell} = \int_{c_1} \vec{\mathbf{A}} \cdot \vec{d\ell} + \int_{c_2} \vec{\mathbf{A}} \cdot \vec{d\ell} + \int_{c_3} \vec{\mathbf{A}} \cdot \vec{d\ell}$$
$$= 2 + 0 + 4 = 6$$

c) Along the path from P to Q, we have $0 \le x \le 1$, $0 \le y \le 1$, and $0 \le z \le 1$. To perform the integration, we can express y and z in terms of x as $y = x$ and $z = x$. Thus, $dy = dx$ and $dz = dx$. Substituting these relations, we get

$$\int_c \vec{\mathbf{A}} \cdot \vec{d\ell} = \int_0^1 (13x - 14x^2 + 8x^3)\,dx = 3.833 \qquad \bullet\bullet\bullet$$

2.7.2 The surface integral

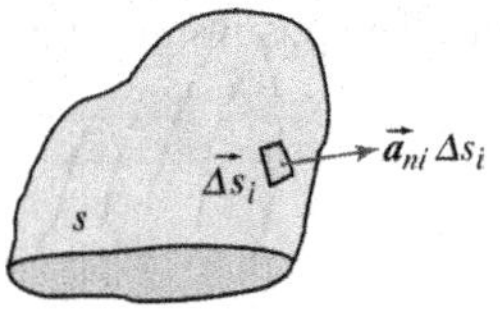

Figure 2.24 Differential surface element

To define the surface integral of a scalar field f or a vector field $\vec{\mathbf{F}}$, we divide the given surface s into a large number of n small surfaces, all of which approach zero in the limit. Each small surface Δs_i has a corresponding vector surface $\vec{\mathbf{\Delta s}}_i$, as indicated in Figure 2.24. To define the surface integral of f, we multiply f by each surface element and sum it for all n elements of s in the limit $\vec{\mathbf{\Delta s}} \to 0$ as $n \to \infty$. This limit is called the **surface integral** of f over s. Thus

$$\int_s f\vec{d\mathbf{s}} = \lim_{\substack{n\to\infty \\ \vec{\Delta \mathbf{s}}_i \to 0}} \sum_{i=1}^{n} f_i \vec{\mathbf{\Delta s}}_i \tag{2.64}$$

where f_i is the value of the scalar function f over the elemental surface $\vec{\mathbf{\Delta s}}_i$. Clearly, the integral in (2.64) is a vector.

By following the same rules, we can define the scalar surface integral by forming the dot product of vector field $\vec{\mathbf{F}}$ with each surface element $\vec{\mathbf{\Delta s}}$ and summing these scalars in the limit. That is,

$$\int_s \vec{\mathbf{F}} \cdot \vec{d\mathbf{s}} = \lim_{\substack{n\to\infty \\ \vec{\Delta \mathbf{s}}_i \to 0}} \sum_{i=1}^{n} \vec{\mathbf{F}}_i \cdot \vec{\mathbf{\Delta s}}_i \tag{2.65}$$

Finally, we define the vector surface integral of a vector field $\vec{\mathbf{F}}$ by taking the cross product as

$$\int_s \vec{\mathbf{F}} \times \vec{d\mathbf{s}} = \lim_{\substack{n\to\infty \\ \vec{\Delta \mathbf{s}}_i \to 0}} \sum_{i=1}^{n} \vec{\mathbf{F}}_i \times \vec{\mathbf{\Delta s}}_i \tag{2.66}$$

EXAMPLE 2.14

Show that over the closed surface of a sphere of radius b, $\oint \vec{d\mathbf{s}} = 0$.

Solution The outward unit normal to the surface of a sphere of radius b is in the direction of the unit vector $\vec{\mathbf{a}}_r$, as shown in Figure 2.25. Therefore,

$$\oint_s \vec{d\mathbf{s}} = \int_{\theta=0}^{\pi} \int_{\phi=0}^{2\pi} \vec{\mathbf{a}}_r b^2 \sin\theta \, d\theta \, d\phi$$

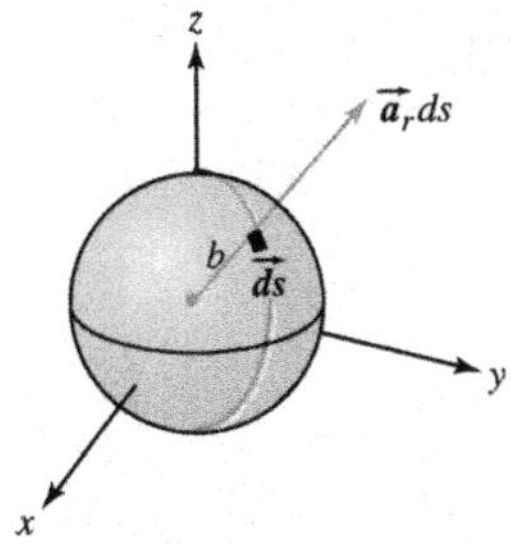

Figure 2.25

Because the unit vector $\vec{\mathbf{a}}_r$ is a function of both θ and ϕ, we must express it in terms of unit vectors in the rectangular coordinate system before integrating. From equation (2.43a,b) we have $\vec{\mathbf{a}}_r = \sin\theta\cos\phi\vec{\mathbf{a}}_x + \sin\theta\sin\phi\,\vec{\mathbf{a}}_y + \cos\theta\,\vec{\mathbf{a}}_z$. Thus,

$$\begin{aligned}\oint_s \vec{d\mathbf{s}} &= \vec{\mathbf{a}}_x b^2 \int_0^{\pi} \sin^2\theta \, d\theta \int_0^{2\pi} \cos\phi \, d\phi + \vec{\mathbf{a}}_y b^2 \int_0^{\pi} \sin^2\theta \, d\theta \\ &\quad \times \int_0^{2\pi} \sin\phi \, d\phi + \vec{\mathbf{a}}_z b^2 \int_0^{\pi} \sin\theta\cos\theta \, d\theta \int_0^{2\pi} d\phi \\ &= 0\end{aligned}$$

$\bullet\bullet\bullet$

EXAMPLE 2.15 Evaluate $\oint \vec{\mathbf{r}} \cdot \overrightarrow{d\mathbf{s}}$ over the closed surface of the cube bounded by $0 \le x \le 1$, $0 \le y \le 1$, and $0 \le z \le 1$, where $\vec{\mathbf{r}}$ is the position vector of any point on the surface of the cube.

Solution The six surfaces bounding the unit cube on which the surface integral is to be evaluated are shown in Figure 2.26. We will evaluate the integral on each surface separately and then sum the results. The position vector of any point P on the surface in general is

$$\vec{\mathbf{r}} = x\vec{\mathbf{a}}_x + y\vec{\mathbf{a}}_y + z\vec{\mathbf{a}}_z$$

Figure 2.26

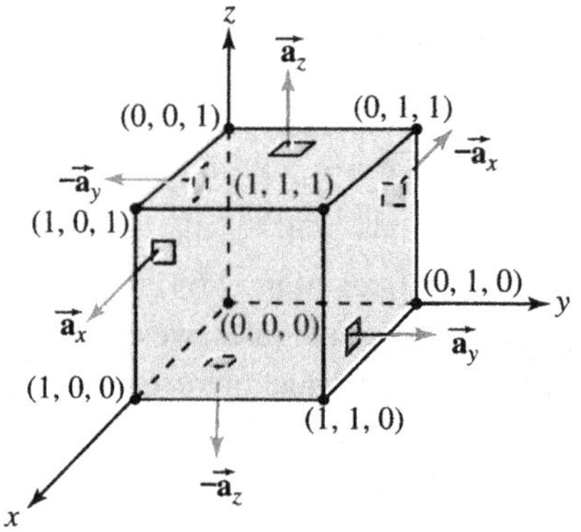

a) Surface at $x = 1$: $\overrightarrow{d\mathbf{s}} = dy\,dz\,\vec{\mathbf{a}}_x$. Thus,

$$\int_{s_1} \vec{\mathbf{r}} \cdot \overrightarrow{d\mathbf{s}} = \int_0^1 dy \int_0^1 dz = 1$$

b) Surface at $x = 0$: $\overrightarrow{d\mathbf{s}} = -dy\,dz\,\vec{\mathbf{a}}_x$ and

$$\int_{s_2} \vec{\mathbf{r}} \cdot \overrightarrow{d\mathbf{s}} = 0$$

c) Surface at $y = 1$: $\overrightarrow{d\mathbf{s}} = dx\,dz\,\vec{\mathbf{a}}_y$ and

$$\int_{s_3} \vec{\mathbf{r}} \cdot \overrightarrow{d\mathbf{s}} = \int_0^1 dx \int_0^1 dz = 1$$

d) Surface at $y = 0$: $\overrightarrow{d\mathbf{s}} = -dx\,dz\,\vec{\mathbf{a}}_y$ and

$$\int_{s_4} \vec{\mathbf{r}} \cdot \overrightarrow{d\mathbf{s}} = 0$$

e) Surface at $z = 1$: $\overrightarrow{d\mathbf{s}} = dx\,dy\,\vec{\mathbf{a}}_z$ and

$$\int_{s_5} \vec{\mathbf{r}} \cdot \overrightarrow{d\mathbf{s}} = \int_0^1 dx \int_0^1 dy = 1$$

f) Surface at $z = 0$: $\vec{ds} = -dx\,dy\,\vec{\mathbf{a}}_z$ and

$$\int_{s_6} \vec{\mathbf{r}} \cdot \vec{ds} = 0$$

Thus,

$$\oint_s \vec{\mathbf{r}} \cdot \vec{ds} = 3$$

• • •

2.7.3 The volume integral

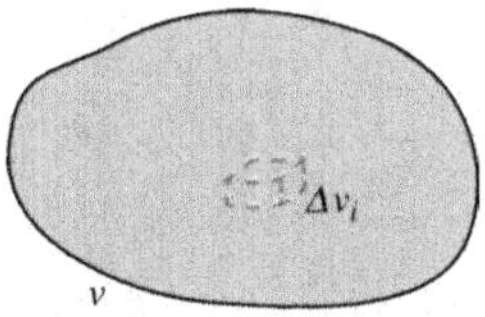

Figure 2.27 A differential volume element

To define a volume integral, we divide a given volume into n small volume elements as shown in Figure 2.27. Each volume element $\Delta v \to 0$ as $n \to \infty$. To define the **scalar volume integral** we multiply each volume element by the scalar field f and sum their products for all volume elements. Then we take the limit for this sum as

$$\int_v f\,dv = \lim_{\substack{n\to\infty \\ \Delta v_i \to 0}} \sum_{i=1}^{n} f_i\,\Delta v_i \tag{2.67}$$

Likewise, we define the **volume integral of a vector field $\vec{\mathbf{F}}$** as

$$\int_v \vec{\mathbf{F}}\,dv = \lim_{\substack{n\to\infty \\ \Delta v_i \to 0}} \sum_{i=1}^{n} \vec{\mathbf{F}}_i\,\Delta v_i \tag{2.68}$$

EXAMPLE 2.16

The electron density distribution within a spherical volume with radius of 2 meters is given as $n_e = (1000/r)\cos(\phi/4)$ electrons/meter^3. Find the charge enclosed if the charge on an electron is -1.6×10^{-19} coulomb.

Solution Let N be the number of electrons in the region bounded by a sphere of 2-meter radius; then

$$\begin{aligned} N &= \int_v n_e\,dv = \int_v \frac{1000}{r}\cos(\phi/4)\,dv \\ &= \int_0^2 \frac{1000}{r} r^2\,dr \int_0^{\pi} \sin\theta\,d\theta \int_0^{2\pi} \cos(\phi/4)\,d\phi \\ &= 16{,}000 \text{ electrons} \end{aligned}$$

Thus, the total charge enclosed is $Q = 16{,}000(-1.6 \times 10^{-19}) = -2.56 \times 10^{-15}$ coulomb.

• • •

2.8 The gradient of a scalar function

Let $f(x, y, z)$ be a real-valued differentiable function of x, y, and z, as shown in Figure 2.28. The differential change in f from point P to Q,

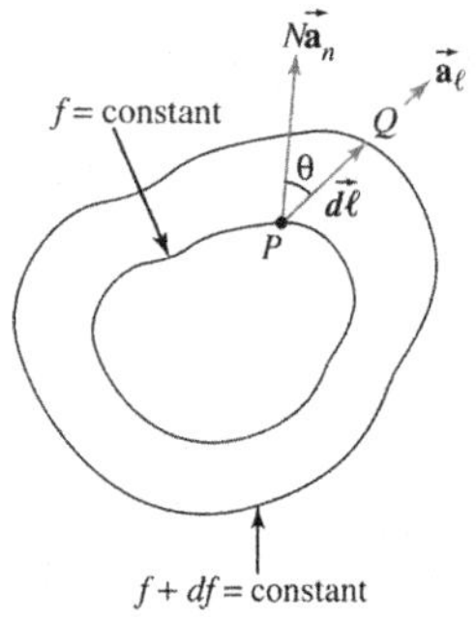

Figure 2.28 Illustration for defining gradient of scalar function

from equation (2.47), can be written as

$$df = \frac{\partial f}{\partial x}dx + \frac{\partial f}{\partial y}dy + \frac{\partial f}{\partial z}dz$$
$$= \left[\frac{\partial f}{dx}\vec{\mathbf{a}}_x + \frac{\partial f}{dy}\vec{\mathbf{a}}_y + \frac{\partial f}{dz}\vec{\mathbf{a}}_z\right] \cdot [dx\,\vec{\mathbf{a}}_x + dy\,\vec{\mathbf{a}}_y + dz\,\vec{\mathbf{a}}_z] \quad (2.69)$$

In terms of the differential length element

$$\vec{d\ell} = dx\,\vec{\mathbf{a}}_x + dy\,\vec{\mathbf{a}}_y + dz\,\vec{\mathbf{a}}_z$$

from P to Q, we can rewrite (2.69) as

$$df = \left[\frac{\partial f}{\partial x}\vec{\mathbf{a}}_x + \frac{\partial f}{\partial y}\vec{\mathbf{a}}_y + \frac{\partial f}{\partial z}\vec{\mathbf{a}}_z\right] \cdot \vec{d\ell} \quad (2.70)$$

or

$$\frac{df}{d\ell} = \left[\frac{\partial f}{\partial x}\vec{\mathbf{a}}_x + \frac{\partial f}{\partial y}\vec{\mathbf{a}}_y + \frac{\partial f}{\partial z}\vec{\mathbf{a}}_z\right] \cdot \frac{\vec{d\ell}}{d\ell}$$
$$= \vec{\mathbf{N}} \cdot \vec{\mathbf{a}}_\ell = N\,\vec{\mathbf{a}}_n \cdot \vec{\mathbf{a}}_\ell \quad (2.71)$$

where $\vec{\mathbf{a}}_\ell = \vec{d\ell}/d\ell$ is a unit vector from P to Q in the direction of $\vec{d\ell}$, and

$$\vec{\mathbf{N}} = \frac{\partial f}{\partial x}\vec{\mathbf{a}}_x + \frac{\partial f}{\partial y}\vec{\mathbf{a}}_y + \frac{\partial f}{\partial z}\vec{\mathbf{a}}_z \quad (2.72)$$

From (2.71) it is evident that the rate of change in the function f is maximum when $\vec{\mathbf{a}}_\ell$ and $\vec{\mathbf{N}}$ are collinear. That is,

$$\left.\frac{df}{d\ell}\right|_{\max} = N \quad (2.73)$$

There exists a surface passing through P on which f is constant. Similarly, there also exists a surface passing through Q on which $f + df$ is constant. For the ratio $df/d\ell$ to be maximum, the distance $d\ell$ from P to Q must be minimum. In other words, $df/d\ell$ is maximum when $\vec{\mathbf{a}}_\ell$ is normal to the surface $f(x, y, z) =$ constant. This, in turn, implies that $\vec{\mathbf{N}}$ is normal to the surface $f(x, y, z) =$ constant. $\vec{\mathbf{N}}$, by definition, is then the **gradient** of $f(x, y, z)$. It is a common practice to write this gradient of $f(x, y, z)$ as ∇f, where ∇, called *del* or *nabla*, is referred to as the gradient operator. The gradient of a scalar function $f(x, y, z)$, from (2.72), is

$$\nabla f = \frac{\partial f}{\partial x}\vec{\mathbf{a}}_x + \frac{\partial f}{\partial y}\vec{\mathbf{a}}_y + \frac{\partial f}{\partial z}\vec{\mathbf{a}}_z \quad (2.74)$$

The gradient operator itself can be written in rectangular coordinates as

$$\nabla = \vec{\mathbf{a}}_x\frac{\partial}{\partial x} + \vec{\mathbf{a}}_y\frac{\partial}{\partial y} + \vec{\mathbf{a}}_z\frac{\partial}{\partial z} \quad (2.75)$$

We stress that the gradient operator is meaningless by itself. Only when applied to a scalar function does it yield a vector. We can now express the differential of a scalar function in terms of the gradient of that function, from (2.70), as

$$df = \nabla f \cdot \vec{d\ell} \tag{2.76a}$$

or

$$df/d\ell = \nabla f \cdot \vec{\mathbf{a}}_\ell \tag{2.76b}$$

We will employ (2.76a) quite frequently in Chapters 3 and 4 to determine the change of a scalar function in a given direction. Equation (2.76b) gives the rate of change of a scalar function f in the direction of the unit vector $\vec{\mathbf{a}}_\ell$. This is called the *directional derivative* of f along $\vec{\mathbf{a}}_\ell$.

We summarize the properties of the gradient of a scalar function at a point as follows

1. It is normal to the surface on which the given function is constant.
2. It points in the direction in which the given function changes most rapidly with position.
3. Its magnitude gives the maximum rate of change of the given function per unit distance.
4. The directional derivative of a function at a point in any direction is equal to the dot product of the gradient of the function and the unit vector in that direction.

We can also obtain expressions for the gradient of a scalar function in the cylindrical coordinate system as

$$\nabla f = \frac{\partial f}{\partial \rho}\vec{\mathbf{a}}_\rho + \frac{1}{\rho}\frac{\partial f}{\partial \phi}\vec{\mathbf{a}}_\phi + \frac{\partial f}{\partial z}\vec{\mathbf{a}}_z \tag{2.77}$$

and in the spherical coordinate system as

$$\nabla f = \frac{\partial f}{\partial r}\vec{\mathbf{a}}_r + \frac{1}{r}\frac{\partial f}{\partial \theta}\vec{\mathbf{a}}_\theta + \frac{1}{r\sin\theta}\frac{\partial f}{\partial \phi}\vec{\mathbf{a}}_\phi \tag{2.78}$$

respectively.

EXAMPLE 2.17

Find the gradient of a scalar field $f(x, y, z) = 6x^2y^3 + e^z$ at the point $P(2, 1, 0)$.

Solution Since $f(x, y, z)$ is given in rectangular coordinates, we use (2.74) to determine its gradient. That is,

$$\begin{aligned}\nabla f &= \frac{\partial}{\partial x}[6x^2y^3 + e^z]\vec{\mathbf{a}}_x + \frac{\partial}{\partial y}[6x^2y^3 + e^z]\vec{\mathbf{a}}_y \\ &\quad + \frac{\partial}{\partial z}[6x^2y^3 + e^z]\vec{\mathbf{a}}_z \\ &= 12xy^3\vec{\mathbf{a}}_x + 18x^2y^2\vec{\mathbf{a}}_y + e^z\vec{\mathbf{a}}_z\end{aligned}$$

At the given point $P(2, 1, 0)$, the gradient of $f(x, y, z)$ is

$$\nabla f = 24\vec{\mathbf{a}}_x + 72\vec{\mathbf{a}}_y + \vec{\mathbf{a}}_z$$

• • •

EXAMPLE 2.18 Find the gradient of r where r is the magnitude of the position vector $\vec{\mathbf{r}} = \rho\vec{\mathbf{a}}_\rho + z\vec{\mathbf{a}}_z$ in the cylindrical coordinate system.

Solution The position vector $\vec{\mathbf{r}}$ is given in the cylindrical coordinate system, so we use (2.77) to determine its gradient. The magnitude of the position vector $\vec{\mathbf{r}}$ is

$$r = [\rho^2 + z^2]^{1/2}$$

The derivatives of r with respect to ρ, ϕ, and z are

$$\frac{\partial r}{\partial \rho} = \frac{\rho}{r}, \quad \frac{\partial r}{\partial \phi} = 0, \quad \text{and} \quad \frac{\partial r}{\partial z} = \frac{z}{r}$$

Thus, from (2.77), the gradient of r is

$$\nabla r = \frac{\rho}{r}\vec{\mathbf{a}}_\rho + \frac{z}{r}\vec{\mathbf{a}}_z = \frac{\vec{\mathbf{r}}}{r} = \vec{\mathbf{a}}_r \tag{2.79}$$

$\nabla r = \vec{\mathbf{a}}_r$ is another important result that we will use from time to time to simplify some of the equations in subsequent chapters. • • •

2.9 Divergence of a vector field

Before defining the divergence of a vector field let us specify a scalar field f at point P in terms of a vector field $\vec{\mathbf{F}}$ as

$$f = \lim_{\Delta v \to 0} \frac{1}{\Delta v} \oint_s \vec{\mathbf{F}} \cdot \overrightarrow{ds} \tag{2.80}$$

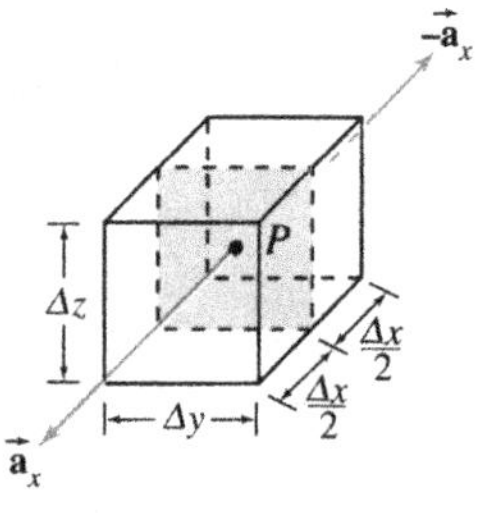

Figure 2.29 A differential volume in the rectangular coordinate system

where the point P is enclosed by volume Δv bounded by surface s. Although Δv can be of any shape, we construct a parallelepiped with sides Δx, Δy, and Δz, as shown in Figure 2.29 in order to evaluate (2.80). Note that $\vec{\mathbf{F}} \cdot \overrightarrow{ds}$ defines the outward flow of the vector field $\vec{\mathbf{F}}$ through the surface $\overrightarrow{ds}$ as the unit normal to ds points away from the volume enclosed. Thus, $\oint \vec{\mathbf{F}} \cdot \overrightarrow{ds}$ gives the net outward flow of flux of a vector field $\vec{\mathbf{F}}$ from the volume Δv. However, the outward flow of a vector field $\vec{\mathbf{F}}$ through the face in the positive x direction, using the Taylor series expansion and neglecting the higher-order terms, is

$$\left[F_x + \frac{\partial F_x}{\partial x}\frac{\Delta x}{2}\right] \Delta y \, \Delta z \tag{2.81}$$

The outward flow of the vector field $\vec{\mathbf{F}}$ through the surface in the negative x direction is

$$-\left[F_x - \frac{\partial F_x}{\partial x}\frac{\Delta x}{2}\right] \Delta y \, \Delta z \tag{2.82}$$

Therefore, the net outward flow of the vector field $\vec{\mathbf{F}}$ through both the surfaces in the x direction is

$$\frac{\partial F_x}{\partial x}\Delta x\ \Delta y\ \Delta z = \frac{\partial F_x}{\partial x}\Delta v \tag{2.83}$$

We can similarly obtain expressions for the net outward flow of the vector field $\vec{\mathbf{F}}$ through the surfaces in the y and z directions. The net outward flow of the vector field $\vec{\mathbf{F}}$ through all the surfaces enclosing the volume Δv then becomes

$$\oint_s \vec{\mathbf{F}} \cdot \vec{d\mathbf{s}} = \left[\frac{\partial F_x}{\partial x} + \frac{\partial F_y}{\partial y} + \frac{\partial F_z}{\partial z}\right]\Delta v \tag{2.84}$$

Comparing (2.80) and (2.84), we get

$$f = \frac{\partial F_x}{\partial x} + \frac{\partial F_y}{\partial y} + \frac{\partial F_z}{\partial z} \tag{2.85}$$

We can express (2.85) in terms of the ∇ operator as

$$f = \left[\vec{\mathbf{a}}_x\frac{\partial}{\partial x} + \vec{\mathbf{a}}_y\frac{\partial}{\partial y} + \vec{\mathbf{a}}_z\frac{\partial}{\partial z}\right]\cdot[F_x\vec{\mathbf{a}}_x + F_y\vec{\mathbf{a}}_y + F_z\vec{\mathbf{a}}_z] \tag{2.86a}$$

$$f = \boldsymbol{\nabla}\cdot\vec{\mathbf{F}} \tag{2.86b}$$

where $\boldsymbol{\nabla}\cdot\vec{\mathbf{F}}$ is called the *divergence of the vector field* $\vec{\mathbf{F}}$. Note that it is a scalar quantity. Equation (2.80) gives us the definition of the divergence of a vector field, and equation (2.85) enables us to compute it. Hence, the **divergence of a vector field** $\vec{\mathbf{F}}$ in the rectangular coordinate system is

$$\boldsymbol{\nabla}\cdot\vec{\mathbf{F}} = \frac{\partial F_x}{\partial x} + \frac{\partial F_y}{\partial y} + \frac{\partial F_z}{\partial z} \tag{2.86c}$$

The physical significance of (2.86) is that by enclosing a point P within any arbitrary small volume we can obtain the net outward flow of a vector field by computing its divergence at that point. The net outward flow is positive at a source point and negative at a sink point. If the vector field is continuous, such as the flow of an incompressible fluid through a pipe or the magnetic lines of field surrounding a magnet, there is no net outward flow. In that case, $\boldsymbol{\nabla}\cdot\vec{\mathbf{F}} = 0$ and $\vec{\mathbf{F}}$ is said to be a *continuous* or *solenoidal* vector field.

We can also obtain the expressions for the divergence of a vector field in cylindrical and spherical coordinates as

$$\boldsymbol{\nabla}\cdot\vec{\mathbf{F}} = \frac{1}{\rho}\frac{\partial}{\partial\rho}[\rho F_\rho] + \frac{1}{\rho}\frac{\partial}{\partial\phi}[F_\phi] + \frac{\partial}{\partial z}[F_z] \tag{2.87}$$

and

$$\boldsymbol{\nabla}\cdot\vec{\mathbf{F}} = \frac{1}{r^2}\frac{\partial}{\partial r}[r^2F_r] + \frac{1}{r\sin\theta}\frac{\partial}{\partial\theta}[\sin\theta\, F_\theta] + \frac{1}{r\sin\theta}\frac{\partial}{\partial\phi}[F_\phi] \tag{2.88}$$

respectively.

EXAMPLE 2.19 Prove that $\nabla \cdot \vec{\mathbf{r}} = 3$ where $\vec{\mathbf{r}}$ is the position vector of any point P in space.

Solution The position vector of any point P in rectangular coordinates is

$$\vec{\mathbf{r}} = x\vec{\mathbf{a}}_x + y\vec{\mathbf{a}}_y + z\vec{\mathbf{a}}_z$$

Thus, the divergence of vector $\vec{\mathbf{r}}$ is

$$\begin{aligned}\nabla \cdot \vec{\mathbf{r}} &= \frac{\partial}{\partial x}[x] + \frac{\partial}{\partial y}[y] + \frac{\partial}{\partial z}[z] \\ &= 1 + 1 + 1 = 3\end{aligned}$$

• • •

2.9.1 The divergence theorem

The definition of the divergence of a vector, (2.80) and (2.86), applies to a point enclosed within an infinitesimal volume Δv – a microscopic operation. If the vector field $\vec{\mathbf{F}}$ is continuously differentiable in a region of volume v bounded by the surface s (see Figure 2.30) the definition of divergence can be extended to cover the entire volume. This is done by subdividing the volume v into n elementary volumes (cells), all of which approach zero in the limit. That is, for an elementary volume Δv_i enclosing a point P_i and bounded by a surface s_i, the divergence of $\vec{\mathbf{F}}$ at P_i is

$$\nabla \cdot \vec{\mathbf{F}}_i = \lim_{\Delta v_i \to 0} \frac{1}{\Delta v_i} \oint_{s_i} \vec{\mathbf{F}} \cdot \vec{ds} \tag{2.89}$$

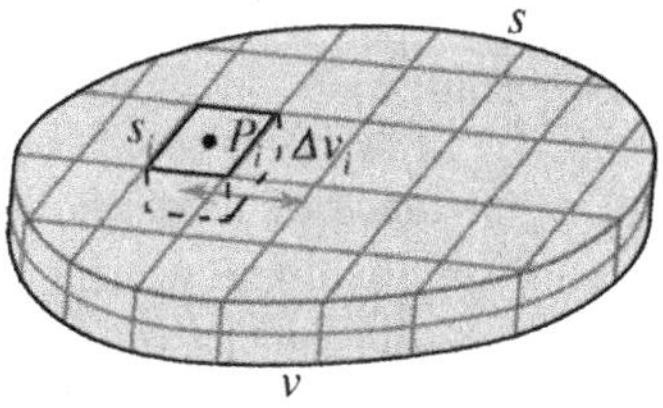

Figure 2.30 Volume *v* subdivided into *n* small volumes to verify divergence theorem

where $\vec{\mathbf{F}}_i$ is the value of the vector field $\vec{\mathbf{F}}$ at P_i. We can rewrite this equation as

$$\oint_{s_i} \vec{\mathbf{F}} \cdot \vec{ds} = \nabla \cdot \vec{\mathbf{F}}_i \, \Delta v_i + \epsilon_i \, \Delta v_i \tag{2.90}$$

where the term $\epsilon_i \, \Delta v_i$ has been introduced because (2.80) is valid strictly for a point enclosed by volume $\Delta v_i \to 0$. Thus, $\epsilon_i \to 0$ as $\Delta v_i \to 0$. Summing for all cells, we obtain

$$\lim_{n\to\infty} \sum_{i=1}^{n} \oint_{s_i} \vec{\mathbf{F}} \cdot \vec{ds} = \lim_{n\to\infty} \sum_{i=1}^{n} \nabla \cdot \vec{\mathbf{F}}_i \, \Delta v_i + \lim_{n\to\infty} \sum_{i=1}^{n} \epsilon_i \, \Delta v_i \tag{2.91}$$

Observe that the surface integrals over the interfaces of the two cells within v vanish as the net flux leaving one cell cancels the net flux leaving the other. Thus, the nonzero terms in the sum correspond to the outermost cells that belong to the surface s. Hence the left-hand side of (2.91) becomes

$$\lim_{n\to\infty} \sum_{i=1}^{n} \oint_{s_i} \vec{\mathbf{F}} \cdot \vec{ds} = \oint_{s} \vec{\mathbf{F}} \cdot \vec{ds}$$

As the number of cells increases, the first term on the right-hand side of (2.91), in the limit, becomes

$$\lim_{n\to\infty} \sum_{i=1}^{n} \nabla \cdot \vec{\mathbf{F}}_i \, \Delta v_i = \int_v \nabla \cdot \vec{\mathbf{F}} \, dv$$

The second term on the right-hand side of (2.91) involves the product of small quantities and vanishes as $n \to \infty$. Therefore, we can write (2.91) in the limit as

$$\int_v \nabla \cdot \vec{\mathbf{F}} \, dv = \oint_s \vec{\mathbf{F}} \cdot \vec{ds} \tag{2.92}$$

Equation (2.92) is a mathematical definition of the *divergence theorem*. It relates the volume integral of the divergence of a vector field to the surface integral of its normal component. *It states that for a continuously differentiable vector field the net outward flux from a closed surface equals the integral of the divergence throughout the region bounded by that surface.*

The divergence theorem is very powerful. It is used extensively in electromagnetic field theory to convert a *closed surface integral* into an equivalent *volume integral* and vice versa.

EXAMPLE 2.20

Verify the divergence theorem for a vector field $\vec{\mathbf{D}} = 3x^2\vec{\mathbf{a}}_x + (3y + z)\vec{\mathbf{a}}_y + (3z - x)\vec{\mathbf{a}}_z$ in the region bounded by the cylinder $x^2 + y^2 = 9$ and the planes $x = 0$, $y = 0$, $z = 0$, and $z = 2$.

Solution Figure 2.31 (page 48) shows five distinct surfaces bounding the volume v. Let us first compute the left-hand side of (2.92).

$$\nabla \cdot \vec{\mathbf{D}} = \frac{\partial}{\partial x}[3x^2] + \frac{\partial}{\partial y}[3y + z] + \frac{\partial}{\partial z}[3z - x]$$
$$= 6x + 6$$

Computing the volume integral in cylindrical coordinates, we get

$$\int_v \nabla \cdot \vec{\mathbf{D}} \, dv = \int_v [6x + 6] \, dv$$
$$= \int_0^3 6\rho^2 \, d\rho \int_0^{\pi/2} \cos\phi \, d\phi \int_0^2 dz + \int_0^3 6\rho \, d\rho$$
$$\times \int_0^{\pi/2} d\phi \int_0^2 dz = 192.82$$

Figure 2.31

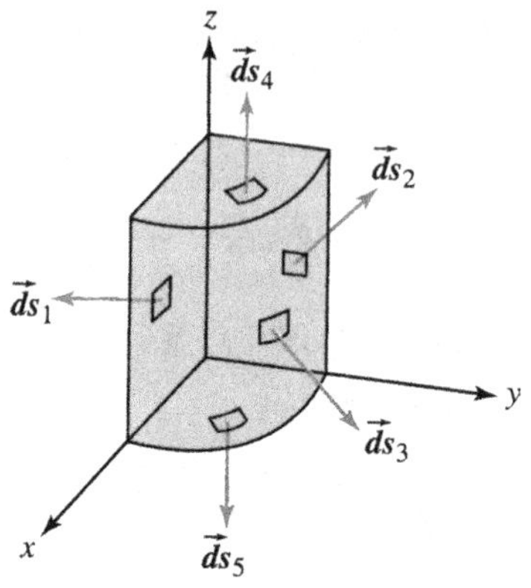

To evaluate the right-hand side of (2.92), we carry out the integration process separately for each of the five surfaces.

For the surface bounded by the plane $y = 0$: $\overrightarrow{d\mathbf{s}_1} = -dx\,dz\,\vec{\mathbf{a}}_y$.

$$\int_{s_1} \vec{\mathbf{D}} \cdot \overrightarrow{d\mathbf{s}_1} = -\int_{x=0}^{3} \int_{z=0}^{2} (3y + z)\,dx\,dz = -6$$

Over the surface bounded by the plane $x = 0$: $\overrightarrow{d\mathbf{s}_2} = -dy\,dz\,\vec{\mathbf{a}}_x$.

$$\int_{s_2} \vec{\mathbf{D}} \cdot \overrightarrow{d\mathbf{s}_2} = -\int_{y=0}^{3} \int_{z=0}^{2} 3x^2\,dy\,dz = 0$$

Over the surface bounded by the radius $\rho = 3$: $\overrightarrow{d\mathbf{s}_3} = 3\,d\phi\,dz\,\vec{\mathbf{a}}_\rho$.

$$\int_{s_3} \vec{\mathbf{D}} \cdot \overrightarrow{d\mathbf{s}_3} = \int_{\phi=0}^{\pi/2} \int_{z=0}^{2} 3D_\rho\,d\phi\,dz$$

However,

$$\begin{aligned} D_\rho &= D_x \cos\phi + D_y \sin\phi \\ &= 3x^2 \cos\phi + (3y + z)\sin\phi \end{aligned}$$

Therefore,

$$\int_{s_3} \vec{\mathbf{D}} \cdot \overrightarrow{d\mathbf{s}_3} = \int_{\phi=0}^{\pi/2} \int_{z=0}^{2} [3x^2 \cos\phi + (3y + z)\sin\phi]3\,d\phi\,dz$$

Substituting $x = 3\cos\phi$ and $y = 3\sin\phi$ in the preceding equation and performing the integration, we get

$$\int_{s_3} \vec{\mathbf{D}} \cdot \overrightarrow{d\mathbf{s}_3} = 156.41$$

Over the surface bounded by the plane $z = 2$: $\overrightarrow{d\mathbf{s}_4} = \rho\,d\rho\,d\phi\,\vec{\mathbf{a}}_z$.

$$\int_{s_4} \vec{\mathbf{D}} \cdot \overrightarrow{d\mathbf{s}_4} = \int_{\rho=0}^{3} \int_{\phi=0}^{\pi/2} (6 - x)\rho\,d\rho\,d\phi$$

Substituting $x = \rho \cos\phi$, the integral yields

$$\int_{s4} \vec{\mathbf{D}} \cdot \overrightarrow{ds_4} = 33.41$$

Finally, over the surface bounded by the plane $z = 0$: $\overrightarrow{ds_5} = -\rho\, d\rho\, d\phi\, \vec{\mathbf{a}}_z$.

$$\int_{s5} \vec{\mathbf{D}} \cdot \overrightarrow{ds_5} = \int_{\rho=0}^{3} \int_{\phi=0}^{\pi/2} x\rho\, d\rho\, d\phi = 9$$

Thus,

$$\oint_{s} \vec{\mathbf{D}} \cdot \overrightarrow{ds} = -6 + 0 + 156.41 + 33.41 + 9 = 192.82$$

which verifies the divergence theorem. • • •

2.10 The curl of a vector field

The line integral of a vector field $\vec{\mathbf{F}}$ around a closed path is called the *circulation* of $\vec{\mathbf{F}}$ and the curl of $\vec{\mathbf{F}}$ is its measure. If we consider a small surface element $\Delta s\, \vec{\mathbf{a}}_n$ bounded by a closed path Δc, we define the component of the curl parallel to the surface normal $\vec{\mathbf{a}}_n$, in the limit $\Delta s \to 0$, as

$$(\mathrm{curl}\, \vec{\mathbf{F}}) \cdot \vec{\mathbf{a}}_n = \lim_{\Delta s \to 0} \frac{1}{\Delta s} \oint_{\Delta c} \vec{\mathbf{F}} \cdot \overrightarrow{d\ell} \tag{2.93}$$

This definition suggests that the curl of a vector field is a vector quantity. The direction of path Δc is determined by the right-hand rule. Equation (2.93) provides a complete definition of **curl** $\vec{\mathbf{F}}$ because it enables us to determine each of the three components of **curl** $\vec{\mathbf{F}}$ in any arbitrary system of orthogonal coordinates.

We begin our evaluation of the z component of curl $\vec{\mathbf{F}}$ in rectangular coordinates by defining the vector field $\vec{\mathbf{F}}$ as

$$\vec{\mathbf{F}} = F_x \vec{\mathbf{a}}_x + F_y \vec{\mathbf{a}}_y + F_z \vec{\mathbf{a}}_z$$

at a point P within the small surface Δs bounded by path Δc, as illustrated in Figure 2.32. The line integral of $\vec{\mathbf{F}}$ along the closed path Δc

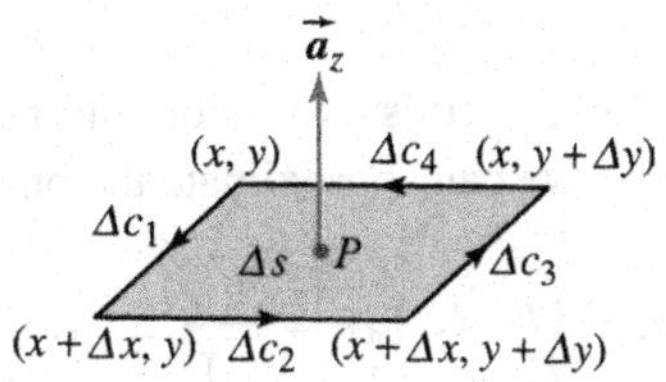

Figure 2.32 A small surface element for defining the curl of a vector field

consists of four separate paths:

$$\oint_{\Delta c} \vec{\mathbf{F}} \cdot \overrightarrow{d\ell} = \int_{\Delta c_1} \vec{\mathbf{F}} \cdot \overrightarrow{d\ell} + \int_{\Delta c_2} \vec{\mathbf{F}} \cdot \overrightarrow{d\ell} + \int_{\Delta c_3} \vec{\mathbf{F}} \cdot \overrightarrow{d\ell} + \int_{\Delta c_4} \vec{\mathbf{F}} \cdot \overrightarrow{d\ell} \tag{2.94}$$

We now evaluate each of the four integrals of (2.94) separately. Along the path Δc_1:

$$\begin{aligned} \int_{\Delta c_1} \vec{\mathbf{F}} \cdot \overrightarrow{d\ell} &= \int_{x}^{x+\Delta x} [F_x \vec{\mathbf{a}}_x + F_y \vec{\mathbf{a}}_y + F_z \vec{\mathbf{a}}_z]_{\text{at } y} \cdot [dx\ \vec{\mathbf{a}}_x] \\ &= [F_x \Delta x]_{\text{at } y} \end{aligned}$$

where $F_x \Delta x$ is to be evaluated at y, and we have made the assumption that the component F_x is approximately constant from x to $x + \Delta x$. This approximation is in accordance with the mean value theorem. We will make similar assumptions for the other components of $\vec{\mathbf{F}}$.

The line integration along Δc_2 is

$$\begin{aligned} \int_{\Delta c_2} \vec{\mathbf{F}} \cdot \overrightarrow{d\ell} &= \int_{y}^{y+\Delta y} [F_x \vec{\mathbf{a}}_x + F_y \vec{\mathbf{a}}_y + F_z \vec{\mathbf{a}}_z]_{\text{at } x+\Delta x} \cdot [dy\ \vec{\mathbf{a}}_y] \\ &= [F_y\ \Delta y]_{\text{at } x+\Delta x} \end{aligned}$$

The line integration along Δc_3 is

$$\begin{aligned} \int_{\Delta c_3} \vec{\mathbf{F}} \cdot \overrightarrow{d\ell} &= \int_{x+\Delta x}^{x} [F_x \vec{\mathbf{a}}_x + F_y \vec{\mathbf{a}}_y + F_z \vec{\mathbf{a}}_z]_{\text{at } y+\Delta y} \cdot [dx\ \vec{\mathbf{a}}_x] \\ &= -[F_x\ \Delta x]_{\text{at } y+\Delta y} \end{aligned}$$

Finally, for the path Δc_4,

$$\begin{aligned} \int_{\Delta c_4} \vec{\mathbf{F}} \cdot \overrightarrow{d\ell} &= \int_{y+\Delta y}^{y} [F_x \vec{\mathbf{a}}_x + F_y \vec{\mathbf{a}}_y + F_z \vec{\mathbf{a}}_z]_{\text{at } x} \cdot [dy\ \vec{\mathbf{a}}_y] \\ &= -[F_y\ \Delta y]_{\text{at } x} \end{aligned}$$

Thus,

$$\oint_{\Delta c} \vec{\mathbf{F}} \cdot \overrightarrow{d\ell} = [F_x\ \Delta x]_{\text{at } y} - [F_x\ \Delta x]_{\text{at } y+\Delta y} + [F_y\ \Delta y]_{\text{at } x+\Delta x} - [F_y\ \Delta y]_{\text{at } x}$$

However, in the limit $\Delta x \to 0$ and $\Delta y \to 0$, we can write

$$[F_x\ \Delta x]_{\text{at } y+\Delta y} - [F_x\ \Delta x]_{\text{at } y} = \frac{\partial F_x}{\partial y} \Delta x\ \Delta y$$

by using the Taylor series expansion and neglecting the higher-order terms. Likewise, we can approximate the other two terms as

$$[F_y\ \Delta y]_{\text{at } x+\Delta x} - [F_y\ \Delta y]_{\text{at } x} = \frac{\partial F_y}{\partial x}\ \Delta x\ \Delta y$$

Therefore,

$$\oint_{\Delta c} \vec{\mathbf{F}} \cdot \vec{d\ell} = \left[\frac{\partial F_y}{\partial x} - \frac{\partial F_x}{\partial y} \right] \Delta x \, \Delta y$$

Dividing both sides of the preceding equation by $\Delta s = \Delta x \, \Delta y$ and taking the limit as $\Delta s \to 0$, we obtain

$$\lim_{\Delta s \to 0} \frac{1}{\Delta s} \oint_{\Delta c \to 0} \vec{\mathbf{F}} \cdot \vec{d\ell} = \frac{\partial F_y}{\partial x} - \frac{\partial F_x}{\Delta y} \tag{2.95}$$

Because the unit vector $\vec{\mathbf{a}}_n = \vec{\mathbf{a}}_z$, we can write $(\text{curl}\,\vec{\mathbf{F}}) \cdot \vec{\mathbf{a}}_z$ as $(\text{curl}\,\vec{\mathbf{F}})_z$, where $(\text{curl}\,\vec{\mathbf{F}})_z$ signifies the z-directed component of curl $\vec{\mathbf{F}}$. Thus, from (2.93) and (2.95), we deduce that the z component of curl $\vec{\mathbf{F}}$ is

$$(\text{curl}\,\vec{\mathbf{F}})_z = \frac{\partial F_y}{\partial x} - \frac{\partial F_x}{\partial y} \tag{2.96a}$$

The other two components of curl $\vec{\mathbf{F}}$ can be obtained similarly by considering vanishingly small surfaces with unit normals in the x and y directions.

The expressions for these components are

$$(\text{curl}\,\vec{\mathbf{F}})_x = \frac{\partial F_z}{\partial y} - \frac{\partial F_y}{\partial z} \tag{2.96b}$$

and

$$(\text{curl}\,\vec{\mathbf{F}})_y = \frac{\partial F_x}{\partial z} - \frac{\partial F_z}{\partial x} \tag{2.96c}$$

Thus, the curl of vector field $\vec{\mathbf{F}}$, curl $\vec{\mathbf{F}}$, is

$$\text{curl}\,\vec{\mathbf{F}} = \left[\frac{\partial F_z}{\partial y} - \frac{\partial F_y}{\partial z} \right] \vec{\mathbf{a}}_x + \left[\frac{\partial F_x}{\partial z} - \frac{\partial F_z}{\partial x} \right] \vec{\mathbf{a}}_y + \left[\frac{\partial F_y}{\partial x} - \frac{\partial F_x}{\partial y} \right] \vec{\mathbf{a}}_z \tag{2.97}$$

in the rectangular coordinate system.

In terms of the cross product we can write equation (2.97) as

$$\begin{aligned} \text{curl}\,\vec{\mathbf{F}} &= \left[\vec{\mathbf{a}}_x \frac{\partial}{\partial x} + \vec{\mathbf{a}}_y \frac{\partial}{\partial y} + \vec{\mathbf{a}}_z \frac{\partial}{\partial z} \right] \times [F_x \vec{\mathbf{a}}_x + F_y \vec{\mathbf{a}}_y + F_z \vec{\mathbf{a}}_z] \\ &= \nabla \times \vec{\mathbf{F}} \end{aligned} \tag{2.98}$$

Thus, we will always write curl $\vec{\mathbf{F}}$ as $\nabla \times \vec{\mathbf{F}}$.

A useful and easy-to-remember expression of $\nabla \times \vec{\mathbf{F}}$ in the Cartesian coordinate system is

$$\nabla \times \vec{\mathbf{F}} = \begin{vmatrix} \vec{\mathbf{a}}_x & \vec{\mathbf{a}}_y & \vec{\mathbf{a}}_z \\ \dfrac{\partial}{\partial x} & \dfrac{\partial}{\partial y} & \dfrac{\partial}{\partial z} \\ F_x & F_y & F_z \end{vmatrix} \tag{2.99}$$

This determinant form of $\nabla \times \vec{\mathbf{F}}$ must be considered as a symbolic abbreviation. It is the expansion of this determinant that leads us to the form given by (2.97).

The expressions for the curl of the vector field $\vec{\mathbf{F}}$ in the cylindrical and spherical coordinate systems, respectively, are

$$\nabla \times \vec{\mathbf{F}} = \frac{1}{\rho}\begin{vmatrix} \vec{\mathbf{a}}_\rho & \rho\vec{\mathbf{a}}_\phi & \vec{\mathbf{a}}_z \\ \dfrac{\partial}{\partial \rho} & \dfrac{\partial}{\partial \phi} & \dfrac{\partial}{\partial z} \\ F_\rho & \rho F_\phi & F_z \end{vmatrix} \tag{2.100}$$

and

$$\nabla \times \vec{\mathbf{F}} = \frac{1}{r^2 \sin\theta}\begin{vmatrix} \vec{\mathbf{a}}_r & r\vec{\mathbf{a}}_\theta & r\sin\theta\,\vec{\mathbf{a}}_\phi \\ \dfrac{\partial}{\partial r} & \dfrac{\partial}{\partial \theta} & \dfrac{\partial}{\partial \phi} \\ F_r & rF_\theta & r\sin\theta\,F_\phi \end{vmatrix} \tag{2.101}$$

The physical significance of the curl of a vector field is that it represents the circulation per unit area of the vector field taken around a small area of any shape. Its direction is normal to the plane of the surface. Stated differently, if the line integral of a vector field about a closed elementary path is nonzero, the curl of the vector field is also nonzero and we say that the vector field is *rotational*. The flow of water out of a tub or a sink provides an excellent example of a rotational velocity field of the flow. On the other hand, if the curl of a vector field is zero, the vector field is said to be *irrotational* or *conservative*. A common example of a conservative field is the work done by a force acting on a body.

EXAMPLE 2.21 If $f(x, y, z)$ is a continuously differentiable scalar function, show that $\nabla \times (\nabla f) = 0$.

Solution The gradient of a scalar function $f(x, y, z)$ from (2.74), is

$$\nabla f = \frac{\partial f}{\partial x}\vec{\mathbf{a}}_x + \frac{\partial f}{\partial y}\vec{\mathbf{a}}_y + \frac{\partial f}{\partial z}\vec{\mathbf{a}}_z$$

From (2.99), the curl of ∇f is

$$\begin{aligned}\nabla \times \nabla f &= \begin{vmatrix} \vec{\mathbf{a}}_x & \vec{\mathbf{a}}_y & \vec{\mathbf{a}}_z \\ \dfrac{\partial}{\partial x} & \dfrac{\partial}{\partial y} & \dfrac{\partial}{\partial z} \\ \dfrac{\partial f}{\partial x} & \dfrac{\partial f}{\partial y} & \dfrac{\partial f}{\partial z} \end{vmatrix} \\ &= \left[\frac{\partial^2 f}{\partial y\,\partial z} - \frac{\partial^2 f}{\partial y\,\partial z}\right]\vec{\mathbf{a}}_x + \left[\frac{\partial^2 f}{\partial z\,\partial x} - \frac{\partial^2 f}{\partial x\,\partial z}\right]\vec{\mathbf{a}}_y \\ &\quad + \left[\frac{\partial^2 f}{\partial x\,\partial y} - \frac{\partial^2 f}{\partial x\,\partial y}\right]\vec{\mathbf{a}}_z\end{aligned}$$

For f to be continuously differentiable,

$$\frac{\partial^2 f}{\partial y\,\partial z} = \frac{\partial^2 f}{\partial z\,\partial y}, \quad \frac{\partial^2 f}{\partial z\,\partial x} = \frac{\partial^2 f}{\partial x\,\partial z}, \quad \text{and} \quad \frac{\partial^2 f}{\partial x\,\partial y} = \frac{\partial^2 f}{\partial y\,\partial x}.$$

Hence $\nabla \times (\nabla f) = 0$. • • •

Because the curl of a gradient of a scalar function is always zero, ∇f is an irrotational or conservative field. Conversely, if the curl of a vector field is zero, the vector field is the gradient of a scalar function. That is, if $\nabla \times \vec{\mathbf{F}} = 0$, then $\vec{\mathbf{F}} = \pm\nabla f$. The choice of plus (+) or minus (−) sign depends upon the physical interpretation of f.

2.10.1 Stokes' theorem

From our definition of $\nabla \times \vec{\mathbf{F}}$, equation (2.93), we can obtain a very important relation, known as Stokes' theorem, for a finite but open surface area s bounded by a closed contour c, as illustrated in Figure 2.33. Let us divide the surface area s into n elementary surface areas (cells) such that an ith cell has an area Δs_i with unit normal $\vec{\mathbf{a}}_{ni}$ and bounded by a closed path Δc_i enclosing a point P_i.

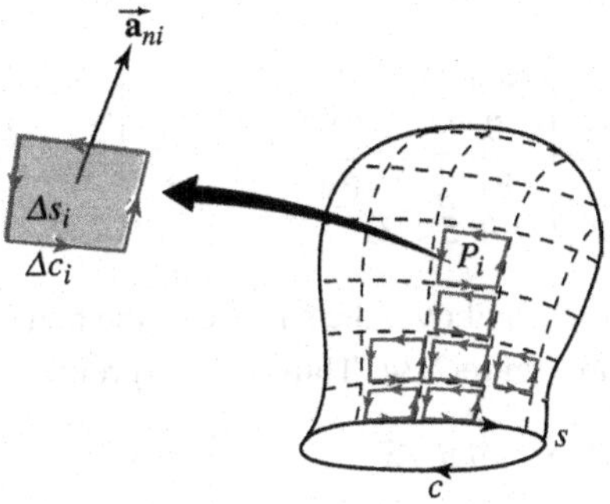

Figure 2.33 Open surface s bounded by a closed contour c to illustrate Stokes' theorem

From (2.93), we can write

$$\int_{\Delta s_i} (\nabla \times \vec{\mathbf{F}}) \cdot \vec{ds_i} = \oint_{\Delta c_i} \vec{\mathbf{F}} \cdot \vec{d\ell} + \epsilon_i\, \Delta s_i$$

where the term $\epsilon_i \Delta s_i$ has been added because (2.93) is strictly true only for a point in the limit when $n \to \infty$, $\epsilon_i \to 0$. Summing over the entire area, we get

$$\sum_{i=1}^{n} \int_{\Delta s_i} (\nabla \times \vec{\mathbf{F}}) \cdot \vec{ds_i} = \sum_{i=1}^{n} \oint_{\Delta c_i} \vec{\mathbf{F}} \cdot \vec{d\ell} + \sum_{i=1}^{n} \epsilon_i\, \Delta s_i \tag{2.102}$$

When $n \to \infty$, the left-hand side of the equation (2.102) becomes

$$\lim_{n\to\infty} \sum_{i=1}^{n} \int_{\Delta s_i} (\nabla \times \vec{\mathbf{F}}) \cdot \vec{ds_i} = \int_s (\nabla \times \vec{\mathbf{F}}) \cdot \vec{ds}$$

where the surface integration is over the open surface s bounded by contour c. The second term on the right-hand side of (2.102) reduces to zero as $n \to \infty$. On the other hand, the line integrals along adjacent elementary areas cancel because the length vectors are directed in opposite directions. The only contribution is from the integration over the path c. Thus,

$$\lim_{n\to\infty} \sum_{i=1}^{n} \oint_{\Delta c_i} \vec{\mathbf{F}} \cdot \overrightarrow{d\ell} = \oint_c \vec{\mathbf{F}} \cdot \overrightarrow{d\ell}$$

Consequently, (2.102) becomes

$$\int_s (\nabla \times \vec{\mathbf{F}}) \cdot \overrightarrow{d\mathbf{s}} = \oint_c \vec{\mathbf{F}} \cdot \overrightarrow{d\ell} \tag{2.103}$$

Equation (2.103) is a statement of *Stokes' theorem. It states that the integral of the normal component of the curl of a vector field over an area is equal to the line integral of the vector field along the curve bounding the area.*

EXAMPLE 2.22

If $\vec{\mathbf{F}} = (2z+5)\vec{\mathbf{a}}_x + (3x-2)\vec{\mathbf{a}}_y + (4x-1)\vec{\mathbf{a}}_z$, verify Stokes' theorem over the hemisphere $x^2 + y^2 + z^2 = 4$ and $z \geq 0$.

Solution

$$\nabla \times \vec{\mathbf{F}} = \begin{vmatrix} \vec{\mathbf{a}}_x & \vec{\mathbf{a}}_y & \vec{\mathbf{a}}_z \\ \dfrac{\partial}{\partial x} & \dfrac{\partial}{\partial y} & \dfrac{\partial}{\partial z} \\ 2z+5 & 3x-2 & 4x-1 \end{vmatrix} = -2\vec{\mathbf{a}}_y + 3\vec{\mathbf{a}}_z$$

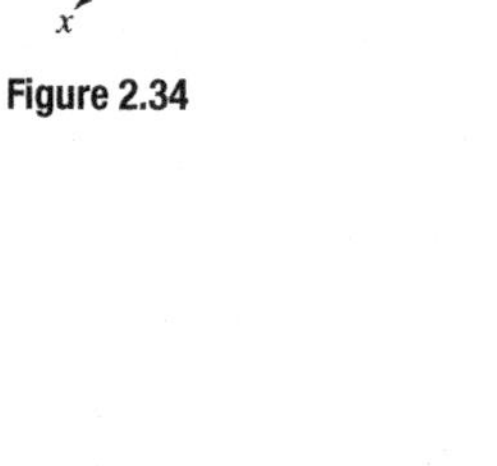

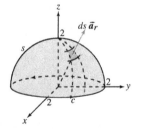

Figure 2.34

The unit normal over the surface of the hemisphere of radius 2 is $\vec{\mathbf{a}}_r$, as shown in Figure 2.34. Thus, the differential surface area is

$$\overrightarrow{d\mathbf{s}} = 4 \sin\theta \, d\theta \, d\phi \, \vec{\mathbf{a}}_r$$

Making the coordinate transformation from rectangular to spherical, the $\vec{\mathbf{a}}_r$ component of $\nabla \times \vec{\mathbf{F}}$ is

$$F_r = -2\sin\theta \sin\phi + 3\cos\theta$$

We can now evaluate the left-hand side of Stokes' theorem as

$$\int_s (\nabla \times \vec{\mathbf{F}}) \cdot \overrightarrow{d\mathbf{s}} = -8 \int_0^{\pi/2} \sin^2\theta \, d\theta \int_0^{2\pi} \sin\phi \, d\phi + 12 \int_0^{\pi/2} \sin\theta\cos\theta \, d\theta \int_0^{2\pi} d\phi = 12\pi$$

The right-hand side of Stokes' theorem involves the line integration over the closed path c in the xy plane with a radius of 2. Because the contour c describes a circle in the xy plane, we can use the cylindrical system to evaluate $\vec{\mathbf{F}} \cdot \overrightarrow{d\ell}$. The length vector of interest is $\overrightarrow{d\ell} = 2\,d\phi\,\vec{\mathbf{a}}_\phi$.

The $\vec{\mathbf{a}}_\phi$ component of $\vec{\mathbf{F}}$, using the rectangular-to-cylindrical coordinate transformation, is

$$F_\phi = -(2z+5)\sin\phi + (3x-2)\cos\phi$$

Substituting $z = 0$ and $x = 2\cos\phi$, we get

$$F_\phi = -5\sin\phi + 6\cos^2\phi - 2\cos\phi$$

Thus,

$$\oint_c \vec{\mathbf{F}}\cdot\vec{d\ell} = -10\int_0^{2\pi}\sin\phi\,d\phi + 12\int_0^{2\pi}\cos^2\phi\,d\phi - 4\int_0^{2\pi}\cos\phi\,d\phi = 12\pi$$

The line integral of $\vec{\mathbf{F}}$ being equal to the surface integral of $\nabla\times\vec{\mathbf{F}}$ verifies Stokes' theorem. • • •

2.11 The Laplacian operator

All of the differential operations discussed so far pertain to first-order differential operators. One second-order differential operator that occurs frequently in the study of field theory is called the **Laplacian operator**, symbolically written as ∇^2. It is defined as the divergence of a gradient of a scalar function. Simply put, if $f(x, y, z)$ is a continuously differentiable scalar function, the Laplacian of $f(x, y, z)$ is

$$\nabla^2 f = \nabla\cdot(\nabla f) \tag{2.104}$$

We can write the divergence of a scalar function f in the rectangular coordinate system as

$$\nabla\cdot(\nabla f) = \left[\vec{\mathbf{a}}_x\frac{\partial}{\partial x} + \vec{\mathbf{a}}_y\frac{\partial}{\partial y} + \vec{\mathbf{a}}_z\frac{\partial}{\partial z}\right]\cdot\left[\vec{\mathbf{a}}_x\frac{\partial f}{\partial x} + \vec{\mathbf{a}}_y\frac{\partial f}{\partial y} + \vec{\mathbf{a}}_z\frac{\partial f}{\partial z}\right]$$

which yields

$$\nabla^2 f = \nabla\cdot(\nabla f) = \frac{\partial^2 f}{\partial x^2} + \frac{\partial^2 f}{\partial y^2} + \frac{\partial^2 f}{\partial z^2} \tag{2.105}$$

From (2.105) it is evident that the Laplacian of a scalar function is a scalar and involves second-order partial differentiation of the function. By simple transformations, we can express the Laplacian of a scalar function in cylindrical coordinates as

$$\nabla^2 f = \frac{1}{\rho}\frac{\partial}{\partial\rho}\left(\rho\frac{\partial f}{\partial\rho}\right) + \frac{1}{\rho^2}\frac{\partial^2 f}{\partial\phi^2} + \frac{\partial^2 f}{\partial z^2} \tag{2.106}$$

A similar transformation from rectangular to spherical coordinates will yield the Laplacian of a scalar in spherical coordinates as

$$\nabla^2 f = \frac{1}{r^2}\frac{\partial}{\partial r}\left(r^2\frac{\partial f}{\partial r}\right) + \frac{1}{r^2 \sin\theta}\frac{\partial}{\partial \theta}\left(\sin\theta\frac{\partial f}{\partial \theta}\right) + \frac{1}{r^2 \sin^2\theta}\frac{\partial^2 f}{\partial \phi^2} \tag{2.107}$$

A scalar function is said to be a *harmonic function* if its Laplacian is zero. That is,

$$\nabla^2 f = 0 \tag{2.108}$$

This equation is routinely referred to as *Laplace's equation.*

In our discussion of electromagnetic fields we will also encounter expressions of the form $\nabla^2\vec{\mathbf{F}}$, where $\vec{\mathbf{F}}$ is a vector field. We call such an expression the Laplacian of a vector field $\vec{\mathbf{F}}$ and define it as

$$\nabla^2\vec{\mathbf{F}} = \nabla(\nabla \cdot \vec{\mathbf{F}}) - \nabla \times (\nabla \times \vec{\mathbf{F}}) \tag{2.109}$$

In the Cartesian coordinate system, equation (2.109) becomes

$$\nabla^2\vec{\mathbf{F}} = \vec{\mathbf{a}}_x \nabla^2 F_x + \vec{\mathbf{a}}_y \nabla^2 F_y + \vec{\mathbf{a}}_z \nabla^2 F_z \tag{2.110}$$

where

$$\nabla^2 = \frac{\partial^2}{\partial x^2} + \frac{\partial^2}{\partial y^2} + \frac{\partial^2}{\partial z^2} \tag{2.111}$$

is the Laplacian operator. *The Laplacian of a vector field is zero if and only if the Laplacian of each of its components is independently zero.*

EXAMPLE 2.23

Show that a scalar function $f = 1/r$, $r \neq 0$, where r is the distance of any point P in space from the origin, is a solution of Laplace's equation.

Solution For a scalar function $f = 1/r$ to be a solution of Laplace's equation, $\nabla^2 f$ must be zero. Working in spherical coordinates, we can write

$$\nabla^2 f = \nabla^2\left[\frac{1}{r}\right] = \frac{1}{r^2}\frac{\partial}{\partial r}\left[r^2\frac{\partial}{\partial r}\left(\frac{1}{r}\right)\right]$$

$$= \frac{1}{r^2}\frac{\partial}{\partial r}\left[r^2\left(-\frac{1}{r^2}\right)\right] = 0$$

• • •

2.12 Some theorems and field classifications

We now examine two more vector identities: Green's first identity and Green's second identity. It is Green's second identity, also known as Green's theorem, that is most useful in our discussion of electromagnetic fields. In the following subsections, using Green's identities we will also prove the uniqueness theorem. We will also show that a vector field must fall into one of four classifications, classes I–IV.

2.12.1 Green's theorem

Let $\vec{\mathbf{A}}$ be a single-valued, continuously differentiable vector field everywhere in volume v and on its surface s. Then according to the divergence theorem,

$$\int_v \nabla \cdot \vec{\mathbf{A}}\, dv = \oint_s \vec{\mathbf{A}} \cdot \overrightarrow{d\mathbf{s}} \tag{2.112}$$

If we define the vector field $\vec{\mathbf{A}}$ to be the product of a scalar field ϕ and a vector field $\nabla\psi$, then

$$\begin{aligned}\nabla \cdot \vec{\mathbf{A}} &= \nabla \cdot (\phi \nabla \psi) \\ &= \nabla\phi \cdot \nabla\psi + \phi \nabla^2 \psi\end{aligned}$$

and its substitution into (2.112) yields

$$\int_v \phi \nabla^2 \psi \, dv + \int_v \nabla\phi \cdot \nabla\psi \, dv = \oint_s \phi \nabla \psi \cdot \overrightarrow{d\mathbf{s}} \tag{2.113}$$

Equation (2.113) is known as *Green's first identity*. By interchanging ϕ and ψ, we can write (2.113) as

$$\int_v \psi \nabla^2 \phi \, dv + \int_v \nabla\psi \cdot \nabla\phi \, dv = \oint_s \psi \nabla \phi \cdot \overrightarrow{d\mathbf{s}} \tag{2.114}$$

Subtracting (2.114) from (2.113), we obtain *Green's second identity* or *Green's theorem* as

$$\int_v [\phi \nabla^2 \psi - \psi \nabla^2 \phi] \, dv = \oint_s [\phi \nabla \psi - \psi \nabla \phi] \cdot \overrightarrow{d\mathbf{s}} \tag{2.115}$$

For the special case of $\phi = \psi$, Green's first identity, (2.113), becomes

$$\int_v \phi \nabla^2 \phi \, dv + \int_v |\nabla \phi|^2 \, dv = \oint_s \phi \nabla \phi \cdot \overrightarrow{d\mathbf{s}} \tag{2.116}$$

We will use this equation to prove the uniqueness theorem.

2.12.2 The uniqueness theorem

The *uniqueness theorem* states that a vector field $\vec{\mathbf{A}}$ is uniquely determined in a region if the following requirements are satisfied:

a) Its divergence is specified throughout the region,
b) Its curl is specified throughout the region, and
c) Its normal component is specified on the closed surface bounding the region.

To prove this theorem, let us consider a volume v bounded by a surface s. Let us also assume that there are two distinct vector fields $\vec{\mathbf{A}}$ and $\vec{\mathbf{B}}$ (apart from an additive constant) that have the same divergence and curl throughout the volume v and the same normal component at the boundary s. In other words, for every point in v,

$$\nabla \cdot \vec{\mathbf{A}} = \nabla \cdot \vec{\mathbf{B}} \quad \text{and} \quad \nabla \times \vec{\mathbf{A}} = \nabla \times \vec{\mathbf{B}}$$

In addition, $\vec{\mathbf{A}} \cdot \overrightarrow{d\mathbf{s}} = \vec{\mathbf{B}} \cdot \overrightarrow{d\mathbf{s}}$ on any differential surface $\overrightarrow{d\mathbf{s}}$.

We now aim to prove that our assumption leads to an inconsistency. Let $\vec{\mathbf{C}}$ be a vector field such that $\vec{\mathbf{C}} = \vec{\mathbf{A}} - \vec{\mathbf{B}}$, then

$$\nabla \cdot \vec{\mathbf{C}} = \nabla \cdot \vec{\mathbf{A}} - \nabla \cdot \vec{\mathbf{B}} = 0$$

and

$$\nabla \times \vec{\mathbf{C}} = \nabla \times \vec{\mathbf{A}} - \nabla \times \vec{\mathbf{B}} = 0$$

throughout the volume v. Furthermore,

$$\vec{\mathbf{C}} \cdot \overrightarrow{d\mathbf{s}} = \vec{\mathbf{A}} \cdot \overrightarrow{d\mathbf{s}} - \vec{\mathbf{B}} \cdot \overrightarrow{d\mathbf{s}} = 0$$

on any differential surface $\overrightarrow{d\mathbf{s}}$. Since $\nabla \times \vec{\mathbf{C}} = 0$, $\vec{\mathbf{C}}$ can be expressed in terms of a gradient of a scalar function f. That is,

$$\vec{\mathbf{C}} = -\nabla f$$
$$\nabla \cdot \vec{\mathbf{C}} = 0 \Rightarrow \nabla \cdot (\nabla f) = 0$$

or

$$\nabla^2 f = 0$$

everywhere in v. Also,

$$\vec{\mathbf{C}} \cdot \overrightarrow{d\mathbf{s}} = 0 \Rightarrow \nabla f \cdot \overrightarrow{d\mathbf{s}} = 0$$

Substituting $\nabla^2 f = 0$ and $\nabla f \cdot \overrightarrow{d\mathbf{s}} = 0$ in (2.116), we obtain

$$\int_v |\nabla f|^2 \, dv = 0 \tag{2.117}$$

Because $|\nabla f|^2$ is a positive quantity, (2.117) can only be satisfied if $\nabla f = 0$ everywhere in v. Therefore, $\vec{\mathbf{C}}$ must be zero and $\vec{\mathbf{A}}$ must be equal to $\vec{\mathbf{B}}$ (apart from an additive constant). Thus, our original assumption that $\vec{\mathbf{A}}$ and $\vec{\mathbf{B}}$ are two distinct vector fields is groundless, which, in turn, establishes the uniqueness of a field.

2.12.3 Classification of fields

The divergence and curl of a vector field are independent operations; therefore, neither one is sufficient to describe a field completely. In fact, in our study of electromagnetic fields, we will find that fields fall into four basic classifications. In solving field problems it is necessary to know which class of field we are working with because this will dictate the procedure we must use to solve the problem. Therefore, let us now examine the traits of the fields pertaining to each class.

Class I fields

We will consider a vector field $\vec{\mathbf{F}}$ to be a class I field everywhere in a region if

$$\nabla \cdot \vec{\mathbf{F}} = 0 \quad \text{and} \quad \nabla \times \vec{\mathbf{F}} = 0$$

However, if the curl of a vector is zero, the vector can be written in terms of a gradient of a scalar function f. That is,

$$\vec{\mathbf{F}} = -\nabla f$$

The reason for the negative sign will become apparent in Chapter 3. From $\nabla \cdot \vec{\mathbf{F}} = 0$, we obtain

$$\nabla \cdot (-\nabla f) = \nabla^2 f = 0$$

which is Laplace's equation. Therefore, to obtain fields of class I, we need to solve Laplace's equation subjected to the conditions at the boundary of the region. Once we know f, we can compute the vector field $\vec{\mathbf{F}}$ as $\vec{\mathbf{F}} = -\nabla f$.

Examples of class I fields are electrostatic fields in charge-free medium and magnetic fields in current-free medium.

Class II fields

We will refer to a vector field $\vec{\mathbf{F}}$ as a class II field in a given region if

$$\nabla \cdot \vec{\mathbf{F}} \neq 0 \quad \text{and} \quad \nabla \times \vec{\mathbf{F}} = 0$$

Once again $\nabla \times \vec{\mathbf{F}} = 0$ implies $\vec{\mathbf{F}} = -\nabla f$. Because $\nabla \cdot \vec{\mathbf{F}} \neq 0$, we can write it as $\nabla \cdot \vec{\mathbf{F}} = \rho$, where ρ is either a constant or a known function within the region. Thus,

$$\nabla^2 f = -\rho$$

which is *Poisson's equation*. Thus, class II fields can be found by solving Poisson's equation within the constraints of the boundary conditions. We can then find the vector field $\vec{\mathbf{F}}$ as $\vec{\mathbf{F}} = -\nabla f$.

An electrostatic field in a charged region is an example of a class II field.

Class III fields

We will consider a vector field $\vec{\mathbf{F}}$ as a class III field in a given region if

$$\nabla \cdot \vec{\mathbf{F}} = 0 \quad \text{and} \quad \nabla \times \vec{\mathbf{F}} \neq 0$$

If the divergence of a vector is zero, then the vector can be expressed in terms of the curl of another vector. For $\nabla \cdot \vec{\mathbf{F}} = 0$, we can express $\vec{\mathbf{F}}$ as

$$\vec{\mathbf{F}} = \nabla \times \vec{\mathbf{A}}$$

where $\vec{\mathbf{A}}$ is another vector field. Because $\nabla \times \vec{\mathbf{F}} \neq 0$, we can write it as

$$\nabla \times \vec{\mathbf{F}} = \vec{\mathbf{J}}$$

where $\vec{\mathbf{J}}$ is a known vector field. Substituting $\vec{\mathbf{F}} = \nabla \times \vec{\mathbf{A}}$, we get

$$\nabla \times \nabla \times \vec{\mathbf{A}} = \vec{\mathbf{J}}$$

Using the vector identity, we can express this equation as

$$\nabla(\nabla \cdot \vec{\mathbf{A}}) - \nabla^2\vec{\mathbf{A}} = \vec{\mathbf{J}}$$

According to the uniqueness theorem, for $\vec{\mathbf{A}}$ to be a unique vector field, we must also define its divergence. If we set an arbitrary constraint that $\nabla \cdot \vec{\mathbf{A}} = 0$, we obtain

$$\nabla^2\vec{\mathbf{A}} = -\vec{\mathbf{J}}$$

which is called *Poisson's vector equation.* Therefore, class III fields require a solution of Poisson's vector equation. The vector field $\vec{\mathbf{F}}$ can be computed from $\vec{\mathbf{A}}$ as $\vec{\mathbf{F}} = \nabla \times \vec{\mathbf{A}}$. The constraint $\nabla \cdot \vec{\mathbf{A}} = 0$ is known as *Coulomb's gauge.*

The magnetic field within a current-carrying conductor falls into class III.

Class IV fields

For a vector field $\vec{\mathbf{F}}$ to be class IV, neither its divergence nor its curl is zero. However, we can decompose $\vec{\mathbf{F}}$ into two vector fields $\vec{\mathbf{G}}$ and $\vec{\mathbf{H}}$ such that $\vec{\mathbf{G}}$ satisfies class III and $\vec{\mathbf{H}}$ satisfies class II requirements. That is,

$$\vec{\mathbf{F}} = \vec{\mathbf{G}} + \vec{\mathbf{H}}$$

$$\nabla \cdot \vec{\mathbf{G}} = 0, \nabla \times \vec{\mathbf{G}} \neq 0, \nabla \times \vec{\mathbf{H}} = 0, \quad \text{and} \quad \nabla \cdot \vec{\mathbf{H}} \neq 0$$

Thus, $\vec{\mathbf{H}} = -\nabla f$ and $\vec{\mathbf{G}} = \nabla \times \vec{\mathbf{A}}$ lead us to conclude that

$$\vec{\mathbf{F}} = \nabla \times \vec{\mathbf{A}} - \nabla f$$

Hydrodynamic fields in a compressible medium are examples of class IV fields.

2.13 Vector identities

There are quite a few vector identities that are important in the study of electromagnetic fields. They are listed below, and we urge the reader to verify them using the rectangular coordinate system.

Zero:

$$\nabla \times (\nabla f) = 0 \tag{2.118}$$

$$\nabla \cdot (\nabla \times \vec{\mathbf{A}}) = 0 \tag{2.119}$$

Notation:

$$\nabla^2 f = \nabla \cdot (\nabla f) \tag{2.120}$$

$$\nabla^2\vec{\mathbf{A}} = \nabla(\nabla \cdot \vec{\mathbf{A}}) - \nabla \times \nabla \times \vec{\mathbf{A}} \tag{2.121}$$

Sums:

$$\nabla(f+g) = \nabla f + \nabla g \tag{2.122}$$

$$\nabla\cdot(\vec{\mathbf{A}}+\vec{\mathbf{B}}) = \boldsymbol{\nabla}\cdot\vec{\mathbf{A}} + \boldsymbol{\nabla}\cdot\vec{\mathbf{B}} \tag{2.123}$$

$$\boldsymbol{\nabla}\times(\vec{\mathbf{A}}+\vec{\mathbf{B}}) = \boldsymbol{\nabla}\times\vec{\mathbf{A}} + \boldsymbol{\nabla}\times\vec{\mathbf{B}} \tag{2.124}$$

Products including a scalar:

$$\nabla(fg) = f\,\nabla g + g\,\nabla f \tag{2.125}$$

$$\boldsymbol{\nabla}\cdot(f\vec{\mathbf{A}}) = f\boldsymbol{\nabla}\cdot\vec{\mathbf{A}} + \vec{\mathbf{A}}\cdot\nabla f \tag{2.126}$$

$$\boldsymbol{\nabla}\times(f\vec{\mathbf{A}}) = f\boldsymbol{\nabla}\times\vec{\mathbf{A}} + \nabla f\times\vec{\mathbf{A}} \tag{2.127}$$

Vector products:

$$\vec{\mathbf{A}}\cdot(\vec{\mathbf{B}}\times\vec{\mathbf{C}}) = \vec{\mathbf{B}}\cdot(\vec{\mathbf{C}}\times\vec{\mathbf{A}}) = \vec{\mathbf{C}}\cdot(\vec{\mathbf{A}}\times\vec{\mathbf{B}}) \tag{2.128}$$

$$\vec{\mathbf{A}}\times(\vec{\mathbf{B}}\times\vec{\mathbf{C}}) = \vec{\mathbf{B}}(\vec{\mathbf{A}}\cdot\vec{\mathbf{C}}) - \vec{\mathbf{C}}(\vec{\mathbf{A}}\cdot\vec{\mathbf{B}}) \tag{2.129}$$

$$\boldsymbol{\nabla}\cdot(\vec{\mathbf{A}}\times\vec{\mathbf{B}}) = \vec{\mathbf{B}}\cdot(\boldsymbol{\nabla}\times\vec{\mathbf{A}}) - \vec{\mathbf{A}}\cdot(\boldsymbol{\nabla}\times\vec{\mathbf{B}}) \tag{2.130}$$

$$\boldsymbol{\nabla}\times(\vec{\mathbf{A}}\times\vec{\mathbf{B}}) = \vec{\mathbf{A}}\boldsymbol{\nabla}\cdot\vec{\mathbf{B}} - \vec{\mathbf{B}}\boldsymbol{\nabla}\cdot\vec{\mathbf{A}} + (\vec{\mathbf{B}}\cdot\boldsymbol{\nabla})\vec{\mathbf{A}} - (\vec{\mathbf{A}}\cdot\boldsymbol{\nabla})\vec{\mathbf{B}} \tag{2.131}$$

Note that f and g are scalar fields and $\vec{\mathbf{A}}$, $\vec{\mathbf{B}}$, and $\vec{\mathbf{C}}$ are vector fields. All fields are single valued and continuously differentiable everywhere in a region and on its bounding surface.

2.14 Summary

In this section we rephrase the definitions used in this chapter and list some of the key equations.

If a physical entity can be completely defined by its magnitude, it is a *scalar*.

A physical entity is called a *vector* if it requires both magnitude and direction for its representation.

A function that characterizes a physical entity at all points in a region is called a *field*. A *scalar field* is specified by a single number at each point in the region. A *vector field* demands the knowledge of both the magnitude and direction for its specification at each point in the region.

To perform the following operations on vector fields in cylindrical coordinates the fields must be defined either at the same point or in a $\phi =$ constant plane. To do the same in a spherical coordinate system, the field must be defined at the intersections of $\theta =$ constant and $\phi =$ constant planes. In other words, the fields must be defined either at the same point or at points on the same radial line. If these conditions are not met, we must first transform the fields into their rectangular components and then perform the required operation.

The dot (scalar) product: $\vec{\mathbf{A}} \cdot \vec{\mathbf{B}} = AB\cos\theta$

Rectangular coordinates	$A_xB_x + A_yB_y + A_zB_z$
Cylindrical coordinates	$A_\rho B_\rho + A_\phi B_\phi + A_zB_z$
Spherical coordinates	$A_rB_r + A_\theta B_\theta + A_\phi B_\phi$

The cross (vector) product: $\vec{\mathbf{A}} \times \vec{\mathbf{B}} = \lvert AB\sin\theta\rvert\, \vec{\mathbf{a}}_n$

Rectangular coordinates	$\begin{vmatrix} \vec{\mathbf{a}}_x & \vec{\mathbf{a}}_y & \vec{\mathbf{a}}_z \\ A_x & A_y & A_z \\ B_x & B_y & B_z \end{vmatrix}$
Cylindrical coordinates	$\begin{vmatrix} \vec{\mathbf{a}}_\rho & \vec{\mathbf{a}}_\phi & \vec{\mathbf{a}}_z \\ A_\rho & A_\phi & A_z \\ B_\rho & B_\phi & B_z \end{vmatrix}$
Spherical coordinates	$\begin{vmatrix} \vec{\mathbf{a}}_r & \vec{\mathbf{a}}_\theta & \vec{\mathbf{a}}_\phi \\ A_r & A_\theta & A_\phi \\ B_r & B_\theta & B_\phi \end{vmatrix}$

The gradient of a scalar function: ∇f

Rectangular coordinates	$\frac{\partial f}{\partial x}\vec{\mathbf{a}}_x + \frac{\partial f}{\partial y}\vec{\mathbf{a}}_y + \frac{\partial f}{\partial z}\vec{\mathbf{a}}_z$
Cylindrical coordinates	$\frac{\partial f}{\partial \rho}\vec{\mathbf{a}}_\rho + \frac{1}{\rho}\frac{\partial f}{\partial \phi}\vec{\mathbf{a}}_\phi + \frac{\partial f}{\partial z}\vec{\mathbf{a}}_z$
Spherical coordinates	$\frac{\partial f}{\partial r}\vec{\mathbf{a}}_r + \frac{1}{r}\frac{\partial f}{\partial \theta}\vec{\mathbf{a}}_\theta + \frac{1}{r\sin\theta}\frac{\partial f}{\partial \phi}\vec{\mathbf{a}}_\phi$

The divergence of a vector field: $\nabla \cdot \vec{\mathbf{A}}$

Rectangular coordinates	$\frac{\partial A_x}{\partial x} + \frac{\partial A_y}{\partial y} + \frac{\partial A_z}{\partial z}$
Cylindrical coordinates	$\frac{1}{\rho}\frac{\partial}{\partial \rho}(\rho A_\rho) + \frac{1}{\rho}\frac{\partial}{\partial \phi}(A_\phi) + \frac{\partial}{\partial z}(A_z)$
Spherical coordinates	$\frac{1}{r^2}\frac{\partial}{\partial r}(r^2 A_r) + \frac{1}{r\sin\theta}\frac{\partial}{\partial \theta}(\sin\theta A_\theta) + \frac{1}{r\sin\theta}\frac{\partial}{\partial \phi}(A_\phi)$

The curl of a vector field: $\nabla \times \vec{\mathbf{B}}$

Rectangular coordinates	$\begin{vmatrix} \vec{\mathbf{a}}_x & \vec{\mathbf{a}}_y & \vec{\mathbf{a}}_z \\ \frac{\partial}{\partial x} & \frac{\partial}{\partial y} & \frac{\partial}{\partial z} \\ B_x & B_y & B_z \end{vmatrix}$
Cylindrical coordinates	$\frac{1}{\rho}\begin{vmatrix} \vec{\mathbf{a}}_\rho & \rho\vec{\mathbf{a}}_\phi & \vec{\mathbf{a}}_z \\ \frac{\partial}{\partial \rho} & \frac{\partial}{\partial \phi} & \frac{\partial}{\partial z} \\ B_\rho & \rho B_\phi & B_z \end{vmatrix}$
Spherical coordinates	$\frac{1}{r^2\sin\theta}\begin{vmatrix} \vec{\mathbf{a}}_r & r\vec{\mathbf{a}}_\theta & r\sin\theta\,\vec{\mathbf{a}}_\phi \\ \frac{\partial}{\partial r} & \frac{\partial}{\partial \theta} & \frac{\partial}{\partial \phi} \\ B_r & rB_\theta & r\sin\theta B_\phi \end{vmatrix}$

The Laplacian of a scalar function: $\nabla^2 f$

Rectangular coordinates	$\dfrac{\partial^2 f}{\partial x^2} + \dfrac{\partial^2 f}{\partial y^2} + \dfrac{\partial^2 f}{\partial z^2}$
Cylindrical coordinates	$\dfrac{1}{\rho}\dfrac{\partial}{\partial \rho}\left(\rho \dfrac{\partial f}{\partial \rho}\right) + \dfrac{1}{\rho^2}\dfrac{\partial^2 f}{\partial \phi^2} + \dfrac{\partial^2 f}{\partial z^2}$
Spherical coordinates	$\dfrac{1}{r^2}\dfrac{\partial}{\partial r}\left(r^2 \dfrac{\partial f}{\partial r}\right) + \dfrac{1}{r^2 \sin\theta}\dfrac{\partial}{\partial \theta}\left(\sin\theta \dfrac{\partial f}{\partial \theta}\right) + \dfrac{1}{r^2 \sin^2\theta}\dfrac{\partial^2 f}{\partial \phi^2}$

The divergence theorem: $$\int_v \nabla \cdot \vec{\mathbf{F}}\, dv = \oint_s \vec{\mathbf{F}} \cdot \vec{ds}$$

Stokes' theorem: $$\int_s (\nabla \times \vec{\mathbf{F}}) \cdot \vec{ds} = \oint_c \vec{\mathbf{F}} \cdot \vec{d\ell}$$

Green's first identity: $$\int_v \phi \nabla^2 \psi\, dv + \int_v \nabla\phi \cdot \nabla\psi\, dv = \oint_s \phi \nabla\psi \cdot \vec{ds}$$

Green's second identity (Green's theorem):

$$\int_v [\phi \nabla^2 \psi - \psi \nabla^2 \phi]\, dv = \oint_s [\phi \nabla\psi - \psi \nabla\phi] \cdot \vec{ds}$$

2.15 Review questions

2.1 What is a scalar quantity? Cite some examples of scalars.

2.2 What is a vector quantity? Give some examples of vectors.

2.3 What do we mean when we say that two vectors are equal?

2.4 Is vector addition "closed"?

2.5 What is the significance of a zero vector?

2.6 Can the dot product be negative? If yes, what must be the condition?

2.7 Can you reason why the dot product of two vectors is known as the scalar product?

2.8 How can you determine if two vectors are dependent or independent?

2.9 Is division of a vector by another vector defined?

2.10 Give some physical examples of dot product and cross product.

2.11 Is the projection of a vector on another vector unique?

2.12 How can you determine the area of a parallelogram using vectors?

2.13 What is the right-hand rule?

2.14 If a vector $\vec{\mathbf{A}}$ is given at point $P(3, \pi/6, 10)$ and vector $\vec{\mathbf{B}}$ is given at $Q(1, \pi/6, 5)$ in cylindrical coordinates, can vector operations be performed without transforming into rectangular coordinates?

2.15 Two vectors $\vec{\mathbf{A}}$ and $\vec{\mathbf{B}}$ are given in the spherical coordinate system at $(2, \pi/2, 2\pi/3)$ and $(10, \pi/2, 2\pi/3)$. Can vector operations be performed without making a transformation from spherical to rectangular coordinates?

2.16 What do we mean by the gradient of a scalar function?

2.17 What does the divergence of a vector signify?

2.18 What is the significance of the curl of a vector?

2.19 Which equations will you use to check if a vector is (a) continuous, (b) solenoidal, (c) rotational, (d) irrotational, and (e) conservative? Give some real-life examples for each case.

2.20 How many vector surfaces does a thin sheet of paper possess if we assume that its thickness $\to 0$?

2.21 If the height of a pillbox approaches zero, how many vector surfaces does it possess?

2.22 If the line integral around a closed loop of vector $\vec{\mathbf{E}}$ is zero, $\vec{\mathbf{E}}$ represents a ________ field.

2.23 If a vector field $\vec{\mathbf{E}}$ can be written in terms of a gradient of a scalar field f, the vector field is ________ in nature.

2.24 If the divergence of a vector field $\vec{\mathbf{B}}$ is zero, the vector field is ________.

2.25 If $\oint \vec{\mathbf{B}} \cdot \vec{ds}$ around a closed surface is zero, the vector field $\vec{\mathbf{B}}$ is said to be ________.

2.26 If the divergence of a vector field $\vec{\mathbf{B}}$ is zero, $\vec{\mathbf{B}}$ can be expressed in terms of another unknown vector $\vec{\mathbf{A}}$ such that $\vec{\mathbf{B}} =$ ________. Is $\vec{\mathbf{A}}$ uniquely defined?

2.27 A thermal field is defined as $\vec{\mathbf{E}} = -\nabla\Phi$ and $\boldsymbol{\nabla} \cdot \vec{\mathbf{E}} = 0$. What is the class of this thermal field?

2.28 State the divergence theorem. What are its advantages and limitations?

2.29 What is Stokes' theorem? What are its advantages and limitations? Can Stokes' theorem be applied to closed surfaces?

2.30 What are Green's identities? Is the uniqueness theorem a consequence of Green's theorem?

2.16 Exercises

2.1 Verify the commutative law for addition of vectors.

2.2 Show that the necessary and sufficient condition for two nonzero vectors $\vec{\mathbf{A}}$ and $\vec{\mathbf{B}}$ to be perpendicular is that $\vec{\mathbf{A}} \cdot \vec{\mathbf{B}} = 0$.

2.3 Prove that vectors obey the distributive law for the scalar product.

2.4 Verify the Pythagorean theorem. In other words, show that $|\vec{\mathbf{A}} + \vec{\mathbf{B}}|^2 = A^2 + B^2$ if and only if $\vec{\mathbf{A}}$ is perpendicular to $\vec{\mathbf{B}}$.

2.5 Prove that vectors obey the distributive law for the cross product.

2.6 Prove that two nonzero vectors are parallel if and only if their cross product is zero.

2.7 Show that $\vec{\mathbf{A}} \cdot (\vec{\mathbf{B}} \times \vec{\mathbf{C}}) = \vec{\mathbf{B}} \cdot (\vec{\mathbf{C}} \times \vec{\mathbf{A}}) = \vec{\mathbf{C}} \cdot (\vec{\mathbf{A}} \times \vec{\mathbf{B}})$.

2.8 Show that $(\vec{\mathbf{A}} \times \vec{\mathbf{B}}) \cdot (\vec{\mathbf{C}} \times \vec{\mathbf{D}}) = (\vec{\mathbf{A}} \cdot \vec{\mathbf{C}})(\vec{\mathbf{B}} \cdot \vec{\mathbf{D}}) - (\vec{\mathbf{A}} \cdot \vec{\mathbf{D}})(\vec{\mathbf{B}} \cdot \vec{\mathbf{C}})$.

2.9 If $\vec{\mathbf{A}} = 2\vec{\mathbf{a}}_x + 0.3\vec{\mathbf{a}}_y - 1.5\vec{\mathbf{a}}_z$ and $\vec{\mathbf{B}} = 10\vec{\mathbf{a}}_x + 1.5\vec{\mathbf{a}}_y - 7.5\vec{\mathbf{a}}_z$, show that $\vec{\mathbf{A}}$ and $\vec{\mathbf{B}}$ are dependent vectors.

2.10 Compute the distance vector from $P(0, -2, 1)$ to $Q(-2, 0, 3)$.

2.11 Show that if $\vec{\mathbf{A}} = 3\vec{\mathbf{a}}_x + 2\vec{\mathbf{a}}_y - \vec{\mathbf{a}}_z$, $\vec{\mathbf{B}} = 4\vec{\mathbf{a}}_x - 8\vec{\mathbf{a}}_y - 4\vec{\mathbf{a}}_z$, and $\vec{\mathbf{C}} = 7\vec{\mathbf{a}}_x - 6\vec{\mathbf{a}}_y - 5\vec{\mathbf{a}}_z$, then $\vec{\mathbf{A}}$, $\vec{\mathbf{B}}$, and $\vec{\mathbf{C}}$ form a right-angle triangle.

2.12 If $\vec{\mathbf{S}} = 3\vec{\mathbf{a}}_x + 5\vec{\mathbf{a}}_y + 17\vec{\mathbf{a}}_z$ and $\vec{\mathbf{G}} = -\vec{\mathbf{a}}_y - 5\vec{\mathbf{a}}_z$, find a unit vector parallel to the sum of $\vec{\mathbf{S}} + \vec{\mathbf{G}}$. Calculate the angle the unit vector makes with the x axis.

2.13 Verify the transformation given in (2.39).

2.14 Calculate $\vec{\mathbf{C}}$ in Example 2.10 using the method of vector projection.

2.15 Express the following vectors in the rectangular coordinate system:

a) $\vec{\mathbf{F}} = \rho \sin\phi\, \vec{\mathbf{a}}_\rho - \rho\cos\phi\, \vec{\mathbf{a}}_\phi$ b) $\vec{\mathbf{H}} = \dfrac{1}{\rho}\vec{\mathbf{a}}_\rho$

2.16 Two points $P(1, \pi, 0)$ and $Q(0, -\pi/2, 2)$ are given in the cylindrical coordinate system. Find the distance vector from P to Q. What is its length? What is the distance vector from Q to P? Express the distance vector from P to Q in terms of the distance vector from Q to P.

2.17 Express the position vector $\vec{\mathbf{r}} = x\vec{\mathbf{a}}_x + y\vec{\mathbf{a}}_y + z\vec{\mathbf{a}}_z$ in the spherical coordinate system.

2.18 If $\vec{\mathbf{F}} = r\vec{\mathbf{a}}_r + r\tan\theta\, \vec{\mathbf{a}}_\theta + r\sin\theta\cos\phi\, \vec{\mathbf{a}}_\phi$, transform $\vec{\mathbf{F}}$ to the rectangular coordinate system.

2.19 Obtain the length of the distance vector from point $P(2, \pi/2, 3\pi/4)$ to $Q(10, \pi/4, \pi/2)$.

2.20 $\vec{\mathbf{S}} = 12\vec{\mathbf{a}}_r + 5\vec{\mathbf{a}}_\theta + \pi\vec{\mathbf{a}}_\phi$ and $\vec{\mathbf{T}} = 2\vec{\mathbf{a}}_r + 0.5\pi\vec{\mathbf{a}}_\theta$ are two vectors at point $(2, \pi, \pi/2)$ and $(5, \pi/2, \pi/2)$, respectively. Determine (a) $\vec{\mathbf{S}} + \vec{\mathbf{T}}$, (b) $\vec{\mathbf{S}} \cdot \vec{\mathbf{T}}$, (c) $\vec{\mathbf{S}} \times \vec{\mathbf{T}}$, (d) the unit vector perpendicular to $\vec{\mathbf{S}} \times \vec{\mathbf{T}}$, and (e) the angle between $\vec{\mathbf{S}}$ and $\vec{\mathbf{T}}$.

2.21 Given a scalar function $g = g[u(t), v(t), s(t)]$, obtain an expression for the derivative of g with respect to $t[dg/dt]$.

2.22 If $G = G(x, y, z, t)$ where x, y, and z are also functions of t, obtain an expression for dG/dt.

2.23 The partial derivative of $\vec{\mathbf{F}}$ with respect to x is given by equation (2.50). What are the expressions for the partial derivatives of $\vec{\mathbf{F}}$ with respect to y and z?

2.24 Obtain equation (2.53) by differentiating the position vector $\vec{\mathbf{r}}$ in the rectangular coordinate system.

2.25 Differentiate the position vector $\vec{\mathbf{r}}$ in the cylindrical coordinate system to obtain equation (2.56).

2.26 Show that equation (2.59) can be obtained by differentiating the position vector $\vec{\mathbf{r}}$ in the spherical coordinate system.

2.27 If $g = 20xy$, evaluate $\int g\,\vec{d\ell}$ from $P(0, 0, 0)$ to $Q(1, 1, 0)$ along (a) the straight line that joins P and Q, and (b) the curve $y = 4x^2$.

2.28 Evaluate $\oint \vec{\rho} \cdot \vec{d\ell}$ along the closed circular path of radius b in the xy plane.

2.29 Determine $\oint \vec{\mathbf{r}} \cdot \vec{d\mathbf{s}}$ over the closed surface of a sphere of radius b.

2.30 Find the volume of a region bounded by the xy plane ($z = 0$) and $z = 4 - x^2 - y^2$.

2.31 Using equation (2.76a) for the differential change in a scalar function f, verify the expressions given in (2.77) and (2.78) for the gradient of f in cylindrical and spherical coordinates, respectively.

2.32 Using the definitions of the position vector $\vec{\mathbf{r}}$ in the rectangular and spherical coordinate systems, show that $\nabla r = \vec{\mathbf{a}}_r$.

2.33 Find the maximum rate of change of a function $f = 12x^2 + yz^2$ with respect to distance at point $P(-1, 0, 1)$. Determine the rate of change of f in the x, y, and z directions. What is the rate of change of f in the direction of point $Q(1, 1, 1)$ from P?

2.34 Using both the cylindrical and spherical coordinate systems, verify that $\boldsymbol{\nabla} \cdot \vec{\mathbf{r}} = 3$.

2.35 If $\vec{\mathbf{F}} = -xy\vec{\mathbf{a}}_x + 3x^2yz\vec{\mathbf{a}}_y + z^3x\vec{\mathbf{a}}_z$, find $\boldsymbol{\nabla} \cdot \vec{\mathbf{F}}$ at $P(1, -1, 2)$.

2.36 If $\vec{\mathbf{r}} = r\vec{\mathbf{a}}_r$, show that $\boldsymbol{\nabla} \cdot (r^n\vec{\mathbf{a}}_r) = (n+2)r^{n-1}$.

2.37 Verify the divergence theorem for a vector field $\vec{\mathbf{F}} = x\vec{\mathbf{a}}_x + xy\vec{\mathbf{a}}_y + xyz\vec{\mathbf{a}}_z$ in the region bounded by a sphere of radius 2.

2.38 Verify equations (2.96b) and (2.96c).

2.39 Verify that equations (2.100) and (2.101) are correct.

2.40 Determine $\nabla \times \vec{\mathbf{F}}$ if $\vec{\mathbf{F}} = (x/r)\vec{\mathbf{a}}_x$, where r is the magnitude of the position vector of a point $P(x, y, z)$ in space.

2.41 Show that the divergence of a curl of a vector field is always zero; i.e., $\nabla \cdot (\nabla \times \vec{\mathbf{F}}) = 0$.

2.42 In Example 2.21 we showed that $\nabla \times (\nabla f) = 0$ using the rectangular coordinate system. Show that it is always true no matter which coordinate system we use.

2.43 Verify Stokes' theorem over the hemisphere shown in Figure 2.34 if the vector field is $\vec{\mathbf{F}} = 10\cos\theta\,\vec{\mathbf{a}}_r - 10\sin\theta\,\vec{\mathbf{a}}_\theta$.

2.44 If $g = 25x^2yz + 12xy^2$, show that $\nabla^2 g = \nabla \cdot (\nabla g)$.

2.45 If $f = 2x^2y^3 + 3yz^3$, show that $\nabla^2 f = \nabla \cdot (\nabla f)$.

2.46 If $h = \rho^2 \sin 2\phi + z^3 \cos\phi$, show that $\nabla^2 h = \nabla \cdot (\nabla h)$.

2.47 A cable with two concentric conductors insulated from one another by a dielectric medium is called a coaxial cable. We are given that the radius of the inner conductor is a and that of the outer conductor is b. The potential distribution within the conductors is given as $\Phi = K \ln(b/\rho)$, where K is a constant. Show that the potential distribution satisfies Laplace's equation.

2.48 Show that the potential distribution as given in Exercise 2.43 also satisfies Green's theorem, equation (2.116). [*Hint:* evaluate each integral on the basis of per unit length of the coaxial cable.]

2.17 Problems

2.1 If A, B, and C form the three sides of a triangle with angle θ opposite to side C, use vectors to prove that $C = [A^2 + B^2 - 2AB\cos\theta]^{1/2}$.

2.2 If $\vec{\mathbf{A}}$, $\vec{\mathbf{B}}$, and $\vec{\mathbf{C}}$ are coplanar vectors, show that $\vec{\mathbf{A}} \cdot (\vec{\mathbf{B}} \times \vec{\mathbf{C}}) = 0$.

2.3 If $P(x, y, z)$ is any point on the surface of a sphere centered at (2, 3, 4), obtain the equation of the sphere using vectors.

2.4 Given $\vec{\mathbf{A}} = \vec{\mathbf{a}}_x \cos\alpha + \vec{\mathbf{a}}_y \sin\alpha$, $\vec{\mathbf{B}} = \vec{\mathbf{a}}_x \cos\beta - \vec{\mathbf{a}}_y \sin\beta$, and $\vec{\mathbf{C}} = \vec{\mathbf{a}}_x \cos\beta + \vec{\mathbf{a}}_y \sin\beta$, show that each is a unit vector. If $\beta < \alpha$, sketch these vectors and show that they are coplanar. Using these vectors obtain the following trigonometric identities: $\sin(\alpha + \beta) = \sin\alpha\cos\beta + \cos\alpha\sin\beta$, and $\sin(\alpha - \beta) = \sin\alpha\cos\beta - \cos\alpha\sin\beta$.

2.5 If $\vec{\mathbf{A}} = \vec{\mathbf{a}}_x + \vec{\mathbf{a}}_y + \vec{\mathbf{a}}_z$ and $\vec{\mathbf{B}} = 4\vec{\mathbf{a}}_x + 4\vec{\mathbf{a}}_y + \vec{\mathbf{a}}_z$, are the two position vectors, what is the distance vector from $\vec{\mathbf{A}}$ to $\vec{\mathbf{B}}$? What is its magnitude?

2.6 If $\vec{\mathbf{A}} = 3\vec{\mathbf{a}}_x + 2\vec{\mathbf{a}}_y - \vec{\mathbf{a}}_z$ and $\vec{\mathbf{B}} = \vec{\mathbf{a}}_x - 2\vec{\mathbf{a}}_y + 3\vec{\mathbf{a}}_z$, determine (a) $\vec{\mathbf{A}} + \vec{\mathbf{B}}$, (b) $\vec{\mathbf{A}} \cdot \vec{\mathbf{B}}$, (c) $\vec{\mathbf{A}} \times \vec{\mathbf{B}}$, (d) the unit normal to $\vec{\mathbf{A}}$ and $\vec{\mathbf{B}}$, (e) the smaller angle between $\vec{\mathbf{A}}$ and $\vec{\mathbf{B}}$, and (f) the scalar and vector projections of $\vec{\mathbf{A}}$ onto $\vec{\mathbf{B}}$.

2.7 The position vectors of points P and Q are given as $5\vec{\mathbf{a}}_x + 12\vec{\mathbf{a}}_y + \vec{\mathbf{a}}_z$ and $2\vec{\mathbf{a}}_x - 3\vec{\mathbf{a}}_y + \vec{\mathbf{a}}_z$, respectively. What is the distance vector from P to Q? What is its length? Is the length segment parallel to the xy plane? What are the coordinates of points P and Q?

2.8 Show that the vectors $\vec{\mathbf{A}} = 5\vec{\mathbf{a}}_x - 5\vec{\mathbf{a}}_y$, $\vec{\mathbf{B}} = 3\vec{\mathbf{a}}_x - 7\vec{\mathbf{a}}_y - \vec{\mathbf{a}}_z$, and $\vec{\mathbf{C}} = -2\vec{\mathbf{a}}_x - 2\vec{\mathbf{a}}_y - \vec{\mathbf{a}}_z$ form the sides of a right-angle triangle. Calculate its area using the vector product.

2.9 Show that $\vec{\mathbf{A}} = 6\vec{\mathbf{a}}_x + 5\vec{\mathbf{a}}_y - 10\vec{\mathbf{a}}_z$ and $\vec{\mathbf{B}} = 5\vec{\mathbf{a}}_x + 2\vec{\mathbf{a}}_y + 4\vec{\mathbf{a}}_z$ are orthogonal vectors.

2.10 Determine the volume of the parallelepiped formed by the lengths of the vectors $\vec{\mathbf{A}} = -2\vec{\mathbf{a}}_x - 3\vec{\mathbf{a}}_y + \vec{\mathbf{a}}_z$, $\vec{\mathbf{B}} = 2\vec{\mathbf{a}}_x - 5\vec{\mathbf{a}}_y + 3\vec{\mathbf{a}}_z$, and $\vec{\mathbf{C}} = 4\vec{\mathbf{a}}_x + 2\vec{\mathbf{a}}_y + 6\vec{\mathbf{a}}_z$.

2.11 Find a unit vector normal to both vectors $\vec{\mathbf{A}} = 4\vec{\mathbf{a}}_x - 3\vec{\mathbf{a}}_y + \vec{\mathbf{a}}_z$ and $\vec{\mathbf{B}} = 2\vec{\mathbf{a}}_x + \vec{\mathbf{a}}_y - \vec{\mathbf{a}}_z$.

2.12 Using vectors, find the area of a triangle formed by the points $P(1, 1, 1)$, $Q(3, 2, 5)$, and $S(5, 7, 9)$.

2.13 Find the smallest angle between the two vectors given in Problem 2.11.

2.14 The components of two vectors at a common point in space are given in cylindrical coordinates as $\vec{\mathbf{A}} = 3\vec{\mathbf{a}}_\rho + 5\vec{\mathbf{a}}_\phi - 4\vec{\mathbf{a}}_z$ and $\vec{\mathbf{B}} = 2\vec{\mathbf{a}}_\rho + 4\vec{\mathbf{a}}_\phi + 3\vec{\mathbf{a}}_z$. Compute (a) $\vec{\mathbf{A}} + \vec{\mathbf{B}}$, (b) $\vec{\mathbf{A}} \cdot \vec{\mathbf{B}}$, (c) $\vec{\mathbf{A}} \times \vec{\mathbf{B}}$, (d) the unit normal to both $\vec{\mathbf{A}}$ and $\vec{\mathbf{B}}$, (e) the smaller angle between $\vec{\mathbf{A}}$ and $\vec{\mathbf{B}}$, and (f) the scalar and vector projections of $\vec{\mathbf{A}}$ onto $\vec{\mathbf{B}}$.

2.15 Calculate the distance between two points given in cylindrical coordinates as $P(5, \pi/6, 5)$ and $Q(2, \pi/3, 4)$.

2.16 Given $\vec{\mathbf{A}} = 2\vec{\mathbf{a}}_\rho + 3\vec{\mathbf{a}}_\phi$ at $P(1, \pi/2, 2)$ and $\vec{\mathbf{B}} = -3\vec{\mathbf{a}}_\rho + 10\vec{\mathbf{a}}_z$ at $Q(2, \pi, 3)$, determine (a) $\vec{\mathbf{A}} + \vec{\mathbf{B}}$, (b) $\vec{\mathbf{A}} \cdot \vec{\mathbf{B}}$, (c) $\vec{\mathbf{A}} \times \vec{\mathbf{B}}$, and (d) the angle between $\vec{\mathbf{A}}$ and $\vec{\mathbf{B}}$.

2.17 Two vectors $\vec{\mathbf{A}} = -7\vec{\mathbf{a}}_r + 2\vec{\mathbf{a}}_\theta + \vec{\mathbf{a}}_\phi$ and $\vec{\mathbf{B}} = \vec{\mathbf{a}}_r - 2\vec{\mathbf{a}}_\theta + 4\vec{\mathbf{a}}_\phi$ are given at the same point in space. Calculate (a) $2\vec{\mathbf{A}} - 3\vec{\mathbf{B}}$, (b) $\vec{\mathbf{A}} \cdot \vec{\mathbf{B}}$, (c) $\vec{\mathbf{A}} \times \vec{\mathbf{B}}$, (d) the unit normal to both $\vec{\mathbf{A}}$ and $\vec{\mathbf{B}}$, and (e) the angle between $\vec{\mathbf{A}}$ and $\vec{\mathbf{B}}$.

2.18 Solve Problem 2.17 if $\vec{\mathbf{A}}$ and $\vec{\mathbf{B}}$ were given at points $P(2, \pi/4, \pi/4)$ and $Q(10, \pi/2, \pi/2)$, respectively.

2.19 Find the distance between two points given in spherical coordinates as $P(10, \pi/4, \pi/3)$ and $Q(2, \pi/2, \pi)$. What is the distance vector from P to Q?

2.20 Given a scalar function $f = 12xy + z$, find (a) $\int f\,\vec{d\ell}$ and (b) $\int f\,d\ell$ along a straight line from (0, 0, 0) to (1, 1, 0).

2.21 The electrical field intensity caused by a charged wire of infinite extent in the z direction is $\vec{\mathbf{E}} = 10/\rho\,\vec{\mathbf{a}}_\rho$ (volts/meter). If the potential of a point at $\rho = a$ with respect to the potential of a point at $\rho = b$ is defined as $V_{ab} = -\int_b^a \vec{\mathbf{E}} \cdot \vec{d\ell}$, calculate the potential difference if $a = 10$ centimeters (cm) and $b = 80$ cm.

2.22 The electron density on the surface of a circular disc of radius 20 m is given as $n_e = 300\rho\cos^2\phi$ electrons/meter^2. Determine the number of electrons residing on the surface of the disc. What is the total charge on the disc?

2.23 If $f = xyz$, evaluate $\int f\,ds$ on the curved surface of a cylinder of radius 2 in the first quadrant and bounded by the planes $z = 0$ and $z = 1$.

2.24 For a vector field $\vec{\mathbf{F}} = x^3\vec{\mathbf{a}}_x + x^2y\vec{\mathbf{a}}_y + x^2z\vec{\mathbf{a}}_z$, determine the total flux $\oint \vec{\mathbf{F}} \cdot \vec{ds}$ passing through the surface of a cylinder of radius 4 and bounded by planes at $z = 0$ and $z = 2$.

2.25 If $\vec{\mathbf{F}} = x\vec{\mathbf{a}}_x$, evaluate $\int \vec{\mathbf{F}} \cdot \vec{d\ell}$ in the xy plane along the x axis from $x = 0$ to $x = 1$, then along the arc of radius 1 from $\phi = 0$ to $\phi = \pi/2$, and finally along the y axis from $y = 1$ to $y = 0$.

2.26 If $\vec{\mathbf{F}} = xy\vec{\mathbf{a}}_x$, find $\int \vec{\mathbf{F}} \cdot \vec{d\ell}$ along the arc of radius 2 from $\theta = 0$ to $\theta = \pi$ when $\phi = \pi/3$.

2.27 If the flux density $\vec{\mathbf{D}}$ in a region is given as $\vec{\mathbf{D}} = (2 + 16\rho^2)\vec{\mathbf{a}}_z$, determine the total flux $\int \vec{\mathbf{D}} \cdot \vec{ds}$ passing through a circular surface of radius $\rho = 2$ in the xy plane.

2.28 If $\vec{\mathbf{D}} = (2 + 16r^2)\vec{\mathbf{a}}_z$, calculate $\int \vec{\mathbf{D}} \cdot \vec{ds}$ over a hemispherical surface bounded by $r = 2$ and $0 \le \theta \le \pi/2$.

2.29 Repeat Problem 2.27 if $\vec{\mathbf{D}} = 10\cos\phi\,\vec{\mathbf{a}}_\rho$.

2.30 Repeat Problem 2.28 if $\vec{\mathbf{D}} = 10\cos\theta\,\vec{\mathbf{a}}_r$.

2.31 A spherical charge distribution centered at the origin is given as $\rho = kr^2$ for $0 \le r \le a$, where k is a constant. Determine the total charge contained within the sphere.

2.32 If $\vec{\mathbf{F}} = xy^2\vec{\mathbf{a}}_x + (x^2y + y)\vec{\mathbf{a}}_y$, evaluate (a) $\oint \vec{\mathbf{F}} \cdot \vec{d\ell}$ along the circumference of a circle of radius 3, and (b) $\int \vec{\mathbf{F}} \cdot \vec{ds}$ over the surface of the same circle.

2.33 If $f = x^3y^2z$, determine (a) ∇f, and (b) $\nabla^2 f$ at $P(2, 3, 5)$.

2.34 Employing the cylindrical coordinate system, prove that (a) $\nabla\phi + \boldsymbol{\nabla} \times [\vec{\mathbf{a}}_z \ln(\rho)] = 0$, and (b) $\boldsymbol{\nabla}[\ln(\rho)] - \boldsymbol{\nabla} \times (\vec{\mathbf{a}}_z\,\phi) = 0$.

2.35 Using spherical coordinates, show that (a) $\nabla(1/r) - \boldsymbol{\nabla} \times (\cos\theta\,\nabla\phi) = 0$, and (b) $\nabla\phi - \boldsymbol{\nabla} \times [(r\nabla\theta)/\sin\theta] = 0$.

2.36 Show that the vector field $\vec{\mathbf{E}} = yz\vec{\mathbf{a}}_x + xz\vec{\mathbf{a}}_x + xy\vec{\mathbf{a}}_z$ is both continuous (solenoidal) and conservative (irrotational).

2.37 Using the rectangular coordinate system, verify that (a) $\nabla \cdot (\nabla \times \vec{\mathbf{A}}) = 0$, and (b) $\nabla \times (\nabla f) = 0$.

2.38 In electrostatic fields, we will define the electric field intensity $\vec{\mathbf{E}}$ as a negative gradient of a scalar function Φ; i.e., $\vec{\mathbf{E}} = -\nabla\Phi$. We will also define the volume charge distribution as $\rho_V = \epsilon_0 \nabla \cdot \vec{\mathbf{E}}$, where ϵ_0 is the permittivity of free space. Determine $\vec{\mathbf{E}}$ and ρ_V if (a) $\Phi = V_0\phi \ln(\rho/a)$ in cylindrical coordinates, where V_0 and a are constants, (b) $\Phi = V_0 r \cos\theta$ in spherical coordinates, and (c) $\Phi = V_0 r \sin\theta$ in spherical coordinates.

2.39 Verify the following identities:

a) $\nabla(fg) = f\nabla g + g\nabla f$

b) $\nabla \cdot (f\vec{\mathbf{A}}) = f\nabla \cdot \vec{\mathbf{A}} + \vec{\mathbf{A}} \cdot \nabla f$

c) $\nabla \times (f\vec{\mathbf{A}}) = f\nabla \times \vec{\mathbf{A}} + \nabla f \times \vec{\mathbf{A}}$

2.40 Show that (a) $\partial/\partial x = \cos\phi\ \partial/\partial\rho - (\sin\phi/\rho)\ \partial/\partial\phi$, and (b) $\partial/\partial y = \sin\phi\ \partial/\partial\rho + (\cos\phi/\rho)\ \partial/\partial\phi$ in the cylindrical coordinate system.

2.41 Show that $\dfrac{\partial^2}{\partial x^2} + \dfrac{\partial^2}{\partial y^2} = \dfrac{1}{\rho^2}\left[\rho\dfrac{\partial}{\partial\rho}\left(\rho\dfrac{\partial}{\partial\rho}\right) + \dfrac{\partial^2}{\partial\phi^2}\right]$ in the cylindrical coordinate system.

2.42 If the electric field intensity in space is given as $\vec{\mathbf{E}} = E_0\cos\theta\ \vec{\mathbf{a}}_r - E_0\sin\theta\ \vec{\mathbf{a}}_\theta$, find $\nabla \cdot \vec{\mathbf{E}}$ and $\nabla \times \vec{\mathbf{E}}$.

2.43 Verify the divergence theorem for the $\vec{\mathbf{E}}$ field given in Problem 2.42 if the region is bounded by a sphere of radius b.

2.44 Given $\vec{\mathbf{F}} = x^3\vec{\mathbf{a}}_x + x^2y\vec{\mathbf{a}}_y + x^2z\vec{\mathbf{a}}_z$, verify the divergence theorem when the region is bounded by a cylinder $x^2 + y^2 = 16$ and the planes at $z = 0$ and $z = 2$.

2.45 If $\vec{\mathbf{A}} = [12 + 6\rho^2]z\vec{\mathbf{a}}_z$, verify the divergence theorem for a region bounded by a cylinder of radius 2 and the planes at $z = -1$ and $z = 1$.

2.46 If $\vec{\mathbf{F}} = 3y^2\vec{\mathbf{a}}_x + 4z\vec{\mathbf{a}}_y + 6y\vec{\mathbf{a}}_z$, verify Stokes' theorem for the open surface $z^2 + y^2 = 4$ in the $x = 0$ plane.

2.47 Verify Stokes' theorem for the function $\vec{\mathbf{F}} = (x/\rho)\vec{\mathbf{a}}_x$ over the first quadrant of a circular region bounded by a radius of 2 in the $z = 0$ plane.

2.48 Verify Stokes' theorem over a hemispherical surface at $r = 2$ and $0 \le \theta \le \pi/2$ for a function $\vec{\mathbf{F}} = 100\cos\theta\ \vec{\mathbf{a}}_r$.

2.49 If $f = x^2$ and $g = y^2$ are two scalar functions in a region bounded by a unit cube centered at the origin, verify Green's first and second identities.

3

Electrostatics

3.1 Introduction

Armed with the necessary tools of vector operations and vector calculus, we are now ready to explore electromagnetic field theory. In this chapter, we study static electric fields (electrostatics), due to charges at rest. A charge can be either concentrated at a point or distributed in some fashion. In any case, the charge is assumed to be constant in time.

We begin our discussion by stating Coulomb's law of electrostatic force between two point charges fixed in space. We define the electric field intensity as the force per unit charge. We then want to establish that

a) The electric field intensity is irrotational or conservative, and
b) The work done in moving a charge from one point to another in an electrostatic field is independent of the path taken and depends only upon the endpoints of the path.

We will express the electric field intensity in terms of electric potential and deduce an expression for the energy required to move a charge from one location to another in an electrostatic field.

We will also explore the influence of the medium on electrostatic fields and define bound charge densities; examine several methods (Gauss's law, Poisson's and Laplace's equations, method of images) of solving electrostatic field problems; and develop the concept of capacitance and obtain an equation for the energy stored in a capacitor.

Some aspects of electrostatic fields discussed in this chapter may appear to be a repetition of what you have already studied in physics. Some repetition is necessary, not only to maintain a continuous link from one section to another, but also to motivate the learning process. We are convinced that a discerning student will find this repetition helpful.

3.2 Coulomb's law

Electrostatics is based upon the quantitative and experimentally verifiable statement of Coulomb's law pertaining to the electric force that one charged particle exerts on another. From his experiments, Charles Augustin de Coulomb, a French physicist, postulated that the electric force between two charged particles is

a) Directly proportional to the product of their charges,
b) Inversely proportional to the square of the distance between them,
c) Directed along the line joining them, and
d) Repulsive (attractive) for like (unlike) charges.

If q_1 and q_2 are two charged particles situated at points $P(x, y, z)$ and $S(x', y', z')$, as shown in Figure 3.1, the electric force acting on q_1 due to q_2 is

$$\vec{\mathbf{F}}_{12} = K \frac{q_1 q_2}{R_{12}^2} \vec{\mathbf{a}}_{12} \tag{3.1}$$

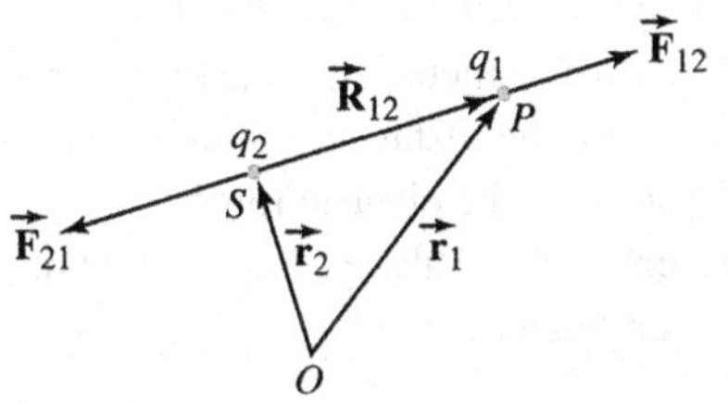

Figure 3.1 Electric force between two point charges

where

a) $\vec{\mathbf{F}}_{12}$ is the force experienced by q_1 due to q_2,
b) K is the constant of proportionality, which depends upon the system of units used,
c) R_{12} is the distance between points P and S, and
d) $\vec{\mathbf{a}}_{12}$ is the unit vector pointing in the direction from point S to point P.

The distance vector from S to P is

$$\vec{\mathbf{R}}_{12} = R_{12}\vec{\mathbf{a}}_{12} = \vec{\mathbf{r}}_1 - \vec{\mathbf{r}}_2 \tag{3.2}$$

where $\vec{\mathbf{r}}_1$ and $\vec{\mathbf{r}}_2$ are the position vectors of points P and S, respectively.

In the SI (International System of Units), the constant of proportionality is

$$K = \frac{1}{4\pi\epsilon_0} \tag{3.3}$$

where $\epsilon_0 = 8.85 \times 10^{-12} \approx 10^{-9}/36\pi$ farad/meter (F/m) is the permittivity of free space (vacuum). Thus,

$$\vec{\mathbf{F}}_{12} = \frac{q_1 q_2}{4\pi\epsilon_0 R_{12}^2} \vec{\mathbf{a}}_{12} \tag{3.4a}$$

or

$$\vec{\mathbf{F}}_{12} = \frac{q_1 q_2 (\vec{\mathbf{r}}_1 - \vec{\mathbf{r}}_2)}{4\pi\epsilon_0 |\vec{\mathbf{r}}_1 - \vec{\mathbf{r}}_2|^3} \tag{3.4b}$$

This equation is valid, not only for the charged particles such as electrons and protons, but also for charged bodies that can be considered *point charges*. Charged bodies are envisioned as point charges as long as their sizes are much less than the distance between them.

If the distance between two point charges, each having a charge of 1 coulomb (C), is 1 meter (m), the magnitude of the force experienced by each charge in free space, from (3.4), is 9×10^9 newtons (N).

From (3.4), it is also clear that the force exerted by q_1 on q_2 is equal in magnitude but opposite in direction to the force that q_2 exerts on q_1. That is,

$$\vec{\mathbf{F}}_{21} = -\vec{\mathbf{F}}_{12} \tag{3.5}$$

Equation (3.5) is in accordance with Newton's third law. We should point out that Coulomb's law has been verified to hold for distances as small as 10^{-14} meter (the distance between the nuclei of atoms). However, when the distance is smaller than 10^{-14} m, the nuclear force tends to dominate the electric force.

In this text, we will always assume that distances are given in meters, unless stated otherwise.

EXAMPLE 3.1

Two point charges of 0.7 mC and 4.9 μC are situated in free space at (2, 3, 6) and (0, 0, 0). Calculate the force acting on the 0.7-mC charge.

Solution The distance vector from the 4.9-μC charge to the 0.7-mC charge is

$$\vec{\mathbf{R}}_{12} = \vec{\mathbf{r}}_1 - \vec{\mathbf{r}}_2 = 2\vec{\mathbf{a}}_x + 3\vec{\mathbf{a}}_y + 6\vec{\mathbf{a}}_z$$

Thus, $R_{12} = [2^2 + 3^2 + 6^2]^{1/2} = 7$ m. The factor $\dfrac{1}{4\pi\epsilon_0} = 9 \times 10^9$. The force acting on the 0.7-mC charge, from (3.4b), is

$$\begin{aligned}\vec{\mathbf{F}}_{0.7\text{mC}} &= \frac{9 \times 10^9 \times 0.7 \times 10^{-3} \times 4.9 \times 10^{-6}}{7^3}[2\vec{\mathbf{a}}_x + 3\vec{\mathbf{a}}_y + 6\vec{\mathbf{a}}_z] \\ &= 0.18\vec{\mathbf{a}}_x + 0.27\vec{\mathbf{a}}_y + 0.54\vec{\mathbf{a}}_z \text{ N}\end{aligned}$$

The magnitude of the force experienced by either charge is 0.63 N.

• • •

Another experimental fact about Coulomb's force is that it obeys the *principle of superposition*. That is, the total force $\vec{\mathbf{F}}_t$ acting on a point charge q due to a system of n point charges is the vector sum of the forces exerted individually by each charge on q, as illustrated in Figure 3.2. That is,

$$\vec{\mathbf{F}}_t = \sum_{i=1}^{n} q \frac{q_i(\vec{\mathbf{r}} - \vec{\mathbf{r}}_i)}{4\pi\epsilon_0|\vec{\mathbf{r}} - \vec{\mathbf{r}}_i|^3} \tag{3.6}$$

where $\vec{\mathbf{r}}$ and $\vec{\mathbf{r}}_i$ are the position vectors of point charges q and q_i.

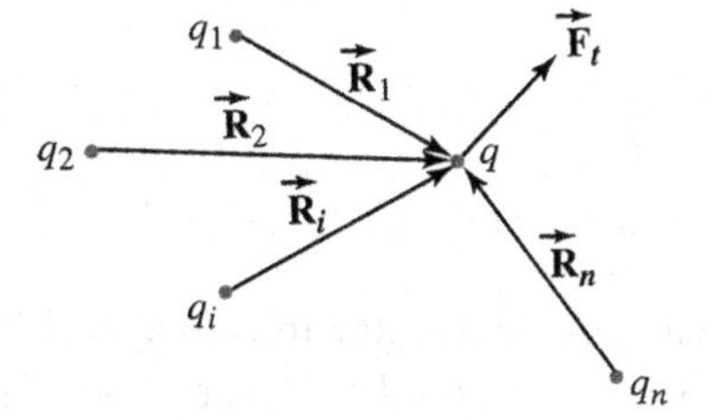

Figure 3.2 Force experienced by a charge q in a system of n charges

EXAMPLE 3.2

Three equal charges of 200 nC are placed in free space at (0, 0, 0), (2, 0, 0), and (0, 2, 0). Determine the total force acting on a charge of 500 nC at (2, 2, 0).

Solution The distance vectors (see Figure 3.3) are

$$\vec{\mathbf{R}}_1 = \vec{\mathbf{r}} - \vec{\mathbf{r}}_1 = 2\vec{\mathbf{a}}_y \Rightarrow R_1 = 2\,\text{m}$$
$$\vec{\mathbf{R}}_2 = \vec{\mathbf{r}} - \vec{\mathbf{r}}_2 = 2\vec{\mathbf{a}}_x \Rightarrow R_2 = 2\,\text{m}$$
$$\vec{\mathbf{R}}_3 = \vec{\mathbf{r}} - \vec{\mathbf{r}}_3 = 2\vec{\mathbf{a}}_x + 2\vec{\mathbf{a}}_y \Rightarrow R_3 = 2.828\,\text{m}$$

Figure 3.3

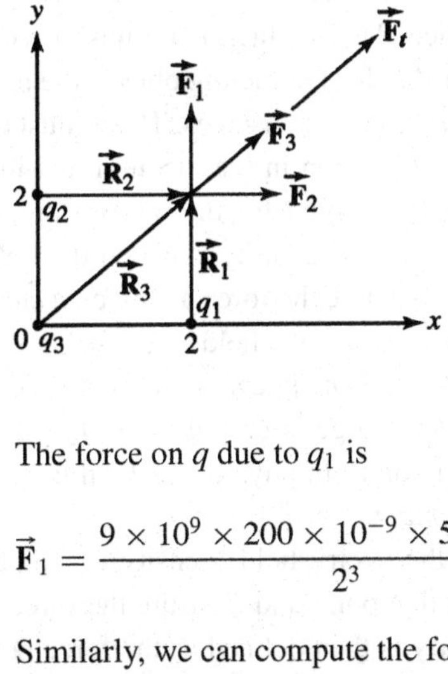

The force on q due to q_1 is

$$\vec{\mathbf{F}}_1 = \frac{9 \times 10^9 \times 200 \times 10^{-9} \times 500 \times 10^{-9}}{2^3}[2\vec{\mathbf{a}}_y] = 225\vec{\mathbf{a}}_y\,\mu\text{N}$$

Similarly, we can compute the forces acting on q due to q_2 and q_3 as

$$\vec{\mathbf{F}}_2 = 225\vec{\mathbf{a}}_x\,\mu\text{N} \quad \text{and} \quad \vec{\mathbf{F}}_3 = 79.6[\vec{\mathbf{a}}_x + \vec{\mathbf{a}}_y]\,\mu\text{N}$$

Thus, the total force experienced by q, from (3.6), is

$$\vec{\mathbf{F}}_t = \vec{\mathbf{F}}_1 + \vec{\mathbf{F}}_2 + \vec{\mathbf{F}}_3 = 304.6[\vec{\mathbf{a}}_x + \vec{\mathbf{a}}_y]\,\mu\text{N}$$

The net force of repulsion exerted by the three charges on q is 430.8 μN and is directed along the line that makes an angle of 45° with the x axis. • • •

3.3 Electric field intensity

Since we already know how to compute the forces among stationary charges, why do we need to define another field quantity? This is a very good question, and the following paragraph attempts to provide an answer.

Coulomb's law states that a charge will always exert a force on another charge, even when the charges are separated by a large distance. In physics, a force acting on one charge due to another is usually referred to as an *action at a distance*. As long as the charges are at rest, the action-at-a-distance point of view satisfies all the necessary requirements. However, if one charge is moved toward the other, the force experienced by the charges must change instantaneously, in accordance with Coulomb's law. In contrast, the theory of relativity requires that the information (or disturbance) about the motion of one charge will take some time to reach the other charge because no signal can travel faster than the speed of light. Thus, the increase in the force acting on the charges cannot be instantaneous, thereby indicating that the energy and momentum associated with the system of charges will be temporarily out of balance. This, in fact, is in unison with the theory of relativity, which states that for interacting objects the momentum and energy cannot be conserved by themselves. There must exist an extra entity, in the form of a perturbation in the medium in which the interacting bodies are situated, to account for the momentum and energy missing from the objects. This extra entity is called the **field**. Therefore, it becomes quite useful to define the force acting on a charge in the presence of another charge in terms of a field. We say that there exists an *electric field* or *electric field intensity* everywhere in space surrounding the charge. When another charge is brought into this electric field, it experiences a force acting on it. In physics, such an interaction is considered as an *action by contact.*

To detect the electric field intensity at a point P we place a positive test charge q_t at that point and measure the force acting on it. The **electric field intensity** is then defined as the force per unit charge. Because q_t also creates its own electric field and distorts the initial electric field, its magnitude must be as small as possible in order to minimize the distortion. In fact, we can make measurements with a continual decrease in the magnitude of q_t and then extrapolate the data to obtain the electric field intensity in the limit $q_t \to 0$, as illustrated in Figure 3.4. Note that the magnitude of the electric field intensity is the slope of the curve at $q_t = 0$.

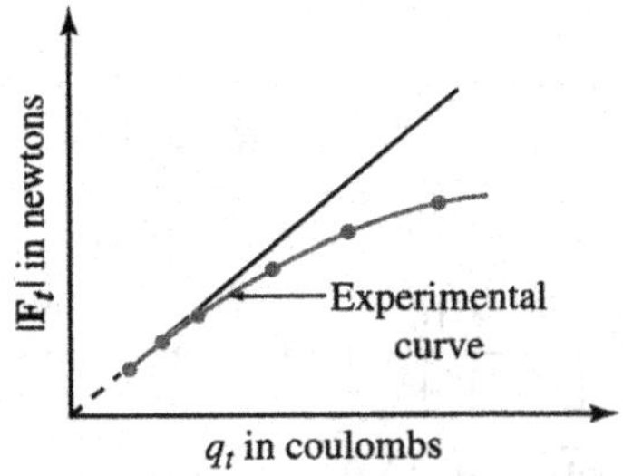

Figure 3.4 Force acting on a test charge

The electric field intensity $\vec{\mathbf{E}}$, the force per unit charge exerted on a test charge q_t as the magnitude of $q_t \to 0$, is

$$\vec{\mathbf{E}} = \lim_{q_t \to 0} \frac{\vec{\mathbf{F}}}{q_t} \tag{3.7}$$

where $\vec{\mathbf{F}}$ is the total force acting on q_t.

The electric field intensity, a vector field, has the units of newtons per coulomb (N/C). As we shall see later, newtons per coulomb are dimensionally equivalent to volts per meter (V/m). Although the electric field intensity is defined as the force per unit charge, it is common to express it in volts per meter.

If $\vec{\mathbf{E}}$ is the electric field intensity at a point P in space, the force acting on a charge q at that point is

$$\vec{\mathbf{F}}_q = q\vec{\mathbf{E}} \tag{3.8}$$

From now on we will use (3.8) to compute the electrostatic force experienced by a charge when placed in an electric field.

From (3.4), we can write the expression for the electric field intensity at any point P due to a point charge q at S as

$$\vec{\mathbf{E}} = \frac{q}{4\pi\epsilon_0} \frac{(\vec{\mathbf{r}}_1 - \vec{\mathbf{r}}_2)}{|\vec{\mathbf{r}}_1 - \vec{\mathbf{r}}_2|^3} = \frac{q}{4\pi\epsilon_0 R^2} \vec{\mathbf{a}}_R \tag{3.9}$$

where the subscript 12 has been dropped from R for brevity and $\vec{\mathbf{a}}_R$ is the unit vector directed from S toward P.

The electric field intensity due to n point charges, from equation (3.6), is

$$\vec{\mathbf{E}} = \sum_{i=1}^{n} \frac{q_i}{4\pi\epsilon_0} \frac{(\vec{\mathbf{r}} - \vec{\mathbf{r}}_i)}{|\vec{\mathbf{r}} - \vec{\mathbf{r}}_i|^3} \tag{3.10}$$

where $\vec{\mathbf{r}}_i$ is the distance vector directed from the location of the charge q_i toward the point of measurement of $\vec{\mathbf{E}}$.

EXAMPLE 3.3

Two point charges of 20 nC and −20 nC are situated at (1, 0, 0) and (0, 1, 0) in free space. Determine the electric field intensity at (0, 0, 1).

Solution The two distance vectors are

$$\vec{\mathbf{R}}_1 = \vec{\mathbf{r}} - \vec{\mathbf{r}}_1 = -\vec{\mathbf{a}}_x + \vec{\mathbf{a}}_z, \quad R_1 = |\vec{\mathbf{r}} - \vec{\mathbf{r}}_1| = 1.414\,\mathrm{m}$$

and

$$\vec{\mathbf{R}}_2 = \vec{\mathbf{r}} - \vec{\mathbf{r}}_2 = -\vec{\mathbf{a}}_y + \vec{\mathbf{a}}_z, \quad R_2 = |\vec{\mathbf{r}} - \vec{\mathbf{r}}_2| = 1.414\,\mathrm{m}$$

Substituting in equation (3.10), we obtain

$$\vec{\mathbf{E}} = 9 \times 10^9 \left[\frac{20 \times 10^{-9}}{1.414^3}(-\vec{\mathbf{a}}_x + \vec{\mathbf{a}}_z) - \frac{20 \times 10^{-9}}{1.414^3}(-\vec{\mathbf{a}}_y + \vec{\mathbf{a}}_z) \right]$$
$$= 63.67[-\vec{\mathbf{a}}_x + \vec{\mathbf{a}}_y]\,\mathrm{V/m}$$

• • •

Thus far, we have assumed that each charge is concentrated at a point. Cases of more complexity involve the continuous distribution of charges on linear elements, on surfaces, and in volumes. Therefore, before proceeding, we first define the charge distributions as follows.

Line charge density

When the charge is distributed over a linear element, we define the line charge density, the charge per unit length, as

$$\rho_\ell = \lim_{\Delta\ell \to 0} \frac{\Delta q}{\Delta \ell} \tag{3.11}$$

where Δq is the charge on a linear element $\Delta \ell$.

Surface charge density

When the charge is distributed over a surface, we define the surface charge density, the charge per unit area, as

$$\rho_s = \lim_{\Delta s \to 0} \frac{\Delta q}{\Delta s} \tag{3.12}$$

where Δq is the charge on a surface element Δs.

Volume charge density

If the charge is confined within a volume, we define the volume charge distribution, the charge per unit volume, as

$$\rho_v = \lim_{\Delta v \to 0} \frac{\Delta q}{\Delta v} \tag{3.13}$$

where Δq is the charge contained in a volume element Δv.

3.3.1 Electric field intensity due to charge distributions

Suppose we are given a line charge distribution (see Figure 3.5a), and our aim is to determine the electric field intensity at some point $P(x, y, z)$. We divide the line into n small sections, all of which approach zero in

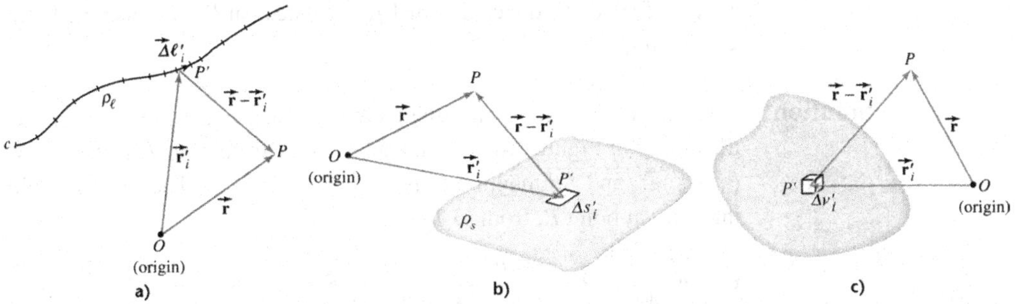

Figure 3.5 Electric field at point P due to (a) line charge distribution, (b) surface charge distribution, and (c) volume charge distribution.

the limit. Then we arbitrarily select a length element $\Delta\ell_i$ on the line that contains a charge $\Delta q_i = \rho_\ell \Delta\ell_i$ and determine its contribution to the electric field intensity. The net electric field intensity is then obtained in the limit of the sum as

$$\vec{\mathbf{E}} = \lim_{n\to\infty} \sum_{i=1}^{n} \frac{\Delta q_i}{4\pi\epsilon_0} \frac{(\vec{\mathbf{r}} - \vec{\mathbf{r}}'_i)}{|\vec{\mathbf{r}} - \vec{\mathbf{r}}'_i|^3}$$

where $\vec{\mathbf{r}}$ is the position vector of the point P and $\vec{\mathbf{r}}'_i$ is the position vector of the point $P'(x', y', z')$ of the charge element $\vec{\Delta\ell}'_i$. We will generally use primed letters for the coordinates of the source point and unprimed letters for points at which the desired quantity is to be determined in order to avoid confusion.

The right-hand side of the preceding equation, in fact, defines the line integral (Section 2.7). Therefore, we can write it as

$$\vec{\mathbf{E}} = \frac{1}{4\pi\epsilon_0} \int_c \frac{\rho_\ell(\vec{\mathbf{r}} - \vec{\mathbf{r}}')}{|\vec{\mathbf{r}} - \vec{\mathbf{r}}'|^3} d\ell' \tag{3.14}$$

where $\vec{\mathbf{r}}$ is the position vector of point $P(x, y, z)$ and $\vec{\mathbf{r}}'$ is the position vector of point $P'(x', y', z')$ at the length element $d\ell'$.

Likewise, we can obtain an expression for the electric field intensity due to a surface charge distribution (see Figure 3.5b) as

$$\vec{\mathbf{E}} = \frac{1}{4\pi\epsilon_0} \int_s \frac{\rho_s(\vec{\mathbf{r}} - \vec{\mathbf{r}}')}{|\vec{\mathbf{r}} - \vec{\mathbf{r}}'|^3} ds' \tag{3.15}$$

Finally, the electric field intensity at point P due to a volume charge distribution, as depicted in Figure 3.5c, is

$$\vec{\mathbf{E}} = \frac{1}{4\pi\epsilon_0} \int_v \frac{\rho_v(\vec{\mathbf{r}} - \vec{\mathbf{r}}')}{|\vec{\mathbf{r}} - \vec{\mathbf{r}}'|^3} dv' \tag{3.16}$$

EXAMPLE 3.4

A semi-infinite line extending from $-\infty$ to 0 along the z axis carries a uniform charge distribution of 100 nC/m. Find the electric field intensity at point $P(0, 0, 2)$. If a charge of 1 μC is placed at P, calculate the force acting on it.

Solution Consider a differential charge element $\rho_\ell\, dz'$ at $z = z'$ from the origin, as illustrated in Figure 3.6. The distance vector from z' to P is $\vec{\mathbf{r}} - \vec{\mathbf{r}}' = (z - z')\vec{\mathbf{a}}_z$ and its magnitude is $|\vec{\mathbf{r}} - \vec{\mathbf{r}}'| = z - z'$. The electric field intensity at point P, from (3.14), is

$$\vec{\mathbf{E}} = \vec{\mathbf{a}}_z \frac{\rho_\ell}{4\pi\epsilon_0} \int_{-\infty}^{0} \frac{dz'}{(z - z')^2} = \frac{\rho_\ell}{4\pi\epsilon_0 z} \vec{\mathbf{a}}_z$$

Figure 3.6

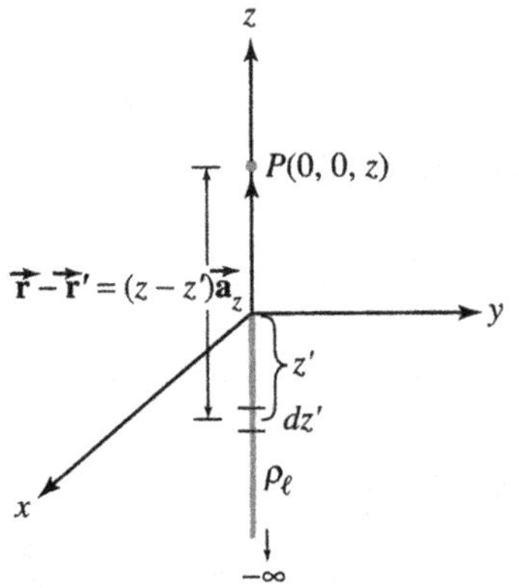

Substituting the values, we obtain

$$\vec{\mathbf{E}} = \frac{9 \times 10^9 \times 100 \times 10^{-9}}{2} \vec{\mathbf{a}}_z = 450\vec{\mathbf{a}}_z \text{ V/m}$$

The force acting on a charge of 1 μC at $z = 2$ m is

$$\vec{\mathbf{F}} = q\vec{\mathbf{E}} = 1 \times 10^{-6} \times 450\vec{\mathbf{a}}_z = 450\vec{\mathbf{a}}_z\ \mu\text{N}$$

• • •

EXAMPLE 3.5 The charge is uniformly distributed in the shape of a ring of radius b, as rendered in Figure 3.7. Determine the electric field intensity at any point on the axis of the ring.

Solution The differential length element in the direction of charge distribution, in cylindrical coordinates, is $b\, d\phi'$. The distance vector from the elemental charge to the point of observation, $P(0, 0, z)$, is

$$\vec{\mathbf{R}} = -b\vec{\mathbf{a}}_\rho + z\vec{\mathbf{a}}_z$$

Thus, from (3.14), we have

$$\vec{\mathbf{E}} = \frac{\rho_\ell}{4\pi\epsilon_0} \int_0^{2\pi} \frac{b\, d\phi'}{[b^2 + z^2]^{3/2}} (-b\vec{\mathbf{a}}_\rho + z\vec{\mathbf{a}}_z)$$

$$= \frac{\rho_\ell b}{4\pi\epsilon_0} \frac{1}{[b^2 + z^2]^{3/2}} \left[-b \int_0^{2\pi} \vec{\mathbf{a}}_\rho\, d\phi' + z \int_0^{2\pi} d\phi'\, \vec{\mathbf{a}}_z \right]$$

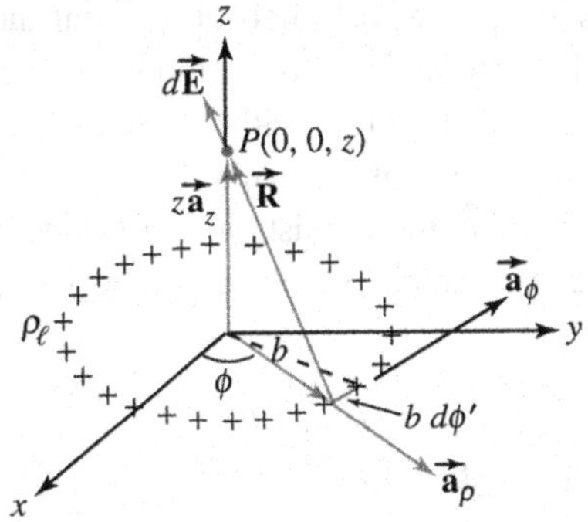

Figure 3.7 Electric field intensity at P due to a ring of uniform charge distribution

Because $\vec{\mathbf{a}}_\rho = \vec{\mathbf{a}}_x \cos\phi' + \vec{\mathbf{a}}_y \sin\phi'$, the first integral on the right-hand side becomes

$$\int_0^{2\pi} \vec{\mathbf{a}}_\rho \, d\phi' = \vec{\mathbf{a}}_x \int_0^{2\pi} \cos\phi' \, d\phi' + \vec{\mathbf{a}}_y \int_0^{2\pi} \sin\phi' \, d\phi' = 0$$

The second integral on the right-hand side is 2π. Thus, the electric field intensity at point P on the axis of the ring is

$$\vec{\mathbf{E}} = \frac{\rho_\ell b z}{2\epsilon_0 [b^2 + z^2]^{3/2}} \vec{\mathbf{a}}_z \tag{3.17}$$

Note that when $z = 0$; i.e., at the center of the loop, the electric field intensity is zero. Do you know why? • • •

EXAMPLE 3.6

A thin annular disc of inner radius a and outer radius b carries a uniform surface charge density ρ_s. Determine the electric field intensity at any point on the z axis when $z \geq 0$.

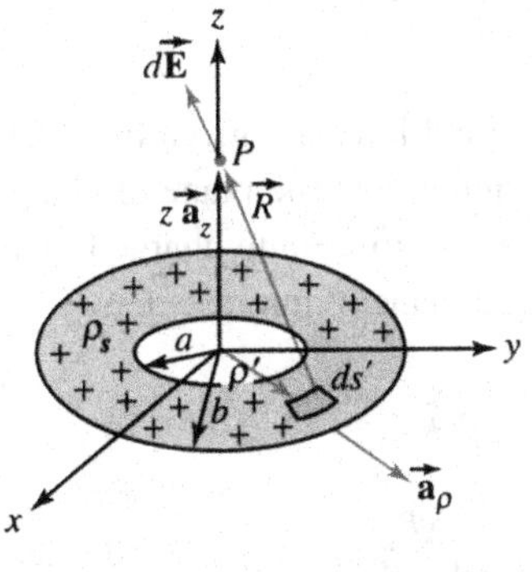

Figure 3.8 Electric field intensity at P due to uniform charge distribution on an annular disc

Solution The charge enclosed by a differential surface element ds', as shown in Figure 3.8, is $\rho_s \rho' \, d\rho' \, d\phi'$. The distance vector from the charge to point P on the z axis is $\vec{\mathbf{R}} = -\rho' \vec{\mathbf{a}}_\rho + z\vec{\mathbf{a}}_z$ and $R = [\rho'^2 + z^2]^{1/2}$. The electric field intensity at point $P(0, 0, z)$, from (3.15), is

$$\vec{\mathbf{E}} = \frac{\rho_s}{4\pi\epsilon_0} \int_a^b \int_0^{2\pi} \frac{\rho' \, d\rho' \, d\phi'}{[\rho'^2 + z^2]^{3/2}} [-\rho' \vec{\mathbf{a}}_\rho + z\vec{\mathbf{a}}_z]$$

Once again, we can show that $\int_0^{2\pi} \vec{\mathbf{a}}_\rho d\phi' = 0.$

At this time, it may be suitable to present another point of view. We can say that there would be no $\vec{\mathbf{a}}_\rho$ component of $\vec{\mathbf{E}}$ because of the symmetry of the charge distribution with respect to the point of observation. For every charge element on one side of P which contributes to the $\vec{\mathbf{a}}_\rho$ component of $\vec{\mathbf{E}}$ there exists a corresponding charge element on the other side of P which exactly cancels it. The net contribution, therefore, is zero. Thus,

$$\vec{\mathbf{E}} = \frac{\rho_s}{4\pi\epsilon_0}\int_a^b\int_0^{2\pi}\frac{\rho'\,d\rho'\,d\phi'}{[\rho'^2+z^2]^{3/2}}z\vec{\mathbf{a}}_z$$
$$= \frac{\rho_s z}{2\epsilon_0}\left[\frac{1}{(a^2+z^2)^{1/2}} - \frac{1}{(b^2+z^2)^{1/2}}\right]\vec{\mathbf{a}}_z \tag{3.18}$$

For an annular disc with very large outer radius $b \to \infty$, as shown in Figure 3.9, the electric field intensity becomes

$$\vec{\mathbf{E}} = \frac{\rho_s z}{2\epsilon_0}\left[\frac{1}{(a^2+z^2)^{1/2}}\right]\vec{\mathbf{a}}_z \tag{3.19}$$

For a solid finite disc of outer radius b, Figure 3.10, the electric field intensity from (3.18), $a = 0$, is

$$\vec{\mathbf{E}} = \frac{\rho_s z}{2\epsilon_0}\left[\frac{1}{z} - \frac{1}{(b^2+z^2)^{1/2}}\right]\vec{\mathbf{a}}_z \tag{3.20}$$

Finally, the electric field intensity at any point due to an infinite plane of charge, Figure 3.11, can be obtained from (3.18) by letting $a \to 0$ and $b \to \infty$. Thus,

$$\vec{\mathbf{E}} = \frac{\rho_s}{2\epsilon_0}\vec{\mathbf{a}}_z \tag{3.21}$$

The electric field intensity given by (3.21) is constant for all values of z. Even though an infinite plane of charge cannot exist, the electric field intensity close to a finite charged plane can be approximated to that of a charged plane of infinte extent.

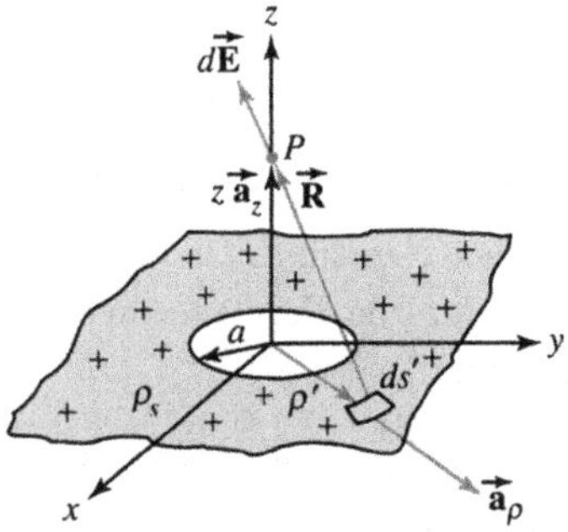

Figure 3.9 Electric field intensity at P due to a hollow circular charged disc of infinite extent

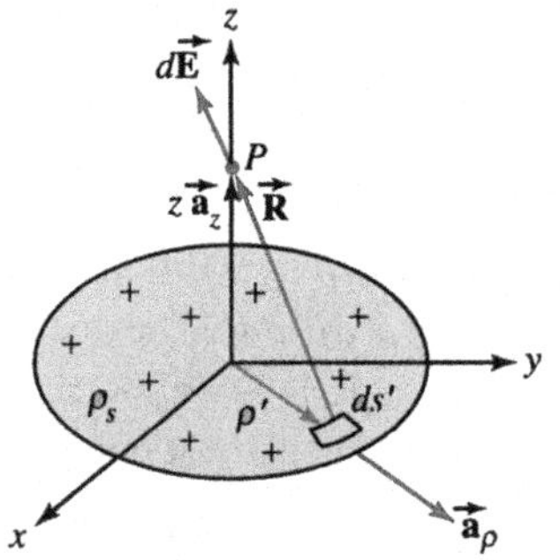

Figure 3.10 Electric field intensity at P due to a finite solid disc of charge

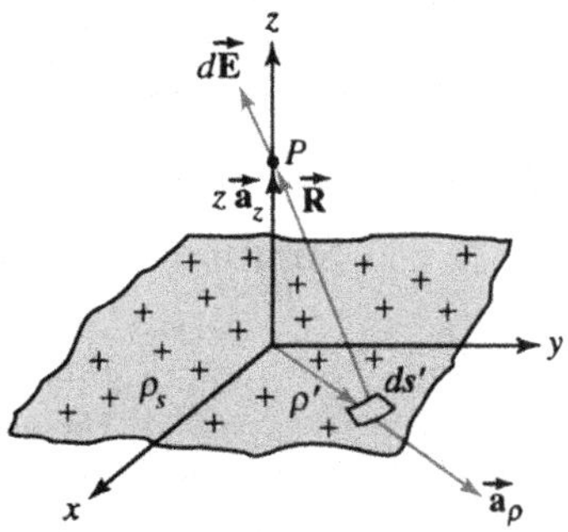

Figure 3.11 Electric field intensity at P due to an infinite plane of charge

• • •

3.4 Electric flux and electric flux density

Let us place a test charge at one point in an electric field and allow it to move. The force acting on the test charge will move it along a certain path. This path is called a *line of force* or a *flux line*. By placing the test charge at a new location we can create another line of force. Therefore, we can create as many lines of force as we desire by repeating the process. In order not to crowd a region with infinite lines of force, it is customary to arbitrarily state that *the number of lines of force due to a charge is equal to the magnitude of the charge in coulombs*. The field lines are then said to represent the *electric flux*. The electric flux lines have no real existence but they are a useful concept in the representation, visualization, and description of electric fields.

For an isolated positive point charge, the electric flux points radially outward, as indicated in Figure 3.12. Figure 3.13 shows the electric flux lines for a pair of equal and opposite point charges and between two positively charged bodies. Lines of electric flux between two oppositely charged parallel planes are shown in Figure 3.14 (page 82). It is indisputable that the electric field intensity at any point is tangential to the lines of electric flux.

Early investigators established the following properties for electric flux:

a) It must be independent of the medium,
b) Its magnitude solely depends upon the charge from which it originates,
c) If a point charge is enclosed in an imaginary sphere of radius R, the electric flux must pass perpendicularly and uniformly through the surface of the sphere, and
d) The electric flux density, the flux per unit area, is then inversely proportional to R^2.

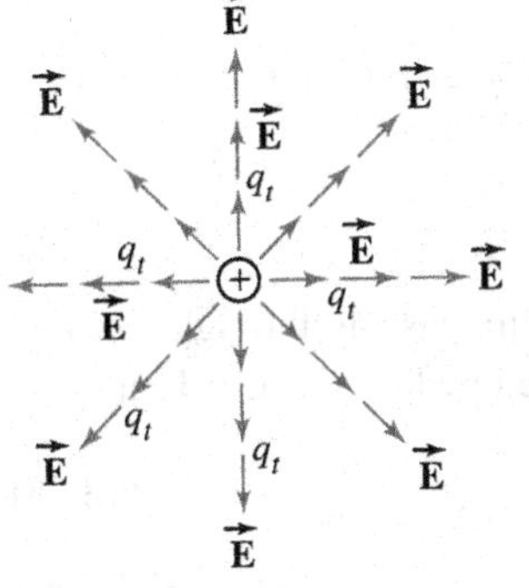

Figure 3.12 Electric flux lines from an isolated positive charge

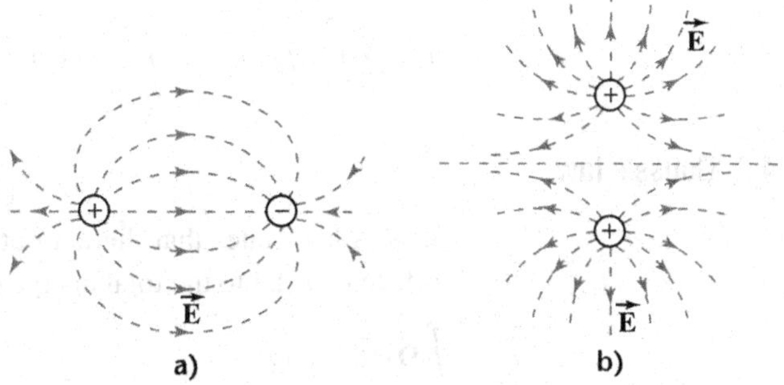

Figure 3.13 Lines of electric flux between (a) a positive and a negative charge, and (b) between two positively charged bodies

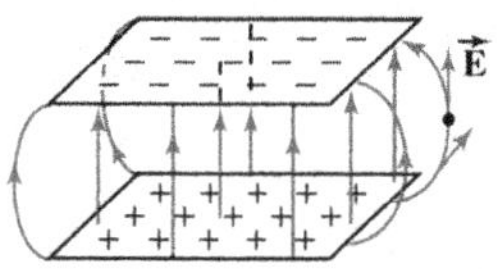

Figure 3.14 Lines of electric flux between oppositely charged parallel planes with fringing

Recall that the electric field intensity also satisfies these constraints, except that its magnitude depends upon the permittivity of the medium [equation (3.9)]. Therefore, you can easily realize that the **electric flux density** $\vec{\mathbf{D}}$ can be defined in terms of electric field intensity $\vec{\mathbf{E}}$ as

$$\vec{\mathbf{D}} = \epsilon_0 \vec{\mathbf{E}} \tag{3.22}$$

where ϵ_0 has been defined earlier as the permittivity of free space (our medium of choice so far).

Substituting for $\vec{\mathbf{E}}$ due to a point charge q in (3.22), the electric flux density at a radius r is

$$\vec{\mathbf{D}} = \frac{q}{4\pi r^2}\vec{\mathbf{a}}_r \tag{3.23}$$

From this equation it is obvious that $\vec{\mathbf{D}}$ has the units of coulombs per square meter (C/m^2).

3.4.1 Definition of electric flux

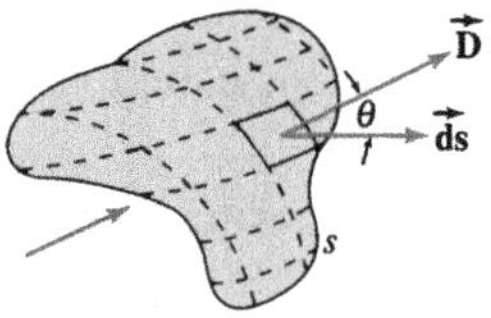

Figure 3.15 Electric flux through a surface

We can now define the **electric flux** Ψ in terms of electric flux density $\vec{\mathbf{D}}$ as

$$\Psi = \int_s \vec{\mathbf{D}} \cdot \vec{d\mathbf{s}} \tag{3.24}$$

where $\vec{d\mathbf{s}}$ is the differential surface element on surface s, as shown in Figure 3.15. The flux passing through s is maximum if $\vec{\mathbf{D}}$ and $\vec{d\mathbf{s}}$ are in the same direction.

EXAMPLE 3.7

The electric flux density in a region is given as $\vec{\mathbf{D}} = 10\vec{\mathbf{a}}_r + 5\vec{\mathbf{a}}_\theta + 3\vec{\mathbf{a}}_\phi$ mC/m^2. Determine the electric flux passing through the surface bounded by the region $z \geq 0$, and $x^2 + y^2 + z^2 = 36$.

Solution The differential surface, in spherical coordinates, at a radius of 6 m is $\vec{d\mathbf{s}} = 36 \sin\theta \, d\theta \, d\phi \, \vec{\mathbf{a}}_r$. The electric flux passing through the surface is

$$\Psi = \int_s \vec{\mathbf{D}} \cdot \vec{d\mathbf{s}} = 360 \int_0^{\pi/2} \sin\theta \, d\theta \int_0^{2\pi} d\phi = 720\pi \text{ mC}$$

• • •

3.4.2 Gauss's law

Gauss's law states that the net outward flux passing through a closed surface is equal to the total charge enclosed by that surface. That is,

$$\oint_s \vec{\mathbf{D}} \cdot \vec{d\mathbf{s}} = Q \tag{3.25}$$

In order to prove Gauss's law, let us enclose a point charge Q at point O by an arbitrary surface s, as shown in Figure 3.16. The electric flux density at point P on the surface s is

$$\vec{\mathbf{D}} = \frac{Q}{4\pi R^2}\vec{\mathbf{a}}_R \tag{3.26}$$

where $\vec{\mathbf{R}} = \vec{\mathbf{r}} - \vec{\mathbf{r}}' = R\vec{\mathbf{a}}_R$ is the distance vector from O to P. The electric flux passing through the closed surface s is

$$\Psi = \oint_s \vec{\mathbf{D}}\cdot\vec{ds} = \frac{Q}{4\pi}\oint_s \frac{\vec{\mathbf{a}}_R\cdot\vec{\mathbf{a}}_n\, ds}{R^2}$$

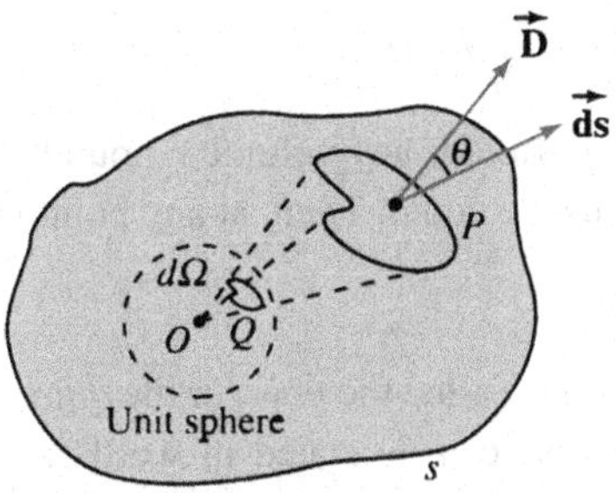

Figure 3.16 Electric flux through a closed surface s from a point charge Q enclosed by s

The integrand in the preceding equation is the solid angle, $d\Omega$, subtended by the surface ds at O as shown in Figure 3.16. Therefore, this equation can be written as

$$\Psi = \oint_s \vec{\mathbf{D}}\cdot\vec{ds} = \frac{Q}{4\pi}\oint_s d\Omega$$

However, from calculus, the solid angle subtended by any closed surface is 4π steradians. Thus, the total flux passing through the surface is

$$\Psi = \oint_s \vec{\mathbf{D}}\cdot\vec{ds} = Q$$

which is what we set out to prove. The surface over which the integral is taken is called a *Gaussian surface*. Equation (3.25) is a mathematical statement of Gauss's law. Gauss's law may be stated in words as follows. *The total electric flux emanating from a closed surface is numerically equal to the net positive charge inside the closed surface.* Gauss's law can also be expressed in terms of electric field intensity in free space as

$$\oint_s \vec{\mathbf{E}}\cdot\vec{ds} = \frac{Q}{\epsilon_0} \tag{3.27}$$

If the charges are distributed in a volume bounded by a surface, (3.25) can then be written as

$$\oint_s \vec{\mathbf{D}}\cdot\vec{ds} = \int_v \rho_v\, dv \tag{3.28}$$

Similar equations can be written if the charges are distributed over a surface or a linear element. Equation (3.28) is known as the *integral form of Gauss's law*. Although it is evident from the preceding development, we point out that a charge outside the closed surface does not contribute to the total charge enclosed. Also, it really does not matter exactly where the charges are located within the closed surface.

Gauss's law can be used to determine the total charge enclosed if the electric field intensity or electric flux density is known at all points on the surface. However, if the charge distribution is symmetric, and a convenient surface can be chosen on which the electric flux density is constant, Gauss's law greatly reduces the complexity of field problems.

By applying the divergence theorem, (3.28) can also be written as

$$\int_v \nabla \cdot \vec{\mathbf{D}} dv = \int_v \rho_v \, dv$$

This must be true for any volume v bounded by a surface s, so the two integrals must be equal. Then, at any point in space, we have

$$\nabla \cdot \vec{\mathbf{D}} = \rho_v \tag{3.29}$$

This equation is called the *point* or the *differential form of Gauss's law*. Equation (3.29) can be stated in words as follows. *Lines of electric flux emanate from any point in space at which there exists a positive charge density*. If the charge density is negative, the lines of electric flux converge toward the point.

Equation (3.29) shows that the electric flux density is always a measure of the free charges present in a region. We will highlight this fact again during our discussion of dielectric materials.

In the examples considered thus far, we have tacitly avoided the calculation of the $\vec{\mathbf{E}}$ field at any point due to a volume charge distribution because of the complexity involved in performing the integration. Some problems of this type can now be solved with much ease by using Gauss's law, as long as the charge distributions are symmetric.

EXAMPLE 3.8

Find $\vec{\mathbf{E}}$ at any point P due to an isolated point charge q using Gauss's law.

Solution

Let us construct a spherical Gaussian surface of radius R passing through P and centered at the charge, as indicated in Figure 3.17. The flux lines are directed radially outward from a positive point charge, so the electric field intensity must be normal to the surface of the sphere (no other direction is unique). That is,

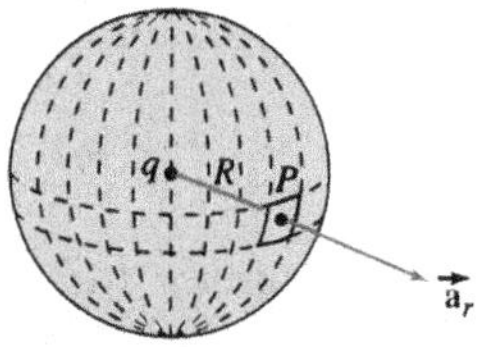

Figure 3.17 A spherical (Gaussian) surface at a radius R enclosing a point charge q at the origin

$$\vec{\mathbf{E}} = E_r \vec{\mathbf{a}}_r$$

Since every point on the sphere is equidistant from its center where q is located, E_r must have the same magnitude at each point on the surface at $r = R$. Thus,

$$\oint_s \vec{\mathbf{E}} \cdot \vec{ds} = E_r \int_0^{\pi} R^2 \sin\theta \, d\theta \int_0^{2\pi} d\phi = 4\pi R^2 E_r$$

The total charge enclosed by the spherical surface is q, thus the electric field intensity at P, from (3.27), is

$$E_r = \frac{q}{4\pi\epsilon_0 R^2}$$

which is identical to the result obtained using Coulomb's law. • • •

EXAMPLE 3.9

A charge is uniformly distributed over a spherical surface of radius a, as illustrated in Figure 3.18. Determine the electric field intensity everywhere in space.

Solution A spherical charge distribution suggests the selection of a spherical Gaussian surface of radius r on which the electric field intensity will be constant. If the surface is of radius $r < a$, the electric field intensity must be zero owing to the absence of charge enclosed. However, for the Gaussian surface when $r > a$, the total charge enclosed is

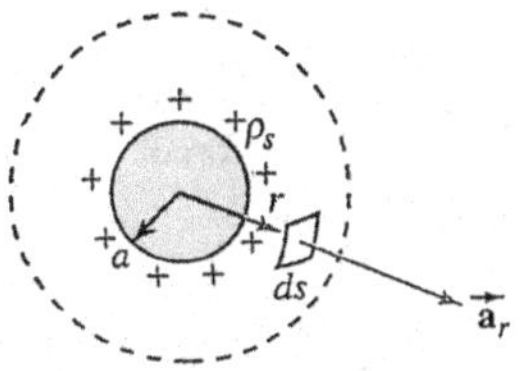

Figure 3.18 A spherical (Gaussian) surface at a radius r enclosing a surface charge distribution ρ_s on a sphere of radius a

$$Q = 4\pi a^2 \rho_s$$

where ρ_s is the uniform surface charge density. Once again,

$$\oint_s \vec{\mathbf{E}} \cdot \vec{ds} = 4\pi r^2 E_r$$

Thus, from Gauss's law, we have

$$E_r = \frac{Q}{4\pi\epsilon_0 r^2} = \frac{\rho_s a^2}{\epsilon_o r^2} \text{ for } r \geq a$$ • • •

3.5 The electric potential

The electrostatic effects we have described so far are in terms of the electric field intensity. In this section we define a *scalar* field, the *electric potential*, because it enables us to simplify a number of otherwise complicated calculations. It is always easier to work with a scalar quantity than a vector quantity.

If we place a positive test charge q in an electric field $\vec{\mathbf{E}}$, there will be a force on the charge given by $\vec{\mathbf{F}} = q\vec{\mathbf{E}}$. Under this force, the charge moves a differential distance $\vec{d\ell}$, as indicated in Figure 3.19a. As the charge moves, work is being done by the electric field. The incremental energy expended by the electric field, or simply the amount of work done by the $\vec{\mathbf{E}}$ field, is

$$dW_e = \vec{\mathbf{F}} \cdot \vec{d\ell} = q\vec{\mathbf{E}} \cdot \vec{d\ell}$$

where the subscript e signifies that the work is done by the $\vec{\mathbf{E}}$ field. Note that a positive test charge always moves in the direction of the $\vec{\mathbf{E}}$ field when the work is done by the $\vec{\mathbf{E}}$ field. However, if the test charge is moved *against* the direction of the $\vec{\mathbf{E}}$ field by an external force $\vec{\mathbf{F}}_{ext}$, the differential work done by the external force is

$$dW = -\vec{\mathbf{F}}_{ext} \cdot \vec{d\ell}$$

where the minus sign indicates that the charge is moved in a direction opposite to the $\vec{\mathbf{E}}$ field. In order to avoid consideration of any kinetic energy that may be acquired by a moving charge, we shall assume that the external force just balances the electric force, as depicted in Figure 3.19b. In that case,

$$dW = -q\vec{\mathbf{E}} \cdot \vec{d\ell}$$

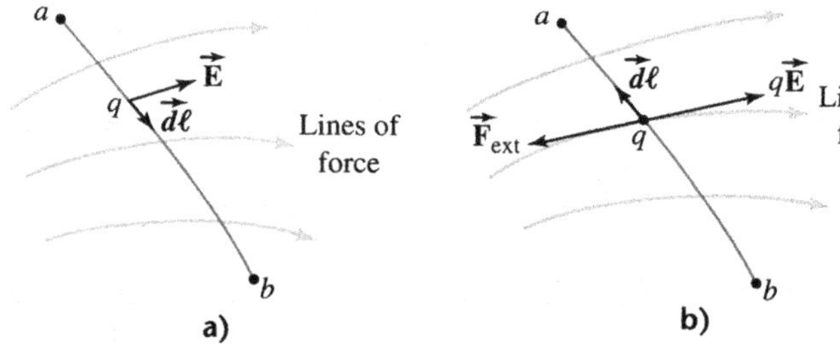

Figure 3.19 Motion of a test charge in an electric field caused by (a) the electric field and (b) the external force

The total work done by the external force in moving the test charge from point b to a is

$$W_{ab} = -q \int_b^a \vec{\mathbf{E}} \cdot \vec{d\ell} \tag{3.30}$$

If we move the charge around a closed path, as indicated in Figure 3.20, the work done must be zero. In other words,

$$\oint_c \vec{\mathbf{E}} \cdot \vec{d\ell} = 0 \tag{3.31}$$

which simply states that the $\vec{\mathbf{E}}$ field under static conditions is irrotational or conservative. However, a field is conservative if its curl is zero. Thus,

$$\nabla \times \vec{\mathbf{E}} = 0 \tag{3.32}$$

If the curl of a vector field is zero, the vector field can be represented in terms of the gradient of a scalar field. Thus, we can express the $\vec{\mathbf{E}}$ field in terms of a scalar field V as

$$\vec{\mathbf{E}} = -\nabla V \tag{3.33}$$

The reason for the minus sign will soon become evident.

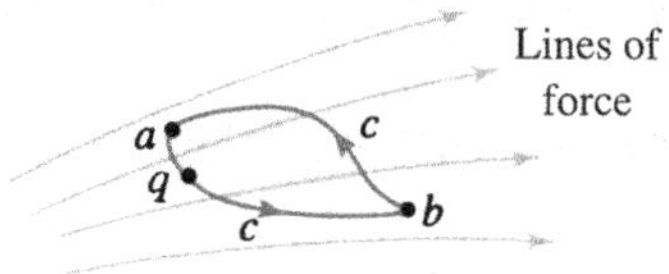

Figure 3.20 Movement of charge q along a closed path c in an electric field

We can now express (3.30) as

$$W_{ab} = -q\int_b^a \vec{\mathbf{E}} \cdot \overrightarrow{d\ell} = q\int_b^a \nabla V \cdot \overrightarrow{d\ell}$$

Substituting $\nabla V \cdot \overrightarrow{d\ell} = dV$ (Section 2.8), we have

$$W_{ab} = -q\int_b^a \vec{\mathbf{E}} \cdot \overrightarrow{d\ell} = q\int_{V_b}^{V_a} dV = q[V_a - V_b] = qV_{ab} \tag{3.34}$$

where V_a and V_b are the values of the scalar field V at points a and b. We speak of V_a and V_b as the electric potentials at points a and b, respectively, with respect to some reference point. It is clear that $V_{ab} = V_a - V_b$ defines the potential of point a with respect to point b (this is called the *potential difference* between the two points).

If the work done is positive, then the potential at point a is higher than that at point b. In other words, when the external force is pushing the positive charge against the $\vec{\mathbf{E}}$ field, the potential energy of the charge is increasing. That is why we have used the negative sign in (3.33). Stated differently, *the work done in moving a positive charge against the electric field is equal to the increase in the potential energy of the charge.*

The **potential difference**, *therefore, is the change in potential energy per unit charge in the limit* $q \to 0$. That is,

$$V_{ab} = \lim_{q \to 0} \frac{W_{ab}}{q} = -\int_b^a \vec{\mathbf{E}} \cdot \overrightarrow{d\ell} \tag{3.35}$$

From (3.35), the unit of potential function is joules per coulomb (J/C) or volts (V). It is now evident from (3.33) or (3.35) why we express the electric field intensity in terms of volts per meter (V/m).

EXAMPLE 3.10 Determine the potential difference between two points due to a point charge q at the origin.

Solution The electric field intensity at a radial distance r from a point charge q is

$$\vec{\mathbf{E}} = \frac{q}{4\pi\epsilon_0 r^2}\vec{\mathbf{a}}_r$$

If the radial distances of two points P and S from the charge q at the origin are r_1 and r_2, respectively, then, from (3.35), we have

$$V_{ab} = -\int_{r_2}^{r_1} \frac{q}{4\pi\epsilon_0 r^2}\,dr = \frac{q}{4\pi\epsilon_0}\left[\frac{1}{r_1} - \frac{1}{r_2}\right]$$

• • •

If we set $r_2 \to \infty$, then the potential of point P with respect to a point S at infinity is known as the *absolute potential*. Thus, the absolute potential of point P at $r_1 = R$ is

$$V_a = \frac{q}{4\pi\epsilon_0 R} \tag{3.36}$$

Equation (3.36) shows that the potential remains unchanged at a surface of constant radius. A surface on which the potential is the same is known as an **equipotential surface**. Thus, for a point charge, the equipotential surfaces are spheres, as shown in Figure 3.21. The reader is urged to verify that the equipotential surfaces are concentric cylinders, as depicted in Figure 3.22 for a uniformly charged line.

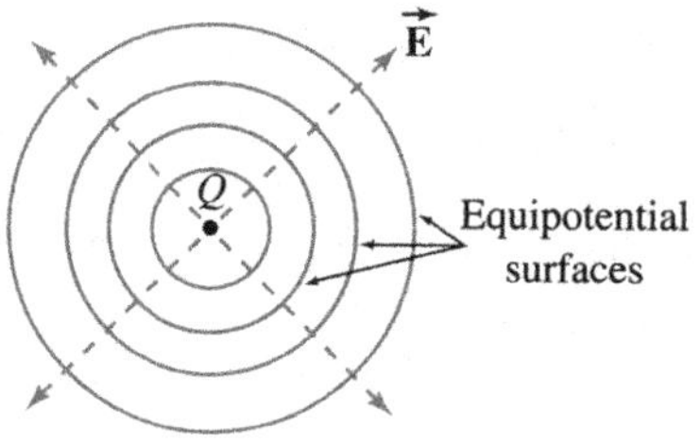

Figure 3.21 Equipotential surfaces of a point charge

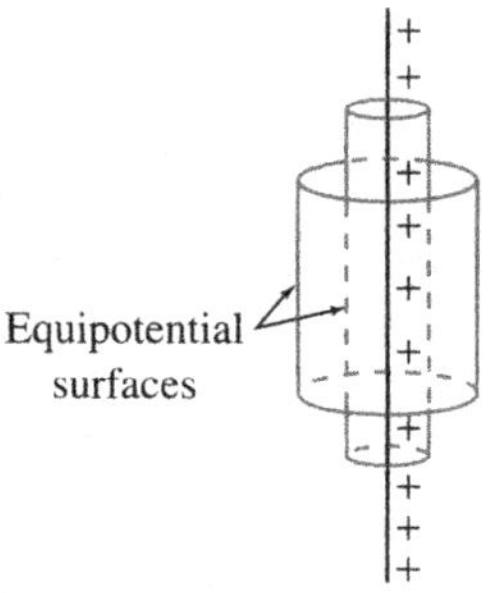

Figure 3.22 Equipotential surfaces for a uniformly charged line

In Section 3.3, we expressed the electric field intensity in terms of the line charge density, surface charge density, and volume charge density. Similar expressions can also be obtained for the potential at any point. We omit the details and cite the equations:

$$V = \frac{1}{4\pi\epsilon_0}\int_v \frac{\rho_v(\mathbf{r}')\,dv'}{|\mathbf{r}-\mathbf{r}'|} \tag{3.37a}$$

for distributions of volume charge density,

$$V = \frac{1}{4\pi\epsilon_0}\int_s \frac{\rho_s(\mathbf{r}')\,ds'}{|\mathbf{r}-\mathbf{r}'|} \tag{3.37b}$$

for distributions of surface charge density, and

$$V = \frac{1}{4\pi\epsilon_0}\int_c \frac{\rho_\ell(\mathbf{r}')\,d\ell'}{|\mathbf{r}-\mathbf{r}'|} \tag{3.37c}$$

for distributions of line charge density.

EXAMPLE 3.11

A charged ring of radius a carries a uniform charge distribution. Determine the potential and the electric field intensity at any point on the axis of the ring.

Solution A charged ring bearing a uniform charge distribution is shown in Figure 3.23. The potential at point $P(0, 0, z)$ on the z axis, from (3.37c), is

$$V(z) = \frac{1}{4\pi\epsilon_0}\int_0^{2\pi} \frac{\rho_\ell a\,d\phi'}{[a^2+z^2]^{1/2}}$$

$$= \frac{\rho_\ell a}{2\epsilon_0\sqrt{a^2+z^2}}$$

which reduces at the center of the ring to

$$V(z=0) = \frac{\rho_\ell}{2\epsilon_0}$$

Figure 3.23 Uniformly charged ring

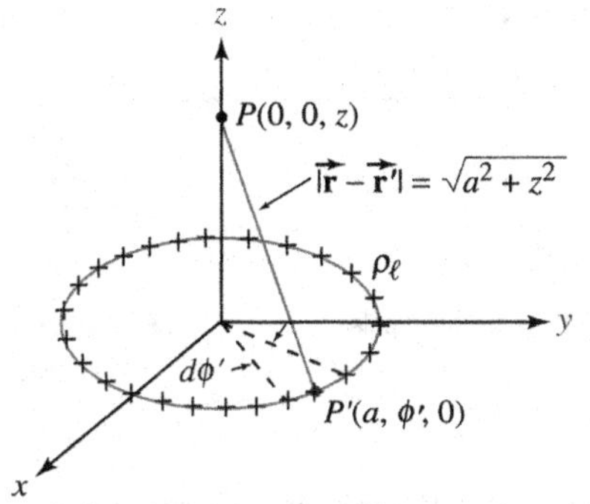

The electric field intensity, from (3.33), is

$$\vec{\mathbf{E}} = -\nabla V = -\frac{\partial V(z)}{\partial z}\vec{\mathbf{a}}_z = \frac{\rho_\ell a}{2\epsilon_0}\left[\frac{z}{(a^2+z^2)^{3/2}}\right]\vec{\mathbf{a}}_z$$

The electric field intensity at the center of the ring, $z = 0$, is zero as expected from the symmetry of the charge distribution. • • •

3.6 Electric dipole

We define an *electric dipole* as a pair of equal charges of opposite signs that are very close together. We will provide a more formal definition at the end of the section. Let us now assume that the magnitude of each charge is q and the separation between them is d, as shown in Figure 3.24. Our aim is to determine the potential and the electric field intensity at any point $P(x, y, z)$ in space established by the dipole. We will assume that the separation between the charges is very small compared with the distance to the point of observation. The total potential at point P is

$$V = \frac{q}{4\pi\epsilon_0}\left[\frac{1}{r_1} - \frac{1}{r_2}\right] = \frac{q}{4\pi\epsilon_0}\left[\frac{r_2 - r_1}{r_1 r_2}\right]$$

where r_1 and r_2 are the distances from the charges to P as shown in the figure.

If the charges are symmetrically placed along the z axis, and the point of observation is quite far away so that $r \gg d$, as illustrated in Figure 3.25, then we can approximate the distances r_1 and r_2 as

$$r_1 \approx r - 0.5d\cos\theta, \quad r_2 \approx r + 0.5d\cos\theta$$

and

$$r_1 r_2 = r^2 - (0.5d\cos\theta)^2 \approx r^2$$

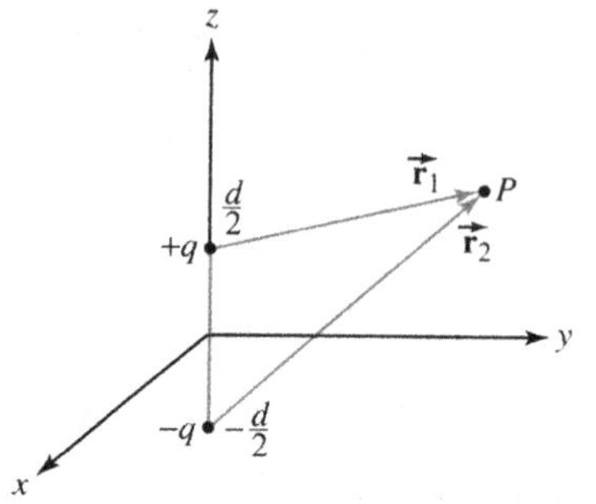

Figure 3.24 An electric dipole

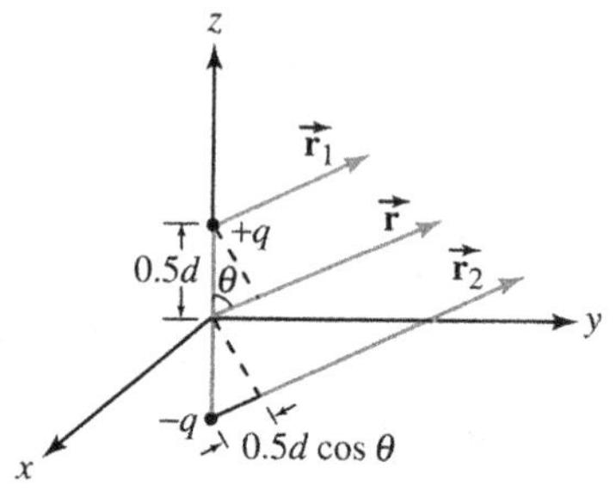

Figure 3.25 Distance approximations when P is far away from the dipole ($r \gg d$)

The potential at P can now be written as

$$V = \frac{q}{4\pi\epsilon_0}\left[\frac{d\cos\theta}{r^2}\right] \tag{3.38}$$

It is interesting to observe that the potential V at any point in the plane bisecting the dipole is zero when $\theta = 90°$. Therefore, there is no expenditure of energy if a charge is moved from one point to another in this plane.

By defining $\vec{\mathbf{p}}$ as a *dipole moment vector* with magnitude $p = qd$ and direction along the line from the negative to the positive charge such that

$$\vec{\mathbf{p}} = qd\vec{\mathbf{a}}_z$$

the potential at point P can now be written as

$$V = \frac{p\cos\theta}{4\pi\epsilon_0 r^2} = \frac{\vec{\mathbf{p}}\cdot\vec{\mathbf{a}}_r}{4\pi\epsilon_0 r^2} \tag{3.39}$$

Note that the potential at a point falls off as the square of the distance for a dipole, whereas it is inversely proportional to distance for a single-point charge.

To obtain equipotential surfaces, we let V in (3.38) assume a series of constant values. Note that the only variables in (3.38) are θ and r. Thus, the equation for an equipotential surface is

$$\frac{\cos\theta}{r^2} = \text{constant} \tag{3.40}$$

The equipotential surfaces of an electric dipole are shown by the dashed lines in Figure 3.26.

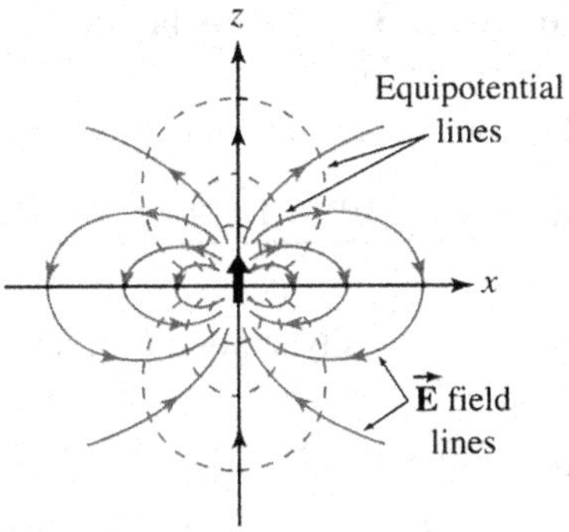

Figure 3.26 Equipotential and electric field lines due to an electric dipole

We can now compute the electric field intensity at point P using (3.33). Evaluating the negative gradient of a scalar potential function V in spherical coordinates, we obtain

$$\vec{\mathbf{E}} = \frac{p}{4\pi\epsilon_0 r^3}[2\cos\theta\,\vec{\mathbf{a}}_r + \sin\theta\,\vec{\mathbf{a}}_\theta] \tag{3.41}$$

However,

$$\begin{aligned} 2\cos\theta\,\vec{\mathbf{a}}_r + \sin\theta\,\vec{\mathbf{a}}_\theta &= 3\cos\theta\,\vec{\mathbf{a}}_r - (\cos\theta\,\vec{\mathbf{a}}_r - \sin\theta\,\vec{\mathbf{a}}_\theta) \\ &= 3\cos\theta\,\vec{\mathbf{a}}_r - \vec{\mathbf{a}}_z \end{aligned} \tag{3.42}$$

Thus, we can write the electric field intensity at point P as

$$\vec{\mathbf{E}} = \frac{3(\vec{\mathbf{p}}\cdot\vec{\mathbf{r}})\vec{\mathbf{r}} - r^2\vec{\mathbf{p}}}{4\pi\epsilon_0 r^5} \tag{3.43}$$

The electric field intensity falls off as the inverse cube of the distance. In the bisecting plane, $\theta = \pm\pi/2$, the field lines are directed along $\vec{\mathbf{a}}_\theta = -\vec{\mathbf{a}}_z$. That is,

$$\vec{\mathbf{E}} = -\frac{\vec{\mathbf{p}}}{4\pi\epsilon_0 r^3} \quad \text{for} \quad \theta = \pm\pi/2 \tag{3.44}$$

However, when $\theta = 0$ or π, the field lines are parallel to the dipole moment $\vec{\mathbf{p}}$. A plot of the electric field of the dipole is shown as solid lines in Figure 3.26.

The concept of an electric dipole is very useful in explaining the behavior of an insulating (dielectric) material when it is placed in an electric field. Therefore, a formal definition is in order.

An **electric dipole** *is defined as two charges of equal strength but of opposite polarity but separated by a small distance.* Associated with each dipole is a vector called the **dipole moment**. If q is the magnitude of each charge and $\vec{\mathbf{d}}$ is the distance vector from the negative to the positive charge, then the dipole moment is $\vec{\mathbf{p}} = q\vec{\mathbf{d}}$.

EXAMPLE 3.12

An electron and a proton separated by a distance of 10^{-11} meter are symmetrically arranged along the z axis with $z = 0$ as its bisecting plane. Determine the potential and $\vec{\mathbf{E}}$ field at $P(3, 4, 12)$.

Solution The position vector: $\vec{\mathbf{r}} = 3\vec{\mathbf{a}}_x + 4\vec{\mathbf{a}}_y + 12\vec{\mathbf{a}}_z \quad r = 13$ m
The dipole moment: $\vec{\mathbf{p}} = 1.6 \times 10^{-19} \times 10^{-11}\vec{\mathbf{a}}_z = 1.6 \times 10^{-30}\vec{\mathbf{a}}_z$
From (3.38), the potential at point P is

$$V = \frac{\vec{\mathbf{p}} \cdot \vec{\mathbf{r}}}{4\pi\epsilon_0 r^3} = \frac{9 \times 10^9 \times 1.6 \times 10^{-30} \times 12}{13^3} = 7.865 \times 10^{-23}\text{V}$$

The electric field intensity at point P, from (3.42), is

$$\begin{aligned}\vec{\mathbf{E}} &= \frac{9 \times 10^9}{13^5}(1.6 \times 10^{-30})[3 \times 12(3\vec{\mathbf{a}}_x + 4\vec{\mathbf{a}}_y + 12\vec{\mathbf{a}}_z) - 13^2\vec{\mathbf{a}}_z] \\ &= [4.189\vec{\mathbf{a}}_x + 5.585\vec{\mathbf{a}}_y + 10.2\vec{\mathbf{a}}_z] \times 10^{-24}\ \text{V/m}\end{aligned}$$

• • •

3.7 Materials in an electric field

We have paid adequate attention to the fields produced by various charge distributions in free space (vacuum), and we are now at a stage when we can discuss materials in order to complete our study of electrostatic fields. We classify materials into three broad categories: conductors, semiconductors, and insulators. Let us look first at electrostatic systems involving conductors.

3.7.1 Conductors in an electric field

A **conductor** is a material, such as a metal, that possesses a relatively large number of free electrons. An electron is said to be a free electron if it (a) is loosely associated with its nucleus, (b) is free to wander throughout the conductor, (c) responds to almost an infinitesimal electric field, and (d) continues to move as long as it experiences a force.

In the space lattice of metal crystals, there are one, two, or three valence electrons per atom that are normally free from the nucleus. Because of thermal agitation, these free electrons move about randomly within the space lattice. There is no net drift in any given direction in an isolated conductor. These are the very same electrons that contribute to the current when an electric field is maintained within a conductor by an external source of energy. Instead of describing a material in terms of the number of free electrons, we prefer to describe it in terms of its conductivity. We will discuss conductivity further in Chapter 4. For now, we simply state that as the number of valence electrons increases the conductivity of the material decreases. In other words, metals with one valence electron have higher conductivities than those with two electrons and so on.

In Chapter 4 we also discuss what really takes place inside a current-carrying conductor. In this subsection, our aim is to investigate the behavior of an isolated conductor when placed in a static electric field.

We remind you that an isolated conductor is electrically neutral. In other words, *the conductor has as many positive charges as it has electrons.*

Let us first raise a question: Can excess charge reside inside a conductor? The answer, of course, is emphatically no because of the force of mutual repulsion among the charges. They will continue to "fly away" due to the repulsive forces until their mutual repulsion is balanced by surface barrier forces. In other words, the excess charge will disappear from inside and redistribute itself on the surface of an isolated conductor. How long does this process take? Again, a quantitative answer is given in Chapter 4; however, the time is extremely small, of the order of 10^{-14} second for a good conductor like copper. This implies that under steady-state (equilibrium) conditions, the net volume charge density within the conductor is zero. That is,

$$\rho_v = 0 \tag{3.45}$$

inside the conductor.

Now let us assume that we place an isolated conductor in an electric field, as shown in Figure 3.27. The externally applied electric field exerts a force on the free electrons and causes them to move in a direction opposite to the $\vec{\mathbf{E}}$ field. One side of the conductor becomes negatively charged, and the other side becomes positively charged. We refer to such a separation of charges as *induced charges* because they are caused without any direct contact with the conductor. The effect of these induced charges is to produce an electric field within the conductor which is finally equal and opposite to the externally applied electric field. In other words, the net electric field inside the conductor is zero when the steady state is reached. Thus,

$$\vec{\mathbf{E}} = 0 \tag{3.46}$$

inside a conductor in the equilibrium state. Equation (3.46) implies that the potential is the same at all points in the conducting material. Thus, *neither volume charge density nor electric field intensity can be maintained within an isolated conductor under static conditions. Each conductor forms an equipotential region of space.*

Figure 3.27 An isolated conductor in an electrostatic field

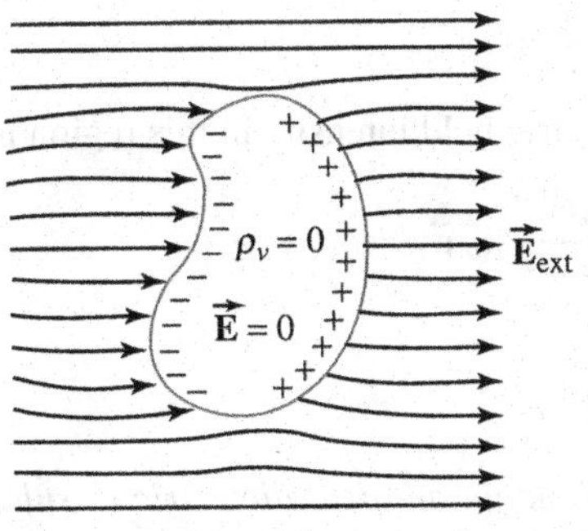

EXAMPLE 3.13

Charge is uniformly distributed within a spherical region of radius a. An isolated conducting spherical shell with inner radius b and outer radius c is placed concentrically, as shown in Figure 3.28. Determine the electric field intensity everywhere in the region.

Solution We can divide the space into four regions, as indicated in the figure.

a) Region I: For any radius $r < a$, the total charge enclosed is

$$Q = \frac{4\pi}{3} r^3 \rho_v$$

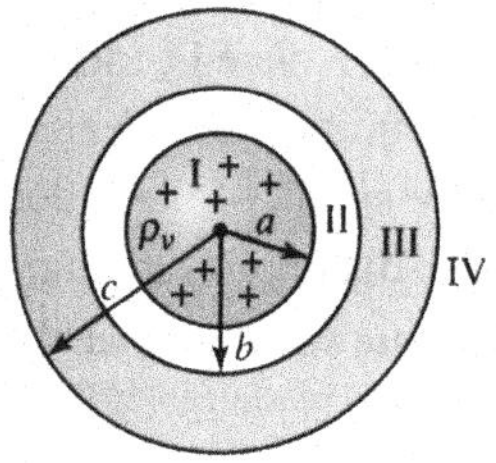

Figure 3.28 A spherical charge distribution enclosed by a conducting shell

Owing to the uniform charge distribution, the $\vec{\mathbf{E}}$ field must be not only in the radial direction but also constant on a spherical (Gaussian) surface. Thus,

$$\oint_s \vec{\mathbf{E}} \cdot \vec{d\mathbf{s}} = 4\pi r^2 E_r$$

Hence, $\vec{\mathbf{E}} = \frac{r}{3\epsilon_0} \rho_v \vec{\mathbf{a}}_r$ for $0 < r < a$

b) Region II: $a \le r < b$. The total charge enclosed is

$$Q = \frac{4\pi}{3} a^3 \rho_v$$

and from Gauss's law,

$$\vec{\mathbf{E}} = \frac{a^3}{3\epsilon_0 r^2} \rho_v \vec{\mathbf{a}}_r \quad \text{for} \quad a \le r \le b$$

c) Region III: $b \le r \le c$. Since $\vec{\mathbf{E}}$ within a conductor must be zero, the surface at $r = b$ must possess a negative charge with magnitude equal to the total charge enclosed. If ρ_{sb} is the surface charge density, then the charge on the surface must be $-4\pi b^2 \rho_{sb}$. Thus,

$$\rho_{sb} = -\frac{a^3}{3b^2} \rho_v$$

d) Region IV: $r \ge c$: If the inner side of an isolated conducting shell acquires a negative charge, the outer side at $r = c$ must acquire an equal amount of positive charge. If ρ_{sc} is the surface charge density on the outer surface, then

$$\rho_{sc} = \frac{a^3}{3c^2} \rho_v$$

The electric field intensity in this region is

$$\vec{\mathbf{E}} = \frac{a^3}{3\epsilon_0 r^2} \rho_v \vec{\mathbf{a}}_r \quad \text{for} \quad r \ge c$$

• • •

3.7.2 Dielectrics in an electric field

Strictly speaking, *an ideal dielectric (insulator) is a material with no free electrons in its lattice structure*. All the electrons associated with an ideal dielectric are strongly bound to its constituent molecules. These electrons experience very strong internal restraining forces that oppose their random movements. Therefore, when an electric field is maintained within a dielectric by an external source of energy, there is no current. Defined formally, *an* **ideal dielectric** *is a material in which positive and negative charges are so sternly bound that they are inseparable.* It has zero conductivity.

Of course, no real substance is an ideal dielectric. However, there are materials with a conductivity 10^{20} times smaller than that of a good conductor. When subjected to electric fields of less than a certain intensity, these materials permit only negligible current to flow. For all practical purposes, these materials can be regarded as ideal (perfect) dielectrics. Under the influence of an electric force the molecules of a dielectric material experience distortion in the sense that the center of a positive charge of a molecule no longer coincides with the center of a negative charge. We speak of such a molecule, and thereby the dielectric material, as being *polarized.* In its polarized state the material contains a large number of dipoles.

A schematic representation of a dielectric slab in its normal state is shown in Figure 3.29a. Figure 3.29b shows the same section under the influence of an electric field.

Figure 3.29

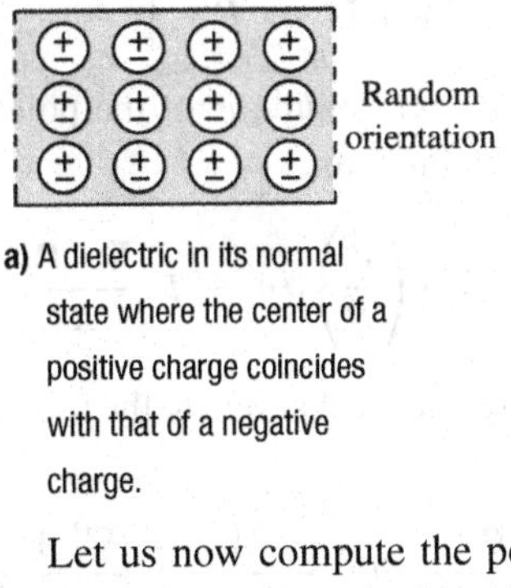

a) A dielectric in its normal state where the center of a positive charge coincides with that of a negative charge.

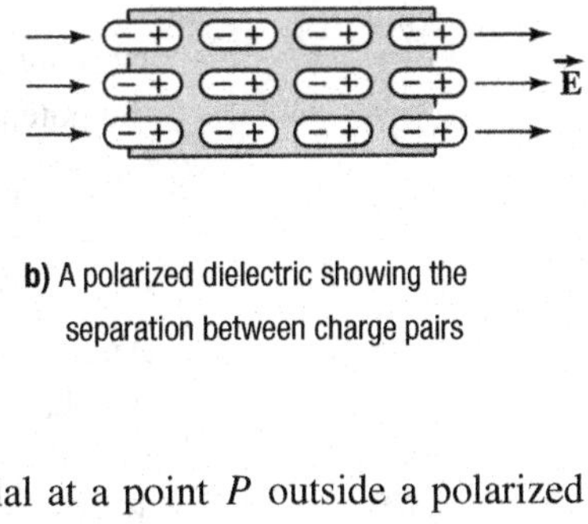

b) A polarized dielectric showing the separation between charge pairs

Let us now compute the potential at a point P outside a polarized dielectric material, as sketched in Figure 3.30. To do so, let us first define a **polarization vector** in terms of the number of dipole moments per unit volume. That is,

$$\vec{\mathbf{P}} = \lim_{\Delta v \to 0} \frac{\Delta \vec{\mathbf{p}}}{\Delta v} \tag{3.47}$$

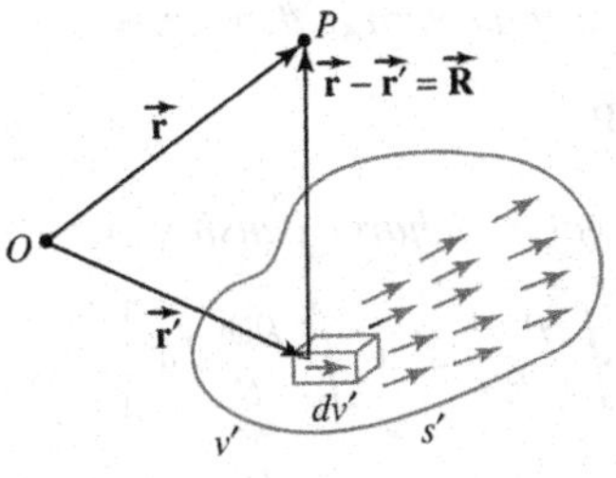

Figure 3.30 Potential at a point due to polarized dielectric material

where $\Delta\vec{\mathbf{p}}$ is the dipole moment of volume Δv in the limit $\Delta v \to 0$. Equation (3.47) is simply a definition of the derivative of $\vec{\mathbf{p}}$ with respect to v. Therefore, for the volume dv' in Figure 3.30, we can express $\overrightarrow{d\mathbf{p}}$ as

$$d\vec{\mathbf{p}} = \vec{\mathbf{P}}\,dv' \tag{3.48}$$

The potential at point P due to $d\vec{\mathbf{p}}$, from (3.38), is

$$dV = \frac{\vec{\mathbf{P}}\cdot\vec{\mathbf{a}}_R}{4\pi\epsilon_0 R^2}dv' \tag{3.49}$$

where $\vec{\mathbf{R}} = \vec{\mathbf{r}} - \vec{\mathbf{r}}' = |\vec{\mathbf{r}} - \vec{\mathbf{r}}'|\vec{\mathbf{a}}_R = R\vec{\mathbf{a}}_R$. Because $\nabla'\left(\frac{1}{R}\right) = \frac{1}{R^2}\vec{\mathbf{a}}_R$, we can write (3.49) as

$$dV = \frac{\vec{\mathbf{P}}\cdot\nabla'(1/R)}{4\pi\epsilon_0}dv' \tag{3.50}$$

Now using the vector identity (Chapter 2),

$$\vec{\mathbf{P}}\cdot\boldsymbol{\nabla}'(1/R) = \boldsymbol{\nabla}'\cdot(\vec{\mathbf{P}}/R) - (\boldsymbol{\nabla}'\cdot\vec{\mathbf{P}})/R$$

we can write (3.50) as

$$dV = \frac{1}{4\pi\epsilon_0}\left[\boldsymbol{\nabla}'\cdot\left(\frac{\vec{\mathbf{P}}}{R}\right) - \frac{\boldsymbol{\nabla}'\cdot\vec{\mathbf{P}}}{R}\right]dv'$$

Now integrating over volume v' of the polarized dielectric, we obtain the potential at point P as

$$V = \frac{1}{4\pi\epsilon_0}\left[\int_{v'}\boldsymbol{\nabla}'\cdot\left(\frac{\vec{\mathbf{P}}}{R}\right)dv' - \int_{v'}\frac{\boldsymbol{\nabla}'\cdot\vec{\mathbf{P}}}{R}dv'\right]$$

Applying the divergence theorem to the first term on the right-hand side, we obtain

$$V = \frac{1}{4\pi\epsilon_0}\oint_{s'}\frac{\vec{\mathbf{P}}\cdot\vec{\mathbf{a}}_n}{R}ds' - \frac{1}{4\pi\epsilon_0}\int_{v'}\frac{\boldsymbol{\nabla}'\cdot\vec{\mathbf{P}}}{R}dv' \tag{3.51}$$

It is obvious from (3.51) that the potential at point P due to a polarized dielectric is the algebraic sum of two terms: a surface term and a volume term. If we define

$$\rho_{sb} = \vec{\mathbf{P}}\cdot\vec{\mathbf{a}}_n \tag{3.52}$$

as the *bound surface charge density*, and

$$\rho_{vb} = -\boldsymbol{\nabla}\cdot\vec{\mathbf{P}} \tag{3.53}$$

as the *bound volume charge density*, then (3.51) can be written as

$$V = \frac{1}{4\pi\epsilon_0}\left[\oint_{s'}\frac{\rho_{sb}}{R}ds' + \int_{v'}\frac{\rho_{vb}}{R}dv'\right] \tag{3.54}$$

Thus, *the polarization of a dielectric material results in bound charge distributions*. These bound charge distributions are not like free charges; they are created by separating the charge pairs as mentioned earlier.

If a dielectric region contains the free charge density in addition to the bound charge density, the contribution due to the free charge density must also be considered to determine the $\vec{\mathbf{E}}$ field in the dielectric region. That is,

$$\nabla\cdot\vec{\mathbf{E}} = \frac{\rho_v + \rho_{vb}}{\epsilon_0} = \frac{\rho_v - \nabla\cdot\vec{\mathbf{P}}}{\epsilon_0}$$

or

$$\nabla\cdot(\epsilon_0\vec{\mathbf{E}} + \vec{\mathbf{P}}) = \rho_v \tag{3.55}$$

The right-hand side of (3.55) is simply the free charge density. When we were discussing fields in free space, we said that the free charge density is equal to $\nabla\cdot\vec{\mathbf{D}}$. This, in fact, is still true and is evident from (3.55) as $\vec{\mathbf{P}} \to 0$ in free space. Therefore, we can now generalize our definition of electric flux density for any medium as

$$\vec{\mathbf{D}} = \epsilon_0\vec{\mathbf{E}} + \vec{\mathbf{P}} \tag{3.56}$$

to include the effect of polarization in a dielectric material. Then $\nabla\cdot\vec{\mathbf{D}}$ *will always represent the free charge density in any medium.*

We know that the dipole moment $\vec{\mathbf{p}}$ in a dielectric medium is induced by an external $\vec{\mathbf{E}}$ field. A material is said to be *linear* if the dipole moment and, thereby, the polarization vector are proportional to $\vec{\mathbf{E}}$. If the electrical properties of the dielectric are independent of the direction, we say that the medium is *isotropic*. A dielectric material is said to be *homogeneous* if all portions of the material are identical. A linear, isotropic, and homogeneous dielectric material is referred to as a *class A* dielectric. Throughout this book, we will always assume a dielectric medium to be of class A type. Thus, we can express the polarization vector $\vec{\mathbf{P}}$ in terms of $\vec{\mathbf{E}}$ as

$$\vec{\mathbf{P}} = \epsilon_0\chi\vec{\mathbf{E}} \tag{3.57}$$

where the proportionality constant χ is called the *electric susceptibility*, and the factor ϵ_0 is included to make it a dimensionless quantity.

Equation (3.56) can now be expressed as

$$\vec{\mathbf{D}} = \epsilon_0(1+\chi)\vec{\mathbf{E}} \tag{3.58a}$$

The quantity $(1+\chi)$ is called the *relative permittivity* or the *dielectric constant* of the medium and is symbolized as ϵ_r. Thus, the general expression for the electric flux density finally becomes

$$\vec{\mathbf{D}} = \epsilon_0\epsilon_r\vec{\mathbf{E}} = \epsilon\vec{\mathbf{E}} \tag{3.58b}$$

where $\epsilon = \epsilon_0\epsilon_r$ is the permittivity of the medium.

Table 3.1. Approximate dielectric constant and dielectric strength of some dielectric materials

Dielectric material	Dielectric constant	Dielectric strength (kV/m)
Air	1.0	3,000
Bakelite	4.5	21,000
Ebonite	2.6	60,000
Epoxy	4	35,000
Glass (Pyrex)	4.5	90,000
Gutta-percha	4	14,000
Mica	6	60,000
Mineral oil	2.5	20,000
Paraffin	2.2	29,000
Polystyrene	2.6	30,000
Paranol	5	20,000
Porcelain	5	11,000
Quartz (fused)	5	30,000
Rubber	2.5–3	25,000
Transformer oil	2–3	12,000
Pure water	81	—

Equation (3.58b) gives us the constitutive relation between $\vec{\mathbf{D}}$ and $\vec{\mathbf{E}}$ in terms of the permittivity of the medium ϵ. For free space, $\vec{\mathbf{D}} = \epsilon_0 \vec{\mathbf{E}}$ because $\epsilon_r = 1$. Therefore, in any medium, the electrostatic fields satisfy the following equations:

$$\nabla \times \vec{\mathbf{E}} = 0 \tag{3.59a}$$

$$\nabla \cdot \vec{\mathbf{D}} = \rho_v \tag{3.59b}$$

$$\vec{\mathbf{D}} = \epsilon \vec{\mathbf{E}} \tag{3.59c}$$

where ρ_v is the free volume charge density in the medium, $\epsilon = \epsilon_0 \epsilon_r$ is the permittivity of the medium, and ϵ_r is the dielectric constant or the relative permittivity of the medium. In fact, we can generalize all the equations we have developed thus far by exchanging ϵ_0 with ϵ.

When we increase the $\vec{\mathbf{E}}$ field to a level at which it pulls electrons completely out of the molecules, at that time the *breakdown* of the dielectric material will take place, and thereafter it will function like a conductor. The maximum $\vec{\mathbf{E}}$ field that a dielectric can withstand prior to breakdown is called the **dielectric strength** of the material. The dielectric constant and the maximum dielectric strength of some materials are given in Table 3.1.

EXAMPLE 3.14

A point charge q is enclosed in a linear, isotropic, and homogeneous dielectric medium of infinite extent. Calculate the $\vec{\mathbf{E}}$ field, the $\vec{\mathbf{D}}$ field, the polarization vector $\vec{\mathbf{P}}$, the bound surface charge density ρ_{sb}, and the bound volume charge density ρ_{vb}.

Solution Since $\vec{\mathbf{E}}$, $\vec{\mathbf{D}}$, and $\vec{\mathbf{P}}$ are all parallel to one another in a linear medium, we still expect that the $\vec{\mathbf{E}}$ field would be in the $\vec{\mathbf{a}}_r$ direction. Thus, from Gauss's law, where q is the only free charge in the medium, we have

$$\oint_s \vec{\mathbf{D}} \cdot \vec{ds} = q$$

or

$$4\pi r^2 D_r = q$$

Therefore,

$$\vec{\mathbf{D}} = \frac{q}{4\pi r^2} \vec{\mathbf{a}}_r$$

The electric field intensity, from (3.59), is

$$\vec{\mathbf{E}} = \frac{q}{4\pi \epsilon_0 \epsilon_r r^2} \vec{\mathbf{a}}_r$$

Thus, the presence of a dielectric material has reduced the $\vec{\mathbf{E}}$ field by a factor of ϵ_r but has left the $\vec{\mathbf{D}}$ field unchanged.

From (3.56), we can compute $\vec{\mathbf{P}}$ as

$$\begin{aligned}\vec{\mathbf{P}} &= \vec{\mathbf{D}} - \epsilon_0 \vec{\mathbf{E}} \\ &= \frac{q}{4\pi \epsilon_r r^2} (\epsilon_r - 1) \vec{\mathbf{a}}_r\end{aligned}$$

Note that $\nabla \cdot \vec{\mathbf{P}} = 0$. Therefore, the bound volume charge density, from (3.53), is zero.

The dielectric medium is bounded by two surfaces: one at $r \to \infty$ and the other around the point charge. The bound surface charge density on the surface at $r \to \infty$ does not contribute to the $\vec{\mathbf{E}}$ field in the region $0 < r < \infty$. However, the surface charge density on the surface around the point charge does contribute to the $\vec{\mathbf{E}}$ field. As $r \to 0$, $\vec{\mathbf{P}} \to \infty$. The singularity in P as $r \to 0$ exists only because of our assumption of a point charge on a macroscopic scale. However, on a molecular scale we can assign it a radius b that in the limit approaches zero. Thus, the total bound charge on the surface of the dielectric next to the charge q is

$$\begin{aligned}Q_{sb} &= \lim_{b \to 0} [4\pi b^2 \rho_{sb}] = \lim_{b \to 0} [4\pi b^2 \vec{\mathbf{P}} \cdot (-\vec{\mathbf{a}}_r)] \\ &= -q(\epsilon_r - 1)/\epsilon_r\end{aligned}$$

Thus, the total charge, which is responsible for the $\vec{\mathbf{E}}$ field,

$$q_t = q + Q_{sb} = q/\epsilon_r$$

has been reduced by a factor of ϵ_r. That is why the $\vec{\mathbf{E}}$ field has been reduced by the same factor. As ϵ_r increases, the electric field intensity in the medium decreases.

• • •

3.7.3 Semiconductors in an electric field

In certain materials, such as silicon and germanium, a small fraction of the total number of valence electrons are free to move about randomly within the space lattice. These free electrons impart some conductivity to the material. This type of material, a **semiconductor**, is a poor conductor. If we place some excess charge inside a semiconductor, it will move toward its outer surface due to the repulsive forces, but at a slower rate than that of a conductor. However, when a state of equilibrium is achieved, there will be no excess charge left within the semiconductor.

If an isolated semiconductor is placed in an electric field, the motion of the free electrons will finally produce an electric field that cancels the externally applied field. Once again, under steady-state conditions, the net electric field inside an isolated semiconductor will be zero. Thus, a semiconductor behaves no differently than a conductor when subjected to electrostatic fields. Therefore, *from the point of view of electrostatic fields, we can group all materials into two categories: conductors and dielectrics.*

3.8 Energy stored in an electric field

In this section we will develop two methods to determine the energy stored in an electric field: one in terms of the sources and the other in terms of the field quantities.

Let us consider a region devoid of electric fields. To have such a region, charges if any, must be located at infinity. We speculate that there are n point charges, each at an infinite distance away from the region under consideration. Let us now bring a point charge q_1 from infinity and place it at point a, as shown in Figure 3.31. The energy required to do so is zero, $W_1 = 0$, because the charge did not experience any force. The presence of point charge q_1 now creates a potential distribution in the region. If we now bring another charge q_2 from infinity to a point b, the energy required to do so is

$$W_2 = q_2 V_{b,a} = \frac{q_1 q_2}{4\pi\epsilon R}$$

where $V_{b,a}$ is the potential at point b established by charge q_1 at point a, and R is the distance between the two charges. In making this statement, we have chosen the reference point for the potential at infinity. The total energy required to bring the two charges from infinity to points a and b is

$$W = W_1 + W_2 = \frac{q_1 q_2}{4\pi\epsilon R} \tag{3.60}$$

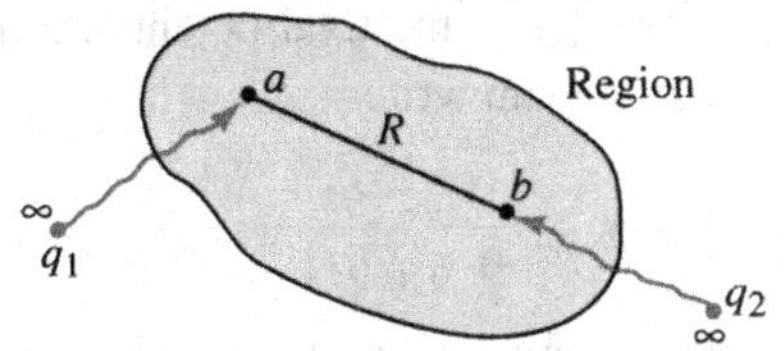

Figure 3.31 Potential energy between two point charges

Equation (3.60) yields the potential energy (strictly speaking, the *mutual potential energy*) of two point charges separated by a distance R in any medium. However, if we had reversed the process and brought charge q_2 first to point b in the region initially devoid of any fields, the energy required to do so would have been zero ($W_2 = 0$). The potential at point a due to q_2 at b would have been

$$V_{a,b} = \frac{q_2}{4\pi\epsilon R}$$

and then the energy expenditure to bring charge q_1 to point a would have been

$$W_1 = q_1 V_{a,b} = \frac{q_1 q_2}{4\pi\epsilon R}$$

The total energy requirements for the reversed process would have been

$$W = W_1 + W_2 = \frac{q_1 q_2}{4\pi\epsilon R} \tag{3.61}$$

The energy required in both cases is the same, so it does not really matter which charge is brought first.

Let us now extend our discussion to a system of three point charges q_1, q_2, and q_3 that are to be brought from infinity to points a, b, and c, respectively (in that order), as shown in Figure 3.32. The energy required to do this would be

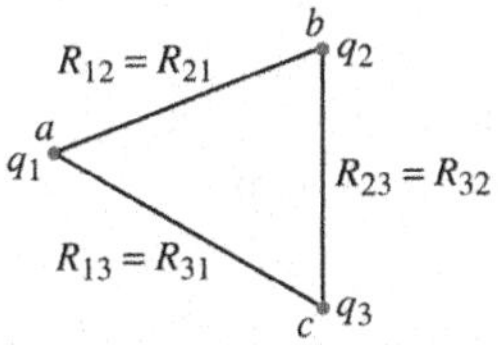

Figure 3.32 Potential energy in a system of three point charges

$$\begin{aligned} W &= W_1 + W_2 + W_3 = 0 + q_2 V_{b,a} + q_3(V_{c,a} + V_{c,b}) \\ &= \frac{1}{4\pi\epsilon}\left[\frac{q_2 q_1}{R_{21}} + \frac{q_3 q_1}{R_{31}} + \frac{q_3 q_2}{R_{32}}\right] \end{aligned} \tag{3.62}$$

However, if the three charges had been brought to their respective positions in the reverse order, the total energy expenditure would have been

$$\begin{aligned} W &= W_3 + W_2 + W_1 = 0 + q_2 V_{b,c} + q_1(V_{a,c} + V_{a,b}) \\ &= \frac{1}{4\pi\epsilon}\left[\frac{q_2 q_3}{R_{23}} + \frac{q_1 q_3}{R_{13}} + \frac{q_1 q_2}{R_{12}}\right] \end{aligned} \tag{3.63}$$

which is the same as (3.62). The work done in each case increases the amount of stored energy in the system of charges.

By adding (3.62) and (3.63), we obtain

$$W = \frac{1}{2}[q_1(V_{a,c} + V_{a,b}) + q_2(V_{b,a} + V_{b,c}) + q_3(V_{c,a} + V_{c,b})]$$

Because $V_{a,c} + V_{a,b}$ is the total potential at point a due to charges at points b and c, we can write

$$V_1 = V_{a,c} + V_{a,b} = \frac{1}{4\pi\epsilon}\left[\frac{q_3}{R_{13}} + \frac{q_2}{R_{12}}\right]$$

Similarly, the potentials at points b and c are

$$V_2 = V_{b,a} + V_{b,c} \quad \text{and} \quad V_3 = V_{c,a} + V_{c,b}$$

The total energy can now be written as

$$W = \frac{1}{2}[q_1 V_1 + q_2 V_2 + q_3 V_3] = \frac{1}{2}\sum_{i=1}^{3} q_i V_i$$

We can generalize this equation for a system of n point charges as

$$W = \frac{1}{2}\sum_{i=1}^{n} q_i V_i \tag{3.64}$$

Equation (3.64) allows us to compute the electrostatic potential energy for a group of point charges in their mutual field.

If the charges are continuously distributed, (3.64) becomes

$$W = \frac{1}{2}\int_v \rho_v V\, dv \tag{3.65}$$

where ρ_v is the volume charge density within v. *This is the standard expression for the energy of a system of charges in terms of the volume charge density and the potential. Surface charge density, line charge density, and point charges represent special cases of this equation.*

EXAMPLE 3.15

A metallic sphere of radius 10 cm has a surface charge density of 10 nC/m^2. Calculate the electric energy stored in the system.

Solution The potential on the surface of the sphere is

$$V = \int_s \frac{\rho_s\, ds}{4\pi\epsilon_0 R} = 9 \times 10^9 \times 10 \times 10^{-9} \times 0.1 \int_0^{\pi} \sin\theta\, d\theta \int_0^{2\pi} d\phi$$
$$= 113.1 \text{ V}$$

Therefore, the energy stored in the system, from (3.65), is

$$W = \frac{1}{2}\int_s \rho_s V\, ds = \frac{1}{2} Q_t V$$

where Q_t is the total charge on the sphere. For uniform charge distribution, the total charge is

$$Q_t = 4\pi R^2 \rho_s = 4\pi(0.1)^2\, 10 \times 10^{-9} = 1.257 \text{ nC}$$

Thus,

$$W = 0.5 \times 1.257 \times 10^{-9} \times 113.1 = 71.08 \times 10^{-9} \text{ joules (J)}$$

• • •

Let us now derive another expression for the energy in an electrostatic system in terms of the field quantities. Using Gauss's law, $\nabla \cdot \vec{\mathbf{D}} = \rho_v$, we can express (3.65) as

$$W = \frac{1}{2} \int_v V(\nabla \cdot \vec{\mathbf{D}}) \, dv$$

However, using the vector identity, equation (2.126),

$$V(\nabla \cdot \vec{\mathbf{D}}) = \nabla \cdot (V\vec{\mathbf{D}}) - \vec{\mathbf{D}} \cdot \nabla V$$

we obtain the expression for the energy as

$$W = \frac{1}{2} \left[\int_v \nabla \cdot (V\vec{\mathbf{D}}) \, dv - \int_v \vec{\mathbf{D}} \cdot (\nabla V) \, dv \right]$$

We now employ the divergence theorem to change the first volume integral into a closed surface integral as

$$\int_v \nabla \cdot (V\vec{\mathbf{D}}) \, dv = \oint_s V\vec{\mathbf{D}} \cdot \vec{ds}$$

The choice of the volume v in the preceding integral is arbitrary. The only constraint is that s bounds v. If we integrate over such a large volume that V and $\vec{\mathbf{D}}$ are negligibly small on the bounding surface, the surface integral vanishes. Therefore, the energy stored in the electrostatic system reduces to

$$W = -\frac{1}{2} \int_v \vec{\mathbf{D}} \cdot (\nabla V) \, dv = \frac{1}{2} \int_v \vec{\mathbf{D}} \cdot \vec{\mathbf{E}} \, dv \tag{3.66}$$

This equation permits us to determine the electrostatic energy in terms of the field quantities. Note that the volume integral in (3.66) is over all space ($R \to \infty$).

If we define the **energy density**, the energy per unit volume, as

$$w = \frac{1}{2} \vec{\mathbf{D}} \cdot \vec{\mathbf{E}} = \frac{1}{2} \epsilon E^2 = \frac{1}{2\epsilon} D^2 \tag{3.67}$$

we can express (3.66) in terms of energy density as

$$W = \int_v w \, dv \tag{3.68}$$

From (3.65), we can obtain another expression for the energy density as

$$w = \frac{1}{2} \rho_v V \tag{3.69}$$

An examination of (3.67) reveals that the energy density may be nonzero all over the space because of the continuity of the fields. However, (3.69) suggests that the energy density is nonzero only where charges exist. Does one equation contradict the other? We declare that there is no contradiction as long as we perceive that *the energy density is merely a quantity that, when integrated over all space, yields the total energy.*

EXAMPLE 3.16 Solve Example 3.15 using (3.66).

Solution Since the charge distribution is on the surface of the sphere, the energy density within the sphere is zero. Using Gauss's law, the electric flux density at any point in space is

$$\oint_s \vec{\mathbf{D}} \cdot \vec{d\mathbf{s}} = Q_t$$

or

$$\vec{\mathbf{D}} = \frac{Q_t}{4\pi r^2} = \frac{0.1 \times 10^{-9}}{r^2} \vec{\mathbf{a}}_r \ \text{C/m}^2$$

From (3.67), the energy density is

$$w = \frac{1}{2} \vec{\mathbf{D}} \cdot \vec{\mathbf{E}} = \frac{(0.1)^2 \times 10^{-18}}{2\epsilon_0 r^4}$$

Thus, the total energy in the system is

$$W = \int_{0.1}^{\infty} \frac{(0.1)^2 \times 10^{-18}}{2\epsilon_0 r^4} r^2 \, dr \int_0^{\pi} \sin\theta \, d\theta \int_0^{2\pi} d\phi$$
$$= 71.06 \text{ nJ}$$

• • •

3.9 Boundary conditions

In this section, we investigate the conditions that govern the behavior of electric fields at the boundary (interface) between two media. The interface may be between a dielectric and a conductor or between two different dielectrics. The equations governing the behavior of electric fields on either side of an interface are known as the **boundary conditions**.

3.9.1 The normal component of $\vec{\mathbf{D}}$

Let us apply Gauss's law to find the boundary condition pertaining to the normal component of the electric flux density at an interface, as exhibited in Figure 3.33a. We have constructed a Gaussian surface in the form of a pillbox, with half in medium 1, and the other half in medium 2. Each flat surface is so small that the electric flux density in each medium is essentially constant over the surface in that medium. We also assume that the area of the curved surface is negligibly small as the height of the pillbox Δh shrinks to zero. We shall also assume that there exists a free surface charge density ρ_s at the interface.

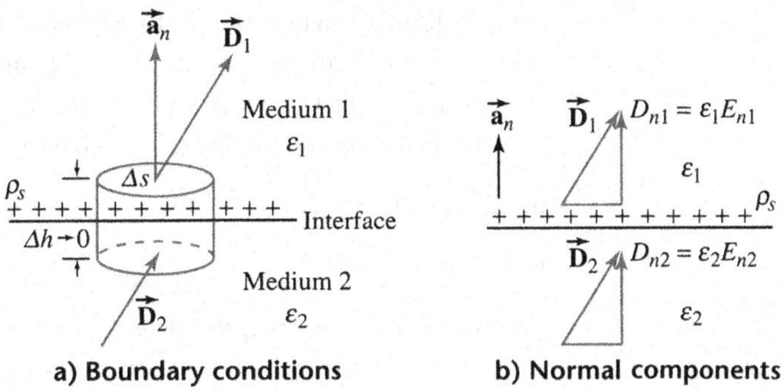

Figure 3.33 Boundary conditions involving normal components of $\vec{\mathbf{D}}$ and $\vec{\mathbf{E}}$ fields

If the surface area is Δs, the total charge enclosed by the pillbox is $\rho_s\,\Delta s$. Applying Gauss's law, we get

$$\vec{\mathbf{D}}_1 \cdot \vec{\mathbf{a}}_n\,\Delta s - \vec{\mathbf{D}}_2 \cdot \vec{\mathbf{a}}_n\,\Delta s = \rho_s\,\Delta s$$

or

$$\vec{\mathbf{a}}_n \cdot (\vec{\mathbf{D}}_1 - \vec{\mathbf{D}}_2) = \rho_s \qquad (3.70a)$$

or

$$D_{n1} - D_{n2} = \rho_s \qquad (3.70b)$$

where $\vec{\mathbf{a}}_n$ is the unit vector normal to the interface pointing from medium 2 to medium 1. D_{n1} and D_{n2} are the components of the $\vec{\mathbf{D}}$ field normal to the interface in medium 1 and medium 2, respectively, as shown in Figure 3.33b. Equation (3.70) states that *the normal components of the electric flux density are discontinuous if a free surface charge density exists at the interface.*

Since $\vec{\mathbf{D}} = \epsilon\vec{\mathbf{E}}$, we can also write (3.70) in terms of the normal components of the $\vec{\mathbf{E}}$ field. That is,

$$\vec{\mathbf{a}}_n \cdot (\epsilon_1 \vec{\mathbf{E}}_1 - \epsilon_2 \vec{\mathbf{E}}_2) = \rho_s \qquad (3.70c)$$

or

$$\epsilon_1 E_{n1} - \epsilon_2 E_{n2} = \rho_s \qquad (3.70d)$$

When the interface is between two different dielectrics, we do not expect any free surface charge density at the boundary unless the charge is deliberately placed there. Ruling out the possibility of such an intentional placement of a charge, we find that *the normal components of the electric flux density are continuous across a dielectric boundary*; i.e.,

$$D_{n1} = D_{n2} \qquad (3.71a)$$

or

$$\epsilon_1 E_{n1} = \epsilon_2 E_{n2} \qquad (3.71b)$$

If medium 2 is a conductor, the electric flux density $\vec{\mathbf{D}}_2$ must be zero under static conditions. For the normal component of the electric flux density $\vec{\mathbf{D}}_1$ to exist in medium 1, there must be a free surface charge density on the conductor's surface in harmony with (3.70). That is,

$$\vec{\mathbf{a}}_n \cdot \vec{\mathbf{D}}_1 = D_{n1} = \rho_s \tag{3.72a}$$

$$\epsilon_1 E_{n1} = \rho_s \tag{3.72b}$$

The normal component of the electric flux density in a dielectric medium just above the surface of a conductor is equal to the surface charge density on the conductor.

3.9.2 The tangential component of $\vec{\mathbf{E}}$

We already know that the electric field is conservative in nature and, accordingly, $\oint \vec{\mathbf{E}} \cdot \vec{d\ell} = 0$. Let us apply this result to the closed path *abcda* lying across the interface, as displayed in Figure 3.34a. The closed path consists of two equal segments *ab* and *cd*, each of length $\vec{\Delta \mathbf{w}}$, parallel to and on opposite sides of the interface, and two shorter segments *bc* and *da*, each of length Δh. As $\Delta h \to 0$, the contributions along the length segments *bc* and *da* to the line integral $\oint \vec{\mathbf{E}} \cdot \vec{d\ell}$ can be neglected. Thus,

$$\vec{\mathbf{E}}_1 \cdot \vec{\Delta \mathbf{w}} - \vec{\mathbf{E}}_2 \cdot \vec{\Delta \mathbf{w}} = 0$$

or

$$(\vec{\mathbf{E}}_1 - \vec{\mathbf{E}}_2) \cdot \vec{\Delta \mathbf{w}} = 0$$

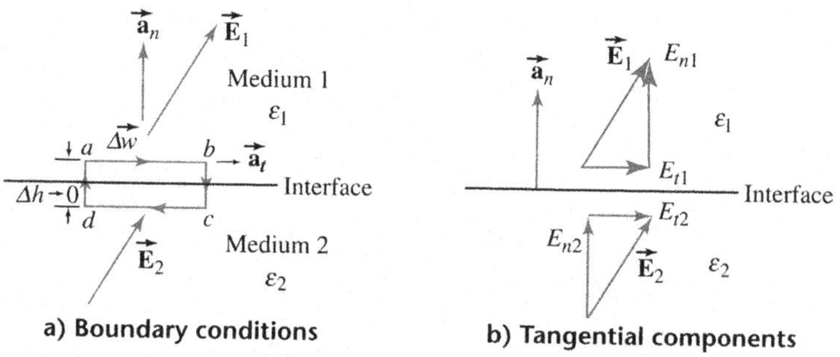

Figure 3.34 Boundary conditions for the tangential components of the $\vec{\mathbf{E}}$ field

If we express $\vec{\Delta \mathbf{w}} = \Delta w\, \vec{\mathbf{a}}_t$, where $\vec{\mathbf{a}}_t$, is the unit vector tangent to the interface as shown in Figure 3.34a, the preceding equation becomes

$$\vec{\mathbf{a}}_t \cdot (\vec{\mathbf{E}}_1 - \vec{\mathbf{E}}_2) = 0$$

or

$$E_{t1} = E_{t2} \tag{3.73a}$$

where E_{t1} and E_{t2} are the tangential components of the $\vec{\mathbf{E}}$ field in medium 1 and medium 2, respectively, as depicted in Figure 3.34b. This equation states that *the tangential components of the electric field intensity are always continuous at the interface.*

Equation (3.73a) can also be written in vector form as

$$\vec{\mathbf{a}}_n \times (\vec{\mathbf{E}}_1 - \vec{\mathbf{E}}_2) = 0 \tag{3.73b}$$

If medium 1 is a dielectric and medium 2 is a conductor, the tangential component of the electric field in medium 1 just next to the conductor must be zero because the electrostatic fields inside a conductor cannot exist. Thus, *the electrostatic field just above a conductor is always normal to the surface of a conductor.*

EXAMPLE 3.17

Charge Q is uniformly distributed over the surface of a metallic sphere of radius R. Determine the $\vec{\mathbf{E}}$ field just above the surface of the sphere.

Solution The surface charge density is

$$\rho_s = \frac{Q}{4\pi R^2}$$

Only the normal component of the $\vec{\mathbf{D}}$ field can exist just above the surface of a conductor, so $\vec{\mathbf{D}} = D_r \vec{\mathbf{a}}_r$. From (3.72a),

$$D_r = \frac{Q}{4\pi R^2}$$

If ϵ is the permittivity of the medium surrounding the sphere, then

$$E_r = D_r/\epsilon = \frac{Q}{4\pi\epsilon R^2}$$

a result which can be easily verified by using Gauss's law. • • •

EXAMPLE 3.18

The plane $z = 0$ marks the boundary between free space and a dielectric medium with a dielectric constant of 40. The $\vec{\mathbf{E}}$ field next to the interface in free space is $\vec{\mathbf{E}} = 13\vec{\mathbf{a}}_x + 40\vec{\mathbf{a}}_y + 50\vec{\mathbf{a}}_z$ V/m. Determine the $\vec{\mathbf{E}}$ field on the other side of the interface.

Solution Let $z > 0$ be the dielectric medium 1 and $z < 0$ be the free space medium 2. Then

$$\vec{\mathbf{E}}_2 = 13\vec{\mathbf{a}}_x + 40\vec{\mathbf{a}}_y + 50\vec{\mathbf{a}}_z$$

The unit vector $\vec{\mathbf{a}}_n$ normal to the interface is $\vec{\mathbf{a}}_z$. Because the tangential components of the $\vec{\mathbf{E}}$ field are continuous, then

$$E_{x1} = E_{x2} = 13 \quad \text{and} \quad E_{y1} = E_{y2} = 40$$

For a dielectric–dielectric interface, the normal components of the $\vec{\mathbf{D}}$ field are also continuous. That is,

$$\epsilon_1 E_{z1} = \epsilon_2 E_{z2}$$

However, $\epsilon_2 = \epsilon_0$ and $\epsilon_1 = 40\epsilon_0$. Therefore,

$$E_{z1} = \frac{E_{z2}}{40} = \frac{50}{40} = 1.25$$

Thus, the $\vec{\mathbf{E}}$ field in medium 1 is

$$\vec{\mathbf{E}} = 13\vec{\mathbf{a}}_x + 40\vec{\mathbf{a}}_y + 1.25\vec{\mathbf{a}}_z \text{ V/m}$$

• • •

3.10 Capacitors and capacitance

Two insulated conductors of any arbitrary shape adjacent to each other, as depicted in Figure 3.35, form a **capacitor**. By applying an external energy, we can transfer charges from one conductor to another. In other words, we are charging the capacitor using an external source. At all times during the charging process, the two conductors will have equal but opposite charges. This separation of charges establishes an electric field in the dielectric medium and thereby a potential difference between the conductors. As we continue the charging process it becomes apparent that the more we transfer charge from one conductor to another, the higher is the potential difference between them. Simply put, the potential difference between the two conductors is proportional to the charge transferred. Such an understanding enables us to define the **capacitance** as the ratio of the charge on one conductor to its potential with respect to the other. Capacitance is expressed mathematically as

$$C = \frac{Q_a}{V_{ab}} \tag{3.74}$$

where C is the capacitance in farads (F), Q_a is the charge on conductor a in coulombs (C), and V_{ab} is the potential of conductor a with respect to conductor b in volts (V).

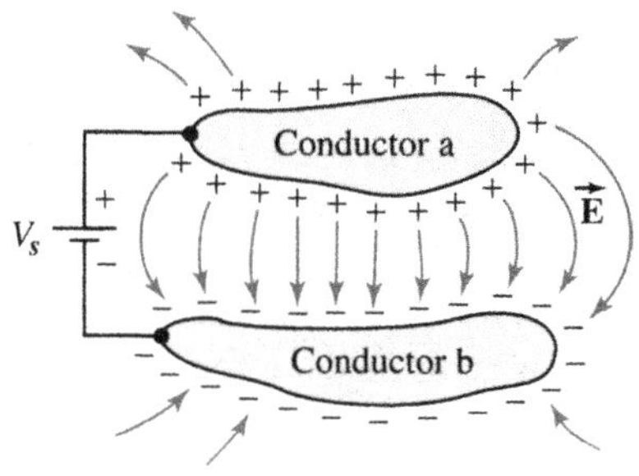

Figure 3.35 A charged capacitor

You may have already used capacitors in the design of tuned circuits in electronics and power factor correction networks in power systems. However, you may not have realized that capacitance also exists, even when it is not being sought, between the conductors of a transmission line and at the pn junction of a diode.

EXAMPLE 3.19

Two parallel conducting plates, each of area A, and separated by a distance d, as shown in Figure 3.36, form a parallel-plate capacitor. The charge on the top is $+Q$ and that on the other plate is $-Q$. What is its capacitance? Also express the energy stored in the medium in terms of the capacitance of the system.

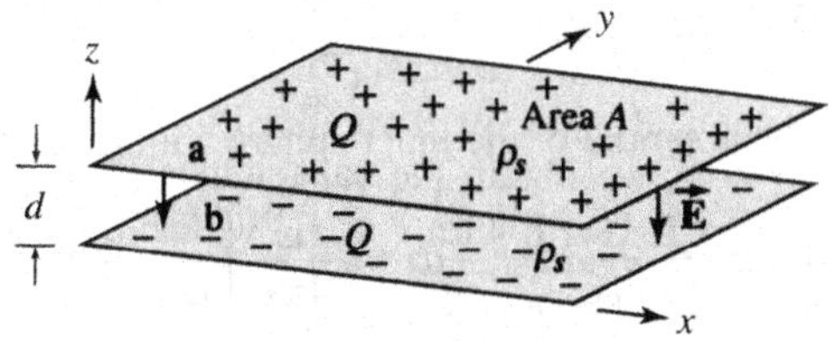

Figure 3.36 Parallel-plate capacitor

Solution Let us assume that the separation between the plates is very small compared to their other dimensions. Therefore, we can neglect the edge effects (fringing) and assume that the charge is uniformly distributed over the inner surface of each plate. The electric field intensity between the conductors is

$$\vec{\mathbf{E}} = -\frac{\rho_s}{\epsilon}\vec{\mathbf{a}}_z \quad \text{and} \quad \rho_s = \frac{Q}{A}$$

where Q is the charge on top plate a at $z = d$, A is the surface area of each plate, and ϵ is the permittivity of the medium. Note that the charge on plate b at $z = 0$ is $-Q$.

The potential of plate a with respect to plate b is

$$V_{ab} = -\int_b^a \vec{\mathbf{E}} \cdot \vec{d\ell} = \frac{\rho_s}{\epsilon}\int_0^d dz = \frac{\rho_s d}{\epsilon} = \frac{Qd}{\epsilon A}$$

Thus, the capacitance of the parallel-plate capacitor is

$$C = \frac{Q}{V_{ab}} = \frac{\epsilon A}{d} \tag{3.75a}$$

The energy stored in the system is

$$W = \frac{1}{2}\int_v \epsilon E^2\, dv = \frac{1}{2}\frac{Ad}{\epsilon}\rho_s^2 = \frac{1}{2}\frac{d}{\epsilon A}Q^2$$

$$= \frac{1}{2C}Q^2 = \frac{1}{2}CV_{ab}^2$$

These are the basic circuit equations for the energy stored in a capacitor.

• • •

EXAMPLE 3.20

A spherical capacitor is formed by two concentric metallic spheres of radii a and b, as shown in Figure 3.37 (see below). The charge on the inner sphere is $+Q$ and that on the outer sphere is $-Q$. Determine the capacitance of the system. What is the capacitance of an isolated sphere? Assuming the earth to be an isolated sphere of radius 6.5×10^6 meters, calculate its capacitance. Deduce an approximate expression for the capacitance when the separation between the spheres is very small as compared to their radii.

Solution

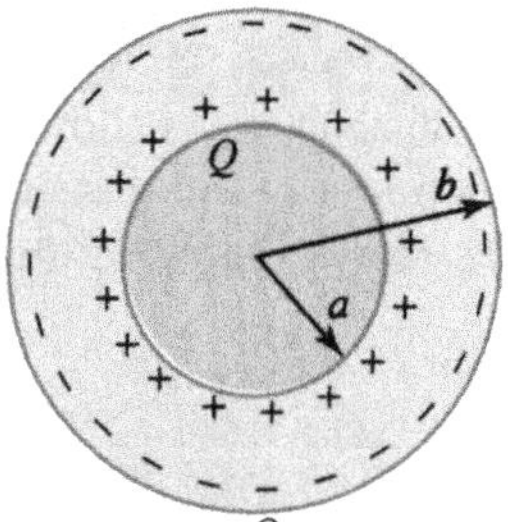

Figure 3.37 A spherical capacitor

For a uniform charge distribution over the spheres, the electric field intensity, from Gauss's law, within the spheres is

$$\vec{\mathbf{E}} = \frac{Q}{4\pi\epsilon r^2}\vec{\mathbf{a}}_r$$

The potential of the inner sphere with respect to the outer sphere is

$$V_{ab} = -\frac{Q}{4\pi\epsilon}\int_b^a \frac{1}{r^2}dr = \frac{Q}{4\pi\epsilon}\left[\frac{1}{a} - \frac{1}{b}\right]$$

Hence, the capacitance of the system is

$$C = \frac{Q}{V_{ab}} = \frac{4\pi\epsilon ab}{b-a} \tag{3.75b}$$

By setting $b \to \infty$, we obtain the capacitance of an isolated sphere as $C = 4\pi\epsilon a$. Substituting the values for earth with $\epsilon = \epsilon_0$, we have

$$C = \frac{6.5 \times 10^6}{9 \times 10^9} = 0.722 \times 10^{-3} \quad \text{or} \quad 722\,\mu\text{F}$$

If the separation between the two spheres is very small; i.e., $d = b - a$ and $d \ll a$, we can approximate $ab \approx a^2$, and the capacitance of the system becomes

$$C = \frac{4\pi\epsilon a^2}{b-a} = \frac{\epsilon A}{d} \tag{3.75c}$$

where $A = 4\pi a^2$ is the surface area of the inner sphere. • • •

From Examples 3.19 and 3.20 it is apparent that the capacitance between two conductors depends upon (a) the sizes and shapes of the conductors, (b) the separation between them, and (c) the permittivity of the medium. As the following example illustrates, (3.75c) is very powerful in the sense that we can always use it to determine the capacitance of a system of two conductors.

EXAMPLE 3.21

Using (3.75c), determine the capacitance of a spherical capacitor.

Solution

Let us divide the region between $a \leq r \leq b$ in Figure 3.38a into n regions so that $\Delta r_i \ll r_i$, where Δr_i is the separation between two surfaces of the ith capacitor at a radius r_i, A_i is its surface area, ϵ_i is its permittivity,

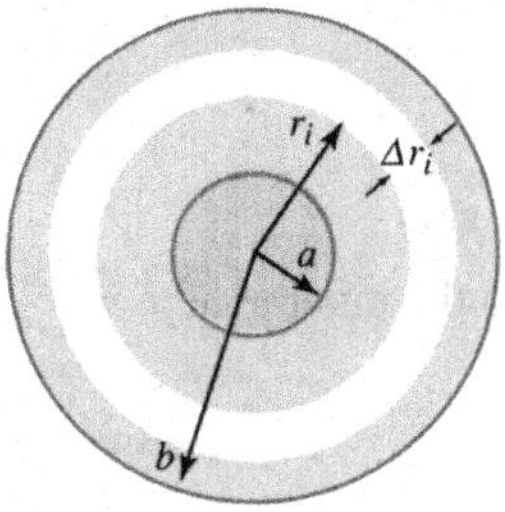

a)

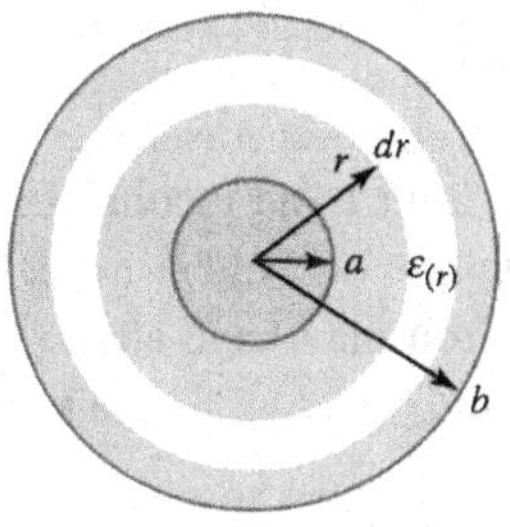

b)

Figure 3.38 Illustrations to determine the capacitance of a spherical capacitor

and C_i is its capacitance, which can be computed from (3.75c). There are n capacitors connected in series between $r = a$ and $r = b$, so the capacitance between the two spheres is

$$\frac{1}{C} = \sum_{i=1}^{n} \frac{\Delta r_i}{\epsilon_i A_i} \tag{3.76}$$

In the limit $n \to \infty$ and $\Delta r_i \to 0$, the summation in equation (3.76) can be replaced by an integral as

$$\frac{1}{C} = \int_a^b \frac{dr}{\epsilon(r) A_r} \tag{3.77}$$

where $\epsilon(r)$ symbolizes that the permittivity, which may be a function of r, and A_r is the cross-sectional area $(4\pi r^2)$ at an arbitrary radius r, as indicated in Figure 3.38b. Equation (3.77), although developed for a spherical capacitor, must be treated as a general equation because we can write similar expressions for parallel-plate and cylindrical capacitors.

If we assume that the permittivity of the medium is constant, then the capacitance between two spherical conductors is

$$\frac{1}{C} = \frac{1}{4\pi\epsilon} \int_a^b \frac{dr}{r^2} = \frac{1}{4\pi\epsilon}\left[\frac{1}{a} - \frac{1}{b}\right]$$

or

$$C = \frac{4\pi\epsilon ab}{b - a}$$

which is exactly the same result as given in (3.75b). • • •

EXAMPLE 3.22

The region between two concentric spherical shells is filled with two different dielectrics, as shown in Figure 3.39. Find the capacitance of the system.

Solution

We expect the $\vec{\mathbf{E}}$ field to be in the radial direction and its tangential components to be continuous at the boundary between the two media. That is,

$$E_{r1} = E_{r2}$$

Since $\vec{\mathbf{D}} = \epsilon \vec{\mathbf{E}}$,

$$D_{r1} = \epsilon_1 E_{r1} \quad \text{and} \quad D_{r2} = \epsilon_2 E_{r2}$$

Therefore,

$$D_{r2} = \frac{\epsilon_2}{\epsilon_1} D_{r1} \tag{3.78}$$

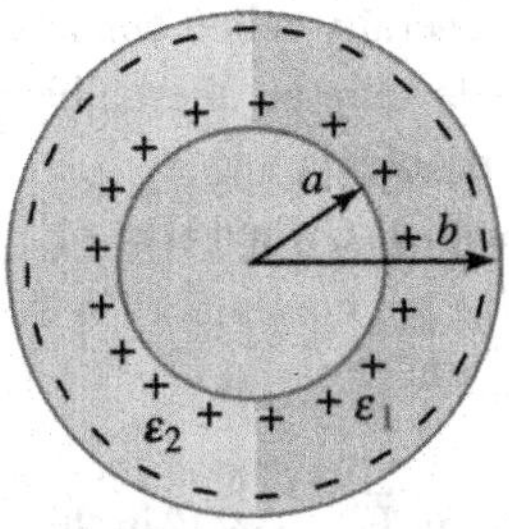

Figure 3.39

From Gauss's law, at any closed surface r, $a \le r \le b$,

$$\oint_s \vec{\mathbf{D}} \cdot \vec{ds} = Q \tag{3.79}$$

Thus,

$$D_{r1} + D_{r2} = \frac{Q}{2\pi r^2} \tag{3.80}$$

From (3.78) and (3.80), we obtain

$$D_{r1} = \frac{Q\epsilon_1}{2\pi r^2(\epsilon_1 + \epsilon_2)} \quad \text{and} \quad E_{r1} = \frac{Q}{2\pi r^2(\epsilon_1 + \epsilon_2)}$$

The potential of the inner sphere with respect to the outer sphere is

$$\begin{aligned} V_{ab} &= -\frac{Q}{2\pi(\epsilon_1 + \epsilon_2)} \int_b^a \frac{1}{r^2}\, dr \\ &= \frac{Q}{2\pi(\epsilon_1 + \epsilon_2)} \left[\frac{b-a}{ab}\right] \end{aligned}$$

Hence, the capacitance of the system is

$$C = 2\pi(\epsilon_1 + \epsilon_2)\left[\frac{ab}{b-a}\right] = C_1 + C_2$$

where

$$C_1 = 2\pi\epsilon_1\left[\frac{ab}{b-a}\right] \quad \text{and} \quad C_2 = 2\pi\epsilon_2\left[\frac{ab}{b-a}\right]$$

Note that C_1 and C_2 are the capacitances of medium 1 and medium 2, respectively. Thus, the capacitance of the system is equivalent to the parallel combination of the two capacitances. You may have already used this result in the analysis of electrical circuits. • • •

3.11 Poisson's and Laplace's equations

In the preceding sections, we have determined electrostatic fields in a medium when the charge distribution is everywhere specified. However, many of the problems encountered in practice are not of this type. Quite often, we must determine the electric fields first, before we can even calculate the charge distribution. We are also confronted with problems that involve boundary surfaces on which either the surface charge density or the potential is specified. Such problems are generally called *boundary value problems*. In this section, we wish to develop an alternative approach for solving such types of electrostatic problems.

We can express Gauss's law in a linear medium; that is, when $\vec{\mathbf{D}} = \epsilon\vec{\mathbf{E}}$, as

$$\nabla \cdot (\epsilon\vec{\mathbf{E}}) = \rho_v$$

where ρ_v is the volume charge density. Substituting $\vec{\mathbf{E}} = -\nabla V$ in this equation, we obtain

$$\nabla \cdot (-\epsilon\nabla V) = \rho_v \tag{3.81}$$

Using the vector identity equation (2.126), we can express (3.81) as

$$\epsilon\nabla \cdot (\nabla V) + \nabla V \cdot \nabla\epsilon = -\rho_v$$

or

$$\epsilon\nabla^2 V + \nabla V \cdot \nabla\epsilon = -\rho_v \tag{3.82}$$

which is a second-order partial differential equation in terms of the potential function V and the volume charge density ρ_v. Equation (3.82) is valid if ϵ is a function of position. This equation can be solved if we know the boundary conditions and the functional dependence of ρ_v and ϵ.

However, if we are interested primarily in solutions to equation (3.82) in a *homogeneous* medium, that is when ϵ is a constant, then $\nabla\epsilon = 0$ and (3.82) reduces to

$$\nabla^2 V = -\rho_v/\epsilon \tag{3.83}$$

This equation, called *Poisson's equation*, states that the potential distribution in a region depends upon the local charge distribution. The solution to (3.83) is, in fact, known and is given by (3.37).

There are some problems in electrostatics that involve charge distributions on the surface of conductors. In these cases, the free volume charge density is zero in the region of interest. Thus, in the region where ρ_v vanishes, (3.83) reduces to

$$\nabla^2 V = 0 \tag{3.84}$$

and is called *Laplace's equation.*

In a charge-free region, we will seek a potential function V that satisfies Laplace's equation subjected to the boundary conditions. Once the potential function in the region is known, the electric field intensity $\vec{\mathbf{E}}$ can be determined as $\vec{\mathbf{E}} = -\nabla V$. In a linear, homogeneous, and charge-free region, $\nabla \cdot \vec{\mathbf{E}} = 0$. These are the fields of class I type, as discussed in Chapter 2. Therefore, the solution to Laplace's equation is unique. Other quantities of interest such as capacitance, charges on the surface of conductors, energy density, and the total energy stored in the system can also be determined.

EXAMPLE 3.23

The two metal plates of Figure 3.40 having an area A and a separation d form a parallel-plate capacitor. The upper plate is held at a potential of V_0, and the lower plate is grounded. Determine (a) the potential distribution, (b) the electric field intensity, (c) the charge distribution on each plate, and (d) the capacitance of the parallel-plate capacitor.

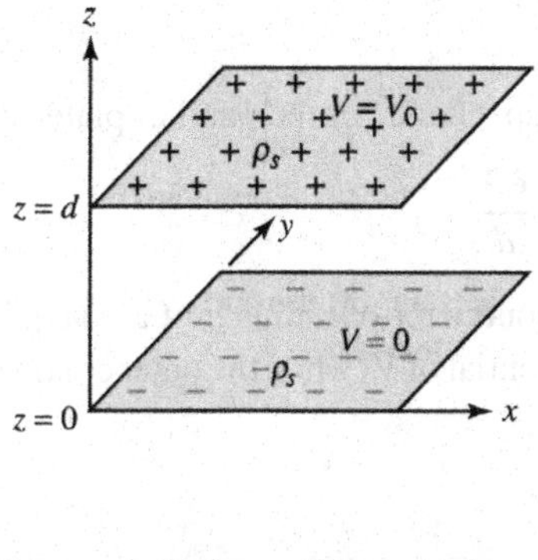

Figure 3.40 A charged parallel-plate capacitor

Solution Since the two metal plates (conductors) form equipotential surfaces in the xy plane at $z = 0$ and $z = d$, we expect that the potential V must be a function of z only. For the charge-free region between the plates, Laplace's equation reduces to

$$\frac{\partial^2 V}{\partial z^2} = 0$$

with a solution

$$V = az + b$$

where a and b are constants to be evaluated from the knowledge of boundary conditions.

When $z = 0$, $V = 0 \Rightarrow b = 0$. The potential distribution within the plates now becomes

$$V = az$$

However, when $z = d$, $V = V_0$ suggests that $a = V_0/d$. Thus, the potential varies linearly in a parallel-plate capacitor as

$$V = \frac{z}{d} V_0$$

We can now compute the electric field intensity as

$$\vec{\mathbf{E}} = -\nabla V = -\vec{\mathbf{a}}_z \frac{\partial V}{\partial z} = -\frac{V_0}{d} \vec{\mathbf{a}}_z$$

and the electric flux density is

$$\vec{\mathbf{D}} = \epsilon \vec{\mathbf{E}} = -\frac{\epsilon V_0}{d} \vec{\mathbf{a}}_z$$

Since the normal component of the $\vec{\mathbf{D}}$ field must be equal to the surface charge density on a conductor, the surface charge density on the lower plate is

$$\rho_s|_{z=0} = -\frac{\epsilon V_0}{d}$$

and that on the upper plate is

$$\rho_s|_{z=d} = \frac{\epsilon V_0}{d}$$

The total charge on the upper plate is

$$Q = \frac{\epsilon V_0 A}{d}$$

Thus, the capacitance of the parallel plate capacitor is

$$C = \frac{Q}{V_0} = \frac{\epsilon A}{d}$$

• • •

EXAMPLE 3.24 The inner conductor of radius a of a coaxial cable (see Figure 3.41) is held at a potential of V_0 while the outer conductor of radius b is grounded.

Figure 3.41

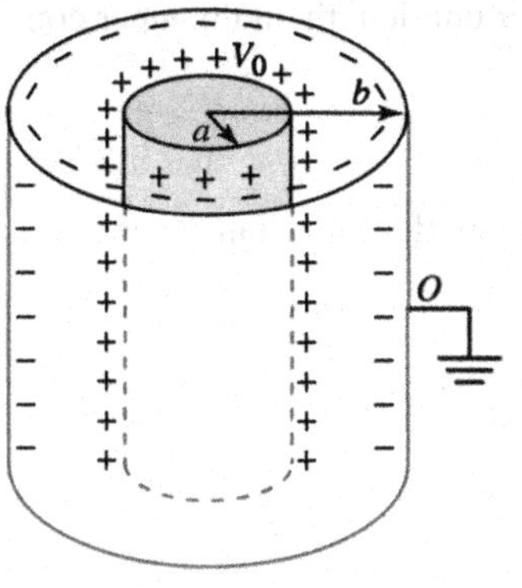

Determine (a) the potential distribution between the conductors, (b) the surface charge density on the inner conductor, and (c) the capacitance per unit length.

Solution Since the two conductors of radii a and b form equipotential surfaces, we expect the potential V to be a function of ρ only. Thus, Laplace's equation reduces to

$$\frac{1}{\rho}\frac{d}{d\rho}\left(\rho\frac{dV}{d\rho}\right) = 0$$

Integrating twice, we obtain

$$V = c\ln\rho + d$$

where c and d are constants of integration.

At $\rho = b$, $V = 0 \Rightarrow d = -c\ln b$. Thus,

$$V = c\ln(\rho/b)$$

At $\rho = a$, $V = V_0 \Rightarrow c = V_0/\ln(a/b)$. Hence, the potential distribution within the region $a \leq \rho \leq b$ is

$$V = V_0\frac{\ln(\rho/b)}{\ln(a/b)}$$

The electric field intensity is

$$\vec{\mathbf{E}} = -\nabla V = -\frac{\partial V}{\partial \rho}\vec{\mathbf{a}}_\rho = \frac{V_0\vec{\mathbf{a}}_\rho}{\rho\ln(b/a)}$$

and the electric flux density is

$$\vec{\mathbf{D}} = \epsilon\vec{\mathbf{E}} = \frac{\epsilon V_0\vec{\mathbf{a}}_\rho}{\rho\ln(b/a)}$$

The normal component of $\vec{\mathbf{D}}$ at $\rho = a$ yields the surface charge density on the inner conductor as

$$\rho_s = \frac{\epsilon V_0}{a\ln(b/a)}$$

The charge per unit length on the inner conductor is

$$Q = \frac{2\pi\epsilon V_0}{\ln(b/a)}$$

Finally, we obtain the capacitance per unit length as

$$C = \frac{2\pi\epsilon}{\ln(b/a)}$$

• • •

3.12 Method of images

Thus far, we have tacitly assumed that charges exist by themselves and that there is nothing else in the region that may influence their fields. Quite often, however, the charges (or the charge distributions) are close to conducting surfaces, and their effects must be taken into consideration in obtaining the total fields in a region. For example, the effect of the earth on the fields from an open-wire transmission line cannot be ignored. Similarly, the field patterns of transmitting and receiving antennas are greatly modified by the presence of the conducting bodies on which they are mounted. To account for the influence of a nearby conductor on a field we must know the charge distribution on the surface of a conductor, which, in turn, depends upon the fields just above its surface. However, in the case of static fields, we know that (a) a conductor forms an equipotential surface, (b) there are no fields inside an isolated conductor, and (c) the fields are normal to the surface of a conductor. These observations will help quantify the charge distribution on the surface of a conductor and its influence on the fields in the region.

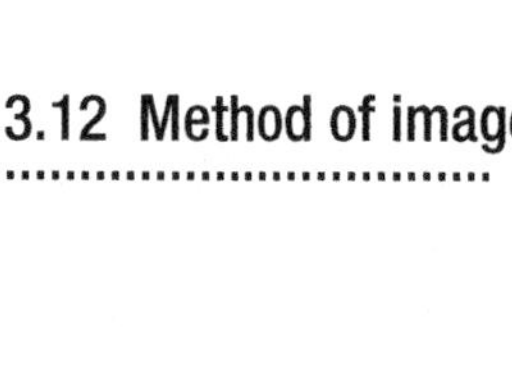

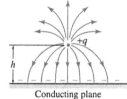

Figure 3.42 The lines of force for a point charge q above an infinite conducting plane

During our discussion of a dipole we stated that the potential at any point on the bisecting plane is zero, and the electric field intensity is normal to the plane. Therefore, the bisecting plane satisfies the requirements of a conducting plane. In other words, if a conducting plane is inserted to coincide with the bisecting plane, the field pattern of the dipole remains unchanged. If the negative charge below the conducting plane is removed, the field distribution in the region above the plane still remains the same, and the total charge induced on the surface of the conductor is $-q$, as shown in Figure 3.42. Conversely, if we are given a point charge q at a distance h above a conducting plane of infinite extent, we can determine the potential and the electric field at any point above the plane by ignoring the plane and imagining a charge $-q$ at the same distance away on the other side of the plane. *The imaginary charge $-q$ is said to be the image of the real charge q. Thus, in the method of images, a conducting plane is temporarily ignored, and an imaginary charge is placed behind the plane. The imaginary charge is equal in magnitude and opposite in polarity to the real charge. The distance between the real and the imaginary charges is twice the distance between the real* *charge and the plane*. However, these statements are only true for a conducting plane of infinite extent and depth. For a curved surface, the imaginary charge is neither equal in magnitude nor as far away on the other side of the conducting surface. We will highlight this fact with an example. The points to remember in the method of images are as follows

a) The image charge is a fictitious charge.
b) The image charge is located in the region of the conducting plane.
c) The conducting plane is an equipotential surface.

When a point charge is enclosed between two parallel conducting planes, the number of images is infinite. However, for the two bisecting planes the number of images will be finite as long as the angle between the planes is a submultiple of 360°. In general, if θ is the angle of intersection of the two planes, the field due to a point charge placed between the planes can be obtained by replacing the planes with $n-1$ image charges as long as $n = 360°/\theta$ is an integer.

EXAMPLE 3.25

A point charge q is located above the surface of a conducting plane of infinite extent and depth. Calculate the potential and electric field intensity at any point P. Show that the total charge induced on the surface of the plane is $-q$.

Solution

Figure 3.43 shows a point charge q at (0, 0, d) above the surface of a conducting plane. To determine the fields, we place an imaginary charge $-q$ at (0, 0, $-d$) and temporarily ignore the existence of the plane. The potential at any point $P(x, y, z)$ and $z \geq 0$ is

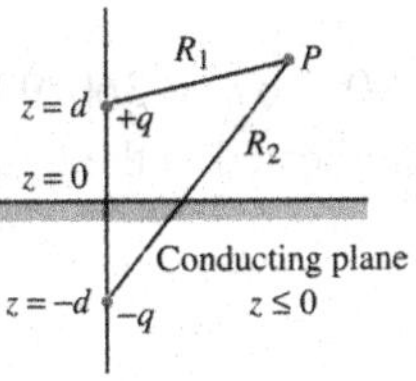

Figure 3.43 A point charge above an infinite conducting plane

$$V = \frac{q}{4\pi\epsilon}\left[\frac{1}{R_1} - \frac{1}{R_2}\right]$$

where $R_1 = [x^2 + y^2 + (z-d)^2]^{1/2}$ and $R_2 = [x^2 + y^2 + (z+d)^2]^{1/2}$. On the surface of the conductor; i.e., the $z = 0$ plane, $V = 0$ as $R_1 = R_2$. The electric field intensity at point P is

$$\vec{\mathbf{E}} = -\nabla V$$

$$= -\frac{q}{4\pi\epsilon}\left[\left(\frac{x}{R_2^3} - \frac{x}{R_1^3}\right)\vec{\mathbf{a}}_x + \left(\frac{y}{R_2^3} - \frac{y}{R_1^3}\right)\vec{\mathbf{a}}_y + \left(\frac{z+d}{R_2^3} - \frac{z-d}{R_1^3}\right)\vec{\mathbf{a}}_z\right]$$

On the surface of the conducting plane, the $\vec{\mathbf{E}}$ field reduces to

$$\vec{\mathbf{E}} = -\frac{2qd}{4\pi\epsilon R^3}\vec{\mathbf{a}}_z \quad \text{where} \quad R = [x^2 + y^2 + d^2]^{1/2}$$

The normal component of the $\vec{\mathbf{D}}$ field must be equal to the surface charge density on the surface of the conductor at $z = 0$, so we have

$$\rho_s = -\frac{2qd}{4\pi R^3}$$

Thus, the total charge induced on the surface of a conductor of an infinite extent is

$$Q = \int_s \rho_s \, ds = -\frac{2qd}{4\pi} \int_0^\infty \frac{\rho \, d\rho}{(\rho^2 + d^2)^{3/2}} \int_0^{2\pi} d\phi = -q$$

Thus, the total charge on the surface of the conductor is $-q$ as expected.

• • •

EXAMPLE 3.26

Two infinite intersecting planes are shown in Figure 3.44. A charge of 100 nC is placed at (3, 4, 0). Find the electric potential and electric field intensity at (3, 5, 0).

Figure 3.44 A point charge in front of two conducting planes intersecting at 90°

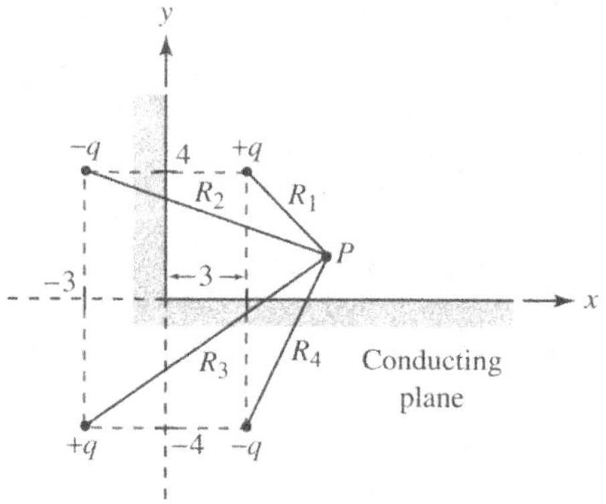

Solution The intersection angle between the two planes is 90°, so $n = 360/90 = 4$. Therefore, we need three fictitious charges as indicated in the figure. If (x, y, z) are the general coordinates of point P, then

$$R_1 = [(x-3)^2 + (y-4)^2 + z^2]^{1/2}$$
$$R_2 = [(x+3)^2 + (y-4)^2 + z^2]^{1/2}$$
$$R_3 = [(x+3)^2 + (y+4)^2 + z^2]^{1/2}$$
$$R_4 = [(x-3)^2 + (y+4)^2 + z^2]^{1/2}$$

Assuming the region to be free space, the potential at $P(x, y, z)$ is

$$V = 9 \times 10^9 \times 100 \times 10^{-9} \left[\frac{1}{R_1} - \frac{1}{R_2} + \frac{1}{R_3} - \frac{1}{R_4} \right]$$

and at $P(3, 5, 0)$,

$$V(3, 5, 0) = 735.2 \text{ V}$$

The electric field intensity is

$$\vec{\mathbf{E}} = -\nabla V$$
$$= -\frac{\partial V}{\partial x} \vec{\mathbf{a}}_x - \frac{\partial V}{\partial y} \vec{\mathbf{a}}_y - \frac{\partial V}{\partial z} \vec{\mathbf{a}}_z$$

However,

$$\frac{\partial V}{\partial x} = 900\left[-\frac{x-3}{R_1^3} + \frac{x+3}{R_2^3} - \frac{x+3}{R_3^3} + \frac{x-3}{R_4^3}\right]$$
$$= 19.8 \text{ at } P(3, 5, 0)$$

Similarly, we obtain at $P(3, 5, 0)$

$$\frac{\partial V}{\partial y} = -891.36 \quad \text{and} \quad \frac{\partial V}{\partial z} = 0$$

Hence, the $\vec{\mathbf{E}}$ field at $P(3, 5, 0)$ is

$$\vec{\mathbf{E}} = -19.8\vec{\mathbf{a}}_x + 891.36\vec{\mathbf{a}}_y \text{ V/m}$$

• • •

EXAMPLE 3.27

A point charge q is placed at a distance d from the center of a grounded conducting sphere of radius a. Calculate the surface charge density on the sphere.

Solution

Because of the curved nature of the surface, we don't expect the image charge to be equal in magnitude to the real charge q. Let the image charge be $-mq$, where m is a constant. The image charge will lie on the line joining the center of the sphere with the real charge, as shown in Figure 3.45. Thus, the potential at any point P is

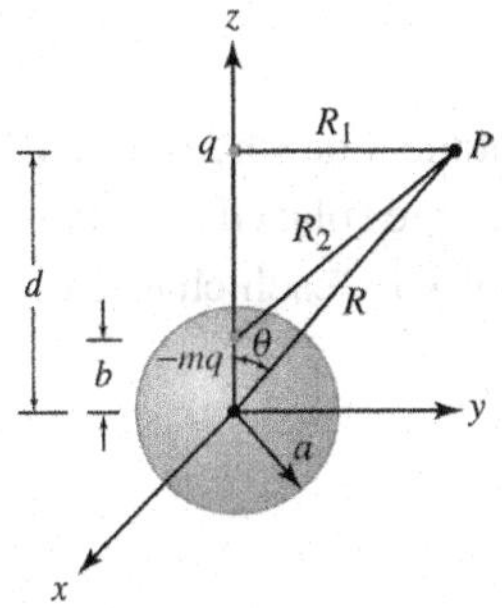

Figure 3.45 A point charge outside a conducting sphere

$$V = \frac{q}{4\pi\epsilon}\left[\frac{1}{R_1} - \frac{m}{R_2}\right]$$

where $R_1 = [r^2 + d^2 - 2rd\cos\theta]^{1/2}$ and $R_2 = [r^2 + b^2 - 2rb\cos\theta]^{1/2}$.

The boundary condition requires that on the surface of the sphere, $r = a$, the potential vanishes. That is,

$$\frac{1}{[a^2 + d^2 - 2ad\cos\theta]^{1/2}} = \frac{m}{[a^2 + b^2 - 2ab\cos\theta]^{1/2}}$$

We need two equations to find m and b. To do so, we square both sides and equate the coefficients of $\cos\theta$ and the remaining terms to obtain

$$(a^2 + d^2)m^2 = a^2 + b^2$$

and

$$2adm^2 = 2ab$$

Solving these equations, we get

$$m^2 = \frac{b}{d}, \quad b = \frac{a^2}{d}, \quad \text{and} \quad m = \frac{a}{d}$$

Hence, the image charge is

$$-mq = -\frac{aq}{d}$$

It is obvious that $m \leq 1$. Only when $d = a$, is $m = 1$. That is, the image charge is equal in magnitude to the real charge only when the real charge is just above the surface of the sphere. The image charge moves toward the center as q moves away from the sphere. Since the surface charge density on the sphere is equal to the normal component of the $\vec{\mathbf{D}}$ field, we can write it as

$$\rho_s = \vec{\mathbf{a}}_r \cdot \vec{\mathbf{D}} = \vec{\mathbf{a}}_r \cdot (-\epsilon \nabla V) = -\epsilon \frac{\partial V}{\partial r}\Big|_{r=a}$$

$$= -\frac{q}{4\pi a}\left[\frac{(d^2 - a^2)}{(d^2 + a^2 - 2ad\cos\theta)^{3/2}}\right] \qquad \bullet\bullet\bullet$$

3.13 Summary

Electrostatic field theory is the study of time-invariant fields established by charges at rest. The entire theory stems from experimental observations made by Coulomb. A quantitative statement of Coulomb's law is

$$\vec{\mathbf{F}} = \frac{q_1 q_2}{4\pi\epsilon R^2}\,\vec{\mathbf{a}}_R$$

Further experimental observations showed that the total force exerted on a point charge q by a number of other point charges placed around it is the vector sum of the forces exerted individually by each charge on q. That is,

$$\vec{\mathbf{F}} = \sum_{i=1}^{n} \frac{q q_i (\vec{\mathbf{r}} - \vec{\mathbf{r}}_i)}{4\pi\epsilon |\vec{\mathbf{r}} - \vec{\mathbf{r}}_i|^3}$$

We defined the electric field intensity in terms of the force experienced by a test charge q_t in the limit $q_t \to 0$. The electric field intensity at a point in any medium due to a point charge is

$$\vec{\mathbf{E}} = \frac{q}{4\pi\epsilon R^2}\,\vec{\mathbf{a}}_R$$

We also obtained the expressions for the electric field intensity at a point due to volume, surface, and line charge distribution as

$$\vec{\mathbf{E}} = \frac{1}{4\pi\epsilon}\int_v \frac{\rho_v'\,dv'}{|\vec{\mathbf{r}} - \vec{\mathbf{r}}'|^3}(\vec{\mathbf{r}} - \vec{\mathbf{r}}')$$

$$\vec{\mathbf{E}} = \frac{1}{4\pi\epsilon}\int_s \frac{\rho_s'\,ds'}{|\vec{\mathbf{r}} - \vec{\mathbf{r}}'|^3}(\vec{\mathbf{r}} - \vec{\mathbf{r}}')$$

$$\vec{\mathbf{E}} = \frac{1}{4\pi\epsilon}\int_c \frac{\rho_\ell'\,d\ell'}{|\vec{\mathbf{r}} - \vec{\mathbf{r}}'|^3}(\vec{\mathbf{r}} - \vec{\mathbf{r}}')$$

We defined the electric flux density as

$$\vec{\mathbf{D}} = \epsilon\vec{\mathbf{E}} = \epsilon_0\vec{\mathbf{E}} + \vec{\mathbf{P}}$$

and the electric flux passing through a surface as

$$\Psi = \int_s \vec{\mathbf{D}} \cdot \overrightarrow{ds}$$

Gauss's law states that the net outward flux from a closed surface is equal to the net positive charge inside the closed surface. That is,

$$\oint_s \vec{\mathbf{D}} \cdot \overrightarrow{ds} = Q$$

in the integral form and

$$\nabla \cdot \vec{\mathbf{D}} = \rho_v$$

in the point or differential form.

As long as the charge distribution is symmetric, we can employ Gauss's law to find the electric flux density. To do so, we must also be able to justify that there exists a Gaussian surface on which the normal component of the $\vec{\mathbf{D}}$ field is constant.

On each surface of an isolated dielectric material placed in an $\vec{\mathbf{E}}$ field we expect a bound surface charge density

$$\rho_{sb} = \vec{\mathbf{P}} \cdot \vec{\mathbf{a}}_n$$

and within the dielectric material we expect a bound volume charge density

$$\rho_{vb} = -\nabla \cdot \vec{\mathbf{P}}$$

We defined the electric potential as work done per unit charge and obtained an equation for it in terms of the $\vec{\mathbf{E}}$ field as

$$V_{ab} = -\int_b^a \vec{\mathbf{E}} \cdot \overrightarrow{d\ell}$$

The absolute potential at point b due to a point charge q is

$$V_b = \frac{q}{4\pi \epsilon R}$$

The electric potential functions due to volume, surface, and line charge distributions are

$$V = \frac{1}{4\pi\epsilon} \int_v \frac{\rho_v' \, dv'}{|\vec{\mathbf{r}} - \vec{\mathbf{r}}'|}$$

$$V = \frac{1}{4\pi\epsilon} \int_s \frac{\rho_s' \, ds'}{|\vec{\mathbf{r}} - \vec{\mathbf{r}}'|}$$

$$V = \frac{1}{4\pi\epsilon} \int_c \frac{\rho_\ell' \, d\ell'}{|\vec{\mathbf{r}} - \vec{\mathbf{r}}'|}$$

From the potential distribution we can find the $\vec{\mathbf{E}}$ field as

$$\vec{\mathbf{E}} = -\nabla V$$

The time-independent electric field intensity is conservative in nature. Thus,

$$\nabla \times \vec{\mathbf{E}} = 0$$

The normal component of the $\vec{\mathbf{D}}$ field, in general, is discontinuous at the boundary. That is,

$$\vec{\mathbf{a}}_n \cdot (\vec{\mathbf{D}}_1 - \vec{\mathbf{D}}_2) = \rho_s$$

The tangential component of the $\vec{\mathbf{E}}$ field is continuous at the boundary. That is,

$$\vec{\mathbf{a}}_n \times (\vec{\mathbf{E}}_1 - \vec{\mathbf{E}}_2) = 0$$

The normal component of $\vec{\mathbf{D}}$ and the tangential component of $\vec{\mathbf{E}}$ are continuous at the interface between two different dielectric materials.

The volume charge density within a conductor in static equilibrium is zero, and so are the fields.

The electrostatic energy due to n point charges is

$$W = \frac{1}{2}\sum_{i=1}^{n} q_i V_i$$

The electrostatic energy for a continuous volume charge distribution is

$$W = \frac{1}{2}\int_v \rho_v V \, dv$$

and in terms of the $\vec{\mathbf{D}}$ and $\vec{\mathbf{E}}$ fields is

$$W = \frac{1}{2}\int_v \vec{\mathbf{D}} \cdot \vec{\mathbf{E}} \, dv$$

The capacitance was defined as the ratio of the charge on one conductor to its potential with respect to the other conductor. The capacitance of a parallel-plate capacitor is

$$C = \frac{\epsilon A}{d}$$

The capacitance of a cylindrical capacitor of length L (coaxial cable) is

$$C = \frac{2\pi\epsilon L}{\ln(b/a)}$$

Finally, the capacitance of a spherical capacitor is

$$C = \frac{4\pi\epsilon ab}{b - a}$$

The general expression to determine the capacitance between two conductors is

$$\frac{1}{C} = \int_a^b \frac{dr}{\epsilon(r) A_r}$$

The general expression for the potential distribution in any medium is

$$\epsilon\nabla^2 V + \nabla V \cdot \nabla\epsilon = -\rho_v$$

which is a second-order differential equation. If the permittivity of the medium is constant, we get Poisson's equation,

$$\nabla^2 V = -\rho_v/\epsilon$$

If the free volume charge density in the region of interest is zero, we obtain Laplace's equation,

$$\nabla^2 V = 0$$

When the charges exist close to a conducting region of infinite extent, we can determine the fields by the method of images. The image charges exist external to the region of interest and the conductor is ignored.

3.14 Review questions

3.1 What do we mean when we say that an object is charged?

3.2 What do we mean when we say that the net charge is conserved in a closed system?

3.3 State Coulomb's law in your own words.

3.4 If we place two positive charges in a region, they will experience a force of ______.

3.5 If we place two negative charges in a medium, they will experience a force of ______.

3.6 If a positive charge is placed near a negative charge, the positive charge will experience a force of ______.

3.7 What is electric field intensity?

3.8 In the strictest sense, what must be the definition of a point charge? What are the other possible charge distributions?

3.9 Prove that newtons per coulomb are dimensionally the same as volts per meter.

3.10 Define surface charge density in terms of volume charge density.

3.11 Define line charge density in terms of surface charge density.

3.12 What is a line of force?

3.13 The number of lines of force from a point charge of 10 C is ______.

3.14 If the work done in moving a positive test charge in an electric field is positive, the work is done by an ______ force. The potential ______ in the direction of motion.

3.15 If the potential decreases in the direction of motion of a positive test charge, the work is done by ______ force.

3.16 What is the significance of the negative sign in the equation $\vec{\mathbf{E}} = -\nabla V$?

3.17 State Gauss's law.

3.18 Can we use Gauss's law in each of the following cases: (a) $\rho_v = k\rho^2$, (b) $\rho_v = k\rho \cos\phi$, (c) $\rho_v = k/\rho$ and $\rho \neq 0$, (d) $\rho_v = kr$, (e) $\rho_v = kr\cos\theta$, (f) $\rho_v = kr\cos\phi$?

3.19 A hollow conductor in the form of a matchbox is placed in a static electric field. What is the electric field inside the conductor? Sketch the charge distribution on the inner and the outer surfaces of the conductor.

3.20 If the hollow conductor in Question 3.19 is held at a potential of 100 V, what must be the potential inside the conductor?

3.21 Is it necessary to find a Gaussian surface on which the normal component of the $\vec{\mathbf{D}}$ field is constant in order to apply Gauss's law? Give reasons to justify your answer.

3.22 A charge of 10 mC is placed inside a conducting shell. What is the induced charge on the inner surface? What is the charge on the outer surface? Does it matter where the charge is placed inside the shell?

3.23 What is the physical significance of $\nabla \cdot \vec{\mathbf{D}}$?

3.24 If the charge is distributed over a thin spherical shell of radius b, what is $\vec{\mathbf{E}}$ inside the shell?

3.25 How much energy is required to assemble a point charge? Do you think that a point charge can really exist?

3.26 Why are the equipotential surfaces perpendicular to the electric flux lines?

3.27 A point charge of 1 coulomb is brought from infinity to point a in free space. How much work is required to do this? If the energy expenditure in bringing a second charge of 1 coulomb from infinity to point b is 1 joule, what is the separation between the charges?

3.28 If the free space in Question 3.27 is replaced by a medium ($\epsilon_r = 4$), what is the separation between the two charges for the same energy expenditure?

3.29 What do we mean by boundary conditions?

3.30 Is the solution of $\nabla^2 V = 0$ unique by itself?

3.31 What is the significance of boundary conditions when solving Laplace's equation?

3.32 What is the definition of a bound charge?

3.33 Show that for any spherical distribution of charge, the field at radius r is the same as if all the charge inside the volume of radius r were concentrated at the center, and that outside of r were removed.

3.34 In a charge-free region, $E_x = \alpha x$, and $E_y = \beta y$. Find E_z.

3.35 If n capacitors are connected in series, what is the effective capacitance?

3.36 If n capacitors are connected in parallel, what is the effective capacitance?

3.37 Why is the charge on each capacitor the same when capacitors are connected in series?

3.38 Is the charge on each capacitor the same when they are connected in parallel?

3.39 If the surface charge density on the surface of a conductor is 10 mC/m^2, what is the electric flux density just above its surface?

3.40 Consider an interface between free space and a dielectric medium with dielectric constant of 5. The normal component of the $\vec{\mathbf{D}}$ field in free space is 10 C/m^2, and the tangential component of the $\vec{\mathbf{E}}$ field in the dielectric medium is 100 V/m. Find their counterparts.

3.41 What is the bound surface charge density at the interface in Question 3.40?

3.42 Do we consider bound charge densities in applying the boundary conditions?

3.43 When we apply Gauss's law in a dielectric medium, should we take into account the bound charge densities?

3.44 An electric field is given by $\vec{\mathbf{E}} = 10\vec{\mathbf{a}}_x + 20\vec{\mathbf{a}}_y + 20\vec{\mathbf{a}}_z$ V/m. Is this a uniform field? Why? What is its magnitude? What are the cosines of the angles it makes with the unit vectors?

3.15 Exercises

3.1 A charge of 5 nC is located at $P(2, \pi/2, -3)$ and another charge of -10 nC is situated at $Q(5, \pi, 0)$. Calculate the force exerted by one charge on the other. What is the nature of this force?

3.2 Three charges of 2 nC, -5 nC, and 0.2 nC are situated at $P(2, \pi/2, \pi/4)$, $Q(1, \pi, \pi/2)$, and $S(5, \pi/3, 2\pi/3)$, respectively. Find the force acting on the 2-nC charge at point P. Is this a force of attraction or repulsion?

3.3 Solve Example 3.6 using the expression for the $\vec{\mathbf{E}}$ field obtained in Example 3.5.

3.4 Two infinite planes with equal and opposite but uniform charge distributions are separated by a distance d. Find the electric field intensity above, below, and in the region between the planes.

3.5 A very thin, finite, and uniformly charged line of length 10 m carries a charge of 10 μC/m. Calculate the electric field intensity in a plane bisecting the line at $\rho = 5$ m.

3.6 The electric flux density passing through a circular region of radius 0.5 m in a $z = 2.5$-m plane is given as $\vec{\mathbf{D}} = 10 \sin\phi\, \vec{\mathbf{a}}_\rho + 12z \cos(\phi/4)\, \vec{\mathbf{a}}_z$ C/m^2. Find the total flux passing through the surface.

3.7 The charge distribution within a spherical region bounded by radii a and b $(a < b)$ is given as $\rho_v = k/r$, where k is a constant. Determine the electric field intensity everywhere in space. What is the total flux passing through a surface at $r = b$?

3.8 A cylindrical conductor of radius a and of infinite length has a uniform charge distribution ρ_s over its surface. Compute the electric field intensity and the electric flux density everywhere in space. Calculate the flux passing through a cylindrical surface of radius b $(b > a)$ and length ℓ.

3.9 Two point charges of 120 nC and 800 nC are separated by a distance of 40 cm. How much energy must be expended to reduce the separation to 30 cm?

3.10 Using (3.9) and vector operations, show that (a) $\vec{\mathbf{E}} = -\nabla V$, and (b) $\nabla \times \vec{\mathbf{E}} = 0$.

3.11 Show that equipotential surfaces for an infinite, uniformly charged line are concentric cylinders.

3.12 Verify equation (3.40).

3.13 Obtain expressions for the lines of force of an electric dipole.

3.14 Show that the magnitude of the electric field intensity of an electric dipole is

$$E = \frac{p}{4\pi\epsilon_0 r^3}[1 + 3\cos^2\theta]^{1/2}$$

3.15 Show that the electric field intensity due to an electric dipole represents a conservative field.

3.16 A very long conducting cylinder of radius a carries a uniform surface charge density ρ_{sa}. It is enclosed by another conducting cylinder of inner and outer radii b and c, respectively. Calculate the electric field intensity everywhere in space. What is the charge density on the inner surface of the outer conductor? What is the charge density on the outer surface of the outer conductor? What happens to the charge and the fields when the outer surface of the outer conductor is grounded?

3.17 If the surface charge density on the inner conductor in Exercise 3.16 is taken to be positive, is the potential of the inner conductor higher or lower than that of the outer conductor? What is the potential difference between the two conductors?

3.18 The charge distribution within a sphere of radius b is given as $\rho_v = (b + r)(b - r)$ C/m^3. Compute the electric field intensity and potential everywhere in free space. Assume that the potential is zero at $r = \infty$.

3.19 Show that $\nabla'(1/R) = (1/R^2)\vec{\mathbf{a}}_R$, where $R = |\vec{\mathbf{r}} - \vec{\mathbf{r}}'|$.

3.20 A dielectric rod of radius 10 mm extends along the z axis from $z = 0$ to $z = 10$ m. The polarization of the rod is given as $\vec{\mathbf{P}} = [2z^2 + 10]\vec{\mathbf{a}}_z$. Calculate the bound volume charge density and the surface polarization charge on each surface. What is the total bound charge?

3.21 The polarization vector in a dielectric cube of side b is given as $\vec{\mathbf{P}} = x\vec{\mathbf{a}}_x + y\vec{\mathbf{a}}_y + z\vec{\mathbf{a}}_z$. If the origin of the coordinates is at the center of the cube, find the bound volume charge density and the bound surface charge density. Does the total bound charge vanish in this case?

3.22 A dielectric cylinder of radius b is polarized along its length and extends along the z axis from $z = -L/2$ to $z = L/2$. If the polarization is uniform and of magnitude P, calculate the electric field resulting from this polarization at a point on the z axis both inside and outside the dielectric cylinder.

3.23 An infinite line charge is enclosed inside a dielectric medium with constant permittivity. Determine the electric field intensity at any point in the medium. What are the bound charge densities when the line is approximated with a cylinder of radius α such that $\alpha \to 0$?

3.24 A uniform volume charge distribution exists in a spherical volume of radius a. Compute the total energy of the system using (3.65).

3.25 Obtain the total energy of the system for Exercise 3.24 using (3.68).

3.26 If the charge distribution in Exercise 3.24 is surrounded by a concentric dielectric shell (inner radius b, outer radius c, permittivity ϵ), determine: (a) the total stored energy in the system, (b) the bound charge densities, and (c) the net bound charge.

3.27 Two dielectric media with permittivities ϵ_1 and ϵ_2 are separated by a plane interface. If θ_1 and θ_2 are the angles that $\vec{\mathbf{E}}_1$ and $\vec{\mathbf{E}}_2$ make with the normal at the interface, find a relationship between θ_1 and θ_2.

3.28 A cylindrical conductor of radius 10 cm carries a uniform surface charge distribution of 200 $\mu C/m^2$. The conductor is embedded in an infinite dielectric medium ($\epsilon_r = 5$). Using boundary conditions, determine $\vec{\mathbf{D}}$ and $\vec{\mathbf{E}}$ in the dielectric medium just above the conductor's surface. What is the bound surface charge density per unit length on the surface of the dielectric next to the conductor?

3.29 Consider a plane interface between free space and a conductor. The x component of the $\vec{\mathbf{E}}$ field in free space is 10 V/m, and it makes an angle of 30° to the normal at the interface. What must be the other components of the $\vec{\mathbf{E}}$ field? What is the surface charge density at the interface?

3.30 Two parallel plates are separated by a dielectric of thickness 2 mm with a dielectric constant of 6. If the area of each plate is 40 cm^2 and the potential difference between them is 1.5 kV, determine (a) the capacitance, (b) the electric field intensity, (c) the electric flux density, (d) the polarization vector, (e) the free surface charge density, (f) the bound charge densities, and (g) the energy stored by the capacitor.

3.31 A cylindrical conductor of radius a is enclosed by another cylindrical conductor of radius b to form a cylindrical capacitor. The permittivity of the medium is ϵ. Obtain an expression for the capacitance per unit length using equation (3.74). If the length of the capacitor is L, what is its total capacitance?

3.32 Solve Exercise 3.31 using equation (3.77).

3.33 The inner spherical shell of radius 10 cm is held at a potential of 1000 V with respect to the outer shell of radius 12 cm. The dielectric constant of the medium is 2.5. Determine $\vec{\mathbf{E}}$, $\vec{\mathbf{D}}$, and $\vec{\mathbf{P}}$ in the medium. What is the surface charge density on each conductor? What are the bound charge densities? What is the capacitance of the system?

3.34 Repeat Exercise 3.30 if the medium is free space. What is the ratio of the two capacitances?

3.35 Repeat Exercise 3.33 if the medium is free space. Calculate the ratio of the two capacitances.

3.36 If the permittivity of the medium increases linearly from ϵ_1 at $z = 0$ to ϵ_2 at $z = d$ in Example 3.19, find the capacitance of the parallel-plate capacitor. What is the capacitance if $\epsilon_2 \to \epsilon_1$?

3.37 A homogeneous dielectric medium fills the region between two concentric spherical shells of radii a and b. The inner shell is held at a potential of V_0, and the outer shell is grounded. Compute (a) the potential distribution, (b) the electric field intensity, (c) the electric flux density, (d) the surface charge density on the inner surface, (e) the capacitance, and (f) the total energy stored in the system.

3.38 The space between the conductors in a coaxial cable is filled with two concentric layers of dielectric, as shown in Figure 3.46. Determine (a) the potential function in each medium, (b) the $\vec{\mathbf{E}}$ and $\vec{\mathbf{D}}$ fields in each region, (c) the charge distribution on the inner conductor, and (d) the capacitance per unit length. Show that the capacitance is equivalent to that of two capacitors connected in series.

Figure 3.46 Cross section of a coaxial cable showing two layers of dielectric

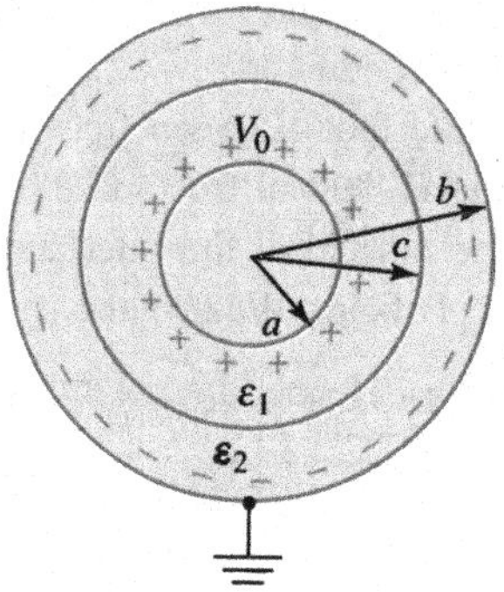

3.39 If $V(x, y, z)$ is a solution to Laplace's equation, show that $\partial V/\partial x$, $\partial^2 V/\partial x^2$, and $\partial^2 V/\partial x\, \partial y$ are also solutions of Laplace's equation.

3.40 Repeat Example 3.23 if the permittivity of the medium is given as $\epsilon = \epsilon_0(1 + mz)$, where m is a constant.

3.41 An infinite, fine line having a uniform charge distribution is at a distance d from an infinite conducting plane. Determine (a) the induced charge per unit length on the conducting plane along the line, and (b) the equation for the equipotential surfaces. Sketch a few of these equipotential surfaces.

3.42 Two conducting planes of infinite extent in the z-direction are at $\phi = 0°$ and $\phi = 60°$. A point charge q is situated at $(2, \pi/6, 0)$ when both plates are at ground potential. Find the potential at a point $(5, \pi/6, 0)$.

3.43 Repeat Exercise 3.42 if the point charge is replaced by a fine line having a uniform charge density.

3.16 Problems

3.1 A point charge of 2 μC is situated at $P(0, 4, 0)$, and a second charge of 10 μC is located at $S(3, 0, 0)$. Calculate the force experienced by each charge.

3.2 Determine the electric force acting on a point charge of 1 μC situated at the origin if a point charge of 200 nC is located at (0.2, 0.3, 0) meter and a charge of -1300 nC is located at (0.5, 0.7, -1.3).

3.3 An infinite, uniformly charged line with a charge density of 100 nC/m extends along the z axis. Find the force experienced by a charge of 500 nC located at (3, 4, 0).

3.4 Two parallel infinite lines with equal and opposite but uniform charge distributions of 100 nC/m are separated by a distance of 1 mm. Determine the force per unit length. What is the nature of this force?

3.5 Calculate the ratio of electric force to the gravitational force between an electron and a proton when the separation between them is 0.05 nm. Assume that the gravitational constant is 6.67×10^{-11} N·m/kg^2 and Newton's law of gravitation holds.

3.6 An electron is circling around the nucleus of a hydrogen atom at a radius of 0.05 nm. Determine the angular velocity and the time period of the electron.

3.7 Two charged particles are suspended by strings each of length L from a common point. If each particle has a mass m and carries a charge q, determine the angle θ made by each string with the vertical.

3.8 The surface charge density on a quarter-disc in the first quadrant is $K \cos\phi$ C/m^2. If the radius of the disc is a, find the $\vec{\mathbf{E}}$ field at a point $P(0, 0, h)$.

3.9 A charged semicircular ring of radius b extending from $\phi = 0$ to $\phi = \pi$ lies in the xy plane and is centered at the origin. If the charge distribution is $k \sin\phi$, compute the electric field intensity at $P(0, 0, h)$.

3.10 The charge distributions on the two isolated surfaces shown in Figure P3.10 are $\rho_{sa} = A\cos\phi$ and $\rho_{sb} = -A\cos\phi$. Obtain the electric field intensity at a point on the z axis at $z = h$.

Figure P3.10 Surface charge distribution

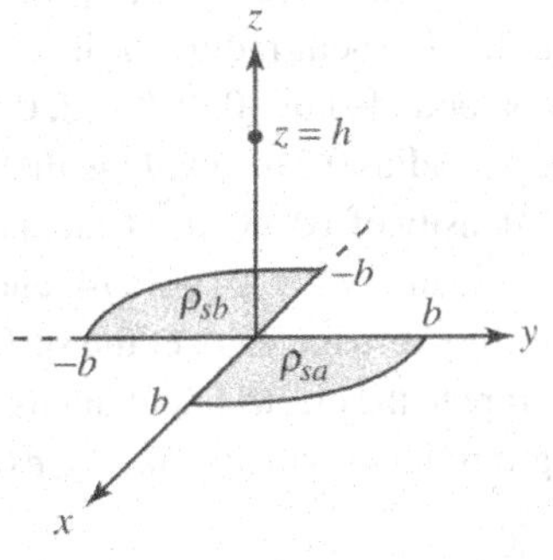

3.11 A straight line with a uniform charge density ρ_ℓ extends from $(x, -L/2, 0)$ to $(x, L/2, 0)$. Find the $\vec{\mathbf{E}}$ field at $P(0, 0, z)$.

3.12 A uniformly charged wire with a charge density of ρ_ℓ extends along the z axis from $z = 0$ to $z = \infty$. Determine the $\vec{\mathbf{E}}$ field at a point $P(\rho, \phi, 0)$.

3.13 Two straight uniformly charged lines are extending along the z axis from $z = -\infty$ to $z = \infty$. One line carries a charge distribution of 1 μC/m and is situated at $y = -3$ m, and the other line is at $y = 3$ m and has a charge distribution of -1 μC/m. Determine the $\vec{\mathbf{E}}$ field at a point on the x axis at $x = 4$ m.

3.14 Three infinite electric sheets carrying uniform charge densities of 2, -3, and 0.5 μC/m^2 are each separated by a 1-mm air gap, as shown in Figure P3.14. Find the electric field intensity everywhere in space.

Figure P3.14

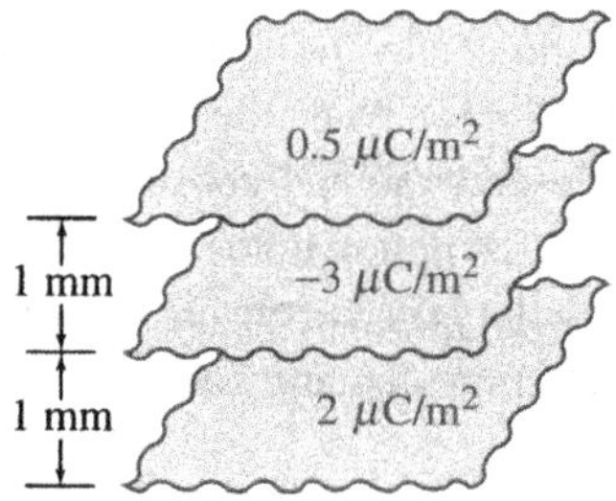

3.15 An arc of radius 0.2 m lies in the $z = 0$ plane and extends from $0 \le \phi \le \pi/2$. It has a charge distribution of 600 sin 2ϕ nC/m. Determine the $\vec{\mathbf{E}}$ field at (a) a point $P(0, 0, 1)$, and (b) the origin.

3.16 A finite line extends from $z = -10$ m to $z = 10$ m and carries a charge distribution of $100z$ nC/m. Determine the $\vec{\mathbf{E}}$ field at a point 2 meters away from the line in the $z = 0$ plane.

3.17 An extremely thin cylinder of radius b carries a uniform surface charge distribution of ρ_s. If the cylinder extends from $Q(0, 0, -L/2)$ to $S(0, 0, L/2)$, find the $\vec{\mathbf{E}}$ field at a point $P(0, 0, h)$. What are the electric field intensities at $h = 0$, $h = L/2$, and $h = -L/2$?

3.18 The electric flux density is given as $\vec{\mathbf{D}} = 6y\vec{\mathbf{a}}_x + 2x\vec{\mathbf{a}}_y + 14xy\vec{\mathbf{a}}_z$ mC/m^2. Determine the electric flux passing through (a) a rectangular window defined by (2, 0, 0), (0, 2, 0), (0, 2, 2), and (2, 0, 2), (b) a circle of 10-cm radius in the xy plane at $z = 0$, and (c) a triangular region bounded by (0, 0, 0), (2, 0, 0), and (0, 2, 0).

3.19 A long cylinder of radius 0.2 m lies along the z axis and carries a uniform surface charge density of 10 mC/m^2. Calculate the flux passing through a window at $\rho = 2$ m, $\pi/4 \le \phi \le 3\pi/4$, and $2 \le z \le 4$.

3.20 Charge is uniformly distributed over the surface of a very long cylinder of radius b. Compute the electric field intensity everywhere. How much flux is passing through a window at $\rho = c$ $(c > b)$, $0 \le \phi \le \pi/2$, and $0 \le z \le h$?

3.21 Using Gauss's law, compute the electric field intensity and electric flux density at any point due to a uniform charge distribution on an infinite plane sheet of charge.

3.22 Four concentric spherical shells of radii 0.2 m, 0.4 m, 0.6 m, and 0.8 m are supporting uniform charge distributions of 10, −2, −0.5, and 0.5 μC/m^2, respectively. Determine the $\vec{\mathbf{E}}$ field at a radius of (a) 0.1 m, (b) 0.3 m, (c) 0.5 m, (d) 0.7 m, and (e) 1 m.

3.23 A point charge Q is located at the origin. Compute the flux passing through the surface at $r = a, 0 \leq \theta \leq \theta_0$, and $0 \leq \phi \leq 2\pi$.

3.24 A very long coaxial cable consists of an inner conductor of radius a and the outer conducting shell with an inner radius of b and outer radius of c. The region between the two conductors carries a volume charge density of k/ρ, where k is a constant. Determine the electric field intensity and the electric flux density at any point in space when the outer conductor is (a) not grounded, and (b) grounded.

3.25 A spherical volume of radius b supports a charge distribution of k/r^2 everywhere except at $r = 0$. Determine the electric flux density in the regions (a) $r < b$, and (b) $r > b$.

3.26 The electric field intensity in free space from a very long, uniformly charged cylinder of radius 5 cm is 100 kV/m at a radius of 1 m. What must be the surface charge density on the cylinder?

3.27 Obtain the expression for the electric potential at a point $P(a, \phi, 0)$ due to a very thin, uniform charged line extending from $z = 0$ to $z = L$. The line charge density is ρ_ℓ C/m. What is the electric field intensity at point P?

3.28 A charge of 500 nC is situated at the origin. Determine the amount of energy released in bringing a charge of −600 nC from infinity to a distance of 1 mm from the fixed charge.

3.29 Find the potential at a point on the axis of a uniformly charged disc of radius b. Also, determine the $\vec{\mathbf{E}}$ field at that point. Assume that the surface charge density on the disc is ρ_s C/m^2.

3.30 Find the potential and the electric field intensity on the axis of a uniformly charged annular disc of inner and outer radii of a and b, respectively.

3.31 The electric potential at 20 cm from a positive point charge is 9 kV. What is the magnitude of the charge? Determine the radius of the equipotential surface at which the potential is (a) 18 kV, and (b) 3 kV.

3.32 An electric field is given by $\vec{\mathbf{E}} = 10\vec{\mathbf{a}}_x + 20\vec{\mathbf{a}}_y + 20\vec{\mathbf{a}}_z$ kV/m. Find the work necessary to a carry a charge of 0.1 nC (a) from the origin to (3, 0, 0), (b) from (3, 0, 0) to (3, 4, 0), and (c) from the origin directly to (3, 4, 0).

3.33 A uniform electric field intensity is given by $\vec{\mathbf{E}} = 10\vec{\mathbf{a}}_x$ kV/m. Find the potential at any point in space if the potential at the origin is zero.

3.34 The electric potential near the origin is given as $V = 10x^2 + 20y^2 + 5z$ V. What is the electric field intensity? Can this potential function exist?

3.35 Charge is uniformly distributed inside an infinitely long cylinder of radius a. If the volume charge density is ρ_v, compute the electric potential and the electric field intensity at all points inside and outside the cylinder.

3.36 Determine the amount of work done in carrying a charge of 0.5 mC along the arc of a circle of radius 2 m from $\phi = 0$ to $\phi = \pi/4$. The electric field intensity in the region is given as $\vec{\mathbf{E}} = 10y\vec{\mathbf{a}}_x + 10x\vec{\mathbf{a}}_y + 2z\vec{\mathbf{a}}_z$ kV/m. Does $\vec{\mathbf{E}}$ represent a viable field? If yes, what is the potential difference between the two ends of the path?

3.37 A uniform line of charge extends from $z = -L/2$ to $z = L/2$. If the charge density is ρ_ℓ, obtain an expression for the electric potential and the electric field intensity at a point $P(0, 0, z)$, where $z > L/2$.

3.38 A dipole is formed by placing a positive charge of 10 nC at $z = 0.5\ \mu$m and a negative charge of -10 nC at $z = -0.5\ \mu$m. Determine the potential and the $\vec{\mathbf{E}}$ field at a point $P(0, 0, 1)$.

3.39 A quadrupole is formed by placing a charge q at $(0, 0, a)$, a charge $-2q$ at $(0, 0, 0)$, and a charge q at $(0, 0, -a)$. Find the potential and electric field intensity at $P(0, 0, z)$, where $z \gg a$.

3.40 Modify the expressions for the potential and the $\vec{\mathbf{E}}$ field of a dipole if an additional charge q is placed at the origin.

3.41 A long coaxial cable consists of an inner conductor of radius a and an outer conductor of inner radius b and outer radius c. The inner conductor carries a uniform surface charge distribution ρ_s. Find the $\vec{\mathbf{E}}$ field at any point in space when (a) the outer conductor is not grounded, and (b) the outer conductor is grounded.

3.42 A uniform charge density ρ_s exists on the surface of an infinitely long metal cylinder of radius a. It is surrounded by a concentric dielectric of inner radius a and outer radius b. Compute (a) $\vec{\mathbf{D}}$, $\vec{\mathbf{E}}$, and $\vec{\mathbf{P}}$ at all points in space, (b) the bound charge densities, and (c) the energy density in the system.

3.43 Find an expression for $\nabla \cdot \vec{\mathbf{E}}$ in a charge-free but nonhomogeneous medium.

3.44 The permittivity of a dielectric medium is given as $\epsilon = \alpha z^n$ where α and n are constants. If the electric field intensity in the medium has only a z component, show that $\nabla \cdot \vec{\mathbf{E}} = -nE/z$, where E is the magnitude of the electric field intensity.

3.45 A metal sphere of radius b has a uniform surface charge distribution. The permittivity of the surrounding region varies as $\epsilon = \epsilon_0(1 + a/r)$. Find (a) $\vec{\mathbf{D}}, \vec{\mathbf{E}}$, and $\vec{\mathbf{P}}$ everywhere in space, (b) the bound charge densities, and (c) the energy density, and (d) show that the potential in the dielectric region is

$$V = \frac{Q}{4\pi\epsilon_0 a}\ln(1 + a/r)$$

where Q is the total charge on the sphere.

3.46 The separation between two point charges in a dielectric medium with $\epsilon_r = 5.5$ is 10 mm. What is the mutual potential energy of the charges if each point charge is equal to 10 μC?

3.47 Two parallel plates, each measuring 20 cm $\times$ 20 cm, are separated by a gap of 1 mm. The plates have equal but opposite uniform surface charge density of 250 nC/m^2. Compute the energy stored if the dielectric constant of the medium is 2.

3.48 Compute the amount of work required to bring two charges of 100 nC and 300 nC from infinity into free space at (0, 3, 3) m and (4, 0, 3) m, respectively.

3.49 Three charges of 100, 200, and 300 nC are so arranged that the separation between any two charges is 5 cm. What is the total energy stored in the system?

3.50 The electric field between two very long concentric cylindrical conductors varies as $\vec{\mathbf{E}} = 100/\rho \vec{\mathbf{a}}_\rho$ V/m. The radius of the inner conductor is 0.2 m and that of the outer conductor is 0.5 m. Determine the energy density and the total energy stored per unit length in the system. Assume $\epsilon_r = 5.5$.

3.51 A metal sphere of radius a is made to acquire a positive charge Q. Show that the electric potential energy of the system is $QV/2$, where V is the electric potential on the surface of the sphere.

3.52 A solid conducting sphere of radius 20 cm is concentrically placed inside a spherical shell of inner radius 30 cm and outer radius 40 cm. A charge of 20 μC is placed on the inner sphere, and a charge of -10 μC is placed on the outer conductor. The dielectric constant of the medium between the two conductors is 5. Find (a) $\vec{\mathbf{D}}$, $\vec{\mathbf{E}}$, and $\vec{\mathbf{P}}$ at all points in space, and (b) the total energy in the system.

3.53 An interface at the plane $x = 5$ separates two dielectrics with dielectric constants of 4 and 16, respectively. The $\vec{\mathbf{E}}$ field on one side of the interface with dielectric constant of 4 is $\vec{\mathbf{E}} = 12\vec{\mathbf{a}}_x + 24\vec{\mathbf{a}}_y - 36\vec{\mathbf{a}}_z$ V/m. Determine the $\vec{\mathbf{E}}$ and $\vec{\mathbf{D}}$ fields on the other side of the interface.

3.54 What must be the charge on the surface of a metallic sphere 20 cm in radius if the magnitude of the electric field just above its surface in free space is 10 MV/m?

3.55 The electric field in the region between a pair of oppositely charged parallel plates is 10 kV/m. If the area of each plate is 25 cm^2 and the separation between them is 1 mm, find the surface charge density and the total charge on each plate. The dielectric constant of the medium is 3.6.

3.56 A plane boundary of infinite extent in the z direction passes through the points (4, 0, 0) and (0, 3, 0), as indicated in Figure P3.56. The electric field intensity in medium 1 ($\epsilon_r = 2.5$) is $\vec{\mathbf{E}} = 25\vec{\mathbf{a}}_x + 50\vec{\mathbf{a}}_y + 25\vec{\mathbf{a}}_z$ V/m. Determine the $\vec{\mathbf{E}}$ field in medium 2 ($\epsilon_r = 5$).

3.57 If medium 2 in Problem 3.56 is a conductor, and the y component of the electric field intensity in medium 1 is 50 V/m, what are the other

Figure P3.56

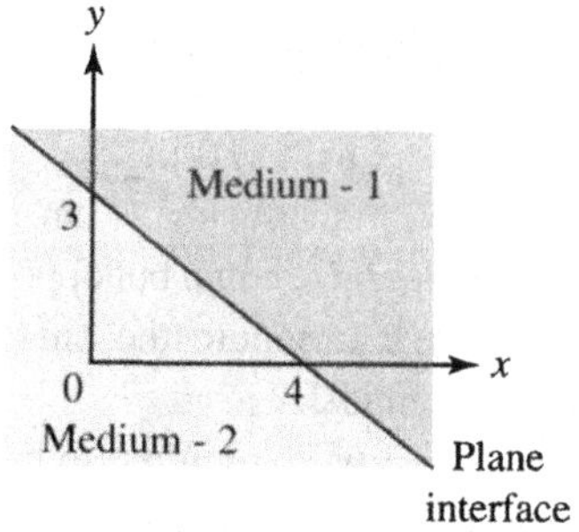

components of the $\vec{\mathbf{E}}$ field? What is the free surface charge density on the conductor?

3.58 A parallel-plate capacitor with three dielectric media is shown in Figure P3.58. What is the total capacitance of the system?

Figure P3.58

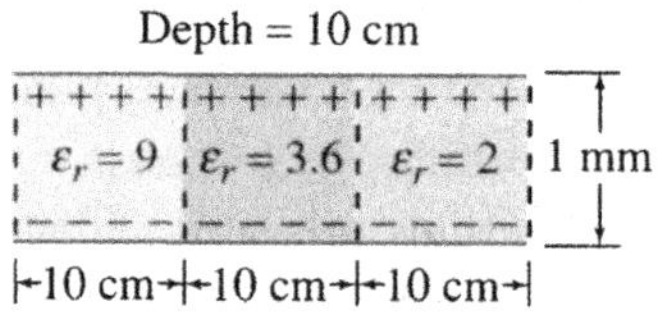

3.59 Find the capacitance of the parallel-plate capacitor shown in Figure P3.59.

Figure P3.59

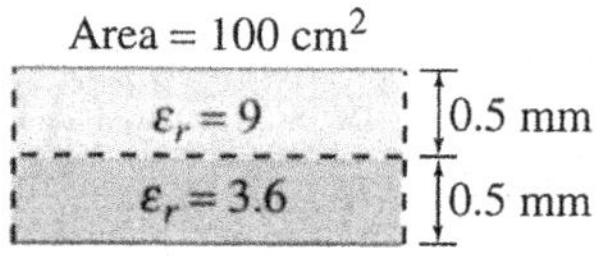

3.60 Determine the capacitance per unit length of a coaxial line, as depicted in Figure P3.60. If $\epsilon_{r1} = 5$, $\epsilon_{r2} = 2.5$, $L = 10$ m, $a = 1$ cm, and $b = 1.5$ cm, what is the total capacitance of the line?

Figure P3.60

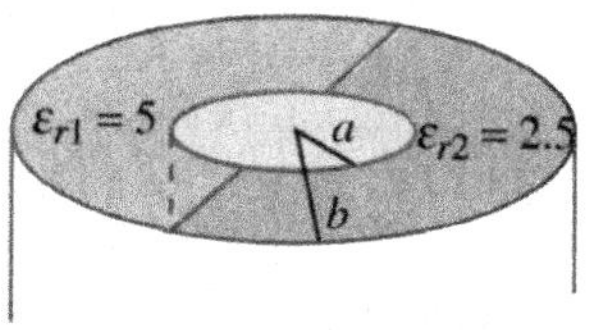

3.61 Two charged conductive plates of a capacitor are at $\rho = 10$ cm and $\rho = 30$ cm, as shown in Figure P3.61. If the medium between the plates has a dielectric constant of 3.6, determine the capacitance of the system.

3.62 The region between two charged concentric spheres is filled with two dielectrics, as shown in Figure P3.62. Determine the capacitance of the system.

Figure P3.61

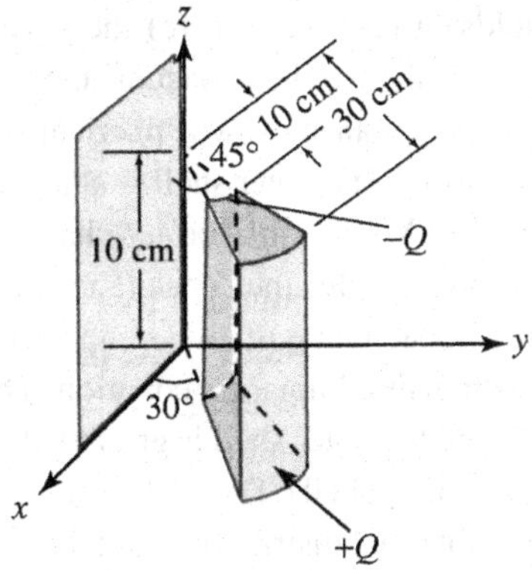

Figure P3.62

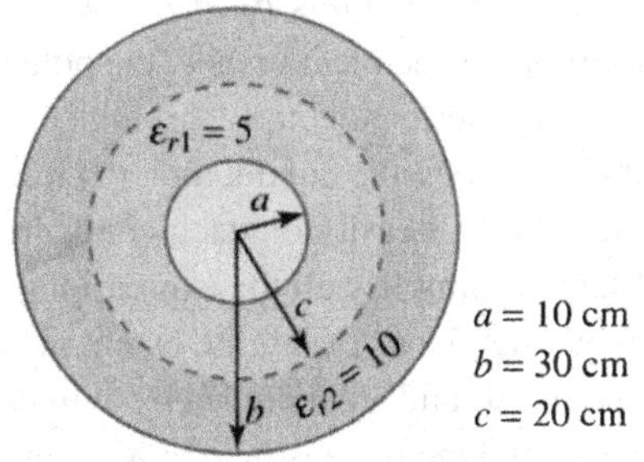

3.63 A large spherical cloud of radius b has a uniform volume charge distribution of ρ_v. Calculate and sketch the potential distribution and the electric field intensity at any point in space using Poisson's and Laplace's equations.

3.64 The upper plate of a parallel-plate capacitor is maintained at a potential of 100 V, and the lower plate at a potential of −100 V. The plates are of infinite extent and are 4 cm apart. When the medium is free space determine (a) the potential distribution between the plates, (b) the $\vec{\mathbf{E}}$ and $\vec{\mathbf{D}}$ fields in the region, and (c) the surface charge density on each plate.

3.65 Determine the potential distribution in a coaxial cable when the inner conductor of radius a is at a potential of V_0 and the outer conductor of radius b is grounded. The space between the conductors is filled with two concentric dielectrics. The permittivity of the inner dielectric is ϵ_1; for the outer dielectric it is ϵ_2. The dielectric interface is at a radius c. Determine (a) the potential distribution, (b) the $\vec{\mathbf{D}}$ and $\vec{\mathbf{E}}$ fields in each region, (c) the surface charge density on the inner conductor, and (d) the capacitance per unit length of the cable.

3.66 Determine the potential distribution in a coaxial cable when the inner conductor of radius 10 cm is at a potential of 100 V and the outer conductor radius 20 cm is grounded. The space between the conductors is filled with two concentric dielectrics. The dielectric constant is 3 for the inner dielectric and 9 for the outer dielectric. The dielectric interface is at a radius of 15 cm. Determine (a) the potential distribution, (b) the

$\vec{\mathbf{D}}$ and $\vec{\mathbf{E}}$ fields in each region, (c) the surface charge density on the inner conductor, and (d) the capacitance of the 100-meter-long cable.

3.67 Calculate the capacitance of two spherical shells of radii a and b using Laplace's equation. The inner shell is at a potential of V_0, and the outer shell is grounded. What is the surface charge density on the inner shell? Obtain the expression for the capacitance of the system.

3.68 Calculate the capacitance of two concentric spherical shells of radii 5 cm and 10 cm using Laplace's equation. The inner shell is at a potential of 500 V, and the outer shell is grounded. What is the surface charge density on the inner shell? The dielectric constant of the medium is 9. Calculate the potential distribution and the electric field in the medium.

3.69 The charge distribution within the region bounded by two conducting cylinders of radii a and b is $\rho_v = A/\rho$ C/m^3, where A is a constant. If the permittivity of the medium is ϵ, the potential of the inner conductor is V_0, and the outer conductor is grounded, determine (a) the potential distribution between the conductors, (b) the electric field intensity, (c) the surface charge density on each conductor, and (d) the capacitance per unit length. Simplify all the expressions if $A \to 0$.

3.70 Two conducting planes of infinite extent in the z direction are arranged at an angle of 30° and are bounded by cylindrical surfaces at $\rho = 0.1$ m and $\rho = 0.2$ m. One plate is held at a potential of 10 kV, and the other is grounded. Find the potential distribution, the $\vec{\mathbf{E}}$ and $\vec{\mathbf{D}}$ fields in the free-space region between the plates, and the capacitance per unit length of the system.

3.71 Two infinite grounded metal plates parallel to the z axis are at $\phi = 0°$ and $\phi = 60°$. A charge of 500 nC is placed at (5, 30°, 10) m. Calculate the potential function and the electric field intensity at (7, 30°, 10) m. Assume that the medium between the plates is free space.

3.72 A charge q is located at a distance d from the center of a grounded metallic sphere of radius R. Show by direct integration that the total charge on the surface of a grounded sphere is $-q\dfrac{R}{d}$. Where does this charge come from?

4

Steady electric currents

4.1 Introduction

In Chapter 3 we were mainly concerned with the forces between charges at rest. In this chapter, we discuss the motion of charges in a conducting medium under the influence of electrostatic fields. More specifically, we will devote most of our attention to the motion of charges in a conductor when an electric field is maintained within the conductor.

We have already discussed that when an isolated conductor is placed in an electric field, the interior of the conductor remains a charge-free region and the charges rearrange themselves on the surfaces of the conductor. The electric field intensity within the isolated conductor vanishes when the conductor attains its electrostatic equilibrium.

Now suppose we suddenly deposit equal but opposite amounts of charged particles on the opposite ends of a conductor. The conductor loses its equilibrium, and an electric field is established within the conductor by these charges. The electric field forces these charges to move toward each other. As soon as the oppositely charged particles come in contact, they cancel. The process continues until all the charges disappear. When that happens the electric field within the conductor also disappears, and the conductor regains its equilibrium.

In this chapter, we intend to show that for conductors the approach to equilibrium is very fast. The time required for a good conductor to achieve equilibrium is a very small fraction of a second. During this very short time the charges rearrange themselves within the conductor. The motion of charges is said to constitute a *current*. Because the charges are in motion only for a short duration, the current is commonly called a *transient current*. We define the **current** as the rate at which the charge is transported past a given point in a conducting medium. That is,

$$i = \frac{dq}{dt} \tag{4.1}$$

where dq is the amount of charge that flows past some given point in time dt. We have used the lowercase letter i for the current to indicate that it is generally a function of time. The SI unit of current is the *ampere*

(A), named in honor of the French physicist, André Marie Ampère. *A current of one ampere corresponds to the transportation of one coulomb of charge in one second.*

In this chapter, we will limit our discussion to *steady currents*; that is, only those currents that are constant in time. We will use the uppercase letter I to represent a steady current. A steady current is also called a *direct current.*

4.2 Nature of current and current density

In this section, we introduce two types of currents: conduction current and convection current. We will also discuss the current per unit area, or current density, for both convection and conduction currents.

4.2.1 Conduction current

In metals, like copper, silver, or gold, the free charge carriers are predominantly electrons. More precisely, it is the valence electron of an atom which is free to contribute to the conduction process. An electron which may be considered as not being attached to any particular atom is called a *free electron.* A free electron has the capability of moving through a whole crystal lattice. However, the heavy, positively charged ions are relatively fixed at their regular positions in the crystal lattice and do not contribute to the current in the metal. Thus, the current in a metal conductor, called **conduction current**, is simply a flow of electrons.

As mentioned in the Introduction, the transitory flow of charges comes to a halt in a very short time in an isolated conductor placed in an electric field. To maintain a steady current within a conductor, a continuous supply of electrons at one end and their removal at the other is necessary. *Even when a steady current is maintained through the conductor, the conductor as a whole is electrostatically neutral.*

In an isolated conductor, the random thermal motion of electrons takes place at high speed, typically in the range of 10^6 m/s, and in all possible directions. Let us assume that the conductor is in the form of a cylindrical wire and extends in the z direction. At any hypothetical plane through the wire perpendicular to its axis, we will find that the rate at which the electrons are passing through it in the positive z direction is the same as the rate at which the electrons are passing through it in the negative z direction. In other words, the net rate is zero, and thus, the net current in an isolated conductor is zero.

Now consider a conductor whose ends are connected to a battery. There now exists a potential difference between the two ends, so there must also exist an electric field inside the conductor, as illustrated in

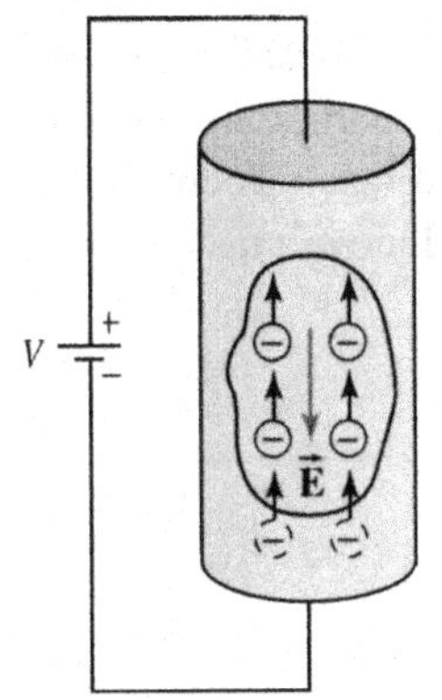

Figure 4.1 (a) A conductor with impressed voltage between its ends

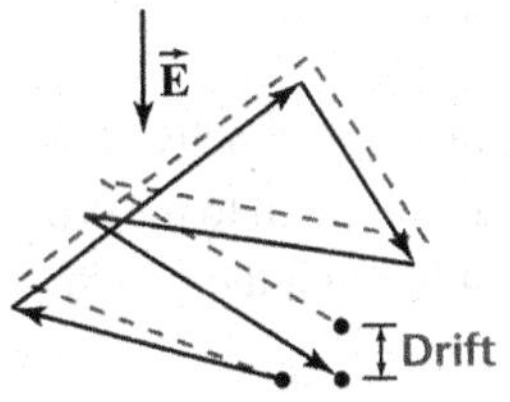

(b) Motion of an electron with (dashed line) and without (solid line) the $\vec{E}$ field

Figure 4.1a. The electric field within the wire exerts a force on the free electron in the z direction. The electron accelerates in response to this force but only for a short period of time because each time an electron moves it eventually suffers a collision with an ion. After each collision the new velocity is almost independent of what the velocity was before the collision. In fact, electrons moving through a piece of copper suffer as many as 10^{14} collisions per second. Each collision slows the electron down, brings it to a stop, or changes its direction of motion. The electric field has to start all over again to change the velocity of the electron. Therefore, the change in velocity produced by the electric force in the z direction is a very small fraction of its random velocity. However, the electric field does produce a systematic component of the random velocity known as the *drift velocity*. The drift velocity causes the electron to drift gradually in the z direction, as illustrated in Figure 4.1b. (The drift has been exaggerated to highlight the concept.) The net drift of the electrons in the z direction now constitutes a current through the conductor.

The conventional direction of the current is taken to be that of the electric field. In other words, *the electrons move in a direction opposite to the conventional direction of the current*. The current through the conductor is the same at all cross sections even though the cross-sectional area may be different at different points. This constancy of the electric current is in accordance with the law of conservation of charge; i.e., the charge must be conserved. Under the assumed steady-state conditions, the charges cannot pile up steadily or drain away steadily from any point in the conductor. Simply stated, a point cannot act as a "source" or a "sink" of charge to maintain a steady current.

4.2.2 Convection current

The motion of charged particles in free space (vacuum) is said to constitute a **convection current**. The motion of electrons from the cathode toward the anode in a vacuum tube is a classic example of convection current. In this case, the electrons which are just emitted by the cathode are moving very slowly. Those electrons that are close to the anode have attained very high velocities. This is so because all along the path from cathode to anode an electron is never involved in any collision. However, for a steady current, the charge crossing a unit area must be the same. Therefore, as the velocity of an electron increases, the charge density decreases, as depicted in Figure 4.2. Thus, a clear distinction between convection current and conduction current is that convection current is not electrostatically neutral and its electrostatic charge must be taken into account. A convection current neither requires a conductor to maintain the charge flow nor obeys Ohm's law.

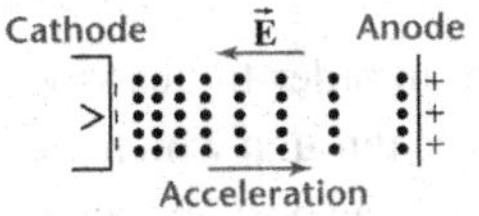

Figure 4.2 Charge density in an electron tube

4.2.3 Convection current density

In order to describe the motion of the charges, let us consider a region with volume charge distribution ρ_v in which the charges are moving under the influence of an electric field with an average velocity $\vec{\mathbf{U}}$, as shown in Figure 4.3. In time Δt these charges will move a distance $\overrightarrow{d\ell}$ such that

$$\overrightarrow{d\ell} = \vec{U}\Delta t$$

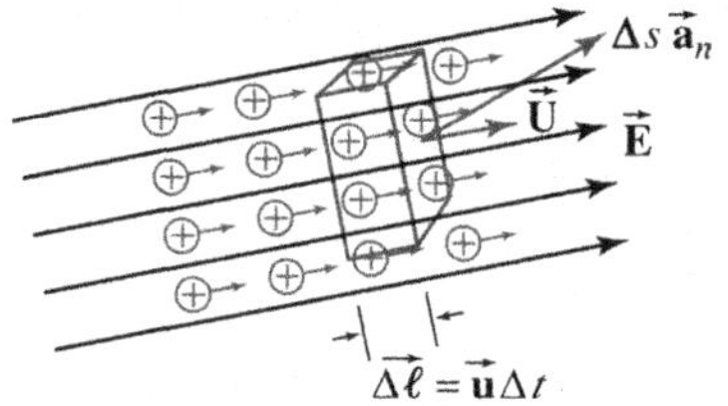

Figure 4.3 Motion of charges in free space under the influence of an $\vec{\mathbf{E}}$ field

The length element is in the same direction as the average velocity. If we consider an imaginary window with surface area $\vec{\mathbf{\Delta s}}$ normal to the drift velocity; i.e., $\vec{\mathbf{\Delta s}} = \Delta s\vec{\mathbf{a}}_n$, the charge movement through the window would be

$$dq = \rho_v \Delta v = \rho_v \vec{\mathbf{\Delta s}} \cdot \overrightarrow{d\ell}$$

The current ΔI through the surface Δs is

$$\Delta I = \rho_v \vec{\mathbf{\Delta s}} \cdot \left(\frac{\overrightarrow{d\ell}}{dt}\right) = \rho_v \vec{\mathbf{\Delta s}} \cdot \vec{\mathbf{U}} \tag{4.2}$$

Let us define the current in terms of the **convection current density**, $\vec{\mathbf{J}}$ (current per unit area), such that

$$\Delta I = \vec{\mathbf{J}} \cdot \vec{\mathbf{\Delta s}}$$

Then from (4.2),

$$\vec{\mathbf{J}} = \rho_v \vec{\mathbf{U}} \tag{4.3}$$

and the current passing through a surface s is

$$I = \int_s \vec{\mathbf{J}} \cdot \overrightarrow{ds} \tag{4.4}$$

It is apparent from the preceding expressions that in order to describe the current in a region we have to specify the surface through which the charges are passing. In contrast, the current density $\vec{\mathbf{J}}$ can be completely described by a vector point function. That is why the current density concept is more useful in describing fields than the current. $\vec{\mathbf{J}}$ is also routinely referred to as the *volume current density*.

If there are positive as well as negative charges with charge densities ρ_{v+} and ρ_{v-} moving with average velocities $\vec{\mathbf{U}}_+$ and $\vec{\mathbf{U}}_-$, respectively, they will move in opposite directions under the influence of an electric field. Positive charges will drift in the direction of the electric field, and the negative charges will move in a direction opposite to the electric field. However, the current due to positive charges will be in the same direction as the current due to the negative charges. Thus, the total current density is

$$\vec{\mathbf{J}} = \rho_{v+}\vec{\mathbf{U}}_+ + \rho_{v-}\vec{\mathbf{U}}_- \tag{4.5}$$

and the total current passing through a surface can be obtained from (4.4). Because we are considering the motion of charges with average velocities, we are, in fact, discussing the steady current in the region; that is, the rate of change of current is zero. *We speak of a steady current in a region if the current density remains constant in time everywhere in that region.*

4.2.4 Conduction current density

Let $\vec{\mathbf{U}}_e$ be the average velocity (or drift velocity) of an electron due to the impressed $\vec{\mathbf{E}}$ field within a conductor, let m_e be its mass, and let τ be the mean time per collision. Then the loss in momentum in time τ of an electron is $m_e\vec{\mathbf{U}}_e$. Thus, the average rate at which an electron loses momentum in collisions is $m_e\vec{\mathbf{U}}_e/\tau$. The rate at which the electron gains momentum from the electric force is $-e\vec{\mathbf{E}}$. Under steady-state conditions, the rate of loss of momentum must match the rate of gain. That is,

$$\frac{m_e\vec{\mathbf{U}}_e}{\tau} = -e\vec{\mathbf{E}}$$

Thus,

$$\vec{\mathbf{U}}_e = -\frac{e\tau\vec{\mathbf{E}}}{m_e}$$

or

$$\vec{\mathbf{U}}_e = -u_e\vec{\mathbf{E}} \tag{4.6}$$

where

$$u_e = \frac{e\tau}{m_e}$$

is called the *electron mobility*. Equation (4.6) states that *the drift velocity of an electron in a conducting medium is proportional to the applied electric field. The constant of proportionality is the electron mobility.*

If there are N electrons per unit volume, the electron charge density is

$$\rho_{v-} = -Ne \tag{4.7}$$

where e is the magnitude of the charge on the electron. Thus, the **conduction current density**, in the conducting medium, is

$$\vec{\mathbf{J}} = \rho_{v-}\vec{\mathbf{U}}_e$$

or

$$\vec{\mathbf{J}} = Neu_e\vec{\mathbf{E}} = \sigma\vec{\mathbf{E}} \tag{4.8}$$

where $\sigma = Neu_e$ is known as the *conductivity* of the medium. The unit of conductivity is siemens per meter (S/m). We refer to equation (4.8) as the microscopic equivalent of Ohm's law. It states that *the current density at any point in a conducting medium is proportional to the electric field intensity. The constant of proportionality is the conductivity of the medium.* For a linear medium (as assumed in this text), $\vec{\mathbf{J}}$ and $\vec{\mathbf{E}}$ are in the same direction.

In electric circuit theory, Ohm's law is valid as long as the resistance does not depend upon the voltage and the current. Similarly, a conducting material obeys Ohm's law if the conductivity of the medium is independent of the electric field intensity. It must, however, be kept in mind that Ohm's law is not a general law of electromagnetism like Gauss's law. Ohm's law is essentially an assertion regarding the electrical properties of certain materials. Materials for which (4.8) holds are called *linear media* or *ohmic media.*

The reciprocal of the conductivity is called the *resistivity.* That is,

$$\rho = \frac{1}{\sigma} \tag{4.9}$$

The unit of resistivity is ohm · meter (Ω · m). Table 4.1 gives the resistivities of a number of common materials.

EXAMPLE 4.1

A potential difference of 10 V is maintained across the ends of a copper wire 2 m in length. If the mean time between collisions is 2.7×10^{-14} s, determine the drift velocity of the free electrons.

Solution Let us assume that the wire extends in the z direction and the upper end is positive with respect to the lower. Then the electric field intensity within the wire is

$$\vec{\mathbf{E}} = -\left(\frac{10}{2}\right)\vec{\mathbf{a}}_z = -5\vec{\mathbf{a}}_z \text{ V/m}$$

The electron mobility is

$$u_e = \frac{e\tau}{m_e} = \frac{1.6 \times 10^{-19} \times 2.7 \times 10^{-14}}{9.1 \times 10^{-31}} = 4.747 \times 10^{-3}$$

Table 4.1. Resistivities of (a) metals, (b) semiconductors, and (c) insulators

(a) Metals	
Material	Resistivity ($\Omega \cdot$ m)
Aluminum	2.83×10^{-8}
Constantan	49×10^{-8}
Copper	1.72×10^{-8}
Gold	2.44×10^{-8}
Iron	8.9×10^{-8}
Mercury	95.8×10^{-8}
Nichrome	100×10^{-8}
Nickel	7.8×10^{-8}
Silver	1.47×10^{-8}
Tungsten	5.51×10^{-8}
(b) Semiconductors	
Material	Resistivity ($\Omega \cdot$ m)
Carbon (graphite)	3.5×10^{-5}
Germanium	0.42
Silicon	2.6×10^{3}
(c) Insulators	
Material	Resistivity ($\Omega \cdot$ m)
Amber	5×10^{14}
Glass	10^{10}–10^{14}
Hard rubber	10^{13}–10^{16}
Mica	10^{11}–10^{15}
Quartz (fused)	7.5×10^{17}
Sulfur	10^{15}

Thus, the drift velocity is

$$\vec{\mathbf{U}}_e = -u_e \vec{\mathbf{E}} = 4.747 \times 10^{-3} \times 5\vec{\mathbf{a}}_z = 23.74 \times 10^{-3} \vec{\mathbf{a}}_z \text{ m/s}$$

Thus, the electron is moving in the z direction at a rate of 23.74 mm/s. It will take nearly 84 seconds (2 m/23.74 mm) for the electron to zigzag from the lower end to the upper end of the wire. However, the current "travels" through the wire at the speed of light. What happens in this case is that one electron that enters at the lower end of the wire pushes on the neighboring electron by means of its electric field and creates a compressional wave within the wire. The compressional wave travels with the speed of light and ejects electrons out of the far end of the wire almost instantaneously.

• • •

4.3 Resistance of a conductor

The resistance of a conductor of length $d\ell$ can be obtained from Ohm's law in terms of the field quantities $\vec{\mathbf{E}}$ and $\vec{\mathbf{J}}$ as

$$dR = \frac{dV}{I} = \frac{-\vec{\mathbf{E}} \cdot \overrightarrow{d\ell}}{\int_s \vec{\mathbf{J}} \cdot \overrightarrow{ds}}$$

where dV is the potential difference between the two ends of the conductor of length $d\ell$, $\vec{\mathbf{E}}$ is the electric field intensity within the conductor, $\vec{\mathbf{J}} = \sigma\vec{\mathbf{E}}$ is the volume current density, and I is the current passing through each surface, as indicated in Figure 4.4. We have assumed that the potential at end a of the conductor is higher than that at end b.

The total resistance of the conductor is

$$R = \int_b^a \frac{-\vec{\mathbf{E}} \cdot \overrightarrow{d\ell}}{\int_s \vec{\mathbf{J}} \cdot \overrightarrow{ds}} \tag{4.10}$$

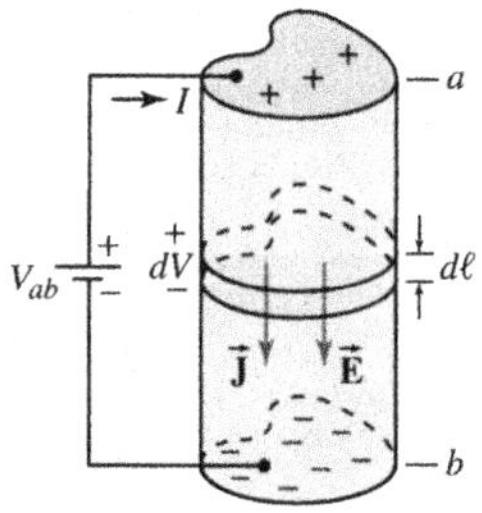

Figure 4.4 A current-carrying conductor

This equation is very general and allows us to determine the resistance of a conducting medium whose conductivity changes in the direction of the current. In the case of a homogeneous medium having constant conductivity, equation (4.10) reduces to

$$R = \frac{-\int_b^a \vec{\mathbf{E}} \cdot \overrightarrow{d\ell}}{\int_s \vec{\mathbf{J}} \cdot \overrightarrow{ds}} = \frac{V_{ab}}{I} \tag{4.11}$$

If the electric field intensity within a homogeneous conducting medium is known, we can use equation (4.11) to determine its resistance. We may not always be able to determine the $\vec{\mathbf{E}}$ field within an arbitrarily shaped conducting material. In that case, we may have to resort to the use of approximate methods or numerical techniques to determine the electric field intensity.

EXAMPLE 4.2

A potential difference of V_0 is maintained across the two ends of a copper wire of length ℓ. If A is the cross-sectional area of the wire, obtain an expression for the resistance of the wire. What is the resistance of the wire if $V_0 = 2$ kV, $\ell = 200$ km, and $A = 40$ mm^2?

Solution Let us assume that the wire extends in the z direction, and the upper end is at a potential of V_0 with respect to the lower end. The electric field intensity within the wire is

$$\vec{\mathbf{E}} = -\frac{V_0}{\ell}\vec{\mathbf{a}}_z$$

If σ is the conductivity of copper, the volume current density at any cross section of the wire is

$$\vec{\mathbf{J}} = \sigma\vec{\mathbf{E}} = -\frac{\sigma V_0}{\ell}\vec{\mathbf{a}}_z$$

The current through the wire is

$$I = \int_s \vec{\mathbf{J}}\cdot\vec{ds} = \frac{\sigma V_0}{\ell}\int_s ds = \frac{\sigma V_0 A}{\ell}$$

Hence, the resistance of the wire, from (4.11), is

$$R = \frac{V_0}{I} = \frac{\ell}{\sigma A} = \frac{\rho\ell}{A}$$

This equation gives us a theoretical expression for the resistance in terms of the physical parameters of the conducting medium. Substituting the values, we obtain

$$R = \frac{1.7\times 10^{-8}\times 200\times 10^{3}}{40\times 10^{-6}} = 85\,\Omega$$

• • •

4.4 The equation of continuity

A conducting region bounded by a closed surface s is shown in Figure 4.5. We assume that the volume charge density in the region is ρ_v, and the current leaving the surface can be described in terms of the volume current density $\vec{\mathbf{J}}$. The total current crossing the closed surface s in the outward direction is

$$i(t) = \oint_s \vec{\mathbf{J}}\cdot\vec{ds} \tag{4.12}$$

Figure 4.5 A conducting region bounded by surface s with an outward charge flow

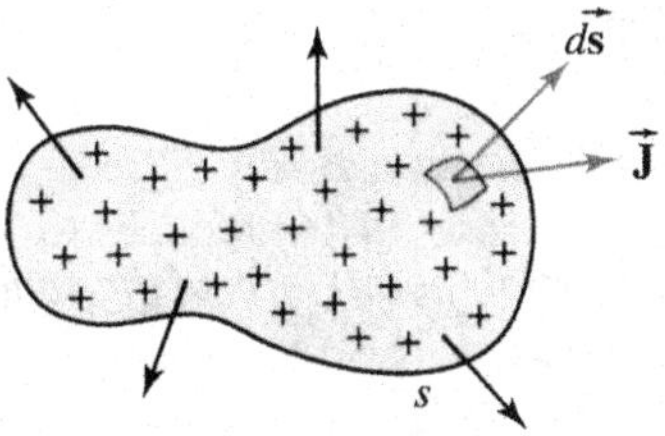

Since current is simply a flow of charge per second, an outward flow of charge must decrease the charge concentration by the same amount within the region bounded by s. Thus, the rate at which the charge is leaving the surface must be equal to the rate at which the charge is diminishing in the bounded region. Therefore, we can also express the current as

$$i(t) = -\frac{dQ}{dt} \tag{4.13}$$

where Q is the total charge enclosed by the surface at any time t. We can write Q in terms of the volume charge density ρ_v as

$$Q = \int_v \rho_v \, dv \tag{4.14}$$

where the integral is taken throughout the region enclosed by s. Combining (4.12), (4.13), and (4.14), we obtain

$$\oint_s \vec{\mathbf{J}} \cdot \overrightarrow{d\mathbf{s}} = -\frac{d}{dt} \int_v \rho_v \, dv \tag{4.15}$$

Equation (4.15) is called *the integral form of the equation of continuity* and is a mathematical expression of the principle of conservation of charge. It states that *any change of charge in a region must be accompanied by a flow of charge across the surface bounding the region.* In other words, the charge can be neither created nor destroyed, but merely transported.

The closed surface integral on the left-hand side of (4.15) can be transformed into a volume integral by applying the divergence theorem. Since the volume under consideration is stationary, the differential with respect to time can be replaced by a partial derivative of volume charge density. We can now rewrite (4.15) as

$$\int_v \nabla \cdot \vec{\mathbf{J}} \, dv = -\int_v \frac{\partial \rho_v}{\partial t} \, dv$$

or

$$\int_v \left(\nabla \cdot \vec{\mathbf{J}} + \frac{\partial \rho_v}{\partial t} \right) dv = 0$$

Since the volume under consideration is arbitrary, the only way for the preceding equation to be true in general is for the integrand to vanish at each point. Hence,

$$\nabla \cdot \vec{\mathbf{J}} + \frac{\partial \rho_v}{\partial t} = 0 \tag{4.16a}$$

This is the differential (point) form of the equation of continuity. We can also write (4.16a) as

$$\nabla \cdot \vec{\mathbf{J}} = -\frac{\partial \rho_v}{\partial t} \tag{4.16b}$$

Equation (4.16) states that *the points of changing charge density ρ_v are sources of volume current density $\vec{\mathbf{J}}$.*

For a conducting medium to sustain a steady (direct) current, there can be no points of changing charge density. In that case, (4.16) reduces to

$$\oint_s \vec{\mathbf{J}} \cdot \overrightarrow{d\mathbf{s}} = 0 \tag{4.17a}$$

or

$$\nabla \cdot \vec{\mathbf{J}} = 0 \tag{4.17b}$$

Equation (4.17a) states that *the net steady current through any closed surface is zero.* If we shrink the closed surface s to a point, we can interpret (4.17a) as

$$\sum I = 0 \tag{4.17c}$$

which is a statement of *Kirchhoff's current law.* That is, *the algebraic sum of the currents at a point (junction or node) is zero.* Equation (4.17b) states that *for a steady current through a conducting medium the current density within the medium is solenoidal or continuous.*

Substituting $\vec{\mathbf{J}} = \sigma\vec{\mathbf{E}}$, where σ is the conductivity of the medium, in (4.17b), we obtain

$$\nabla \cdot (\sigma\vec{\mathbf{E}}) = 0$$

or

$$\sigma\nabla \cdot \vec{\mathbf{E}} + \vec{\mathbf{E}} \cdot \nabla\sigma = 0 \tag{4.18}$$

For a homogeneous medium, $\nabla\sigma = 0$ and (4.18) becomes

$$\nabla \cdot \vec{\mathbf{E}} = 0$$

Substituting $\vec{\mathbf{E}} = -\nabla V$, where V is the potential at any point within the conducting medium, we can write the preceding equation as

$$\nabla^2 V = 0 \tag{4.19}$$

Equation (4.19) asserts that *the potential distribution within a conducting medium satisfies Laplace's equation as long as the medium is homogeneous and the current distribution is time invariant.*

EXAMPLE 4.3

Two infinitely conducting parallel plates, each of cross-sectional area A, are separated by a distance ℓ. The potential difference between the plates is V_{ab}, as shown in Figure 4.6. If the medium between the plates is homogeneous and has a finite conductivity σ, determine the resistance of the region between the plates.

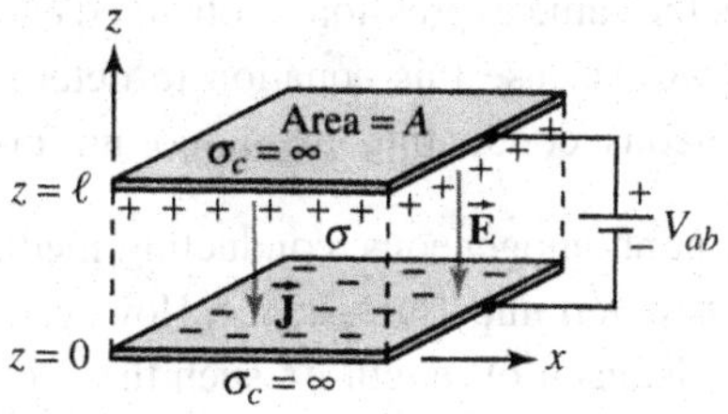

Figure 4.6 Two parallel plates separated by a conducting medium

Solution The resistance of each parallel plate is zero because of its infinite conductivity. We can use (4.19) to determine the potential distribution in the homogeneous, conducting medium. We expect the potential distribution to be a function of z only. From (4.19), it follows that

$$\frac{d^2V}{dz^2} = 0$$

Integrating twice, we obtain

$$V = az + b$$

where a and b are constants of integration. Applying the boundary conditions,

$$V = 0 \text{ at } z = 0 \Rightarrow b = 0$$

and

$$V = V_{ab} \text{ at } z = \ell \Rightarrow a = V_{ab}/\ell$$

Thus, the potential distribution in the conducting medium between the plates is

$$V = \frac{z}{\ell} V_{ab}$$

The electric field intensity in the conducting medium is

$$\vec{\mathbf{E}} = -\nabla V = -\frac{\partial V}{\partial z} \vec{\mathbf{a}}_z = -\frac{V_{ab}}{\ell} \vec{\mathbf{a}}_z$$

The volume current density in the medium is

$$\vec{\mathbf{J}} = \sigma \vec{\mathbf{E}} = -\frac{\sigma V_{ab}}{\ell} \vec{\mathbf{a}}_z$$

The current through a surface normal to $\vec{\mathbf{J}}$ is

$$I = \int_s \vec{\mathbf{J}} \cdot \vec{ds} = \frac{\sigma A V_{ab}}{\ell}$$

Finally, the resistance of the conducting medium is

$$R = \frac{V_{ab}}{I} = \frac{\ell}{\sigma A} \tag{4.20}$$

which is the same expression as obtained earlier for a conducting wire. In fact, we can use this equation to determine the resistance of any homogeneous, conducting medium of uniform cross section. • • •

For a nonhomogeneous, conducting medium, we cannot use (4.20) directly to determine its resistance. However, if we subdivide the region into n cells, each of length $d\ell$ such that $d\ell \to 0$ as $n \to \infty$, then we can assume that each cell has a constant conductivity, as illustrated in

Figure 4.7. The resistance of cell i from (4.20) is

$$R_i = \frac{d\ell_i}{\sigma_i A_i}$$

where $d\ell_i$, σ_i, and A_i are the length, conductivity, and the area of cell i, respectively. Thus, the total resistance of n cells connected in series is

$$R = \sum_{i=1}^{n} R_i = \sum_{i=1}^{n} \frac{d\ell_i}{\sigma_i A_i} \tag{4.21a}$$

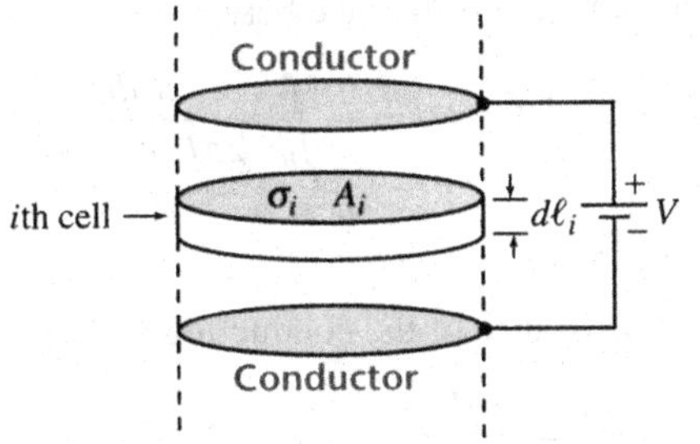

Figure 4.7 A nonhomogeneous, conducting medium divided into n cells (only the ith cell is shown)

In the limit $d\ell \to 0$ as $n \to \infty$, (4.21a) becomes

$$R = \int_c \frac{d\ell}{\sigma A} \tag{4.21b}$$

If the changes in the conductivity are discrete, we can use (4.21a) to determine the total resistance of a conducting medium. If the conductivity is given as a function of length, we can employ (4.21b), (4.18), or (4.10) to calculate the resistance of a nonhomogeneous, conducting medium.

EXAMPLE 4.4

A material with conductivity $\sigma = m/\rho + k$, where m and k are constants, fills the space between two concentric, cylindrical conductors of radii a and b, as shown in Figure 4.8. If V_0 is the potential difference between the two conductors, and L is the length of each conductor, obtain expressions for the resistance of the material, the current density, and the electric field intensity in the material.

Solution a) Method 1: Let us calculate the resistance using (4.21b). At any radius ρ the differential length is $d\rho$, and the cross-sectional area is $2\pi\rho L$. The resistance of the material is

$$R = \int_a^b \frac{d\rho}{(m + k\rho)2\pi L} = \frac{1}{2\pi Lk} \ln\left[\frac{m + kb}{m + ka}\right]$$

Let $M = \ln\left[\dfrac{m + kb}{m + ka}\right]$; then $R = \dfrac{M}{2\pi Lk}$.

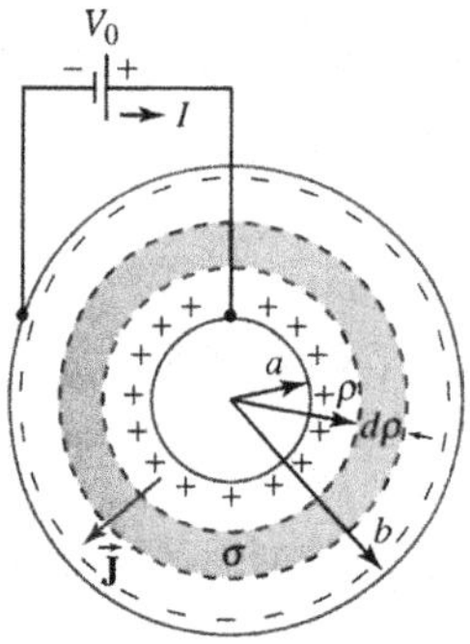

Figure 4.8 Current through a nonhomogeneous, conducting medium between two concentric, cylindrical conductors

b) Method 2: Let us now use (4.10) to determine the resistance of the material. The total current I through any cross-sectional area must be the same, so the current density $\vec{\mathbf{J}}$ in the material is

$$\vec{\mathbf{J}} = \frac{I}{2\pi\rho L}\vec{\mathbf{a}}_\rho$$

where the current I is unknown at this time. The electric field intensity in the medium is

$$\vec{\mathbf{E}} = \frac{\vec{\mathbf{J}}}{\sigma} = \frac{I}{2\pi L(m+k\rho)}\vec{\mathbf{a}}_\rho$$

The potential difference between the two conductors is

$$V_0 = -\int_c \vec{\mathbf{E}}\cdot\vec{d\ell} = -\int_b^a \frac{I\,d\rho}{2\pi L(m+k\rho)} = \frac{I}{2\pi Lk}\ln\left[\frac{m+kb}{m+ka}\right]$$
$$= \frac{IM}{2\pi Lk}$$

The resistance of the conducting material is

$$R = \frac{V_0}{I} = \frac{M}{2\pi Lk}$$

The current through the conducting material is

$$I = \frac{2\pi Lk}{M}V_0$$

Hence, the electric field intensity and the current density in the medium are, respectively,

$$\vec{\mathbf{E}} = \frac{k}{(m+k\rho)M}V_0\vec{\mathbf{a}}_\rho$$

and

$$\vec{\mathbf{J}} = \sigma\vec{\mathbf{E}} = \frac{k}{M\rho}V_0\vec{\mathbf{a}}_\rho$$

This method, lengthy as it may be, is useful when a steady current is maintained throughout the medium and we are interested in determining (a) the $\vec{\mathbf{E}}$ and $\vec{\mathbf{D}}$ fields in the conducting medium, (b) the surface charge densities on the conductors, (c) the volume charge density within the medium, and (d) the total charge in the conducting medium. • • •

4.5 Relaxation time

Let us consider an isolated, linear, homogeneous, and isotropic medium with permittivity ϵ and conductivity σ having an excess charge density of ρ_v. As mentioned earlier, the electrostatic forces of repulsion among the charges will move the excess charge onto the bounding surface of the medium in order to attain its electrostatic equilibrium. During this charge migration process, the equation of continuity must be satisfied. That is, at any point in the medium

$$\nabla \cdot \vec{\mathbf{J}} + \frac{\partial \rho_v}{\partial t} = 0$$

Substituting $\vec{\mathbf{J}} = \sigma \vec{\mathbf{E}}$, in this equation, we get

$$\sigma \nabla \cdot \vec{\mathbf{E}} + \frac{\partial \rho_v}{\partial t} = 0$$

Replacing $\nabla \cdot \vec{\mathbf{E}}$ with ρ_v/ϵ, we obtain

$$\frac{\partial \rho_v}{\partial t} + \frac{\sigma}{\epsilon} \rho_v = 0$$

which is a first-order differential equation in terms of the volume charge density ρ_v. The solution is

$$\rho_v = \rho_0 \, e^{-(\sigma/\epsilon)t} \tag{4.22}$$

where ρ_0 is the volume charge density at $t = 0$. Equation (4.22) highlights the fact that equilibrium is reached exponentially, and theoretically it will take forever for all of the excess charge to vanish from the interior of the conducting medium.

We can show that the ratio of ϵ to σ has the dimensions of time and is called the *relaxation time* τ. That is,

$$\tau = \frac{\epsilon}{\sigma} \tag{4.23}$$

The relaxation time is a measure of how fast a conducting medium approaches electrostatic equilibrium. In fact, it is the time required for the charge in any conducting medium to decay to $1/e$ (36.8%) of its initial value. By the time $t = 5\tau$, the charge density within the medium will fall below 1% of its initial value. *We can generally state that the conducting medium has attained its equilibrium state after five relaxation times.*

The relaxation time is inversely proportional to the conductivity of the medium—the larger the conductivity, the smaller the time required for the conducting medium to achieve electrostatic equilibrium. For copper, with $\sigma = 5.8 \times 10^7$ S/m and $\epsilon \approx \epsilon_0$, the relaxation time is $\tau = 1.52 \times 10^{-19}$ s. We can say that copper attains electrostatic equilibrium almost immediately. On the other hand, the relaxation time for pure water is 40 ns and that for amber is about 70 minutes.

EXAMPLE 4.5

A certain amount of charge is placed within an isolated conductor. The current through a closed surface bounding the charge is observed to be $i(t) = 0.125e^{-25t}$ A. Determine (a) the relaxation time, (b) the initial charge, and (c) the charge transported through the surface in time $t = 5\tau$.

Solution The relaxation time is $\tau = 1/25 = 0.04$ s.

The charge passing through the surface in time t is

$$Q = \int_0^t i\,dt = 0.125 \int_0^t e^{-25t}\,dt$$
$$= 5[1 - e^{-25t}] \text{ mC} \tag{4.24}$$

By setting $t = 5\tau = 0.2$ s, we obtain the charge passing through a closed surface as

$$Q = 4.97 \text{ mC}$$

However, by setting $t = \infty$ in (4.24) we obtain the total charge that has passed through a closed surface as 5 mC. Because there is no further transportation of the charge, the current through the surface is zero. Therefore, the total charge placed inside the conductor at $t = 0$ must be 5 mC. • • •

4.6 Joule's law

Consider a medium in which the charges are moving with an average velocity $\vec{\mathbf{U}}$ under the influence of an electric field. If ρ_v is the volume charge density, the force experienced by the charge in a volume dv is

$$\vec{d\mathbf{F}} = \rho_v\, dv \vec{\mathbf{E}}$$

If the charges move a distance $\vec{d\ell}$ in time dt such that $\vec{d\ell} = \vec{\mathbf{U}}\,dt$, the work done by the electric field is

$$dW = \vec{d\mathbf{F}} \cdot \vec{d\ell} = \rho_v \vec{\mathbf{U}} \cdot \vec{\mathbf{E}}\, dv\, dt$$
$$= \vec{\mathbf{J}} \cdot \vec{\mathbf{E}}\, dv\, dt$$

where $\vec{\mathbf{J}} = \rho_v \vec{\mathbf{U}}$.

Since power is the work done per unit time, the power supplied by the electric field is

$$dp = \frac{dW}{dt} = \vec{\mathbf{J}} \cdot \vec{\mathbf{E}}\, dv$$

If we define the power density p as the power per unit volume such that $dp = p\,dv$, then we can write the preceding equation as

$$p = \vec{\mathbf{J}} \cdot \vec{\mathbf{E}} \tag{4.25a}$$

Equation (4.25a) is called the point (or differential) form of Joule's law. It follows that the power delivered per unit volume by the electric field is a scalar product of electric field intensity and the volume current density.

The power associated with a volume v is

$$P = \int_v p\,dv = \int_v \vec{\mathbf{J}} \cdot \vec{\mathbf{E}}\, dv \tag{4.25b}$$

We refer to this equation as the integral form of Joule's law.

If the motion of free charges takes place within a conducting medium, the forces exerted by the $\vec{\mathbf{E}}$ field are balanced by the loss in momentum during the collision processes. In this case, the power supplied by the electric field is dissipated in the form of **heat**, a process that yields the ohmic or Joule heating of resistors. Then, the power density p represents the time rate at which heat is being generated per unit volume.

For a linear conductor, $\vec{\mathbf{J}} = \sigma\vec{\mathbf{E}}$, the power dissipation per unit volume is

$$p = \sigma\vec{\mathbf{E}}\cdot\vec{\mathbf{E}} = \sigma E^2 \tag{4.26a}$$

and the total power dissipation is

$$P = \int_v \sigma E^2\, dv \tag{4.26b}$$

If V is the potential difference between the two ends of a conductor of length L and uniform cross section A, then the power density is

$$p = \sigma\left[\frac{V}{L}\right]^2 \text{ W/m}^3$$

and the total power lost by the conductor as heat is

$$P = \frac{\sigma A V^2}{L} = \frac{V^2}{R}\text{W} \tag{4.27a}$$

where $R = \dfrac{L}{\sigma A}$ is the resistance of the conductor.

Equation (4.27a) is an equivalent form of Joule's law that is commonly used in electrical circuit theory to determine the power dissipated as heat by a resistor. The reader is urged to verify that

$$P = I^2 R \tag{4.27b}$$

is another equivalent form of Joule's law. It states that *the rate of heat dissipation varies as the square of the current in a linear conductor.*

EXAMPLE 4.6

A parallel-plate capacitor whose plates are 10 cm square and 0.2 cm apart contains a medium with $\epsilon_r = 2$ and $\sigma = 4 \times 10^{-5}$ S/m. To maintain a steady current through the medium a potential difference of 120 V is applied between the plates. Determine the electric field intensity, the volume current density, the power density, the power dissipation, the current, and the resistance of the medium.

Solution If the potential of the lower plate at $z = 0$ is 0 V, then the potential of the upper plate at $z = 0.2$ cm is 120 V. The electric field intensity in the medium is

$$\vec{\mathbf{E}} = -\frac{120}{0.002}\vec{\mathbf{a}}_z = -60\vec{\mathbf{a}}_z \text{ kV/m}$$

For the medium with $\sigma = 4 \times 10^{-5}$ S/m, the current density is

$$\vec{\mathbf{J}} = -4 \times 10^{-5} \times 60 \times 10^3\vec{\mathbf{a}}_z = -2.4\vec{\mathbf{a}}_z \text{ A/m}^2$$

Thus, the current through the medium is

$$I = \int_s \vec{\mathbf{J}} \cdot \overrightarrow{ds} = 2.4 \times 100 \times 10^{-4} = 24 \text{ mA}$$

The power density in the medium is

$$p = \vec{\mathbf{J}} \cdot \vec{\mathbf{E}} = 2.4 \times 60 \times 10^3 = 144 \text{ kW/m}^3$$

The total power dissipation in the medium is

$$P = \int_v p\, dv = 144 \times 10^3 \times 100 \times 10^{-4} \times 0.2 \times 10^{-2} = 2.88 \text{ W}$$

Because $P = I^2 R$, the resistance of the medium is

$$R = \frac{2.88}{[24 \times 10^{-3}]^2} = 5000\,\Omega \quad \text{or} \quad 5\,\text{k}\Omega$$

• • •

4.7 Steady current in a diode

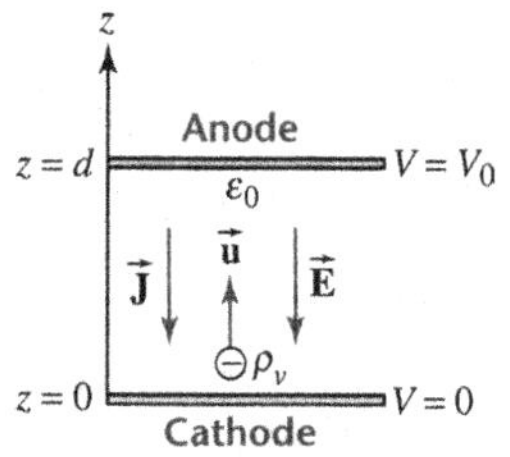

Figure 4.9 A vacuum tube diode

Let us consider a diode consisting of two parallel plates, as shown in Figure 4.9. One plate is the cathode; the other is the anode. The anode is held at a positive potential V_0, and the cathode is grounded. If we assume that the area of the plates is large compared to their separation, the potential distribution will be a function of z only. Thus, the electric field intensity at any point in the region is

$$\vec{\mathbf{E}} = -\frac{dV}{dz}\vec{\mathbf{a}}_z$$

The cathode is heated to emit electrons, which travel toward the anode under the influence of the $\vec{\mathbf{E}}$ field. Let $\vec{\mathbf{U}} = U\vec{\mathbf{a}}_z$ be the velocity of an electron at any time t. If m is the mass of the electron and $-e$ is its charge, then the force acting on the electron is $-e\vec{\mathbf{E}}$. This force imparts acceleration to the electron such that

$$m\frac{dU}{dt}\vec{\mathbf{a}}_z = e\frac{dV}{dz}\vec{\mathbf{a}}_z$$

$$mU\frac{dU}{dz} = e\frac{dV}{dz}$$

or

$$\frac{d}{dz}\left[\frac{1}{2}mU^2 - eV\right] = 0$$

Integrating both sides, we obtain

$$\frac{1}{2}mU^2 = eV + c$$

where c is the constant of integration. If the electron begins at rest

($U = 0$) at the cathode ($V = 0$), then $c = 0$. Thus,

$$\frac{1}{2} m U^2 = eV \tag{4.28}$$

Equation (4.28) states that the *potential energy supplied by the electric field is transformed into the kinetic energy of the electron.* The velocity of the electron at any point between the plates is

$$U = \left[\frac{2eV}{m}\right]^{1/2} \tag{4.29}$$

To determine U we must know the potential distribution V between the plates. However, the potential distribution between the plates must satisfy Poisson's equation. That is,

$$\frac{d^2V}{dz^2} = -\frac{\rho_v}{\epsilon_0} \tag{4.30}$$

where $\rho_v = -Ne$ and N is the number of electrons per unit volume in the region. In addition, $\vec{\mathbf{J}} = \rho_v \vec{\mathbf{U}} = \rho_v \mathbf{U} \vec{\mathbf{a}}_z = J \vec{\mathbf{a}}_z$, where $J = \rho_v U$. For steady current, $\nabla \cdot \vec{\mathbf{J}} = 0 \Rightarrow \vec{\mathbf{J}} = \rho_v \vec{\mathbf{U}} =$ constant. Therefore, as U increases, ρ_v decreases; a fact we alluded to in Section 4.2. Thus, we can express the volume charge density as

$$\rho_v = \frac{J}{\sqrt{\dfrac{2eV}{m}}} = \frac{K}{\sqrt{V}} \tag{4.31}$$

where

$$K = \frac{J}{\sqrt{\dfrac{2e}{m}}}$$

We eliminate ρ_v from (4.30) and obtain

$$\frac{d^2V}{dz^2} = -\frac{K}{\epsilon_0 \sqrt{V}}$$

Integrating, we obtain

$$\left(\frac{dV}{dz}\right)^2 = -4\frac{K}{\epsilon_0}\sqrt{V} + k$$

where k is another constant of integration. At the cathode, $z = 0$, $V = 0$, and $dV/dz = 0$, thus the constant of integration $k = 0$. Thus,

$$\frac{dV}{V^{1/4}} = \sqrt{-\frac{4K}{\epsilon_0}} dz$$

Integrating again, we have

$$\left(\frac{4}{3}\right)V^{3/4} = \sqrt{-\frac{4K}{\epsilon_0}}\,z + k_1$$

where k_1 is another constant of integration and is also zero because $V = 0$ at $z = 0$. Hence

$$\frac{16}{9}V^{3/2} = -4\frac{J}{\epsilon_0}\sqrt{\frac{m}{2e}}\,z^2 \tag{4.32}$$

Because $V = V_0$ for $z = d$, we finally have

$$J = -\left(\frac{4}{9}\right)\left[\frac{\epsilon_0}{d^2}\right]\sqrt{\frac{2e}{m}}\,(V_0)^{3/2} \tag{4.33}$$

This equation is called the Child–Langmuir relation. It is a nonlinear relation because the current density and thereby the current are proportional to $V_0^{3/2}$. Also note that the current density $\vec{\mathbf{J}}$ is negative as it must be because a stream of electrons moving in the $+z$ direction constitutes a current in the $-z$ direction.

We can also show that the potential distribution within the parallel plates is

$$V = V_0\left[\frac{z}{d}\right]^{4/3} \tag{4.34}$$

and the electric field intensity in the region between the plates is

$$\vec{\mathbf{E}} = -\frac{dV}{dz}\vec{\mathbf{a}}_z = -\frac{4V_0}{3d}\left[\frac{z}{d}\right]^{1/3}\vec{\mathbf{a}}_z \tag{4.35}$$

which clearly indicates that the $\vec{\mathbf{E}}$ field at $z = 0$ is zero. In practice, it will have a small but finite value. Finally, the space charge density from Poisson's equation is

$$\rho_v = -\frac{4\epsilon_0 V_0}{9d^{4/3}}\,z^{-2/3} \tag{4.36}$$

EXAMPLE 4.7

WORKSHEET 1

Mathcad

The anode of a vacuum tube diode is at 1000 V, and the cathode is grounded. The plates are 5 cm apart. Determine (a) the potential distribution, (b) the electric field intensity, (c) the volume current density, and (d) the charge density in the diode.

Solution From (4.34), the potential distribution between the plates is

$$V = 1000\left[\frac{z}{0.05}\right]^{4/3} = 54.288z^{4/3}\ \text{kV}$$

The electric field intensity, from (4.35), is

$$\vec{\mathbf{E}} = -\frac{4\times 1000}{3\times 0.05}\left[\frac{z}{0.05}\right]^{1/3}\vec{\mathbf{a}}_z = -72.384z^{1/3}\,\vec{\mathbf{a}}_z\ \text{kV/m}$$

The current density, from (4.33) is

$$\vec{\mathbf{J}} = -\frac{4}{9}\left[\frac{10^{-9}}{36\pi(0.05)^2}\right]\left[\frac{2\times 1.6\times 10^{19}}{9.11\times 10^{-31}}\right]^{1/2} \times 1000^{3/2}\,\vec{\mathbf{a}}_z$$
$$= -29.46\,\vec{\mathbf{a}}_z \text{ A/m}^2$$

From (4.36), the space charge density is

$$\rho_v = -\frac{4\times 10^{-9}\times 1000}{36\pi\times 9\times (0.05)^{4/3}} z^{-2/3} = -213.34 z^{-2/3} \text{ nC/m}^3$$

• • •

4.8 Boundary conditions for current density

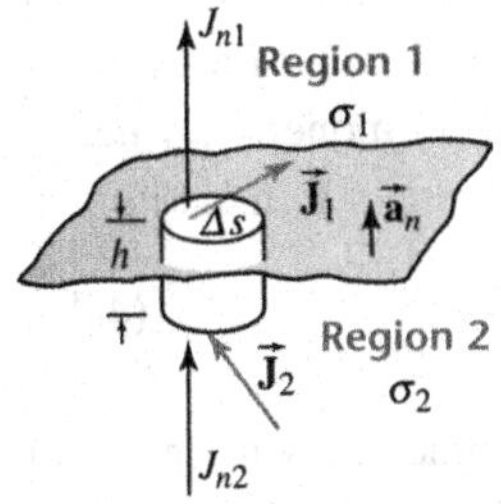

Figure 4.10 Boundary conditions for normal component of $\vec{\mathbf{J}}$

In this section we study how the current density vector $\vec{\mathbf{J}}$ changes when passing through an interface between media of conductivities σ_1 and σ_2. Let us construct a closed surface in the form of a pillbox, as illustrated in Figure 4.10. The height of the pillbox is so small that the contribution from the radial surface to the current can be neglected. Computing the integral, (4.17a),

$$\oint_s \vec{\mathbf{J}}\cdot\overrightarrow{ds} = 0$$

over the closed surface s of the pillbox when $h \to 0$, we find

$$\vec{\mathbf{a}}_n\cdot\vec{\mathbf{J}}_1\,\Delta s - \vec{\mathbf{a}}_n\cdot\vec{\mathbf{J}}_2\,\Delta s = 0$$
$$\vec{\mathbf{a}}_n\cdot(\vec{\mathbf{J}}_1 - \vec{\mathbf{J}}_2) = 0 \tag{4.37a}$$

or

$$J_{n1} = J_{n2} \tag{4.37b}$$

where the subscript n stands for the normal component of the field quantity. Equation (4.37) states that *the normal component of electric current density* $\vec{\mathbf{J}}$ *is continuous across the boundary.*

Since the tangential component of the $\vec{\mathbf{E}}$ field is continuous across the boundary; that is,

$$\vec{\mathbf{a}}_n \times [\vec{\mathbf{E}}_1 - \vec{\mathbf{E}}_2] = 0$$

and $\vec{\mathbf{J}} = \sigma\vec{\mathbf{E}}$, we can now write an equation for the tangential component of $\vec{\mathbf{J}}$ at the interface (see Figure 4.11) as

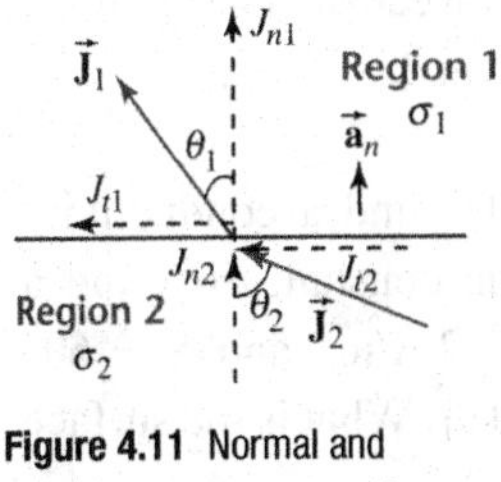

Figure 4.11 Normal and tangential components of current densities just above and below the interface

$$\vec{\mathbf{a}}_n \times \left[\frac{\vec{\mathbf{J}}_1}{\sigma_1} - \frac{\vec{\mathbf{J}}_2}{\sigma_2}\right] = 0 \tag{4.38a}$$

or

$$\frac{J_{t1}}{J_{t2}} = \frac{\sigma_1}{\sigma_2} \tag{4.38b}$$

where the subscript t implies the tangential component of the field quantity. Equation (4.38) signifies that *the ratio of the tangential components of the current densities at the interface is equal to the ratio of the conductivities.*

From (4.37), (4.38), and Figure 4.11, we have

$$\frac{J_{n1}\sigma_1}{J_{t1}} = \frac{J_{n2}\sigma_2}{J_{t2}}$$

or

$$\frac{\tan\theta_1}{\tan\theta_2} = \frac{\sigma_1}{\sigma_2} \tag{4.39}$$

As a special case, let us consider an interface between a poorly conducting region (medium 1) and a highly conducting region (medium 2). If θ_2 is an angle between 0° and 90°, then from (4.39), θ_1 will be a very small angle because $\sigma_2 \gg \sigma_1$. In other words, $\vec{\mathbf{J}}$ and $\vec{\mathbf{E}}$ in medium 1 are almost normal to the interface. Therefore, their tangential components are negligibly small. On the other hand, the normal component of $\vec{\mathbf{E}}$ in medium 2,

$$E_{n2} = \frac{\sigma_1}{\sigma_2}E_{n1} \tag{4.40}$$

is also very small. This means that the $\vec{\mathbf{E}}$ field is practically nonexistent in a highly conducting medium. Therefore, there must exist a free surface charge density at the interface. We can compute the free surface charge density from the normal components of the $\vec{\mathbf{D}}$ field as

$$\begin{aligned}\rho_s &= D_{n1} - D_{n2}\\ &= D_{n1}\left[1 - \frac{\sigma_1\epsilon_2}{\sigma_2\epsilon_1}\right]\\ &= E_{n1}\left[\frac{\epsilon_1\sigma_2 - \epsilon_2\sigma_1}{\sigma_2}\right]\\ &= J_{n1}\left[\frac{\epsilon_1}{\sigma_1} - \frac{\epsilon_2}{\sigma_2}\right]\end{aligned} \tag{4.41}$$

Equation (4.41) gives the surface charge density in terms of the normal component of $\vec{\mathbf{J}}$ in medium 1. A similar expression can also be obtained in terms of the normal component of $\vec{\mathbf{J}}$ in medium 2.

EXAMPLE 4.8

Medium 1 ($z \geq 0$) has a dielectric constant of 2 and a conductivity of 40 μS/m. Medium 2 ($z \leq 0$) has a dielectric constant of 5 and a conductivity of 50 nS/m. If $\vec{\mathbf{J}}_2$ has a magnitude of 2 A/m², and $\theta_2 = 60°$ with the normal to the interface, compute $\vec{\mathbf{J}}_1$ and θ_1. What is the surface charge density at the interface?

Solution From the given information,

$$J_{n2} = 2\cos 60° = 1\text{ A/m}^2 \text{ and } J_{t2} = 2\sin 60° = 1.732\text{ A/m}^2$$

From the boundary condition (4.37b), $J_{n1} = 1$ A/m^2.

Applying the boundary condition (4.38b), we obtain

$$J_{t1} = \frac{40 \times 10^{-6}}{50 \times 10^{-9}} \times 1.732 = 1385.6 \text{ A/m}^2$$

Thus, $J_1 = [1^2 + 1385.6^2]^{1/2} \approx 1385.6$ A/m^2 and $\theta_1 = \tan^{-1}[1385.6] = 89.96°$.

Finally, from (4.41), the surface charge density is

$$\rho_s = 1\left[\frac{2}{40 \times 10^{-6}} - \frac{5}{50 \times 10^{-9}}\right]\frac{10^{-9}}{36\pi}$$
$$= -0.88 \text{ mC/m}^2$$

• • •

4.9 Analogy between $\vec{D}$ and $\vec{J}$

At this time it is advantageous to emphasize that there exists an analogy between the $\vec{\mathbf{D}}$ and $\vec{\mathbf{J}}$ fields under static (time-invariant) conditions. By this we mean that the two fields can be described by equations of the same mathematical form. For example, for steady current,

$$\nabla \cdot \vec{\mathbf{J}} = 0 \tag{4.42a}$$

and in a charge-free region,

$$\nabla \cdot \vec{\mathbf{D}} = 0 \tag{4.42b}$$

Because

$$\vec{\mathbf{J}} = \sigma \vec{\mathbf{E}} \tag{4.43a}$$
$$\vec{\mathbf{D}} = \epsilon \vec{\mathbf{E}} \tag{4.43b}$$

and

$$\nabla \times \vec{\mathbf{E}} = 0$$

for a linear medium with constant permittivity ϵ and conductivity σ, we have

$$\nabla \times \vec{\mathbf{J}} = 0 \tag{4.44a}$$
$$\nabla \times \vec{\mathbf{D}} = 0 \tag{4.44b}$$

From the continuity of the normal components at the interface between two conducting media, we have

$$J_{n1} = J_{n2} \tag{4.45a}$$

and

$$D_{n1} = D_{n2} \tag{4.45b}$$

between two dielectric media. From the boundary conditions on the tangential components, we have

$$\frac{J_{t1}}{J_{t2}} = \frac{\sigma_1}{\sigma_2} \tag{4.46a}$$

and

$$\frac{D_{t1}}{D_{t2}} = \frac{\epsilon_1}{\epsilon_2} \tag{4.46b}$$

It is evident from the preceding expressions that an equation in terms of $\vec{\mathbf{J}}$ can be obtained from an equation in terms of $\vec{\mathbf{D}}$ by relating $\vec{\mathbf{D}}$ with $\vec{\mathbf{J}}$ and ϵ with σ. In order to exploit the analogy between the $\vec{\mathbf{D}}$ and $\vec{\mathbf{J}}$ fields in a charge-free medium, we can first determine the $\vec{\mathbf{D}}$ field, assuming that the medium is dielectric. We can then obtain the current density by substituting σ for ϵ. The following example illustrates this procedure.

EXAMPLE 4.9

The potential difference between the two plates in a parallel-plate capacitor is V_0. If the area of each plate is A, the separation between them is d, and the conducting medium is characterized by permittivity ϵ and conductivity σ, determine the current through the medium using the analogy between the $\vec{\mathbf{D}}$ and $\vec{\mathbf{J}}$ fields.

Solution The electric field intensity in the parallel-plate capacitor is

$$\vec{\mathbf{E}} = -\frac{V_0}{d}\vec{\mathbf{a}}_z$$

where we have assumed that the separation between the plates is in the z direction and the upper plate at $z = d$ is positive with respect to the lower plate at $z = 0$. The electric flux density in the medium is

$$\vec{\mathbf{D}} = -\frac{\epsilon}{d}V_0\vec{\mathbf{a}}_z$$

Using the analogy between $\vec{\mathbf{D}}$ and $\vec{\mathbf{J}}$ for a charge-free medium, we can obtain the volume current density in the medium by substituting σ for ϵ as

$$\vec{\mathbf{J}} = -\frac{\sigma}{d}V_0\vec{\mathbf{a}}_z$$

Hence, the current through the medium is

$$I = \int_s \vec{\mathbf{J}} \cdot \vec{ds} = \frac{\sigma A}{d}V_0 = \frac{V_0}{R}$$

where $R = \dfrac{d}{\sigma A}$ is the resistance of the medium. • • •

In Chapter 3 we defined the capacitance as

$$C = \frac{Q}{V_{ab}} = \frac{\int_s \rho_s\, ds}{-\int_b^a \vec{\mathbf{E}} \cdot \vec{d\ell}} = \frac{\int_s \epsilon E_n\, ds}{-\int_b^a \vec{\mathbf{E}} \cdot \vec{d\ell}} \tag{4.47}$$

where $\rho_s = \epsilon E_n$ is the surface charge density at the surface of conductor a of the capacitor. If we define the conductance G as the reciprocal of the resistance, then from (4.11), we have

$$G = \frac{I}{V_{ab}} = \frac{\int_s \vec{\mathbf{J}} \cdot \vec{ds}}{-\int_b^a \vec{\mathbf{E}} \cdot \vec{d\ell}} = \frac{\int_s \sigma E_n \, ds}{-\int_b^a \vec{\mathbf{E}} \cdot \vec{d\ell}} \tag{4.48}$$

Comparing (4.47) and (4.48), we find that

$$G = \frac{\sigma}{\epsilon} C \tag{4.49}$$

This equation enables us to find the conductance and thereby the resistance of any configuration whose capacitance is known. The converse is also true.

EXAMPLE 4.10

Two parallel plates, each of area A, are separated by a distance d. The medium between the plates has a conductivity σ and a permittivity ϵ. Find the resistance of the parallel-plate capacitor.

Solution We know that the capacitance of a parallel-plate capacitor is

$$C = \frac{\epsilon A}{d}$$

From (4.49), the conductance is

$$G = \frac{\sigma}{\epsilon} C = \frac{\sigma A}{d}$$

Therefore, the resistance between the plates of the capacitor is

$$R = \frac{1}{G} = \frac{d}{\sigma A}$$

• • •

EXAMPLE 4.11

The conductivity of the medium between two concentric metal spheres is σ and the permittivity is ϵ. If the radius of the inner sphere is a and the inner radius of the outer sphere is b, determine the resistance of the medium between the spheres.

Solution In Chapter 3 we obtained an expression for the capacitance between two concentric spheres as

$$C = \frac{4\pi\epsilon ab}{b - a}$$

Thus, the conductance, from (4.49), is

$$G = \frac{4\pi\sigma ab}{b - a}$$

Finally, the resistance of the region between the spheres is

$$R = \frac{b - a}{4\pi\sigma ab}$$

• • •

EXAMPLE 4.12 The region between a very long coaxial cable is filled with a material of conductivity σ and permittivity ϵ. If the radii of the inner and outer conductors are a and b, respectively, find the resistance per unit length between the conductors.

Solution The capacitance per unit length of a coaxial cable was obtained in Chapter 3 as

$$C = \frac{2\pi\epsilon}{\ln(b/a)}$$

Replacing ϵ with σ, we obtain the conductance per unit length as

$$G = \frac{2\pi\sigma}{\ln(b/a)}$$

Hence, the resistance per unit length between the two concentric conductors is

$$R = \frac{\ln(b/a)}{2\pi\sigma}$$

• • •

4.10 The electromotive force

In Chapter 3 it was stated that the integral of the tangential component of the electric field intensity around any closed path vanishes; i.e.,

$$\oint_c \vec{\mathbf{E}} \cdot \vec{d\ell} = 0$$

for an electrostatic field. In this chapter, we have shown that the volume current density in a conducting medium is $\vec{\mathbf{J}} = \sigma\vec{\mathbf{E}}$ and the current through the conductor is

$$I = \int_s \vec{\mathbf{J}} \cdot \vec{ds} = \int_s \sigma\vec{\mathbf{E}} \cdot \vec{ds}$$

From the preceding equations, it follows that a purely electrostatic field cannot cause a current to circulate in a closed path (loop). In addition to the electrostatic field, there must also exist a source of energy to maintain the steady current in a closed loop. The external source of energy may be nonelectrical, such as a chemical reaction (battery), a mechanical drive (direct-current generator), a light-activated source (solar cell), or a temperature-sensitive device (thermocouple). Because these devices convert nonelectrical energy into electrical energy, we consider them to be nonconservative elements in the electrical circuit, and they will set up nonconservative electric fields $\vec{\mathbf{E}}'$.

The total electric field in the closed loop is then $\vec{\mathbf{E}} + \vec{\mathbf{E}}'$. The total power associated with the loop is

$$P = \int_v (\vec{\mathbf{E}} + \vec{\mathbf{E}}') \cdot \vec{\mathbf{J}} \, dv$$

If we assume that the steady current in the loop is uniformly distributed, then $\vec{\mathbf{J}} \, dv$ can be replaced by I $\overrightarrow{d\ell}$, and the volume integral reduces to

$$P = \oint_c I(\vec{\mathbf{E}} + \vec{\mathbf{E}}') \cdot \overrightarrow{d\ell} = I \oint_c \vec{\mathbf{E}}' \cdot \overrightarrow{d\ell}$$

By defining the *electromotive force (emf)* in the closed loop as

$$\mathscr{E} = \oint_c \vec{\mathbf{E}}' \cdot \overrightarrow{d\ell} \tag{4.50}$$

we obtain the power delivered to the loop as

$$P = \mathscr{E} I \tag{4.51}$$

Thus, *the power delivered to the loop is equal to the product of the emf and the current.*

For a part of the loop (branch) of the circuit between two points a and b, we can write

$$\begin{aligned} \int_a^b \frac{1}{\sigma} \vec{\mathbf{J}} \cdot \overrightarrow{d\ell} &= \int_a^b [\vec{\mathbf{E}} + \vec{\mathbf{E}}'] \cdot \overrightarrow{d\ell} \\ &= \int_a^b \vec{\mathbf{E}} \cdot \overrightarrow{d\ell} + \int_a^b \vec{\mathbf{E}}' \cdot \overrightarrow{d\ell} \\ &= -[V_b - V_a] + \mathscr{E}_{ab} \end{aligned} \tag{4.52}$$

where $\mathscr{E}_{ab}$ is the emf of the source(s) between points a and b. If $\mathscr{E}_{ab}$ is zero, the circuit branch between a and b is said to be a *passive branch.* If $\mathscr{E}_{ab}$ is not zero, we say that the branch contains a *seat of emf (active source).*

The left-hand side of (4.52) is simply IR. The simplest possible way to show that this is so is to consider that the current is uniformly distributed in a cylindrical conductor of area A and length L between points a and b, as depicted in Figure 4.12. Then $J = I/A$ and the left-hand side of (4.52) becomes

$$\frac{IL}{\sigma A} = IR$$

where

$$R = \frac{L}{\sigma A}$$

is the resistance of the conductor between points a and b. Equation (4.52) can now be written as

$$-[V_b - V_a] + \mathscr{E}_{ab} = IR \tag{4.53}$$

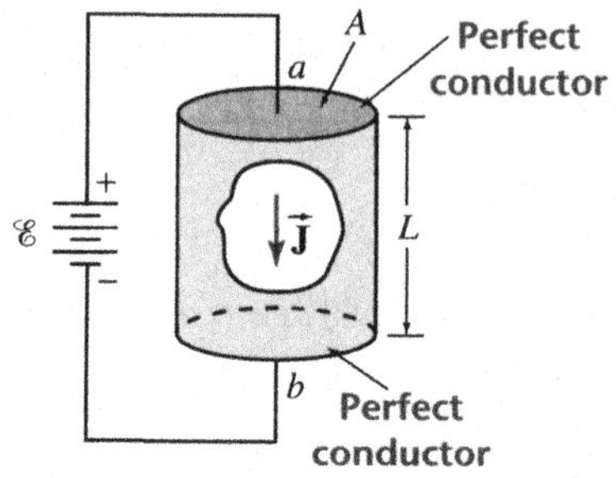

Figure 4.12 Uniform current flow through a cylindrical conductor

If the branch from a to b does not contain an active source, then this equation becomes

$$V_a - V_b = IR \tag{4.54}$$

Equation (4.54) gives the relation between the voltage drop across the resistor and the current passing through it. For I to be positive, V_a must be greater than V_b; that is, point a must be at a higher potential than point b. The current enters at point a and leaves at point b.

If we consider a closed loop such that point b is the same as point a, then $V_b = V_a$, and (4.53) becomes

$$\mathscr{E} = IR \tag{4.55a}$$

In this case, R represents the total resistance of the circuit, and $\mathscr{E}$ is the entire emf in the circuit. If the closed loop contains m seats of emf (sources) and n resistances carrying the same current, then (4.55a) becomes

$$\sum_{k=1}^{m} \mathscr{E}_k = \sum_{j=1}^{n} IR_j \tag{4.55b}$$

This equation is a mathematical statement of *Kirchhoff's voltage law*, which states that *the algebraic sum of the emfs in any closed loop is equal to the sum of the voltage drops in the same loop.*

The following example shows how to apply Kirchhoff's current law, (4.17c), and Kirchhoff's voltage law, (4.55b), to a simple electric circuit. You have already used these laws in a previous course on circuits, so we will not discuss them further in this text.

EXAMPLE 4.13

WORKSHEET 2

Find the current through each element of the circuit shown in Figure 4.13. Calculate the total power supplied by the sources.

Figure 4.13

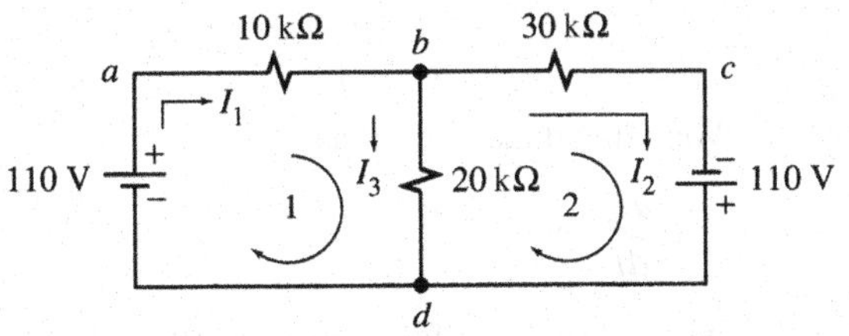

Solution We have arbitrarily marked the directions of the currents through each element of the circuit in the figure. Because the algebraic sum of the currents at any junction (node) is zero, from (4.17c), the sum of the currents entering a junction must be equal to the sum of the currents leaving the junction. Thus, at junction b, we have

$$I_1 = I_2 + I_3$$

or

$$I_3 = I_1 - I_2 \tag{4.56}$$

Equating the sum of the emfs with appropriate sign to the sum of the IR in loops 1 ($abda$) and 2 ($bcdb$), we obtain

$$10 \times 10^3 I_1 + 20 \times 10^3 I_3 = 110$$
$$30 \times 10^3 I_2 - 20 \times 10^3 I_3 = 110$$

or

$$I_1 + 2I_3 = 0.011$$
$$3I_2 - 2I_3 = 0.011$$

Substituting for I_3 from (4.56), we can write the preceding equations as

$$3I_1 - 2I_2 = 0.011$$
$$-2I_1 + 5I_2 = 0.011$$

Solving by any standard method, we get

$$I_1 = 7\,\text{mA}, \quad I_2 = 5\,\text{mA}, \quad \text{and} \quad I_3 = 2\,\text{mA}.$$

The total power delivered to the circuit, from (4.51), is

$$P_s = 110 \times 7 \times 10^{-3} + 110 \times 5 \times 10^{-3} = 1.32\ \text{W}$$

The total power dissipated as heat by the resistors, from (4.27b), is

$$P_d = (0.007)^2 10 \times 10^3 + (0.005)^2 30 \times 10^3 + (0.002)^2 20 \times 10^3$$
$$= 1.32\ \text{W}$$

Thus, the power supplied by the sources equals the power lost by the resistors in accordance with the law of conservation of energy. • • •

4.11 Summary

We defined the current as

$$i = \frac{dq}{dt}$$

where dq is the differential charge transferred across a finite cross section in time dt.

We discussed two types of currents: the convection current and the conduction current. The convection current is due to the flow of charges in a vacuum. The conduction current results from the flow of electrons in a conductor.

The current per unit area was defined as the current density. The convection current density is

$$\vec{\mathbf{J}} = \rho_v \vec{\mathbf{U}}_v$$

where ρ_v is the volume charge density and $\vec{\mathbf{U}}_v$ is the average velocity of these charges. The conduction current density is

$$\vec{\mathbf{J}} = \sigma \vec{\mathbf{E}}$$

where σ is the conductivity of the medium and $\vec{\mathbf{E}}$ is the electric field intensity within the conductor.

The resistance of a conductor, in terms of the field quantities, is

$$R = \int_b^a \frac{-\vec{\mathbf{E}} \cdot \overrightarrow{d\ell}}{\int_s \vec{\mathbf{J}} \cdot \overrightarrow{d\mathbf{s}}}$$

For a linear medium,

$$R = \frac{\ell}{\sigma A}$$

where ℓ is the length, σ is the conductivity, and A is the area of the conductor.

We obtained a general expression for the equation of continuity as

$$\nabla \cdot \vec{\mathbf{J}} = -\frac{\partial \rho_v}{\partial t}$$

which reduced to

$$\nabla \cdot \vec{\mathbf{J}} = 0$$

for steady current through a medium.

The potential distribution in a linear, isotropic, and homogeneous conducting medium satisfies Laplace's equation. That is,

$$\nabla^2 V = 0$$

The solution of this equation enables us to determine the $\vec{\mathbf{E}}$ field, the $\vec{\mathbf{D}}$ field, the current density $\vec{\mathbf{J}}$, the current I, and the resistance of the medium.

If we place an excess charge within a conductor, it will redistribute over the surface of the conductor according to the following equation:

$$\rho_v = \rho_0 e^{-t/\tau}$$

where $\tau = \epsilon/\sigma$ is the relaxation time. For all practical purposes, the charge will vanish from the interior of the conductor in a time $t = 5\tau$. The power dissipated by a conductor as heat is

$$P_d = I^2 R = \frac{V^2}{R}$$

and the power associated with a conducting medium, in terms of field quantities, is

$$P = \int_v \vec{\mathbf{E}} \cdot \vec{\mathbf{J}}\, dv$$

From the boundary conditions between two conducting media, we found that the normal components of $\vec{\mathbf{J}}$ are continuous at the interface. That is,

$$J_{n1} = J_{n2}$$

and the ratio of the tangential components of $\vec{\mathbf{J}}$ at the interface is equal to the ratio of the conductivities:

$$\frac{J_{t1}}{J_{t2}} = \frac{\sigma_1}{\sigma_2}$$

From the knowledge of capacitance for a given arrangement of conductors, we can determine the conductance as

$$G = \frac{\sigma}{\epsilon} C$$

An electromotive force (emf) is a device that maintains a potential difference between its output terminals. It supplies energy to the electric circuit when the current flows through it from the negative to the positive terminal. The power delivered by an emf source is

$$P = \mathscr{E} I$$

Kirchhoff's current law states that the algebraic sum of electric currents at a node is zero. That is,

$$\sum_{k=1}^{n} I_k = 0$$

Kirchhoff's voltage law states that the algebraic sum of the emfs in a closed loop is equal to the algebraic sum of the voltage drops across the resistors. That is,

$$\sum_{j=1}^{m} \mathscr{E}_j = \sum_{k=1}^{n} IR_k$$

4.12 Review questions

4.1 Why is the electric field intensity not zero in a steady-current-carrying conductor?

4.2 A copper wire and an aluminum wire of the same length and the same cross-sectional area have the same potential difference across them. Do they carry the same current?

4.3 A potential difference of V_0 volts is maintained between the two terminals of a wire of diameter d and length L. How is the drift velocity of an electron affected if (a) the potential difference is doubled, (b) the diameter is doubled, (c) the length is doubled?

4.4 A wire is carrying a steady current I. How is the current density $\vec{\mathbf{J}}$ affected if (a) its length is doubled, (b) its area is doubled, (c) its length is doubled, but its area is halved?

4.5 A wire of area A and length L has a resistance of R ohms. It is drawn out through a die so that its new length is $3L$. What is its new resistance?

4.6 When can we use Laplace's equation to determine the potential distribution in a conducting medium carrying steady current?

4.7 What is the difference between emf and potential difference?

4.8 The conductivity and the permittivity of the medium between the plates of a parallel-plate capacitor are σ and ϵ, respectively. At time $t = 0$, the capacitor is charged to V_0 volts. Will the charge on the capacitor remain the same?

4.9 Someone suggests that the capacitor in Question 4.8 will discharge with a time constant $\tau = \sigma/\epsilon$. What do you think?

4.10 A 10-Ω, 0.5-W resistor is connected in series with a 10-Ω, 5-W resistor. What is the effective resistance? What is its power rating? What is the maximum current that can safely flow through the series combination?

4.11 The two resistors in Question 4.10 are connected in parallel. Find (a) the effective resistance, (b) its power rating, and (c) the maximum current that can flow safely through the parallel combination.

4.12 Does $\mathscr{E} = IR$ apply to a medium that does not obey Ohm's law?

4.13 What do we really mean when we say that the potential at a point p is V_p?

4.14 A very thin conductor carries a current of 10 A. Determine the number of electrons that pass a point on the conductor each second.

4.15 Is a wire carrying steady current in electrostatic equilibrium?

4.16 Can there be a current in a conductor if the electric field intensity within the conductor is zero?

4.17 Can there be an electric field within a conductor if the net current through it is zero?

4.18 State Kirchhoff's current law.

4.19 State Kirchhoff's voltage law.

4.20 What is Joule's law?

4.13 Exercises

4.1 One end of an aluminum wire of 0.125 cm in diameter is welded to a copper wire of 0.25 cm in diameter. The composite wire carries a current of 8 mA. What is the current density in each wire?

4.2 A 100-km-long, high-voltage transmission line uses a copper cable of diameter 3 cm. If the cable carries a steady current of 1000 A, determine (a) the electric field intensity inside the cable, (b) the drift velocity of the free electrons, (c) the current density in the cable, and (d) the time taken by an electron to travel the full length of the cable. Assume the average time between collisions is 2.7×10^{-14} s.

4.3 The average velocity of electrons in a vacuum tube is 1.5×10^6 m/s. If the current density is 5 A/mm^2, determine the number of electrons per unit area passing through a hypothetical plane normal to their flow.

4.4 An aluminum conductor has a radius of 2 cm and carries a current of 100 A. If the length of the conductor is 100 km, determine (a) the current density in the conductor, (b) the $\vec{\mathbf{E}}$ field within the conductor, (c) the potential drop across the conductor, and (d) the resistance of the conductor.

4.5 A hollow circular iron cylinder has an inner diameter of 2 cm and an outer diameter of 5 cm. The length of the cylinder is 200 m. The magnitude of the electric field intensity within the cylinder is 10 mV/m. Determine (a) the potential drop across the cylinder, (b) the current through the cylinder, and (c) the resistance of the cylinder.

4.6 The hollow circular cylinder in Exercise 4.5 is replaced by a solid copper cylinder of the same length. For the same resistance, find the radius of the copper cylinder. For the same applied voltage, compute the electric field intensity and the current density within the copper cylinder.

4.7 Show that the volume current density $\vec{\mathbf{J}}$ in Example 4.4 satisfies (4.17b).

4.8 If ϵ is the permittivity of the conducting material in Example 4.4, determine (a) the electric flux density in the medium, (b) the surface charge density and the total charge on the inner conductor, (c) the surface charge density and the total charge on the outer conductor, and (d) the volume charge density and the total charge in the medium.

4.9 Repeat Example 4.4 using (4.18).

4.10 The conductivity of a homogeneous conducting medium, bounded by 10 cm $\leq r \leq$ 20 cm, $30^\circ \leq \theta \leq 45^\circ$, and $\pi/6 \leq \phi \leq \pi/3$, is 0.4 S/m. The surface at $\theta = 45^\circ$ is at a ground potential, and the surface at

$\theta = 30°$ is at 100 V. Using Laplace's equation, determine the resistance of the medium, neglecting the edge effects.

4.11 If $\epsilon = 5\epsilon_0$ is the permittivity of the conducting medium in Exercise 4.10, determine (a) the electric flux density in the medium, (b) the surface charge density and the total charge on the conductor at $\theta = 45°$, (c) the surface charge density and the total charge on the conductor at $\theta = 30°$, and (d) the volume charge density and the total charge in the medium.

4.12 Determine the time rate of change of the volume charge density if the volume current density in the medium is $\vec{\mathbf{J}} = \sin(10x)\vec{\mathbf{a}}_x + y\vec{\mathbf{a}}_y + e^{-3z}\vec{\mathbf{a}}_z$ A/m^2.

4.13 Repeat Exercise 4.12 for $\vec{\mathbf{J}} = e^{-\beta\rho}\cos\phi\vec{\mathbf{a}}_\rho + \ln(\cos\beta z)\vec{\mathbf{a}}_z$ A/m^2.

4.14 A cylinder of radius 10 cm with constant permittivity ϵ and conductivity σ is uniformly charged up to radius 2 cm with a charge density of 10 μC/m^3 at time $t = 0$. Determine (a) the charge distribution for all time, (b) the electric field intensity everywhere in space, (c) the charge density on the outer surface, and (d) the conduction current density. Assuming that the charge transfer process takes five relaxation times, how long will it take for the charge to build up on the outer surface of (i) a copper cylinder, (ii) an aluminum cylinder, (iii) a carbon cylinder, and (iv) a quartz cylinder?

4.15 Using the concept of resistance, show that the resistance of the medium in Example 4.6 is 5 kΩ. Also, verify the current and the power lost as heat in the medium.

4.16 Two circular metal plates, each of radius 5 cm, are held 5 mm apart to form a parallel-plate capacitor. The space between them is filled with two slabs, one 3 mm thick with a conductivity of 40 μS/m and a dielectric constant of 5, and the other 2 mm thick with a conductivity of 60 μS/m and a dielectric constant of 2. A steady current is maintained through the medium by applying a potential difference of 200 V across the plates. What are the electric field intensity, the current density, the power density, and the power dissipation in each region? Find the resistance of each region. What is the total resistance? Find the charge density on each plate and the free charge density at the interface between the two slabs.

4.17 If each plate in Example 4.7 is 10 cm square, obtain the current in the diode. What is the maximum velocity of the electron? What is the maximum energy supplied by the $\vec{\mathbf{E}}$ field?

4.18 Repeat Example 4.7 and Exercise 4.17 if the potential at the anode is (a) 500 V and (b) 5000 V.

4.19 Obtain the expression for the surface charge density at the interface between two conducting media with different conductivities in terms of the normal component of the current density in medium 2.

4.20 The volume current density in medium $1(x \geq 0, \epsilon_{r1} = 1$, and $\sigma_1 = 20$ μS/m) is $\vec{\mathbf{J}}_1 = 100\vec{\mathbf{a}}_x + 20\vec{\mathbf{a}}_y - 50\vec{\mathbf{a}}_z$ A/m^2. Obtain the volume current density in medium $2(x \leq 0, \epsilon_{r2} = 5, \sigma_2 80$ μS/m). Also compute θ_1, θ_2, and ρ_s at the interface. What are the $\vec{\mathbf{E}}$ and $\vec{\mathbf{D}}$ fields on both sides of the interface?

4.21 Two conducting spherical shells with inner radius a and outer radius b are held at a potential difference of V_0. The medium between the shells is characterized by permittivity ϵ and conductivity σ. Using the analogy between the $\vec{\mathbf{D}}$ and $\vec{\mathbf{J}}$ fields, determine the current through the medium.

4.22 If $V_0 = 1000$ V, $a = 2$ cm, $b = 5$ cm, $\epsilon_r = 1$, and $\sigma = 4$ μS/m, what is the current through the medium in Exercise 4.21? Also calculate the capacitance and the resistance of the medium. What is the power dissipated in the medium?

4.23 The radii of the inner and outer conductors of a coaxial cable of length 100 m are 2 cm and 5 cm, respectively. A potential difference of 5 kV is maintained between the conductors. Using the analogy between the $\vec{\mathbf{D}}$ and $\vec{\mathbf{J}}$ fields, determine the current through the medium. The medium is characterized by $\epsilon_r = 2$ and $\sigma = 10$ μS/m. Calculate the power dissipated in the medium. Represent the medium by an equivalent circuit.

4.24 A copper wire 10 km in length and 1.3 mm in diameter is connected to an emf source of 24 V. Determine (a) the resistance of the wire, (b) the current through the wire, (c) the current density in the wire, (d) the power dissipated as heat by the wire, and (e) the power supplied by the source.

4.25 The copper wire in Exercise 4.24 is replaced by a Nichrome wire of the same length. What must be the diameter of the Nichrome wire if the power dissipated is to be the same?

4.26 Show that the equivalent resistance of n resistors connected in series is

$$R = \sum_{i=1}^{n} R_i$$

where R_i is the resistance of the ith resistor.

4.27 Show that the equivalent conductance of n resistors connected in parallel is

$$G = \sum_{i=1}^{n} G_i$$

where $G_i = 1/R_i$ is the conductance of the ith resistor.

4.28 Find the current through each element of the circuit shown in Figure E4.28. Calculate (a) the power dissipated by each resistor, (b) the power supplied by each source, and (c) the potential difference between a and b.

Figure E4.28

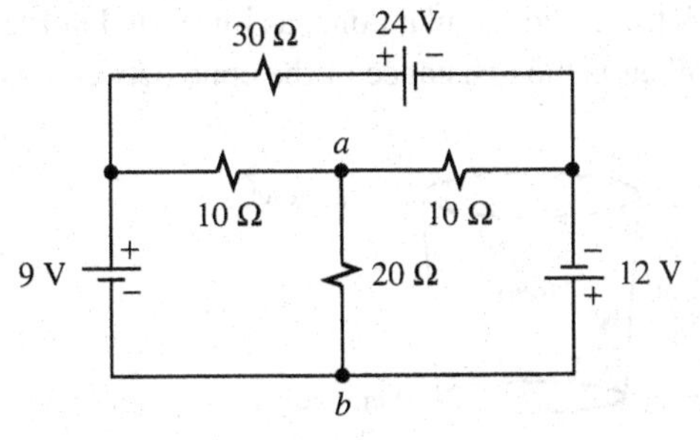

4.14 Problems

4.1 There are approximately 8.5×10^{28} free electrons per cubic meter in copper. If a copper wire having a cross-sectional area of 10 cm^2 carries a current of 1500 A, calculate (a) the average drift velocity of the electrons, (b) the current density, (c) the electric field intensity, and (d) the electron mobility.

4.2 A potential difference of 100 V is maintained across the two ends of a nickel wire of radius 2 mm and length 10 m. Calculate (a) the electric field intensity, (b) the current density, and (c) the electric current through the wire.

4.3 The average velocity of the electrons in a vacuum under the influence of an electric field is 3×10^5 m/s. If the current density is 10 A/cm^2, determine the number of electrons per unit area passing through a hypothetical plane perpendicular to the flow of electrons.

4.4 In an electrolyte, positive and negative ions of the same mass are responsible for a current density of 0.2 nA/m^2. If the mean charge density of either ion has a magnitude of $25e$ per cubic meter, where e is the charge on an electron, what is the mean ion velocity?

4.5 The space between two parallel conducting plates is filled with pure silicon ($\epsilon_r \approx 1$). If the leakage current through the capacitor is 100 A, determine the charge on each plate.

4.6 A 30-km-long wire has a diameter of 2.58 mm. What is the resistance of the wire if it is made of (a) copper, (b) aluminum, (c) silver, (d) Nichrome?

4.7 A section of copper ring is shown in Figure P4.7. Find the resistance between the boundaries A and B.

Figure P4.7

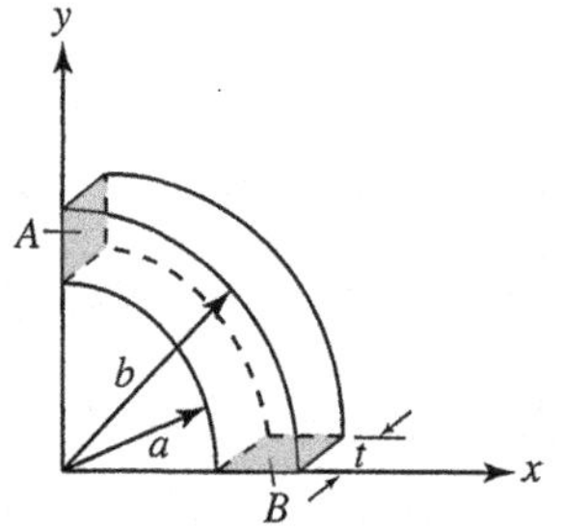

4.8 A truncated circular cone, as illustrated in Figure P4.8, is made of iron. What is the resistance of the cone between its flat ends?

Figure P4.8

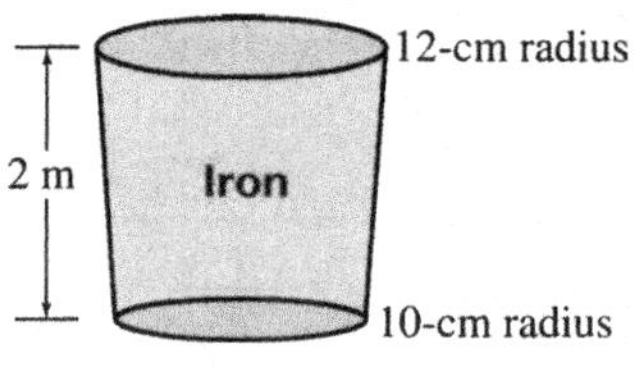

4.9 A 10-km-long solid cable is made of two contacting materials. The inner material is copper and has a radius of 2 cm. The outer material is constantan and has an outer radius of 3 cm. If the cable carries a current of 100 A, determine (a) the resistance of each material, (b) the current in each material, (c) the current density in each material, and (d) the electric field intensity in each material.

4.10 A copper wire of area 1 cm^2 carries a uniformly distributed current of 200 A. If the electron density in copper is 8.5×10^{28} electrons per cubic meter, determine the average drift velocity of the electrons. What is the electric field intensity in the wire? If the wire is 100 km long, calculate the potential difference between its ends. What is the resistance of the wire?

4.11 The potential difference between the flat ends of the truncated circular cone shown in Figure P4.8 is 2 mV. Calculate (a) the potential distribution within the cone, (b) the electric field intensity, (c) the volume current density, and (d) the current through the cone.

4.12 The space between two conducting concentric spheres of radii a and $b(b > a)$ is filled with inhomogeneous material with conductivity $\sigma = m/r + k$, where $a \leq r \leq b$, and m and k are constants. The inner sphere is held at a potential of V_0 volts, and the outer sphere is grounded. Compute (a) the resistance of the medium, (b) the surface charge density on each sphere, (c) the volume charge density in the medium, (d) the total charge on each sphere, (e) the current density in the region, and (f) the current through the region. What is the resistance when $m \to 0$?

4.13 An electromagnet is constructed of 200 turns of copper wire around a cylindrical core. The diameter of the copper wire is 0.45 mm, and the average radius of a turn is 8 mm. What is the resistance of the wire?

4.14 A wire of length 10 m and radius 0.5 mm carries a current of 2 A when a potential difference of 12 V is applied between its ends. What is the resistance of the wire? What is its conductivity?

4.15 How long a piece of carbon rod is needed to make a resistance of 10 Ω if its radius is 0.25 mm?

4.16 A solid conductor of radius 2 mm and length 10 m is to be replaced with a hollow conductor of inner radius 2 mm. If the two conductors are made of the same material and have the same length and resistance, what must be the outer radius of the hollow conductor?

4.17 A high-voltage line has a diameter of 4 cm and a length of 200 km. The cable carries a current of 1.2 kA. If the resistance of the cable is 4.5 Ω, determine (a) the potential drop between its ends, (b) the electric field intensity, (c) the current density, and (d) the resistivity of the material. Can you identify the material?

4.18 An excess charge of 500 trillion electrons is dumped at $t = 0$ within an aluminum sphere of radius 10 cm. What is the amount of excess charge? Assuming the dielectric constant of aluminum is unity, what is the relaxation time? How long does it take for the charge to decrease to 80% of its initial value?

4.19 The excess charge in a conducting medium decays to one-half of its initial amount in 100 ns. If the dielectric constant of the medium is 2.5, what is its conductivity? What is the relaxation time? What fraction of the charge will still be present in the medium after 200 ns?

4.20 A large quantity of charge is placed within an isolated conducting region. The current through a closed surface bounding the charge is observed to be $i(t) = 0.2e^{-50t}$ A. Determine (a) the relaxation time, (b) the initial charge, (c) the total charge transported through the surface in $t = 2\tau$, and (d) the time when the current will be 10% of its initial value.

4.21 Find the time rate of change of volume charge density in a medium where the volume current density is $\vec{\mathbf{J}} = e^{-x} \sin \omega x \vec{\mathbf{a}}_x$ A/m^2.

4.22 If ρ_v is the volume charge density of charges in motion and $\vec{\mathbf{U}}$ is their average velocity, show that $\rho_v \nabla \cdot \vec{\mathbf{U}} + (\vec{\mathbf{U}} \cdot \nabla)\rho_v + \partial\rho/\partial t = 0$.

4.23 Find the power dissipated by the wire in Problem 4.2 using Equation (4.25). Verify your answer using (4.27).

4.24 Obtain the power lost as heat in Problem 4.14.

4.25 What is the power dissipated by the carbon resistor in Problem 4.15 when a 12-V potential difference is applied between its ends?

4.26 The medium between the conductors of a coaxial cable has a dielectric constant of 2 and a conductivity of 6.25 μS/m. The radii of the inner and outer conductors of the cable are 8 mm and 10 mm, respectively. What is the resistance per unit length of the cable? If a potential difference of 230 V is applied between the conductors, and the cable is 100 m in length, compute the total power supplied to the cable.

4.27 The belt of an electrostatic generator is 30 cm wide and travels at 20 m/s. The belt carries the charge at a rate corresponding to 50 μA. Compute the surface charge density on the belt.

4.28 The region between two parallel metal plates, each of area 1 m^2, is filled with three conducting media of thicknesses 0.5 mm, 0.2 mm, and 0.3 mm, and of conductivities 10 kS/m, 500 S/m, and 0.2 MS/m, respectively. What is the effective resistance between the two plates? If a potential difference of 10 mV is maintained between the plates, calculate the $\vec{\mathbf{J}}$ and $\vec{\mathbf{E}}$ fields in each region. How much power is dissipated in each medium? What is the total power dissipation?

4.29 The anode of a diode is at a potential of 10 kV with respect to the cathode. The separation between the plates is 10 cm. If the initial velocity of an electron at the cathode is zero, what is its final velocity when it hits the anode? Determine the voltage distribution, the electric field intensity, and the current density in the diode.

4.30 If each plate in Problem 4.29 is 4 cm square, calculate the current in the diode.

4.31 Using boundary conditions, determine the surface charge densities at the two interfaces between the three conducting media in Problem 4.28 if the dielectric constant of each region is unity.

4.32 The current density in medium $1 (\sigma_1 = 100$ S/m, $\epsilon_{r1} = 9.6)$ is 50 A/m^2. It makes an angle of 30° with respect to the normal at the interface. If medium 2 has a conductivity of 10 S/m and a dielectric constant of 4, what is the current density in medium 2? What angle does it make with the normal? What is the surface charge density at the interface?

4.33 A plane interface between two conducting media is shown in Figure P4.33. If the current density in medium 1 just above the interface ($\sigma_1 =$ 100 S/m, $\epsilon_{r1} = 2$) is $\vec{\mathbf{J}}_1 = 20\vec{\mathbf{a}}_x + 30\vec{\mathbf{a}}_y - 10\vec{\mathbf{a}}_z$ A/m^2, what is the current density in medium 2 just below the interface ($\sigma_2 = 1000$ S/m, $\epsilon_{r2} = 9$)? What are the corresponding components of the $\vec{\mathbf{E}}$ and $\vec{\mathbf{D}}$ fields on both sides? What is the surface charge density at the interface?

Figure P4.33

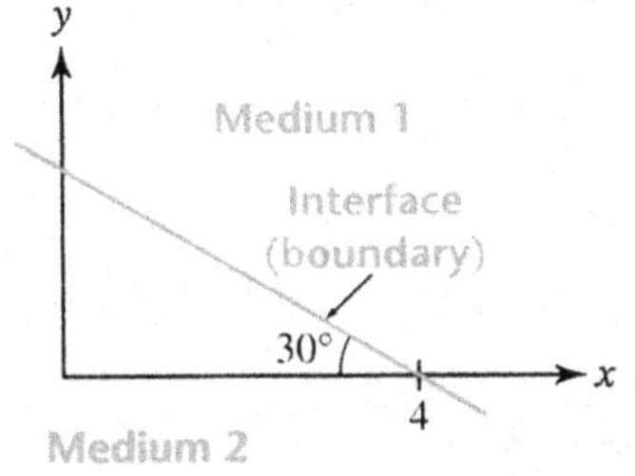

4.34 The radius of the inner conductor of a coaxial cable is 10 cm, and the radius of the outer conductor is 40 cm. There are two media. The inner one, extending from 10 cm to 20 cm, has a conductivity of $\sigma_1 = 50\,\mu$S/m and a permittivity of $\epsilon_1 = 2\epsilon_0$; the outer one, extending from 20 cm to 40 cm, has a conductivity of $\sigma_2 = 100\,\mu$S/m and a permittivity of $\epsilon_2 = 4\epsilon_0$. Using the analogy between $\vec{\mathbf{D}}$ and $\vec{\mathbf{J}}$, determine on a per-unit-length basis (a) the capacitance of each region, (b) the resistance of each region, (c) the total capacitance, and (d) the total resistance.

4.35 Using boundary conditions, determine the surface charge density at the interface in Problem 4.34 when the length is 100 m and the potential difference between the conductors is 10 V.

4.36 Draw the equivalent circuit for Problem 4.34. A voltage of V_0 is applied to charge the capacitor. If the voltage is removed at $t = 0$, determine the time when the voltage across the capacitor will be half of its initial value. For all practical purpose, how long will it take for the capacitor to discharge completely?

4.37 The radius of an inner spherical conductor is 3 cm, and the radius of the outer spherical conductor is 9 cm. There are two media: the inner one, extending from 3 cm to 6 cm, has a conductivity of 50 μS/m and a permittivity of $3\epsilon_0$; the outer one, extending from 6 cm to 9 cm, has a conductivity of 100 μS/m and a permittivity of $4\epsilon_0$. Using the analogy between $\vec{\mathbf{D}}$ and $\vec{\mathbf{J}}$, determine on a per-unit-length basis (a) the capacitance of each region, (b) the resistance of each region, (c) the total capacitance, and (d) the total resistance.

4.38 Using boundary conditions, determine the surface charge density at the interface in Problem 4.37 when the potential difference between the conductors is 50 V.

4.39 Draw the equivalent circuit for Problem 4.37. A voltage of 50 V is applied to charge the capacitor. If the voltage is removed at $t = 0$, determine the time when the voltage across the capacitor will be half of its initial value. For all practical purposes, how long will it take for the capacitor to discharge completely?

4.40 What is the equivalent resistance of the circuit shown in Figure P4.40?

Figure P4.40

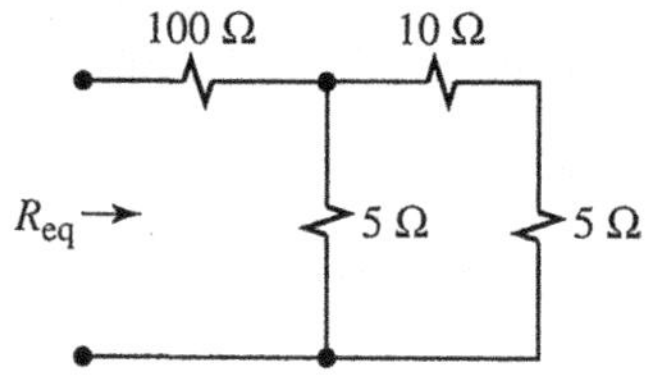

4.41 Determine V_{ab} in the circuit of Figure P4.41.

Figure P4.41

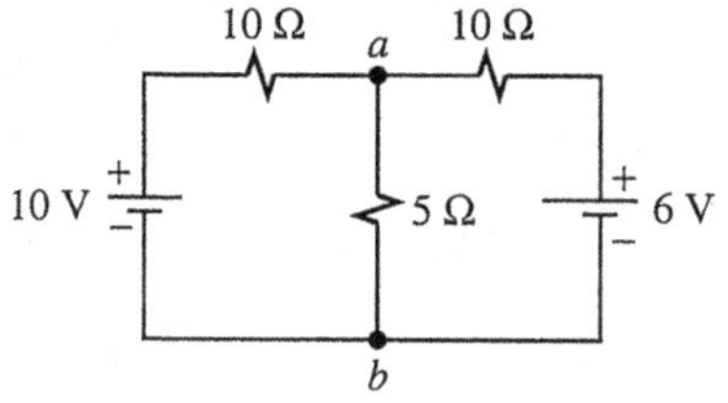

4.42 What is the power supplied by each source in Figure P4.42? Is the power supplied equal to the power dissipated in the circuit?

Figure P4.42

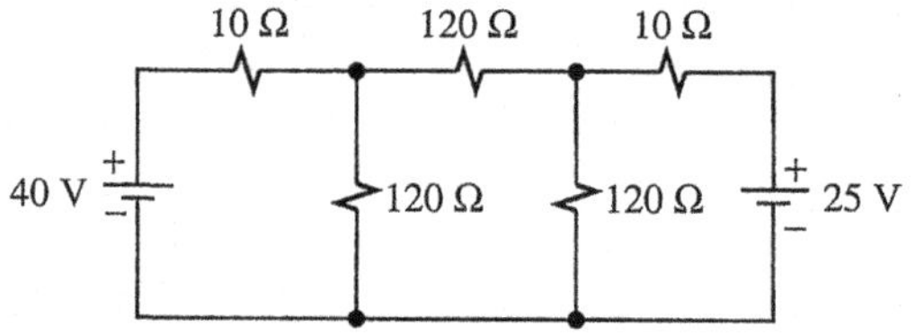

5

Magnetostatics

5.1 Introduction

The discovery of a permanently magnetized iron ore, lodestone, generated interest among scientists in an area of study called *magnetism.* Lodestone's ability to orient itself in the north and south directions led to the postulation of the existence of another force, which is now referred to as the *magnetic force*. A material that can be influenced (magnetized) by the magnetic force is called a *magnetic material*. Included in the family of magnetic materials are iron, cobalt, and nickel. A magnetized material is called a *magnet.* The end of a freely suspended magnet that points toward north is named as the north (seeking) pole; the other end is the south (seeking) pole. That a north pole of a magnet always points toward north had a profound influence on early navigation and exploration.

A *magnetic field* is associated with each magnet in the same way as an electric field is associated with a charge. *Magnetic lines of force* (outside the magnet) are said to emanate from the north pole and terminate at the south pole, as indicated in Figure 5.1. If another magnet is placed in the magnetic field, it will experience a force of attraction or repulsion. From experimental observations, it was found that *like poles repel and unlike poles attract.*

Experience with magnets has also shown that the two ends (north pole and south pole) cannot be separated, no matter in how many small pieces the magnet is divided. In other words, an isolated magnetic pole is not a physical reality.

Except for the fact that magnetic materials can be magnetized and used as magnets, magnetism was the least understood and exploited field of science until the early nineteenth century. A major breakthrough came in 1820 when Hans Christian Oersted discovered experimentally that a magnetic needle was deflected by a current in a wire. This event bridged the gap between the science of electricity and magnetism. Scientists immediately realized that electric currents are also sources of magnetic fields.

Within a short time after Oersted's discovery, Biot and Savart experimentally formulated an equation to determine the *magnetic flux density*

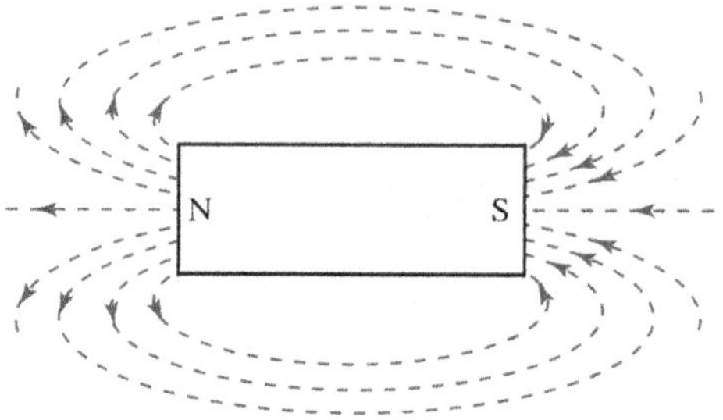

Figure 5.1 Magnetic lines of flux surrounding a bar magnet

at a point produced by a current-carrying conductor. We now view the *Biot–Savart law* as the magnetic equivalent of Coulomb's law. By 1825 André Marie Ampère had discovered the existence of magnetic force between current-carrying conductors and formulated a set of qualitative relationships based upon a series of experiments. These discoveries lead to the development of electric machines we use in our daily lives.

This chapter is devoted to the study of **magnetostatics**; i.e., the magnetic fields produced by steady currents. We begin our discussion with the Biot-Savart law and use it as a basic tool to calculate the magnetic field set up by any given distribution of currents.

5.2 The Biot–Savart law

It has been found experimentally that the *magnetic flux density* produced at a point P from an element of length $\vec{d\ell}$ of a filamentary wire carrying a steady current I, as shown in Figure 5.2, is

$$\vec{d\mathbf{B}} = k\frac{I\vec{d\ell} \times \vec{\mathbf{a}}_R}{R^2}$$

In this equation, $\vec{d\mathbf{B}}$ is the elemental magnetic flux density in teslas (T), where one tesla is equal to one weber per square meter (Wb/m^2), $\vec{d\ell}$ is an element of length in the direction of the current, $\vec{\mathbf{a}}_R$ is the unit vector pointing from $\vec{d\ell}$ to point P, the point P is at a distance R from the current element $\vec{d\ell}$, and k is the constant of proportionality.

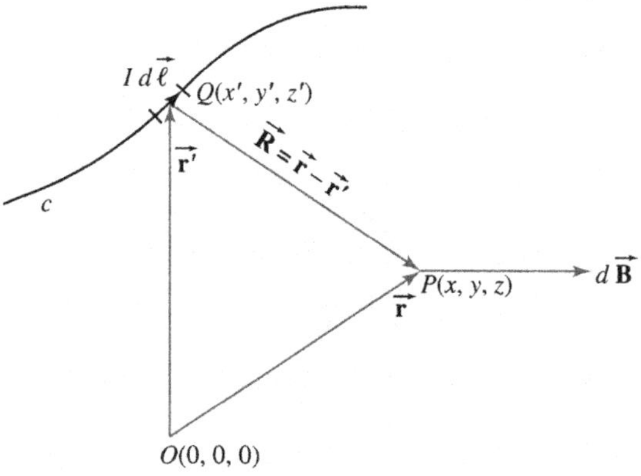

Figure 5.2 Magnetic flux density at point P produced by current element at Q

The distance vector $\vec{\mathbf{R}}$ from $Q(x', y', z')$ to point $P(x, y, z)$ is

$$\vec{\mathbf{R}} = \vec{\mathbf{r}} - \vec{\mathbf{r}}'$$

In the SI system of units, the constant k is expressed as

$$k = \frac{\mu_0}{4\pi}$$

where $\mu_0 = 4\pi 10^{-7}$ H/m is the free space permeability. Substituting for k, we can express $d\vec{\mathbf{B}}$ as

$$d\vec{\mathbf{B}} = \frac{\mu_0 I d\vec{\ell} \times \vec{\mathbf{R}}}{4\pi R^3} \tag{5.1}$$

Integrating (5.1), we obtain

$$\vec{\mathbf{B}} = \frac{\mu_0}{4\pi} \int_c \frac{I d\vec{\ell} \times \vec{\mathbf{R}}}{R^3} \tag{5.2}$$

where $\vec{\mathbf{B}}$ is the magnetic flux density at point $P(x,y,z)$ due to a filamentary wire carrying steady current I. Note that the direction of $\vec{\mathbf{B}}$ is perpendicular to the plane containing $d\vec{\ell}$ and $\vec{\mathbf{R}}$.

The integrand in (5.2) involves six variables: x, y, z, x', y', and z'. The unprimed variables x, y, and z are the coordinates of point P and are not involved in the integration process. However, the primed variables (also known as the dummy variables) x', y', and z' are the coordinates of point Q and are involved in the integration process. The integration process eliminates the primed variables. Thus, $\vec{\mathbf{B}}$ is a function of unprimed variables only. Point $P(x, y, z)$ is routinely referred to as the *field point* and point $Q(x', y', z')$ as the *source point.* We could have written $d\vec{\ell}$ as $d\vec{\ell}\,'$ in (5.2) but we chose not to do so to avoid using an extra superscript. We will institute the use of primed coordinates whenever it is necessary to distinguish between the primed and the unprimed variables.

As explained in Chapter 4, we can express the current element $I d\vec{\ell}$ in terms of the volume current density $\vec{\mathbf{J}}_v$ as

$$I d\vec{\ell} = \vec{\mathbf{J}}_v \, dv$$

and obtain an expression for $\vec{\mathbf{B}}$ in terms of $\vec{\mathbf{J}}_v$ (see Figure 5.3) as

$$\vec{\mathbf{B}} = \frac{\mu_0}{4\pi} \int_v \frac{\vec{\mathbf{J}}_v \times \vec{\mathbf{R}}}{R^3} dv \tag{5.3}$$

We can also obtain a similar expression in terms of the surface current density $\vec{\mathbf{J}}_s$(A/m); i.e., when the current flows over the surface of a conductor as illustrated in Figure 5.4, as

$$\vec{\mathbf{B}} = \frac{\mu_0}{4\pi} \int_s \frac{\vec{\mathbf{J}}_s \times \vec{\mathbf{R}}}{R^3} ds \tag{5.4}$$

Because current is simply a flow of charge, we can also express (5.2) in terms of a charge q moving with an average velocity $\vec{\mathbf{U}}$. If ρ_v is the volume charge density, A is the cross section of the wire, and $d\ell$ is the element of length of the wire, then $dq = \rho_v A\, d\ell$ and $\vec{\mathbf{J}}_v\, dv = dq\vec{\mathbf{U}}$. Thus, from (5.3), we have

$$\vec{\mathbf{B}} = \frac{\mu_0}{4\pi}\left[\frac{q\vec{\mathbf{U}}\times\vec{\mathbf{R}}}{R^3}\right] \tag{5.5}$$

This equation gives the magnetic flux density produced by a charge q moving with an average velocity $\vec{\mathbf{U}}$ at a distance $\vec{\mathbf{R}}$ away from the charge.

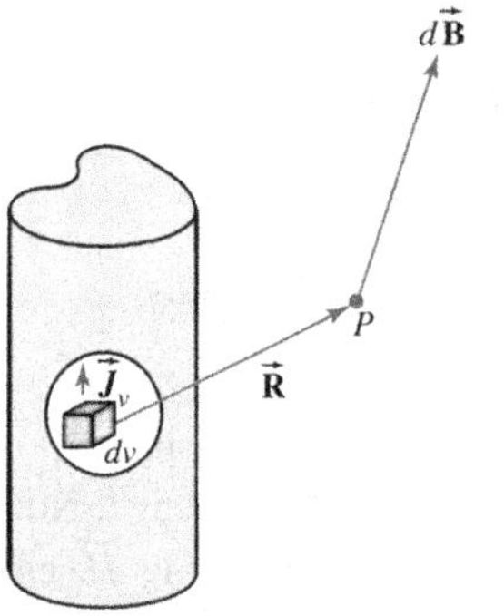

Figure 5.3 Magnetic flux density at point P due to a volume current density

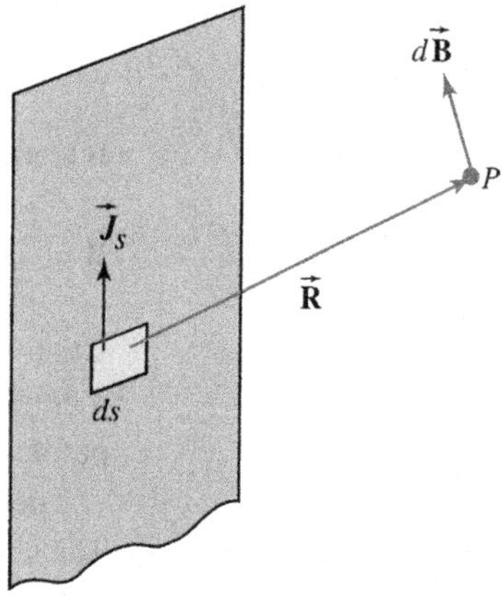

Figure 5.4 Magnetic flux density at a point P due to surface current distribution

The following examples illustrate the applications of the Biot–Savart law to determine the magnetic flux density at a point due to current-carrying conductors.

EXAMPLE 5.1

A filamentary wire of finite length extends from $z = a$ to $z = b$, as shown in Figure 5.5a. Determine the magnetic flux density at a point P in the xy plane. What is the magnetic flux density at P if $a \to -\infty$ and $b \to \infty$?

Solution As $I\vec{d\ell} = I\, dz\vec{\mathbf{a}}_z$ and $\vec{\mathbf{R}} = \rho\vec{\mathbf{a}}_\rho - z\vec{\mathbf{a}}_z$, so

$$I\vec{d\ell}\times\vec{\mathbf{R}} = I\rho\, dz\vec{\mathbf{a}}_\phi$$

Substituting in (5.2), we have

$$\vec{\mathbf{B}} = \frac{\mu_0 I\rho}{4\pi}\int_a^b \frac{dz}{[\rho^2+z^2]^{3/2}}\,\vec{\mathbf{a}}_\phi$$

$$= \frac{\mu_0 I}{4\pi\rho}\left[\frac{b}{\sqrt{\rho^2+b^2}} - \frac{a}{\sqrt{\rho^2+a^2}}\right]\vec{\mathbf{a}}_\phi$$

This result shows that $\vec{\mathbf{B}}$ has a nonzero component only in the $\vec{\mathbf{a}}_\phi$ direction. This is expected because the current is in the z direction, and $\vec{\mathbf{B}}$ must be normal to it.

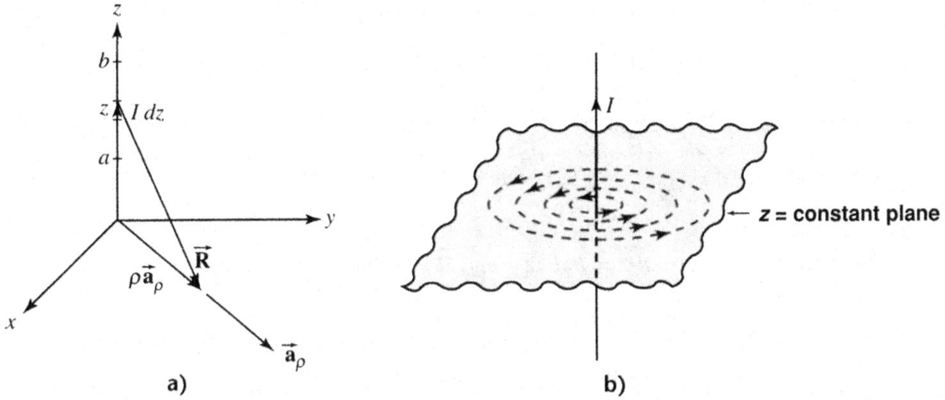

Figure 5.5 (a) Magnetic flux density produced by a finite current-carrying conductor (Example 5.1); (b) Magnetic flux lines are concentric circles in a plane perpendicular to an infinitely long current-carrying conductor

By setting $a = -\infty$ and $b = \infty$ in the preceding expression, we obtain the $\vec{\mathbf{B}}$ field produced at a point by a wire of infinite extent as

$$\vec{\mathbf{B}} = \frac{\mu_0 I}{2\pi\rho}\vec{\mathbf{a}}_\phi \tag{5.6}$$

• • •

As you can see from equation (5.6), the magnetic flux density varies inversely as a function of ρ. In a plane perpendicular to the wire, the magnetic flux lines are circles surrounding the wire, as illustrated in Figure 5.5b. As a memory aid, we can say that if we grasp the current-carrying conductor in the right hand and extend the thumb in the direction of the current, the lines of magnetic flux are in the direction of the curled fingers.

It is also interesting to note that the magnitude of the $\vec{\mathbf{B}}$ field varies with ρ in the same way as the magnitude of the $\vec{\mathbf{E}}$ field varies due to a long line of uniform charge density. This shows that under certain circumstances there may be a correspondence between the electrostatic and magnetostatic fields, even though the directions of the fields are different. We will have more to say about this correspondence in later sections.

EXAMPLE 5.2

A circular loop of radius b is in the xy plane and carries a current I, as depicted in Figure 5.6. Obtain an expression for the magnetic flux density at a point on the positive z axis. What is the approximate expression for the magnetic flux density at a point far away from the loop?

Figure 5.6 Magnetic flux density at a point P on the z axis produced by a current-carrying loop

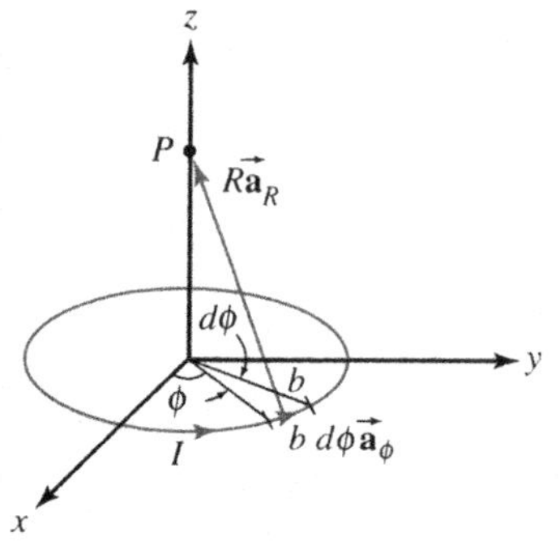

Solution Because $\vec{d\ell} = b\,d\phi\vec{\mathbf{a}}_\phi$ and $\vec{\mathbf{R}} = -b\vec{\mathbf{a}}_\rho + z\vec{\mathbf{a}}_z$, then

$$\vec{d\ell} \times \vec{R} = (b^2\vec{\mathbf{a}}_z + bz\vec{\mathbf{a}}_\rho)\,d\phi$$

From (5.2), the magnetic flux density is

$$\begin{aligned}\vec{\mathbf{B}} &= \frac{\mu_0 I b^2}{4\pi}\int_0^{2\pi}\frac{\vec{\mathbf{a}}_z\,d\phi}{(b^2+z^2)^{3/2}} + \frac{\mu_0 I bz}{4\pi}\int_0^{2\pi}\frac{\vec{\mathbf{a}}_\rho\,d\phi}{(b^2+z^2)^{3/2}}\\ &= \frac{\mu_0 I b^2}{2(b^2+z^2)^{3/2}}\vec{\mathbf{a}}_z \end{aligned} \tag{5.7}$$

Thus, on the axis of a current-carrying loop, the magnetic flux density has only a z-directed component. By setting $z = 0$, we obtain the magnetic flux density at the center of the loop as

$$\vec{\mathbf{B}} = \frac{\mu_0 I}{2b}\vec{\mathbf{a}}_z \tag{5.8}$$

When the point of observation is far away from the loop, we can approximate the term in the denominator of (5.7) as

$$(b^2 + z^2)^{3/2} \approx z^3$$

and obtain the expression for the magnetic flux density as

$$\vec{\mathbf{B}} = \frac{\mu_0 I b^2}{2z^3}\vec{\mathbf{a}}_z \tag{5.9}$$

• • •

When the point of observation is far away from the loop, the size of the loop is very small in comparison with the distance z. In this case, we refer to the current-carrying loop as a *magnetic dipole*. If we define the *magnetic dipole moment* as

$$\vec{\mathbf{m}} = I\pi b^2\vec{\mathbf{a}}_z = IA\vec{\mathbf{a}}_z \tag{5.10}$$

where A is the area of the loop, then the $\vec{\mathbf{B}}$ field, from (5.9), is

$$\vec{\mathbf{B}} = \frac{\mu_0\vec{\mathbf{m}}}{2\pi z^3}$$

In Example 5.2, we chose the point P on the z axis to determine the $\vec{\mathbf{B}}$ field. The calculation of the $\vec{\mathbf{B}}$ field at any arbitrary point in space is quite involved. However, a general sketch of magnetic lines of flux due to a current-carrying loop is given in Figure 5.7. From the direction of the lines of flux, it is evident that the region above the loop acts like a north pole. Likewise, the region under the loop behaves like a south pole. Therefore, a current-carrying coil forms what is commonly termed an *electromagnet*.

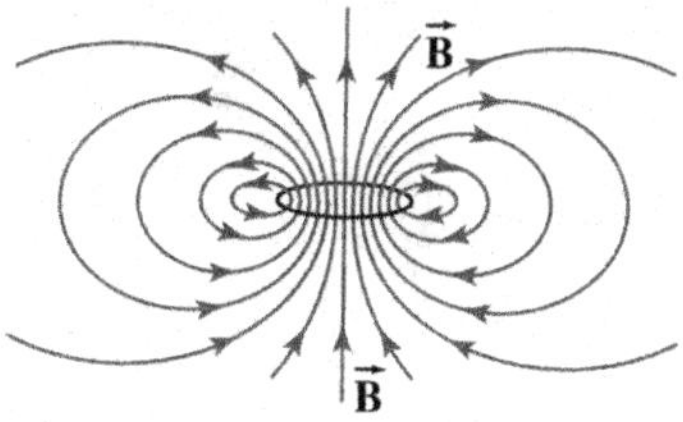

Figure 5.7 Magnetic lines of flux associated with a current-carrying loop

5.3 Ampère's force law

Most of the experiments conducted by Ampère were related to the determination of the force that one current-carrying conductor experiences in the presence of another current-carrying conductor. From his experiments, Ampère was able to demonstrate that when two current-carrying elements $I_1\overrightarrow{d\ell}_1$ and $I_2\overrightarrow{d\ell}_2$ interact, the elemental magnetic force exerted by element 1 upon element 2 is

$$\overrightarrow{d\mathbf{F}}_2 = \frac{\mu_0 I_2\overrightarrow{d\ell}_2}{4\pi} \times \left[\frac{I_1\overrightarrow{d\ell}_1 \times \vec{\mathbf{R}}_{21}}{R_{21}^3}\right]$$

where $\vec{\mathbf{R}}_{21}$ is the distance vector from $I_1\overrightarrow{d\ell}_1$ to $I_2\overrightarrow{d\ell}_2$ as depicted in Figure 5.8. If each current-carrying element is a part of a current-carrying conductor, as illustrated in Figure 5.9, the magnetic force exerted by current-carrying conductor 1 upon current-carrying conductor 2 is

$$\vec{\mathbf{F}}_2 = \frac{\mu_0}{4\pi}\int_{c_2} I_2\overrightarrow{d\ell}_2 \times \int_{c_1} \frac{I_1\overrightarrow{d\ell}_1 \times \vec{\mathbf{R}}_{21}}{R_{21}^3} \tag{5.11a}$$

This equation is referred to as *Ampère's force law*.

Using (5.2), we can write equation (5.11a) as

$$\vec{\mathbf{F}}_2 = \int_{c_2} I_2\overrightarrow{d\ell}_2 \times \vec{\mathbf{B}}_1 \tag{5.11b}$$

where $\vec{\mathbf{B}}_1$, the magnetic flux density produced by a current-carrying conductor 1 at the location of current-carrying element $I_2\,\overrightarrow{d\ell}_2$, is given as

$$\vec{\mathbf{B}}_1 = \frac{\mu_0}{4\pi}\int_{c_1}\frac{I_1\vec{d\ell}_1 \times \vec{\mathbf{R}}_{21}}{R_{21}^3} \tag{5.11c}$$

In general, when a current-carrying conductor is placed in an external magnetic field $\vec{\mathbf{B}}$, the magnetic force experienced by the conductor is

$$\vec{\mathbf{F}} = \int_c I\vec{d\ell} \times \vec{\mathbf{B}} \tag{5.12a}$$

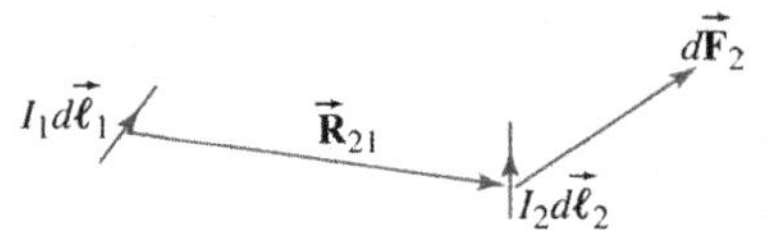

Figure 5.8 Elemental magnetic force on current element 2 due to current element 1

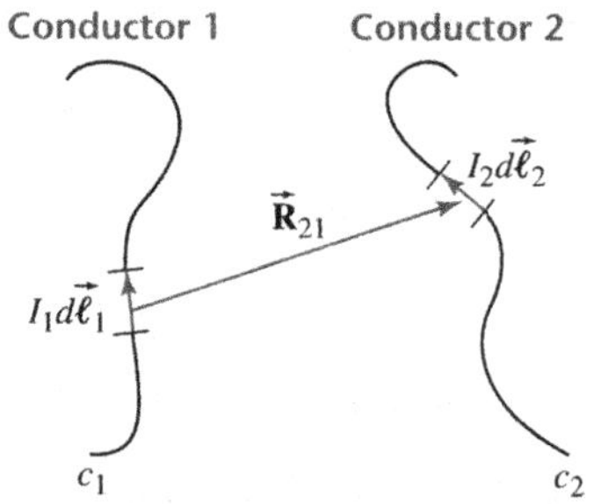

Figure 5.9 Magnetic force on conductor 2 exerted by conductor 1

In terms of the volume current density, we can express equation (5.12a) as

$$\vec{\mathbf{F}} = \int_v \vec{\mathbf{J}}_v \times \vec{\mathbf{B}}\,dv \tag{5.12b}$$

Equation (5.12b) can be treated as a general form of Ampère's force law. By replacing $\vec{\mathbf{J}}_v\,dv$ with $\vec{\mathbf{J}}_s\,ds$ we can obtain an expression for the magnetic force experienced by a surface current distribution in an external magnetic field.

If ρ_{v1} is the volume charge density, $\vec{\mathbf{U}}_1$ is the average velocity of the charge, and A_1 is the cross section of current-carrying conductor 1, then $dq_1 = \rho_{v1}A\,d\ell_1$ and $\vec{\mathbf{J}}_{v1}\,dv_1 = dq_1\,\vec{\mathbf{U}}_1$. If $\vec{\mathbf{B}}$ is the magnetic flux density in the region, then the magnetic force experienced by the charge q_1 is

$$\vec{\mathbf{F}}_1 = q_1\vec{\mathbf{U}}_1 \times \vec{\mathbf{B}} \tag{5.13}$$

If $\vec{\mathbf{B}}$ is also created by the motion of charges, then from (5.5), the magnetic force experienced by charge q_1 in a magnetic field produced by q_2 moving with an average velocity $\vec{\mathbf{U}}_2$ is

$$\vec{\mathbf{F}}_1 = \frac{\mu_0}{4\pi R_{12}^3}[q_1\vec{\mathbf{U}}_1 \times q_2\vec{\mathbf{U}}_2 \times \vec{\mathbf{R}}_{12}] \tag{5.14}$$

We could have postulated (5.14) as the fundamental law of magnetic force and then used it to obtain expressions for Ampère's force law and the Biot–Savart law. We mention that, like the electric force and the gravitational force, the magnetic force between two charges in motion varies as the inverse square of the distance.

EXAMPLE 5.3

A wire bent as shown in Figure 5.10 lies in the xy plane and carries a current I. If the magnetic flux density in the region is $\vec{\mathbf{B}} = B\vec{\mathbf{a}}_z$, determine the magnetic force acting on the wire.

Figure 5.10

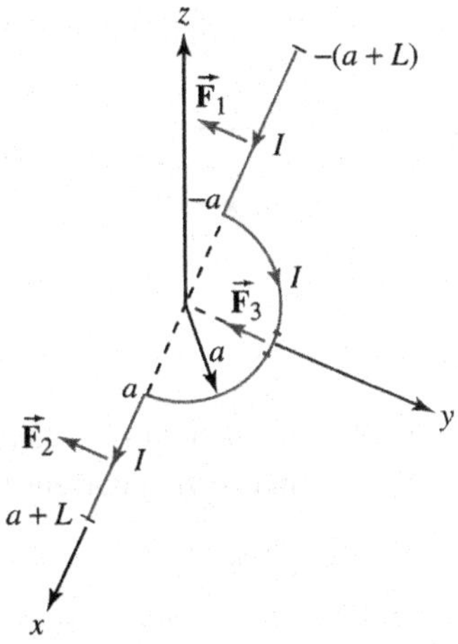

Solution The magnetic force acting on the section of the wire from $x = -(a + L)$ to $x = -a$, from (5.12a), is

$$\vec{\mathbf{F}}_1 = \int_{-(a+L)}^{-a} IB(\vec{\mathbf{a}}_x \times \vec{\mathbf{a}}_z)\, dx = -\,BIL\vec{\mathbf{a}}_y$$

Similarly, the magnetic force experienced by the section of the wire from $x = a$ to $x = a + L$ is

$$\vec{\mathbf{F}}_2 = -BIL\vec{\mathbf{a}}_y$$

The magnetic force acting on the semicircular arc of radius a is

$$\vec{\mathbf{F}}_3 = \int_{\pi}^{0} IB(-\vec{\mathbf{a}}_\phi \times \vec{\mathbf{a}}_z)a\, d\phi = -\int_{\pi}^{0} \vec{\mathbf{a}}_\rho BIa\, d\phi$$

$$= BIa \int_{0}^{\pi} [\vec{\mathbf{a}}_x \cos\phi + \vec{\mathbf{a}}_y \sin\phi]\, d\phi = -2IBa\vec{\mathbf{a}}_y$$

The resultant magnetic force on the whole wire is

$$\vec{\mathbf{F}} = \vec{\mathbf{F}}_1 + \vec{\mathbf{F}}_2 + \vec{\mathbf{F}}_3 = -2IB(a + L)\vec{\mathbf{a}}_y$$

It is interesting to note that the total magnetic force acting on the bent wire is the same as that acting on a straight wire of length $2(L + a)$.

• • •

EXAMPLE 5.4

Figure 5.11 shows a current-carrying conductor of finite length L placed at a distance b from another current-carrying conductor of infinite extent. Determine the magnetic force per unit length acting on the finite conductor.

Figure 5.11

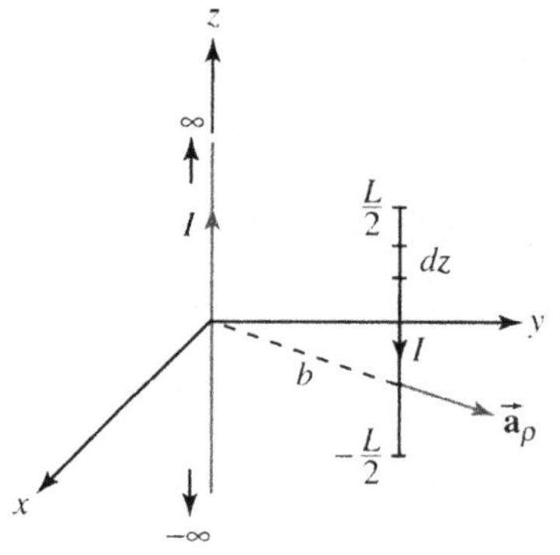

Solution From Example 5.1, the magnetic flux density produced at a distance b by a wire of infinite extent carrying current I is

$$\vec{\mathbf{B}} = \frac{\mu_0 I}{2\pi b}\,\vec{\mathbf{a}}_\phi$$

From (5.12), the magnetic force acting on the finite conductor is

$$\begin{aligned}\vec{\mathbf{F}} &= -\frac{\mu_0 I^2}{2\pi b}\int_{-L/2}^{L/2}(\vec{\mathbf{a}}_z \times \vec{\mathbf{a}}_\phi)\,dz \\ &= \frac{\mu_0}{2\pi b}\,I^2 L\,\vec{\mathbf{a}}_\rho\end{aligned}$$

Hence the magnetic force per unit length experienced by the finite current-carrying conductor is

$$\vec{\mathbf{F}}_{\text{per unit length}} = \frac{\vec{\mathbf{F}}}{L} = \frac{\mu_0}{2\pi b}I^2\vec{\mathbf{a}}_\rho \text{ N/m} \qquad (5.15)$$

• • •

Since the magnetic force $\vec{\mathbf{F}}$ is directed along $\vec{a}_\rho$ and points in the direction away from the infinitely long wire, it is repulsive in nature. If both wires were carrying currents in the same direction, the magnetic force between them would be attractive.

Equation (5.15) is, in fact, used to define the unit of current, the ampere. When two current-carrying parallel conductors, each of length one meter, and separated by a distance of one meter, experience a force of 2×10^{-7} newton, the current through each conductor is 1 ampere.

For the two current-carrying conductors, (5.11a) can also be written as

$$\vec{\mathbf{F}}_2 = \frac{\mu_0 I_1 I_2}{4\pi}\int_{c_2}\int_{c_1}\frac{1}{R_{21}^3}[\vec{d\ell}_2 \times \vec{d\ell}_1 \times \vec{\mathbf{R}}_{21}]$$

Using the vector identity

$$\vec{\mathbf{A}} \times (\vec{\mathbf{B}} \times \vec{\mathbf{C}}) = \vec{\mathbf{B}}(\vec{\mathbf{A}} \cdot \vec{\mathbf{C}}) - \vec{\mathbf{C}}(\vec{\mathbf{A}} \cdot \vec{\mathbf{B}})$$

we can write the previous equation as

$$\vec{\mathbf{F}}_2 = \frac{\mu_0 I_1 I_2}{4\pi}\left[\int_{c_2}\int_{c_1} \frac{\vec{d\ell}_2 \cdot \vec{\mathbf{R}}_{21}}{R_{21}^3}\vec{d\ell}_1 - \int_{c_2}\int_{c_1} \frac{\vec{d\ell}_1 \cdot \vec{d\ell}_2}{R_{21}^3}\vec{\mathbf{R}}_{21}\right]$$

Because $\vec{\mathbf{R}}_{21}/R_{21}^3 = -\nabla(1/R_{21})$, the first integral on the right-hand side can be written as

$$-\int_{c_2}\int_{c_1}\left[\nabla\left(\frac{1}{R_{21}}\right) \cdot \vec{d\ell}_2\right]\vec{d\ell}_1$$

If the conductor carrying current I_2 forms a closed loop, then we can apply Stokes' theorem and transform the line integral to a surface integral as

$$-\int_{c_1}\int_{s_2}\left[\nabla \times \nabla\left(\frac{1}{R_{21}}\right) \cdot \vec{d\mathbf{s}}_2\right]\vec{d\ell}_1$$

This integral is, however, zero because the curl of a gradient of a scalar function is always zero.

Thus, the magnetic force experienced by a *current-carrying loop* of arbitrary shape is

$$\vec{\mathbf{F}}_2 = -\frac{\mu_0 I_1 I_2}{4\pi}\int_{c_1}\oint_{c_2} \frac{\vec{d\ell}_1 \cdot \vec{d\ell}_2}{R_{21}^3}\vec{\mathbf{R}}_{21} \tag{5.16}$$

The following example applies equation (5.16) to determine the magnetic force experienced by a current-carrying loop.

EXAMPLE 5.5

A rectangular loop carrying current I_2 is placed close to a straight conductor carrying current I_1, as shown in Figure 5.12. Obtain an expression for the magnetic force experienced by the loop.

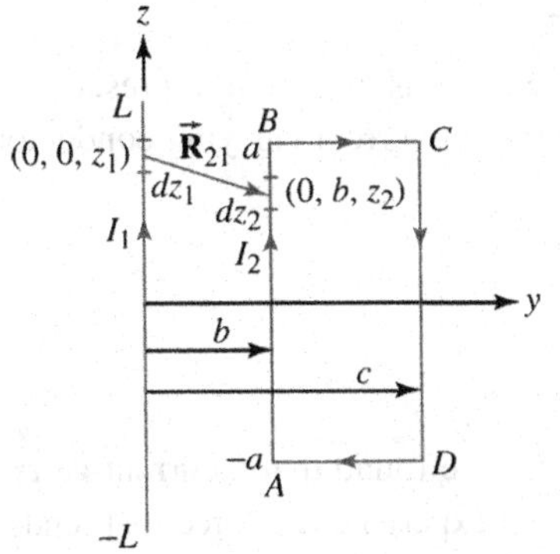

Figure 5.12 A current-carrying loop placed in a magnetic field created by a straight conductor of finite length

Solution The total magnetic force acting on the loop can be expressed as the sum of the forces acting on four sections of the loop, *AB*, *BC*, *CD*, and

DA. The differential length segment for section AB or CD of the loop is $\vec{d\ell}_2 = dz_2\vec{\mathbf{a}}_z$, and that for section BC or DA is $\vec{d\ell}_2 = dy_2\vec{\mathbf{a}}_y$. The differential length segment for the straight conductor is $\vec{d\ell}_1 = dz_1\vec{\mathbf{a}}_z$. Equation (5.16) involves the dot product of differential length segments $\vec{d\ell}_1$ and $\vec{d\ell}_2$, and for sections BC and DA of the loop, the dot product is zero. Therefore, only the two sections AB and CD contribute to the total magnetic force acting on the closed loop.

Let us first determine the magnetic force acting on segment AB of the loop. The distance vector is

$$\vec{\mathbf{R}}_{21} = b\vec{\mathbf{a}}_y + (z_2 - z_1)\vec{\mathbf{a}}_z$$

From (5.16), the magnetic force on section AB is

$$\begin{aligned}\vec{\mathbf{F}}_{AB} &= -\frac{\mu_0 I_1 I_2}{4\pi}\int_{-L}^{L} dz_1 \int_{-a}^{a} \frac{b\vec{\mathbf{a}}_y + (z_2 - z_1)\vec{\mathbf{a}}_z}{\left[b^2 + (z_2 - z_1)^2\right]^{3/2}}\,dz_2 \\ &= -\frac{\mu_0 I_1 I_2}{2\pi b}\left[\sqrt{(L+a)^2 + b^2} - \sqrt{(L-a)^2 + b^2}\right]\vec{\mathbf{a}}_y \end{aligned} \tag{5.17}$$

The quantity within the brackets is positive; thus, the negative sign indicates that the magnetic force acting on segment AB is attractive in nature.

Similarly, we can obtain the following expression for the magnetic force acting on segment CD of the loop:

$$\vec{\mathbf{F}}_{CD} = \frac{\mu_0 I_1 I_2}{2\pi c}\left[\sqrt{(L+a)^2 + c^2} - \sqrt{(L-a)^2 + c^2}\right]\vec{\mathbf{a}}_y \tag{5.18}$$

$\vec{\mathbf{F}}_{CD}$ is clearly a force of repulsion. Thus, the total magnetic force acting on the rectangular loop is

$$\begin{aligned}\vec{\mathbf{F}} = -\vec{\mathbf{a}}_y\frac{\mu_0 I_1 I_2}{2\pi}\Bigg[&\frac{1}{b}\left[\sqrt{(L+a)^2 + b^2} - \sqrt{(L-a)^2 + b^2}\right] \\ &- \frac{1}{c}\left[\sqrt{(L+a)^2 + c^2} - \sqrt{(L-a)^2 + c^2}\right]\Bigg]\end{aligned}$$

Because $c > b$, this equation represents a force of attraction between the straight current-carrying conductor and the current-carrying loop. • • •

5.4 Magnetic torque

In Section 5.3 we found that a current-carrying conductor placed in a magnetic field experiences a force that tends to move the conductor in a direction perpendicular to both the magnetic field and the conductor. However, if a current-carrying coil is placed in a magnetic field, the magnetic force acting on the coil may impart a rotation to the coil. This,

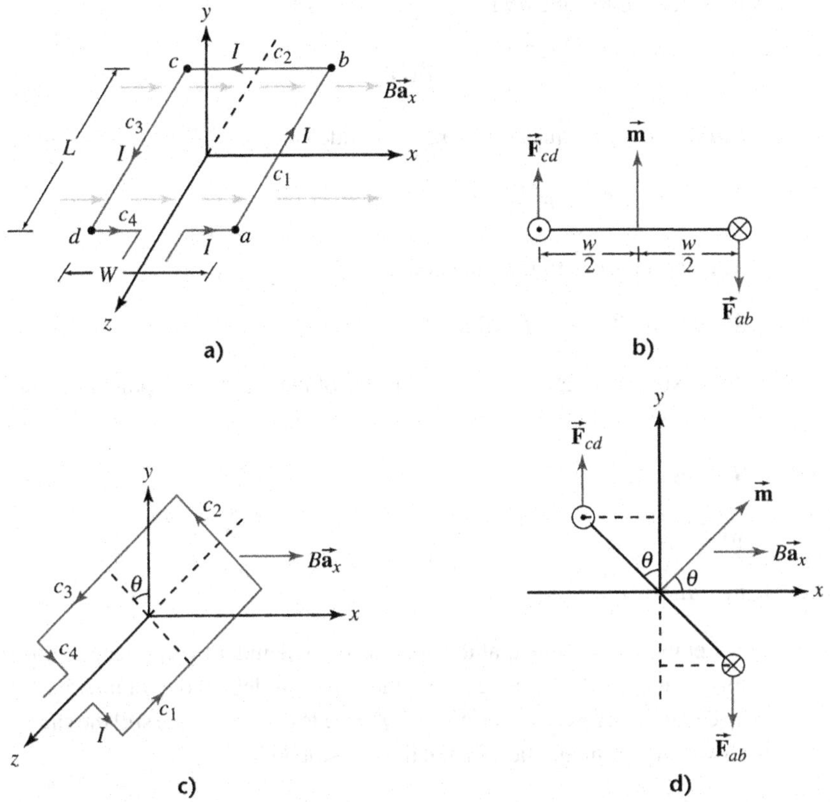

Figure 5.13 Torque exerted on a current-carrying loop by a magnetic field

in fact, is the underlying principle of operation of electric motors and the D' Arsonval type of electric meters.

A single-turn rectangular coil carrying current I is placed in a magnetic field $\vec{\mathbf{B}}$, as shown in Figure 5.13a. The plane of the coil is parallel to the magnetic field, and the coil is free to rotate about the z axis as shown. The cross-sectional area of the coil is LW. In accordance with Ampère's force law, there will be no force acting on sides bc and da. However, the force exerted by the $\vec{\mathbf{B}}$ field on side ab is

$$\vec{\mathbf{F}}_{ab} = -BIL\vec{\mathbf{a}}_y$$

and the force experienced by the side cd is

$$\vec{\mathbf{F}}_{cd} = BIL\vec{\mathbf{a}}_y$$

Figure 5.13b shows the forces acting on the sides of the coil. Because the lines of action of the two forces do not coincide, these forces exert a torque, which tends to rotate the coil about its axis. The moment arm of side ab is $W/2\vec{\mathbf{a}}_x$ and that for side cd is $-W/2\vec{\mathbf{a}}_x$. The torque experienced by side *ab* is

$$\vec{\mathbf{T}}_{ab} = \frac{W}{2}\vec{\mathbf{a}}_x \times \vec{\mathbf{F}}_{ab} = -\frac{1}{2}BILW\vec{\mathbf{a}}_z$$

Similarly, the torque experienced by side *cd* is

$$\vec{\mathbf{T}}_{cd} = -\frac{W}{2}\vec{\mathbf{a}}_x \times \vec{\mathbf{F}}_{cd} = -\frac{1}{2}BILW\vec{\mathbf{a}}_z$$

The net torque acting on the coil is

$$\vec{\mathbf{T}} = \vec{\mathbf{T}}_{ab} + \vec{\mathbf{T}}_{cd} = -BILW\vec{\mathbf{a}}_z$$

This expression can be written in terms of the magnetic dipole moment, as

$$\vec{\mathbf{T}} = \vec{\mathbf{m}} \times \vec{\mathbf{B}} \tag{5.19}$$

where

$$\vec{\mathbf{m}} = ILW\vec{\mathbf{a}}_y = IA\vec{\mathbf{a}}_y \tag{5.20}$$

Let us now assume that the coil has rotated under the influence of the torque and makes an angle θ with the y axis, as depicted in Figure 5.13c. The magnetic forces experienced by the sides *ab* and *cd* are still the same. However, the magnetic force acting on side *bc* is

$$\vec{\mathbf{F}}_{bc} = \int_{c_2} I(dx\vec{\mathbf{a}}_x + dy\vec{\mathbf{a}}_y) \times \vec{\mathbf{a}}_x B = -BIW\cos\theta\,\vec{\mathbf{a}}_z$$

Likewise, the magnetic force acting on side *da* is

$$\vec{\mathbf{F}}_{da} = BIW\cos\theta\,\vec{\mathbf{a}}_z$$

Since the line of action of the forces $\vec{\mathbf{F}}_{bc}$ and $\vec{\mathbf{F}}_{da}$ is the same, the resultant force in the z direction is zero. Thus, the only forces that contribute to the torque are $\vec{\mathbf{F}}_{ab}$ and $\vec{\mathbf{F}}_{cd}$ as shown in Figure 5.13d. The torques acting on sides *ab* and *cd* are

$$\vec{\mathbf{T}}_{ab} = \frac{W}{2}[\vec{\mathbf{a}}_x\sin\theta + \vec{\mathbf{a}}_y\cos\theta] \times (-\vec{\mathbf{a}}_y BIL) = -\frac{1}{2}BILW\sin\theta\,\vec{\mathbf{a}}_z$$

and

$$\vec{\mathbf{T}}_{cd} = \frac{W}{2}[-\vec{\mathbf{a}}_x\sin\theta + \vec{\mathbf{a}}_y\cos\theta] \times (\vec{\mathbf{a}}_y BIL) = -\frac{1}{2}BILW\sin\theta\,\vec{\mathbf{a}}_z$$

The resultant torque is

$$\vec{\mathbf{T}} = \vec{\mathbf{T}}_{ab} + \vec{\mathbf{T}}_{cd} = -BILW\sin\theta\,\vec{\mathbf{a}}_z = \vec{\mathbf{m}} \times \vec{\mathbf{B}} \tag{5.21}$$

As the torque experienced by the coil varies sinusoidally, the torque is maximum when the plane of the coil is parallel to the magnetic field.

By the same token, the torque is zero when the magnetic field is normal to the plane of the coil. If the plane of the coil is not normal to the magnetic field, the direction of the torque is such that it tends to rotate the coil until its plane becomes normal to the field. In other words, $\vec{\mathbf{m}}$ tends to align with $\vec{\mathbf{B}}$. Once the plane of the coil is perpendicular to the magnetic field, the coil is said to be locked in that position, as no further rotation can be imparted. We point out that (5.21), although developed for a rectangular coil, is valid for a coil of any arbitrary shape.

If the coil has N closely wound turns, the net torque exerted by the magnetic field on the coil will be N times as much as that experienced by a single-turn coil.

EXAMPLE 5.6

A circular coil of 200 turns has a mean area of 10 cm^2, and the plane of the coil makes an angle of 30° with the uniform magnetic flux density of 1.2 T, as depicted in Figure 5.14a. Determine the torque experienced by the coil if it carries a current of 50 A.

Solution The side view of the coil indicating the direction of the dipole moment is shown in Figure 5.14b. The magnetic dipole moment lies in the xy plane and has a magnitude of

$$m = NIA = 200 \times 50 \times 10 \times 10^{-4} = 10 \text{ At} \cdot \text{m}^2$$

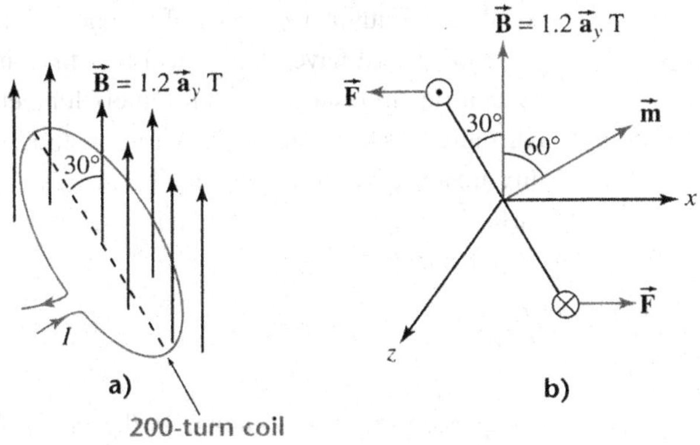

Figure 5.14 Circular coil for Example 5.6. (a) Coil immersed in $\vec{\mathbf{B}}$ field, Example 5.6; (b) Side view

The torque experienced by the coil is

$$\vec{\mathbf{T}} = \vec{\mathbf{m}} \times \vec{\mathbf{B}} = \vec{\mathbf{a}}_z 10 \times 1.2 \sin 60° = 10.39 \vec{\mathbf{a}}_z \text{ N} \cdot \text{m}$$

• • •

5.5 Magnetic flux and Gauss's law for magnetic fields

Figure 5.15a shows the lines of a magnetic field passing through an open surface s bounded by a contour c. The magnetic flux density $\vec{\mathbf{B}}$ may or may not be uniform over the entire surface. If we subdivide the surface into n very small elementary surfaces, as illustrated in Figure 5.15b, and assume that the $\vec{\mathbf{B}}$ field passing through each elementary surface is of uniform strength, then the elemental magnetic flux passing through surface $\Delta\vec{\mathbf{s}}_i$ is defined as

$$\Delta\Phi_i = \vec{\mathbf{B}}_i \cdot \Delta\vec{\mathbf{s}}_i$$

where $\vec{\mathbf{B}}_i$ is the magnetic flux density passing through surface $\Delta\vec{\mathbf{s}}_i$. The total magnetic flux passing through the surface s is

$$\Phi = \sum_{i=1}^{n} \vec{\mathbf{B}}_i \cdot \Delta\vec{\mathbf{s}}_i$$

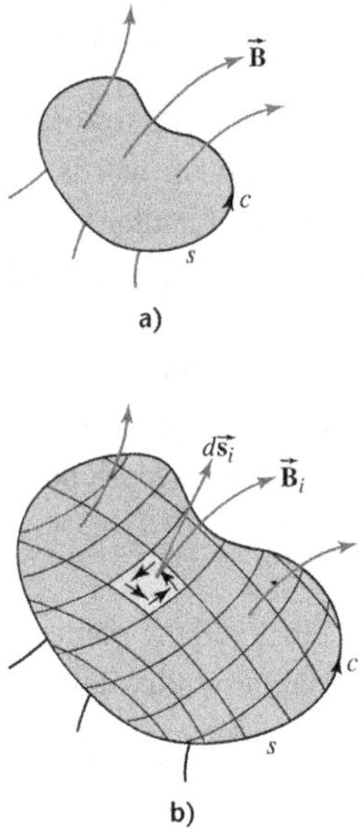

Figure 5.15 (a) Magnetic lines of flux passing through an open surface; (b) The same open surface divided into n elementary surfaces

The summation in this equation can be replaced by a definite integral as the elementary surface area approaches zero. Thus, *the magnetic flux* passing through an open surface s is

$$\Phi = \int_s \vec{\mathbf{B}} \cdot \vec{d\mathbf{s}} \tag{5.22}$$

The magnetic flux is measured in webers (Wb). If the magnetic flux density is tangential to the surface, the total flux passing through (or linking) the surface is zero.

The north and south poles of a magnet cannot be separated; therefore, the number of lines of magnetic flux leaving the north pole is exactly equal to the number of lines of magnetic flux entering the south pole. In addition, we have also shown that the lines of magnetic flux form concentric circles around an infinitely long current-carrying conductor. All these observations lead us to conclude that the lines of magnetic flux are always continuous. In other words, the flux penetrating a closed surface is equal to the flux leaving the closed surface. Therefore, for a closed surface,

$$\oint_s \vec{\mathbf{B}} \cdot \vec{d\mathbf{s}} = 0 \tag{5.23a}$$

Equation (5.23a) is known as the *integral form of Gauss's law for magnetic fields*. The closed surface integral can, however, be converted into a volume integral by the direct application of the divergence theorem. That is,

$$\int_v \nabla \cdot \vec{\mathbf{B}}\, dv = 0$$

where v is the volume bounded by the closed surface s. Because the volume under consideration is generally not zero, this equation implies that

$$\nabla \cdot \vec{\mathbf{B}} = 0 \tag{5.23b}$$

Equation (5.23b) is known as the *point form* or the *differential form of Gauss's law for magnetic fields*. Since the divergence of $\vec{\mathbf{B}}$ is always zero, the magnetic flux density is *solenoidal*. Although we have been discussing the magnetic fields produced by steady currents, (5.23a) and (5.23b) are completely general and are valid even when the currents vary with time in any fashion. Equation (5.23b) will appear later as one of Maxwell's four equations.

EXAMPLE 5.7

Two very long, identical, and parallel conductors carrying 1000 A in opposite directions are strung on poles 100 m apart. If the radius of each conductor is 2 cm and the separation between their axes is 1 m, determine the flux passing through the region bounded by the conductors and the two consecutive poles.

Solution Two parallel conductors, each of radius a, separated by a distance b, and carrying currents in opposite directions are shown in Figure 5.16. The distance between two consecutive poles is L. The magnetic flux density at a point y in the plane of the conductors is

$$\vec{\mathbf{B}} = -\frac{\mu_0 I}{2\pi}\left[\frac{1}{y} + \frac{1}{b-y}\right]\vec{\mathbf{a}}_x$$

The elementary surface is $\overrightarrow{ds} = -dy\, dz\vec{\mathbf{a}}_x$. Thus, the required flux is

$$\Phi = \frac{\mu_0 I}{2\pi}\int_a^{b-a}\left[\frac{1}{y} + \frac{1}{b-y}\right]dy \int_0^L dz$$
$$= \frac{\mu_0 IL}{\pi}\ln\left[\frac{b-a}{a}\right]$$

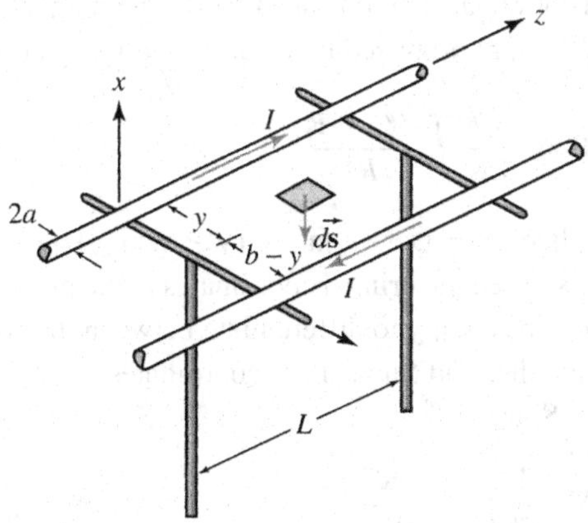

Figure 5.16 Two-wire transmission line

For our example, $a = 0.02$ m, $b = 1$ m, $L = 100$ m, and $I = 1000$ A. Substituting these values in the preceding expression, we obtain

$$\Phi = 155.67 \text{ mWb}$$

• • •

EXAMPLE 5.8

If $\vec{\mathbf{B}} = B\vec{\mathbf{a}}_z$, compute the magnetic flux passing through a hemisphere of radius R centered at the origin and bounded by the plane $z = 0$.

Solution

The hemisphere and the circular disc of radius R form a closed surface, as illustrated in Figure 5.17; therefore, the flux passing through the hemisphere must be exactly equal to the flux passing through the disc. The flux passing through the disc is

$$\Phi = \int_s \vec{\mathbf{B}} \cdot \vec{ds} = \int_0^R \int_0^{2\pi} B\rho \, d\rho \, d\phi = \pi R^2 B$$

The reader is encouraged to verify this result by integrating over the surface of the hemisphere. • • •

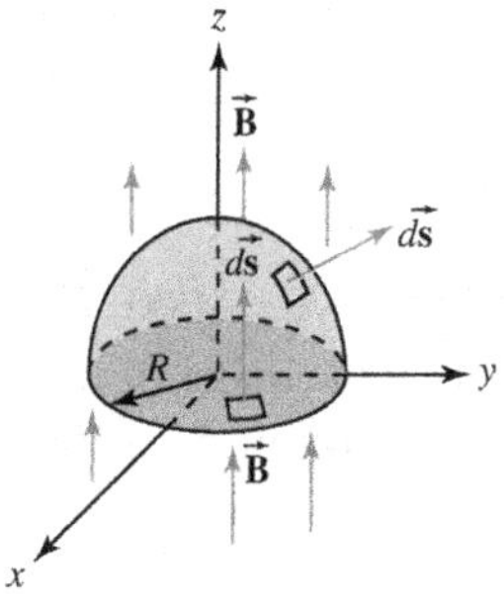

Figure 5.17 Magnetic flux passing through a hemisphere

5.6 Magnetic vector potential

As mentioned is Section 5.5, the magnetic flux density is always solenoidal (continuous) because its divergence is zero. A vector whose divergence is zero can be expressed in terms of the curl of another vector quantity as

$$\vec{\mathbf{B}} = \nabla \times \vec{\mathbf{A}} \tag{5.24}$$

where $\vec{\mathbf{A}}$ is called the *magnetic vector potential* and is expressed in webers per meter (Wb/m). Quite often, we find it expedient to work with the magnetic vector potential $\vec{\mathbf{A}}$ and then obtain $\vec{\mathbf{B}}$ using (5.24).

To obtain an expression for $\vec{\mathbf{A}}$, we begin our discussion with the Biot-Savart law for the $\vec{\mathbf{B}}$ field. The magnetic flux density at any point $P(x, y, z)$ produced by a current-carrying conductor is

$$\vec{\mathbf{B}} = \frac{\mu_0 I}{4\pi} \int_c \frac{\vec{d\ell}' \times \vec{\mathbf{R}}}{R^3}$$

where $\vec{\mathbf{R}} = (x - x')\vec{\mathbf{a}}_x + (y - y')\vec{\mathbf{a}}_y + (z - z')\vec{\mathbf{a}}_z$. Note that we have also used the primed coordinates in the preceding equation. In this case, it is necessary to differentiate between the source (primed) coordinates and the field (unprimed) coordinates.

Since

$$\nabla\left(\frac{1}{R}\right) = -\frac{\vec{\mathbf{R}}}{R^3}$$

the magnetic flux density can also be written as

$$\vec{\mathbf{B}} = \frac{\mu_0 I}{4\pi} \int_c \nabla\left(\frac{1}{R}\right) \times \vec{d\ell}' \tag{5.25}$$

where the negative sign has been eliminated by reversing the terms in the vector product. Using the vector identity, (2.127), the integrand in equation (5.25) can be expressed as

$$\nabla\left(\frac{1}{R}\right) \times \vec{d\ell}' = \nabla \times \left[\frac{\vec{d\ell}'}{R}\right] - \frac{1}{R}[\nabla \times \vec{d\ell}']$$

Because the curl operation is with respect to the unprimed coordinates of point $P(x, y, z)$, $\nabla\times\vec{d\ell}' = 0$. Thus, from (5.25), we have

$$\vec{\mathbf{B}} = \frac{\mu_0 I}{4\pi}\int_c \nabla \times \left[\frac{\vec{d\ell}'}{R}\right]$$

The integration and the differentiation are with respect to two different sets of variables, so we can interchange the order and write the preceding equation as

$$\vec{\mathbf{B}} = \nabla \times \left[\frac{\mu_0 I}{4\pi}\int_c \frac{\vec{d\ell}'}{R}\right] \tag{5.26}$$

Comparing (5.24) and (5.26), we obtain an expression for the magnetic vector potential $\vec{\mathbf{A}}$ as

$$\vec{\mathbf{A}} = \frac{\mu_0}{4\pi}\int_c \frac{I\vec{d\ell}'}{R} \tag{5.27a}$$

If the current-carrying conductor forms a closed loop, this equation becomes

$$\vec{\mathbf{A}} = \frac{\mu_0}{4\pi}\oint_c \frac{I\vec{d\ell}'}{R} \tag{5.27b}$$

We can now generalize (5.27a) by expressing it in terms of the volume current density as

$$\vec{\mathbf{A}} = \frac{\mu_0}{4\pi}\int_v \frac{\vec{\mathbf{J}}_v\, dv'}{R} \tag{5.27c}$$

We have defined the vector quantity $\vec{\mathbf{A}}$ as the magnetic vector potential whose curl yields the magnetic flux density $\vec{\mathbf{B}}$. As discussed in Chapter 2, a vector field is uniquely defined if and only if both its curl and divergence are defined. Therefore, we must still define the divergence of $\vec{\mathbf{A}}$. In magnetostatics, we define $\nabla \cdot \vec{\mathbf{A}} = 0$ and refer to this constraint as Coulomb's gauge.

We can also express the magnetic flux Φ in terms of $\vec{\mathbf{A}}$ as

$$\Phi = \int_s \vec{\mathbf{B}} \cdot \vec{ds} = \int_s (\nabla \times \vec{\mathbf{A}}) \cdot \vec{ds}$$

A direct application of Stokes' theorem yields

$$\Phi = \oint_c \vec{\mathbf{A}} \cdot \vec{d\ell} \tag{5.28}$$

where c is the contour bounding the open surface s.

EXAMPLE 5.9

A very long, straight conductor located along the z axis carries a current I in the z direction. Obtain an expression for the magnetic vector potential at a point in the bisecting plane of the conductor. What is the magnetic flux density at that point?

Solution

A current-carrying conductor extending in the z direction from $z = -L$ to $z = L$ is shown in Figure 5.18. The distance vector $\vec{\mathbf{R}}$ of point P from the current element $I\, dz\vec{\mathbf{a}}_z$ is $\vec{\mathbf{R}} = \rho\vec{\mathbf{a}}_\rho - z\vec{\mathbf{a}}_z$. Thus, the magnetic vector potential at point P is

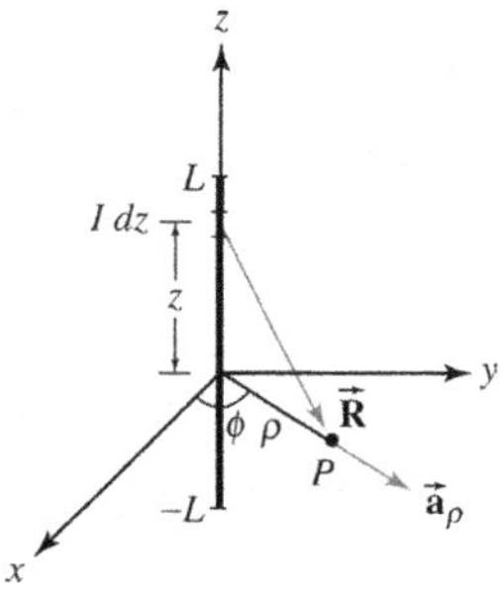

Figure 5.18 Magnetic vector potential at P produced by a finite current-carrying conductor

$$\vec{\mathbf{A}} = \frac{\mu_0 I \vec{\mathbf{a}}_z}{4\pi} \int_{-L}^{L} \frac{dz}{[\rho^2 + z^2]^{1/2}}$$

$$= \frac{\mu_0 I}{4\pi} \left[\ln[L + \sqrt{L^2 + \rho^2}] - \ln[-L + \sqrt{L^2 + \rho^2}] \right] \vec{\mathbf{a}}_z \tag{5.29}$$

This is an exact expression for the magnetic vector potential in the bisecting plane of the current-carrying conductor. For a very long conductor, $L \gg \rho$, we can make the following approximations:

$$L + \sqrt{L^2 + \rho^2} \approx L + L\left[1 + \left(\frac{\rho}{2L}\right)^2\right] \approx 2L$$

and

$$-L + \sqrt{L^2 + \rho^2} \approx -L + L\left[1 + \left(\frac{\rho}{2L}\right)^2\right] \approx \frac{\rho^2}{2L}$$

Using these approximations, we obtain

$$\vec{\mathbf{A}} = \frac{\mu_0 I}{2\pi} \ln\left[\frac{2L}{\rho}\right] \vec{\mathbf{a}}_z \tag{5.30}$$

The magnetic flux density at point P, from (5.24), is

$$\vec{\mathbf{B}} = \nabla \times \vec{\mathbf{A}} = -\frac{\partial A_z}{\partial \rho} \vec{\mathbf{a}}_\phi$$

$$= \frac{\mu_0 I L}{2\pi\rho} \left[\frac{1}{\sqrt{L^2 + \rho^2}} \right] \vec{\mathbf{a}}_\phi$$

Once again, making the approximation that $L \gg \rho$, we get

$$\vec{\mathbf{B}} = \frac{\mu_0 I}{2\pi\rho} \vec{\mathbf{a}}_\phi \tag{5.31}$$

which is the same expression for the $\vec{\mathbf{B}}$ field as was obtained earlier using the Biot–Savart law for an infinitely long, current-carrying conductor. • • •

EXAMPLE 5.10

The inner conductor of a 100-meter-long coaxial cable has a radius of 1 cm and carries a current of 80 A in the z direction, as depicted in Figure 5.19. The outer conductor is very thin and has a radius of 10 cm. Calculate the total flux enclosed within the conductors.

Solution The separation between the conductors is very small in comparison with the length of the cable, so we can use the approximate expression, (5.30), for the magnetic vector potential at any point within the cable. The total flux enclosed, from (5.28), is

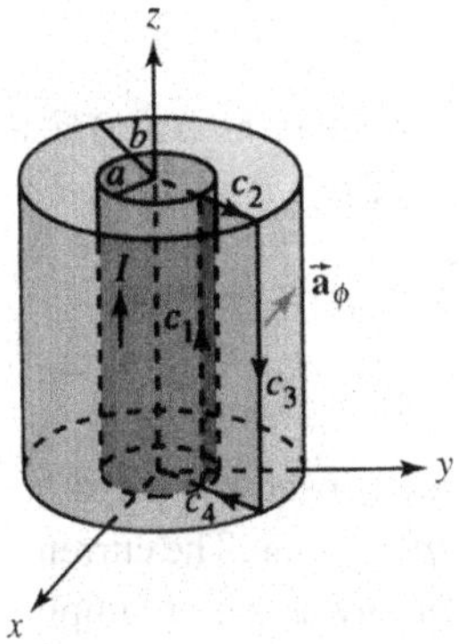

Figure 5.19 A coaxial cable carrying current

$$\Phi = \oint_c \vec{\mathbf{A}} \cdot \vec{d\ell} = \int_{c_1} \vec{\mathbf{A}} \cdot \vec{d\ell} + \int_{c_2} \vec{\mathbf{A}} \cdot \vec{d\ell} + \int_{c_3} \vec{\mathbf{A}} \cdot \vec{d\ell} + \int_{c_4} \vec{\mathbf{A}} \cdot \vec{d\ell}$$

To obtain a general expression for the flux enclosed by the coaxial cable, let a and b be the radii of the inner and the outer conductors, respectively. Because the magnetic vector potential has only a z component, there will be no contribution from the integration along the c_2 and c_4 paths. Thus,

$$\begin{aligned}\Phi &= \int_{c_1} \vec{\mathbf{A}} \cdot \vec{d\ell} + \int_{c_3} \vec{\mathbf{A}} \cdot \vec{d\ell} \\ &= \frac{\mu_0 I}{2\pi} \int_{-L}^{L} \ln(2L/a)\, dz - \frac{\mu_0 I}{2\pi} \int_{-L}^{L} \ln(2L/b)\, dz \\ &= \frac{\mu_0 IL}{\pi} \ln\left(\frac{b}{a}\right) \qquad (5.32)\end{aligned}$$

Substituting the values $I = 80$ A, $L = 50$ m, $a = 1$ cm, and $b = 10$ cm in equation (5.32), we obtain

$$\Phi = 3.68\,\text{mWb}$$

• • •

5.7 Magnetic field intensity and Ampère's circuital law

In the study of electrostatic fields we defined the electric flux density in terms of the electric field intensity as $\vec{\mathbf{D}} = \epsilon\vec{\mathbf{E}}$ so that $\vec{\mathbf{D}}$ was independent of the permittivity of the medium. We shall now define the *magnetic field intensity* $\vec{\mathbf{H}}$ in free space as

$$\vec{\mathbf{H}} = \frac{\vec{\mathbf{B}}}{\mu_0} \qquad (5.33a)$$

or

$$\vec{\mathbf{B}} = \mu_0 \vec{\mathbf{H}} \qquad (5.33b)$$

From (5.33) it is clear that the magnetic field intensity is independent of the permeability and the connection between $\vec{\mathbf{B}}$ and $\vec{\mathbf{H}}$ is analogous to that between $\vec{\mathbf{D}}$ and $\vec{\mathbf{E}}$. In Section 5.8, we will define $\vec{\mathbf{H}}$ for a material medium and discuss its behavior in more detail. We will also show that in regions where $\vec{\mathbf{J}}$ is zero, $\vec{\mathbf{H}}$ is conservative; that is, $\vec{\mathbf{H}}$ can be expressed in terms of the gradient of another field quantity called the *magnetic scalar potential*. From (5.33) it is also obvious that in free space $\vec{\mathbf{B}}$ and $\vec{\mathbf{H}}$ are in the same direction. We can now state Ampère's circuital law in terms of the magnetic field intensity.

5.7.1 Ampère's circuital law

Ampère's circuital law, which we will hereafter refer to as Ampère's law, states that *the line integral of the magnetic field intensity around a closed path equals the current enclosed*. That is,

$$\oint_c \vec{\mathbf{H}} \cdot \vec{d\ell} = I \tag{5.34a}$$

where I is the net current intercepted by the area enclosed by the path. We will refer to (5.34a) as the *integral form of Ampère's law*. The current in (5.34a) can either be carried by a conductor of any shape or simply be a flow of charges (a beam of electrons in a vacuum tube).

In electrostatics we exploited Gauss's law for sufficiently symmetric charge distributions in order to compute the electric fields in a region. In magnetostatics, Ampère's law facilitates the determination of magnetic fields without the laborious process of integration associated with the Biot-Savart law. The only restriction is that the current or the current distribution must possess a high degree of symmetry.

Since the current can be expressed in terms of volume current density as

$$I = \int_s \vec{\mathbf{J}}_v \cdot \vec{d\mathbf{s}}$$

the integral form of Ampère's law, from (5.34a), becomes

$$\oint_c \vec{\mathbf{H}} \cdot \vec{d\ell} = \int_s \vec{\mathbf{J}}_v \cdot \vec{d\mathbf{s}}$$

Stokes' theorem allows us to express the line integral in terms of the surface integral as

$$\int_s (\nabla \times \vec{\mathbf{H}}) \cdot \vec{d\mathbf{s}} = \int_s \vec{\mathbf{J}}_v \cdot \vec{d\mathbf{s}}$$

As s can be any arbitrary open surface bounded by a closed contour c, the preceding equation can be written in the general form as

$$\nabla \times \vec{\mathbf{H}} = \vec{\mathbf{J}}_v \tag{5.34b}$$

Equation (5.34b) will always be referred to as *the point (differential) form of Ampère's law for static magnetic fields.*

The following examples illustrate some applications of Ampère's law to determine the magnetic fields for the current distributions that satisfy the symmetry requirements.

EXAMPLE 5.11

A very long, very thin, straight wire located along the z axis carries a current I in the z direction. Find the magnetic field intensity at any point in free space using Ampère's law.

Solution

The symmetry arguments dictate that the magnetic field lines must be concentric circles, as shown in Figure 5.20. The magnetic field intensity will have a constant magnitude along each circle. Thus, at any radius ρ, we have

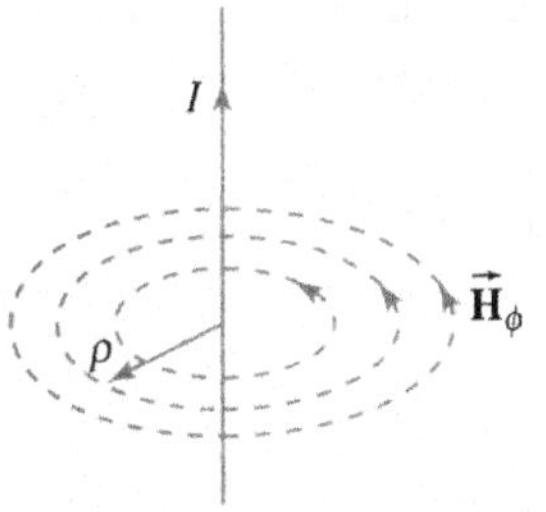

Figure 5.20 Magnetic field surrounding a very long current-carrying conductor

$$\oint_c \vec{\mathbf{H}} \cdot \vec{d\ell} = \int_0^{2\pi} H_\phi \rho \, d\phi = 2\pi\rho H_\phi$$

Since the current enclosed by the closed path is I, Ampère's law gives us

$$\vec{\mathbf{H}} = \frac{I}{2\pi\rho} \vec{\mathbf{a}}_\phi$$

Thus, Ampère's law yields the same result that was obtained earlier using the Biot–Savart law.

• • •

EXAMPLE 5.12

A very long, hollow conductor of inner radius a and outer radius b is located along the z axis and carries a current I in the z direction, as depicted in Figure 5.21a. If the current distribution is uniform, determine the magnetic field intensity at any point in space.

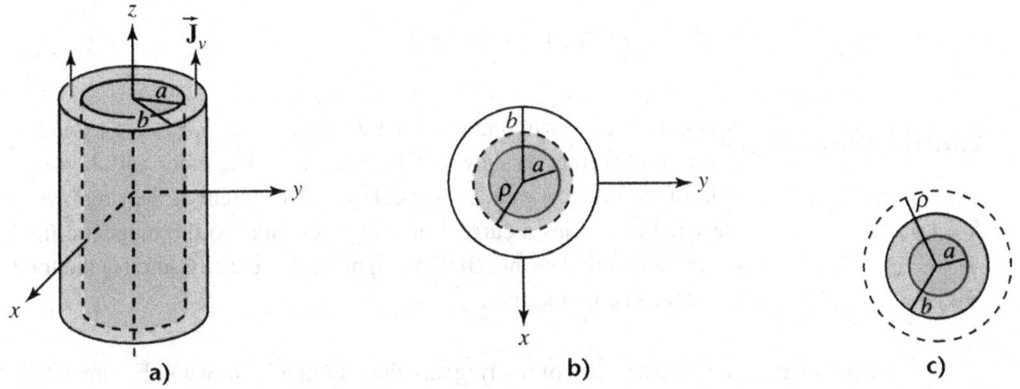

Figure 5.21 Hollow conductor for Example 5.12. (a) A hollow conductor carrying current; (b) Cross-sectional view shows the closed circular path of $a \leq \rho \leq b$; (c) The cross-sectional view shows the closed path at $\rho \geq b$

Solution As the current is uniformly distributed, we can express it in terms of the volume current density as

$$\vec{\mathbf{J}}_v = \frac{I}{\pi(b^2 - a^2)}\vec{\mathbf{a}}_z$$

Following the symmetry arguments, we expect that the field lines must be concentric circles, the magnetic field intensity must be in the ϕ direction, and H_ϕ has a constant magnitude along each circle. There are three regions of interest; we will separately calculate the magnetic field intensity in each region.

a) Region 1, $\rho \le a$: For any closed path within the hollow region of the cylinder the current enclosed is zero. Therefore, the magnetic field intensity is identically zero in this region. That is, $\vec{\mathbf{H}} = 0$ for $\rho \le a$.

b) Region 2, $a \le \rho \le b$: The cross-sectional view of the current-carrying conductor with a closed circular contour at a radius ρ is shown in Figure 5.21b. The net current enclosed is

$$I_{\text{enc}} = \int_s \vec{\mathbf{J}}_v \cdot \vec{d\mathbf{s}} = \frac{I}{\pi(b^2 - a^2)} \int_a^\rho \rho\, d\rho \int_0^{2\pi} d\phi$$
$$= \frac{I(\rho^2 - a^2)}{b^2 - a^2}$$

On the other hand,

$$\oint_c \vec{\mathbf{H}} \cdot \vec{d\ell} = 2\pi\rho H_\phi$$

Thus, from Ampère's law, we have

$$\vec{\mathbf{H}} = \frac{I}{2\pi\rho}\left[\frac{\rho^2 - a^2}{b^2 - a^2}\right]\vec{\mathbf{a}}_\phi \quad \text{for} \quad a \le \rho \le b$$

c) Region 3, $\rho \ge b$: In this case, the point of observation is outside the conductor. (See Figure 5.21c.) Therefore, the net current enclosed is I. Thus, the magnetic field intensity in this region is

$$\vec{\mathbf{H}} = \frac{I}{2\pi\rho}\vec{\mathbf{a}}_\phi \quad \text{for} \quad \rho \ge b$$

• • •

EXAMPLE 5.13 A closely spaced winding (toroidal winding) with N turns is wound in the form of a ring, as shown in Figure 5.22a. The inner and the outer radii of the ring are a and b, respectively. The height of the ring is h. If the winding carries a current of I amperes, find (a) the magnetic field intensity within the ring, (b) the magnetic flux density, and (c) the total flux enclosed by the ring.

Solution A cut-away view of the ring and the winding is shown in Figure 5.22b. The application of Ampère's law reveals that the magnetic field intensity exists only inside the ring. At any radius ρ within the ring the magnetic

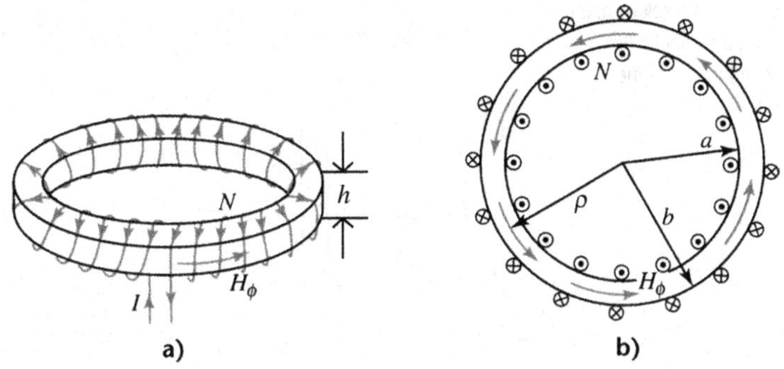

Figure 5.22 (a) A toroidal winding; (b) Cut-away view to show the total current enclosed by the circular loop at radius $a \le \rho \le b$

field intensity is in the ϕ direction and its magnitude is constant. The total current enclosed is NI; therefore, from Ampère's law, the magnetic field intensity inside the ring is

$$\vec{\mathbf{H}} = \frac{NI}{2\pi\rho}\vec{\mathbf{a}}_\phi \quad \text{for} \quad a \le \rho \le b$$

The magnetic flux density at any radius ρ within the ring is

$$\vec{\mathbf{B}} = \mu_0\vec{\mathbf{H}} = \frac{\mu_0 NI}{2\pi\rho}\vec{\mathbf{a}}_\phi \quad \text{for} \quad a \le \rho \le b$$

The total flux enclosed inside the ring is

$$\begin{aligned}\mathbf{\Phi} &= \int \vec{\mathbf{B}} \cdot \vec{d\mathbf{s}} = \frac{\mu_0 NI}{2\pi}\int_a^b \frac{d\rho}{\rho}\int_0^h dz \\ &= \frac{\mu_0 NIh}{2\pi}\ln(b/a)\end{aligned}$$

• • •

5.8 Magnetic materials

Let us now extend our theory of magnetic fields to regions containing magnetic materials. To some extent our discussion will run parallel to that of the electric fields in dielectric materials, but there are some important differences, which we will emphasize from time to time.

Let us perform an experiment with a cylindrical coil of length L, usually called a solenoid, carrying current I, as depicted in Figure 5.23a. It is assumed that the wire is closely wound and uniformly spaced. In Exercise 5.3, you found that the magnetic flux density at the center of the solenoid is twice as much as that at either end, as shown in Figure 5.23b. If we place samples of various substances into this field we realize that the magnetic force experienced by these samples is maximum near the ends of the solenoid, where the gradient dB_z/dz is large. In order to continue the experiment, let us now assume that we will always place a sample at the upper end of the solenoid and observe the force experienced by it. Our observations will reveal that the force

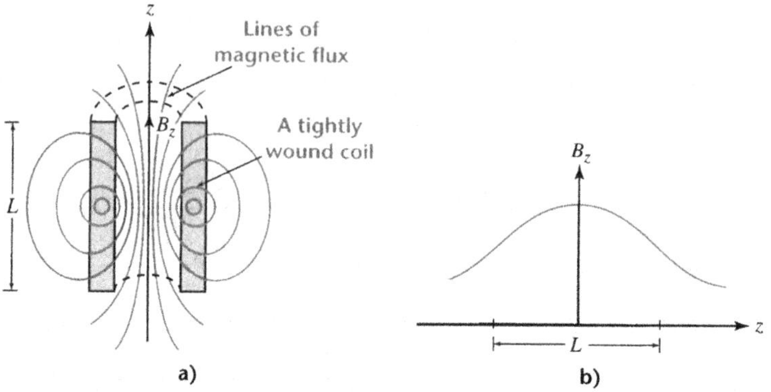

Figure 5.23 (a) A solenoid; (b) Graph of the magnetic flux density on the axis of the solenoid

on a particular substance is proportional to the mass of the sample and is independent of its shape as long as the sample is not too large. We will also observe that some samples are attracted toward the region of stronger field, whereas other samples are repelled.

Those substances that experience a feeble force of repulsion are called *diamagnetic*. Practically all organic compounds and the majority of inorganic compounds are diamagnetic. In fact, we now consider diamagnetism to be a property of every atom and molecule.

There are two distinct types of substances that experience a force of attraction. Those substances that are pulled toward the center with a feeble force are called *paramagnetic*. Paramagnetism in metals such as aluminum, copper, and many others is not much stronger than diamagnetism. However, some substances, like iron and magnetite, are literally sucked in by the magnetic force. These substances are called **ferromagnetic**. The magnetic force experienced by ferromagnetic materials may be 5000 times as much as that experienced by paramagnetic materials.

As the force experienced by paramagnetic and diamagnetic substances is quite feeble, for all practical purposes we can group them together and refer to them as nonmagnetic materials. *We also assume that the permeability of all nonmagnetic materials is the same as that of free space.*

To fully describe the magnetic properties of materials we need the concept of *quantum mechanics*, which is considered to be beyond the scope of this book. However, we can use a simple and easily visualized model of an atom to explain some of the magnetic properties. We know that the electrons orbit around the nucleus at constant speed, as illustrated in Figure 5.24a. Since the current is the amount of charge that passes through a given point per second, an orbiting electron produces a ring current of magnitude

$$I = \frac{eU_e}{2\pi\rho} \tag{5.35}$$

where e is the magnitude of the charge on the electron, U_e is its speed, and ρ is the radius. The orbiting electron gives rise to an orbital magnetic moment

$$\vec{\mathbf{m}} = \frac{eU_e\rho}{2}\vec{\mathbf{a}}_z \qquad (5.36)$$

as depicted in Figure 5.24b.

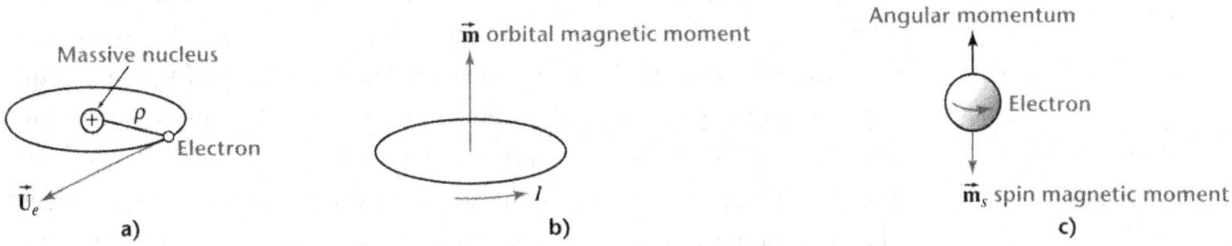

Figure 5.24 (a) A model of an atom showing an electron moving in a circular orbit; (b) Orbital magnetic moment; (c) Spin magnetic moment

A fundamental tenet of quantum mechanics is that the magnitude of the orbital angular momentum is always some integral multiple of $h/2\pi$, where h is Planck's constant ($h = 6.63 \times 10^{-34}$ J · s). An electron also possesses angular momentum that has, of course, nothing to do with its orbital motion. Think of it as if the electron is continually rotating (spinning) around its own axis at a fixed rate. The spinning motion involves circulating charge and it gives an electron a spin magnetic moment, as shown in Figure 5.24c. The spin magnetic moment has a fixed magnitude

$$m_s = \frac{he}{8\pi m_e} = 9.27 \times 10^{-24}\ \text{A} \cdot \text{m}^2 \qquad (5.37)$$

where m_e is the mass of the electron.

The net magnetic moment of the atom is obtained by combining both the orbital and spin moments of all the electrons, taking into account the directions of these moments. The net magnetic moment produces a

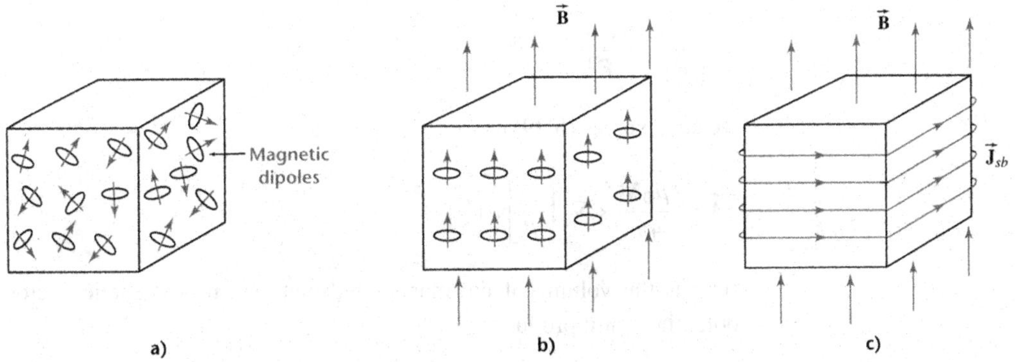

Figure 5.25 (a) A piece of magnetic material with randomly oriented magnetic dipoles; (b) An external $\vec{\mathbf{B}}$ field causes the magnetic dipoles to align with it; (c) The small aligned current loops of (b) are equivalent to a current along the surface of the material

far field similar to that produced by a current loop (magnetic dipole). In the absence of an external magnetic field, the magnetic dipoles in a piece of material are oriented at random, as shown in Figure 5.25a. Thus, the net magnetic moment is nearly zero. In the presence of an external magnetic field, each magnetic dipole experiences a torque (see Section 5.4) that tends to align it with the magnetic field, as illustrated in Figure 5.25b. The figure shows an ideal case of perfect alignment, but in reality, the alignment is only partial. The alignment of the magnetic dipoles is analogous to the alignment of the electric dipoles in a dielectric medium – however, there is a remarkable difference. The alignment of the electric dipoles always decreases the original electric field, whereas the alignment of the magnetic poles in paramagnetic and ferromagnetic materials increases the original magnetic field. The alignment of the magnetic dipoles within the material is equivalent to the current along the surface of the material, as depicted in Figure 5.25c. This current results in an additional magnetic field within the material. Let us now quantitatively prove that this is so.

If there are n atoms in an elemental volume Δv, and $\vec{\mathbf{m}}_i$ is the magnetic moment of the ith atom, then the *magnetic moment per unit volume* is defined as

$$\vec{\mathbf{M}} = \lim_{\Delta v \to 0} \frac{\sum_{i=1}^{n} \vec{\mathbf{m}}_i}{\Delta v} \tag{5.38}$$

A material is said to be magnetized if $\vec{\mathbf{M}} \neq 0$. The magnetic dipole moment $\overrightarrow{d\mathbf{m}}$ for an elemental volume dv' is $\overrightarrow{d\mathbf{m}} = \vec{\mathbf{M}}\, dv'$. The magnetic vector potential set up by $\overrightarrow{d\mathbf{m}}$ is

$$\overrightarrow{d\mathbf{A}} = \frac{\mu_0 \vec{\mathbf{M}} \times \vec{\mathbf{a}}_R}{4\pi R^2}\, dv' \tag{5.39}$$

Using the vector identity

$$\nabla' \left[\frac{1}{R} \right] = \frac{\vec{\mathbf{a}}_R}{R^2}$$

we can express (5.39) as

$$\overrightarrow{d\mathbf{A}} = \frac{\mu_0 \vec{\mathbf{M}}}{4\pi} \times \nabla' \left[\frac{1}{R} \right] dv'$$

If v' is the volume of the magnetized material, the magnetic vector potential that it produces is

$$\vec{\mathbf{A}} = \frac{\mu_0}{4\pi} \int_{v'} \vec{\mathbf{M}} \times \nabla' \left[\frac{1}{R} \right] dv'$$

We now use the vector identity

$$\vec{\mathbf{M}} \times \nabla' \left[\frac{1}{R}\right] = \frac{1}{R}\nabla' \times \vec{\mathbf{M}} - \nabla' \times \left[\frac{\vec{\mathbf{M}}}{R}\right]$$

and write the magnetic vector potential as

$$\vec{\mathbf{A}} = \frac{\mu_0}{4\pi}\int_{v'} \frac{\nabla' \times \vec{\mathbf{M}}}{R} dv' - \frac{\mu_0}{4\pi}\int_{v'} \nabla' \times \left[\frac{\vec{\mathbf{M}}}{R}\right] dv'$$

Employing the vector identity

$$\int_{v'} \nabla' \times \vec{\mathbf{M}} dv' = -\oint_{s'} \vec{\mathbf{M}} \times ds'$$

We can rewrite $\vec{\mathbf{A}}$ as

$$\vec{\mathbf{A}} = \frac{\mu_0}{4\pi}\int_{v'} \frac{\nabla' \times \vec{\mathbf{M}}}{R} dv' + \frac{\mu_0}{4\pi}\oint_{s'} \frac{\vec{\mathbf{M}} \times \vec{\mathbf{a}}_n}{R} ds'$$

$$\vec{\mathbf{A}} = \frac{\mu_0}{4\pi}\int_{v'} \frac{\vec{\mathbf{J}}_{vb}}{R} dv' + \frac{\mu_0}{4\pi}\int_{s'} \frac{\vec{\mathbf{J}}_{sb}}{R} ds' \tag{5.40}$$

where

$$\vec{\mathbf{J}}_{vb} = \nabla \times \vec{\mathbf{M}} \tag{5.41}$$

is the *bound volume current density*, and

$$\vec{\mathbf{J}}_{sb} = \vec{\mathbf{M}} \times \vec{\mathbf{a}}_n \tag{5.42}$$

is the *bound surface current density*. In (5.41) and (5.42) we have dropped the primes with an understanding that both the curl and the cross product operations refer to the coordinates of the source point. Equation (5.40) reveals that the bound volume current density within a magnetized material and the bound surface current density on the surface of the material can be used to determine the magnetic vector potential due to the magnetized material. In addition, there may be a free volume current density $\vec{\mathbf{J}}_{vf}$ and a free surface current density $\vec{\mathbf{J}}_{sf}$ that contribute to the magnetic vector potential. The total volume current density in the medium is $\vec{\mathbf{J}}_v = \vec{\mathbf{J}}_{vf} + \vec{\mathbf{J}}_{vb}$. However, from (5.34), $\vec{\mathbf{J}}_{vf} = \nabla \times \vec{\mathbf{H}}$. The magnetic flux density in free space is $\vec{\mathbf{B}} = \mu_0 \vec{\mathbf{H}}$ or $\vec{\mathbf{H}} = \vec{\mathbf{B}}/\mu_0$. Thus, in free space, we have

$$\nabla \times \left[\frac{\vec{\mathbf{B}}}{\mu_0}\right] = \vec{\mathbf{J}}_{vf}$$

To account for the contribution due to $\vec{\mathbf{J}}_{vb}$, the enhanced $\vec{\mathbf{B}}$ field in the magnetic medium is

$$\nabla \times \left[\frac{\vec{\mathbf{B}}}{\mu_0}\right] = \vec{\mathbf{J}}_{vf} + \vec{\mathbf{J}}_{vb} = \nabla \times \vec{\mathbf{H}} + \nabla \times \vec{\mathbf{M}}$$

or

$$\vec{\mathbf{B}} = \mu_0[\vec{\mathbf{H}} + \vec{\mathbf{M}}] \tag{5.43}$$

Equation (5.43) is so general that it is valid for any medium, linear or not. For a linear homogeneous and isotropic medium, however, we can express $\vec{\mathbf{M}}$ in terms of $\vec{\mathbf{H}}$ as

$$\vec{\mathbf{M}} = \chi_m \vec{\mathbf{H}} \tag{5.44}$$

where χ_m is a constant of proportionality and is called the *magnetic susceptibility*. Substituting (5.44) in (5.43), we obtain

$$\vec{\mathbf{B}} = \mu_0[1 + \chi_m]\vec{\mathbf{H}} = \mu_0\mu_r\vec{\mathbf{H}} = \mu\vec{\mathbf{H}} \tag{5.45}$$

The quantity $\mu = \mu_0\mu_r$ is the permeability of the medium, and the parameter μ_r is called the *relative permeability* of the medium. For a linear, isotropic, and homogeneous medium both χ_m and μ_r are constants.

It is a common practice to assume $\mu_r = 1$ for paramagnetic and diamagnetic materials (we referred to these earlier as nonmagnetic materials). However, for ferromagnetic materials the relative permeability can be as high as 5000 at a flux density of 1 T. We remind you that (5.44) is only valid for linear homogeneous and isotropic materials. For anisotropic materials, $\vec{\mathbf{B}}$, $\vec{\mathbf{H}}$, and $\vec{\mathbf{M}}$ are no longer parallel. A detailed discussion of anisotropic materials is beyond the scope of this book; however, some insight into the behavior of ferromagnetic materials is needed as a basis for our discussion of magnetic circuits in Section 5.12.

5.8.1 Ferromagnetism

The behavior of a ferromagnetic material such as iron, cobalt, or nickel is explained in terms of magnetic domains. A *magnetic domain* is a very small region in which all the magnetic dipoles are perfectly aligned, as depicted in Figure 5.26. The direction of alignment of the magnetic dipoles varies from one domain to the next; thus this virgin material exists in a nonmagnetized state.

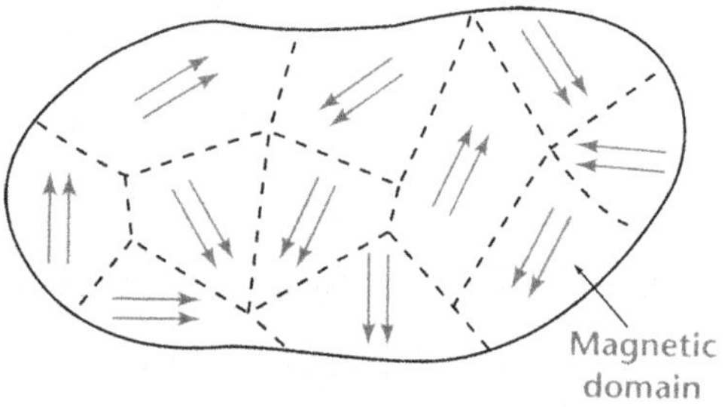

Figure 5.26 Random orientation of magnetic dipoles in a nonmagnetized ferromagnetic material

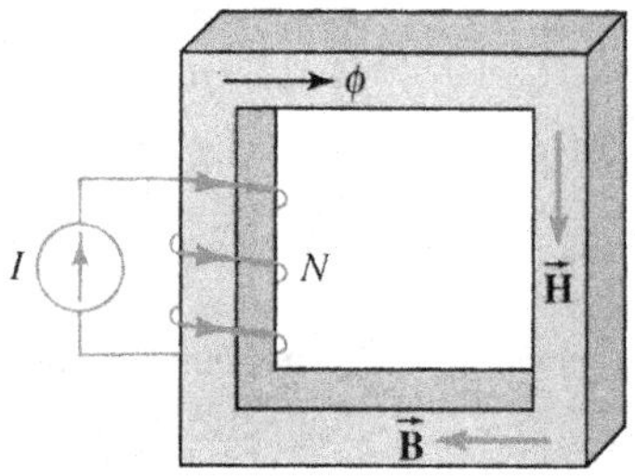

Figure 5.27 A wrapped coil establishes flux in a magnetic material

When the magnetic material is placed in an external magnetic field, all dipoles will tend to align along this magnetic field. One way to place the magnetic material in an external magnetic field is to wind a current-carrying wire around it, as shown in Figure 5.27. We can expect that there are some domains in the material that are already more or less aligned in the direction of the field. These domains have the tendency to grow in size at the expense of the neighboring domains. The growth of a domain merely changes its boundaries. The movement of domain boundaries, however, depends upon the grain structure of the material. We can also expect that some domains will rotate their dipoles in the direction of the applied field. As a result, the magnetic flux density within the material increases.

The current in the coil establishes the $\vec{\mathbf{H}}$ field in the magnetic material (medium) which we can consider as an independent variable. The applied $\vec{\mathbf{H}}$ field creates a $\vec{\mathbf{B}}$ field within the medium. As long as the $\vec{\mathbf{B}}$ field in the medium is weak, the movement of the domain walls is reversible. As we continue increasing the $\vec{\mathbf{H}}$ field by increasing the current through the coil, the $\vec{\mathbf{B}}$ field within the medium becomes stronger and stronger as more and more magnetic dipoles align themselves with the $\vec{\mathbf{B}}$ field. If we measure the $\vec{\mathbf{B}}$ field within the magnetic material (Chapter 7 explains how) we will find that $\vec{\mathbf{B}}$ increases slowly at first, then more rapidly, then very slowly, and finally flattens off, as illustrated in Figure 5.28. The solid curve in Figure 5.28 is commonly referred to as the *magnetization characteristic* of the magnetic material. Each magnetic material has a different magnetization characteristic. The changes in $\vec{\mathbf{B}}$ are due to the changes in $\vec{\mathbf{M}}$. The flattening-off region indicates that almost all the magnetic dipoles in the magnetic material have aligned themselves in the direction of the $\vec{\mathbf{B}}$ field. Knowing $\vec{\mathbf{B}}$ and $\vec{\mathbf{H}}$, we can actually determine $\vec{\mathbf{M}}$ because $\vec{\mathbf{M}} = \vec{\mathbf{B}}/\mu_0 - \vec{\mathbf{H}}$.

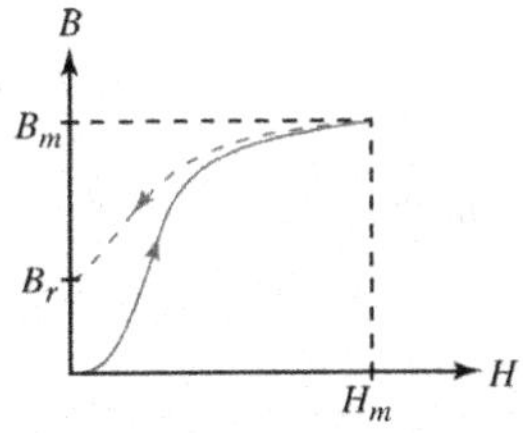

Figure 5.28 Magnetization characteristic of a magnetic material

If we now start lowering the $\vec{\mathbf{H}}$ field by decreasing the current in the coil, we find that the $\vec{\mathbf{B}}$ field does not decrease as fast, as indicated by the dashed line in Figure 5.28. This irreversibility is called **hysteresis**. The dashed curve shows that even when the $\vec{\mathbf{H}}$ field is reduced to zero, there is still some magnetic flux density in the material. We refer to this as the *residual* or *remanent flux density*, $\vec{\mathbf{B}}_r$. The magnetic material has been magnetized and acts like a permanent magnet because once the magnetic domains have been aligned in a certain direction in response to an external magnetic field, some of them tend to stay that way. The higher the residual magnetic flux density, the better suited is the magnetic material for applications requiring permanent magnets. The direct-current machines fall into that category. A magnetic material that retains high residual flux density is referred to as a *hard* magnetic material.

If we reverse the direction of the current through the coil of Figure 5.27, we find that the flux density in the material becomes zero at a certain $\vec{\mathbf{H}}$ in the opposite direction. This value of the $\vec{\mathbf{H}}$ field is called the *coercive force*, $\vec{\mathbf{H}}_c$. By increasing and then decreasing the $\vec{\mathbf{H}}$ field in both directions, we can trace a loop known as the *hysteresis loop*, shown in Figure 5.29. The area of the hysteresis loop determines the loss in energy per cycle (hysteresis loss). We need this energy to align the magnetic domains in one direction and then realign them in the opposite direction once in each cycle. For alternating-current applications such as transformers, induction motors, etc., we need magnetic materials with as low hysteresis loss as possible. In other words, the residual flux density for these materials should be as low as possible. Those materials that exhibit such properties are called *soft* magnetic materials.

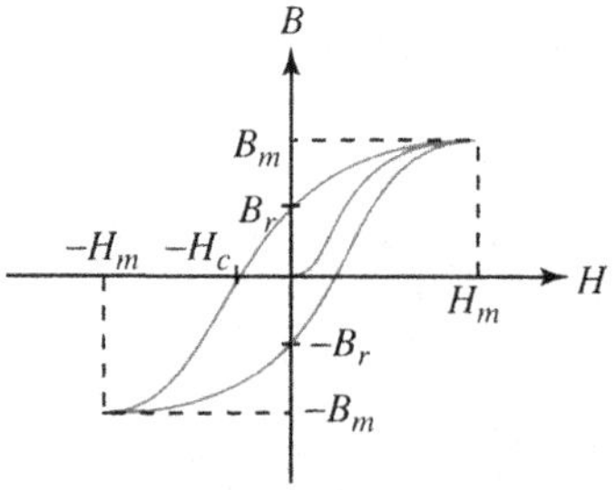

Figure 5.29 Hysteresis loop

EXAMPLE 5.14

If the winding in Example 5.13 is wound over a magnetic material with relative permeability μ_r, find (a) the magnetic moment per unit volume, (b) the bound volume current density, and (c) the bound surface current density.

Solution In terms of relative permeability, the magnetic susceptibility is

$$\chi_m = \mu_r - 1$$

Thus, the magnetic moment per unit volume or the *magnetization vector*, as it is sometimes called, is

$$\vec{\mathbf{M}} = \frac{(\mu_r - 1)NI}{2\pi\rho}\vec{\mathbf{a}}_\phi$$

The bound volume current density, from (5.41), is

$$\vec{\mathbf{J}}_{vb} = \nabla \times \vec{\mathbf{M}} = 0$$

The volume is bounded by four surfaces; let us compute the bound surface current density at each surface separately.

The bound surface current density on the top surface, from (5.42), is

$$\vec{\mathbf{J}}_{sb}\big|_{\text{top surface}} = \vec{\mathbf{M}} \times \vec{\mathbf{a}}_z = \frac{(\mu_r - 1)NI}{2\pi\rho}\vec{\mathbf{a}}_\rho$$

The bound surface current density on the bottom surface is

$$\vec{\mathbf{J}}_{sb}|_{\text{bottom surface}} = \vec{\mathbf{M}} \times (-\vec{a}_z) = -\frac{(\mu_r - 1)NI}{2\pi\rho}\vec{\mathbf{a}}_\rho$$

The bound surface current density on the surface at $\rho = a$ is

$$\vec{\mathbf{J}}_{sb}|_{\rho=a} = \vec{\mathbf{M}} \times (-\vec{\mathbf{a}}_\rho)|_{\rho=a} = \frac{(\mu_r - 1)NI}{2\pi a}\vec{\mathbf{a}}_z$$

Finally, the bound surface current density on the surface at $\rho = b$ is

$$\vec{\mathbf{J}}_{sb}|_{\rho=b} = -\frac{(\mu_r - 1)NI}{2\pi b}\vec{\mathbf{a}}_z$$

• • •

5.9 Magnetic scalar potential

In Section 5.6, we defined the magnetic flux density $\vec{\mathbf{B}}$ in terms of the magnetic vector potential $\vec{\mathbf{A}}$ as

$$\vec{\mathbf{B}} = \nabla \times \vec{\mathbf{A}}$$

and obtained a general expression for $\vec{\mathbf{A}}$ in terms of the volume current density $\vec{\mathbf{J}}$ as

$$\vec{\mathbf{A}} = \frac{\mu_0}{4\pi}\int_v \frac{\vec{\mathbf{J}}_v\, dv}{R}$$

We also obtained an expression for the magnetic field intensity $\vec{\mathbf{H}}$ at a point due to the volume current density $\vec{\mathbf{J}}_v$ as

$$\nabla \times \vec{\mathbf{H}} = \vec{\mathbf{J}}_v \tag{5.46}$$

and we referred to this equation as Ampère's law. From this equation, it is obvious that in a current-carrying region the magnetic field intensity is not conservative in nature. *In general, the $\vec{\mathbf{H}}$ field is rotational.* In contrast, the electric field intensity $\vec{\mathbf{E}}$ at any point due to fixed charges always represents a conservative field because $\nabla \times \vec{\mathbf{E}} = 0$.

In a source-free region; i.e., a region devoid of currents, (5.46) becomes

$$\nabla \times \vec{\mathbf{H}} = 0 \tag{5.47a}$$

which also implies that

$$\oint_c \vec{\mathbf{H}} \cdot \vec{d\ell} = 0 \tag{5.47b}$$

when the closed path c does not enclose any current.

In electrostatic fields we represented the conservative field $\vec{\mathbf{E}}$ in terms of an electric potential V as

$$\vec{\mathbf{E}} = -\nabla V$$

and obtained the potential at point a with respect to point b (the potential difference between points a and b) as

$$V_{ab} = -\int_b^a \vec{\mathbf{E}} \cdot \overrightarrow{d\ell}$$

Because (5.47a) claims that the $\vec{\mathbf{H}}$ field is conservative as long as the region is devoid of currents, we can express $\vec{\mathbf{H}}$ in terms of a scalar field as

$$\vec{\mathbf{H}} = -\nabla \mathscr{F} \tag{5.48a}$$

where $\mathscr{F}$ is called the *magnetic scalar potential* or *magnetostatic potential*. The SI unit of magnetic scalar potential is the ampere. If $\mathscr{F}_a$ and $\mathscr{F}_b$ are the magnetic scalar potentials of points a and b, then the magnetic potential (difference) of point a with respect to point b is

$$\mathscr{F}_{ab} = \mathscr{F}_a - \mathscr{F}_b = -\int_b^a \vec{\mathbf{H}} \cdot \overrightarrow{d\ell} \tag{5.48b}$$

The term *magnetomotive force* or *mmf* is commonly used to describe the difference in magnetic potential between any two points.

From $\nabla \cdot \vec{\mathbf{B}} = 0$ and $\vec{\mathbf{B}} = \mu \vec{\mathbf{H}}$, we obtain

$$\nabla \cdot \vec{\mathbf{H}} = 0$$

as long as the medium is linear, isotropic, and homogeneous. Substituting for $\vec{\mathbf{H}}$ from (5.47) in the preceding equation, we get

$$\nabla^2 \mathscr{F} = 0 \tag{5.49}$$

which represents Laplace's equation for the magnetic scalar potential in a current-free region. Equation (5.49) can be solved in exactly the same way as we solved $\nabla^2 V = 0$ in Chapter 3. We will exploit the preceding equations extensively in the study of magnetic circuits in Section 5.12.

EXAMPLE 5.15

A very long, straight conductor lies along the z axis. It carries a uniform current I in the z direction. Obtain an expression for the magnetic potential difference between two points in space.

Solution The region surrounding the conductor satisfies (5.46). The magnetic field intensity in this region, from Ampère's law, is

$$\vec{\mathbf{H}} = \frac{I}{2\pi\rho} \vec{\mathbf{a}}_\phi$$

and

$$\begin{aligned} \vec{\mathbf{H}} \cdot \overrightarrow{d\ell} &= H_\phi \vec{\mathbf{a}}_\phi \cdot [d\rho \vec{\mathbf{a}}_\rho + \rho\, d\phi \vec{\mathbf{a}}_\phi + dz \vec{\mathbf{a}}_z] = \rho H_\phi\, d\phi \\ &= \frac{I}{2\pi}\, d\phi \end{aligned}$$

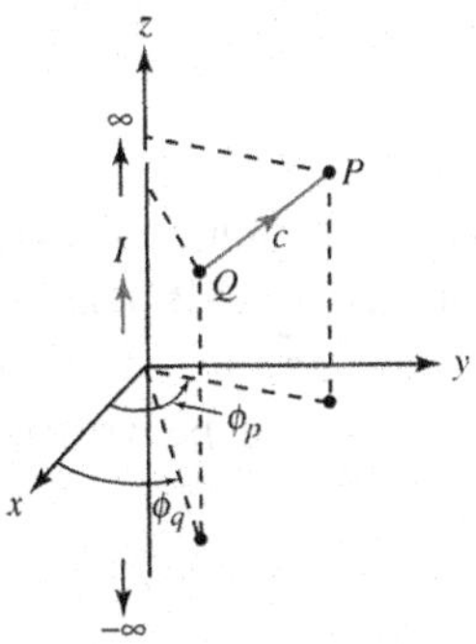

Figure 5.30 Magnetic potential of point P with respect to point Q

If the two points in space are $P(\rho_p, \phi_p, z_p)$ and $Q(\rho_q, \phi_q, z_q)$ as shown in Figure 5.30, then the magnetic potential of point P with respect to Q is

$$\mathscr{F}_{PQ} = -\int_{\phi_q}^{\phi_p} \frac{I}{2\pi}\, d\phi = -\frac{I}{2\pi}[\phi_p - \phi_q]$$

$$= \frac{I}{2\pi}[\phi_q - \phi_p]$$

Figure 5.30 shows that $\phi_p > \phi_q$; therefore, this expression yields the mmf drop in traversing the path from point Q to point P. • • •

5.10 Boundary conditions for magnetic fields

Prior to considering either the applications of magnetic fields or the analysis of magnetic circuits, we must know the behavior of magnetic fields at the boundary between two media (regions) having different permeabilities. A *boundary*, also known as an interface, is an area of infinitesimally small thickness that marks the end of one region and the beginning of the other.

5.10.1 Boundary condition for normal components of $\vec{\mathbf{B}}$ field

To determine the boundary condition for the normal components of the magnetic flux density at the interface between the two regions, let us construct a Gaussian surface in the form of a flat cylinder (a pillbox) with vanishingly small thickness, as shown in Figure 5.31. Since the magnetic flux lines are continuous, we have

$$\oint_s \vec{\mathbf{B}} \cdot \vec{d\mathbf{s}} = 0$$

where s is the entire surface of the pillbox. Neglecting the flux that flows through the vanishingly small thickness of the pillbox, this equation

Figure 5.31 Boundary condition for the normal components of the $\vec{\mathbf{B}}$ field

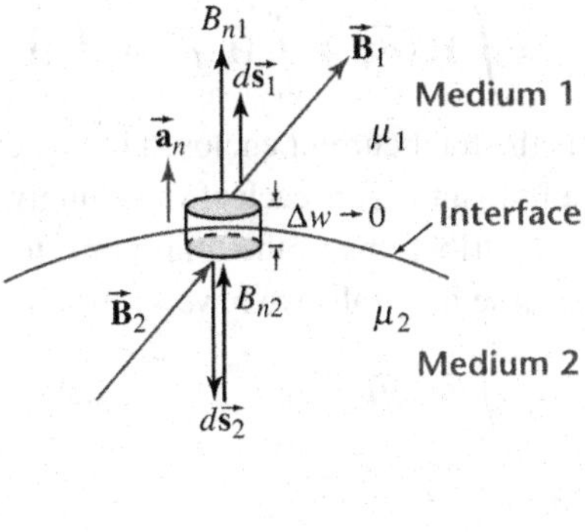

becomes

$$\int_{s_1} \vec{\mathbf{B}} \cdot \overrightarrow{d\mathbf{s}} + \int_{s_2} \vec{\mathbf{B}} \cdot \overrightarrow{d\mathbf{s}} = 0$$

If $\vec{\mathbf{a}}_n$ is the unit normal to the interface pointing into region 1, $B_{n1} = \vec{\mathbf{a}}_n \cdot \vec{\mathbf{B}}_1$ and $B_{n2} = \vec{\mathbf{a}}_n \cdot \vec{\mathbf{B}}_2$ are the normal components of the $\vec{\mathbf{B}}$ field in the two regions at the interface, and $\overrightarrow{d\mathbf{s}}_1 = \vec{\mathbf{a}}_n\, ds_1$ and $\overrightarrow{d\mathbf{s}}_2 = -\vec{\mathbf{a}}_n\, ds_2$ are the differential surfaces, then the preceding equation can be written as

$$\int_{s_1} B_{n1}\, ds_1 - \int_{s_2} B_{n2}\, ds_2 = 0$$

or

$$\int_{s_1} B_{n1}\, ds_1 = \int_{s_2} B_{n2}\, ds_2 \tag{5.50}$$

which simply states that the total flux leaving the boundary is equal to the amount of flux entering the boundary. We will use this equation in the analysis of magnetic circuits.

For the two surfaces to be equal, which is true for a pillbox of vanishingly small thickness, we have

$$\int_{s} (B_{n1} - B_{n2})\, ds = 0$$

Since the surface under consideration is arbitrary, we can express this result in scalar form as

$$B_{n1} = B_{n2} \tag{5.51a}$$

which states that *the normal components of the magnetic flux density are equal at the boundary*. Equation (5.51a) can also be expressed in vector form as

$$\vec{\mathbf{a}}_n \cdot (\vec{\mathbf{B}}_1 - \vec{\mathbf{B}}_2) = 0 \tag{5.51b}$$

5.10.2 Boundary condition for tangential components of $\vec{\mathbf{H}}$ field

To obtain the boundary condition for the tangential components of the $\vec{\mathbf{H}}$ field, consider the closed path shown in Figure 5.32. Applying Ampère's law to the closed path, we obtain

$$\oint_{c} \vec{\mathbf{H}} \cdot \overrightarrow{d\ell} = \int_{c_1} \vec{\mathbf{H}} \cdot \overrightarrow{d\ell} + \int_{c_2} \vec{\mathbf{H}} \cdot \overrightarrow{d\ell} + \int_{c_3} \vec{\mathbf{H}} \cdot \overrightarrow{d\ell} + \int_{c_4} \vec{\mathbf{H}} \cdot \overrightarrow{d\ell} = I$$

where I is the total current enclosed by the closed path c.

The paths c_2 and c_4 are each of vanishingly small thickness, $\Delta w \to 0$, and their contributions to the total mmf drop can be neglected. Thus, dropping these integrals, we have

$$\int_{c_1} \vec{\mathbf{H}} \cdot \overrightarrow{d\ell} + \int_{c_3} \vec{\mathbf{H}} \cdot \overrightarrow{d\ell} = I$$

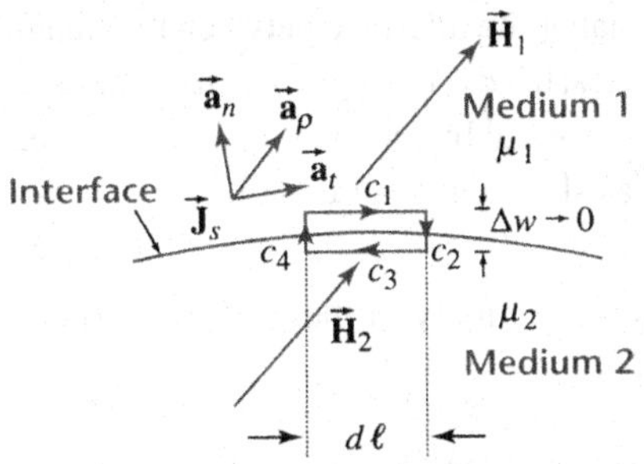

Figure 5.32 Boundary condition for the tangential components of the $\vec{\mathbf{H}}$ field

If $\vec{\mathbf{a}}_n$, $\vec{\mathbf{a}}_t$, and $\vec{\mathbf{a}}_\rho$ are three mutually perpendicular unit vectors, as shown in the figure, and I can be expressed in terms of the volume current density, the preceding equation can then be written as

$$\int_{c_1} (\vec{\mathbf{H}}_1 - \vec{\mathbf{H}}_2) \cdot \vec{\mathbf{a}}_t \, d\ell = \int_{s} \vec{\mathbf{J}}_v \cdot \vec{\mathbf{a}}_\rho \, d\ell \, \Delta w \tag{5.52}$$

However, in the limit $\Delta w \to 0$,

$$\lim_{\Delta w \to 0} \vec{\mathbf{J}}_v \, \Delta w = \vec{\mathbf{J}}_s$$

where $\vec{\mathbf{J}}_s$ is the surface current density (in A/m). Furthermore, in accordance with the right-hand rule, $\vec{\mathbf{a}}_t = \vec{\mathbf{a}}_\rho \times \vec{\mathbf{a}}_n$. Therefore, we can express (5.52) as

$$\int_{c_1} (\vec{\mathbf{H}}_1 - \vec{\mathbf{H}}_2) \cdot (\vec{\mathbf{a}}_\rho \times \vec{\mathbf{a}}_n) \, d\ell = \int_{c_1} \vec{\mathbf{J}}_s \cdot \vec{\mathbf{a}}_\rho \, d\ell$$

Using the vector identity, (2.128), this expression transforms as

$$\int_{c_1} [\vec{\mathbf{a}}_n \times (\vec{\mathbf{H}}_1 - \vec{\mathbf{H}}_2)] \cdot \vec{\mathbf{a}}_\rho \, d\ell = \int_{c_1} \vec{\mathbf{J}}_s \cdot \vec{\mathbf{a}}_\rho \, d\ell$$

from which we obtain

$$\vec{\mathbf{a}}_n \times (\vec{\mathbf{H}}_1 - \vec{\mathbf{H}}_2) = \vec{\mathbf{J}}_s \tag{5.53a}$$

which states that *the tangential components of the* $\vec{\mathbf{H}}$ *field at the boundary are discontinuous.* Equation (5.53a) can also be written in scalar form as

$$H_{t1} - H_{t2} = J_s \tag{5.53b}$$

In applying (5.53b) we must keep in mind that H_{t1} will be greater than H_{t2} when the surface current density is in the $\vec{\mathbf{a}}_\rho$ direction. Also note that $\vec{\mathbf{a}}_\rho$ is the unit normal to the plane containing the tangential component of $\vec{\mathbf{H}}$. We would like to mention here that for two magnetic media with finite conductivities the surface current density $\vec{\mathbf{J}}_s$ is zero; if there is any current flow in either medium it will be in terms of the volume current density $\vec{\mathbf{J}}_v$. If one of the media is a perfect conductor, $\vec{\mathbf{J}}_s$ exists on the surface of the perfect conductor because there is no magnetic field inside the perfect conductor.

EXAMPLE 5.16

Show that at the interface between two magnetic media with finite conductivities, $\tan\phi_1/\tan\phi_2 = \mu_1/\mu_2$, where ϕ_1 and ϕ_2 are the angles made with the normal by the magnetic fields in region 1 and region 2, respectively, as shown in Figure 5.33.

Solution From the continuity of the normal components of the $\vec{\mathbf{B}}$ field, (5.51a), we obtain

$$B_1 \cos\phi_1 = B_2 \cos\phi_2 \tag{5.54}$$

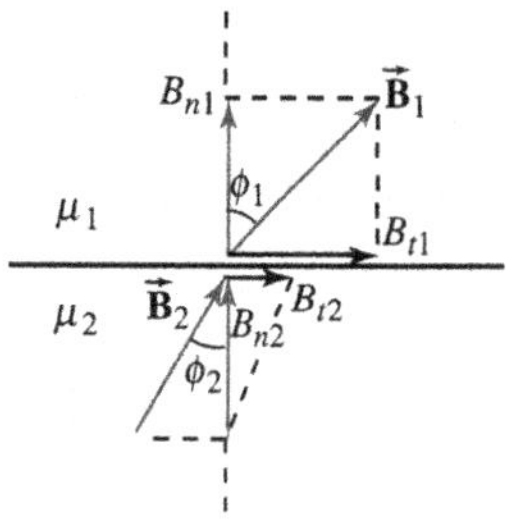

Figure 5.33 Interface between two magnetic media

Since each medium has a finite conductivity, $\vec{\mathbf{J}}_s = 0$. Therefore, the tangential components of the $\vec{\mathbf{H}}$ field, from (5.53b), are also continuous. That is,

$$H_{t1} = H_{t2}$$

or

$$\frac{B_{t1}}{\mu_1} = \frac{B_{t2}}{\mu_2}$$

or

$$B_1 \sin\phi_1 = \frac{\mu_1}{\mu_2} B_2 \sin\phi_2 \tag{5.55}$$

From (5.54) and (5.55), we obtain

$$\frac{\tan\phi_1}{\tan\phi_2} = \frac{\mu_1}{\mu_2}$$

which is the required relation between the angles and the permeabilities of the two regions. However, we want to make the following remarks.

a) If $\phi_1 = 0$, then ϕ_2 is also zero. In other words, the magnetic field lines are normal to the boundary in each region and have the same magnitude.
b) If the permeability of region 2 is very high in comparison with the permeability of region 1 and ϕ_2 is less than 90°, the angle ϕ_1 will be quite small. In other words, the magnetic field lines are normal when they enter a highly permeable magnetic region. For example, if region 1 is free space and region 2 is steel with a relative permeability of 2400 and $\phi_2 = 45^\circ$, then $\phi_1 = 0.02^\circ$. This fact is exploited in the shaping of the magnetic paths in electrical machines. • • •

EXAMPLE 5.17

The magnetic flux density in a finitely conducting cylinder of radius 10 cm and with a relative permeability of 5 is found to vary as $0.2/\rho\vec{\mathbf{a}}_\phi$ T.

If the region surrounding the cylinder is characterized by free space, determine the magnetic flux density just outside the cylinder.

Solution The interface is at a radius of 10 cm; thus, the magnetic flux density in the cylinder just beneath the boundary is

$$\vec{\mathbf{B}}_c = \frac{0.2}{0.1}\vec{\mathbf{a}}_\phi = 2\vec{\mathbf{a}}_\phi \text{ T}$$

Note that the $\vec{\mathbf{B}}$ field is tangential to the boundary. In addition, the finite conductivity of the cylinder suggests that $\vec{\mathbf{J}}_s = 0$. Therefore, the tangential components of the $\vec{\mathbf{H}}$ field must be continuous. However, the tangential component of $\vec{\mathbf{H}}$ in the cylinder at $\rho = 10$ cm is

$$\vec{\mathbf{H}}_c = \frac{2}{5 \times 4\pi \times 10^{-7}}\vec{\mathbf{a}}_\phi = 318.31\vec{\mathbf{a}}_\phi \text{ kA/m}$$

Thus, the magnetic field intensity just above the surface of the cylinder in free space is

$$\vec{\mathbf{H}}_a = \vec{\mathbf{H}}_c = 318.31\vec{\mathbf{a}}_\phi \text{ kA/m}$$

Finally, the magnetic flux density just above the surface of the cylinder in free space is

$$\vec{\mathbf{B}}_a = \mu_0\vec{\mathbf{H}}_a = 4\pi \times 10^{-7} \times 318.31 \times 10^3\vec{\mathbf{a}}_\phi = 0.4\vec{\mathbf{a}}_\phi \text{ T}$$

• • •

5.11 Energy in a magnetic field

We conceive of the magnetic field as storing magnetic energy in the same sense that we conceive of the electric field as storing electric energy. The magnetic energy should come from the circuit containing a current-carrying conductor, or the coil, which establishes the magnetic field. If the field is established in free space, all the magnetic energy is recoverable and returned to the circuit when the current ceases and the magnetic field collapses. In any other medium, a part of the energy is lost in that medium and is not recoverable. We will discuss in detail the loss of magnetic energy in a magnetic medium in Chapter 7.

In Chapter 3, we went to great lengths to derive an expression for the energy density in an electric field,

$$w_e = \frac{1}{2}\vec{\mathbf{D}} \cdot \vec{\mathbf{E}}$$

and the total electric energy stored in a medium,

$$W_e = \frac{1}{2}\int_v \vec{\mathbf{D}} \cdot \vec{\mathbf{E}}\,dv$$

It turns out that we cannot easily obtain an expression for the energy density in a static magnetic field. However, we will obtain an expression for the energy density in a time-varying magnetic field in Chapter 7. Until then, we will just state that the energy density w_m in a magnetic field is

$$w_m = \frac{1}{2}\vec{\mathbf{B}} \cdot \vec{\mathbf{H}} \tag{5.56a}$$

based upon the similar expression for the electric field. Because $\vec{\mathbf{B}} = \mu\vec{\mathbf{H}}$, we can also express (5.56a) as

$$w_m = \frac{1}{2}\,\mu H^2 = \frac{1}{2\mu}\,B^2 \tag{5.56b}$$

The total magnetic energy in any finite volume is simply the integral of the magnetic energy density over the volume. That is,

$$W_m = \int_v w_m\,dv \tag{5.57}$$

where W_m is the total magnetic energy in joules.

EXAMPLE 5.18

Calculate the energy stored in the magnetic field of the toroidal winding discussed in Example 5.13.

Solution We obtained an expression for the magnetic field intensity inside the toroid as

$$\vec{\mathbf{H}} = \frac{NI}{2\pi\rho}\,\vec{\mathbf{a}}_\phi, \quad \text{for} \quad a \le \rho \le b$$

where a and b are the inner and the outer radii of the toroid. Thus, the magnetic energy density within the toroid, from (5.56b), is

$$w_m = \frac{1}{2}\,\mu_0 H^2 = \frac{1}{8}\,\mu_0\left[\frac{NI}{\pi\rho}\right]^2$$

The total magnetic energy within the toroid is

$$\begin{aligned} W_m &= \frac{N^2I^2}{8\pi^2}\,\mu_0 \int_a^b \frac{1}{\rho}\,d\rho \int_0^{2\pi} d\phi \int_0^h dz \\ &= \frac{\mu_0}{4\pi}\,N^2I^2h\ln[b/a] \end{aligned}$$

• • •

5.12 Magnetic circuits

As the magnetic flux lines form a closed path, and the magnetic flux entering a boundary is the same as the magnetic flux leaving a boundary, we can draw an analogy between the magnetic flux and the current in a closed conducting circuit. In a conducting circuit the current flows exclusively through the conductor without any leakage through the region surrounding the conductor. The magnetic flux cannot be completely confined to follow a given path in a magnetic material. However, if the permeability of the magnetic material is very high compared to that of the material surrounding it, such as free space, most of the flux will be confined to the highly permeable material. This causes the magnetic flux to be concentrated within a magnetic material with almost negligible flux existing in the region surrounding it. Magnetic shielding is based upon such a behavior of the magnetic flux. The channeling of the flux through a highly permeable material is very similar to the current flow through a conductor. For that reason, we call the closed path followed by the flux in a magnetic material a *magnetic circuit*. Magnetic circuits form an integral part of such devices as rotating machines, transformers, electromagnets, and relays.

We have tacitly considered a simple magnetic circuit in the form of a toroid wound with a closely spaced helical winding. We have stated that the magnetic flux existed only within the core of the toroid. Let us now extend and generalize that observation. When the core of the toroid is made of a very highly permeable magnetic material with the winding concentrated only over its small portion, a large portion of the magnetic flux will still circulate through the core of the toroid. A fraction of the total flux produced by the coil does complete its path through the medium surrounding the magnetic circuit and is referred to as the *leakage flux*. In the design of magnetic circuits, an attempt is always made to keep the leakage flux to a minimum value that is economically possible. For this reason, in the analysis of magnetic circuits, we will disregard the leakage flux.

In the case of a toroid, we found that the magnetic field intensity and thereby the magnetic flux density were inversely proportional to the radius of the circular path. In other words, the magnetic flux density is maximum at the inner radius of the toroid and minimum at the outer radius. In the analysis of magnetic circuits, we usually assume that the magnetic flux density is uniform within the magnetic material, and its magnitude is equal to the magnetic flux density at the mean radius.

The toroid that we studied formed a continuous closed path for the magnetic circuit. However, in applications such as rotating machines, the closed path is broken by an air gap. The magnetic circuit now consists of a highly permeable magnetic material in series with an air gap, as depicted in Figure 5.34. Because it is a series circuit, the magnetic flux in the magnetic material is equal to the magnetic flux in the air gap. The spreading of the magnetic flux in the air gap, known as *fringing*, is inevitable, as shown in the figure. However, if the length of the air gap is very small compared to its other dimensions, most of the flux lines are well confined between the opposite surfaces of the magnetic core at the air gap, and the fringing effect is negligible.

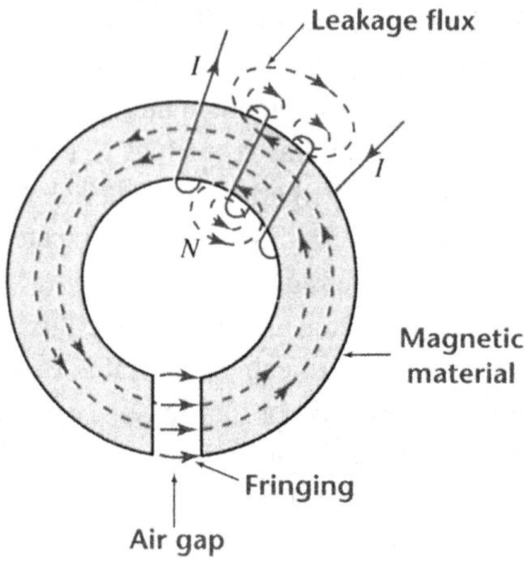

Figure 5.34 A magnetic circuit with an air gap

In summary, we will always assume that

a) The magnetic flux is restricted to flow through the magnetic material with no leakage,
b) There is no spreading or fringing of the magnetic flux in the air gap regions, and
c) The magnetic flux density is uniform within the magnetic material.

Let us now consider a magnetic circuit, as shown in Figure 5.35a. If the coil has N turns and carries a current I, the applied magnetomotive force (mmf) is NI. Even though in the SI system of units, the turn is a dimensionless quantity, we will still use the ampere-turn (A.t) as the unit of mmf in order to differentiate it from the basic unit of current. Thus,

$$\mathscr{F} = NI = \oint_c \vec{\mathbf{H}} \cdot \vec{d\ell}$$

If the magnetic field intensity is considered to be uniform within the magnetic material, then this equation becomes

$$HL = NI \tag{5.58}$$

where L is the mean length of the magnetic path, as shown in the figure. The magnetic flux density in the magnetic material is

$$B = \mu H = \frac{\mu NI}{L}$$

where μ is the permeability of the magnetic material. The flux in the magnetic material is

$$\Phi = \int_s \vec{\mathbf{B}} \cdot \vec{d\mathbf{s}} = BA = \frac{\mu NIA}{L}$$

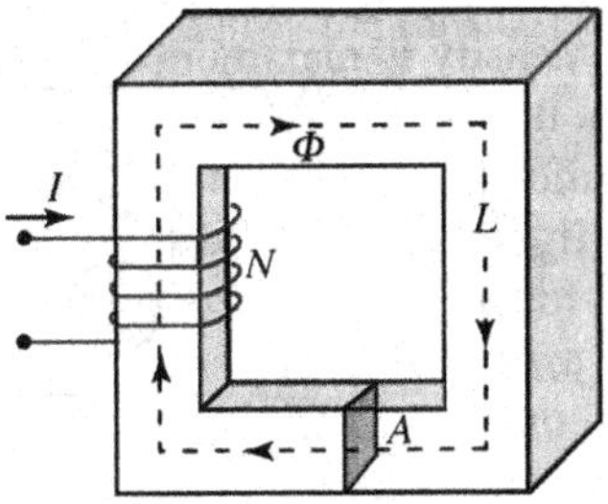

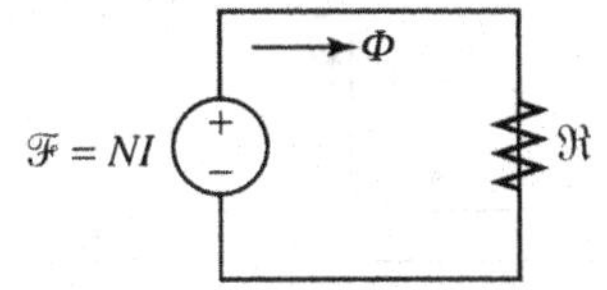

Figure 5.35 (a) Magnetic circuit with mean length L and cross-sectional area A; (b) Its equivalent circuit

where A is the cross-sectional area of the magnetic material. This equation can also be written as

$$\Phi = \frac{NI}{L/\mu A} = \frac{\mathscr{F}}{L/\mu A} \tag{5.59}$$

By considering the magnetic flux and the applied magnetomotive force (mmf) in the magnetic circuit analogous to the current and the applied electromotive force (emf) in the electric circuit, the quantity in the denominator of (5.59) must be like the resistance in an electric circuit. This quantity is defined as the *reluctance* of the magnetic circuit. It is denoted by $\mathscr{R}$ and has the units of ampere-turns per weber (A.t/Wb). Thus,

$$\mathscr{R} = \frac{L}{\mu A} \tag{5.60}$$

In terms of the reluctance $\mathscr{R}$, we can rearrange (5.59) as

$$\Phi\mathscr{R} = NI \tag{5.61}$$

Equation (5.61) is known as *Ohm's law for the magnetic circuit.* Since the resistance of a conductor is

$$R = L/\sigma A$$

the permeability of a magnetic material is similar to the conductivity of a conductor. The higher the permeability of the magnetic material, the lower is its reluctance. For the same applied mmf, the flux in a highly permeable material will be higher than that in a material of low permeability. This result should not surprise us because it is in accordance with our assumptions. We can now represent the magnetic circuit by an equivalent circuit as shown in Figure 5.35b.

When the magnetic circuit consists of two or more sections of magnetic material, as portrayed in Figure 5.36a, it can be represented in terms of reluctances, as shown in Figure 5.36b. The total reluctance can be obtained from series and parallel combinations of reluctances of individual sections because the reluctances obey the same rules as resistances.

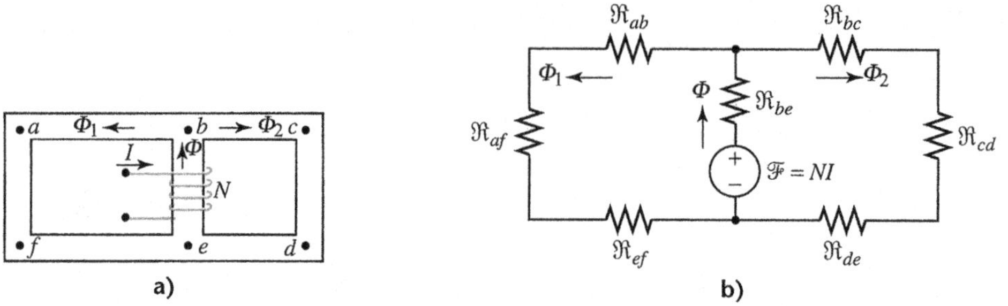

Figure 5.36 (a) A series-parallel magnetic circuit of uniform thickness; (b) Its equivalent circuit

If H_i is the magnetic field intensity in the ith section of a magnetic circuit and L_i is the mean length, then the total mmf drop in the magnetic circuit must be equal to the applied mmf. That is,

$$\sum_{i=1}^{n} H_i L_i = NI \tag{5.62}$$

Equation (5.62) is analogous to Kirchhoff's voltage law for an electric circuit.

It appears from (5.62) that each magnetic circuit can always be analyzed using an analogous circuit. However, this is only true for linear magnetic materials; i.e., those with constant permeability. For a ferromagnetic material, the permeability is a function of the magnetic flux density, as shown in Figure 5.37. The curve describes a relation between the applied mmf and the magnetic flux density in the magnetic material. It is referred to as the *magnetization characteristic* or simply the *B–H curve*. When the permeability of a magnetic material varies with the flux density, the magnetic circuit is said to be *nonlinear*. All devices

Figure 5.37 Magnetization characteristic (*B–H* curve) for a magnetic material

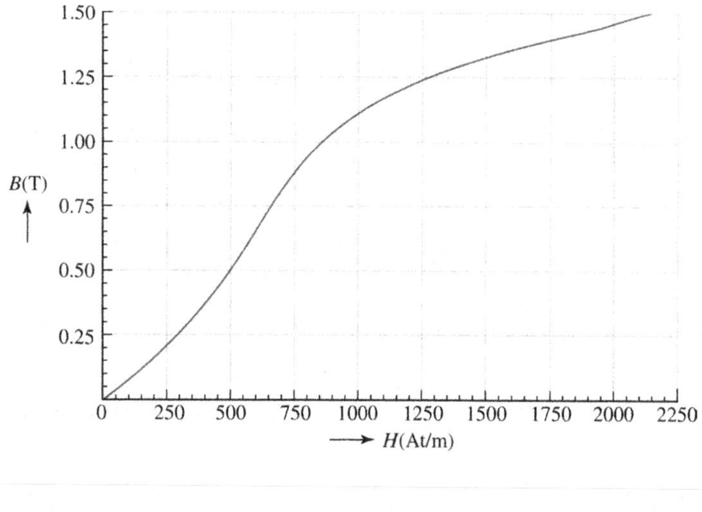

using ferromagnetic materials such as iron form nonlinear magnetic circuits.

There are basically two types of problems pertaining to the analysis of magnetic circuits. The first type of problem requires the determination of the applied mmf to establish a given flux density in a magnetic circuit. The other problem deals with the calculation of magnetic flux density and thereby the flux in a magnetic circuit when the applied mmf is given.

For a linear magnetic circuit, the solution to either problem can be obtained using the equivalent circuit approach because the permeability of the magnetic material is constant. In a nonlinear magnetic circuit, it is relatively simple and straightforward to determine the required mmf in order to maintain a certain flux density in the magnetic circuit. In this case, we can calculate the flux density in each magnetic section and then obtain H from the B–H curve. Knowing H we can determine the mmf drops across each magnetic section. The required mmf is simply the sum of the individual mmf drops in accordance with (5.62).

The second type of problem in a nonlinear circuit may be solved using an iterative technique. In this case, we make an educated guess for the mmf drop in one of the magnetic regions and then obtain the total mmf requirements. At this time, we compare the results with the given mmf and make another educated guess if we are quite off. By iterating this way, we soon arrive at a situation when the error between the calculated mmf and the applied mmf is within permissible limits. What constitutes a permissible limit is another debatable point. In our discussion, we will use $\pm 2\%$ as the permissible limit for the error if it is not specified. A computer program can be written to reduce the error even further. We next provide some examples of linear and nonlinear magnetic circuits.

EXAMPLE 5.19

An electromagnet of square cross section similar to the one shown in Figure 5.34 has a tightly wound coil with 1500 turns. The inner and the outer radii of the magnetic core are 10 cm and 12 cm, respectively. The length of the air gap is 1 cm. If the current in the coil is 4 A and the relative permeability of the magnetic material is 1200, determine the flux density in the magnetic circuit.

Solution Since the permeability of the magnetic material is given to be a constant and the applied mmf is known, we can use the reluctance method to determine the flux density in the core.

The mean radius is 11 cm and the mean length of the magnetic path is

$$L_m = 2\pi \times 11 - 1 = 68.12 \text{ cm}$$

Neglecting the effect of fringing, the cross-sectional area of the magnetic path is the same as the air gap. That is,

$$A_m = A_g = 2 \times 2 = 4 \text{ cm}^2$$

The reluctance of each region is

$$\mathcal{R}_m = \frac{68.12 \times 10^{-2}}{1200 \times 4\pi \times 10^{-7} \times 4 \times 10^{-4}} = 1.129 \times 10^6 \text{A.t/Wb}$$

$$\mathcal{R}_g = \frac{1 \times 10^{-2}}{4\pi \times 10^{-7} \times 4 \times 10^{-4}} = 19.894 \times 10^6 \text{A.t/Wb}$$

The total reluctance in the series circuit is

$$\mathcal{R} = \mathcal{R}_m + \mathcal{R}_g = 21.023 \times 10^6 \text{A.t/Wb}$$

Thus, the flux in the magnetic circuit is

$$\Phi = \frac{1500 \times 4}{21.023 \times 10^6} = 285.402 \times 10^{-6} \text{ Wb}$$

The flux density either in the air gap or in the magnetic region is

$$B_m = B_g = \frac{285.402 \times 10^{-6}}{4 \times 10^{-4}} = 0.714 \text{ T}$$

• • •

EXAMPLE 5.20

A series-parallel magnetic circuit with its pertinent dimensions in centimeters is given in Figure 5.38. If the flux density in the air gap is 0.05 T, and the relative permeability of the magnetic region is 500, calculate the current in the 1000-turn coil using the fields approach.

Figure 5.38

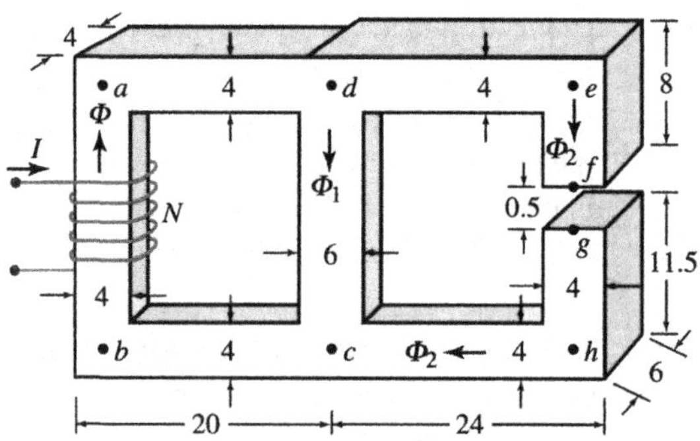

Solution Since the flux density in the air gap is given, we can calculate the flux in the air gap. The magnetic sections *def* and *chg* are in series with the air

Section	Flux (mWb)	Area (cm^2)	*B* (T)	*H* (At/m)	*L* (cm)	mmf (At)
fg	0.12	24	0.05	39,788.74	0.5	198.94
def	0.12	24	0.05	79.58	28.0	22.28
chg	0.12	24	0.05	79.58	31.5	25.07
Total mmf drop for magnetic sections *fg*, *def*, and *chg*						246.29

gap; therefore, they carry the same flux. Thus, we can compute the mmf drop for each of these sections as tabulated here. Since the mmf drop for the region *dc* is the same as that for the combined regions *fg*, *def*, and *chg*, we can determine the flux in region *dc* by working backward. The flux in the region *dabc* is the sum of the fluxes in regions *dc* and *fg*. The mmf drop for each of these regions is tabulated below.

Section	Flux (mWb)	Area (cm^2)	*B* (T)	*H* (A.t/m)	*L* (cm)	mmf (A.t)
dc	3.48	36	0.967	1539.31	16	246.29
ad	3.60	16	2.25	3580.99	18	644.58
ab	3.60	16	2.25	3580.99	16	572.96
bc	3.60	16	2.25	3580.99	18	644.58
Total mmf drop for the magnetic circuit						2108.41

The current in the coil: $I = 2108.41/1000 = 2.108$ A. • • •

EXAMPLE 5.21

A magnetic circuit with its pertinent dimensions in millimeters is given in Figure 5.39. The magnetization characteristic of the magnetic material is shown in Figure 5.37. If the magnetic circuit has a uniform thickness of 20 mm and the flux density in the air gap is 1.0 T, find the current in the 500-turn coil.

Solution The permeability of the magnetic material depends upon the flux density, thus we cannot compute the reluctance unless the flux density is known. Problems of this type can be easily solved using the fields approach.

The flux density in the air gap is known; thus we compute the flux in the air gap as

$$\Phi_{ab} = 1.0 \times 6 \times 20 \times 10^{-6} = 0.12 \times 10^{-3} \text{ Wb}$$

The flux in each section of the circuit is the same because the given magnetic structure forms a series magnetic circuit. We can now compute

Figure 5.39

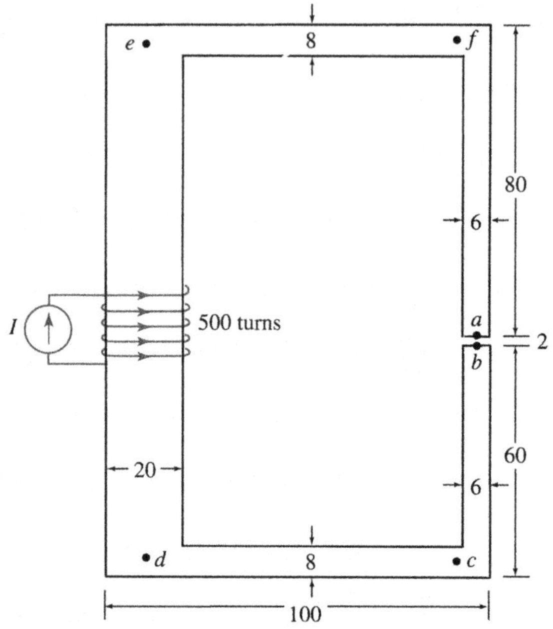

the mmf drop for each section using the fields approach as tabulated below.

Section	Flux (mWb)	Area (mm^2)	*B* (T)	*H* (A.t/m)	*L* (mm)	mmf (A.t)
ab	0.12	120	1.00	795,774.72	2	1591.55
bc	0.12	120	1.00	850.00	56	47.60
cd	0.12	160	0.75	650.00	87	56.55
de	0.12	400	0.30	350.00	134	46.90
ef	0.12	160	0.75	650.00	87	56.55
fa	0.12	120	1.00	850.00	76	64.60
	Total mmf drop in the magnetic circuit					1863.75

Therefore, the current in the 500-turn coil is

$$I = \frac{1863.75}{500} = 3.73\text{A}.$$

• • •

EXAMPLE 5.22

WORKSHEET 3

Mathcad

A magnetic circuit with its mean lengths and cross-sectional areas is shown in Figure 5.40. If a 600-turn coil carries a current of 10 A, what is the flux in the series magnetic circuit? Use the magnetization curve given in Figure 5.37 for the magnetic material.

Solution The applied mmf $= 600 \times 10 = 6000$ A.t. Since the magnetic circuit is nonlinear, we have to use the iterative method to determine the flux. In

Figure 5.40

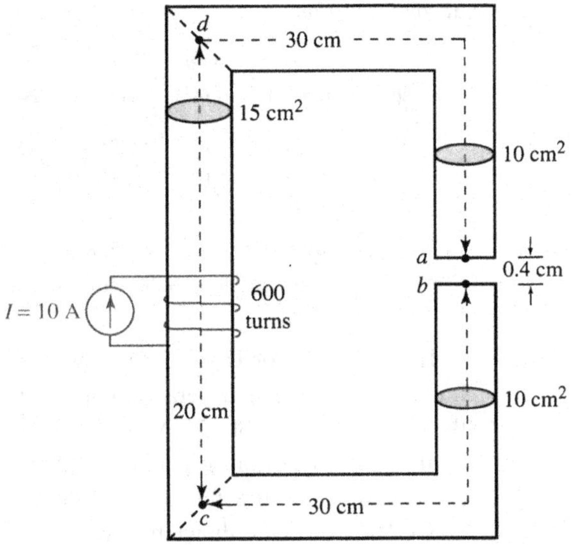

the absence of any other information, let us assume that 50% of the total mmf drop takes place in the air gap. We can now calculate the total mmf drop using the fields approach. The results are given in the following table.

First iteration

Section	Flux (mWb)	Area (cm^2)	B (T)	H (A.t/m)	L (cm)	mmf (A.t)
ab	0.942	10	0.942	750,000	0.4	3000
bc	0.942	10	0.942	780	30.0	234
cd	0.942	15	0.628	570	20.0	114
da	0.942	10	0.942	780	30.0	234
	Total mmf drop in the series magnetic circuit					3582

It is now evident that most of the applied mmf appears as a drop across the air gap. The ratio of the mmf drop across the air gap to the total mmf drop is 0.837 (3000/3582); that is, the mmf drop across the air gap seems to be 83.7% of the applied mmf. However, any increase in the mmf drop in the air gap increases the flux density in each magnetic region. Nonlinear magnetic behavior may also cause the increase in mmf drop in each magnetic section to be considerable. So, instead of 83.7% of the total mmf drop across the air gap, let us assume that the mmf drop is only 80%. Hence, we begin our second iteration with an mmf drop of 4800 At (0.8×6000) across the air gap. The results are shown below.

Second iteration

Section	Flux (mWb)	Area (cm^2)	B (T)	H (A.t/m)	L (cm)	mmf (A.t)
ab	1.508	10	1.508	1,200,000	0.4	4800.0
bc	1.508	10	1.508	2,175	30.0	652.5
cd	1.508	15	1.005	850	20.0	170.0
da	1.508	10	1.508	2,175	30.0	652.5
	Total mmf drop in the series magnetic circuit					6275.0

The percent error is still 4.58%, which is not within the desirable limit. From the second iteration table, we can conclude that most of the extra mmf drop of 275 A.t is across the air gap. If we reduce the mmf drop across the air gap to 4600 A.t or so, it is possible to bring the percent error to well within $\pm 2\%$. Let us perform one more iteration to do so. The results are shown below.

Third iteration

Section	Flux (mWb)	Area (cm^2)	B (T)	H (A.t/m)	L (cm)	mmf (A.t)
ab	1.445	10	1.445	1,150,000	0.4	4600
bc	1.445	10	1.445	1,950	30.0	585
cd	1.445	15	0.963	820	20.0	164
da	1.445	10	1.445	1,950	30.0	585
	Total mmf drop in the series magnetic circuit					5934

The percent error is now -1.1%, which is well within the desirable limit. Therefore, no further iteration is necessary. The flux in the magnetic structure is 1.445 mWb. • • •

5.13 Summary

Magnetostatic field theory is the study of time-independent fields established by moving charges. Since a moving charge constitutes a current flow, we defined the magnetic flux density in a medium in terms of the current (Biot–Savart law) as

$$\vec{\mathbf{B}} = \frac{\mu}{4\pi} \int_c \frac{I d\vec{\ell} \times \vec{\mathbf{R}}}{R^3}$$

We also defined the magnetic flux density in terms of the magnetic field intensity as

$$\vec{\mathbf{B}} = \mu \vec{\mathbf{H}}$$

The force experienced by a current-carrying conductor in a magnetic field is given by Ampère's force law,

$$\vec{\mathbf{F}} = \int_c I\overrightarrow{d\ell} \times \vec{\mathbf{B}}$$

where $\vec{\mathbf{B}}$ is the magnetic flux density produced by a source other than the current I given in the equation.

If a current-carrying loop is placed in a magnetic field, it experiences a torque

$$\vec{\mathbf{T}} = \vec{\mathbf{m}} \times \vec{\mathbf{B}}$$

that tends to rotate it until its plane becomes normal to the $\vec{\mathbf{B}}$ field. The magnitude of the magnetic dipole moment ($\vec{\mathbf{m}}$) is simply the product of the current and the cross-sectional area of the loop. Its direction is given by the right-hand rule.

The magnetic flux through an open surface is

$$\Phi = \int_s \vec{\mathbf{B}} \cdot \overrightarrow{d\mathbf{s}}$$

The net flux leaving a closed surface is zero. That is,

$$\oint_s \vec{\mathbf{B}} \cdot \overrightarrow{d\mathbf{s}} = 0 \quad \text{or} \quad \nabla \cdot \vec{\mathbf{B}} = 0$$

This equation is known as *Gauss's law for the magnetic field.* It states that the magnetic flux density is solenoidal.

Ampère's circuital law, given as

$$\oint_c \vec{\mathbf{H}} \cdot \overrightarrow{d\ell} = I \quad \text{or} \quad \nabla \times \vec{\mathbf{H}} = \vec{\mathbf{J}}_v$$

enables us to compute the $\vec{\mathbf{H}}$ field with much ease as long as the current or the current distribution possesses a high degree of symmetry.

We can also compute $\vec{\mathbf{B}}$ from the magnetic vector potential $\vec{\mathbf{A}}$ as

$$\vec{\mathbf{B}} = \nabla \times \vec{\mathbf{A}}$$

where

$$\vec{\mathbf{A}} = \frac{\mu}{4\pi} \oint_c \frac{I\overrightarrow{d\ell}}{R}$$

For surface or volume current distribution, $I\,\overrightarrow{d\ell}$ in this equation can be replaced by $\vec{\mathbf{J}}_s\,ds$ or $\vec{\mathbf{J}}_v\,dv$.

The flux passing through an open surface can also be given in terms of $\vec{\mathbf{A}}$ as

$$\Phi = \oint_c \vec{\mathbf{A}} \cdot \overrightarrow{d\ell}$$

where c is the contour bounding the open surface s.

We also obtained expressions for the bound volume current density and the bound surface current density in a magnetic region as

$$\vec{\mathbf{J}}_{vb} = \nabla \times \vec{\mathbf{M}}$$

and

$$\vec{\mathbf{J}}_{sb} = \vec{\mathbf{M}} \times \vec{\mathbf{a}}_n$$

where $\vec{\mathbf{M}}$ is the polarization vector.

The two boundary conditions for magnetic fields are

$$\vec{\mathbf{a}}_n \cdot (\vec{\mathbf{B}}_1 - \vec{\mathbf{B}}_2) = 0$$

and

$$\vec{\mathbf{a}}_n \times (\vec{\mathbf{H}}_1 - \vec{\mathbf{H}}_2) = \vec{\mathbf{J}}_s$$

The energy density in a magnetic field is

$$w_m = \frac{1}{2}\vec{\mathbf{B}} \cdot \vec{\mathbf{H}} = \frac{1}{2}\mu H^2 = \frac{1}{2\mu}B^2$$

We analyzed magnetic circuits using the magnetic scalar potential. We defined the magnetic potential difference between any two points a and b as

$$\mathscr{F}_{ab} = \mathscr{F}_a - \mathscr{F}_b = -\int_b^a \vec{\mathbf{H}} \cdot \overrightarrow{d\ell}$$

We obtained an expression for Ohm's law for the magnetic field as

$$NI = \Phi\mathscr{R}$$

where

$$\mathscr{R} = \frac{L}{\mu A}$$

where $\mathscr{R}$ is the reluctance of the magnetic circuit.

We also mentioned that the mmf drop in a closed magnetic circuit is equal to the total applied mmf. That is,

$$\sum_{i=1}^{n} H_i L_i = NI$$

We used the above equation to analyze magnetic circuits. With the help of examples we illustrated the techniques to analyze two types of magnetic circuit problems.

5.14 Review questions

5.1 A charge q is moving with a velocity $\vec{\mathbf{U}}$ in free space. Write an expression for the magnetic field produced at any point by this charge.

5.2 Does a charge in motion establish an electric field?

5.3 Does a charge at rest establish a magnetic field?

5.4 If a stationary charge is placed in a magnetic field, what is the force experienced by it?

5.5 State the Biot-Savart law in your own words.

5.6 Two charges are moving in the same direction. What is the force experienced by either one of them?

5.7 A charge is moving in the direction of the $\vec{\mathbf{B}}$ field. What is the force experienced by the charge?

5.8 A charged particle passes through a magnetic field without experiencing any force. What can you conclude about the magnetic field?

5.9 An electron moving in the x direction enters a region in which the magnetic field is in the y direction. What is the direction of motion of the electron in the region?

5.10 What is a magnetic dipole? How does a magnetic dipole differ from an electric dipole?

5.11 When can we use Ampère's circuital law to determine the magnetic field?

5.12 Is the magnetic field intensity, in general, a conservative field?

5.13 A wire 10 cm in diameter carries a current of 100 A. Find the magnetic field intensity just outside the wire if (a) the current is uniformly distributed within the wire, and (b) the current flows just over the surface of the wire.

5.14 What is the magnetic field intensity just inside the wire in Question 5.13?

5.15 Explain Ampère's force law.

5.16 Two parallel conductors carry currents in the same direction. Is the force experienced by them attractive or repulsive?

5.17 What do we mean by magnetic vector potential?

5.18 What is the difference between magnetic vector potential and magnetic scalar potential?

5.19 Why is there no such thing as electric vector potential in the case of electrostatic fields?

5.20 What must be the condition for $\vec{\mathbf{J}}_s$ to exist at the boundary between free space and a magnetic material?

5.21 Can you derive the Biot–Savart law from the magnetic vector potential?

5.22 What is the significance of $\nabla \cdot \vec{\mathbf{B}} = 0$?

5.23 What do we mean by bound volume current density and bound surface current density?

5.24 What is the significance of the polarization vector?

5.25 What is the relation between mmf and reluctance?

5.26 What is the relationship between flux and reluctance? Is it a linear relation?

5.27 Can we use the force equation to define current? If yes, how?

5.28 Is any energy required to maintain a steady magnetic flux after it is established? Explain.

5.29 In the design of magnetic circuits we always strive to set up the required flux density with as few mmf as possible. What does this mean in terms of (a) the permeability of the magnetic material and (b) the reluctance of the magnetic material?

5.30 Is it possible for the mmf to be numerically equal to the current producing it?

5.31 What happens to the reluctance of a ferromagnetic material if the flux density is increased?

5.32 Does the permeability of a ferromagnetic material increase as the flux density in the material decreases?

5.33 Does the permeability of a nonmagnetic material change with the flux density? How about the reluctance?

5.34 Can we apply Ohm's law for the magnetic field to ferromagnetic materials?

5.35 What do we mean when we say that a magnetic material is saturated? What happens to the permeability of a magnetic material when it is saturated?

5.36 What is hysteresis? What is meant by hysteresis loss? Is the hysteresis loss more for hard steel than for soft steel?

5.37 A nickel-cobalt-manganese alloy known as *Perminvar* is found to have constant permeability over a wide range of flux densities. Why is it important to find such a material?

5.38 Bismuth is not a ferromagnetic material; however, it changes its resistance when placed in a magnetic field. Do you think we can use it to measure the magnetic flux density in a region? Explain.

5.39 There are some alloys for which the Curie point is quite low. Each of these alloys has been found to show appreciable change in its permeability for a small change in temperature. Do you think we can use such alloys to measure temperature? Explain.

5.40 What is residual flux density?

5.15 Exercises

5.1 A very thin wire extends from $z = 0$ to $z = \infty$ and carries a current I. Obtain an expression for the magnetic flux density at any point in the $z = 0$ plane.

5.2 A straight wire extends from $z = -L$ to $z = L$ and carries a current I. What is the $\vec{\mathbf{B}}$ field in a plane bisecting the wire?

5.3 A cylinder of radius b and length L is closely and tightly wound with N turns of a very fine wire. If the wire carries a steady current I, find the magnetic flux density at any point on the axis of the cylinder. What is the magnetic flux density at the center of the cylinder? Also obtain expressions for the $\vec{\mathbf{B}}$ field at the ends of the cylinder.

5.4 Assuming (5.14) as the fundamental law of magnetic force, obtain expressions for the Biot-Savart law and Ampère's force law.

5.5 Verify (5.17) and (5.18).

5.6 If $L = 10$ m, $b = 2$ cm, $c = 10$ cm, $a = 5$ cm, and $I_1 = I_2 = 10$ A, find the magnetic force exerted on the loop in Example 5.5.

5.7 A 10-turn coil, 10 cm by 20 cm, is placed in a magnetic field of 0.8 T. The coil carries a current of 15 A and is free to rotate about its long axis. Plot the torque experienced by the coil against the displacement angle of the coil for the one complete rotation.

5.8 A D'Arsonval meter is designed to have a coil of 25 turns mounted in a magnetic field of 0.2 T. The coil is 4 cm long and 2.5 cm broad. The restoring torque of the meter is applied by a spring and is proportional to the deflection angle θ. The spring constant is 50 micronewton meters per degree. The scale covers 50° of arc and is divided into 100 equal parts. The meter design is such that the magnetic field is always in the radial direction with respect to the axis of the coil. Calculate the current through the coil (a) per degree of deflection, (b) per scale division, and (c) for full-scale deflection.

5.9 Show that the $\vec{\mathbf{B}}$ field set up by an infinitely long current-carrying conductor satisfies Gauss's law.

5.10 Repeat Example 5.7 when both the conductors carry currents in the z direction.

5.11 A cube of edge $2b$ is centered at the origin. A very long, straight wire located along the z axis carries a current I in the z direction. Find the flux passing through the surface at $x = b$.

5.12 If $\vec{\mathbf{B}} = 12x\vec{\mathbf{a}}_x + 25y\vec{\mathbf{a}}_y + cz\vec{\mathbf{a}}_z$, find c.

5.13 Verify the result in Example 5.8 by integrating over the surface of the hemisphere.

5.14 Determine the total flux enclosed in Example 5.10 using the magnetic flux density in the region within the conductors.

5.15 A short, straight conductor of length L carries a current I in the z direction. Show that the magnetic vector potential at a point far away from the conductor is

$$\vec{\mathbf{A}} = \frac{\mu_0 IL}{4\pi R}\vec{\mathbf{a}}_z$$

where R is the distance of the point of observation from the origin. What is the magnetic flux density at that point?

5.16 A very long, straight conductor located along the z axis has a circular cross section of radius 10 cm. The conductor carries 100 A in the z

direction which is uniformly distributed over its cross section. Find the magnetic field intensity (a) inside the conductor and (b) outside the conductor. Sketch the magnetic field intensity as a function of the distance from the center of the conductor.

5.17 Find the magnetic field intensity, the magnetic flux density, and the total flux inside the ring of Example 5.13 if $N = 500$ turns, $a = 15$ cm, $b = 20$ cm, $h = 5$ cm, and $I = 2$ A. If the magnetic field intensity is assumed to be uniform within the ring and has a magnitude equal to that at the mean radius, compute the magnetic flux density and the total flux in the ring. What is the error introduced by this assumption?

5.18 A fine wire wound in the form of a tight helical coil is said to form a solenoid. If the inner radius of the coil is b, and the solenoid is very long, show that the magnitude of the magnetic field intensity within the coil is nI, where I is the current in the coil and n is the number of turns per unit length. Also compute the magnetic flux density within the coil and the total flux within (linking) the coil.

5.19 Repeat Exercise 5.18 if the material on which the coil is wound has a relative permeability of μ_r. Also calculate the magnetization vector $\vec{\mathbf{M}}$, the bound volume current density $\vec{\mathbf{J}}_{vb}$, and the bound surface current density $\vec{\mathbf{J}}_{sb}$.

5.20 Find the magnetic flux density, bound volume current density, bound surface current density, and the total flux inside the ring in Example 5.14 if $\mu_r = 1200$, $N = 500$ turns, $I = 2$ A, $a = 15$ cm, $= 20$ cm, and $h = 5$ cm.

5.21 Consider the toroidal winding discussed in Example 5.13. Obtain an expression for the mmf drop between any two points on the mean radius of the ring.

5.22 What is the effect of the permeability on the magnetic potential difference between any two points in a region devoid of currents when the flux density in the region is known?

5.23 A circular loop of radius b lies in the xy plane and carries a current I in the ϕ direction. Obtain an expression for the magnetic potential difference between any two points on the z axis. What is the magnetic potential difference between the points $P(0, 0, 0)$ and $Q(0, 0, \infty)$?

5.24 The relative permeability and the magnetic flux density in a finitely conducting magnetic region bounded by a plane $2y - x + 4 \leq 0$ are 10 and $\vec{\mathbf{B}} = 2\vec{\mathbf{a}}_x + 3\vec{\mathbf{a}}_y + 5\vec{\mathbf{a}}_z$ T, respectively. If the other region is characterized by free space, compute (a) the $\vec{\mathbf{H}}$ field in both regions, (b) the magnetization vector in the magnetic region, and (c) the $\vec{\mathbf{B}}$ field in free space.

5.25 Consider a plane interface at $x = 0$ between air and a magnetic material. The magnitude of the flux density in air is 0.5 T and it makes an angle of 5° with the x axis. If the flux density in the magnetic region is 1.2 T, determine (a) the angle it makes with the x axis and (b) the permeability of the magnetic material. What are the corresponding $\vec{\mathbf{H}}$ fields in both regions?

5.26 The inner radius of a toroid is 10 cm, the outer radius is 14 cm, and the height is 4 cm. A uniformly wound coil carries a current of 0.5 A. If the magnetic field intensity at the mean radius is 79.578 A/m, determine the number of turns in the coil. Calculate the energy stored in the magnetic field if the permeability of the core is 500.

5.27 The current is restricted to flow on the outer surface of the inner conductor ($\rho = a$) and the inner surface of the outer conductor ($\rho = b$) in a coaxial cable. If the coaxial cable carries a current I, determine the energy stored per unit length in the magnetic field in the region between the two conductors. Assume that the medium is nonmagnetic.

5.28 Repeat Exercise 5.27 if the current through the inner conductor is distributed uniformly.

5.29 Find the current in the 1000-turn coil for the magnetic circuit given in Figure 5.38 using the reluctance method. Also draw the analogous magnetic circuit.

5.30 Find the mmf necessary to establish a flux of 10 mWb in a magnetic ring of circular cross section. The inner diameter of the ring is 20 cm, and the outer diameter is 30 cm. The permeability of the magnetic material is 1200. What is the reluctance of the magnetic path?

5.31 A magnetic core with its pertinent dimensions is shown in Figure 5.41. The thickness of the magnetic material is 10 cm. What must be the current in a 500-turn coil in order to establish a flux of 7 mWb in leg c? Use the magnetization curve given in Figure 5.37.

5.32 If the 500-turn coil in Figure 5.41 carries a current of 2 A, determine the flux in each leg of the magnetic circuit.

Figure 5.41

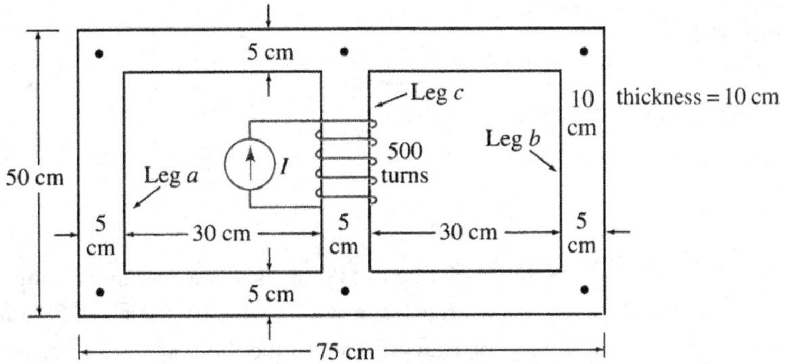

5.16 Problems

5.1 A circular loop of radius 2 cm carries a current of 10 A. Using exact and approximate expressions, find the magnetic flux density at (a) the center of the loop, (b) 10 cm on the axis of the loop, and (c) 10 m on the axis of the loop. What is the magnetic dipole moment of the loop?

5.2 A solenoid has a radius of 2 mm and a length of 1.2 cm. If the number of turns per unit length is 200 and the current is 12 A, calculate the magnetic flux density at (a) the center and (b) the ends of the solenoid.

5.3 A long solenoid has 2 turns per millimeter. Determine the current in its windings needed to produce a magnetic field of 0.5 T in its interior.

5.4 A square loop 10 cm on a side has 500 turns that are closely and tightly wound and carries a current of 120 A. Determine the magnetic flux density at the center of the loop.

5.5 Find the magnetic flux density at point P in Figure P5.5.

Figure P5.5

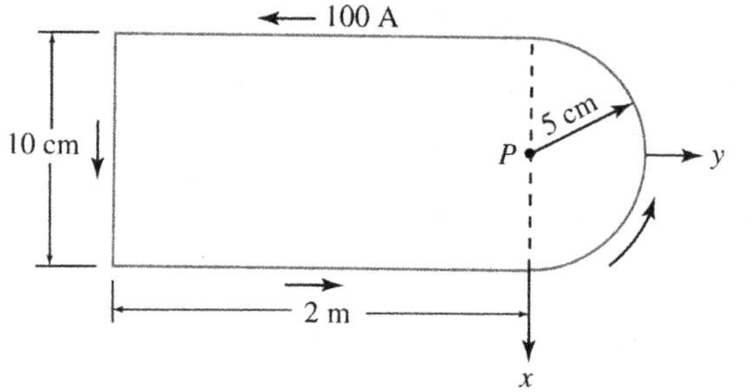

5.6 Two semicircular conductors of radii a and b are joined by straight segments to form a complete loop. If the current in the loop is I, determine the magnetic flux density at any point on the axis of the loop.

5.7 A wire bent as shown in Figure P5.7 lies in the xy plane and carries a current of 20 A. The magnetic field in the region is $1.25\vec{\mathbf{a}}_z$ T. Determine the force experienced by the wire.

Figure P5.7

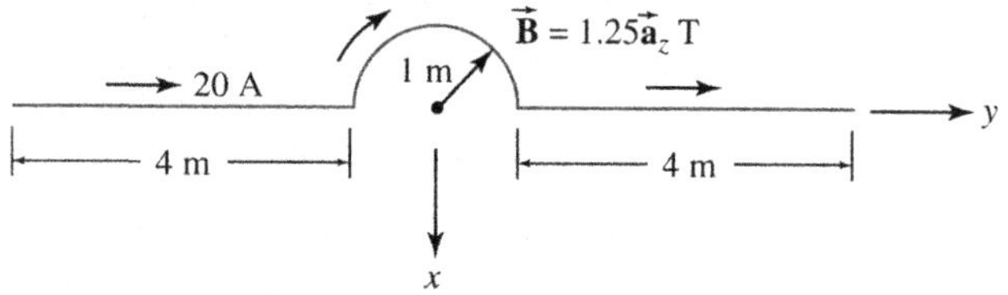

5.8 A charge of 500 nC is at one instant passing through a point (3, 4, 5) m in free space with a velocity $500\vec{\mathbf{a}}_x + 2000\vec{\mathbf{a}}_y$ m/s in a magnetic field of $1.2\vec{\mathbf{a}}_z$ T. Determine the magnetic force experienced by the charge.

5.9 A linear conductor with its ends at $(-3, -4, 0)$ m and $(5, 12, 0)$ m carries a current of 250 A. If the magnetic flux density in free space is $0.2\vec{\mathbf{a}}_z$ T, determine the magnetic force acting on the conductor.

5.10 A metallic rod 1.2 m in length and having a mass of 500 grams is suspended by a pair of flexible leads in a magnetic field of 0.9 T, as shown in Figure P5.10. Determine the current needed to remove the tension in the supporting leads. What must be the direction of the current?

5.11 Two long straight wires carrying equal currents of 15 A in the same direction are separated by a distance of 15 mm. What is the magnetic

Figure P5.10

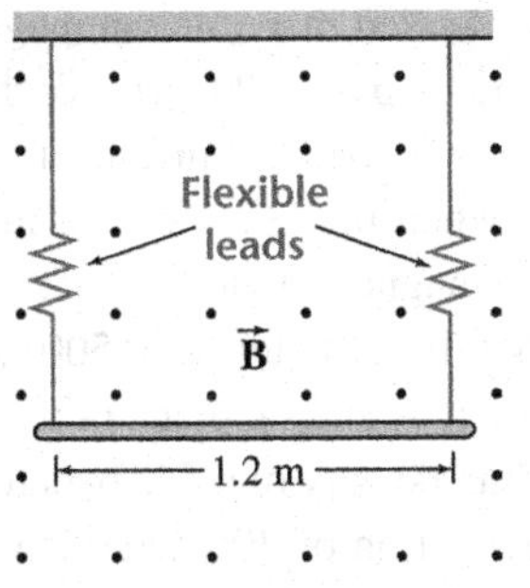

force experienced by a 0.5-m segment of each wire due to the entire length of the other?

5.12 Two point charges of 800 nC and 400 nC are moving in a plane with velocities of 20,000 km/s and 50,000 km/s, respectively, in the same direction. If at one instant the 400-nC charge is at (0, 0, 0) and 800-nC charge is at (0, 0.1, 0), determine the ratio of the electric force to the magnetic force acting on the 800-nC charge at that instant. What is the total force experienced by the 800-nC charge? What is the ratio of electric force to magnetic force experienced by the 400-nC charge? What is the total force acting on the 400-nC charge? Is the total force acting on the two charges the same?

5.13 Repeat Problem 5.12 when the charges are moving in opposite directions.

5.14 In a hydrogen atom, an electron revolves in a circular orbit of radius 5.3×10^{-11} m. The velocity of the electron is 2200 km/s. What is the magnetic field produced by the electron at the center of its orbit?

5.15 Show that $\vec{\mathbf{B}} = \mu_0 \epsilon_0 \vec{\mathbf{U}} \times \vec{\mathbf{E}}$ for a point charge q moving with a velocity $\vec{\mathbf{U}}$. $\vec{\mathbf{E}}$ is the electric field intensity produced by the point charge.

5.16 Two parallel conductors of infinite extent are carrying currents of 10 A and 20 A in opposite directions. If the separation between the conductors is 10 cm, calculate the force per unit length experienced by either conductor.

5.17 A very long, straight wire carries a current of 500 A. An 80 cm × 20 cm rectangular loop carries a current of 20 A. If the 80-cm side of the loop is parallel to the wire, as shown in Figure P5.17, what is the magnetic force acting on the loop?

Figure P5.17

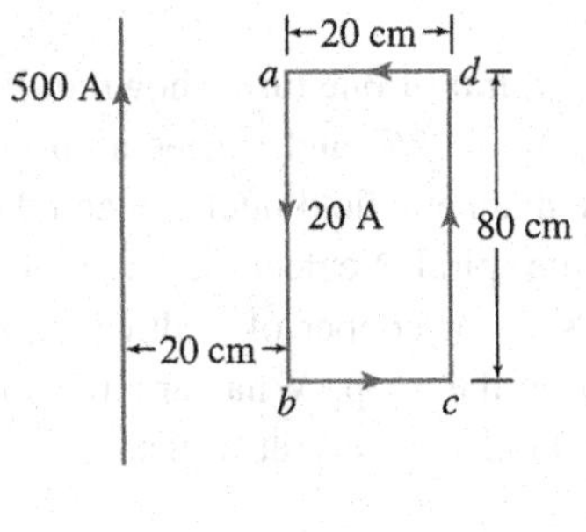

5.18 A square loop is suspended in a uniform magnetic field of $1.2\vec{\mathbf{a}}_y$ T and is free to rotate. If the loop has 400 turns, each side of the loop is 50 cm in length, and the loop carries a current of 8 A, determine the torque acting on the loop when the plane of the loop is (a) parallel and (b) perpendicular to the magnetic field.

5.19 A circular coil in a galvanometer has 1500 turns, very closely wound, and a mean radius of 1.2 cm. The plane of the coil is held parallel to the uniform magnetic flux density of 0.5 T by a torsion wire. If a current of 10 A produces a deflection of 30°, what is the torsion constant or the restoring torque of the torsion wire per radian deflection?

5.20 A square coil 10 cm on a side has 1200 turns and carries a current of 25 A. Calculate the amount of work required to rotate it in a magnetic field of 1.2 T from $\phi = 0°$ to $\phi = 180°$, where ϕ is the angle between the magnetic dipole moment and the magnetic field.

5.21 A long straight wire carries a current of 100 A. Calculate the total flux passing through a plane bounded by $\rho = 1$ cm, $\rho = 10$ cm, $z = 5$ cm, and $z = 50$ cm.

5.22 Two very long, straight conductors, each of radius 1 mm, are separated by a distance of 10 cm. If they carry equal and opposite currents of 200 A, determine the flux per unit length passing through the region between the two conductors. What will the flux be if the currents are in the same direction?

5.23 An incomplete circular loop with very long leads carries a current of 10 A, as shown in Figure P5.23. Calculate the magnetic field intensity and the magnetic flux density at the center of the loop.

Figure P5.23

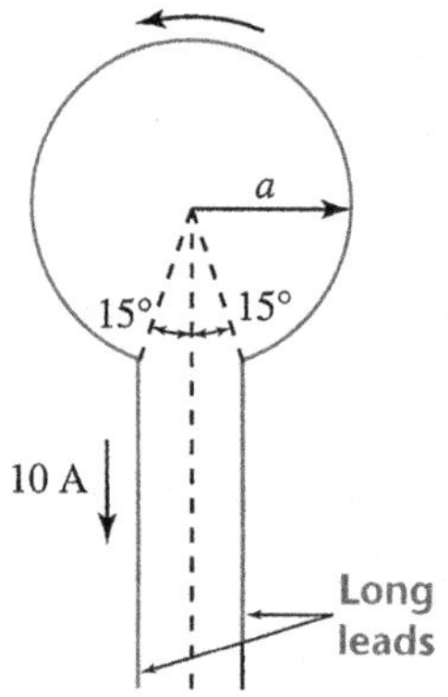

5.24 A logarithmic spiral of one turn, shown in Figure P5.24, is defined by an equation $\rho = ae^{-\phi/\pi}$ and carries a current of 5 A. If $a = 10$ cm, determine the magnetic field intensity and the magnetic flux density at the origin of the spiral. Neglect the effect of the leads.

5.25 A very long strip of copper of width b carries a current I uniformly distributed over the strip. What are the magnetic flux density and the magnetic field intensity at a distance z above the midline of the strip?

Figure P5.24

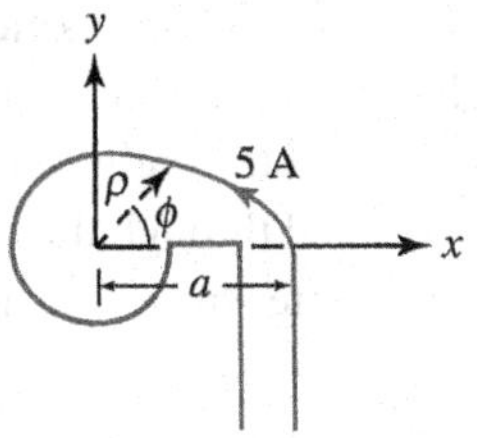

5.26 The area of a tightly wound 50-turn coil is 20 cm^2. It carries a current of 10 A and lies in a plane $3x + 4y + 12z = 26$. Determine the magnetic moment of the coil if it is directed away from the origin.

5.27 Two straight, parallel conductors, each of length 10 m, carry equal and opposite currents of 10 A. The separation between the conductors is 2 m, as shown in Figure P5.27. Calculate the magnetic vector potential, magnetic flux density, and the magnetic field intensity at a point $P(3, 4, 0)$ m.

Figure P5.27

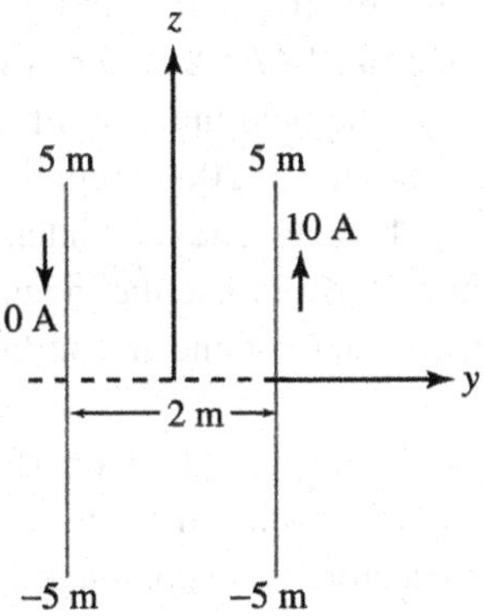

5.28 A current element of length L carrying a current I is directed along the z direction. Find the magnetic vector potential and the magnetic field intensity at a very distant point.

5.29 An iron ring with inner and outer radii of 30 cm and 40 cm has a rectangular cross section with a height of 5 cm. A uniformly wound toroidal winding of 1000 turns carries a current of 1 A. If the relative permeability of iron is 500, calculate the minimum and the maximum flux densities in the ring. What are the corresponding magnetic field intensities? What is the flux in the ring?

5.30 A long cylinder of nonmagnetic material with finite conductivity has a diameter of 20 cm and carries a current of 100 A. For a uniform current distribution within the cylinder, determine the magnetic field intensity at any point inside and outside of the cylinder. What is the curl of the magnetic field intensity at any point in space? What will happen to the fields if the conductivity of the cylinder is infinite?

5.31 The current density in a very long, cylindrical conductor of radius 10 cm is given as $\vec{\mathbf{J}}_v = 200e^{-0.5\rho}\vec{\mathbf{a}}_z$ A/m^2. Calculate the magnetic field intensity at any point in space.

5.32 A coaxial cable consists of a long cylindrical conductor of radius a surrounded by a cylindrical shell of inner radius b and outer radius c. The inner conductor and the outer shell each carry equal and opposite currents I uniformly distributed through the conductors. Obtain expressions for the magnetic field intensity in each of the regions (a) $\rho \leq a$, (b) $a \leq \rho \leq b$, (c) $b \leq \rho \leq c$, and (d) $\rho \geq c$.

5.33 Refer to Problem 5.32. Compute the flux per unit length enclosed in the region bounded by $a \leq \rho \leq b$. What is the energy stored per unit length in this region?

5.34 A current I flows in a thin wire bent into a circle of radius b. The axis of the circular loop coincides with the z axis. Compute $\int \vec{\mathbf{H}} \cdot \vec{d\ell}$ along the z axis from $z = -\infty$ to $z = \infty$. What do you conclude from your answer?

5.35 The magnetic flux density in a region 1, $z > 0$, is $\vec{\mathbf{B}} = 1.5\vec{\mathbf{a}}_x + 0.8\vec{\mathbf{a}}_y + 0.6\vec{\mathbf{a}}_z$ mT. If $z = 0$ marks the boundary between regions 1 and 2, determine the magnetic flux density in region 2. Consider region 1 as free space; region 2 has a relative permeability of 100.

5.36 A current sheet of $12\vec{\mathbf{a}}_y$ kA/m separates two regions at $z = 0$. Region 2, $z > 0$, has a magnetic field intensity of $40\vec{\mathbf{a}}_x + 50\vec{\mathbf{a}}_y + 12\vec{\mathbf{a}}_z$ kA/m and a relative permeability of 200. If region 1, $z < 0$, has a permeability of 1000, determine the magnetic field intensity in this region.

5.37 Refer to Problem 5.35. Compute the magnetization vector in each region. What are the bound volume and surface current densities in each region?

5.38 The magnetic field intensity around a perfect cylindrical conductor of radius 10 cm is $10/\rho\vec{\mathbf{a}}_\phi$ A/m. What is the surface current density on the surface of the conductor? Also compute the current on the surface of the conductor.

5.39 In a magnetic material the B field is 1.2 T when $H = 300$ A/m. When H is increased to 1500 A/m, the B field is 1.5 T. What is the change in the magnetization vector?

5.40 A series magnetic circuit with a uniform thickness of 2 cm is shown in Figure P5.40 with all its pertinent dimensions in centimeters. If

Figure P5.40

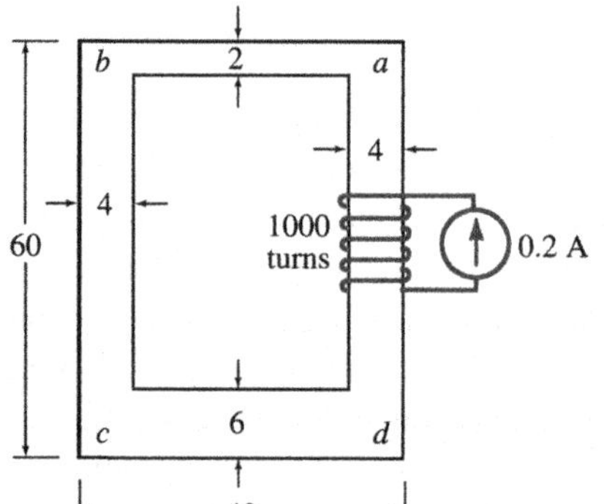

the current in the 1000-turn coil is 0.2 A, determine the flux in the magnetic circuit. The relative permeability of the magnetic material is 2000.

5.41 Refer to Problem 5.40. Compute (a) the magnetic energy density and the magnetic energy stored in each section, (b) the total energy stored in the magnetic medium, and (c) the equivalent inductance.

5.42 Using the reluctance concept, determine the inductance and the energy stored in the magnetic circuit of Problem 5.40.

5.43 A series magnetic circuit with a uniform thickness of 6 cm is shown in Figure P5.43 with all dimensions in centimeters. If the current through the 500-turn coil is 0.8 A, determine the current in the 700-turn coil in order to maintain a flux of 1.44 mWb in the air gap. Assume the permeability of the magnetic material is 500. Compute the ratio of the mmf drop across the air gap to the applied mmf.

Figure P5.43

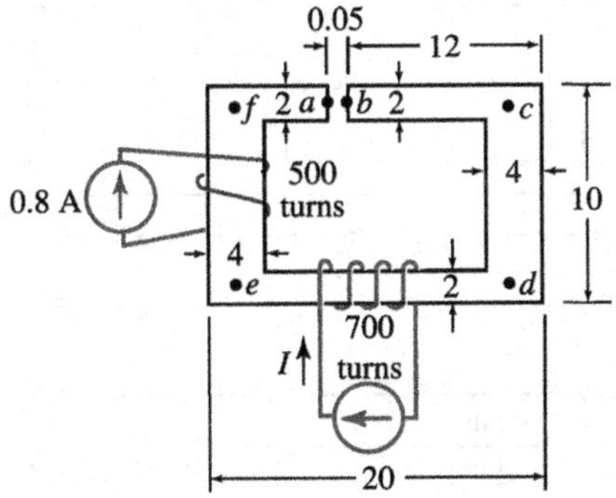

5.44 Repeat Problem 5.43 but use the *B–H* curve for the magnetic material given in Figure 5.37.

5.45 Refer to Problem 5.43. Compute (a) the magnetic energy density and the magnetic energy stored in each section, (b) the total energy stored in the magnetic medium, (c) the inductance using the energy concept, and (d) the inductance using the reluctance concept.

5.46 A magnetic circuit with all the pertinent dimensions in centimeters is shown in Figure P5.46. Using the *B–H* curve for the magnetic material

Figure P5.46

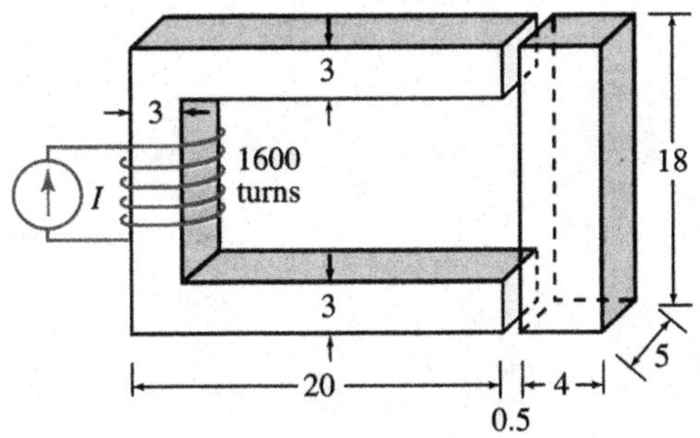

given in Figure 5.37, determine the current in the 1600-turn coil to establish a flux density of 0.75 T in each air gap.

5.47 Refer to Problem 5.46. What is the flux density in the air gap if the current in the coil is increased by 50%?

5.48 Refer to Problem 5.46. Compute (a) the magnetic energy density and the magnetic energy stored in each section, (b) the total energy stored in the magnetic medium, (c) the inductance, and (d) the total reluctance.

5.49 A series-parallel magnetic circuit is shown in Figure P5.49. A coil with 200 turns is wound over the center leg. If the flux density in the air gap is 0.2 T, find the current in the coil. Use the *B–H* curve given in Figure 5.37. Also compute the inductance and the energy stored in the magnetic field.

Figure P5.49

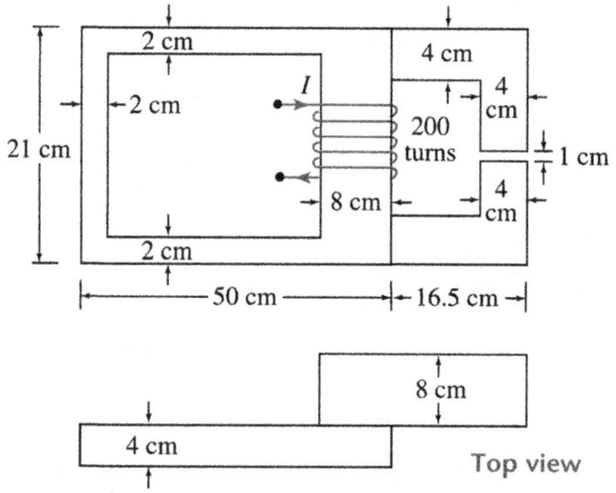

5.50 If the current in the coil in Problem 5.49 is increased by 20%, what will the flux density be in the air gap?

6

Applications of static fields

6.1 Introduction

Now that we have discussed the fundamentals of electrostatic and magnetostatic fields, we can explain some of the applications of static fields. It may appear unusual that we have devoted a complete chapter to discussing the applications of static fields when some of these applications could easily have been included in the preceding chapters. We have several reasons for doing so:

1. To discuss some of the applications in their entirety requires knowledge of both electrostatic and magnetostatic fields. For instance, the acceleration of a charged particle in a cyclotron is accomplished by an electric field, whereas the rotation is imparted by a magnetic field.
2. By presenting the major applications of static fields in one chapter we hope to convince the reader of their importance. We have seen some recently published textbooks that tend to skip over the subject of static fields as if they are of no significance.
3. If there is not enough time to discuss the applications of static fields in the classroom, we presume that this chapter epitomizes a very good reading assignment for the student.

6.2 Deflection of a charged particle

One of the most common applications of electrostatic fields is the deflection of a charged particle such as an electron or a proton in order to control its trajectory. Devices such as the cathode-ray oscilloscope, cyclotron, ink-jet printer, and velocity selector are based on this principle. Whereas the charge of an electron beam in a cathode-ray oscilloscope is constant, the charge on the fine particles of ink in an ink-jet printer varies with the character to be printed. In any case, the deflection of a charged particle is accomplished by maintaining a potential difference between a pair of parallel plates.

Consider a charged particle with charge q and mass m moving in the x direction with a velocity u_x, as shown in Figure 6.1. At time $t = 0$

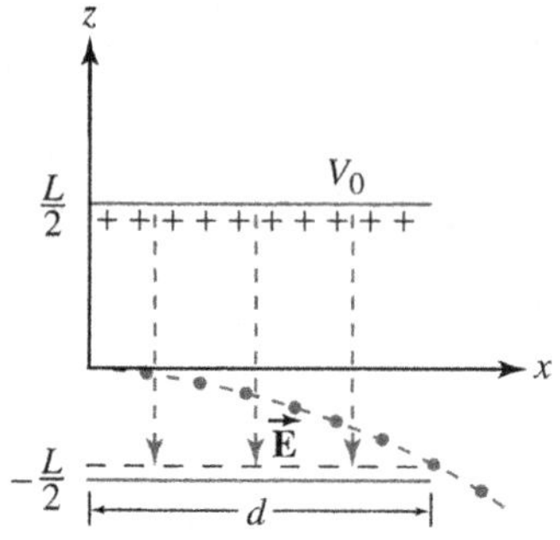

Figure 6.1 Trajectory of a charged particle in a uniform $\vec{\mathbf{E}}$ field

the charged particle enters the region between the pair of parallel plates held at a potential difference of V_0. Ignoring the effects of fringing of the electric field lines, the electric field intensity within the parallel plates is

$$\vec{\mathrm{E}} = -\frac{V_0}{L}\vec{\mathbf{a}}_z$$

where L is the separation between the two parallel plates. The force acting on the charged particle due to the electric field is

$$\vec{\mathbf{F}} = q\vec{\mathbf{E}}$$

which is in the downward direction. Neglecting the effect of gravitational force on the charged particle, the acceleration in the z direction is

$$a_z = -\frac{qV_0}{mL} \tag{6.1}$$

Thus, the velocity of the charged particle within the parallel plates in the z direction is

$$u_z = a_z t \tag{6.2}$$

because $u_z = 0$ at time $t = 0$.

The displacement of the charged particle in the z direction is

$$z = \tfrac{1}{2}a_z t^2 \tag{6.3}$$

because $z = 0$ at time $t = 0$. However, the displacement of the charged particle in the x direction in time t is

$$x = u_x t \tag{6.4}$$

The time taken by the charged particle to exit the region between the parallel plates is

$$T = \frac{d}{u_x} \tag{6.5}$$

Thus, the trajectory of the charged particle within the parallel plates, from (6.3) and (6.4), is

$$z = -\frac{qV_0}{2mL}\left[\frac{x}{u_x}\right]^2 \tag{6.6}$$

which is an equation for a parabola.

EXAMPLE 6.1

A potential difference of 1.5 kV is maintained between two parallel plates that are held 10 cm apart. An electron with a kinetic energy of 2 keV enters the deflection plates at right angles to the electric field. If the plates are 20 cm long, determine (a) the time taken by the electron to exit the plates and (b) the deflection of the electron as it exits the plates.

Solution From the kinetic energy of the electron, we can determine the velocity of the electron at time $t = 0$ in the x direction as

$$\tfrac{1}{2}mu_x^2 = 2 \times 10^3 \times 1.6 \times 10^{-19}$$

Substituting $m = 9.11 \times 10^{-31}$ kg for the electron, we obtain

$$u_x = 26.52 \times 10^6 \text{ m/s}$$

a) The time taken by the electron to exit the parallel plates is

$$T = \frac{20 \times 10^{-2}}{26.52 \times 10^6} = 7.54 \times 10^{-9} \text{ s} \quad \text{or } 7.54 \text{ ns}$$

b) The deflection of the electron, from (6.6), is

$$z = \frac{1.6 \times 10^{-19} \times 1.5 \times 10^3}{2 \times 9.1 \times 10^{-31} \times 0.1}\left[\frac{20 \times 10^{-2}}{26.52 \times 10^6}\right]^2$$

$$= 74.97 \times 10^{-3} \text{ m} \quad \text{or} \quad 74.97 \text{ mm}$$

• • •

6.3 Cathode-ray oscilloscope

The essential features of a *cathode-ray oscilloscope* are illustrated in Figure 6.2. The tube itself is of glass and is highly evacuated. The cathode emits electrons when it is heated by a heating filament. These electrons are then accelerated toward an anode that is held at a potential of several hundred volts with respect to the cathode. The anode has a small hole that allows a narrow beam of electrons to pass through it. The accelerated electrons then enter a region where they can be deflected in both the horizontal and vertical directions in a similar manner as discussed in Section 6.2. Finally, the electron beam bombards the inner surface of a screen coated with a substance (phosphor) that emits visible light.

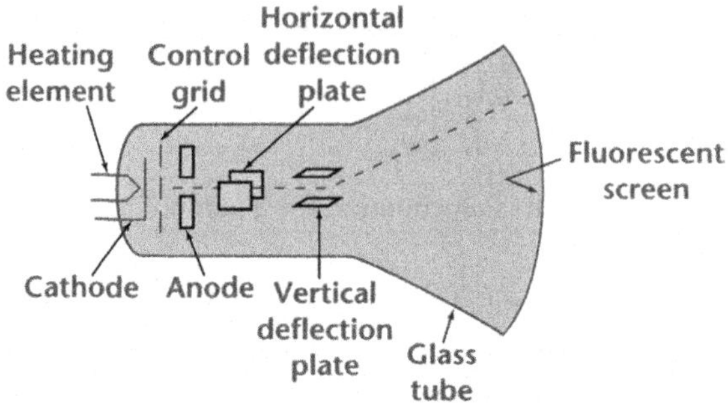

Figure 6.2 Basic elements of a cathode-ray oscilloscope

Let us assume that the initial velocity of the electron as it is being emitted from the cathode's surface is zero. If V_1 is the potential difference between the anode and the cathode, the velocity of the electron as it exits the anode can be obtained from the gain in its kinetic energy as

$$\tfrac{1}{2}mu_x^2 = eV_1$$

or

$$u_x = \left[\frac{2e}{m}V_1\right]^{1/2} \tag{6.7}$$

Let us assume that there exists no potential difference between the horizontal deflection plates and that the top vertical deflection plate is held at a potential of V_0 with respect to the lower plate. The electron passes undisturbed through the horizontal deflection plates and experiences a force in the positive z direction in the region between the vertical deflection plates. The vertical displacement as the electron exits the vertical deflection region $x = d$, as shown in Figure 6.3, from (6.6), is

$$z_1 = \frac{eV_0}{2mL}\left[\frac{d}{u_x}\right]^2 \tag{6.8}$$

The corresponding velocity in the z direction for $x = d$ is

$$u_z = \frac{edV_0}{mLu_x} \tag{6.9}$$

whereas the velocity in the x direction remains unchanged. As the electron exits the vertical deflection region, it moves in a straight-line path as u_x and u_z are both constant. The velocity $\vec{\mathbf{u}}$ now makes an angle θ with the x axis, where

$$\tan\,\theta = \frac{u_z}{u_x} \tag{6.10}$$

The time required by the electron to travel a distance D on emerging from the deflection plates to the screen is

$$t_2 = \frac{D}{u_x}$$

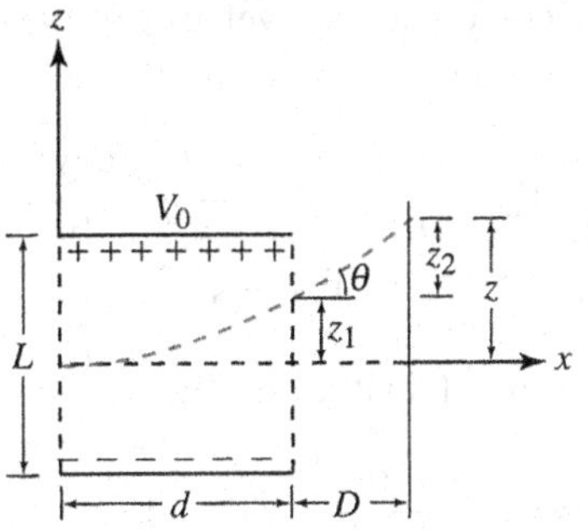

Figure 6.3 Operation of a cathode-ray oscilloscope

Thus,

$$z_2 = u_z t_2 = \frac{edD}{mL} V_0 \left[\frac{1}{u_x}\right]^2 \qquad (6.11)$$

Therefore, after substituting for u_x in this equation, the total vertical displacement of the electron as it strikes the screen is

$$z = z_1 + z_2 = \frac{d}{2L}\,[0.5d + D]\left[\frac{V_0}{V_1}\right] \qquad (6.12)$$

It is evident from (6.12) that if the potential difference between the anode and the cathode is held constant, the deflection of the electron is proportional to the potential difference between the vertical deflection plates. By applying a potential difference between the horizontal deflection plates, we can cause the electron to move in the y direction. Therefore, the point at which the electron beam will strike the screen depends upon the vertical and the horizontal deflecting voltages.

EXAMPLE 6.2

WORKSHEET 4

Mathcad

The potential difference between the anode and the cathode of a cathode-ray oscilloscope is 1000 V. For the vertical deflection plates, $L = 5$ mm, $d = 1.5$ cm, and $V_0 = 200$ V. The distance D is 15 cm. For an electron released from the anode with a zero initial velocity, find (a) the velocity in the x direction as it enters the vertical deflection plates, (b) the acceleration and the velocity in the z direction within the plates, (c) the exit velocity in the z direction, and (d) the total displacement of the electron on the screen.

Solution

a) From (6.7), the velocity in the x direction as the electron leaves the anode is

$$u_x = \left[\frac{2 \times 1.6 \times 10^{-19} \times 1000}{9.1 \times 10^{-31}}\right]^{1/2} = 18.75 \times 10^6 \text{ m/s}$$

which is small enough to ignore the relativistic effects because the velocity is less than 10% of the speed of light. At this speed the mass of the electron is almost the same as that at rest.

b) The acceleration and the velocity of the electron in the z direction, from (6.1) and (6.2), are, respectively,

$$a_z = \frac{e}{mL}V_0 = \frac{1.6 \times 10^{-19} \times 200}{9.1 \times 10^{-31} \times 5 \times 10^{-3}} = 7.03 \times 10^{15} \text{ m/s}^2$$
$$u_z = a_z t = 7.03 \times 10^{15} t \text{ m/s}$$

c) The electron will exit the deflection plates in time $t = T$ such that

$$T = \frac{d}{u_x} = \frac{1.5 \times 10^{-2}}{18.75 \times 10^6} = 8 \times 10^{-10} \text{ s}, \quad \text{or} \quad 0.8 \text{ ns}$$

Thus, the exit velocity in the z direction is

$$u_z = 7.03 \times 10^{15} \times 8 \times 10^{-10} = 5.624 \times 10^6 \text{ m/s}$$

d) The total deflection of the electron, from (6.12), is

$$z = \frac{1.5 \times 10^{-2} \times 200}{2 \times 5 \times 10^{-3} \times 1000}[0.75 + 15] \times 10^{-2} = 4.725 \text{ cm}$$

• • •

6.4 Ink-jet printer

A novel printing technique based upon the electrostatic deflection principle has been developed to increase the speed of the printing process and enhance the print quality. The resulting printer is called an *ink-jet printer.* In an ink-jet printer, a nozzle vibrating at ultrasonic frequency sprays ink in the form of very fine, uniformly sized droplets separated by a certain spacing. These droplets acquire charge proportional to the character to be printed while passing through a set of charged plates, as depicted in Figure 6.4. With a fixed potential difference between the vertical deflection plates, the vertical displacement of an ink droplet is-proportional to its charge. A blank space between characters is achieved by having no charge imparted to the ink droplets (in this case, the ink droplets are collected by the ink receptor). In a cathode-ray oscilloscope the horizontal deflection of the electron is obtained by constantly changing the potential difference between the horizontal deflection plates. However, in an ink-jet printer the printer head is moved horizontally at a constant speed, and the characters can be formed at the rate of 100 characters per second (cps).

As there are very few moving parts, ink-jet printers are very quiet and reliable in operation compared with impact printers. Also, impact printers limit printing to only those characters that are on the print-wheel, whereas any character can be formed with ink-jet printers, making them very versatile. As you may have guessed, the equations that determine

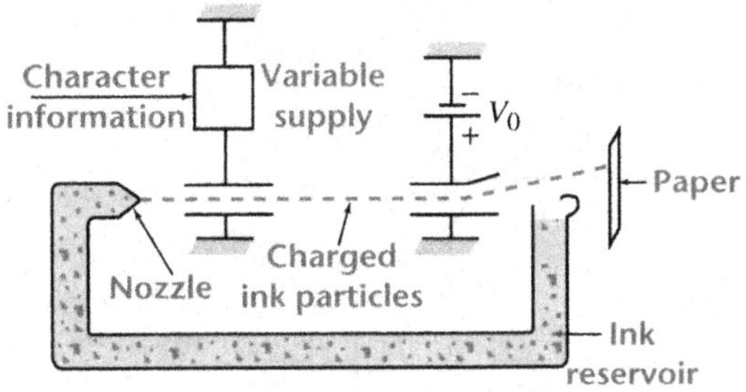

Figure 6.4 Basic constructional details and operation of an ink-jet printer

the trajectory of an ink droplet are exactly the same as those for an electron in a cathode-ray oscilloscope.

EXAMPLE 6.3

An ink droplet of diameter 0.02 mm attains a charge of −0.2 pC as it passes through the charging plates at a speed of 25 m/s. The potential difference between the vertical deflection plates held 2 mm apart is 2 kV. If the length of each deflection plate is 2 mm, and the distance from the exit end of the deflection plate to the paper is 8 mm, determine the vertical displacement of the ink droplet. Assume that the density of the ink is 2 grams per cubic centimeter.

Solution The mass of the ink droplet is

$$m = \frac{4\pi}{3}\left[\frac{1}{2} \times 0.02 \times 10^{-3}\right]^3 \times 2 \times 10^3 = 8.38 \times 10^{-12}\ \text{kg}$$

The total vertical deflection is

$$\begin{aligned} z &= \frac{-qd}{mL} V_0 \left[\frac{1}{u_x}\right]^2 [0.5d + D] \\ &= \frac{2 \times 10^{-13} \times 2 \times 10^{-3} \times 2000}{8.38 \times 10^{-12} \times 2 \times 10^{-3}} \left[\frac{1}{25^2}\right] [0.5 \times 2 + 8] \times 10^{-3} \\ &= 0.69\ \text{mm} \end{aligned}$$

• • •

6.5 Sorting of minerals

The principle of electrostatic deflection is also employed by the mining industry to sort oppositely charged minerals. For example, in an *ore separator*, phosphate ore containing granules of phosphate rock and quartz is dropped onto a vibrating feeder, as illustrated in Figure 6.5. The vibrations cause the granules of phosphate rock to rub against the particles of quartz. During the rubbing process each quartz granule acquires a positive charge and each phosphate particle acquires a negative charge. The sorting of the oppositely charged particles is accomplished by passing them through an electric field set up by a parallel-plate capacitor.

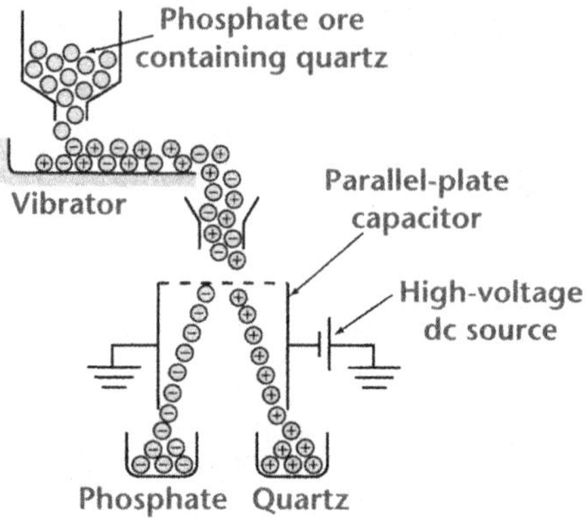

Figure 6.5 Ore separator

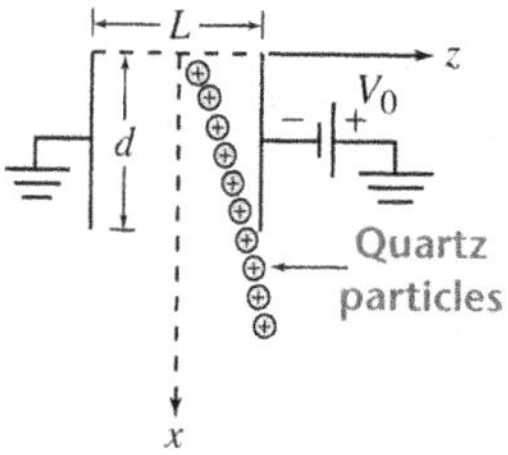

Figure 6.6 Trajectory of a quartz particle in the parallel-plane region

To develop an expression for the trajectory of the charged particle within the parallel-plate capacitor region, let us assume that the mass and the charge of the quartz particle are m and q, respectively. Let the initial velocity of each particle be zero at the instant it enters the charged region between the parallel plates, as shown in Figure 6.6. Then $u_x = 0$ and $u_z = 0$ at $t = 0$. The force of gravity will impart acceleration in the x direction. At any time t, the velocity and the distance traveled in the x direction are

$$u_x = \frac{dx}{dt} = gt \tag{6.13}$$

and

$$x = \tfrac{1}{2}gt^2 \tag{6.14}$$

The motion of the charged particle in the z direction can be described as

$$a_z = \frac{q}{mL} V_0 \tag{6.15}$$

$$u_z = a_z t \tag{6.16}$$

and

$$z = \tfrac{1}{2}a_z t^2 \tag{6.17}$$

From (6.14) and (6.17), we obtain the trajectory of each charged particle as

$$z = a_z \frac{x}{g} \tag{6.18}$$

This equation reveals that the trajectory of a charged particle is a straight line within the parallel-plate region. The time taken by the charged particle to exit the parallel-plate region, $x = d$, is

$$T = \left[\frac{2d}{g}\right]^{1/2} \tag{6.19}$$

For any time $t \geq T$, the velocity of the charged particle in the z direction is constant and, from (6.16), is given as

$$u_z = a_z T = \frac{qV_0}{mL}\left[\frac{2d}{g}\right]^{1/2} \quad \text{for } t \geq T \tag{6.20}$$

and

$$z = u_z t \quad \text{for } t \geq T \tag{6.21}$$

From (6.14) and (6.21), we can express z in terms of x as

$$z^2 = \frac{2}{g}u_z^2 x \quad \text{for } t \geq T \tag{6.22}$$

which is an equation for a parabola. Thus, a charged particle follows a straight-line path within the parallel plates and a parabolic path thereafter.

EXAMPLE 6.4

A quartz particle with a mass of 2 grams acquires a charge of 100 nC on the vibrating feeder. The particle then falls freely at the middle of the top edge of the parallel plates, which are held at a potential difference of 10 kV. If the plates are 2 m in length and are 50 cm apart, determine the position and the velocity of the particle at the end of the plates.

Solution The time taken by the quartz particle to leave the plates, from (6.19), is

$$T = \left[\frac{2 \times 2}{9.81}\right]^{1/2} = 638.55 \text{ ms}$$

From (6.15), the acceleration of the quartz particle within the parallel-plate region is

$$a_z = \frac{100 \times 10^{-9} \times 10 \times 10^3}{2 \times 10^{-3} \times 0.5} = 1.0 \text{ m/s}^2$$

The distance traveled in the z direction in time $t = T$ is

$$z = \tfrac{1}{2} \times 1.0 \times [638.55 \times 10^{-3}]^2 = 0.204 \text{ m} \quad \text{or} \quad 20.4 \text{ cm}$$

At the time of exit the velocities of the charged particle in the x and z directions are

$$u_x = 9.81 \times 638.55 \times 10^{-3} = 6.264 \text{ m/s}$$

$$u_z = 1.0 \times 638.55 \times 10^{-3} = 0.639 \text{ m/s}$$

Thus, the exit velocity of the quartz particle is

$$\vec{\mathbf{u}} = 6.264\vec{\mathbf{a}}_x + 0.639\vec{\mathbf{a}}_z \text{ m/s}$$

• • •

6.6 Electrostatic generator

An electrostatic generator conceived by Lord Kelvin was put in practice by Robert J. Van de Graaff and since then has been called the *Van de Graaff generator*. It consists of a hollow spherical conductor (dome) supported on an insulated hollow column, as shown in Figure 6.7a. A belt passes over the pulleys. The lower pulley is driven by a motor, and the upper is an idler. A number of sharp points projecting from a rod are maintained at a very high positive potential, and the air around the points becomes ionized. The positive ions are repelled from the sharp points and some of these ions attach themselves to the surface of the moving belt. A similar process takes place at the metal brush inside the dome. As the charge builds up, the potential of the dome rises. With the Van de Graaff generator a potential difference as high as several million volts can be realized. Its chief application is to accelerate the charged particles to acquire high kinetic energies, which are then used in atom-smashing experiments.

In order to understand this generator's basic principle of operation, consider a hollow, uncharged, conducting sphere (dome) with a small opening, as shown in Figure 6.7b. Let us now introduce a positively charged small sphere with charge q through the opening into the cavity. As soon as the equilibrium state is reached, the inner surface of the dome acquires a net negative charge, while a positive charge q is induced on its outer surface. If the small sphere is now made to touch the inner surface of the dome, the positive charge of the small sphere will be completely neutralized by the negative charge on the inner surface of the dome. However, the outer surface will still maintain the positive charge q. If the small sphere is now withdrawn, charged again to q, and reinserted into the dome, the inner surface will again acquire a negative charge, causing the charge on the outer surface to increase by the same amount. By

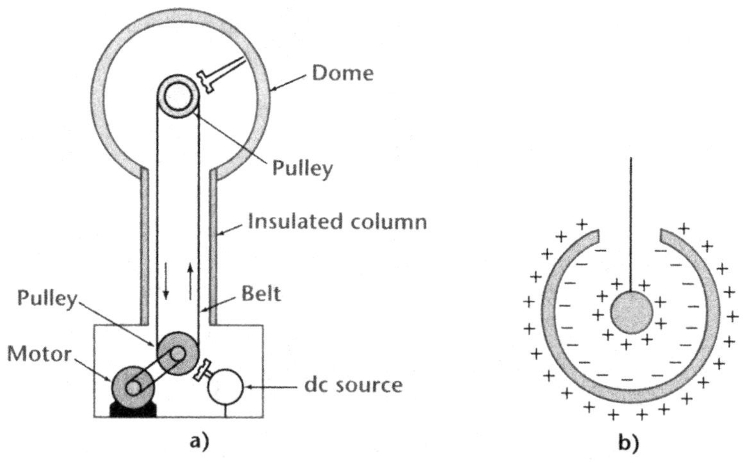

Figure 6.7 (a) Schematic of Van de Graaff generator; (b) Its principle of operation

touching the small sphere to the inner surface of the dome, both the inner surface of the dome and the small sphere are again free of charge. But there is now twice as much charge on the outer surface of the dome. In other words, by bringing in a charged body and touching it to the inner surface of the dome, all the charge on the charged body can be transferred to the outer surface of the dome. This process is, of course, independent of the initial charge on the outer surface of the dome.

Let us suppose that at any instant, after the equilibrium state has been reached, the charge on the smaller sphere inside the dome is q and that on the outer surface of the dome is Q. If the radii of the inner and outer spheres are r and R, respectively, the potential at any point on the dome is

$$V_R = \frac{1}{4\pi\epsilon_0}\left[\frac{Q}{R} + \frac{q}{R}\right]$$

The first term inside the brackets is the contribution to the potential on the dome by its own charge Q, and the second term is due to the equipotential surface at a radius R created by the charge q on the small sphere. The potential of the small sphere is

$$V_r = \frac{1}{4\pi\epsilon_0}\left[\frac{q}{r} + \frac{Q}{R}\right]$$

where the first term is due to the charge on the small sphere, and the second term takes into account that the small sphere is inside the large sphere.

Thus, the potential difference between the spheres is

$$V = V_r - V_R = \frac{q}{4\pi\epsilon_0}\left[\frac{1}{r} - \frac{1}{R}\right] \tag{6.23}$$

For a positive charge q the potential of the inner sphere will always be higher than that of the dome. If the two spheres are electrically connected, the entire charge on the inner sphere will flow toward the outer surface of the dome regardless of the charge Q on it. This is another way that we can explain the charge transfer from the inner sphere to the outer surface of the dome. Note that the potential difference is zero only if $q = 0$.

EXAMPLE 6.5

How much charge is required to raise an isolated metallic sphere of 45-cm radius to a potential of 900 kV? What is the electric field intensity on the surface of the sphere?

Solution If Q is the total charge required, then

$$Q = 4\pi\epsilon_0 RV = \frac{4\pi \times 10^{-9}}{36\pi} \times 0.45 \times 900 \times 10^3$$
$$= 45 \times 10^{-6}\ \text{C} \quad \text{or} \quad 45\ \mu\text{C}$$

The magnitude of the electric field intensity on the surface of the sphere is

$$E_r = \frac{V}{R} = \frac{900 \times 10^3}{0.45} = 2 \times 10^6 \text{ V/m} \quad \text{or} \quad 2 \text{ MV/m}$$ • • •

6.7 Electrostatic voltmeter

There are numerous other applications of the basic principles of electrostatics. In our final application, let us explore the operation of an *electrostatic voltmeter*, which is used to measure true rms voltage. The voltmeter is equally useful for measuring both dc and ac voltages. Figure 6.8 shows an electrostatic voltmeter in its simplest form. Plates a and b form a capacitor whose capacitance increases as the pointer moves to the right when a voltage is applied between terminals 1 and 2. The helical spring as shown not only controls the motion of the pointer but also establishes an electrical contact between the movable plate b and the external point 2. When the applied voltage is held constant and the pointer takes up its final position at an angle θ, the increase in electrostatic energy is equal to the amount of mechanical work done.

Any change in the potential difference across the electrostatic voltmeter can be expressed as

$$dV = d\left(\frac{Q}{C}\right) = \frac{1}{C}dQ - \frac{Q}{C^2}dC$$

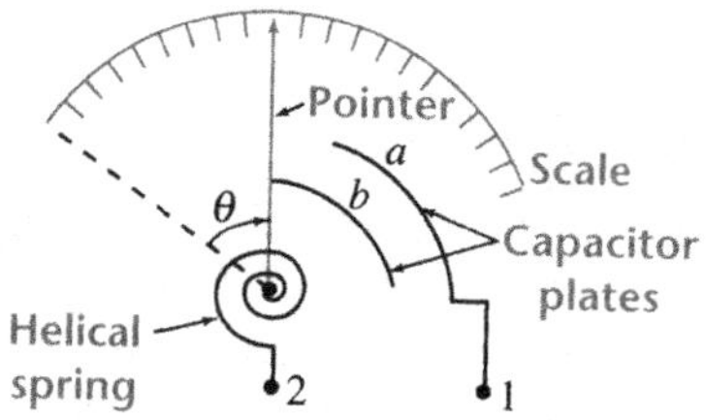

Figure 6.8 Schematic of an electrostatic voltmeter

When the potential is held constant, $dV = 0$ and

$$\frac{1}{C}dQ = \frac{Q}{C^2}dC \tag{6.24}$$

The change in the electrostatic energy is

$$dW_e = d\left[\frac{Q^2}{2C}\right] = \frac{Q}{C}dQ - \frac{Q^2}{2C^2}dC$$

$$= \frac{Q^2}{2C^2}dC \tag{6.25}$$

and the mechanical work done is

$$dW = T d\theta \tag{6.26}$$

Equating (6.25) and (6.26), we obtain

$$T = \frac{1}{2}\left[\frac{Q}{C}\right]^2 \frac{dC}{d\theta} = \frac{1}{2}V^2\frac{dC}{d\theta} \tag{6.27}$$

In the equilibrium position,

$$T = \tau\theta \tag{6.28}$$

where τ is the torsional constant of the spring. Thus, the deflection of the pointer

$$\theta = \frac{1}{2\tau} V^2 \frac{dC}{d\theta} \tag{6.29}$$

is proportional to the square of the applied voltage if $dC/d\theta$ is constant. In real electrostatic voltmeters, the value of $dC/d\theta$ depends upon θ and, therefore, must be properly calibrated by the manufacturer.

6.8 Magnetic separator

An important application of magnetostatic fields is a device called a *magnetic separator* (see Figure 6.9 below), which is designed to separate magnetic from nonmagnetic materials. A mix of magnetic and nonmagnetic materials is fed on an endless belt running at a constant speed. The belt passes over a magnetic pulley. The magnetic pulley consists of an iron shell containing an exciting coil that produces the magnetic field. The nonmagnetic material immediately drops off into a bin while the magnetic material is held by the pulley until the belt leaves the pulley. The magnetic material is therefore carried further round the pulley and then dropped into a second bin as shown in the figure.

Figure 6.9 A magnetic separator

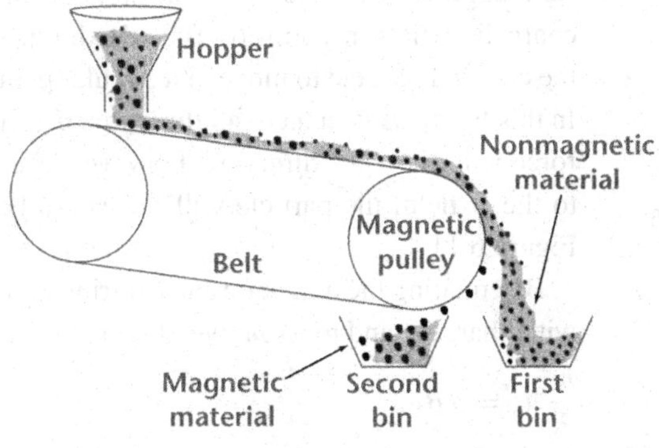

6.9 Magnetic deflection

Most of the applications of a steady magnetic field are based upon the magnetic force it exerts either on a charged particle in motion or on a current-carrying conductor. If a positive charge q is moving with a velocity $\vec{\mathbf{u}}$ in a steady magnetic field $\vec{\mathbf{B}}$, the magnetic force experienced by the charge is

$$\vec{\mathbf{F}} = q\vec{\mathbf{u}} \times \vec{\mathbf{B}}$$

The magnetic force is not only at right angles to the magnetic field $\vec{\mathbf{B}}$ but is also normal to the direction of motion of the charge. The direction of motion being perpendicular to the direction of the force emphasizes the fact that the magnetic force does no work on the charged particle. In other words, the kinetic energy of the charged particle remains the same. However, the magnetic field does influence the direction of the motion of the charged particle, as we will now explain.

If we express the velocity of the charged particle as

$$\vec{\mathbf{u}} = \vec{\mathbf{u}}_p + \vec{\mathbf{u}}_n$$

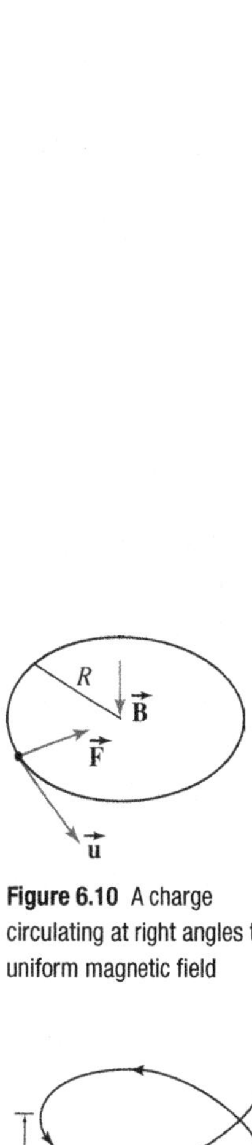

Figure 6.10 A charge circulating at right angles to a uniform magnetic field

where $\vec{\mathbf{u}}_p$ and $\vec{\mathbf{u}}_n$ are the components of $\vec{\mathbf{u}}$ parallel and normal to the $\vec{\mathbf{B}}$ field, respectively. The parallel component $\vec{\mathbf{u}}_p$ does not contribute to the magnetic force. Thus, the magnitude of the magnetic force on the charged particle can be expressed in terms of the normal component of the velocity $\vec{\mathbf{u}}_n$ as

$$F = qu_nB$$

and the direction of the force can be ascertained by the right-hand rule. In the absence of the parallel component $\vec{\mathbf{u}}_p$, the magnetic force will keep the particle in the plane perpendicular to the $\vec{\mathbf{B}}$ field. If the magnetic field is uniform, the force acting on the charged particle will be constant; that is, the normal component of $\vec{\mathbf{u}}$ will have constant magnitude at any point in a plane perpendicular to the $\vec{\mathbf{B}}$ field. This is similar to a situation where a stone tied to one end of a rope and whirled in a plane to create a circular motion results in a constant tension on the rope which is at right angles to the velocity of the stone at any instant. The motion of a charged particle in a uniform $\vec{\mathbf{B}}$ field exhibits similar traits, so we expect the charged particle to move in a circular path, as shown in Figure 6.10. In this figure, $\vec{\mathbf{B}}$ is in the $-\vec{\mathbf{a}}_z$ direction, $\vec{\mathbf{u}}$ is in the $\vec{\mathbf{a}}_\phi$ direction, and the force is in the $-\vec{\mathbf{a}}_\rho$ direction. However, if $\vec{\mathbf{u}}$ has a component parallel to the $\vec{\mathbf{B}}$ field, the particle will traverse a helical path, as depicted in Figure 6.11.

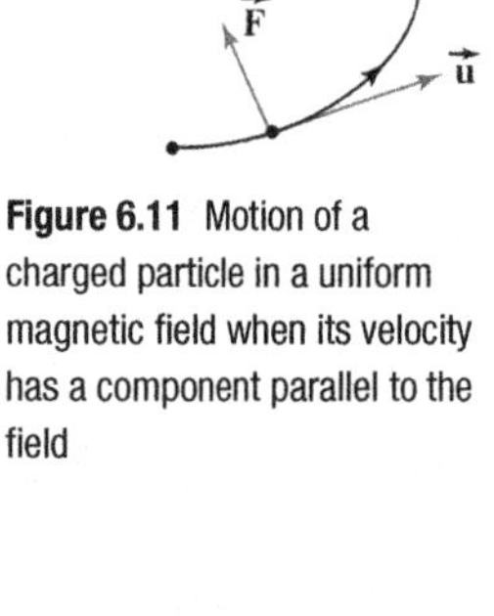

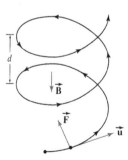

Figure 6.11 Motion of a charged particle in a uniform magnetic field when its velocity has a component parallel to the field

By equating the magnetic and centripetal forces acting on the particle with charge q and mass m, we obtain

$$\frac{m}{R} u_n^2 = qBu_n$$

which gives the radius of the circular orbit as

$$R = \frac{m}{qB} u_n \tag{6.30}$$

This equation simply states that the radius of a circular orbit of a charged particle is directly proportional to the component of its velocity normal to the $\vec{\mathbf{B}}$ field and inversely proportional to the strength of the $\vec{\mathbf{B}}$ field. Also, the heavier the particle, the larger the radius of the circular orbit. This fact is used in the separation of isotopes as discussed in Section 6.11.

The time required for the charged particle to complete one cycle is known as the *time period* and is denoted by T. Thus,

$$T = \frac{2\pi R}{u_n} = \frac{2\pi m}{qB} \tag{6.31}$$

We can now define the frequency, often referred to as the *cyclotron frequency*, as

$$f = \frac{1}{T} = \frac{qB}{2\pi m} \tag{6.32}$$

and the angular frequency as

$$\omega = 2\pi f = \frac{qB}{m} \tag{6.33}$$

The time period and the frequency of the charged particle are constant as long as the particle is orbiting in a uniform magnetic field. This observation led to the development of a particle accelerator known as the cyclotron (see Section 6.10).

For a charged particle with a component of velocity parallel to the $\vec{\mathbf{B}}$ field, we can calculate the distance traveled in one period as

$$d = u_p T = \frac{2\pi m}{qB} u_p \tag{6.34}$$

As you can see from Figure 6.11, d is the spacing between two adjacent turns of the helical path and is, therefore, the pitch of the helix.

EXAMPLE 6.6

A proton is revolving in a uniform magnetic field of $1.75\vec{\mathbf{a}}_z$ T with a velocity of $3000\vec{\mathbf{a}}_z - 4000\vec{\mathbf{a}}_\phi$ km/s. Determine (a) the force acting on the proton, (b) the direction of rotation, (c) the radius of the orbit, (d) the time period, (e) the cyclotron frequency, and (f) the pitch of the helix.

Solution a) The magnetic force acting on the proton is

$$\begin{aligned}\vec{\mathbf{F}} &= q\vec{\mathbf{u}} \times \vec{\mathbf{B}} = 1.6 \times 10^{-19} \times 10^6 \times 1.75 \times [3\vec{\mathbf{a}}_z - 4\vec{\mathbf{a}}_\phi] \times \vec{\mathbf{a}}_z \\ &= -1.12 \times 10^{-12}\vec{\mathbf{a}}_\rho \text{ N}\end{aligned}$$

b) Because the normal component of the proton's velocity to the $\vec{\mathbf{B}}$ field is in the $-\vec{\mathbf{a}}_\phi$ direction, the proton revolves clockwise as viewed from the z axis.

c) The radius of the orbit is

$$R = \frac{m}{F} u_n^2 = \frac{1.7 \times 10^{-27}}{1.12 \times 10^{-12}} [4 \times 10^6]^2$$
$$= 24.286 \times 10^{-3} \text{ m} \quad \text{or} \quad 24.286 \text{ mm}$$

d) The time period for one complete revolution is

$$T = \frac{2\pi R}{u_n} = \frac{2\pi \times 24.286 \times 10^{-3}}{4 \times 10^6}$$
$$= 38.148 \times 10^{-9} \text{ s} \quad \text{or} \quad 38.148 \text{ ns}$$

e) The frequency is

$$f = \frac{1}{T} = \frac{1}{38.148 \times 10^{-9}} = 26.21 \times 10^6 \text{ Hz} \quad \text{or} \quad 26.21 \text{ MHz}$$

f) The pitch of the helix is

$$d = u_p T = 3 \times 10^6 \times 38.148 \times 10^{-9} = 0.1144 \text{ m} \quad \text{or} \quad 11.44 \text{ cm}$$

• • •

6.10 Cyclotron

Beams of high-energy charged particles, such as protons or deuterons, are required for the so-called atom-smashing experiments that are used to investigate the subatomic structure of an atom. To impart high energy to a charged particle, an electric field is used to accelerate it to a very high speed. For this reason, the device imparting high energy to a charged particle is known as an *accelerator*.

The most common type of accelerator is the electron gun used in a cathode-ray tube. A very high potential difference is required for the particle to achieve the desired high speed in a one-shot operation. However, many such guns with modest potential difference can be arranged in a line and the particle made to pass through each one of them. In this way, the particle gains energy each time it passes through a gun. Such a device consisting of an array of guns is called a *linear accelerator*. As you can expect, a linear accelerator tends to be quite long.

A *cyclotron*, on the other hand, requires one electron gun through which the charged particle is made to pass again and again. In its simplest form, a cyclotron consists of two D-shaped cavities made of copper, as shown in Figure 6.12. A high-frequency oscillator is connected across the two cavities. As expected, the electric field will exist only within the gap between the cavities, and the charged particle will gain energy

Figure 6.12 The elements of a cyclotron

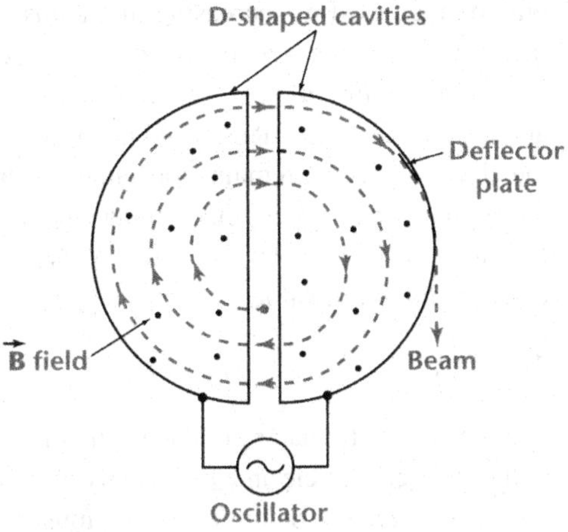

only while passing through the gap. The two cavities are sealed in a vacuum chamber to minimize the loss in energy due to collisions with air molecules. The whole structure is immersed in a uniform magnetic field.

The action begins when the charged particle is accelerated by the electric field in the gap and enters one of the two D-shaped cavities. Once the charged particle is inside the cavity, it follows a semicircular path in accordance with (6.30). There is no electric field within the cavity, so the velocity of the charged particle remains the same. If the frequency of the oscillator is the same as the cyclotron frequency, (6.32), the applied voltage will reverse its polarity by the time the charged particle reaches the gap. The reversal of the applied voltage changes the direction of the electric field within the gap and accelerates the charged particle into the other D-shaped cavity, where the particle describes another circular path of somewhat larger radius. Thus, the particle gains kinetic energy each time it crosses the gap, thereby moving into an orbit of larger radius. This process continues until the charged particle reaches the outer edge of the D-shaped cavity, where it is ejected out. If u is the velocity of the charged particle in a plane normal to the $\vec{\mathbf{B}}$ field at the time of exit and R is the radius, then the exit velocity, from (6.30), is

$$u = \frac{qBR}{m} \tag{6.35}$$

where q and m are the charge and the mass of the charged particle, respectively. The kinetic energy of the charged particle is

$$W_k = \frac{1}{2}mu^2 = \frac{q^2B^2R^2}{2m} \tag{6.36}$$

From this equation, it is evident that the kinetic energy of the charged particle depends upon the radius of the D-shaped cavity. Thus, for a given magnetic flux density the kinetic energy of a charged particle can only be increased by increasing the radius of the cavity. As the radius increases, so do the size and the cost of the electromagnet.

In order to limit the cost, we can vary the frequency of the oscillator and the magnetic flux density in unison so that the radius of the orbit of the charged particle remains the same. Such a design allows us to use an annular (ring-shaped) electromagnet at a tremendous saving in cost. Because $u = \omega R_c = 2\pi f R_c$, where R_c is the fixed radius, the frequency of the oscillator must be

$$f = \frac{u}{2\pi R_c} \tag{6.37}$$

Note that the frequency of the oscillator is proportional to the velocity of the charged particle and the velocity changes at every half-revolution, thus the frequency of the oscillator must be adjusted accordingly. In order to satisfy (6.32), the magnetic flux density must also follow a similar adjustment. A device that embodies this principle is known as a *synchrotron*. One such synchrotron, which is about 175 meters in diameter, has been built at Geneva, Switzerland, by the European Organization for Nuclear Research (CERN) at a cost of $28 million. A proton traveling a distance of 80 kilometers within the two cavities of this synchrotron can gain a kinetic energy of 4.5 nJ [28 billion electron volts, 1 eV $= 1.6 \times 10^{-19}$ J].

EXAMPLE 6.7

The radius of a D-shaped cavity of a cyclotron is 53 cm, and the frequency of the applied voltage source is 12 MHz. What value of B is needed to accelerate deuterons? What is the kinetic energy of a deuteron as it exits the cavity? A deuteron has the same charge as a proton but almost twice the mass.

Solution The magnetic flux density B must be

$$B = \frac{2\pi f m}{q} = \frac{2\pi \times 12 \times 10^6 \times 3.4 \times 10^{-27}}{1.6 \times 10^{-19}} = 1.6\text{ T}$$

The kinetic energy of a deuteron at the time of exit, from (6.36), is

$$W_k = \left(\frac{1.6 \times 10^{-19} \times 1.6 \times 0.53}{2 \times 3.4 \times 10^{-27}}\right)^2$$
$$= 2.707 \times 10^{-12}\text{ J} \quad \text{or} \quad 16.92\text{ MeV}$$

• • •

6.11 The velocity selector and the mass spectrometer

When a neutral gas is bombarded with high-energy particles, it acts as a source of positive ions. The positively charged particles can be collimated into a beam by passing them through an electron gun. There may be a wide spread in the speed of these particles in the beam. In order to create a beam of charged particles having the same speed, a *velocity selector* is used.

The operation of a velocity selector is based upon the Lorentz force equation:

$$\vec{\mathbf{F}} = q\vec{\mathbf{E}} + q\vec{\mathbf{u}} \times \vec{\mathbf{B}} \tag{6.38}$$

where q is the charge of a positive ion and $\vec{\mathbf{u}}$ is its velocity. In the design of a velocity selector, the $\vec{\mathbf{E}}$ and $\vec{\mathbf{B}}$ fields are arranged at right angles to the incoming beam of positively charged particles, as depicted in Figure 6.13. This is done to ensure that the electrostatic force experienced by the positive ion is in opposition to the magnetostatic force. The electrostatic force is the same on each ion but the magnetostatic force varies directly with its speed. Therefore, the net force experienced by a charged particle is zero at one particular velocity $\vec{\mathbf{u}}_0$ such that

$$\vec{\mathbf{E}} = -\vec{\mathbf{u}}_0 \times \vec{\mathbf{B}} \tag{6.39}$$

Thus the speed of the positive ion is

$$u_0 = \frac{E}{B} \tag{6.40}$$

A positive ion having a speed of u_0 will pass through the region without experiencing any force. A positive ion with a speed that is less than u_0 will be deflected upward. When the speed of a positive ion is greater than u_0, it will follow a downward trajectory. Therefore, all the positive ions passing through the fine aperture will have a speed of u_0; thus the device acts as a velocity selector (Figure 6.14). The velocity selector is also referred to as a *velocity filter*.

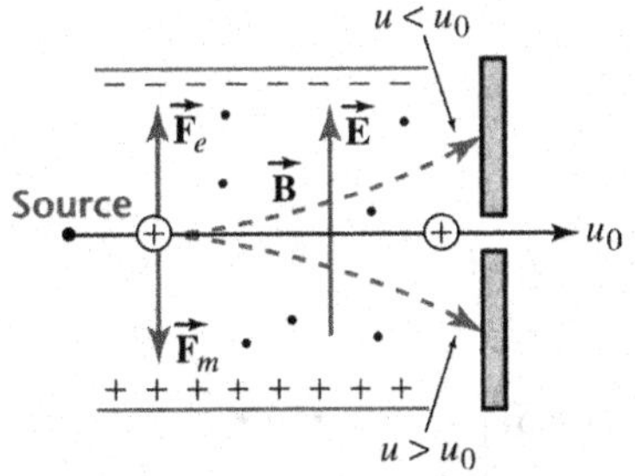

Figure 6.13 Elements of a velocity selector

We can also separate ions according to their masses using a *mass spectrometer*. A mass spectrometer consists of four basic parts: an ion source, a velocity selector, a deflection region with uniform magnetic field, and an ion detector such as an electron multimeter, electrometer, or photographic plate, as illustrated in Figure 6.15. The ion source produces the positively charged particles, and the velocity selector produces a beam of these charged particles moving with the same speed. These

Figure 6.14 Velocity selector (Courtesy of National Electrostatics Corp.)

charged particles of different masses now enter a region where there exists a uniform $\vec{\mathbf{B}}'$ field normal to their motion. In accordance with (6.30), each charged particle will follow a semicircular trajectory before being detected by the ion detector. The radius of the orbit depends upon the mass of each charged particle, so by measuring its radius we can determine a particle's mass. From (6.30) and (6.40), we obtain

$$m = \frac{qRBB'}{E} \tag{6.41}$$

Figure 6.15 Elements of a mass spectrometer

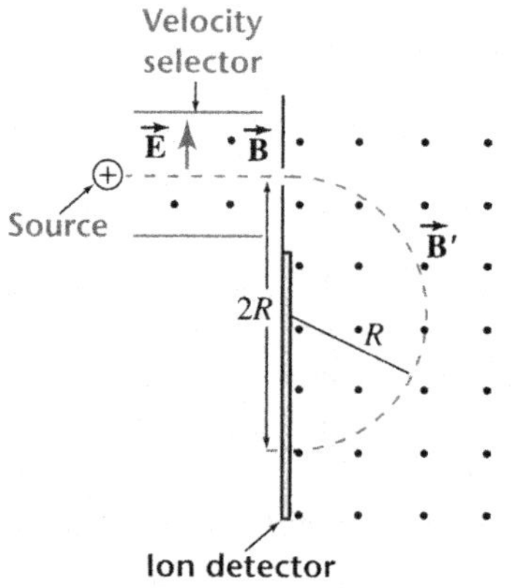

The mass spectrometer is also used in the study of **isotopes**. Isotopes are different forms of the same element having different masses but exhibiting the same behavior chemically. For that reason, they cannot be separated using chemical reactions. By replacing the ion detector by an ion collector we can transform a mass spectrometer into an *isotope separator*.

EXAMPLE 6.8

WORKSHEET 5

Mathcad

A velocity selector is used to select alpha particles of energy 200 keV from a beam containing particles of several energies. The electric field strength is 800 kV/m. What must be the magnetic field strength?

Solution The mass of an alpha particle is 6.68×10^{-27} kg. Thus, the velocity of the alpha particle is

$$u_0 = \left[\frac{2 \times 200 \times 10^3 \times 1.6 \times 10^{-19}}{6.68 \times 10^{-27}}\right]^{1/2} = 3.095 \times 10^6 \text{ m/s}$$

Thus, the magnetic field strength, from (6.40), must be

$$B = \frac{800 \times 10^3}{3.095 \times 10^6} = 0.258 \text{ T} \quad \text{or} \quad 258 \text{ mT}$$

• • •

6.12 The Hall effect

In 1879 Edwin Herbert Hall devised an experiment to determine the sign of the predominant charge carrier in a given conducting material. He placed a current-carrying strip in a plane perpendicular to a uniform magnetic field, as shown in Figure 6.16. If the current through the strip is due to the positive charges, the motion of the positive charges will be in the direction of the current, as depicted in Figure 6.16a. A positive charge moving with a velocity $\vec{\mathbf{u}}$ at right angles to a magnetic field $\vec{\mathbf{B}}$ will experience a force that will tend to move it toward side b of the strip. Therefore, there will be an excess of positive charges on side b, whereas side a will experience a deficiency of these charges. This results in a potential difference, known as the **Hall-effect voltage**, between the two sides. In this case, the potential of side b is higher than that of side a. The build of potential is not without limit. The potential difference between the two sides creates a transverse $\vec{\mathbf{E}}$ field within the strip which exerts a force on the positive charge in a direction opposite to the force created by the magnetic field. When the electrostatic force is equal and opposite to the magnetostatic force, the positive charges will move along the length of the strip without being deflected. The potential of side b with respect to side a can be expressed as

$$V_{ba} = V_b - V_a = -\int_a^b \vec{\mathbf{E}}_H \cdot \vec{d\ell} = E_H w$$

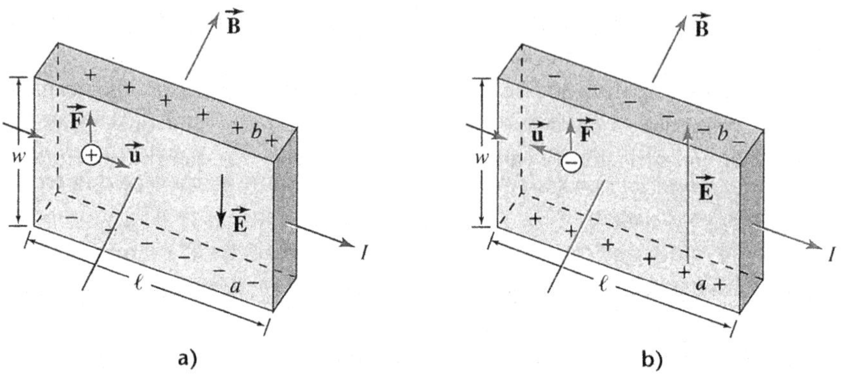

Figure 6.16 Hall-effect voltage in (a) a p-type material and (b) n-type material

where w is the width of the strip and E_H is the electric field intensity due to the Hall effect. Under equilibrium conditions, the electric field, from (6.40), is $E_H = uB$. Therefore, the Hall-effect voltage is

$$V_{ba} = uBw \tag{6.42}$$

In a p-type semiconductor, the Hall-effect voltage V_{ba} is positive, indicating that the current in a p-type semiconductor is due to the positive charges. It is now believed that a p-type semiconductor conducts by a process known as *hole conduction*.

Let us now consider that the current is due to the motion of electrons in a conducting material. The motion of an electron is in a direction opposite to that of the current, as illustrated in Figure 6.16b. Once again, the magnetic force acting on an electron pushes it toward side b. Soon there will be a concentration of electrons on side b and a deficiency of electrons on side a. Thus, the potential of side b with respect to side a will be negative. The direction of the $\vec{\mathbf{E}}_H$ field will be from side a toward side b. Under equilibrium conditions, the magnitude of the Hall-effect voltage can be obtained from (6.42). However, in this case, it will be negative. From experiments it has been found that the current in conductors such as copper, aluminum, silver, gold, etc., is due to the flow of electrons. This is also true for all n-type semiconductors. Therefore, using a static magnetic field we can determine whether a given wafer of a semiconducting material is a p-type or an n-type.

If A is the cross-sectional area of the strip, n is the number of the predominant charge carriers per unit volume, and $\Delta\ell$ is the length the charge carriers have advanced in time Δt, the current in the strip is

$$I = \frac{qnA\Delta\ell}{\Delta t} \tag{6.43}$$

However,

$$\Delta t = \frac{\Delta\ell}{u} \tag{6.44}$$

where u is the average velocity of the charge carriers in the medium. From these equations, we obtain

$$u = \frac{I}{qnA} \tag{6.45}$$

The Hall-effect voltage can now be expressed in terms of the current in the strip as

$$V_{ba} = \frac{BIw}{qAn} \tag{6.46}$$

In practice, the Hall effect is used (a) to determine the density of free electrons in a metal and (b) to measure the magnetic flux density in the air gap of an electric machine.

EXAMPLE 6.9

A 10-cm-wide, 10-cm-long, and 1-cm-thick copper strip carrying a current of 100 A is placed at right angles in a uniform magnetic field of 1.75 T. Determine the Hall-effect voltage. What is the Hall-effect electric field intensity? What is the electric field intensity responsible for the current in the copper strip? The conductivity of copper is 5.8×10^7 S/m and there are 8.5×10^{28} free electrons per cubic meter in copper.

Solution The electron is the predominant charge carrier in copper. Therefore, the Hall-effect voltage, from (6.46), is

$$V_{ba} = -\frac{1.75 \times 100 \times 10 \times 10^{-2}}{1.6 \times 10^{-19} \times 8.5 \times 10^{28} \times 10 \times 1 \times 10^{-4}}$$
$$= -1.287 \times 10^{-6} \text{ V}, \quad \text{or} \quad 1.287\ \mu\text{V}$$

The Hall-effect electric field is

$$E_H = \frac{V_{ba}}{w} = -\frac{1.287 \times 10^{-6}}{0.1} = -12.87 \times 10^{-6} \text{ V/m}$$

Since $J = \sigma E = I/A$, the electric field intensity responsible for the current is

$$E = \frac{I}{A\sigma} = \frac{100}{10 \times 1 \times 10^{-4} \times 5.8 \times 10^{7}}$$
$$= 1.72 \times 10^{-3} \text{ V/m} \quad \text{or} \quad 1.72 \text{ mV/m}$$

• • •

6.13 Magnetohydrodynamic generator

We have already observed that a current-carrying strip immersed in a magnetic field gives rise to the Hall-effect voltage between the opposite sides in the transverse plane. A *magnetohydrodynamic (MHD) generator* also employs this principle. In this case, hot ionized gas or plasma is made to flow through a rectangular channel in a plane perpendicular to the uniform magnetic field, as depicted in Figure 6.17. Because the

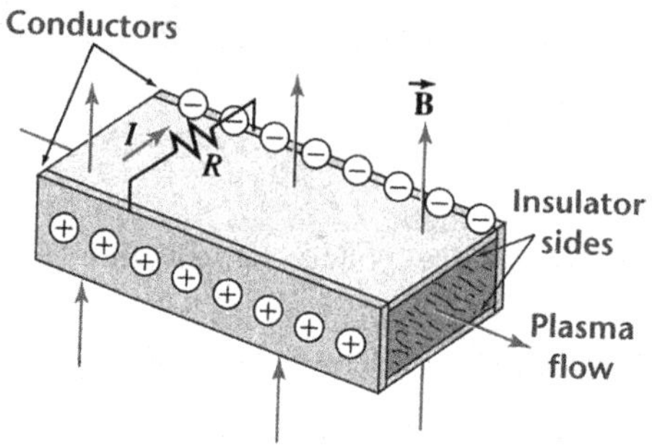

Figure 6.17 Elements of an MHD generator

plasma contains positively charged ions, these ions will be deflected to one side in accordance with the Hall effect. This produces a Hall-effect voltage between two sides of the plasma. For the directions of the magnetic field and the plasma flow as indicated in the figure, the right side of the stream of plasma acts as a positive terminal of a source of emf and the left side as the negative terminal. This source of emf will drive a current through an externally connected resistance as shown. In order to establish a good electrical contact, the right and the left sides of the channel must be made of a conductor. However, the top and the bottom sides of the channel must be made of an insulator in order to stop the circulating currents in the channel.

As auxiliary generators, MHDs can play a major role in the development of electrical energy from the burning of fossil fuel. However, the burning of the fossil fuel must take place in a special chamber, like a jet engine, so that the exhaust ionized gas entering the MHD can be further utilized in heating a boiler that provides steam for a conventional power generator. The channel can also be used as an *electromagnetic flowmeter* to measure the rate of flow of a conducting liquid such as beer, sewage, detergent, etc., by detecting the Hall-effect voltage induced by the motion of the liquid in a magnetic field.

EXAMPLE 6.10

A 75-cm-wide and 3-cm-thick rectangular channel with its transverse sides properly insulated is designed to carry a flow of exhaust gas at a speed of 1000 m/s. If the strength of the magnetic field is 1.5 T, what is the Hall-effect voltage between the two sides of the channel?

Solution From (6.42), the Hall-effect voltage is

$$V = uBw = 1000 \times 1.5 \times 0.75 = 1125 \text{ V}$$

6.14 An electromagnetic pump

The magnetic force exerted by the magnetic field on a moving charge has also led to the development of a pumping device without any moving parts, commonly referred to as an *electromagnetic pump*. The only thing

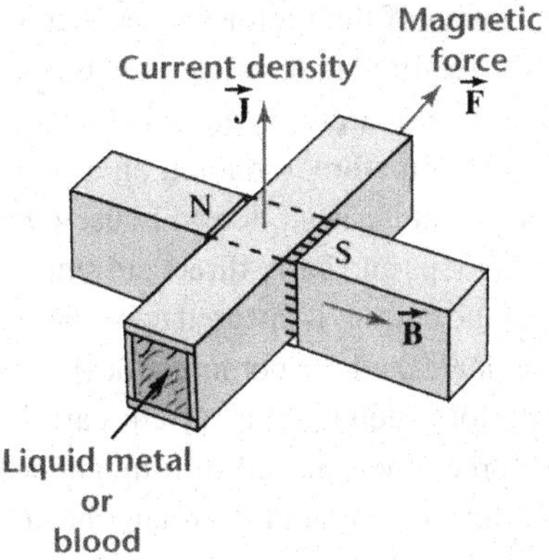

Figure 6.18 Elements of an electromagnetic pump

that moves in this device is the liquid itself that is to be circulated from one place to another. It has been used in the transfer of heat from the core of a nuclear reactor to a place where it can be utilized via liquid metals such as bismuth, lithium, sodium, etc. It has also been found useful in pumping blood without causing any damage to blood cells in the heart-lung and artificial kidney machines.

In its simplest form, an electromagnetic pump consists of a channel that is placed in a magnetic field, as shown in Figure 6.18. The channel may carry liquid metal or blood, depending upon its application. When a current is passed in the transverse direction as indicated in the figure, the resulting magnetic force drives the liquid along the channel.

6.15 A direct-current motor

The cross section of a bipolar *direct-current (dc) motor* is shown in Figure 6.19a. The field winding wound on the two poles of the stationary

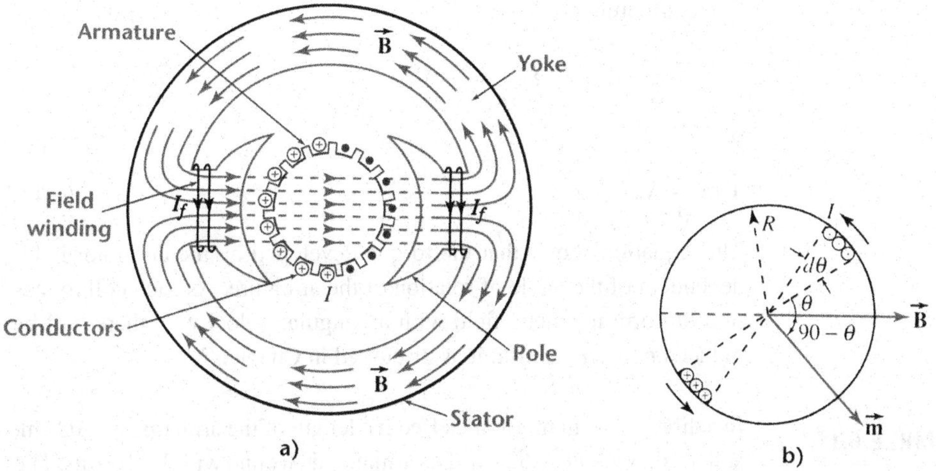

Figure 6.19 (a) Cross section of a dc motor; (b) Torque as experienced by the conductors on the armature

member (stator) of the motor carries a constant current I_f in order to establish the required magnetic flux in the machine. A cylindrical structure, called the armature, is placed concentrically in the region within the poles. The armature, mounted on a shaft, is free to rotate. The surface of the armature is slotted to house conductors. These conductors form closed loops and carry direct current I as shown in the figure. The direction of the current is ensured to be the same at all times in a conductor by a device called the commutator (not shown). In accordance with the Lorentz force equation, a current-carrying conductor experiences a magnetic force when placed in a magnetic field. The magnetic force produces a torque that tends to rotate the armature as we now explain.

Let us assume that all the conductors in the armature are uniformly distributed over its surface and form N turns. The turns enclosed by a differential angle $d\theta$ are $N\,d\theta/2\pi$. For the direction of current as shown in Figure 6.19b, the differential magnetic moment is

$$dm = \frac{NIA}{2\pi}\,d\theta$$

where I is the current in each turn, and A is the cross-sectional area of the turn. The direction of the magnetic moment is indicated in Figure 6.19b in accordance with the right-hand rule. The differential torque developed is

$$\begin{aligned}\vec{dT} &= \vec{dm} \times \vec{\mathbf{B}} = \frac{1}{2\pi}\,NIAB\,\sin(\pi/2-\theta)\,d\theta\vec{\mathbf{a}}_z\\ &= \frac{1}{2\pi}\,NIAB\,\cos\theta\,d\theta\vec{\mathbf{a}}_z\end{aligned}$$

where we have assumed that $\vec{\mathbf{a}}_z$ is the unit normal to the plane containing $\vec{\mathbf{m}}$ and $\vec{\mathbf{B}}$. In other words, both the vectors $\vec{\mathbf{m}}$ and $\vec{\mathbf{B}}$ are assumed to be in the xy plane. As you can see from the cross product, the torque experienced by the conductors will tend to rotate the armature in the counterclockwise direction. The total torque exerted on the conductors in the armature is

$$\vec{\mathbf{T}} = \vec{\mathbf{a}}_z\frac{1}{2\pi}\,NIAB\int_{-\pi/2}^{\pi/2}\cos\theta\,d\theta$$

or

$$\vec{\mathbf{T}} = \frac{1}{\pi}\,NIAB\,\vec{\mathbf{a}}_z \tag{6.47}$$

This equation shows that the torque developed by the dc motor is independent of the angle of rotation of the armature. As each coil rotates in a uniform magnetic field with an angular velocity ω, there will be induced motional emf in it as discussed in Chapter 7.

EXAMPLE 6.11

In a direct-current motor the effective length of the armature is 5.08 cm, and the diameter is 7.62 cm. The armature is wound with 1000 turns. The magnetic flux density is 1.08 T. If the current in the armature windings is 2.5 A, determine the torque developed by the motor.

Solution The cross-sectional area of each turn is

$$A = 5.08 \times 7.62 = 38.71 \text{ cm}^2$$

Thus, the torque developed by the motor is

$$T = \frac{1}{\pi} \times 1000 \times 2.5 \times 38.71 \times 10^{-4} \times 1.08$$
$$= 3.327 \text{ N·m}$$

• • •

6.16 Summary

We have devoted this chapter to explaining some of the applications of electrostatic and magnetostatic fields, but we have just scratched the surface. There are numerous other applications of static fields which embody the same principles as outlined in this chapter. Almost all of the applications fall into at least one of the following categories.

1. Force exerted on a charged particle by an electric field. If the charged particle is free to move, the electric force increases the kinetic energy of the charged particle.
2. Deflection of a charged particle by an electric field.
3. Deflection of a charged particle by a magnetic field.
4. Force experienced by a current-carrying conductor when placed in a magnetic field.

If a charged particle of mass m and charge q moves through a potential difference of V, the change in its kinetic energy is

$$\tfrac{1}{2}m\left[u_2^2 - u_1^2\right] = qV$$

where u_1 and u_2 are its initial and final velocities. We used this equation in the discussion of the cathode-ray oscilloscope. The ink-jet printer also employs the same principle of operation.

The ore separator for sorting minerals is based on the deflection experienced by a charged particle when it moves through an electric field. The force of gravity is used to move the charged particle in the downward direction. The trajectory of the charged particle within the electric field region was found to be a straight line. That is,

$$z = a_z \frac{x}{g}$$

Using the concept of change in energy in a variable capacitor, we explained the operation of an electrostatic voltmeter. We showed that the deflection of the pointer is proportional to the square of the applied voltage:

$$\theta = \frac{1}{2\tau} V^2 \frac{dC}{d\theta}$$

If a charged particle of charge q has a velocity $\vec{\mathbf{u}}$ in a plane perpendicular to a uniform magnetic field $\vec{\mathbf{B}}$, the charged particle moves in a circular orbit. Its radius is

$$R = \frac{mu}{qB} = \frac{u}{\omega}$$

where $\omega = (qB)/m$ is the cyclotron frequency. The cyclotron, a high-energy particle accelerator, uses this basic equation for its operation.

It was interesting to find out that by simply arranging electric and magnetic fields at right angles to each other we can select only those charged particles that move with a desired velocity. The desired velocity is simply a ratio of the electric field to magnetic field:

$$u_0 = \frac{E}{B}$$

The velocity selector becomes an integral part of a mass spectrometer. The mass spectrometer can be used to determine the mass of a charged particle by measuring the radius of its orbit. That is,

$$m = \frac{qRBB'}{E}$$

A current in a wafer placed in a magnetic field gives rise to the Hall-effect voltage. If the motion of charges in the wafer is at right angles to the uniform $\vec{\mathbf{B}}$ field, the Hall-effect voltage is

$$V = uBw = \frac{BIw}{qAn}$$

The polarity of the Hall-effect voltage enables us to determine whether the wafer is of p-type or n-type material.

A direct-current motor utilizes the force experienced by a current-carrying conductor when placed in a magnetic field. The magnetic force results in a torque, which in turn tends to rotate the armature of the motor. The magnitude of the torque is

$$T = \frac{1}{\pi} NIAB$$

We also briefly discussed other applications such as the electromagnetic pump and magnetohydrodynamic generator. You can find books on these subjects in your library, and you will be amazed at how simple ideas can be transformed into useful applications.

6.17 Review questions

6.1 What is the force on a test charge q in an electric field $\vec{\mathbf{E}}$?

6.2 What is the force on a test charge q moving through a magnetic field $\vec{\mathbf{B}}$ with a velocity $\vec{\mathbf{u}}$? When is the force maximum? When is the force minimum?

6.3 What is the net force on a charge q moving with a velocity $\vec{\mathbf{u}}$ in a region where $\vec{\mathbf{E}}$ and $\vec{\mathbf{B}}$ fields are at right angles to each other?

6.4 Devise a simple statement of how a current-carrying loop will orient itself in a magnetic field.

6.5 What is the trajectory of a charged particle after it exits a parallel-plate region?

6.6 A current-carrying conductor has zero net charge. Why, then, does a magnetic field exert a force on it?

6.7 What are the primary functions of the electric and the magnetic fields in a cyclotron?

6.8 What is the major difference between a cyclotron and a synchrotron?

6.9 What is the relation between the torque and the force on a conductor in a dc motor?

6.10 Can you use a cathode-ray oscilloscope to measure the ratio of the charge and the mass of an electron? Explain.

6.11 If a charged particle moves in an electric field, there is a change in its kinetic energy. If the same charged particle moves in a magnetic field, the kinetic energy remains unchanged. Why?

6.12 What is the Hall-effect voltage? Is Hall-effect voltage similar to motional emf?

6.13 Explain the operation of a cathode-ray tube.

6.14 Explain the operation of an electromagnetic pump.

6.15 Explain the principle of operation of a magnetohydrodynamic generator.

6.16 Explain the principle of operation of a dc motor.

6.17 A uniform $\vec{\mathbf{B}}$ field is along the z direction. A proton is rotating in a circular orbit with velocity $\vec{\mathbf{u}}$. What is its direction of rotation as viewed from the z axis?

6.18 An electron is deflected upward when it passes through an electric field. What is the direction of the electric field?

6.19 A proton is deflected upward when it passes through a magnetic field. What is the direction of the magnetic field?

6.20 An electric and a magnetic field are at right angles to each other. An electron moving from left to right passes through this region without experiencing any force. What are the directions of the $\vec{\mathbf{E}}$ and $\vec{\mathbf{B}}$ fields?

6.21 Nuclear physicists like to measure energies in electron volts (eV). Why?

6.22 The energy of a charged particle is 2 MeV. What is its energy in joules?

6.23 Why is a velocity selector called a velocity filter?

6.24 Show that the potential energy of a current loop in a magnetic field is $-\vec{\mathbf{m}} \cdot \vec{\mathbf{B}}$. What is the significance of the potential energy?

6.25 When is the potential energy of a loop maximum? Minimum?

6.26 Is it possible to define magnetic potential energy for a charged particle moving in a magnetic field?

6.18 Exercises

6.1 If the separation between the plates in Example 6.1 is reduced to 5 cm, what must be the applied voltage to cause a deflection of 0.5 cm? Sketch the velocity vs. time and velocity vs. distance curves for the electron during its travel between the parallel plates.

6.2 How long must the plates be in Example 6.1 so that the deflection is 2 cm? Compute the time now taken by the electron to exit the parallel plates. What is its velocity in the direction of its deflection at the time of exit?

6.3 Two parallel plates are held 5 cm apart. An electron is released at the surface of the negatively charged plate and strikes the surface of the opposite plate in 12.5 ns. Find (a) the velocity of the electron at the instant it strikes the positively charged plate, (b) the acceleration of the electron, (c) the potential of the positively charged plate with respect to the negatively charged plate, and (d) the electric field intensity within the plates.

6.4 An electron moving with a velocity of $60 \times 10^6 \vec{\mathbf{a}}_z$ m/s enters an electric field of $100\vec{\mathbf{a}}_z$ kV/m. Find (a) the distance traveled by the electron in the z direction before it temporarily comes to rest and (b) the time taken by the electron to travel that distance. How far will the electron travel before it loses 80% of its kinetic energy?

6.5 Compute the velocity of the ink droplet in Example 6.3 as it strikes the paper. What is the total time taken by the ink droplet from the instant it enters the deflection plate region and strikes the paper?

6.6 An ink droplet of diameter 0.01 mm attains a charge of -2 pC as it passes through the charging plates at a speed of 20 m/s. The potential difference between the vertical deflection plates held 5 mm apart is 200 V. If the length of each deflection plate is 1.5 mm, and the distance from the exit end of the deflection plate to the paper is 12 mm, determine the vertical displacement of the ink droplet. Assume that the density of the ink is 2 grams per cubic centimeter.

6.7 A phosphate granule with a mass of 1.2 grams acquires a charge of -100 nC on the vibrating feeder. The particle falls freely at the middle of the top edge of the parallel plates, which are held at a potential of 5 kV. If the plates are 1.5 m in length, find the separation between the plates so that the phosphate granule barely touches the bottom of the positively charged plate. What is the velocity of the charged particle at the time of its exit?

6.8 The mass of a phosphate granule in Exercise 6.7 varies from 0.5 gram to 2.5 grams. The charge acquired by each granule also varies from -80 nC to -120 nC. What must be the minimum separation between the plates if the particles fall freely at the middle of the top edge of the parallel plates?

6.9 The breakdown electric field intensity of air is 3 MV/m. If the factor of safety is assumed to be 10, what must be the radius of an isolated metallic sphere if it is held at a potential of 240 kV? What is the charge on the sphere?

6.10 In a Van de Graaff generator the radius of the inner sphere is 1 cm and that of the dome is 1 m. If the charges on the inner sphere and the dome are 10 nC and 10 μC, what is the potential difference between them?

6.11 The deflection on an electrostatic voltmeter is 30° when a potential difference of 100 V is maintained across its terminals. What is the change in capacitance per radian if the torsional constant of the spring is 1.5 N•m/rad?

6.12 If the deflection of the electrostatic voltmeter in Exercise 6.11 is 60°, what is the applied voltage?

6.13 Verify equation (6.30) using $m\vec{\mathbf{a}} = q\vec{\mathbf{u}} \times \vec{\mathbf{B}}$, where $\vec{\mathbf{a}}$ is the acceleration of the charged particle of charge q and mass m moving in a uniform magnetic field $\vec{\mathbf{B}} = B\vec{\mathbf{a}}_z$ with a velocity $\vec{\mathbf{u}}$. Assume at $t = 0$, $\vec{\mathbf{u}} = u_0\vec{\mathbf{a}}_x$, and the particle enters the magnetic field region at the origin.

6.14 An electron having energy of 5 electron volts (eV) is orbiting in a plane at right angles to a uniform magnetic field of 1.2 mT. Calculate (a) its velocity, (b) the radius of the orbit, (c) the cyclotron frequency, and (d) the period of oscillation. [*Note*: 1 eV $= 1.6 \times 10^{-19}$ J.]

6.15 An electron is accelerated from rest through a potential difference of 20 kV. It enters a uniform magnetic field that is at right angles to its direction of motion. If the strength of the magnetic field is 50 mT, calculate the distance between the electron's entry and exit points. How much time does the electron take to exit the field region?

6.16 The radius of a D-shaped cavity of a cyclotron is 75 cm, and the frequency of the applied voltage source is 10 MHz. What value of B is needed to accelerate protons? What is the kinetic energy of a proton as it exits the cavity?

6.17 In a certain cyclotron an electron moves in a circle of radius 25 cm. The magnitude of the $\vec{\mathbf{B}}$ field is 1.2 mT. Calculate (a) the cyclotron frequency and (b) the kinetic energy of the electron.

6.18 In a mass spectrometer, ions of one isotope of oxygen having a mass of 26.72×10^{-27} kg are detected at a distance of 20 cm from the point of entry. If an ion of another isotope of oxygen is detected at a distance of 22 cm, what is its mass?

6.19 The magnitude of magnetic flux density in a velocity selector is 0.5 T. If the electric field intensity is 1 MV/m, what must be the speed of a proton that passes through the region without being deflected? What is its kinetic energy? [Proton mass $= 1.67 \times 10^{-27}$ kg.]

6.20 The proton in Exercise 6.19 enters a uniform $\vec{\mathbf{B}}$ field of 0.5 T, where it moves in a semicircular path. What is the distance between its point of entry and the point of exit? How long does it take to exit?

6.21 Show that the number of charge carriers per unit volume in a conducting medium is

$$n = \frac{JB}{qE_H}$$

where J is the current density.

6.22 If E is the electric field intensity responsible for the current in a conducting medium and E_H is the Hall-effect electric field, show that

$$\frac{E_H}{E} = \frac{B\sigma}{nq}$$

where σ is the conductivity of the medium.

6.23 The width of a rectangular channel is 50 cm, and the strength of the magnetic field is 1.6 T. What must be the rate of flow of plasma in the channel so that the current through a 1150-Ω resistance connected between its sides is 2 A?

6.24 A plastic pipe of diameter 20 cm carries sewage. It is immersed in a transverse magnetic field of 0.5 T. If the maximum induced voltage between the opposite sides of the column of sewage is 0.25 V, what is the rate of flow of sewage in the pipe? Assume that the sewage flows at a uniform rate in the pipe.

6.25 In a direct-current motor the effective length of the armature is 2.54 cm, and the diameter is 5.08 cm. The armature is wound with 1500 turns. The magnetic flux density is 1.5 T. If the current in the armature windings is 12.5 A, determine the torque developed by the motor.

6.26 Refer to Exercise 6.25. Since a turn is made of two conductors, what is the force on each conductor? What is the force on all the conductors? What is the torque on each conductor? Compute the torque developed by the motor.

6.19 Problems

6.1 The separation between two plates of a parallel-plate capacitor is 5 cm. The potential difference between the plates is 10 kV. The plates are 10 cm on each side. A proton with a kinetic energy of 2 keV enters the region at right angles to the electric field. Determine (a) the time taken by the proton to exit the region, (b) the deflection of the proton as it exits the region, (c) the initial and final velocities of the proton, and (d) the change in its kinetic energy.

6.2 A deuteron moving with a velocity of $2 \times 10^6 \vec{\mathbf{a}}_x$ m/s enters an electric field of $-50\vec{\mathbf{a}}_x$ kV/m. Find (a) the distance traveled by the deuteron before it temporarily comes to rest and (b) the time taken by the deuteron to travel that distance. How far does the deuteron travel before it loses 50% of its kinetic energy? The mass of a deuteron is 3.4×10^{-27} kg.

6.3 An electron moving with a velocity of $0.8 \times 10^6 \vec{\mathbf{a}}_z$ enters a region where the electric field is $-10\vec{\mathbf{a}}_z$ kV/m. What is the velocity of the electron after it travels a distance of 3 cm?

6.4 The potential difference across two parallel plates 2 cm apart is 200 V. An electron is released from the negatively charged plate at $t = 0$ from rest. Determine (a) the velocity of the electron as it reaches the positively charged plate, (b) the time required to do so, and (c) the kinetic energy of the electron.

6.5 If the electron in Problem 6.4 has an initial velocity of 2 m/μs when it leaves the negatively charged plate, what is its velocity when it just touches the positively charged plate? How much energy has the electron gained?

6.6 At $t = 0$ a particle of mass 2 grams and a charge of 100 nC is passing through the origin with a velocity of 141.4 m/s at 45° with respect to the x axis in the xy plane. Determine its velocity and position at $t = 10$ s if the electric field intensity in the region is $200\vec{\mathbf{a}}_x$ kV/m.

6.7 The electric field intensity in a region is $150\vec{\mathbf{a}}_x$ kV/m. A proton enters that region with a velocity of $-32.48 \times 10^6 \vec{\mathbf{a}}_x$ m/s. What is the kinetic energy of the proton as it enters the region? How far will the proton travel before it comes to rest temporarily?

6.8 An ink droplet of diameter 0.025 mm attains a charge of −0.25 pC as it passes through the charging plates at a speed of 10 m/s. The electric field intensity between the vertical plates is 100 kV/m when the plates are 2.5 cm apart. If the length of each deflection plate is 1.5 mm and the distance from the end of the deflection plate and the paper is 5 mm, determine the vertical displacement of the ink droplet. The density of the ink is 2 grams per cubic centimeter.

6.9 Refer to Problem 6.8. What must be the electric field intensity to produce a total deflection of 5 mm on the paper?

6.10 Refer to Problem 6.8. Compute the velocity of the ink droplet as it strikes the paper. What is the total time taken by the ink droplet from the instant it enters the deflection plate region and strikes the paper?

6.11 A graphite particle of 1.2 grams acquires a charge of −50 nC on the vibrating feeder. The particle then falls freely in the middle of the top edge of a parallel-plate capacitor. The potential difference between the plates is 15 kV. If the plates are 1 m in length and 40 cm apart, determine the position and the velocity of the particle at the end of the plates.

6.12 Refer to Problem 6.11. If the plates are held at a distance of 2 m above ground, determine the velocity when the particle hits the ground. What is the total deflection of the particle?

6.13 How much charge is required to raise an isolated metallic sphere of 25-cm radius to a potential of 200 kV? What is the electric field intensity on the surface of the sphere? Can this electric field exist without breakdown when the surrounding medium is air?

6.14 In a Van de Graaff generator the radius of the inner sphere is 2 cm and that of the outer sphere is 20 cm. The charge on the outer sphere is 5 nC. What must be the charge on the inner sphere so that the potential difference between them is 500 V?

6.15 A particle of mass m and charge q is projected from infinity with an initial kinetic energy k toward another heavy particle of charge Q and mass M. What is the distance between the two particles when the particle in motion comes instantaneously to rest?

6.16 The capacitance in an electrostatic voltmeter varies as $b\theta$. The torsional constant of the spring is 1.2 N•m/rad. When a potential difference of 100 V is applied, the angle of deflection is 30°. What is b? What is the applied voltage if the angle of deflection is 45°?

6.17 Find the radius of curvature of an electron in a magnetic field of 1.5 T if an electron has a kinetic energy of (a) 100 eV and (b) 10 keV, in a plane perpendicular to the magnetic field.

6.18 The frequency of rotation of a proton in a $\vec{\mathbf{B}}$ field is 10 MHz. What is the $\vec{\mathbf{B}}$ field? If the radius of the orbit is 10 cm, what is the kinetic energy of the proton in (a) joules and (b) electron volts?

6.19 An electron moving along the z direction with a kinetic energy of 20 keV enters a region of uniform magnetic field where $\vec{\mathbf{B}} = 1.25\vec{\mathbf{a}}_x$ T. Determine the magnetic force acting on the electron. If the magnetic field is confined in the region $z \geq 0$, what is the radius of its trajectory? What is its direction at the time of exit? What is the distance between the point of entry and the point of exit?

6.20 A copper strip 20 cm wide and 0.2 cm thick carrying a current of 500 A is placed normally in a magnetic field of 1.2 T. Calculate the Hall-effect voltage that appears across the strip. The density of free electrons within copper is 8.5×10^{28} electrons per cubic meter.

6.21 An electric field of $\vec{\mathbf{E}} = 20\vec{\mathbf{a}}_z$ kV/m and a magnetic field of $\vec{\mathbf{B}} = 0.5\vec{\mathbf{a}}_x$ T exist in a region where a proton is moving without being deflected. What is the kinetic energy of the proton?

6.22 When a proton reaches its maximum energy in a cyclotron, its orbit has a radius of 12 cm. If the magnetic flux density is 1.5 T, determine the momentum and the kinetic energy of the proton.

6.23 A cyclotron is used to accelerate a proton to attain an energy of 8 MeV. If the radius of the orbit is 0.5 m, what must be the magnetic flux density? What is the cyclotron frequency? What is the exit velocity of the proton?

6.24 The radius of the orbit of a deuteron is 50 cm. The oscillator frequency is 10 MHz. Compute the magnetic field and the kinetic energy of the deuteron.

6.25 The armature of a dc motor has a diameter of 12 cm and a length of 30 cm, and is uniformly wound with 1200 turns. The stator winding produces a flux density of 0.8 T. If the armature current is 120 A, determine the torque developed by the motor.

6.26 A rectangular loop with 25 turns and a surface area of 10 cm^2 is rotating in a constant magnetic field of 0.5 T. If the current in the loop is 2 A, determine the torque developed. Assume that the angle between the lines of the magnetic field and the normal to the coil is 30°.

6.27 How much work is required to turn the coil in Problem 6.26 from $\theta = 0°$ to $\theta = 180°$?

7

Time-varying electromagnetic fields

7.1 Introduction

In the study of static fields we concluded that (a) static electric fields are created by charges, (b) static magnetic fields are produced by charges in motion or steady currents, (c) the static electric field is a conservative field because it has no curl, (d) the static magnetic field is continuous because its divergence is zero, and (e) the static electric field can exist even when there is no static magnetic field and vice versa.

In this chapter, we show that a time-varying electric field can be produced by a time-varying magnetic field. We will refer to an electric field created by a magnetic field as an *induced electric field* or an *emf-producing electric field*. We will also highlight the fact that the induced electric field is not a conservative field. The line integral of an induced electric field around a closed path is, in fact, called the *induced emf (electro-motive force)*. We will also discover that a time-varying electric field gives rise to a time-varying magnetic field. Simply stated, if there exists a time-varying electric (magnetic) field in a region, there also exists a time-varying magnetic (electric) field in that region. The equations describing the relations between electric and magnetic fields are known as *Maxwell's equations* because they were concisely formulated by James Clerk Maxwell. During the formulation of these equations, it will also become evident that Maxwell's equations are extensions of the known works of Gauss, Faraday, and Ampère.

We can begin our study either by stating Faraday's law of induction as an experimental fact and then developing the corresponding Maxwell equation or by looking at the magnetic force acting on a charged particle. The effects of the magnetic force on a charged particle are already known to us, so let us begin with this latter approach.

7.2 Motional electromotive force

Let us consider a conductor moving with a uniform velocity $\vec{\mathbf{u}}$ in the x direction, as depicted in Figure 7.1. In the region there also exists a

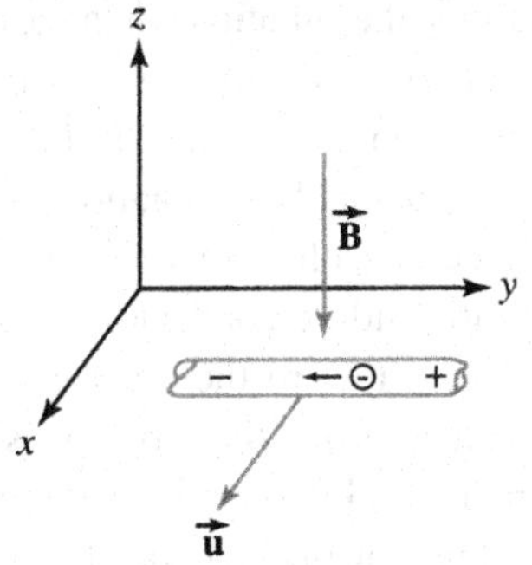

Figure 7.1 A conductor moving in a uniform magnetic field

uniform flux density $\vec{\mathbf{B}}$ such that $\vec{\mathbf{B}} = -B\vec{\mathbf{a}}_z$. The magnetic force acting on each of the free electrons in the conductor is

$$\begin{aligned}\vec{\mathbf{F}} &= q_e\vec{\mathbf{u}} \times \vec{\mathbf{B}} \\ &= q_e\, uB\vec{\mathbf{a}}_y\end{aligned} \tag{7.1}$$

where q_e is the magnitude of the charge on an electron. Under the influence of this force the free electrons within the conductor will move from right to left. Such a migration of electrons will result in a net negative charge at the left end of the conductor and a net positive charge at the right end. Barnett was able to demonstrate that such a separation of charges does take place in a conductor moving in a magnetic field. He managed to cut the conductor in the middle while it was still in motion. When the two pieces of the conductor were brought to rest, one was found to be positively charged while the other was negatively charged.

The force per unit charge is the electric field intensity $\vec{\mathbf{E}}$, thus we obtain an expression for the $\vec{\mathbf{E}}$ field from (7.1) as

$$\vec{\mathbf{E}} = \vec{\mathbf{u}} \times \vec{\mathbf{B}} = uB\vec{\mathbf{a}}_y \tag{7.2}$$

Since the electric field as given by (7.2) is established by a magnetic field, it is referred to as the *induced electric field.* As the $\vec{\mathbf{E}}$ field is a result of the motion of a conductor in the magnetic field, it is also called the *motional electric field.* Note that the induced electric field is perpendicular to the plane containing $\vec{\mathbf{u}}$ and $\vec{\mathbf{B}}$. We will later show that the *induced electric field is a nonconservative field.*

The induced electric field is tangential to the surface of the conductor. As the tangential component of the electric field just above the surface of the conductor is zero, so the electric field just beneath the surface of the conductor must also be zero. In order to satisfy this boundary condition, the flow of electrons inside the conductor from right to left will cease as soon as the electric field intensity within the conductor due to the separation of charges is equal and opposite to the induced electric field. When that happens, the conductor will be in its equilibrium state, and the net force acting on the free electrons will cease to exist.

Let us now examine the situation when the conductor is sliding freely over a pair of stationary conductors, as illustrated in Figure 7.2. A resistor is connected between the far ends of the two stationary conductors. Thus, the sliding conductor, the two stationary conductors, and the resistor form a closed electric path. In this case, accumulation of the electrons on the left side of the sliding conductor is no longer possible. Instead, the electrons will flow toward the other end of the conductor via the two stationary conductors and the series resistor. This flow of electrons results in a current in the closed circuit and is called the *induced current*. The conventional direction of the induced current, which is in opposition to the flow of electrons, is shown in the figure. The sliding conductor, therefore, acts as a source of *induced emf* (electromotive force). The induced current is merely a consequence of the induced emf in a closed electric circuit. Because the induced current in the resistor is from right to left, the right end (at $y = b$) of the sliding conductor is positive with respect to its left end (at $y = a$). Note that *the conventional direction of the induced current is the same as that of the induced electric field in the conductor.*

Figure 7.2 A sliding conductor

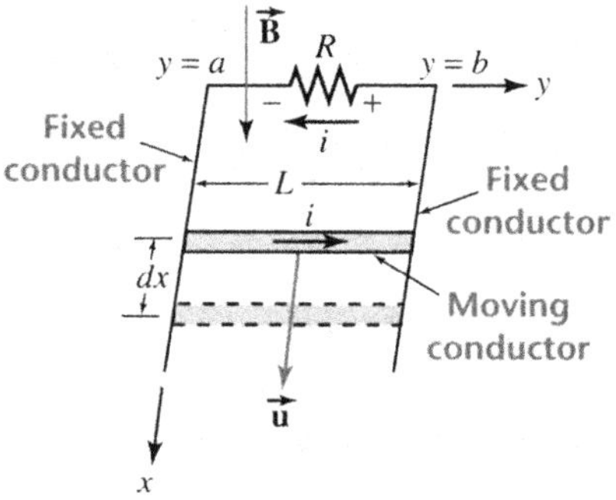

In accordance with the Lorentz force equation, the magnetic force experienced by the sliding conductor is

$$\vec{\mathbf{F}}_m = i\vec{\mathbf{L}} \times \vec{\mathbf{B}} = -BiL\vec{\mathbf{a}}_x \tag{7.3}$$

where i is the current in the sliding conductor and L is its effective length. As expected, the magnetic force is in a direction that opposes the motion of the conductor. Therefore, we must apply an external force in the x direction to keep the conductor moving in that direction. The external force that must be applied to keep the conductor in motion with a uniform velocity must be

$$\vec{\mathbf{F}}_{\text{ext}} = -\vec{\mathbf{F}}_m = BiL\vec{\mathbf{a}}_x \tag{7.4}$$

When the conductor moves a distance dx in a time interval dt, the work done by the external force is

$$dW = BLi\,dx = BLiu\,dt$$

where $dx = u\,dt$. Because $i\,dt$ represents the amount of charge dq that has been transferred in time dt, this expression can also be written as

$$dW = BLu\,dq$$

We now define the *electromotive force* or the *induced emf* as the amount of work done per unit positive charge by the external force:

$$e = \frac{dW}{dq} = BLu \tag{7.5}$$

In this case e is the *induced emf* between the two ends of the sliding conductor. It is also referred to as the *motional emf* because it is due to the motion (flux cutting action) of a conductor in a magnetic field. In the SI system of units, B is in teslas (T, Wb/m^2), L is in meters (m), u is in meters per second (m/s), and e is in joules per coulomb (J/C) or volts (V).

Equation (7.5) is valid only for a linear conductor moving in a plane normal to the magnetic field and along a direction at right angles to its length. We tacitly incorporated these assumptions in our development in order to explain the concept of motional emf. We now develop a general expression for motional emf.

7.2.1 General expression for motional emf

The general expression for the force that is required to keep a conductor in motion is

$$\vec{\mathbf{F}}_{\text{ext}} = -\int_c i\,\vec{d\ell}_c \times \vec{\mathbf{B}}$$

where $\vec{d\ell}_c$ is an element of length of the conductor in the direction of current i and c indicates the path of integration in the direction of the induced current in the conductor. The work done by the external force in moving the conductor a length $\vec{d\ell}$ in time dt is

$$dW = \vec{\mathbf{F}}_{\text{ext}} \cdot \vec{d\ell} = -i\vec{d\ell} \cdot \int_c \vec{d\ell}_c \times \vec{\mathbf{B}}$$

Substituting

$$i = \frac{dq}{dt}$$

and

$$\vec{\mathbf{u}} = \frac{\vec{d\ell}}{dt}$$

in this equation, we obtain a general expression for the motional emf in the conductor as

$$e = \frac{dW}{dq} = -\vec{\mathbf{u}} \cdot \int_c \vec{d\ell}_c \times \vec{\mathbf{B}}$$

As $\vec{\mathbf{u}}$ does not vary along the length of the conductor, we can also write the above equation as

$$e = -\int_c \vec{\mathbf{u}} \cdot (\vec{d\ell}_c \times \vec{\mathbf{B}}) = \int_c \vec{\mathbf{u}} \cdot (\vec{\mathbf{B}} \times \vec{d\ell}_c)$$

Finally, using the vector identity, we can write this equation as

$$e = \int_c (\vec{\mathbf{u}} \times \vec{\mathbf{B}}) \cdot \vec{d\ell}_c \tag{7.6}$$

which reduces to (7.5) when $\vec{\mathbf{u}}$, $\vec{\mathbf{B}}$, and the length of the conductor are mutually perpendicular to each other. *Equation (7.6) is the one we will use to determine the motional emf in a conductor moving in a magnetic field.*

We mention again that $\vec{\mathbf{u}} \times \vec{\mathbf{B}}$ is the induced electric field intensity and its direction is the same as that of the induced current in the conductor. This fact enables us to establish the polarity of the motional emf across the two ends of a conductor. In a nutshell, *the induced current due to the motional emf is in the direction of the induced electric field.*

EXAMPLE 7.1

A copper strip of length L pivoted at one end is rotating freely with an angular velocity ω in a uniform magnetic field, as shown in Figure 7.3. What is the induced emf between the two ends of the strip?

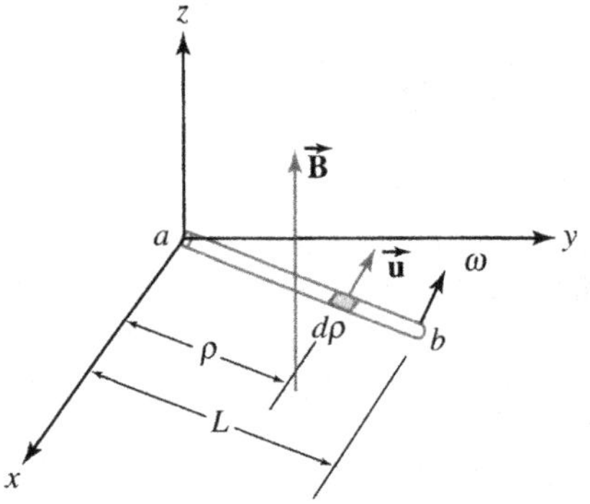

Figure 7.3 A copper strip rotating in a uniform magnetic field

Solution The velocity at any radius ρ of the strip is

$$\vec{\mathbf{u}} = \rho\omega\vec{\mathbf{a}}_\phi$$

The induced electric field intensity is

$$\vec{\mathbf{E}} = \vec{\mathbf{u}} \times \vec{\mathbf{B}} = \rho\omega B(\vec{\mathbf{a}}_\phi \times \vec{\mathbf{a}}_z) = \rho\omega B\vec{\mathbf{a}}_\rho$$

Because the induced emf is in the radial direction, end b of the strip is positive with respect to the pivoted end a. Hence, the induced emf, from (7.6), is

$$e_{ba} = \omega B \int_0^L \rho \, d\rho$$
$$= \tfrac{1}{2} B\omega L^2$$

• • •

EXAMPLE 7.2 A copper strip of length $2L$ pivoted at the midpoint is rotating with an angular velocity ω in a uniform magnetic field, as illustrated in Figure 7.4. Determine the induced emf between the midpoint and one of the ends of the strip. What is the induced emf between the two ends?

Solution Let us imagine that the copper strip of length $2L$ is made of two copper strips each of length L but joined together at the pivoted end. From Example 7.1, it is then evident that the far end of each strip is at a higher potential with respect to its pivoted end. As both strips are of equal length and are rotating with the same angular velocity in a common uniform magnetic field, the magnitude of the induced emf between the free and the pivoted end of each strip must be the same. Thus, the induced emf between one end of the strip and the midpoint is

$$e_{ba} = \tfrac{1}{2} B\omega L^2$$

The induced emf between the two far ends of the copper strip is zero.

• • •

If the strip shown in Figure 7.3 can supply a maximum current I, the two strips shown in Figure 7.4 have the capability of supplying a total current of $2I$ as long as the strips are identical. Thus, by joining many strips in this fashion, the current through R can be increased considerably. A device based upon this principle is called a *homopolar generator* and is shown in Figure 7.5. It is simply a thin circular disc, known as *Faraday's disc*, made of a conducting material such as copper. When the disc rotates with a constant angular velocity in a uniform

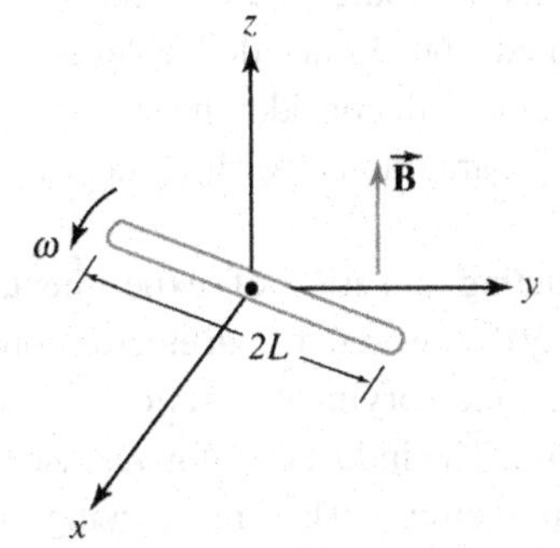

Figure 7.4 A copper strip pivoted at the midpoint and rotating in a uniform magnetic field

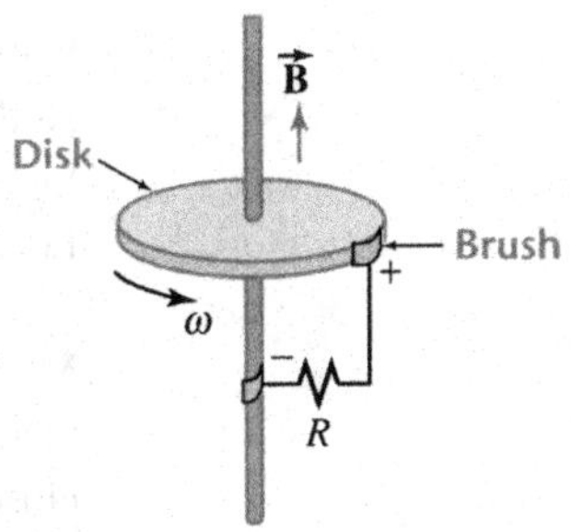

Figure 7.5 A homopolar generator

magnetic field, it acts as a constant (dc) voltage source. Thus its name – homopolar generator.

7.3 Faraday's law of induction

The induced emf in the closed circuit of Figure 7.2 can also be studied from another point of view. As the sliding conductor moves in the x direction, the increase in the cross-sectional area of the closed circuit formed by the sliding conductor, the two stationary conductors, and the resistor is

$$\vec{d\mathbf{s}} = L\,dx\vec{\mathbf{a}}_z$$

where L is the length of the conductor and dx is the distance it has moved in time dt. The change in the magnetic flux passing through the plane of the closed loop is

$$d\Phi = \vec{\mathbf{B}} \cdot \vec{d\mathbf{s}} = -BL\,dx$$

The rate of change of the magnetic flux passing through the loop is

$$\frac{d\Phi}{dt} = -BL\frac{dx}{dt} = -BLu$$

From (7.5), we can write this equation as

$$e = -\frac{d\Phi}{dt} \tag{7.7}$$

Equation (7.7) is, in fact, a mathematical definition of *Faraday's law of induction*. It states that *the induced emf around a closed path is equal to the negative rate of change of the magnetic flux with respect to time passing through the area enclosed by the path.*

Although we have derived equation (7.7) for a special case, it is valid in general. Strictly speaking, (a) this equation is based upon experimental observations and is usually stated as such in most textbooks, and (b) the negative sign is attributed to Lenz and is called *Lenz's law*. We could have simply stated it but we decided to develop it from the motional emf. At this time, we can still consider the statement as an experimental fact. In the following paragraphs we shed more light on Faraday's law and Lenz's law.

After conducting a series of experiments with stationary coils, Michael Faraday discovered that an electromotive force (emf) is induced in a coil when a time-varying magnetic flux passes through the area enclosed by the coil. The induced emf gives rise to an induced current in a closed conducting circuit. The time-varying flux may be created either by moving a magnet in the vicinity of the coil, as shown in Figure 7.6, or by opening or closing a switch in another coil, as shown in Figure 7.7.

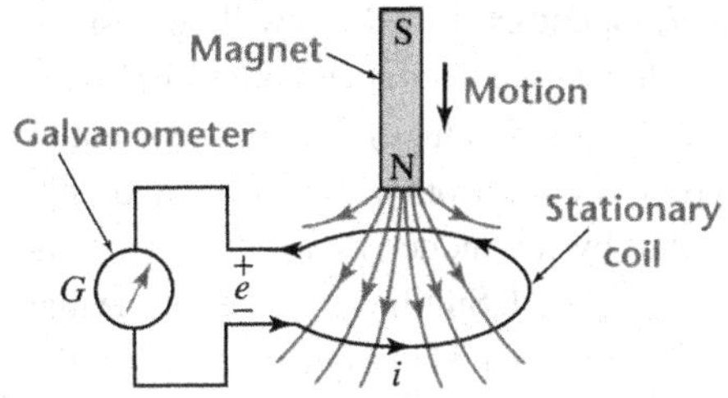

Figure 7.6 Induced emf and current in a coil due to increase in the magnetic flux

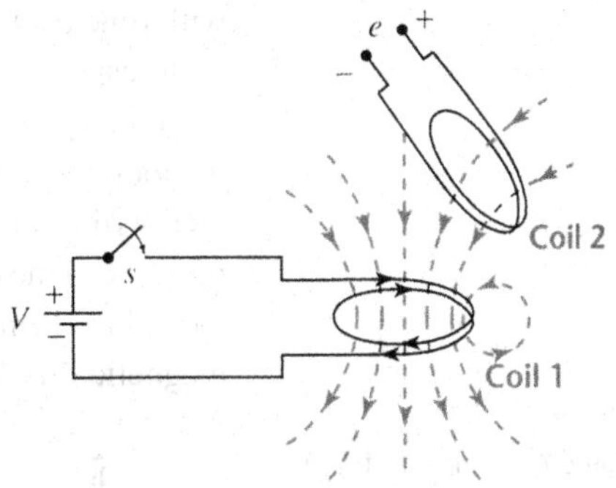

Figure 7.7 Induced emf in coil 2 at the time of closing of switch *s* in coil 1

The process of inducing an emf in a coil (also called a loop) by placing it in a time-varying magnetic field is now commonly referred to as an *electromagnetic induction*. In fact, the electromagnetic induction will take place as long as one of the following conditions holds.

1. The magnetic flux passing through (linking) a stationary coil varies as a function of time.
2. The coil continuously changes its shape or position when the magnetic flux distribution in the region is uniform. The shape can be changed by squeezing or stretching the coil.
3. The magnetic flux linking the coil changes with time, and the coil is in motion and/or changes its shape.

The induced current in a closed conducting path is a consequence of the induced emf in the loop. As far as the induced emf is concerned, the closed path forming a loop does not have to be conductive. The induced emf will still exist if the closed path is in free space or an insulating medium.

As mentioned earlier, the negative sign of equation (7.7) was introduced by Heinrich Friedrich Emil Lenz in order to comply with the polarity of the induced emf and is now known as **Lenz's law**. It states that *the current induced in a closed conducting loop by a change in magnetic flux through the loop is in a direction such that the flux generated by the induced current tends to counterbalance the change in the original magnetic flux*. Later in this chapter we will realize that Lenz's law is simply a consequence of the principle of conservation of energy in electrical terms. It helps us to determine the direction of the induced current in a closed loop. The polarity of the induced emf in an open loop can be determined by visualizing the direction of the induced current if the loop were a closed path.

Let us now consider an open loop placed in a magnetic field. When the magnetic flux density, and thereby the magnetic flux linking the loop, is of uniform strength, as shown in Figure 7.8a, the induced emf in the loop will be zero. When the magnetic flux density is increasing with time as depicted in Figure 7.8b, end b of the loop will be positive with respect to the other end a. This is in accordance with Lenz's law. If we imagine that the loop forms a closed path, the induced current in the loop must be in the clockwise direction from a to b. Then and only then will the magnetic flux created by the induced current *oppose the increase* in the original flux linking the coil. Similarly, we can argue that the induced emf will have the polarity indicated in Figure 7.8c when the magnetic flux linking the coil is decreasing with time.

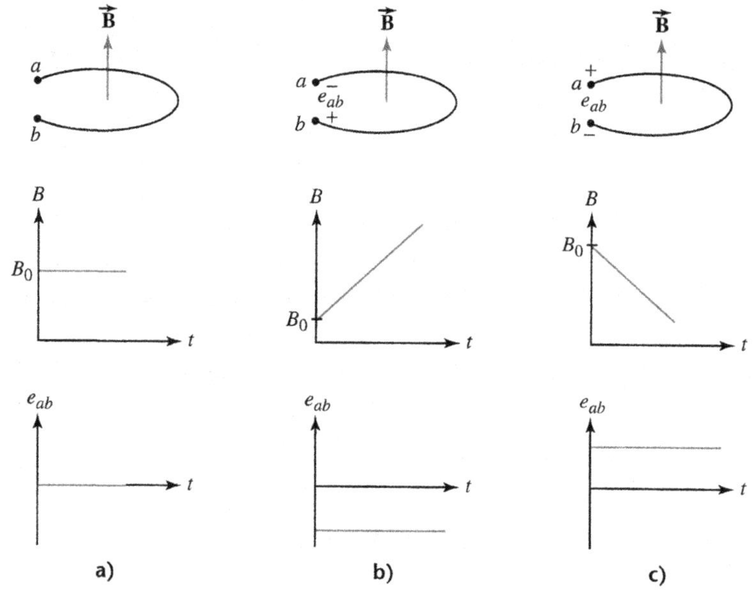

Figure 7.8 Induced emf in an open loop resulting from the flux linking the loop is (a) constant with time, (b) increasing with time, and (c) decreasing with time

7.3.1 Induced emf equation

In a coil with N tightly wound turns, the change in the magnetic flux linking the coil induces an emf in each turn of the coil. The total emf induced in the coil is the sum of the induced emfs of the individual turns connected in series. That is,

$$e = -N\frac{d\Phi}{dt} \tag{7.8}$$

By defining the *flux linkages* as

$$\lambda = N\Phi \tag{7.9}$$

we can express (7.8) as

$$e = -\frac{d\lambda}{dt} \tag{7.10}$$

We can either use (7.8) or (7.10) to determine the induced emf in a tightly

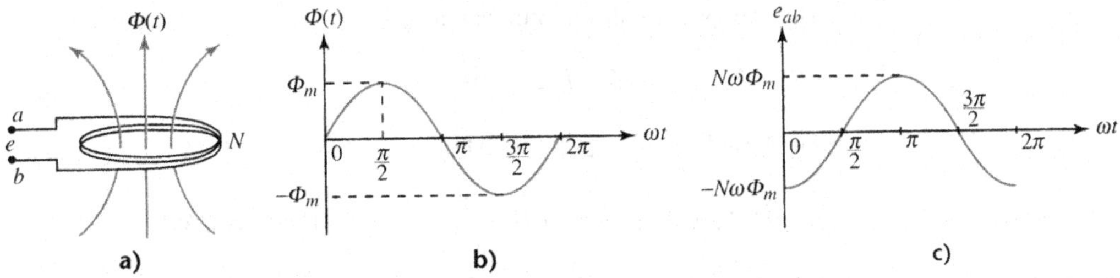

Figure 7.9 (a) An open coil with *N* turns; (b) Magnetic flux linking the coil; (c) Induced emf in the coil

wound coil having N turns. However, if the flux linking the N-turn coil (Figure 7.9a) varies sinusoidally (Figure 7.9b) as

$$\Phi = \Phi_m \sin \omega t$$

then the induced emf in the coil is

$$e_{ab} = -N\omega\Phi_m \cos \omega t$$

The instantaneous value of the induced emf is sketched in Figure 7.9c. The maximum value of the induced emf is

$$E_m = N\Phi_m\omega$$

and the effective (or rms) value is

$$E = \frac{1}{\sqrt{2}} E_m = \sqrt{2}\, \pi f N \Phi_m \tag{7.11}$$

It is a common practice to approximate $\sqrt{2}\pi$ as 4.44 and state equation (7.11) as

$$E = 4.44 f N \Phi_m \tag{7.12}$$

We developed this equation on the assumption that the coil is stationary and the magnetic flux linking the coil is changing sinusoidally with time. This is basically the operating principle of a transformer (discussed in detail in Section 7.14). Thus equation (7.12) is routinely referred to as the *transformer equation*, and the induced emf is commonly called the *transformer emf.* We will discuss the transformer emf and the transformer equation further in Section 7.4.

EXAMPLE 7.3

A circular conducting loop of radius 40 cm lies in the xy plane and has a resistance of 20 Ω. If the magnetic flux density in the region is given as $\vec{\mathbf{B}} = 0.2 \cos 500t\, \vec{\mathbf{a}}_x + 0.75 \sin 400t\, \vec{\mathbf{a}}_y + 1.2 \cos 314t\, \vec{\mathbf{a}}_z$ T, determine the effective value of the induced current in the loop.

Solution As the loop is in the xy plane, the unit normal to the loop is in the z direction. Thus, the differential surface area of the loop is

$$\overrightarrow{ds} = \rho\, d\rho\, d\phi\, \vec{\mathbf{a}}_z$$

The flux passing through this area is

$$d\Phi = \vec{\mathbf{B}} \cdot \vec{d\mathbf{s}} = 1.2\rho\, d\rho\, d\phi\, \cos 314t$$

The total flux linking the loop at any time is

$$\Phi = 1.2 \cos 314t \int_0^{0.4} \rho\, d\rho \int_0^{2\pi} d\phi = 0.603 \cos 314t \text{ Wb}$$

Because the flux varies sinusoidally with $\omega = 314$ rad/s, the frequency of the induced emf is 50 Hz. The maximum value of the flux is 0.603 Wb. Hence the effective value of the induced emf, from (7.12), is

$$\begin{aligned} E &= 4.44 \times 50 \times 1 \times 0.603 \\ &= 133.866\, V \end{aligned}$$

The effective value of the induced current in the closed loop having a resistance of 20 Ω is

$$I = \frac{133.866}{20} = 6.693 \text{ A}$$

• • •

7.4 Maxwell's equation (Faraday's law)

We already know that there must exist an electric field inside a conductor to sustain a current within the conductor. Such an understanding enables us to define the induced emf in a conductor in terms of the induced electric field intensity inside the conductor as

$$e = \oint_c \vec{\mathbf{E}} \cdot \vec{d\ell} \tag{7.13}$$

where the path of integration along c is in the direction of the induced current that will circulate if the path is conductive. If we express the total flux enclosed by contour c as

$$\Phi = \int_s \vec{\mathbf{B}} \cdot \vec{d\mathbf{s}}$$

then we can express (7.7) as

$$\oint_c \vec{\mathbf{E}} \cdot \vec{d\ell} = -\frac{d}{dt} \int_s \vec{\mathbf{B}} \cdot \vec{d\mathbf{s}} \tag{7.14}$$

The direction of the surface $\vec{d\mathbf{s}}$ is defined by the direction of contour c and the right-hand rule. When we curl the fingers of the right hand in the direction of contour c, the thumb points in the direction of a unit normal to the surface ds.

If we consider the surface to be fixed in space, then the time derivative in equation (7.14) applies only to the time-varying magnetic field $\vec{\mathbf{B}}$. In that case, we can express the equation as

$$\oint_c \vec{\mathbf{E}} \cdot \vec{d\ell} = -\int_s \frac{\partial \vec{\mathbf{B}}}{\partial t} \cdot \vec{d\mathbf{s}} \tag{7.15}$$

In order to indicate that the differentiation of $\vec{\mathbf{B}}$ is with respect to time only, we have expressed it as a partial derivative. *Equation (7.15) is also a definition of Faraday's law of induction in the integral form as applied to a stationary loop immersed in a time-varying magnetic field.* As the line integration of the induced electric field around a closed path is equal to the induced emf, *the induced electric field is not a conservative field.*

Using Stokes' theorem, we can transform the line integral around a closed path c into a surface integral over the surface s bounded by c as

$$\int_s (\nabla \times \vec{\mathbf{E}}) \cdot \vec{d\mathbf{s}} = -\int_s \frac{\partial \vec{\mathbf{B}}}{\partial t} \vec{d\mathbf{s}}$$

The integration on either side of this equation is over the same surface s bounded by closed path c; therefore, the equation is valid if and only if the two integrands are equal. That is,

$$\nabla \times \vec{\mathbf{E}} = -\frac{\partial \vec{\mathbf{B}}}{\partial t} \tag{7.16}$$

Equation (7.16) is also a statement of Faraday's law of induction for a fixed point of observation in a stationary medium. This equation is one of the four well-known *Maxwell equations*, and we will refer to it as *Maxwell's equation (Faraday's law) in the point, or differential, form.* This equation enables us to compute the electric field intensity at a fixed point in space when the magnetic field is a function of time. Note that for a static field, it reduces to $\nabla \times \vec{\mathbf{E}} = 0$.

We also identify (7.15) as *Maxwell's equation (Faraday's law) in the integral form.* We can use this equation to calculate the induced emf around a stationary closed path. As mentioned in Section 7.3, this equation is then the integral form of the transformer equation. We can also express it as

$$e_t = -\int_s \frac{\partial \vec{\mathbf{B}}}{\partial t} \cdot \vec{d\mathbf{s}} \tag{7.17}$$

where the subscript t is added to indicate that it is the transformer emf only.

7.4.1 General equations

The motion of a closed loop (circuit) in a magnetic field produces a motional emf in that loop as stated by (7.6). If we use the subscript m

to indicate the motional emf, we can write (7.6) for the closed circuit as

$$e_m = \oint_c (\vec{\mathbf{u}} \times \vec{\mathbf{B}}) \cdot \vec{d\ell} \tag{7.18}$$

When the loop is moving in a time-varying magnetic field, the total induced emf will be

$$\begin{aligned} e &= e_t + e_m \\ &= -\int_s \frac{\partial \vec{\mathbf{B}}}{\partial t} \cdot \vec{ds} + \oint_c (\vec{\mathbf{u}} \times \vec{\mathbf{B}}) \cdot \vec{d\ell} \end{aligned} \tag{7.19}$$

The direction of contour c in this equation defines the direction of the unit normal to the surface ds in accordance with the right-hand rule. *This equation is another general statement of Faraday's law of induction.* In terms of the induced electric field, equation (7.19) can also be written as

$$\oint_c \vec{\mathbf{E}} \cdot \vec{d\ell} = -\int_s \frac{\partial \vec{\mathbf{B}}}{\partial t} \vec{ds} + \oint_c (\vec{\mathbf{u}} \times \vec{\mathbf{B}}) \cdot \vec{d\ell}$$

The application of Stokes' theorem yields

$$\oint_s (\nabla \times \vec{\mathbf{E}}) \cdot \vec{ds} = -\int_s \frac{\partial \vec{\mathbf{B}}}{\partial t} \vec{ds} + \int_x [\nabla \times (\vec{\mathbf{u}} \times \vec{\mathbf{B}})] \cdot \vec{ds}$$

Since the surface s is bounded by an arbitrary contour c, and for the equation to be valid in general, we can equate the integrands and obtain

$$\nabla \times \vec{\mathbf{E}} = -\frac{\partial \vec{\mathbf{B}}}{\partial t} + \nabla \times (\vec{\mathbf{u}} \times \vec{\mathbf{B}}) \tag{7.20}$$

This equation is the most general form of Maxwell's equation (Faraday's law) in the point form. It enables us to determine the electric field at a point of observation moving with a velocity $\vec{\mathbf{u}}$ in a magnetic field $\vec{\mathbf{B}}$.

EXAMPLE 7.4

A tightly wound rectangular coil having N turns is rotating in a uniform magnetic field, as shown in Figure 7.10. Determine the induced emf in the coil using (a) the concept of motional emf and (b) Faraday's law of induction.

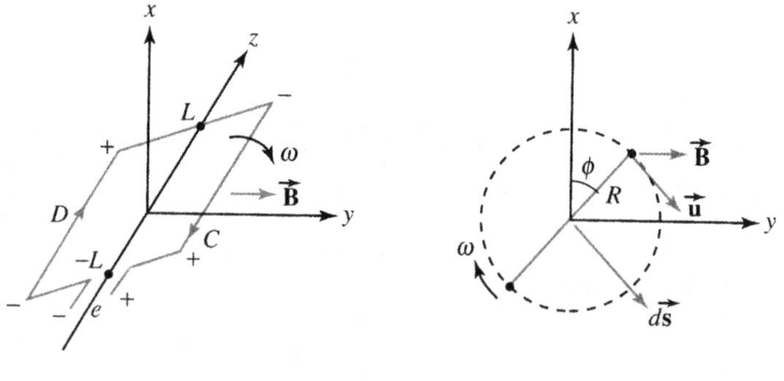

Figure 7.10 Two views of a coil rotating in a magnetic field

Solution a) Motional emf: The magnetic flux density is uniform, so the induced emf in the coil will only be due to its motion. In addition, only the conductors at radius R are responsible for the induced emf. The magnitude of the induced emf in N conductors at C is the same as that in N conductors at D. However, they will exhibit a 180° phase shift. With this understanding, let us compute the induced emf in N conductors at C. The velocity with which these conductors are moving is

$$\vec{\mathbf{u}} = \omega R \vec{\mathbf{a}}_\phi$$

As $\phi = \omega t$ and $\vec{\mathbf{B}} = B\vec{\mathbf{a}}_y$, $\vec{\mathbf{u}} \times \vec{\mathbf{B}} = -\omega RB \sin\omega t\, \vec{\mathbf{a}}_z$
Thus, the motional emf in N conductors at C is

$$e_m = \int_c N(\vec{\mathbf{u}} \times \vec{\mathbf{B}}) \cdot \vec{d\ell} = -NBR\omega \sin\omega t \int_L^{-L} dz$$
$$= 2NLRB\omega \sin\omega t$$

Hence, the induced emf in the coil with N tightly wound turns is

$$e = 2e_m = 4LRNB\omega \sin\omega t = NBA\omega \sin\omega t$$

where $A = 4LR$ is the area of the coil.

b) Faraday's law: For the direction of the differential surface, as shown in Figure 7.10, we have

$$\Phi = \int_s \vec{\mathbf{B}} \cdot \vec{d\mathbf{s}} = \int_s (\vec{\mathbf{a}}_y \cdot \vec{\mathbf{a}}_\rho) B\, ds = B\cos\omega t \int_s ds = BA\cos\omega t$$

The induced emf in an N-turn coil is

$$e = -N\frac{d\Phi}{dt} = BAN\omega \sin\omega t$$

• • •

EXAMPLE 7.5 If the magnetic flux density in Example 7.4 varies as $B_m \sin\omega t$, determine the induced emf using (a) the concepts of motional and transformer emfs and (b) Faraday's law of induction.

Solution a) Motional emf:

$$e = 2e_m = 2N \int_c (\vec{\mathbf{u}} \times \vec{\mathbf{B}}) \cdot \vec{d\ell} = 2N\omega B_m R \sin^2\omega t \int_L^{-L} dz$$
$$= B_m AN\omega \sin^2\omega t$$

where $A = 4LR$.
Transformer emf:

$$e_t = -N \int_s \frac{\partial \vec{\mathbf{B}}}{\partial t} \cdot \vec{d\mathbf{s}} = -N\omega B_m \cos\omega t(\vec{\mathbf{a}}_y \cdot \vec{\mathbf{a}}_\rho) \int_s ds$$
$$= -B_m AN\omega \cos^2\omega t$$

where $\vec{\mathbf{a}}_y \cdot \vec{\mathbf{a}}_\rho = \cos \omega t$. Thus, the induced emf in the coil is

$$\begin{aligned} e &= e_m + e_t \\ &= -B_m A N \omega (\cos^2 \omega t - \sin^2 \omega t) \\ &= -B_m A N \omega \cos 2\omega t \end{aligned}$$

b) Faraday's law:

$$\Phi = \int_s \vec{\mathbf{B}} \cdot \overrightarrow{ds} = B_m \sin \omega t \cos \omega t \int_s ds = \frac{1}{2} B_m A \sin 2\omega t$$

Hence, the induced emf in an N-turn coil is

$$e = -\frac{1}{2} B_m A N \frac{d}{dt} (\sin 2\omega t) = -B_m A N \omega \cos 2\omega t$$

• • •

7.5 Self-inductance

Consider a coil with N closely wound turns connected to a time-varying source and carrying a current $i(t)$, as shown in Figure 7.11a. The current generates a time-varying magnetic flux, which in turn induces an emf in the coil, as indicated in the figure. The induced emf causes an induced current in the coil which tends to oppose the change in the original current $i(t)$. If $\Phi(t)$ is the flux at any instant linking all the turns in the coil, the induced emf $e(t)$ in the coil at that instant is

$$e = N \frac{d\Phi}{dt}$$

Note that we have dropped the negative sign in this equation because we have marked the polarity of the induced emf in the figure.

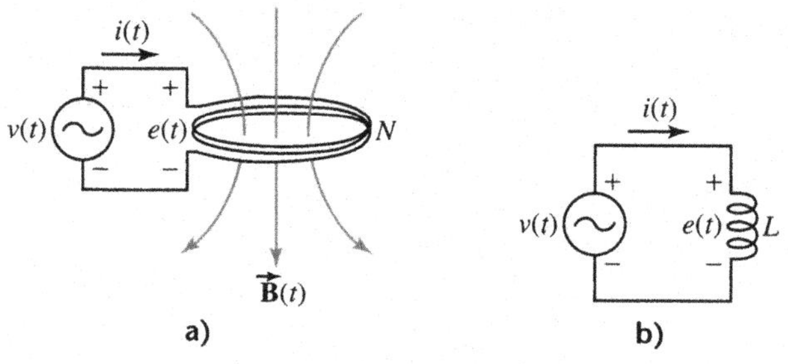

Figure 7.11 (a) A coil carrying time-varying current; (b) The coil modeled as an inductor

Since the induced emf in the coil opposes the applied voltage, it is also known as the *induced voltage, back emf*, or *counter emf*. Thus, we can write the preceding equation in terms of the applied voltage as

$$v = e = N \frac{d\Phi}{dt} \tag{7.21}$$

The product $N\Phi$ is commonly called the *number of flux linkages* and is denoted by λ as follows:

$$\lambda = N\Phi \tag{7.22}$$

The rate of change of flux linkages per unit change in the current is called the *self-inductance* or *inductance* of the coil and is usually symbolized by L. Thus

$$L = \frac{d\lambda}{di} = N\frac{d\Phi}{dt} \tag{7.23}$$

This equation defines the unit of inductance as the weber-turn per ampere. However, we refer to one weber-turn per ampere as one henry in honor of Joseph Henry.

Equation (7.23) can also be written as

$$L\,di = N\,d\Phi \tag{7.24a}$$

In the development that follows we will assume that the flux $\Phi(t)$ is directly proportional to the current $i(t)$. In other words, the coil is wound over a linear (constant permeability) magnetic material, and the inductance L of the coil is constant. In this case, we can also express equation (7.24a) as

$$Li = N\Phi \tag{7.24b}$$

This equation states that the product Li is equal to the total flux linkages when the flux linking each turn is the same.

From (7.24b), we obtain an expression for the inductance of a coil wound over a linear magnetic material as

$$L = \frac{N\Phi}{i} \tag{7.25}$$

This equation is commonly used to determine the inductance of a magnetic circuit for a steady current in the coil. Differentiating both sides of (7.24a) with respect to t, we get

$$L\frac{di}{dt} = N\frac{d\Phi}{dt} \tag{7.26}$$

Comparing (7.21) and (7.26), we get

$$v = L\frac{di}{dt} \tag{7.27}$$

This is a well-known circuit equation that yields the voltage drop across an inductance L. It helps us to model a current-carrying coil by its inductance, as shown in Figure 7.11b. An element such as the coil in this case, that has an inductance is called an *inductor*. The inductance of an element is one henry if the voltage drop across the element is one volt when the current in the element changes at the rate of one ampere per second.

We can also define inductance in terms of the reluctance or the permeance of the magnetic circuit. We do this to show that the inductance essentially depends upon the parameters of the magnetic circuit. We can rewrite (7.25) by multiplying the numerator and the denominator by N as

$$L = \frac{N^2\Phi}{Ni} = \frac{N^2\Phi}{\mathscr{F}} = \frac{N^2}{\mathscr{R}} = \mathscr{P}N^2 \tag{7.28}$$

where $\mathscr{F} = Ni$ is the applied mmf to the magnetic circuit (coil in our case), $\mathscr{R}$ is the reluctance, and $\mathscr{P}$ is the permeance of the magnetic circuit. Thus, each of the magnetic circuits discussed in Chapter 5 can now be represented by an equivalent electric circuit in terms of its inductance.

EXAMPLE 7.6

A very long cylinder of radius 20 cm is closely and tightly wound with 200 turns per unit length to form an air-core inductor (solenoid). If the current in the coil is constant, determine its inductance.

Solution The magnetic flux density inside a very long cylinder is

$$\vec{\mathbf{B}} = \mu_0 n I \vec{\mathbf{a}}_z$$

where n is the number of turns per unit length. The flux enclosed by a cylinder of radius b is

$$\Phi = \int_s \vec{\mathbf{B}} \cdot \vec{d\mathbf{s}} = \mu_0 n I \int_0^b \rho\, d\rho \int_0^{2\pi} d\phi = \mu_0 n I \pi b^2$$

The inductance of the solenoid per unit length, from (7.25), is

$$L = \mu_0 \pi n^2 b^2$$

Substituting the values, we get

$$L = 4\pi \times 10^{-7} \times \pi \times 200^2 \times 0.2^2$$
$$= 6.32 \text{ mH/m}$$

• • •

EXAMPLE 7.7

Obtain an expression for the self-inductance per unit length of a coaxial cable of inner conductor a and outer conductor b. The outer conductor has negligible thickness, and the current is uniformly distributed inside the inner conductor.

Solution The current density inside the inner conductor (Figure 7.12) is

$$\vec{\mathbf{J}} = \frac{I}{\pi a^2} \vec{\mathbf{a}}_z$$

The magnetic flux density within the inner conductor at any radius ρ such that $0 \leq \rho \leq a$, from Ampère's law, is

$$\vec{\mathbf{B}}_i = \frac{\mu_0 I \rho}{2\pi a^2} \vec{\mathbf{a}}_\phi$$

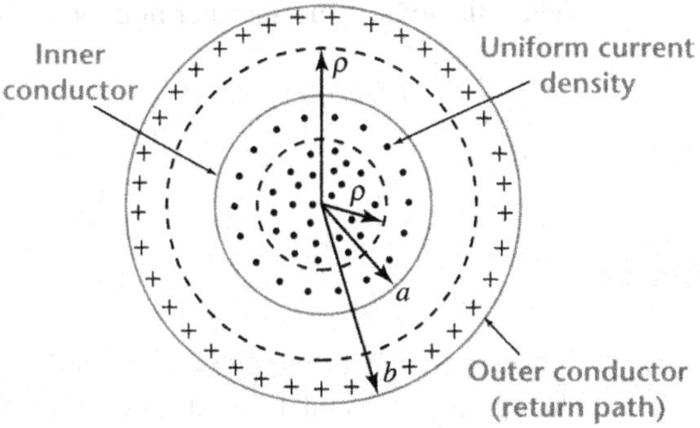

Figure 7.12 A coaxial cable with a uniform current distribution

The flux enclosed in the region between ρ and $\rho + d\rho$ and the unit length in the z direction is

$$d\Phi_i = \frac{\mu_0 I \rho \, d\rho}{2\pi a^2}$$

As only a fraction of the total current is enclosed by the contour at radius ρ, the flux linkages are

$$d\lambda_i = \frac{\mu_0 I \rho \, d\rho}{2\pi a^2} \left(\frac{\rho}{a}\right)^2$$

Thus, the total flux linkages within the inner conductor are

$$\lambda_i = \frac{\mu_0 I}{2\pi a^4} \int_0^a \rho^3 \, d\rho = \frac{\mu_0 I}{8\pi}$$

Hence, the inductance per unit length of the inner conductor due to the flux inside it is

$$L_i = \frac{\lambda_i}{I} = \frac{\mu_0}{8\pi} \text{ H/m}$$

Let us now determine the inductance due to the flux between the two conductors (Figure 7.12). The flux density in the region $a \le \rho \le b$ is

$$\vec{\mathbf{B}}_e = \frac{\mu_0 I}{2\pi\rho} \vec{\mathbf{a}}_\phi$$

The flux passing through the region between ρ and $\rho + d\rho$ and the unit length in the z direction is

$$d\Phi_e = \frac{\mu_0 I \, d\rho}{2\pi\rho}$$

The total flux linkages are

$$\lambda_e = \frac{\mu_0 I}{2\pi} \int_a^b \frac{1}{\rho} \, d\rho = \frac{\mu_0 I}{2\pi} \ln(b/a)$$

Thus, the contribution to the self-inductance by these flux linkages is

$$L_e = \frac{\lambda_e}{I} = \frac{\mu_0}{2\pi} \ln(b/a) \text{ H/m}$$

Hence, the total inductance per unit length of the coaxial cable is

$$L = \frac{\mu_0}{2\pi}\left[\frac{1}{4} + \ln(b/a)\right] \text{ H/m} \tag{7.29}$$

• • •

7.6 Mutual Inductance

Consider a magnetic circuit with two coils, as illustrated in Figure 7.13. The current i_1 in coil 1 establishes a flux Φ_1, and coil 2 is open. The self-inductance of coil 1 is

$$L_{11} = N_1 \frac{d\Phi_1}{di_1} \tag{7.30}$$

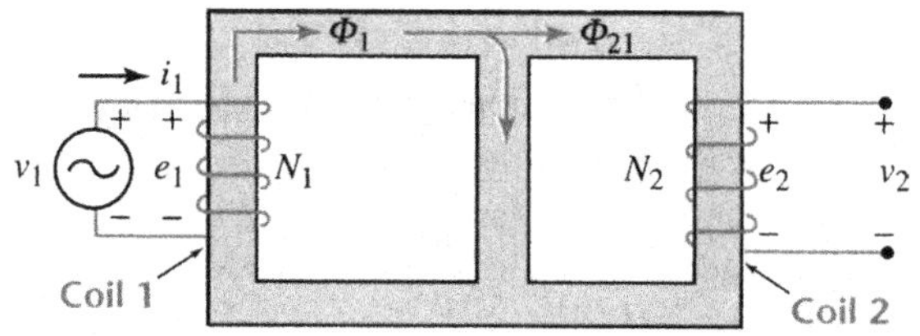

Figure 7.13 A magnetic circuit with two current-carrying coils

The first subscript for the inductance in this equation refers to the coil whose inductance is being calculated. The second subscript is for the coil that is carrying the current and producing the flux. Let Φ_{21} be the fraction of the flux Φ_1 that links the coil 2; then, the induced emf in coil 2 is

$$e_2 = N_2 \frac{d\Phi_{21}}{dt} \tag{7.31}$$

with the polarity as indicated in the figure. If v_2 is the open-circuit voltage across coil 2, then

$$v_2 = N_2 \frac{d\Phi_{21}}{dt} \tag{7.32}$$

If we define the *mutual inductance* such that

$$v_2 = L_{21} \frac{di_1}{dt} \tag{7.33}$$

then

$$L_{21} = N_2 \frac{d\Phi_{21}}{di_1} \tag{7.34}$$

where L_{21} is the mutual inductance of coil 2 due to the flux generated by coil 1. This equation lets us define the mutual inductance in terms of the flux in the magnetic circuit. The **mutual inductance** between two coils is defined as the total flux linking one coil per unit change in current in the other coil.

If we now excite coil 2 with a current i_2 such that it produces a flux Φ_2 while coil 1 is open, the self-inductance of coil 2 is

$$L_{22} = N_2 \frac{d\Phi_2}{di_2} \tag{7.35}$$

Let Φ_{12} be the fraction of the flux Φ_2 that links coil 1; then the induced emf in coil 1 is

$$e_1 = N_1 \frac{d\Phi_{12}}{dt} \tag{7.36}$$

If v_1 is the open-circuit voltage across coil 1, then

$$v_1 = N_1 \frac{d\Phi_{12}}{dt} \tag{7.37}$$

If we define the mutual inductance such that

$$v_1 = L_{12} \frac{di_2}{dt} \tag{7.38}$$

then

$$L_{12} = N_1 \frac{d\Phi_{12}}{di_2} \tag{7.39}$$

where L_{12} is the mutual inductance of coil 1 due to the flux generated by coil 2.

Because the geometry of the relative arrangement of the coils is the same in both cases, we will prove later that

$$L_{12} = L_{21} = M \tag{7.40}$$

where M is the mutual inductance between the two coils. Since a part of the flux produced by one coil links the other, we can express these fluxes as

$$\Phi_{21} = k_1 \Phi_1 \tag{7.41a}$$

and

$$\Phi_{12} = k_2 \Phi_2 \tag{7.41b}$$

where k_1 is the fraction of the flux linking coil 2 produced by coil 1. Similarly, k_2 is the fraction of the flux that links coil 1 when produced by coil 2. It is evident that $0 \leq k_1 \leq 1$ and $0 \leq k_2 \leq 1$.

From (7.34), (7.39), and (7.40), we can show that the mutual inductance is

$$M = k\sqrt{L_{11} L_{22}} \tag{7.42a}$$

where

$$k = \sqrt{k_1 k_2} \tag{7.42b}$$

is called the *coefficient of coupling* between the two coils and $0 \leq k \leq 1$. Under ideal conditions, the coefficient of coupling between the two coils is unity, and the coils are said to be *perfectly coupled*. If more than two coils are magnetically coupled together, we can determine the mutual inductance for each pair of coils separately. Because L_{11} and L_{22} can be defined in terms of the reluctance of the magnetic circuit, we can also obtain an expression for the mutual inductance in terms of the reluctance as

$$M = k\frac{N_1 N_2}{\mathcal{R}} \tag{7.43}$$

This equation is quite useful in determining the mutual inductance between any two coils for a linear magnetic circuit.

Let us now verify the assertion that the mutual inductance between the two coils depends upon the geometry, size and shape, and relative arrangement of the coils. Figure 7.14 shows two tightly and closely wound coils when coil 1 is carrying a current and coil 2 is left open. The total flux linkages with coil 2 are

$$\lambda_{21} = N_2 \int_{s_2} \vec{\mathbf{B}}_1 \cdot \overrightarrow{d\mathbf{s}}_2$$

where $\vec{\mathbf{B}}_1$ is the flux density in the plane of coil 2 due to the current $i_1(t)$ in coil 1. The magnetic flux density can be expressed in terms of the magnetic vector potential $\vec{\mathbf{A}}$; therefore the preceding equation can be written as

$$\lambda_{21} = N_2 \int_{s_2} (\nabla \times \vec{\mathbf{A}}_1) \cdot \overrightarrow{d\mathbf{s}}_2$$

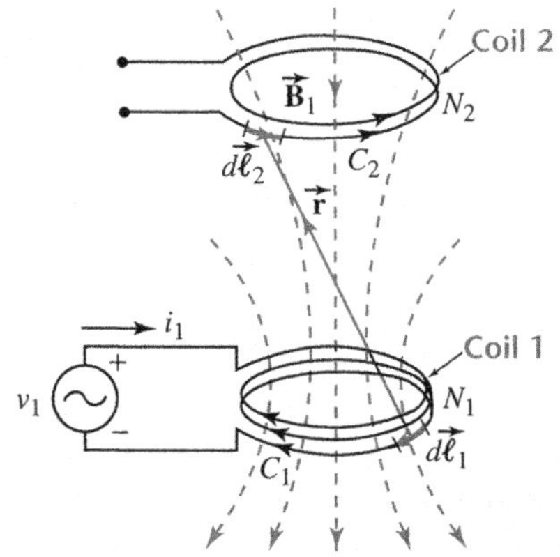

Figure 7.14 Mutual flux linkages between two coils

The application of Stokes' theorem yields

$$\lambda_{21} = N_2 \oint_{c_2} \vec{\mathbf{A}}_1 \cdot \overrightarrow{d\ell}_2 \tag{7.44}$$

However, the magnetic vector potential at any point on coil 2 due to the current in coil 1 is

$$\vec{\mathbf{A}}_1 = \frac{\mu_1 N_1 i_1}{4\pi} \oint_{c_1} \frac{\overrightarrow{d\ell}_1}{r}$$

Substituting for $\vec{\mathbf{A}}_1$ in (7.44), we obtain

$$\lambda_{21} = \frac{\mu_1 N_1 N_2 i_1}{4\pi} \oint_{c_1} \oint_{c_2} \frac{\vec{d\ell}_1 \cdot \vec{d\ell}_2}{r}$$

Thus, the mutual inductance of coil 2 due to the current in coil 1 is

$$L_{21} = \frac{\mu_1 N_1 N_2}{4\pi} \oint_{c_1} \oint_{c_2} \frac{\vec{d\ell}_1 \cdot \vec{d\ell}_2}{r} \tag{7.45}$$

Similarly, when coil 2 carries a current $i_2(t)$ while coil 1 is open, we can obtain an expression for the mutual inductance of coil 1 due to the current in coil 2 as

$$L_{12} = \frac{\mu_2 N_1 N_2}{4\pi} \oint_{c_1} \oint_{c_2} \frac{\vec{d\ell}_1 \cdot \vec{d\ell}_2}{r} \tag{7.46}$$

Equations (7.45) and (7.46) are known as *Neumann's formulas* for the inductance between two current-carrying coils. These equations state that the mutual inductance depends upon the geometrical arrangement of the two coils and the permeability of the magnetic region. For a linear magnetic medium such as free space, the two equations are identical. These equations are rarely used to determine the mutual inductance between any two coils because of the double integration. It is much easier to calculate the self- and mutual inductances on the basis of flux linking the coils.

EXAMPLE 7.8

A toroidal coil of 2000 turns is wound over a magnetic ring with inner radius of 10 mm, outer radius of 15 mm, height of 10 mm, and relative permeability of 500. A very long, straight conductor passing through the center of the toroid carries a time-varying current. Determine the mutual inductance between the toroid and the straight conductor.

Solution

A toroid with inner radius a, outer radius b, and height h with a very long conductor carrying current $i(t)$ passing through its center is shown in Figure 7.15. From the application of Ampère's law, the magnetic flux density at any radius ρ within the toroid is

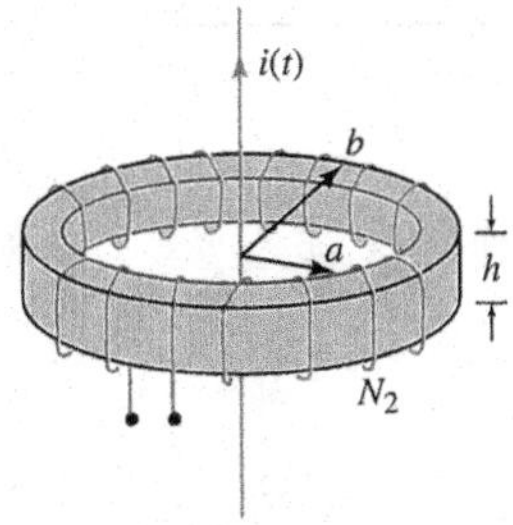

Figure 7.15 Mutual inductance between a straight current-carrying conductor and a toroid

$$\vec{\mathbf{B}}_1 = \frac{\mu i}{2\pi\rho} \vec{\mathbf{a}}_\phi$$

Thus, the flux linking the toroid is

$$\Phi_{21} = \int_{s_2} \vec{\mathbf{B}}_1 \cdot \vec{d\mathbf{s}}_2$$

$$= \frac{\mu i}{2\pi} \int_a^b \frac{1}{\rho} d\rho \int_0^h dz = \frac{\mu i}{2\pi} \ln(b/a)\, h$$

Hence, the mutual inductance is

$$L_{21} = N_2 \frac{d\Phi_{21}}{di} = \frac{\mu}{2\pi} h N_2 \ln(b/a)$$

Substituting the values, we obtain

$$L_{21} = \frac{500 \times 4\pi \times 10^{-7}}{2\pi} \times 0.01 \times 2000 \times \ln(15/10)$$
$$= 0.81 \text{ mH}$$

• • •

7.7 Inductance of coupled coils

Two magnetically coupled coils can be connected either in series or in parallel. In each case, the effective inductance of the coupled coils depends upon the orientation of the coils and the direction of the flux produced by each coil. We now discuss series and parallel connections and in turn their aiding and opposing connections.

7.7.1 Series connection

When the two coils are connected in tandem (end to end), they are said to be connected in series. When two series-connected coils produce a flux in the same direction (Figure 7.16a) they are said to be connected in *series aiding*. On the other hand, the coils are connected in *series opposing* when they produce fluxes in opposite directions (Figure 7.16b). When the flux set up by one coil is normal to the flux produced by the other, the two coils behave independently of each other, and the mutual inductance between them is zero. In this case, the magnetic axes of the two coils are said to be normal to each other.

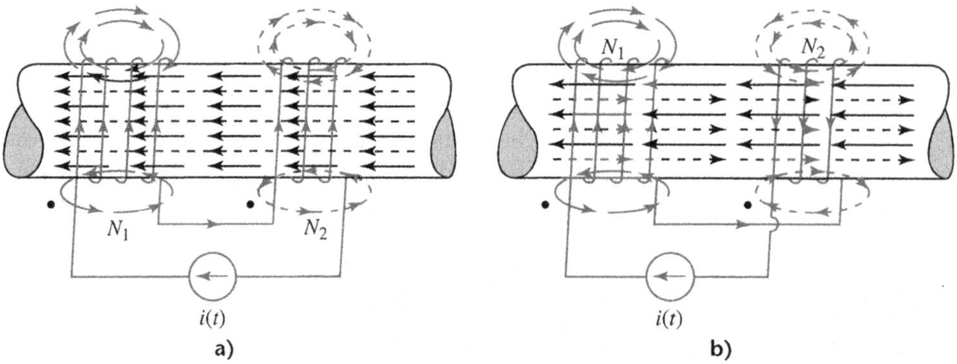

Figure 7.16 (a) Series-aiding connections and (b) series-opposing connections

Consider a series circuit with two magnetically coupled coils, as shown in Figure 7.17. Let L_1 and L_2 be the inductances of the two coils, let M be the mutual inductance, and let R_1 and R_2 be their internal

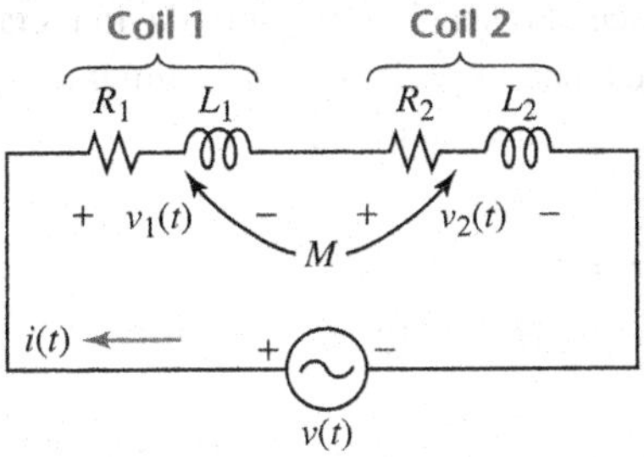

Figure 7.17 Magnetically coupled coils connected in series

resistances. When the current in the series circuit is $i(t)$, the voltage drop across each coil is

$$v_1 = L_1\frac{di}{dt} + iR_1 \pm M\frac{di}{dt}$$

and

$$v_2 = L_2\frac{di}{dt} + iR_2 \pm M\frac{di}{dt}$$

The sign in these equations is plus when the two coils are connected in series aiding and minus when they are in series opposing. From Kirchhoff's voltage law, we have

$$v = (L_1 + L_2 \pm 2M)\frac{di}{dt} + i(R_1 + R_2)$$

If L is the effective inductance and R is the effective resistance of the two coils, then

$$L = L_1 + L_2 \pm 2M \tag{7.47a}$$

and

$$R = R_1 + R_2 \tag{7.47b}$$

From (7.47a), it is evident that the effective inductance is maximum when the two coils are connected in series aiding and minimum when they are in series opposing.

Whether the fluxes produced by coupled coils aid or oppose each other can be easily identified by marking one terminal of each coil with a dot (•) as we have done in Figure 7.16. The understanding here is that when all the currents in coupled coils enter (or leave) the dotted terminals (Figure 7.16a), the fluxes aid, and the sign is plus for the mutual inductance M in (7.47a). However, when the current in one coil enters the dotted terminal and that in the other coil leaves the dotted terminal (Figure 7.16b), the fluxes oppose, and the sign is minus for the mutual inductance M in (7.47a).

EXAMPLE 7.9

The effective inductances when two coils are connected in series aiding and series opposing are 2.38 H and 1.02 H, respectively. If the inductance of one coil is 16 times the inductance of the other, determine the inductance of each coil, the mutual inductance, and the coefficient of coupling between them.

Solution From the given data, we have

$$L_1 + L_2 + 2M = 2.38$$

and

$$L_1 + L_2 - 2M = 1.02$$

Thus, $L_1 + L_2 = 1.7$ H and $M = 0.34$ H. If we set $L_1 = 16L_2$, we obtain $L_1 = 1.6$ H and $L_2 = 0.1$ H. The coefficient of coupling between the two coils is

$$k = \frac{M}{\sqrt{L_1 L_2}} = \frac{0.34}{\sqrt{1.6 \times 0.1}} = 0.85$$

• • •

7.7.2 Parallel connection

When two coils are connected in parallel, as depicted in Figure 7.18, it can be shown that the effective inductance is

$$L = \frac{L_1 L_2 - M^2}{L_1 + L_2 \pm 2M} \tag{7.48}$$

The term in the denominator has a minus sign when the two coils are connected in *parallel aiding* and a plus sign for *parallel opposing*. We leave the derivation of this equation as an exercise for the student.

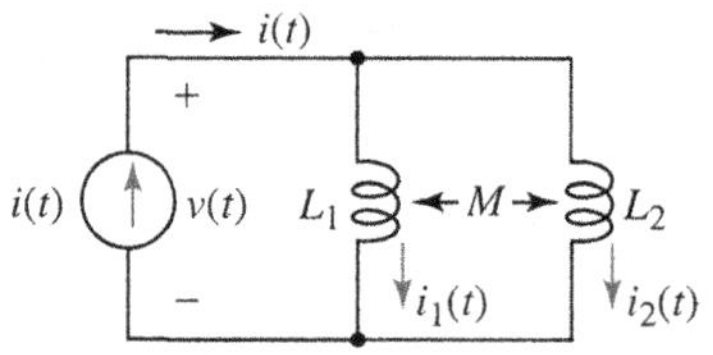

Figure 7.18 Magnetically coupled coils connected in parallel

EXAMPLE 7.10

The self-inductances of two coils are 800 mH and 200 mH. The coefficient of coupling is 0.8. Calculate the effective inductance when the coils are connected in (a) parallel aiding and (b) parallel opposing.

Solution Let $L_1 = 0.8$ H and $L_2 = 0.2$ H. Then the mutual inductance is

$$M = k\sqrt{L_1 L_2} = 0.8\sqrt{0.8 \times 0.2} = 0.32 \text{ H}$$

a) For the parallel-aiding connection, the effective inductance from (7.48) is

$$L = \frac{0.8 \times 0.2 - 0.32^2}{0.8 + 0.2 - 2 \times 0.32} = 0.16 \text{ H} \quad \text{or} \quad 160 \text{ mH}$$

b) Similarly, for the parallel-opposing connection, we have

$$L = \frac{0.8 \times 0.2 - 0.32^2}{0.8 + 0.2 + 2 \times 0.32} = 0.035 \text{ H} \quad \text{or} \quad 35 \text{ mH}$$

• • •

7.8 Energy in a magnetic field

In this section we obtain an expression for the energy stored in a magnetic field established by (a) a single coil and (b) coupled coils.

7.8.1 Single coil

Let us consider an N-turn coil connected to an electric source and carrying a current $i(t)$. When the current is increasing in the coil, the induced emf across its terminals is

$$e = -N\frac{d\Phi}{dt}$$

where $d\Phi$ is a function of the current $i(t)$ in the loop. In order to maintain an increase in the current, the electric source must supply energy. The amount of work done in time dt is

$$dW = -ei\,dt = iN\,d\Phi \tag{7.49}$$

where the minus sign indicates that the source is supplying energy or the coil is absorbing energy as the current in the coil increases. The total work done is

$$W = N\int i\,d\Phi \tag{7.50}$$

In order to integrate equation (7.50), we must know how the flux varies with the current. For a linear magnetic circuit, however, we know that

$$N\,d\Phi = L\,di \tag{7.51}$$

where L is the self-inductance of the coil. Substituting (7.51) in (7.49), we obtain

$$dW = Li\,di$$

If W_0 is the initial energy in the coil corresponding to the initial current I_0, and W_f is the final energy in the coil when the current is I_f, the increase in the energy in the coil is

$$\int_{W_0}^{W_f} dW = \int_{I_0}^{I_f} Li\,di$$

Thus, the increase in the energy is

$$W = W_f - W_i = \tfrac{1}{2}LI_f^2 - \tfrac{1}{2}LI_0^2 \tag{7.52}$$

If the initial current in the coil is zero, and the current at any time t is $i(t)$, the energy stored in the magnetic circuit, from (7.52), is

$$W = \tfrac{1}{2}Li^2 \tag{7.53}$$

For a linear magnetic circuit, this equation can also be written as

$$W = \tfrac{1}{2}N\Phi I = \tfrac{1}{2}\lambda I \tag{7.54}$$

where $\lambda = N\Phi$ represents the total flux linkages with the coil.

The energy stored in the coil can also be expressed in terms of the field quantities as follows:

$$\Phi = \int_s \vec{\mathbf{B}} \cdot \vec{d\mathbf{s}} = BA$$

and

$$Ni = \oint_c \vec{\mathbf{H}} \cdot \vec{d\ell} = H\ell$$

where Ni is the total current enclosed by contour c, A is the cross-sectional area of the coil, and ℓ is its length. Thus, $A\ell$ is the volume enclosed by the coil. Equation (7.54) can now be written as

$$W = \tfrac{1}{2}BHA\ell$$

which yields the energy stored per unit volume w_m, the *magnetic energy density*, as

$$w_m = \frac{1}{2}BH = \frac{1}{2}\mu H^2 = \frac{1}{2\mu}B^2 \tag{7.55a}$$

Equation (7.55a) can also be written in vector form as

$$w_m = \tfrac{1}{2}\vec{\mathbf{B}} \cdot \vec{\mathbf{H}} \tag{7.55b}$$

Equation (7.55) highlights the fact that the magnetic energy associated with a coil is distributed throughout its magnetic field region. However, it is a common practice to state that the energy is stored in the inductor.

EXAMPLE 7.11

The outer radius of the inner conductor and the inner radius of the outer conductor of a coaxial transmission line are a and b, respectively. Using the stored magnetic energy concept, determine the inductance per unit length of the line.

Solution In the region $a \le \rho \le b$, when the inner conductor carries a current $i(t)$ in the z direction, the magnetic field intensity is

$$H = \frac{i}{2\pi\rho}$$

Thus, the energy density at any point within the region is

$$w_m = \frac{1}{2}\mu\left[\frac{i}{2\pi\rho}\right]^2$$

Hence, the energy stored per unit length is

$$W_m = \frac{\mu}{8\pi^2}i^2\int_a^b \frac{1}{\rho}\,d\rho\int_0^{2\pi} d\phi$$

$$= \frac{\mu}{4\pi}i^2 \ln\left(\frac{b}{a}\right) \text{ J/m}$$

Comparing this expression with equation (7.53), we obtain the inductance per unit length as

$$L = \frac{\mu}{2\pi}\ln\left(\frac{b}{a}\right) \text{ H/m}$$

• • •

7.8.2 Coupled coils

We now proceed to determine the energy stored in two coupled coils, as depicted in Figure 7.14. If Φ_1 is the total flux linking coil 1 carrying current i_1, then

$$\Phi_1 = \Phi_{11} + \Phi_{12}$$

where Φ_{11} is the flux created by coil 1 when there is no current in coil 2, and Φ_{12} is the flux created by current i_2 in coil 2 and linking coil 1 when there is no current in coil 1. The use of a plus sign in the preceding equation indicates that Φ_{11} and Φ_{12} are in the same direction at any time. The sign must be changed to minus when the coils produce opposing fluxes. Similarly, the total flux linking the second coil is

$$\Phi_2 = \Phi_{22} + \Phi_{21}$$

Thus, the magnetic energy stored in the region is

$$W = \tfrac{1}{2}N_1\Phi_1 i_1 + \tfrac{1}{2}N_2\Phi_2 i_2$$
$$= \tfrac{1}{2}N_1\Phi_{11}i_1 + \tfrac{1}{2}N_1\Phi_{12}i_1 + \tfrac{1}{2}N_2\Phi_{22}i_2 + \tfrac{1}{2}N_2\Phi_{21}i_2$$

Using the definitions of self- and mutual inductance as

$$L_{11} = \frac{N_1\Phi_{11}}{i_1}, \quad L_{22} = \frac{N_2\Phi_{22}}{i_2}$$
$$L_{12} = \frac{N_1\Phi_{12}}{i_2}, \quad \text{and} \quad L_{21} = \frac{N_2\Phi_{21}}{i_2}$$

we can write the expression for stored magnetic energy as

$$W = \tfrac{1}{2}L_{11}i_1^2 + \tfrac{1}{2}L_{12}i_1 i_2 + \tfrac{1}{2}L_{22}i_2^2 + \tfrac{1}{2}L_{21}i_1 i_2$$

If we substitute $M = L_{12} = L_{21}$ for a linear magnetic circuit in this equation, we obtain

$$W = \frac{1}{2}\left[L_{11}i_1^2 + L_{22}i_2^2 + 2M\, i_1 i_2\right] \tag{7.56}$$

as the stored magnetic energy in two coupled coils when the *fluxes aid each other*. When the *fluxes oppose each other*, the stored magnetic energy is

$$W = \frac{1}{2}\left[L_{11}i_1^2 + L_{22}i_2^2 - 2M\, i_1 i_2\right] \tag{7.57}$$

When the two coils are connected in series such that $i_1 = i_2 = i$, the total energy stored for *series aiding* is

$$W = \frac{1}{2}[L_{11} + L_{22} + 2M]i^2 \tag{7.58a}$$

Likewise, the total energy stored in two coupled coils connected in *series opposing* is

$$W = \frac{1}{2}[L_{11} + L_{22} - 2M]i^2 \tag{7.58b}$$

From (7.58a) and (7.58b) it is evident that the energy stored in a coupled magnetic circuit can be computed in terms of the equivalent inductance of the coupled circuit.

EXAMPLE 7.12

The current in the coupled coils in Example 7.9 varies from an initial value of 2 A to a final value of 5 A. If the coils are connected in series aiding, compute (a) the initial energy, (b) the final energy, and (c) the change in the stored energy.

Solution

a) The initial energy in the coupled coils is

$$W_i = \frac{1}{2} \times 2.38 \times 2^2 = 4.76 \text{ J}$$

b) The final energy in the coupled coils is

$$W_f = \frac{1}{2} \times 2.38 \times 5^2 = 29.75 \text{ J}$$

c) Hence, the increase in the energy stored in the coupled coils is

$$W = 29.75 - 4.76 = 24.99 \text{ J}$$

• • •

7.9 Maxwell's equation from Ampère's law

During our study of magnetostatics we formulated Ampère's law in integral form as

$$\oint_c \vec{\mathbf{H}} \cdot \vec{d\ell} = I$$

where $\vec{\mathbf{H}}$ is the magnetic field intensity and I is the uniform current enclosed by contour c. By describing I in terms of the volume current density $\vec{\mathbf{J}}$ over surface s bounded by closed contour c as

$$I = \int_s \vec{\mathbf{J}} \cdot \vec{ds}$$

we obtain Ampère's law in point (differential) form such that

$$\nabla \times \vec{\mathbf{H}} = \vec{\mathbf{J}} \tag{7.59}$$

Taking the divergence of both sides, we get

$$\nabla \cdot \vec{\mathbf{J}} = 0 \tag{7.60}$$

because $\nabla \cdot (\nabla \times \vec{\mathbf{H}}) = 0$.

However, for time-varying fields, $\nabla \cdot \vec{\mathbf{J}}$ is not necessarily zero. In fact, the equation of continuity, derived in Chapter 4, states that

$$\nabla \cdot \vec{\mathbf{J}} = -\frac{\partial \rho_v}{\partial t} \tag{7.61}$$

where $\rho_v(t)$ is the volume charge density. As the presence of time-varying charge cannot always be ruled out in (7.61), $\vec{\mathbf{J}}$ cannot be, in general, a continuous (nondiverging) time-varying field. Therefore, (7.59) leads to a contradiction in the time-varying case. Let us elaborate this point further.

Imagine a capacitor connected to a time-varying voltage source as shown in Figure 7.19. The rise and fall of the applied voltage in time is indicative of the amount of charge that is transferred from the source to each electrode of the capacitor. In other words, accumulation of charge on each electrode of the capacitor is a time-dependent process. Because the rate of change of charge constitutes a current, there must be a time-varying current $i(t)$ in the circuit. This current must also establish a time-varying magnetic field in the region. Thus, if we select an open surface s bounded by a closed contour c, Ampère's law suggests that

$$\oint_c \vec{\mathbf{H}} \cdot \vec{d\ell} = i \tag{7.62}$$

where $\vec{\mathbf{H}}$ is the time-varying magnetic field intensity.

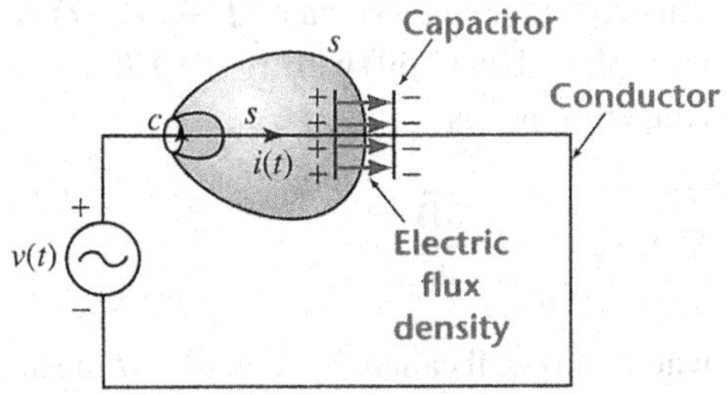

Figure 7.19 The displacement current in a capacitor establishes the continuity of the conduction current in the conductor

However, if we consider another surface S as the open surface bounded by the same closed contour c as shown in the figure, the current passing through this surface is zero. In other words, we now have

$$\oint_c \vec{\mathbf{H}} \cdot \vec{d\ell} = 0 \tag{7.63}$$

Once again, (7.63) is in contradiction of (7.62). If we eliminate the uncertainty in these equations by setting $i(t)$ to zero, then we cannot justify the existence of either the current in the circuit or the magnetic field associated with it.

These inconsistencies led Maxwell to predict that there must exist a current in the capacitor. As the current cannot be due to conduction, he referred to it as *displacement current*. In order to account for the displacement current, Maxwell added another term to Ampère's law and ensured its validity for the time-varying case as well. The additional term is, in fact, a consequence of conservation of charge. We can obtain this term from Gauss's law

$$\nabla \cdot \vec{\mathbf{D}} = \rho_v \tag{7.64}$$

and the equation of continuity. Substituting for ρ_v from (7.64) in (7.61), we obtain

$$\nabla \cdot \vec{\mathbf{J}} = -\frac{\partial}{\partial t}(\nabla \cdot \vec{\mathbf{D}})$$

As space and time are independent variables, we can change the sequence of differentiation in the preceding equation and obtain

$$\nabla \cdot \vec{\mathbf{J}} = -\nabla \cdot \left(\frac{\partial \vec{\mathbf{D}}}{\partial t}\right)$$

or

$$\nabla \cdot \left(\vec{\mathbf{J}} + \frac{\partial \vec{\mathbf{D}}}{\partial t}\right) = 0 \tag{7.65}$$

This equation suggests that $(\vec{\mathbf{J}} + \partial\vec{\mathbf{D}}/\partial t)$ is a continuous field. When we replace $\vec{\mathbf{J}}$ in (7.59) with $\vec{\mathbf{J}} + \partial\vec{\mathbf{D}}/\partial t$, we obtain the modified form of Ampère's law as

$$\nabla \times \vec{\mathbf{H}} = \vec{\mathbf{J}} + \frac{\partial \vec{\mathbf{D}}}{\partial t} \tag{7.66}$$

where Maxwell called the term $\partial\vec{\mathbf{D}}/\partial t$ the *displacement current density* (measured in A/m^2). The name is still in use, although there is not a real physical current. When we use (7.66) as a statement for Ampère's law we find that all the contradictions disappear.

The right-hand side of (7.66) also states that at any point in a medium there exists a *total current density* which is the sum of the *conduction current density* and the *displacement current density*:

$$\text{total current density} = \vec{\mathbf{J}} + \frac{\partial \vec{\mathbf{D}}}{\partial t} \tag{7.67}$$

The modification of Ampère's law is one of the most significant contributions by Maxwell and it led to the development of a unified electromagnetic field theory. At this stage of your study you may not fully appreciate the significance of displacement current density, but the very presence of this term enabled Maxwell to predict that electromagnetic fields should propagate through space as waves. A few years thereafter, Hertz was able to experimentally show that such waves do exist. All our modern means of communications are based on this modification of Ampère's law.

From now on we refer to (7.66) as *Maxwell's equation in the point or differential form*. For any arbitrary open surface s bounded by a closed contour c we can rewrite (7.66) in the integral form as

$$\oint_c \vec{\mathbf{H}} \cdot \vec{d\ell} = \int_s \vec{\mathbf{J}} \cdot \vec{ds} + \int_s \frac{\partial \vec{\mathbf{D}}}{\partial t} \cdot \vec{ds} \tag{7.68}$$

where the first term on the right-hand side corresponds to the conduction current, and the second term represents the displacement current. For the capacitor circuit discussed earlier, we now conclude that it is the displacement current through the capacitor that produces the time-varying magnetic field. On the other hand, for the current to be continuous in the circuit, the displacement current through the capacitor must be equal to the conduction current in the wire.

We now make the following important observations:

a) The displacement current density is simply the rate at which the electric flux density $\vec{\mathbf{D}}$ varies with time.
b) Because $\partial \vec{\mathbf{D}}/\partial t$ acts as a *source* for the magnetic field, a time-varying electric field creates a time-varying magnetic field.
c) The addition of the term $\partial \vec{\mathbf{D}}/\partial t$ in (7.67) has not changed the fact that the magnetic field ($\vec{\mathbf{H}}$ and $\vec{\mathbf{B}}$) is solenoidal.
d) The time-varying magnetic field establishes a time-varying electric field (Faraday's law).
e) The time-varying electric and magnetic fields are interdependent.

EXAMPLE 7.13

The magnetic field intensity in free space is given as $\vec{\mathbf{H}} = H_0 \sin\theta \vec{\mathbf{a}}_y$ A/m, where $\theta = \omega t - \beta z$, and β is a constant quantity. Determine (a) the displacement current density and (b) the electric field intensity.

Solution The conduction current density in free space is zero. Thus, from (7.67), the displacement current density is equal to $\nabla \times \vec{\mathbf{H}}$. That is,

$$\frac{\partial \vec{\mathbf{D}}}{\partial t} = \begin{vmatrix} \vec{\mathbf{a}}_x & \vec{\mathbf{a}}_y & \vec{\mathbf{a}}_z \\ \frac{\partial}{\partial x} & \frac{\partial}{\partial y} & \frac{\partial}{\partial z} \\ 0 & H_0 \sin\theta & 0 \end{vmatrix}$$

$$= -\frac{\partial}{\partial z}[H_0 \sin\theta]\vec{\mathbf{a}}_x + \frac{\partial}{\partial x}[H_0 \sin\theta]\vec{\mathbf{a}}_z$$

$$= \beta H_0 \cos\theta\, \vec{\mathbf{a}}_x \text{ A/m}^2$$

Thus, the displacement current density has an amplitude of βH_0 A/m^2. By integrating the displacement current density with respect to time, we obtain the electric flux density as

$$\vec{\mathbf{D}} = \frac{\beta}{\omega} H_0 \sin\theta\, \vec{\mathbf{a}}_x \text{ C/m}^2$$

Finally, the electric field intensity in free space is

$$\vec{\mathbf{E}} = \frac{\vec{\mathbf{D}}}{\epsilon_0} = \frac{\beta}{\omega\epsilon_0} H_0 \sin\theta\, \vec{\mathbf{a}}_x \text{ V/m}$$

• • •

7.10 Maxwell's equations from Gauss's laws

In our study of electrostatics, we obtained a mathematical statement for Gauss's law in the point (differential) form as

$$\nabla \cdot \vec{\mathbf{D}} = \rho_v \tag{7.69}$$

where $\vec{\mathbf{D}}$ is the electric flux density, and ρ_v is the free (or real) volume charge density in the medium. We remind you that ρ_v is zero within a dielectric medium because the effect of polarized charges is already included in the definition of relative permittivity ϵ_r (or dielectric constant). The arguments we used to derive this equation are equally applicable in the time-varying case. The only difference is that both $\vec{\mathbf{D}}$ and ρ_v are now time-dependent field quantities. Equation (7.69) is one of the four Maxwell equations. It can also be expressed in integral form as

$$\oint_s \vec{\mathbf{D}} \cdot \vec{ds} = \int_v \rho_v \, dv = q \tag{7.70}$$

where $q(t)$ is the total charge present at any time t within the volume v bounded by a closed surface s.

Because the magnetic flux is always continuous, Gauss's law derived earlier for magnetostatic fields is also valid when the fields vary with time. Thus,

$$\nabla \cdot \vec{\mathbf{B}} = 0 \tag{7.71}$$

where the magnetic flux density $\vec{\mathbf{B}}$ is now a time-varying field. This equation completes the set of four Maxwell's equations. It can also be written in integral form as

$$\oint_s \vec{\mathbf{B}} \cdot \vec{ds} = 0 \tag{7.72}$$

7.11 Maxwell's equations and boundary conditions

Before we proceed any further, for clarity we will (a) group the four Maxwell equations and highlight the significance of each equation, (b) state the constitutive relationships, and (c) review the boundary conditions. We do this now because the rest of the text is devoted to the solution of Maxwell's equations under various boundary conditions.

7.11.1 Maxwell's equations

The four **Maxwell equations** in the point (differential) and integral forms are

$$\nabla \times \vec{\mathbf{E}} = -\frac{\partial \vec{\mathbf{B}}}{\partial t} \Rightarrow \oint_c \vec{\mathbf{E}} \cdot \vec{d\ell} = -\int_s \frac{\partial \vec{\mathbf{B}}}{\partial t} \cdot \vec{ds} \tag{7.73}$$

$$\nabla \times \vec{\mathbf{H}} = \vec{\mathbf{J}} + \frac{\partial \vec{\mathbf{D}}}{\partial t} \Rightarrow \oint_c \vec{\mathbf{H}} \cdot \vec{d\ell} = \int_s \vec{\mathbf{J}} \cdot \vec{ds} + \int_s \frac{\partial \vec{\mathbf{D}}}{\partial t} \cdot \vec{ds} \tag{7.74}$$

$$\nabla \cdot \vec{\mathbf{D}} = \rho_v \Rightarrow \oint_s \vec{\mathbf{D}} \cdot \vec{ds} = \int_v \rho_v \, dv \tag{7.75}$$

$$\nabla \cdot \vec{\mathbf{B}} = 0 \Rightarrow \oint_s \vec{\mathbf{B}} \cdot \vec{ds} = 0 \tag{7.76}$$

where

$\vec{\mathbf{E}}$ = electric field intensity (vector), in V/m
$\vec{\mathbf{H}}$ = magnetic field intensity (vector), in A/m
$\vec{\mathbf{D}}$ = electric flux density (vector), in C/m^2
$\vec{\mathbf{B}}$ = magnetic flux density (vector), in Wb/m^2 (T)
ρ_v = free volume charge density (scalar), in C/m^3
$\vec{\mathbf{J}}$ = volume current density (vector), in A/m^2

The integrals involving the conduction current density $\vec{\mathbf{J}}$ and the volume charge density ρv can also be written as

$$I = \int_s \vec{\mathbf{J}} \cdot \vec{ds} \tag{7.77}$$

$$q = \int_v \rho_v \, dv \tag{7.78}$$

where

$I =$ electric current through surface s (scalar), in A
$q =$ free charge enclosed in volume v (scalar), in C

Equation (7.73) states that a time-varying magnetic field produces a time-varying electric field. This is the principle on which transformers and induction motors operate.

Equation (7.74) asserts that a time-varying magnetic field can be produced not only by conduction current but also by displacement current. The displacement current represents the rate of change of electric flux density, thus this equation suggests that a time-varying electric field creates a time-varying magnetic field, which in turn produces a time-varying electric field. That is, the energy from the electric field can be transferred to the magnetic field, which then transfers it back to the electric field. Knowledge of the continuation of transfer of energy from one field to another enabled Maxwell to predict the propagation of electromagnetic energy in any medium. The realization that the electromagnetic fields travel as waves helped Maxwell to predict the velocities and other peculiar characteristics of these waves. That the velocity of these waves in free space is equal to the velocity of light led Maxwell to conclude that light has the same nature as electromagnetic waves. In the 1880s Heinrich Rudolf Hertz showed experimentally the existence of electromagnetic waves and also confirmed that these waves exhibit the properties as predicted by Maxwell.

Equation (7.75) affirms that the total electric flux emanating from a closed volume at any time is equal to the charge enclosed. The electric flux lines are continuous if the charge enclosed is zero.

Equation (7.76) validates the fact that the magnetic flux lines are always continuous: the net flux emerging from any closed surface at any time is zero.

Further links to Maxwell's equations are established by the equation of continuity (in differential and integral form)

$$\nabla \cdot \vec{\mathbf{J}} = -\frac{\partial \rho_v}{\partial t} \quad \Rightarrow \quad \oint_s \vec{\mathbf{J}} \cdot \overrightarrow{d\mathbf{s}} = -\int_v \frac{\partial \rho_v}{\partial t}\, dv \tag{7.79}$$

and the Lorentz force equation

$$\vec{\mathbf{F}} = q(\vec{\mathbf{E}} + \vec{\mathbf{u}} \times \vec{\mathbf{B}}) \text{ [N]} \tag{7.80a}$$

where $\vec{\mathbf{F}}$ is the force experienced by a charge q moving with a velocity $\vec{\mathbf{u}}$ in a region of electromagnetic fields. Since the drifting charges produce a current, we can also express the force acting on charge and current per unit volume as

$$\vec{\mathbf{f}} = \rho_v \vec{\mathbf{E}} + \vec{\mathbf{J}} \times \vec{\mathbf{B}} \ \text{[N/m}^3\text{]} \tag{7.80b}$$

where

$$\vec{\mathbf{J}} = \rho_v \vec{\mathbf{u}} \ \text{[A/m}^2\text{]} \tag{7.80c}$$

Maxwell's equations in concert with the equation of continuity and the Lorentz force equation completely describe the interactions among charges, currents, and electric and magnetic fields. By manipulating these equations properly, we can obtain the characteristics of electromagnetic fields in any medium. We will usually employ the equations in the point (differential) form.

7.11.2 The constitutive equations

The *constitutive equations*, which define the relationships between the field quantities in a linear, homogeneous, and isotropic medium are

$$\vec{\mathbf{D}} = \epsilon \vec{\mathbf{E}} \tag{7.81}$$

$$\vec{\mathbf{J}} = \sigma \vec{\mathbf{E}} \tag{7.82}$$

$$\vec{\mathbf{B}} = \mu \vec{\mathbf{H}} \tag{7.83}$$

where

$$\begin{aligned}
\epsilon &= \text{permittivity (scalar), in F/m} \\
\epsilon_0 &= 8.854 \times 10^{-12} \text{ F/m for free space} \\
\mu &= \text{permeability (scalar), in H/m} \\
\mu_0 &= 4\pi \times 10^{-7} \text{ H/m for free space} \\
\sigma &= \text{conductivity (scalar), in S/m} \\
\sigma_{\text{Cu}} &= 5.8 \times 10^{7} \text{ S/m for copper}
\end{aligned}$$

Equation (7.82), known as *Ohm's law*, states that the motion of a charge in a conductor under the influence of an electric field produces a current in the conductor.

7.11.3 Boundary conditions

The electromagnetic fields obtained from the solution of Maxwell's equations must also satisfy the boundary conditions at the interface between different media. As it turns out, the boundary conditions for time-varying fields are exactly the same as those for static fields. We leave the proof as an exercise for the student and state the boundary conditions as follows:

Scalar form	Vector form	
$E_{t1} = E_{t2}$	$\vec{\mathbf{a}}_n \times (\vec{\mathbf{E}}_1 - \vec{\mathbf{E}}_2) = 0$	(7.84)
$H_{t1} - H_{t2} = J_s$	$\vec{\mathbf{a}}_n \times (\vec{\mathbf{H}}_1 - \vec{\mathbf{H}}_2) = \vec{\mathbf{J}}_s$	(7.85)
$B_{n1} = B_{n2}$	$\vec{\mathbf{a}}_n \cdot (\vec{\mathbf{B}}_1 - \vec{\mathbf{B}}_2) = 0$	(7.86)
$D_{n1} - D_{n2} = \rho_s$	$\vec{\mathbf{a}}_n \cdot (\vec{\mathbf{D}}_1 - \vec{\mathbf{D}}_2) = \rho_s$	(7.87)
$J_{n1} = J_{n2}$	$\vec{\mathbf{a}}_n \cdot (\vec{\mathbf{J}}_1 - \vec{\mathbf{J}}_2) = 0$	(7.88)
$\dfrac{J_{t1}}{\sigma_1} = \dfrac{J_{t2}}{\sigma_2}$	$\vec{\mathbf{a}}_n \times \left[\dfrac{\vec{\mathbf{J}}_1}{\sigma_1} - \dfrac{\vec{\mathbf{J}}_2}{\sigma_2}\right] = 0$	(7.89)

The subscripts $t1$ and $t2$ refer to the components of fields tangential to the boundary in media 1 and 2, respectively. Similarly, the subscripts $n1$ and $n2$ indicate the field components normal to the boundary. Note that the unit vector $\vec{\mathbf{a}}_n$ at the interface points into medium 1, ρ_s is the free surface charge density, and $\vec{\mathbf{J}}_s$ is the free surface current density.

Equation (7.84) states that the tangential components of $\vec{\mathbf{E}}_1$ and $\vec{\mathbf{E}}_2$ are equal at the interface (boundary). However, (7.85) asserts that the tangential components of $\vec{\mathbf{H}}_1$ and $\vec{\mathbf{H}}_2$ at any point on the interface are discontinuous by an amount equal to the surface current density at that point.

Equation (7.86) states that the normal components of $\vec{\mathbf{B}}_1$ and $\vec{\mathbf{B}}_2$ at the interface are continuous. However, (7.87) states that the normal components of $\vec{\mathbf{D}}_1$ and $\vec{\mathbf{D}}_2$ are discontinuous at any point on the interface by an amount equal to the surface charge density at that point.

Equation (7.88) states that the normal components of $\vec{\mathbf{J}}_1$ and $\vec{\mathbf{J}}_2$ are equal at the interface. Equation (7.89) states that the ratio of the tangential components of current densities at the interface is equal to the ratio of the conductivities.

When we apply the boundary conditions we must keep in mind the following conditions.

a) The electromagnetic fields inside a perfect conductor ($\sigma = \infty$) are zero. Thus, on the surface of a perfect conductor, both ρ_s and $\vec{\mathbf{J}}_s$ can exist.
b) Time-varying fields can exist inside a conductor ($\sigma < \infty$). Hence, $\vec{\mathbf{J}}_s$ is zero, but ρ_s can exist at the boundary between a conductor and a perfect dielectric.
c) At the interface between two perfect dielectrics $\vec{\mathbf{J}}_s$ is zero. However, ρ_s is zero unless the charge is physically placed at the interface.

Electromagnetic fields existing in any medium must satisfy Maxwell's equations. When we seek solutions of Maxwell's equations in two or more media, we must ascertain that the fields are matched at the boundaries. The following example illustrates how we can check whether or not the assumed fields satisfy Maxwell's equations.

EXAMPLE 7.14 An electric field intensity in a source-free dielectric medium is given by $\vec{\mathbf{E}} = C\cos(\omega t - \beta z)\vec{\mathbf{a}}_x$ V/m, where C is the amplitude of the field, ω is the frequency, and β is a constant quantity. Under what condition can this field exist? What are the other field quantities?

Solution A field can exist if and only if it satisfies all of Maxwell's equations. Let us assume that the given electric field intensity can exist in a source-free ($\rho_v = 0, \sigma = 0,$ and $\vec{\mathbf{J}} = 0$) dielectric medium. Then, using Maxwell's equation based upon Faraday's law, we obtain the magnetic flux density as

$$\nabla \times \vec{\mathbf{E}} = -\frac{\partial \vec{\mathbf{B}}}{\partial t}$$

$$\frac{\partial \vec{\mathbf{B}}}{\partial t} = -\frac{\partial}{\partial z}[E_x]\vec{\mathbf{a}}_y = -C\beta \sin(\omega t - \beta z)\,\vec{\mathbf{a}}_y$$

Integrating this expression, we obtain an expression for the magnetic flux density as

$$\vec{\mathbf{B}} = \frac{C\beta}{\omega}\cos(\omega t - \beta y)\,\vec{\mathbf{a}}_y \text{ T}$$

Using the constitutive relationship $\vec{\mathbf{B}} = \mu\vec{\mathbf{H}}$, we obtain the magnetic field intensity as

$$\vec{\mathbf{H}} = \frac{C\beta}{\mu\omega}\cos(\omega t - \beta z)\,\vec{\mathbf{a}}_y \text{ A/m}$$

Finally, we obtain the electric flux density from the $\vec{\mathbf{E}}$ field, using the constitutive equation $\vec{\mathbf{D}} = \epsilon\vec{\mathbf{E}}$, as

$$\vec{\mathbf{D}} = \epsilon C\cos(\omega t - \beta z)\,\vec{\mathbf{a}}_x \text{ C/m}^2$$

We now check if Maxwell's equation based upon Gauss's law for the electric field is satisfied or not. For a source-free dielectric medium

$$\nabla \cdot \vec{\mathbf{D}} = 0$$

or

$$\frac{\partial D_x}{\partial x} + \frac{\partial D_y}{\partial y} + \frac{\partial D_z}{\partial z} = 0$$

As the $\vec{\mathbf{D}}$ field has only one component D_x in the x direction and it is not a function of x, $\nabla \cdot \vec{\mathbf{D}} = 0$ is satisfied.

Similarly, we can show that $\nabla \cdot \vec{\mathbf{B}} = 0$ is also satisfied because the $\vec{\mathbf{B}}$ field has only one component B_y in the y direction and it is not a function of y. For the fields to exist, the last Maxwell equation that must be satisfied is

$$\begin{aligned}\nabla \times \vec{\mathbf{H}} &= \frac{\partial \vec{\mathbf{D}}}{\partial t}\\ &= -\frac{\partial}{\partial z}\left[\frac{C\beta}{\mu\omega}\cos(\omega t - \beta z)\right]\vec{\mathbf{a}}_x\\ &= \frac{\partial}{\partial t}[\epsilon C\cos(\omega t - \beta z)]\,\vec{\mathbf{a}}_x\\ &= -\frac{C}{\mu\omega}\beta^2\sin(\omega t - \beta z)\,\vec{\mathbf{a}}_x\\ &= -C\omega\epsilon\sin(\omega t - \beta z)\,\vec{\mathbf{a}}_x\end{aligned}$$

Hence,

$$\beta^2 = \omega^2\mu\epsilon$$

or

$$\beta = \pm\omega\sqrt{\mu\epsilon} \tag{7.90}$$

Thus, the given electric field, along with the other field components, can exist in a source-free dielectric medium as long as (7.90) is satisfied.

• • •

7.12 Poynting's theorem

In the preceding chapters we obtained expressions for the energy density associated with static electric and magnetic fields. In this section, our intention is to show that these expressions are also valid for time-varying fields. In addition, we will obtain an expression that represents the propagation of energy in a medium.

Let us consider a charged particle q moving with a velocity $\vec{\mathbf{u}}$ in a region where there exist time-varying electromagnetic fields. At any instant, the force experienced by the charged particle is

$$\vec{\mathbf{F}} = q(\vec{\mathbf{E}} + \vec{\mathbf{u}} \times \vec{\mathbf{B}})$$

where $\vec{\mathbf{E}}$ and $\vec{\mathbf{B}}$ are the time-varying electric field intensity and magnetic flux density, respectively. When the charge moves a distance $\overrightarrow{d\ell}$ in time dt under the influence of the force $\vec{\mathbf{F}}$, the work dW done by the force on the charged particle is

$$dW = q(\vec{\mathbf{E}} + \vec{\mathbf{u}} \times \vec{\mathbf{B}}) \cdot \overrightarrow{d\ell}$$

Because $\overrightarrow{d\ell} = \vec{\mathbf{u}}\,dt$ and $dW = P\,dt$, this equation can be written in terms of the power P supplied by the field as

$$P = q(\vec{\mathbf{E}} + \vec{\mathbf{u}} \times \vec{\mathbf{B}}) \cdot \vec{\mathbf{u}} \\ = q\vec{\mathbf{u}} \cdot \vec{\mathbf{E}} \tag{7.91}$$

because $(\vec{\mathbf{u}} \times \vec{\mathbf{B}}) \cdot \vec{\mathbf{u}} = 0$.

From (7.91) it is evident that the *time-varying magnetic field does not supply any energy to the charged particle.* Only the electric field intensity supplies power to the charged particle as it moves through the region. In order to generalize our development, let us now consider a volume distribution of charge with charge density ρ_v moving with an average velocity $\vec{\mathbf{u}}$. The power supplied by the field to an infinitesimal charge $\rho_v\, dv$ contained in volume dv is

$$dP = \rho_v\, dv\, \vec{\mathbf{E}} \cdot \vec{\mathbf{u}} = \vec{\mathbf{E}} \cdot \rho_v \vec{\mathbf{u}}\, dv \tag{7.92}$$

As $\vec{\mathbf{J}} = \rho_v \vec{\mathbf{u}}$, we now obtain the power in terms of the *power density* p (power per unit volume) as

$$p = \frac{dP}{dv} = \vec{\mathbf{J}} \cdot \vec{\mathbf{E}} \tag{7.93}$$

We obtained a similar expression for static fields on the basis of conservation of energy, and we have now shown that it is true for time-varying fields. We use this equation to derive relationships for (a) the energy stored in a time-varying electric field, (b) the energy stored in a time-varying magnetic field, and (c) the instantaneous power flow out of or into a given region. From Maxwell's equation (Ampère's law), we have

$$\vec{\mathbf{J}} = \nabla \times \vec{\mathbf{H}} - \frac{\partial \vec{\mathbf{D}}}{\partial t}$$

Substituting for $\vec{\mathbf{J}}$ in (7.93), we obtain

$$\vec{\mathbf{J}} \cdot \vec{\mathbf{E}} = \vec{\mathbf{E}} \cdot (\nabla \times \vec{\mathbf{H}}) - \vec{\mathbf{E}} \cdot \frac{\partial \vec{\mathbf{D}}}{\partial t} \tag{7.94}$$

Using the vector identity

$$\vec{\mathbf{E}} \cdot (\nabla \times \vec{\mathbf{H}}) = \vec{\mathbf{H}} \cdot (\nabla \times \vec{\mathbf{E}}) - \nabla \cdot (\vec{\mathbf{E}} \times \vec{\mathbf{H}})$$

we can write (7.94) as

$$\vec{\mathbf{J}} \cdot \vec{\mathbf{E}} = \vec{\mathbf{H}} \cdot (\nabla \times \vec{\mathbf{E}}) - \nabla \cdot (\vec{\mathbf{E}} \times \vec{\mathbf{H}}) - \vec{\mathbf{E}} \cdot \frac{\partial \vec{\mathbf{D}}}{\partial t}$$

Substituting for $\nabla \times \vec{\mathbf{E}}$ from (7.73), we get

$$\nabla \cdot (\vec{\mathbf{E}} \times \vec{\mathbf{H}}) + \vec{\mathbf{J}} \cdot \vec{\mathbf{E}} + \vec{\mathbf{H}} \cdot \frac{\partial \vec{\mathbf{B}}}{\partial t} + \vec{\mathbf{E}} \cdot \frac{\partial \vec{\mathbf{D}}}{\partial t} = 0 \tag{7.95}$$

This equation is known as the *point (differential) form of Poynting's theorem*, in honor of John H. Poynting, who is credited with its development. This equation is, in fact, a statement of the law of conservation of energy. The vector product $\vec{\mathbf{E}} \times \vec{\mathbf{H}}$ has the units of power density, watts per square meter (W/m^2), and is called the *Poynting vector.* The Poynting vector, therefore, yields the instantaneous flow of power per unit area. The direction of power flow is normal to the plane containing $\vec{\mathbf{E}}$ and $\vec{\mathbf{H}}$. In this text, we will denote the Poynting vector by $\vec{\mathbf{S}}$ such that

$$\vec{\mathbf{S}} = \vec{\mathbf{E}} \times \vec{\mathbf{H}} \tag{7.96}$$

For the time-varying fields in a linear, homogeneous, and isotropic medium, $\vec{\mathbf{B}} = \mu\vec{\mathbf{H}}$ and $\vec{\mathbf{D}} = \epsilon\vec{\mathbf{E}}$. In addition,

$$\vec{\mathbf{H}} \cdot \frac{\partial \vec{\mathbf{B}}}{\partial t} = \frac{1}{2}\frac{\partial}{\partial t}[\vec{\mathbf{B}} \cdot \vec{\mathbf{H}}] = \frac{1}{2}\frac{\partial}{\partial t}[\mu H^2]$$

$$\vec{\mathbf{E}} \cdot \frac{\partial \vec{\mathbf{D}}}{\partial t} = \frac{1}{2}\frac{\partial}{\partial t}[\vec{\mathbf{D}} \cdot \vec{\mathbf{E}}] = \frac{1}{2}\frac{\partial}{\partial t}[\epsilon E^2]$$

We can now express (7.95) as

$$\nabla \cdot \vec{\mathbf{S}} + \vec{\mathbf{J}} \cdot \vec{\mathbf{E}} + \frac{\partial}{\partial t}\left[\frac{1}{2}\mu H^2\right] + \frac{\partial}{\partial t}\left[\frac{1}{2}\epsilon E^2\right] = 0 \tag{7.97}$$

The third term in this equation represents the rate of change of energy density in the magnetic field; the fourth term yields the rate of change of energy density in the electric field. Thus, the instantaneous expressions for the energy densities in the magnetic and electric fields are, respectively,

$$w_m = \tfrac{1}{2}\vec{\mathbf{B}} \cdot \vec{\mathbf{H}} = \tfrac{1}{2}\mu H^2 \tag{7.98}$$

$$w_e = \tfrac{1}{2}\vec{\mathbf{D}} \cdot \vec{\mathbf{E}} = \tfrac{1}{2}\epsilon E^2 \tag{7.99}$$

Note that these expressions for the energy densities for time-varying fields are the same as those obtained for static fields. The only difference is that the fields now vary in time. In order to interpret what equation (7.97) really represents we have to write it in integral form as

$$\int_v \nabla \cdot \vec{\mathbf{S}}\, dv + \int_v \vec{\mathbf{J}} \cdot \vec{\mathbf{E}}\, dv + \int_v \frac{\partial}{\partial t} w_m\, dv + \int_v \frac{\partial}{\partial t} w_e\, dv = 0$$

or

$$\oint_s \vec{\mathbf{S}} \cdot \vec{ds} + \int_v \vec{\mathbf{J}} \cdot \vec{\mathbf{E}}\, dv + \frac{d}{dt}\int_v w_m\, dv + \frac{d}{dt}\int_v w_e\, dv = 0 \tag{7.100}$$

where the volume v is bounded by a surface s.

Equation (7.100) is the *integral form of Poynting's theorem.* The first term represents the power crossing the closed surface s bounding the volume v. If this integral is positive, the net power is flowing out of the volume. The power is flowing into volume v when the integral is negative.

The second integral represents the power supplied to the charged particles by the field. When this integral is positive, the field is doing work on the charged particles. However, an external force is doing the work in making the charged paticles move against the field when the integral is negative. In a conductive medium, $\vec{\mathbf{J}} = \sigma \vec{\mathbf{E}}$, this term represents power dissipation or ohmic power loss.

The third term represents the rate of change of stored magnetic energy. When this integral is positive, an external source is supplying energy to the magnetic field, resulting in an increase in the magnetic field. When the integral is negative, the magnetic energy is being extracted from the magnetic field, causing the field to decay.

The final term represents the rate of change of stored energy in the electric field, and it is interpreted exactly like the third term. Quite often, (7.100) is written as

$$-\oint_s \vec{\mathbf{S}} \cdot \vec{ds} = \int_v \vec{\mathbf{J}} \cdot \vec{\mathbf{E}}\, dv + \frac{d}{dt}\int_v (w_m + w_e)\, dv \tag{7.101}$$

The negative sign on the left-hand side of this equation indicates that the net power must flow into volume v in order to account for (a) the power dissipation in the region as heat and (b) the increase in the energy stored in both the electric and magnetic fields. For *static fields*, equation (7.101) becomes

$$-\oint_s \vec{\mathbf{S}} \cdot \vec{ds} = \int_v \vec{\mathbf{J}} \cdot \vec{\mathbf{E}}\, dv \tag{7.102}$$

which simply states that the net power flowing through surface s into volume v is equal to the power dissipation in that volume.

EXAMPLE 7.15

The electric field intensity in a dielectric (perfect) medium is given as $\vec{\mathbf{E}} = E\cos(\omega t - kz)\vec{\mathbf{a}}_x$ V/m, where E is its peak value, and k is a constant quantity. Determine (a) the magnetic field intensity in the region, (b) the direction of power flow, and (c) the average power density.

Solution

a) Let us first check to see if the given electric field intensity can exist in the dielectric medium. The x-directed component of the $\vec{\mathbf{E}}$ field is

$$E_x = E\cos(\omega t - kz)\ \text{V/m}$$

If ϵ is the permittivity of the dielectric medium, the electric flux density is

$$D_x = \epsilon E\cos(\omega t - kz)\ \text{C/m}^2$$

The time-varying charge density in the medium is

$$\rho_v = \nabla \cdot \vec{\mathbf{D}} = \frac{\partial}{\partial x}[\epsilon E\cos(\omega t - kz)] = 0$$

As expected, the free charge density in the dielectric medium is zero. Thus, Maxwell's equation $\nabla \cdot \vec{\mathbf{D}} = 0$ in a dielectric medium is satisfied. Let us use Maxwell's equation to determine the $\vec{\mathbf{B}}$ field as

$$\begin{aligned}\frac{\partial \vec{\mathbf{B}}}{\partial t} &= -\nabla \times \vec{\mathbf{E}} \\ &= -\frac{\partial}{\partial z}[E_x]\vec{\mathbf{a}}_y = -Ek\sin(\omega t - kz)\vec{\mathbf{a}}_y\end{aligned}$$

Integrating with respect to time, we obtain the y component of the $\vec{\mathbf{B}}$ field as

$$B_y = \frac{Ek}{\omega}\cos(\omega t - kz)\ \mathrm{T}$$

As $\vec{\mathbf{B}} = \mu\vec{\mathbf{H}}$, where μ is the permeability of the dielectric medium, the magnetic filed intensity is

$$H_y = \frac{Ek}{\omega\mu}\cos(\omega t - kz)\ \mathrm{A/m}$$

Let us now verify if the $\vec{\mathbf{B}}$ or $\vec{\mathbf{H}}$ field can exist:

$$\nabla \cdot \vec{\mathbf{B}} = \frac{\partial}{\partial y}[B_y] = \frac{\partial}{\partial y}\left[\frac{Ek}{\omega}\cos(\omega t - kz)\right] = 0$$

As $\nabla \cdot \vec{\mathbf{B}} = 0$, the $\vec{\mathbf{B}}$ field can exist.

Let us now compute the volume current density $\vec{\mathbf{J}}$ from the remaining Maxwell equation as

$$\begin{aligned}\vec{\mathbf{J}} &= \nabla \times \vec{\mathbf{H}} - \frac{\partial \vec{\mathbf{D}}}{\partial t} \\ &= -\frac{\partial}{\partial z}[H_y]\vec{\mathbf{a}}_x - \frac{\partial}{\partial t}[D_x]\vec{\mathbf{a}}_x \\ &= \left[\omega\epsilon - \frac{1}{\omega\mu}k^2\right]E\sin(\omega t - kz)\vec{\mathbf{a}}_x\end{aligned}$$

As $\vec{\mathbf{J}}$ must be zero in a perfectly dielectric medium, and this equation is zero only when

$$\omega\epsilon - \frac{1}{\omega\mu}k^2 = 0$$

or

$$k = \pm\omega\sqrt{\mu\epsilon}$$

Thus, k is not an arbitrary constant. It depends upon the frequency of the time-varying fields, and the permittivity and permeability of the medium. We now say that the fields can exist because they satisfy all the Maxwell equations.

b) The instantaneous power density, or the Poynting vector, is

$$\vec{\mathbf{S}} = \vec{\mathbf{E}} \times \vec{\mathbf{H}}$$
$$= \frac{k}{\omega\mu} E^2 \cos^2(\omega t - kz)\,\vec{\mathbf{a}}_z \text{ W/m}^2$$

Since $\vec{\mathbf{S}}$ has only a z component, the power flow is in the z direction.

c) The average power density in the z direction can now be obtained as

$$\langle S_z \rangle = \frac{1}{T} \int_0^T \frac{k}{\omega\mu} E^2 \cos^2(\omega t - kz)\, dt$$

where T is the time period such that $\omega T = 2\pi$. Using the trigonometric identity

$$\cos(2\omega t - 2kz) = 2\cos^2(\omega t - kz) - 1$$

we can express the average power density as

$$\langle S_z \rangle = \frac{1}{2T} \int_0^T \frac{k}{\omega\mu} E^2\, dt + \frac{1}{2T} \int_0^T \frac{k}{\omega\mu} E^2 \cos(2\omega t - 2kz)\, dt$$
$$= \frac{k}{2\omega\mu} E^2 \text{ W/m}^2$$

• • •

7.13 Time-harmonic fields

One of the most important cases of time-varying electromagnetic fields is the time-harmonic (sinusoidal) field. In this type of field, the excitation source varies sinusoidally in time with a single frequency. In a linear system, a sinusoidally varying source generates fields that also vary sinusoidally in time at all points in the system. For time-harmonic fields we can employ phasor analysis to obtain the single-frequency (monochromatic) steady-state response. When fields are examined in this manner, there is no loss in generality as (a) any time-varying periodic function can be represented by a Fourier series in terms of sinusoidal functions, and (b) the principle of superposition can be applied under linear conditions. In other words, we can obtain the complete response of time-varying periodic fields by using linear combinations of monochromatic responses.

In circuit theory, you have already used the phasor notation to represent voltages and currents varying sinusoidally in time. In this section we extend these definitions to encompass vector quantities. Any vector quantity can be represented in terms of its components along three mutually perpendicular axes, so each component can be treated as a scalar quantity. For example, if the $\vec{\mathbf{E}}$ field is given as

$$\vec{\mathbf{E}}(x, y, z, t) = E_x(x, y, z, t)\vec{\mathbf{a}}_x + E_y(x, y, z, t)\vec{\mathbf{a}}_y + E_z(x, y, z, t)\vec{\mathbf{a}}_z \tag{7.103}$$

then $E_x(x, y, z, t)$, $E_y(x, y, z, t)$, and $E_z(x, y, z, t)$ are the scalar components of the $\vec{\mathbf{E}}$ field along the $\vec{\mathbf{a}}_x$, $\vec{\mathbf{a}}_y$, and $\vec{\mathbf{a}}_z$ directions, respectively. The time-harmonic variations of these components may be written as

$$E_x(x, y, z, t) = E_x(r, t) = E_{x0}(r)\cos[\omega t + \alpha(r)] \tag{7.104a}$$

$$E_y(x, y, z, t) = E_y(r, t) = E_{y0}(r)\cos[\omega t + \beta(r)] \tag{7.104b}$$

$$E_z(x, y, z, t) = E_z(r, t) = E_{z0}(r)\cos[\omega t + \gamma(r)] \tag{7.104c}$$

where E_{x0}, E_{y0}, and E_{z0} are the amplitudes of the components of the $\vec{\mathbf{E}}$ field along the $\vec{\mathbf{a}}_x$, $\vec{\mathbf{a}}_y$, and $\vec{\mathbf{a}}_z$ directions, respectively. We have also used a shorthand notation (r) to imply that the fields are functions of space coordinates x, y, and z. In addition, $\alpha(r)$, $\beta(r)$, and $\gamma(r)$ are the phase shifts of the x, y, and z components of the $\vec{\mathbf{E}}$ field at a given point (x, y, z) in space. The amplitude of each component is now a function of space only. We can also write each component as

$$E_x(r, t) = \text{Re}[E_{x0}(r)e^{j\alpha(r)}e^{j\omega t}] \tag{7.105a}$$

$$E_y(r, t) = \text{Re}[E_{y0}(r)e^{j\beta(r)}e^{j\omega t}] \tag{7.105b}$$

$$E_z(r, t) = \text{Re}[E_{z0}(r)e^{j\gamma(r)}e^{j\omega t}] \tag{7.105c}$$

where Re stands for the real part of the complex function enclosed in the brackets. If we define

$$\tilde{E}_x(r) = E_{x0}(r)e^{j\alpha(r)} \tag{7.106a}$$

$$\tilde{E}_y(r) = E_{y0}(r)e^{j\beta(r)} \tag{7.106b}$$

$$\tilde{E}_z(r) = E_{z0}(r)e^{j\gamma(r)} \tag{7.106c}$$

then (7.105) can be written as

$$E_x(r, t) = \text{Re}[\tilde{E}_x(r)e^{j\omega t}] \tag{7.107a}$$

$$E_y(r, t) = \text{Re}[\tilde{E}_y(r)e^{j\omega t}] \tag{7.107b}$$

$$E_z(r, t) = \text{Re}[\tilde{E}_z(r)e^{j\omega t}] \tag{7.107c}$$

In equations (7.106) and (7.107), $\tilde{E}_x(r)$, $\tilde{E}_y(r)$, and $\tilde{E}_z(r)$ are said to be the phasor equivalents of $E_x(r, t)$, $E_y(r, t)$, and $E_z(r, t)$. Thus, the **phasor representation** of a time-harmonic field is a complex quantity that is a function of space only. The time dependency is completely embodied in the term $e^{j\omega t}$. We have used the tilde (˜) symbol to denote a phasor quantity and will continue its use in this text.

Equation (7.103) can now be written as

$$\begin{aligned}\vec{\mathbf{E}}(r, t) &= \text{Re}\{[\tilde{E}_x(r)\vec{\mathbf{a}}_x + \tilde{E}_y(r)\vec{\mathbf{a}}_y + \tilde{E}_z(r)\vec{\mathbf{a}}_z]e^{j\omega t}\} \\ &= \text{Re}[\tilde{\vec{\mathbf{E}}}(r)e^{j\omega t}]\end{aligned} \tag{7.108}$$

where

$$\tilde{\vec{\mathbf{E}}}(r) = \tilde{E}_x(r)\vec{\mathbf{a}}_x + \tilde{E}_y(r)\vec{\mathbf{a}}_y + \tilde{E}_z(r)\vec{\mathbf{a}}_z \tag{7.109}$$

is the phasor representation of the $\vec{\mathbf{E}}$ field (vector) at any point in the region where the time-harmonic $\vec{\mathbf{E}}$ field exists. Once again, the space dependency is included in $\tilde{\vec{\mathbf{E}}}(r)$, and the time dependency is retained in the implicit form. It is evident from (7.108) that we can always express a field in the time domain by multiplying its counterpart in the phasor or frequency domain by $e^{j\omega t}$ and taking its real part only. From (7.108), the time rate of change of the $\vec{\mathbf{E}}$ field is

$$\frac{\partial \vec{\mathbf{E}}(r,t)}{\partial t} = \mathrm{Re}[j\omega \tilde{\vec{\mathbf{E}}}(r) e^{j\omega t}]$$

which states that the differentiation with respect to time in the time domain yields a factor $j\omega$ in the phasor domain. Similarly, we can show that the integration with respect to time becomes division by $j\omega$.

7.13.1 Maxwell's equations in phasor form

Substituting (7.109) for the $\vec{\mathbf{E}}$ field and similar expressions for the $\vec{\mathbf{D}}$, $\vec{\mathbf{B}}$, and $\vec{\mathbf{H}}$ fields into Maxwell's equations, we obtain the phasor forms of these equations in point (differential) and integral form as

$$\nabla \times \tilde{\vec{\mathbf{E}}} = -j\omega \tilde{\vec{\mathbf{B}}} \Rightarrow \oint_c \tilde{\vec{\mathbf{E}}} \cdot \vec{d\ell} = -j\omega \int_s \tilde{\vec{\mathbf{B}}} \cdot \vec{ds} \tag{7.110}$$

$$\nabla \times \tilde{\vec{\mathbf{H}}} = \tilde{\vec{\mathbf{J}}} + j\omega \tilde{\vec{\mathbf{D}}} \Rightarrow \oint_c \tilde{\vec{\mathbf{H}}} \cdot \vec{d\ell} = \int_s \tilde{\vec{\mathbf{J}}} \cdot \vec{ds} + j\omega \int_s \tilde{\vec{\mathbf{D}}} \cdot \vec{ds} \tag{7.111}$$

$$\nabla \cdot \tilde{\vec{\mathbf{D}}} = \tilde{\rho}_v \Rightarrow \oint_s \tilde{\vec{\mathbf{D}}} \cdot \vec{ds} = \int_v \tilde{\rho}_v \, dv \tag{7.112}$$

$$\nabla \cdot \tilde{\vec{\mathbf{B}}} = 0 \Rightarrow \oint_s \tilde{\vec{\mathbf{B}}} \cdot \vec{ds} = 0 \tag{7.113}$$

$$\nabla \cdot \tilde{\vec{\mathbf{J}}} = -j\omega \tilde{\rho}_v \Rightarrow \oint_s \tilde{\vec{\mathbf{J}}} \cdot \vec{ds} = -j\omega \int_v \tilde{\rho}_v \, dv \tag{7.114}$$

The constitutive relationships in the phasor form are

$$\tilde{\vec{\mathbf{D}}} = \epsilon \tilde{\vec{\mathbf{E}}} \tag{7.115}$$

$$\tilde{\vec{\mathbf{B}}} = \mu \tilde{\vec{\mathbf{H}}} \tag{7.116}$$

where μ and ϵ are the permeability and the permittivity of the medium.

7.13.2 Boundary conditions in phasor form

The general boundary conditions at the interface between two media in the phasor form are

$$\vec{\mathbf{a}}_n \cdot (\tilde{\vec{\mathbf{B}}}_1 - \tilde{\vec{\mathbf{B}}}_2) = 0 \tag{7.117}$$

$$\vec{\mathbf{a}}_n \cdot (\tilde{\vec{\mathbf{D}}}_1 - \tilde{\vec{\mathbf{D}}}_2) = \tilde{\rho}_s \tag{7.118}$$

$$\vec{\mathbf{a}}_n \times (\tilde{\vec{\mathbf{E}}}_1 - \tilde{\vec{\mathbf{E}}}_2) = 0 \tag{7.119}$$

$$\vec{\mathbf{a}}_n \times (\tilde{\vec{\mathbf{H}}}_1 - \tilde{\vec{\mathbf{H}}}_2) = \tilde{\vec{\mathbf{J}}}_s \tag{7.120}$$

where $\vec{\mathbf{a}}_n$ is the unit vector normal to the interface and pointing into medium 1, $\tilde{\rho}_s$ is the surface charge density, and $\vec{\mathbf{J}}_s$ is the surface current density at the interface.

7.13.3 Poynting theorem in phasor form

The scalar product of (7.110) with $\vec{\mathbf{H}}^*$, where $*$ represents the conjugate of a field quantity, and of the conjugate of (7.111) with $\vec{\mathbf{E}}$ yield

$$(\nabla \times \vec{\mathbf{E}}) \cdot \vec{\mathbf{H}}^* = -j\omega \vec{\mathbf{B}} \cdot \vec{\mathbf{H}}^* \tag{7.121}$$

$$(\nabla \times \vec{\mathbf{H}}^*) \cdot \vec{\mathbf{E}} = (\vec{\mathbf{J}}^* - j\omega \vec{\mathbf{D}}^*) \cdot \vec{\mathbf{E}} \tag{7.122}$$

Subtracting (7.121) from (7.122), we obtain

$$\vec{\mathbf{E}} \cdot (\nabla \times \vec{\mathbf{H}}^*) - \vec{\mathbf{H}}^* \cdot (\nabla \times \vec{\mathbf{E}}) = \vec{\mathbf{E}} \cdot \vec{\mathbf{J}}^* + j\omega (\vec{\mathbf{B}} \cdot \vec{\mathbf{H}}^* - \vec{\mathbf{E}} \cdot \vec{\mathbf{D}}^*) \tag{7.123}$$

Using the vector identity

$$\nabla \cdot (\vec{\mathbf{E}} \times \vec{\mathbf{H}}^*) = \vec{\mathbf{H}}^* \cdot (\nabla \times \vec{\mathbf{E}}) - \vec{\mathbf{E}} \cdot (\nabla \times \vec{\mathbf{H}}^*)$$

and the definition of the *complex Poynting vector* or *complex power density*

$$\hat{\mathbf{S}} = \tfrac{1}{2}[\vec{\mathbf{E}} \times \vec{\mathbf{H}}^*] \tag{7.124}$$

we can write (7.123) as

$$-\nabla \cdot \hat{\mathbf{S}} = \tfrac{1}{2}\vec{\mathbf{E}} \cdot \vec{\mathbf{J}}^* + j\omega[\tfrac{1}{2}\vec{\mathbf{B}} \cdot \vec{\mathbf{H}}^* - \tfrac{1}{2}\vec{\mathbf{E}} \cdot \vec{\mathbf{D}}^*] \tag{7.125}$$

Equation (7.125) is known as the *complex Poynting theorem in the point (differential) form.* Integrating over volume v bounded by surface s and applying the divergence theorem, we obtain the *complex Poynting theorem in integral form as*

$$-\oint_s \hat{\mathbf{S}} \cdot \overrightarrow{ds} = \int_v \frac{1}{2}\vec{\mathbf{E}} \cdot \vec{\mathbf{J}}^* dv + j2\omega \int_v \frac{1}{4}\vec{\mathbf{B}} \cdot \vec{\mathbf{H}}^* dv - j2\omega \int_v \frac{1}{4}\vec{\mathbf{E}} \cdot \vec{\mathbf{D}}^* dv \tag{7.126}$$

The term on the left-hand side of this equation ($-\oint_s \hat{\mathbf{S}} \cdot \overrightarrow{ds}$) represents the complex power flow into volume v. If there are sources inside v, then $\oint_s \hat{\mathbf{S}} \cdot \overrightarrow{ds}$ yields the complex power radiated or transmitted from the region.

For a conductive medium, $\vec{\mathbf{J}} = \sigma \vec{\mathbf{E}}$, the first term on the right-hand side yields the time-average power dissipated within the conductive medium. If $\vec{\mathbf{J}}$ is the convection current due to the motion of charges in a medium such that $\vec{\mathbf{J}} = \tilde{\rho}_{v+}\vec{\mathbf{u}}_+ + \tilde{\rho}_{v-}\vec{\mathbf{u}}_-$, then this term represents the rate of expenditure of energy by the $\vec{\mathbf{E}}$ field in moving these charges.

The second term on the right-hand side represents the magnetic energy stored within the region. Note that

$$\langle w_m \rangle = \tfrac{1}{4}\vec{\mathbf{B}} \cdot \vec{\mathbf{H}}^* = \tfrac{1}{4}\mu H^2 \tag{7.127}$$

is the time-average magnetic energy density in the region. The last term on the right-hand side yields the time-average electric energy within the region. In this case,

$$\langle w_e \rangle = \tfrac{1}{4}\tilde{\mathbf{E}} \cdot \tilde{\mathbf{D}}^* = \tfrac{1}{4}\epsilon E^2 \tag{7.128}$$

is the time-average electric energy density in the region.

The time average power density can be obtained from (7.124) as

$$\langle \hat{\mathbf{S}} \rangle = \text{Re}[\hat{\mathbf{S}}] \tag{7.129}$$

and the average power flow through a surface s is

$$\langle P \rangle = \int_s \langle \hat{\mathbf{S}} \rangle \cdot \vec{ds} \tag{7.130}$$

EXAMPLE 7.16

The $\vec{\mathbf{E}}$ field in a source-free dielectric region is given as $\vec{\mathbf{E}} = C \sin \alpha x \cos(\omega t - kz)\vec{\mathbf{a}}_y$ V/m. Using phasors, determine (a) the magnetic field intensity, (b) the necessary condition for the fields to exist, and (c) the time-average power flow per unit area.

Solution a) The phasor equivalent of the $\vec{\mathbf{E}}$ field is

$$\tilde{E}_y = C \sin \alpha x \, e^{-jkz}$$

Let us determine the $\tilde{\mathbf{H}}$ field using Maxwell's equation

$$\nabla \times \tilde{\mathbf{E}} = -j\omega\mu\tilde{\mathbf{H}}$$

However,

$$\begin{aligned}\nabla \times \tilde{\mathbf{E}} &= -\frac{\partial}{\partial z}[\tilde{E}_y]\vec{\mathbf{a}}_x + \frac{\partial}{\partial x}[\tilde{E}_y]\vec{\mathbf{a}}_z \\ &= jkC \sin \alpha x \, e^{-jkz}\vec{\mathbf{a}}_x + \alpha C \cos \alpha x \, e^{-jkz}\vec{\mathbf{a}}_z\end{aligned}$$

Hence, the $\tilde{\mathbf{H}}$ field is

$$\tilde{\mathbf{H}} = -\frac{kC}{\omega\mu} \sin \alpha x \, e^{-jkz}\vec{\mathbf{a}}_x + j\frac{\alpha C}{\omega\mu} \cos \alpha x \, e^{-jkz}\vec{\mathbf{a}}_z$$

b) In the time domain, the x and z components of the $\vec{\mathbf{H}}$ field are

$$H_x(r, t) = -\frac{kC}{\omega\mu} \sin \alpha x \cos(\omega t - kz)$$

and

$$\begin{aligned}H_z(r, t) &= -\frac{\alpha C}{\omega\mu} \cos \alpha x \cos(\omega t - kz + 90°) \\ &= -\frac{\alpha C}{\omega\mu} \cos \alpha x \sin(\omega t - kz)\end{aligned}$$

Note that in the source-free region ($\tilde{\rho}_v = 0$) the fields do satisfy the two divergence equations. That is, $\nabla \cdot \tilde{\mathbf{D}} = 0$ and $\nabla \cdot \tilde{\mathbf{B}} = 0$.

Because $\vec{\mathbf{J}} = 0$ in a source-free dielectric region, the last Maxwell equation becomes

$$\nabla \times \vec{\mathbf{H}} = j\omega\epsilon\vec{\mathbf{E}}$$

Substituting for $\vec{\mathbf{E}}$ and $\vec{\mathbf{H}}$ in this equation and performing some simplifications, we obtain

$$k^2 = \omega^2\mu\epsilon - \alpha^2$$

as the necessary condition for the fields to exist.

c) The complex power density in the region is

$$\begin{aligned}\hat{\mathbf{S}} &= \tfrac{1}{2}[\vec{\mathbf{E}} \times \vec{\mathbf{H}}^*] \\ &= \tfrac{1}{2}[\tilde{E}_y\tilde{H}_z^*\vec{\mathbf{a}}_x - \tilde{E}_y\tilde{H}_x^*\vec{\mathbf{a}}_z] \\ &= -j\frac{\alpha}{2\omega\mu}C^2 \sin\alpha x \cos\alpha x\, \vec{\mathbf{a}}_x + \frac{k}{2\omega\mu}C^2 \sin^2 \alpha x\, \vec{\mathbf{a}}_z\end{aligned}$$

Finally, the time-average real (active) power flow per unit area is

$$\langle\hat{\mathbf{S}}\rangle = \mathrm{Re}[\hat{\mathbf{S}}] = \frac{k}{2\omega\mu}C^2 \sin^2 \alpha x\, \vec{\mathbf{a}}_z$$

Although the complex power density has components in the x and z directions, the time-average real power flow per unit area is in the z direction. To sustain such a power flow, there must be appropriate components of the $\vec{\mathbf{E}}$ and $\vec{\mathbf{H}}$ fields at each point along the z direction. This is a clear indication of the propagation of fields as waves.

• • •

7.14 Applications of electromagnetic fields

In this chapter we have learned that all electromagnetic fields must satisfy Maxwell's equations as the condition for their existence. In fact, when Maxwell's equations are satisfied, each field quantity ($\vec{\mathbf{E}}$ or $\vec{\mathbf{H}}$) separately satisfies a *wave equation* or *Helmholtz equation*. Among the many possible solutions of a wave equation is a *uniform plane-wave solution*, in which the field components are in a plane normal to the direction of propagation of the wave, and the magnitude of each field component is constant (uniform) in that plane. We will derive the wave equation and discuss uniform plane waves, or *transverse electromagnetic (TEM)* waves, as they are also called, in Chapter 8. We will also show that a uniform plane wave propagates with the velocity of light in free space.

A TEM wave can also exist where there are two or more conductors insulated from one another. This type of TEM wave is also called a *guided wave* because it propagates along the length of the conductors. Parallel-wire transmission lines and coaxial cables usually support the TEM mode, also called the *principal mode*, to transmit power or information (signals) from one end to the other. The guided TEM wave, which can exist at any frequency, is the subject of Chapter 9.

We can also use hollow conductors to guide waves through the dielectric region that they bound. When a single conductor is used to guide the wave, it is called a **waveguide**. There are two types of commonly used waveguides: cylindrical and rectangular. We discuss the rectangular waveguide in Chapter 10. We will study two types of waves that can exist in a rectangular waveguide: the *transverse electric* wave (*TE* mode), and the *transverse magnetic* wave (*TM* mode). A single conductor cannot support the TEM wave. The striking feature of both TE and TM waves is that each wave can only exist above a certain limiting frequency, called the *cutoff frequency*.

We use *antennas* to generate the electromagnetic fields that can propagate either through an unbounded region, along a transmission line, or through the medium enclosed by a waveguide. An antenna can simply be a straight cylindrical conductor or it can be a properly arranged combination of such conductors of various dimensions. Waveguides with flares and parabolic dishes are also used as antennas. We discuss the principles of operation of some antennas in Chapter 11.

As you can see, the succeeding chapters in this text present nothing but the applications of Maxwell's equations under various boundary conditions. The wave solution in each case is unique. Among the numerous other applications of Maxwell's equations, we have chosen three for discussion in this section: the *transformer*, the *autotransformer*, and the *betatron*.

7.14.1 The transformer

When two electrically isolated coils are arranged so that the (time-varying) magnetic flux produced by one coil links the other coil and induces an emf in it, the two coils are said to be magnetically coupled and they form a two-winding *transformer*. Figure 7.20 shows a two-winding transformer in its simplest form. The coil connected to the

Figure 7.20 Two magnetically coupled coils form a transformer

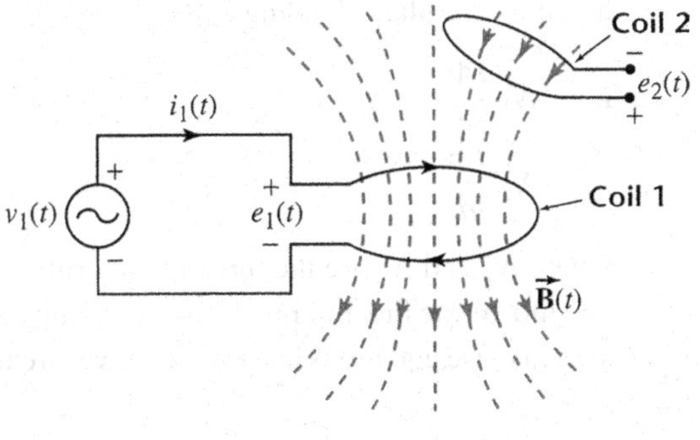

source is called the *primary winding*; the other coil is referred to as the *secondary winding*. When the two coils are isolated from each other, either in free space or wound over a nonmagnetic material (known as the core), the transformer is usually called an *air-core transformer*. The total flux linking the secondary coil depends upon its proximity and orientation with respect to the primary coil. To ensure maximum flux linkages between the two windings, they can be wound over a magnetic material with high permeability to form a common magnetic circuit. This arrangement is then called an *iron-core transformer*.

When the permeability of the magnetic core is high and the secondary side of the transformer is left open (no-load condition), as shown in Figure 7.21, a small amount of current $i_m(t)$, called the *magnetization current*, circulates in the primary winding to (a) establish a time-varying flux $\Phi(t)$ in the core, (b) account for the mmf drop in the magnetic circuit due to its reluctance, and (c) justify the power loss in the primary winding and the magnetic loss in the core.

Figure 7.21 Transformer under no-load condition

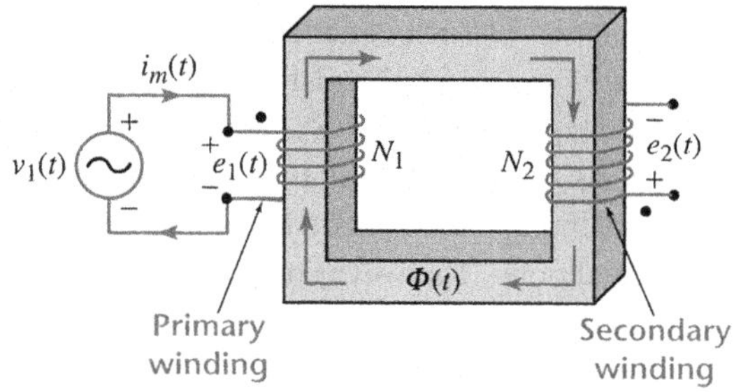

An ideal transformer

Let us now consider an ideal transformer; that is, one (a) of infinite permeability, (b) whose winding resistance is zero, and (c) whose magnetic losses are nonexistent. These assumptions mandate that the magnetization current under the no-load condition will be vanishingly small and can be ignored, and all the time-varying flux produced by the primary winding will follow the magnetic path without any leakages. Under these ideal conditions, the emfs induced in the primary winding $e_1(t)$ and the secondary winding $e_2(t)$ are

$$e_1 = N_1 \frac{d\Phi}{dt} \tag{7.131}$$

$$e_2 = N_2 \frac{d\Phi}{dt} \tag{7.132}$$

where N_1 and N_2 are the turns in the primary and secondary windings, and $\Phi(t)$ is the flux linking the two windings. We have omitted the minus sign in these equations because we have already marked the polarities of these induced emfs in Figure 7.21. We have also placed a dot next to one end of each winding. It is understood that the induced emf is positive at the dotted end of a winding with respect to its other end when the flux linking the winding is increasing with time. We will use this *dot convention* to represent a transformer by its equivalent circuit.

We can express the ratio of induced emfs as

$$\frac{e_1}{e_2} = \frac{N_1}{N_2} \tag{7.133}$$

That is, *the ratio of the induced emfs in the two windings is equal to the ratio of their turns.* We would have obtained the same expression had we impressed the voltage source on the secondary winding and left the primary winding open. Under ideal conditions, the induced emf in either winding must be equal to the voltage rating of that winding. Thus,

$$\frac{v_1}{v_2} = \frac{N_1}{N_2} = a \tag{7.134}$$

where $v_1(t)$ and $v_2(t)$ are the voltage ratings of the primary and secondary windings, and a is the ratio of the number of turns in the primary winding to that in the secondary and is called the *a-ratio* or the *ratio of transformation.*

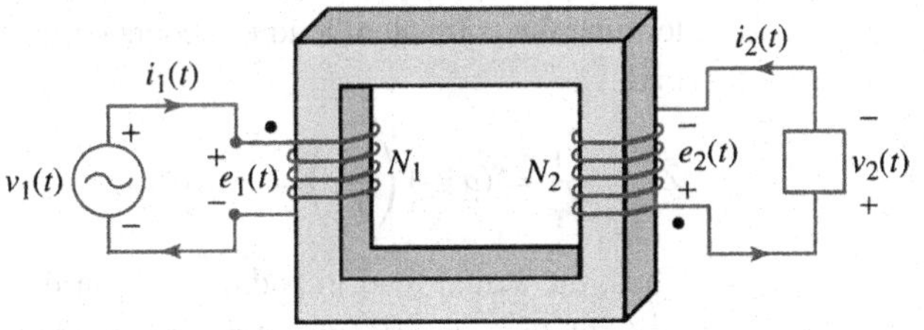

Figure 7.22 Transformer under load condition

When the secondary winding is connected to a load (see Figure 7.22), the current in the secondary winding produces its own flux, which opposes the original flux. The net flux in the core and thereby the induced emf in each winding will tend to decrease from its no-load value. As soon as the induced emf in the primary winding tends to decrease, it causes a current in the primary winding to nullify the decrease in the flux and the induced emf. The increase in the current continues until the flux in the core and thereby the induced emfs in the two windings are restored to their no-load values. Hence the source supplies power to the primary winding, and the secondary winding delivers that power to the load. The magnetic flux acts as a medium in the power transfer process. Under ideal conditions, the power input must be equal to the power output. That is,

$$v_1 i_1 = v_2 i_2$$

or

$$\frac{i_2}{i_1} = \frac{v_1}{v_2} = \frac{e_1}{e_2} \tag{7.135}$$

From this equation we conclude that *the ratio of the currents is inversely proportional to the ratio of the induced emfs.*

From (7.134) and (7.135) we can show that

$$N_1 i_1 - N_2 i_2 = 0 \tag{7.136}$$

This expression simply emphasizes the fact that under ideal conditions the net mmf needed to excite the transformer is zero. This is another way of saying that the magnetic material has infinite permeability, or that the reluctance of the magnetic circuit is zero.

For sinusoidally varying sources the preceding relations can be expressed in phasor form as

$$\frac{\tilde{V}_1}{\tilde{V}_2} = \frac{\tilde{I}_2}{\tilde{I}_1} = \frac{N_1}{N_2} = a \tag{7.137}$$

If we define

$$\hat{Z}_2 = \frac{\tilde{V}_2}{\tilde{I}_2}$$

as the load impedance connected to the secondary winding, we can determine the equivalent load impedance referred to the primary winding (side) as

$$\hat{Z}_1 = \frac{\tilde{V}_1}{\tilde{I}_1} = (a\tilde{V}_2)\left(\frac{a}{\tilde{I}_2}\right) = a^2\frac{\tilde{V}_2}{\tilde{I}_2} = a^2\hat{Z}_2 \tag{7.138}$$

Thus, the actual load impedance $\hat{Z}_2$ on the secondary side appears as $a^2\hat{Z}_2$ on the primary side, and the ideal transformer can be represented by an equivalent circuit, as depicted in Figure 7.23.

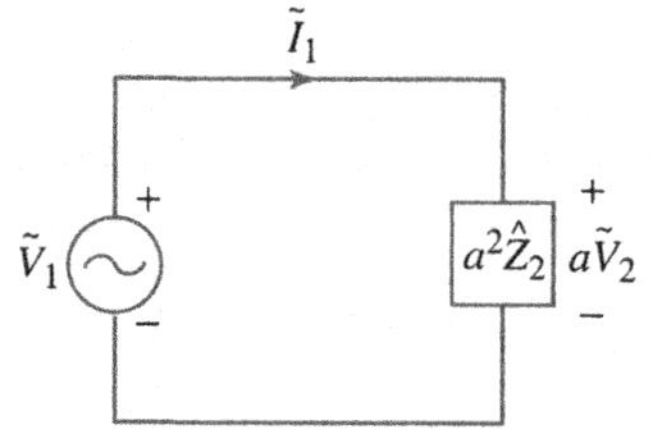

Figure 7.23 An equivalent circuit of an ideal transformer as referred to the primary side

A real transformer

When the magnetic core has a finite permeability, each coil has a finite self-inductance, and the magnetic coupling between the two coils contributes to the mutual inductance. In addition, each winding must have its own winding resistance. Taking these factors into consideration, we can represent a two-winding transformer by its inductively coupled equivalent circuit, as shown in Figure 7.24 (see below). In this circuit, R_1 and L_1 are the resistance and the self-inductance of the primary winding, R_2 and L_2 are the resistance and the self-inductance of the secondary winding, and M is the mutual inductance between them. If the core is made of a highly permeable magnetic material, we expect the flux to confine itself within the core with negligibly small leakage flux.

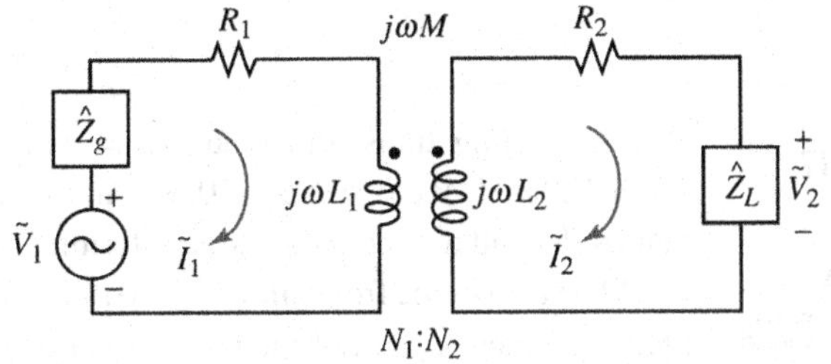

Figure 7.24 Inductively coupled equivalent circuit of a transformer

Thus, we can assume that the two coils are perfectly coupled and the mutual inductance between them is

$$M = \sqrt{L_1 L_2} \tag{7.139}$$

Let $\tilde{I}_1$ and $\tilde{I}_2$ be the primary and secondary winding currents when a load impedance $\hat{Z}_L$ is connected to the secondary winding, and a voltage source $\tilde{V}_1$ with internal impedance $\hat{Z}_g$ is impressed upon the primary. We can express the two coupled equations as

$$(R_1 + j\omega L_1 + \hat{Z}_g)\tilde{I}_1 - j\omega M \tilde{I}_2 = \tilde{V}_1 \tag{7.140}$$

$$-j\omega M \tilde{I}_1 + (R_2 + j\omega L_2 + \hat{Z}_L)\tilde{I}_2 = 0 \tag{7.141}$$

The solution of these equations yields $\tilde{I}_1$ and $\tilde{I}_2$.

We can now compute the load voltage as

$$\tilde{V}_2 = \tilde{I}_2 \hat{Z}_L$$

The power delivered to the load is

$$P_o = \mathrm{Re}[\tilde{V}_2 \tilde{I}_2^*]$$

Finally, the power input to the transformer is

$$P_i = \mathrm{Re}[\tilde{V}_1 \tilde{I}_1^*]$$

The efficiency of the transformer is simply the ratio of the power output to the power input. Note that we have not included magnetic losses in our analysis. Information on how to include these losses is available in many textbooks on electrical machines. In this section, however, we will assume that these losses are negligible.

Based upon (7.140) and (7.141) we can also represent the transformer by a conductively coupled equivalent circuit. Figure 7.25 shows one of the equivalent circuits that are commonly used to analyze the performance of a transformer.

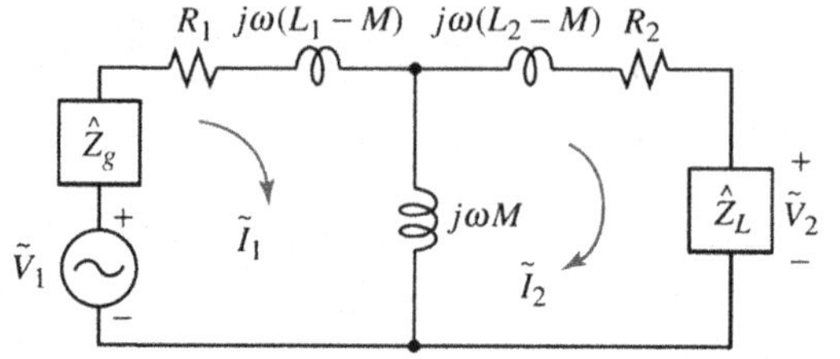

Figure 7.25 Conductively coupled equivalent circuit of a transformer

EXAMPLE 7.17

WORKSHEET 6

Mathcad

A two-winding transformer has $R_1 = 4\,\Omega$, $R_2 = 1\,\Omega$, $L_1 = 30$ mH, $L_2 = 120$ mH, and $k = 1$. A 120-V (rms) source is connected to the primary side, and the secondary side is terminated into a load resistance of 100 Ω. If the source frequency is 1000 rad/s, draw the conductively coupled equivalent circuit of the transformer and determine its efficiency.

Solution The mutual inductance between the two windings is

$$M = \sqrt{0.03 \times 0.12} = 0.06\,\text{H}$$

Using Figure 7.25 as a reference, we can draw the conductively coupled equivalent circuit as shown in Figure 7.26. Using the mesh-current method, the two equations are

$$(4 + j30)\tilde{I}_1 - j60\tilde{I}_2 = 120$$
$$-j60\tilde{I}_1 + (101 + j120)\tilde{I}_2 = 0$$

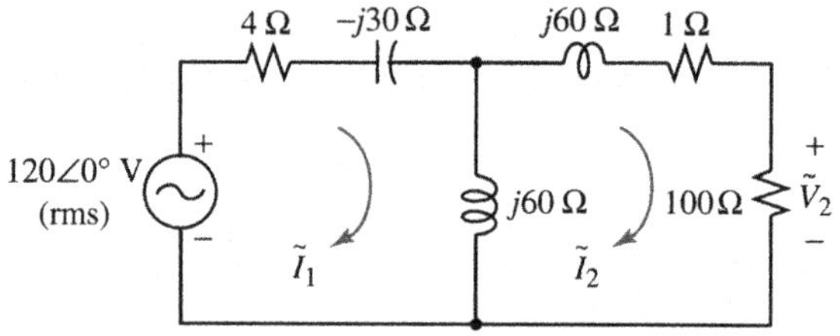

Figure 7.26

The simultaneous solution of these equations yields

$$\tilde{I}_1 = 5.33\underline{/-33.52^\circ}\ \text{A}$$

and

$$\tilde{I}_2 = 2.04\underline{/6.57^\circ}\ \text{A}$$

Thus, the load voltage, power output, and power input are

$$\tilde{V}_2 = 100 \times 2.04\underline{/6.57^\circ} = 204\underline{/6.57^\circ}\ \text{V}$$
$$P_o = \text{Re}[204\underline{/6.57^\circ} \times 2.04\underline{/-6.57^\circ}] = 416.16\,\text{W}$$
$$P_i = \text{Re}[120 \times 5.33\underline{/33.52^\circ}] = 533.23\,\text{W}$$

Finally, the efficiency of the transformer is

$$\eta = \frac{P_o}{P_i} = \frac{416.16}{533.23} = 0.78 \quad \text{or} \quad 78\%$$

• • •

7.14.2 The autotransformer

In the two-winding transformer we have considered thus far the primary winding is electrically insulated from the secondary winding. The two windings are coupled together magnetically by a common core. Thus, it is magnetic induction that is responsible for the energy transfer from the primary to the secondary winding.

When the two windings of a transformer are also interconnected electrically, it is called an *autotransformer*. An autotransformer may have a single continuous winding serving as both primary and secondary winding, or it can consist of two or more distinct coils wound on the same magnetic core. In either case, the principle of operation is the same. The direct electrical connection between the windings ensures that a part of the energy is transferred from the primary to the secondary winding by *conduction*. The magnetic coupling between the windings guarantees that some of the energy is also delivered by *induction*.

Autotransformers may be used for almost all applications where we would employ a two-winding transformer. The only disadvantage is the loss of electrical isolation between the high- and low-voltage sides of the autotransformer. Some of the advantages of an autotransformer over a two-winding transformer are as follows.

1. An autotransformer is less expensive than a conventional two-winding transformer of similar voltage and power ratings.
2. An autotransformer delivers more power than a two-winding transformer of similar physical dimensions.
3. For a similar power rating, an autotransformer is more efficient than a two-winding transformer.
4. An autotransformer requires a lower excitation current to establish the same flux in the core than a two-winding transformer.

We begin our discussion of an autotransformer by connecting an ideal two-winding transformer as an autotransformer. In fact, there are four possible ways to do this, as shown in Figure 7.27. Let us consider the circuit shown in Figure 7.27a, where the two-winding transformer is connected as a step-down autotransformer. Note that the secondary winding of the two-winding transformer is now the common winding of the autotransformer. Under ideal conditions,

$$\begin{aligned}
\tilde{V}_{1a} &= \tilde{E}_{1a} = \tilde{E}_1 + \tilde{E}_2 \\
\tilde{V}_{2a} &= \tilde{E}_{2a} = \tilde{E}_2 \\
\frac{\tilde{V}_{1a}}{\tilde{V}_{2a}} &= \frac{\tilde{E}_{1a}}{\tilde{E}_{2a}} = \frac{\tilde{E}_1 + \tilde{E}_2}{\tilde{E}_2} = \frac{N_1 + N_2}{N_2} = 1 + a = a_T
\end{aligned} \tag{7.142}$$

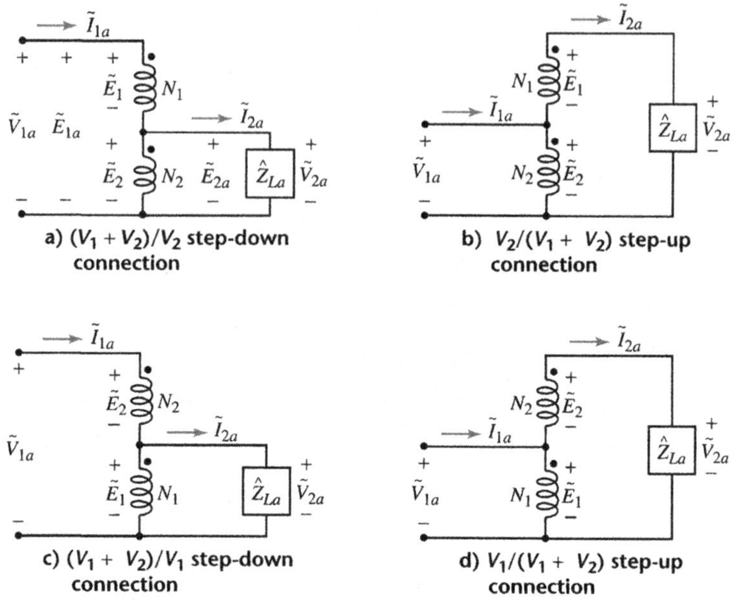

Figure 7.27 Possible ways to connect a two-winding transformer as an autotransformer

where $a = N_1/N_2$ is the a-ratio of a two-winding transformer, and $a_T = 1 + a$ is the a-ratio of the autotransformer under consideration. The a-ratio for the other connections should also be computed in the same way because a_T is not the same for all connections.

In an ideal autotransformer the primary mmf must be equal and opposite to the secondary mmf:

$$(N_1 + N_2)I_{1a} = N_2 I_{2a}$$

From this equation, we obtain

$$\frac{I_{2a}}{I_{1a}} = \frac{N_1 + N_2}{N_2} = 1 + a = a_T \tag{7.143}$$

Thus, the apparent power supplied by an ideal transformer to the load, S_{oa}, is

$$\begin{aligned} S_{oa} &= V_{2a} I_{2a} \\ &= \left[\frac{V_{1a}}{a_T}\right][a_T I_{1a}] \\ &= V_{1a} I_{1a} \\ &= S_{\text{ina}} \end{aligned} \tag{7.144}$$

where S_{ina} is the apparent power input to the autotransformer. This equation simply highlights the fact that the power input is equal to the power output under ideal conditions.

Let us now express the apparent output power in terms of the parameters of a two-winding transformer. For the configuration under consideration,

$$V_{2a} = V_2$$

and

$$I_{2a} = a_T I_{1a} = (a + 1)I_{1a}$$

However, for the rated load, $I_{1a} = I_1$. Thus,

$$\begin{aligned} S_{oa} &= V_2 I_1 (a + 1) \\ &= V_2 I_2 \frac{a + 1}{a} = S_o \left[1 + \frac{1}{a} \right] \end{aligned}$$

where $S_o = V_2 I_2$ is the apparent power output of a two-winding transformer. This power is associated with the common winding of the autotransformer and is, therefore, the power transferred to the load by induction. The rest of the power, S_o/a in this case, is conducted directly from the source to the load and is called the *conduction power*. Hence, a two-winding transformer delivers more power when connected as an autotransformer.

EXAMPLE 7.18

A 24-kVA, 2400/240-V, distribution transformer is to be connected as an autotransformer. For each possible combination, determine (a) the primary winding voltage, (b) the secondary winding voltage, (c) the ratio of transformation, and (d) the nominal rating of the autotransformer.

Solution From the given information for the two-winding transformer, we conclude that $V_1 = 2400$ V, $V_2 = 240$ V, $S_o = 24$ kVA, $I_1 = 10$ A, and $I_2 = 100$ A.

a) For the autotransformer connection shown in Figure 7.27a,

$$\begin{aligned} V_{1a} &= 2400 + 240 = 2640\text{ V} \\ V_{2a} &= 240\text{ V} \\ a_T &= 2640/240 = 11 \\ S_{oa} &= V_{2a} I_{2a} = V_{1a} I_{1a} = V_{1a} I_1 \\ &= 2640 \times 10 = 26{,}400\text{ VA or } 26.4\text{ kVA} \end{aligned}$$

The nominal rating of the autotransformer: 26.4-kVA, 2640/240-V.

b) For the autotransformer connection shown in Figure 7.27b,

$$\begin{aligned} V_{1a} &= 240\text{ V} \\ V_{2a} &= 2400 + 240 = 2640\text{ V} \\ a_T &= 240/2640 = 0.091 \\ S_{oa} &= V_{2a} I_{2a} = V_{2a} I_1 \\ &= 2640 \times 10 = 26{,}400\text{ VA or } 26.4\text{ kVA} \end{aligned}$$

The nominal rating of the autotransformer: 26.4-kVA, 240/2640-V

c) For the autotransformer connection shown in Figure 7.27c,

$$\begin{aligned} V_{1a} &= 240 + 2400 = 2640\,\text{V} \\ V_{2a} &= 2400\,\text{V} \\ a_T &= 2640/240 = 1.1 \\ S_{oa} &= V_{2a}I_{2a} = V_{1a}I_{1a} = V_{1a}I_2 \\ &= 2640 \times 100 = 264{,}000\,\text{VA or } 264\,\text{kVA} \end{aligned}$$

The nominal rating of the autotransformer: 264-kVA, 2640/2400-V

d) Finally, for the autotransformer connection shown in Figure 7.27d,

$$\begin{aligned} V_{1a} &= 2400\,\text{V} \\ V_{2a} &= 2400 + 240 = 2640\,\text{V} \\ a_T &= 2400/2640 = 0.91 \\ S_{oa} &= V_{2a}I_{2a} = V_{2a}I_2 \\ &= 2640 \times 100 = 264{,}000\,\text{VA or } 264\,\text{kVA} \end{aligned}$$

The nominal power rating of the autotransformer: 264-kVA, 2400/2640-V

Note the 10-fold increase in the power ratings when the two-winding transformer is connected either as a step-down autotransformer in Figure 7.27c or a step-up autotransformer in Figure 7.27d. • • •

7.14.3 The betatron

In a cyclotron, a charged particle is accelerated by applying a time-varying electric field in the region between the two conducting D-shaped cavities. A uniform magnetic field is used to force the charged particle to follow a circular orbit within each D-shaped region. The radius of the charged particle increases each time it enters the cavity. In a *betatron*, the charged particle revolves in an evacuated glass chamber called a **torus**, at a constant radius. An electromagnet creates a time-varying magnetic field, and the gap between the pole faces of the electromagnet increases radially outward to control the strength of the magnetic field as shown in Figure 7.28.

Let us assume that a charged particle (an electron) is at rest, and the magnetic field is zero. As the magnetic field increases in the z direction, it induces an electric field, Figure 7.29, which forms closed circular loops in the plane of the torus. The electric field intensity, from Maxwell's equation, is

$$\oint_c \vec{\mathbf{E}} \cdot \vec{d\ell} = -\int_s \frac{\partial \vec{\mathbf{B}}}{\partial t} \cdot \vec{ds}$$

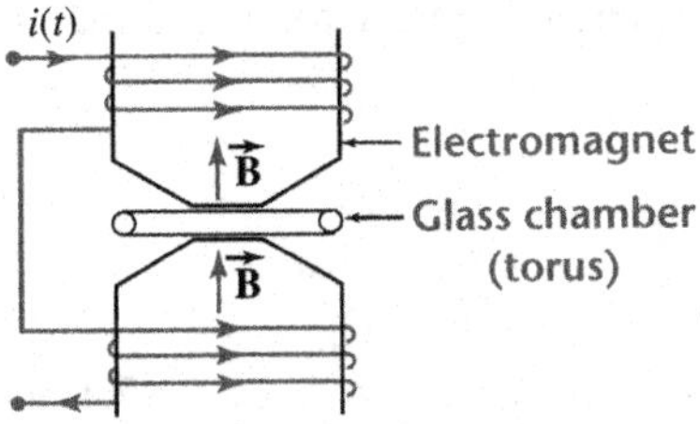

Figure 7.28 Schematic of a betatron

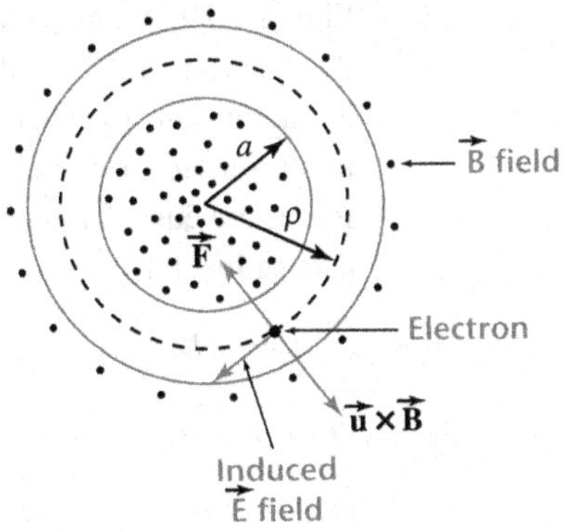

Figure 7.29 Forces experienced by an electron revolving with a velocity u at radius ρ

where $\vec{\mathbf{B}}(r, t)$, the magnetic flux density, is a function of space and time. The symmetric design of the electromagnet ensures that the magnitude of the $\vec{\mathbf{B}}$ field is the same at a constant radius from the center. Thus, at the same radius, the strength of the $\vec{\mathbf{E}}$ field is also constant. Thus, for a loop of constant radius a, the preceding equation yields

$$E_\phi = -\frac{1}{2\pi a}\frac{d\Phi}{dt} \tag{7.145}$$

where

$$\Phi = \int_0^a B\rho\, d\rho \int_0^{2\pi} d\phi \tag{7.146}$$

is the total flux passing through the surface bounded by the circular loop of radius a.

The force exerted by the $\vec{\mathbf{E}}$ field on the electron is

$$F_\phi = -eE_\phi = \frac{e}{2\pi a}\frac{d\Phi}{dt} \tag{7.147}$$

where e is the magnitude of the charge on the electron (1.602×10^{-19} C). According to Newton's second law of motion, the rate of change of momentum is equal to the impressed force. That is

$$\frac{dp}{dt} = \frac{e}{2\pi a}\frac{d\Phi}{dt}$$

The electron is at rest at $t = 0$; therefore, the gain in momentum at any time t is

$$p = \frac{e\Phi}{2\pi a} \tag{7.148}$$

As soon as the electron starts revolving in the circular path at a radial distance a from the center, it experiences the Lorentz force, $-e(\vec{\mathbf{u}} \times \vec{\mathbf{B}})$. This force tends to move the electron toward its center as indicated in

Figure 7.29. At the same time, however, the centripetal force acting on the electron tends to make it move to escape the device. As the two forces are acting on the electron at the same time but in opposite directions, it is possible for the electron to maintain a circular orbit when the two forces are equal in magnitude. For the circular orbit at a constant radius a, we must have

$$\frac{m}{a} u^2 = eBu$$

or

$$mu = eBa \tag{7.149}$$

Here m is the mass of the electron, and it may be many times its rest mass (9.1×10^{-31} kg) because the velocity attained by the electron is comparable with the velocity of light. Thus, we must treat m as a variable.

Because $p = mu$ is the momentum of the electron, equating (7.148) and (7.149) we get

$$B = \frac{\Phi}{2\pi a^2} \tag{7.150}$$

When we define the space average flux density (over the surface area bounded by the orbit) as

$$B_0 = \frac{\Phi}{\pi a^2} \tag{7.151}$$

and compare it with (7.150), we find that the magnetic flux density at radius a must be exactly equal to one-half of its average value:

$$B = \tfrac{1}{2} B_0 \tag{7.152}$$

For this reason, the poles of the electromagnet are tapered to create a $\vec{\mathbf{B}}$ field that decreases in the outward radial direction.

The first betatron was constructed in 1940 by D. W. Kerst at the University of Illinois, but the idea was suggested as early as 1928 by R. Wideroe. Since then, betatrons that can accelerate electrons and impart energies in excess of 400 MeV have been successfully built.

7.15 Summary

The motional emf in a conductor moving in a magnetic field is given by

$$e_m = \int_c (\vec{\mathbf{u}} \times \vec{\mathbf{B}}) \cdot \vec{d\ell}$$

where $\vec{\mathbf{u}}$ is the velocity of the conductor and $\vec{\mathbf{B}}$ is the magnetic flux density. The induced current in a closed conductor due to the motional emf is in the direction of the induced electric field $(\vec{\mathbf{u}} \times \vec{\mathbf{B}})$.

A time-varying flux passing through the area enclosed by an N-turn coil fixed in space produces a transformer emf, which is given as

$$e_t = -\int_s \frac{\partial \vec{\mathbf{B}}}{\partial t} \cdot \vec{ds}$$

For time-harmonic variations in field quantities, the effective (rms) value of the transformer emf is

$$E = 4.44\, f N \Phi_m$$

where f is the frequency of oscillation and Φ_m is the amplitude (maximum) value of the flux. When a closed conductor (loop) is moving in a time-varying magnetic field, the total induced emf is

$$e = e_m + e_t = -\frac{d\Phi}{dt}$$

where Φ is the total flux passing through surface s inside a closed path c. This equation is a mathematical statement of Faraday's law of induction. When expressed in terms of $\vec{\mathbf{E}}$ and $\vec{\mathbf{B}}$ fields, it gives the most general form of one of the four Maxwell equations as

$$\nabla \times \vec{\mathbf{E}} = -\frac{\partial \vec{\mathbf{B}}}{\partial t} + \nabla \times (\vec{\mathbf{u}} \times \vec{\mathbf{B}})$$

We defined the self-inductance of a coil as

$$L = N\frac{d\Phi}{di} = \frac{N^2}{\mathscr{R}} = \mathscr{P}N^2$$

where $i(t)$ is the current in the coil, $\mathscr{R}$ is the reluctance, and $\mathscr{P}$ is the permeance of the magnetic circuit. The mutual inductance between two coils is

$$M = k\frac{N_1 N_2}{\mathscr{R}} = k\sqrt{L_1 L_2}$$

where L_1 and L_2 are the self-inductances of coils 1 and 2, respectively. The effective inductance of two magnetically coupled coils connected in series is

$$L = L_1 + L_2 \pm 2M$$

where the plus sign is used for series aiding and the minus sign is used for series opposing. The effective inductance of two magnetically coupled coils connected in parallel is

$$L = \frac{L_1 L_2 - M^2}{L_1 + L_2 \pm 2M}$$

where the minus sign is for parallel aiding and the plus sign is for parallel opposing. Maxwell's equation from the modified Ampère law is

$$\nabla \times \vec{\mathbf{H}} = \vec{\mathbf{J}} + \frac{\partial \vec{\mathbf{D}}}{\partial t}$$

where $\vec{\mathbf{J}}$ represents (a) the volume current density due to the sources in a region, (b) the conduction current density in a conducting medium ($\vec{\mathbf{J}} = \sigma\vec{\mathbf{E}}$), or (c) the convection current density due to drifting charges ($\vec{\mathbf{J}} = \rho_v\vec{\mathbf{u}}$). The other two Maxwell equations are

$$\nabla \cdot \vec{\mathbf{D}} = \rho_v$$
$$\nabla \cdot \vec{\mathbf{B}} = 0$$

Fields can exist in a medium if and only if they satisfy the four Maxwell equations.

The equation of continuity satisfied by Maxwell's equations is given as

$$\nabla \cdot \vec{\mathbf{J}} = -\frac{\partial \rho_v}{\partial t}$$

The force experienced by a charge q moving with a velocity $\vec{\mathbf{u}}$ in a region where there exist time-varying fields is given as

$$\vec{\mathbf{F}} = q(\vec{\mathbf{E}} + \vec{\mathbf{u}} \times \vec{\mathbf{B}})$$

The instantaneous power density, or Poynting's vector, is

$$\vec{\mathbf{S}} = \vec{\mathbf{E}} \times \vec{\mathbf{H}}$$

When the fields vary sinusoidally, we can compute the average power per unit area as

$$\langle\hat{\mathbf{S}}\rangle = \tfrac{1}{2}\mathrm{Re}[\tilde{\vec{\mathbf{E}}} \times \tilde{\vec{\mathbf{H}}}^*]$$

where $\tilde{\vec{\mathbf{E}}}$ and $\tilde{\vec{\mathbf{H}}}$ are the phasor representations of time-harmonic $\vec{\mathbf{E}}$ and $\vec{\mathbf{H}}$ fields in terms of their maximum values.

The average power flow through a surface s is

$$\langle P\rangle = \int_s \langle\hat{\mathbf{S}}\rangle \cdot \vec{d\mathbf{s}}$$

The voltage and current relationships in a two-winding transformer are

$$\frac{e_1}{e_2} = \frac{i_2}{i_1} = \frac{N_1}{N_2} = a$$

A two-winding transformer can be connected in four different ways as an autotransformer. In each case, the autotransformer has a higher power rating because part of the power is now being delivered by conduction. The time-varying magnetic field is used to accelerate a charged particle in a device called a betatron. The electromagnet is tapered in such a way that the maximum value of the $\vec{\mathbf{B}}$ field at the stable orbit of the electron is one-half the space average flux density.

7.16 Review questions

7.1 Derive an expression for the motional emf and state all the assumptions.

7.2 Explain transformer emf.

7.3 State Faraday's law of induction.

7.4 What is Lenz's law? What is its significance?

7.5 Explain the following terms: series aiding, series opposing, parallel aiding, and parallel opposing.

7.6 Determine the effective inductance when three magnetically coupled coils are connected in (a) series aiding, (b) series opposing, (c) parallel aiding, and (d) parallel opposing.

7.7 Why is the induced electric field not a conservative field?

7.8 What is meant by a perfectly coupled coil? What is its coefficient of coupling?

7.9 What is the significance of polarity marking?

7.10 Derive the expression for magnetic energy density.

7.11 Why was it necessary to modify Ampère's law for time-varying fields?

7.12 Write Maxwell's equations in (a) point form and (b) integral form. Explain the significance of each equation.

7.13 Can the equation of continuity be derived from Maxwell's equations? If yes, derive it. If not, why not?

7.14 Derive boundary conditions for time-varying fields.

7.15 Is it necessary for the fields to satisfy Maxwell's equations in order to exist?

7.16 What is $\vec{\mathbf{J}}$ in a source-free dielectric medium?

7.17 What is $\vec{\mathbf{J}}$ in a source-free conductive medium?

7.18 Can a surface current exist at the interface between a dielectric and a conducting medium? Explain.

7.19 Can a surface charge exist at the interface between a dielectric and a conducting medium? Explain.

7.20 Can fields exist in a conducting medium? Justify your answer.

7.21 Can fields exist in a perfectly conducting medium? Justify your answer.

7.22 State the boundary conditions at the interface between a dielectric and a perfect conductor.

7.23 Obtain expressions for the average energy densities for time-harmonic fields.

7.24 State Poynting's theorem. What is Poynting's vector?

7.25 What is average power density?

7.26 What is an ideal transformer?

7.27 The primary and secondary windings of a 120-VA, two-winding transformer are rated at 120 V and 60 V, respectively. What is the current rating of each winding?

7.28 A 240-VA, 240/12-V, two-winding transformer is connected as an autotransformer. Determine the voltage, current, and power rating for each possible connection.

7.29 Explain the operation of a betatron.

7.30 How does a time-varying magnetic field accelerate a charged particle?

7.31 The electric field intensity in a medium is given as $E_0 \cos \omega t \cos \beta z \vec{\mathbf{a}}_x$ (V/m), where ω is the frequency and β is a constant. Can this field exist in free space?

7.17 Exercises

7.1 A 2-meter-long copper strip is rotating in a plane perpendicular to the uniform magnetic flux density of 12.5 mT. Determine the induced emf between the two ends of the strip when it is pivoted at one end and the other end is rotating with an angular velocity of 188 rad/s. When a 2-Ω resistance is connected as shown in Figure 7.30a, what is the current in the strip? What is the power supplied by the strip? Compute the magnetic force acting on the strip. What is the significance of this force?

7.2 If the copper strip in Exercise 7.1 is pivoted in the middle as shown in Figure 7.30b, determine (a) the induced emf between the midpoint and one of its freely rotating ends, (b) the current in each section of the strip, and (c) the total power supplied by the strip.

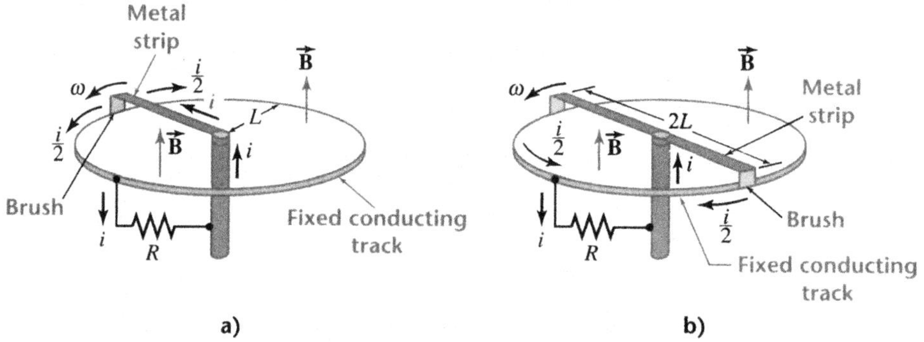

Figure 7.30 Copper strips rotating in a magnetic field

7.3 A rectangular loop of dimensions 20 cm × 10 cm is made of aluminum wire of radius 1.2 mm. The loop is placed in a magnetic field that is increasing at the rate of 40 T/s. What is the induced current in the loop? The conductivity of aluminum is 3.57×10^7 S/m. Draw a sketch and show the direction of the induced current.

7.4 Using Faraday's law, determine the induced emf in the copper strip of Exercise 7.1.

7.5 A circular conducting loop of radius 10 cm is located in a region in which the magnetic field intensity with a peak value of 10 A/m is changing sinusoidally with time at a frequency of 200 kHz. If a voltmeter is connected in series with the loop, and the plane of the loop is normal to the magnetic field intensity, determine the reading on the voltmeter using (a) Faraday's law of induction and (b) the transformer equation.

7.6 An N-turn rectangular coil is situated midway between a parallel-wire transmission line, as shown in Figure 7.31. What is the induced emf in the coil?

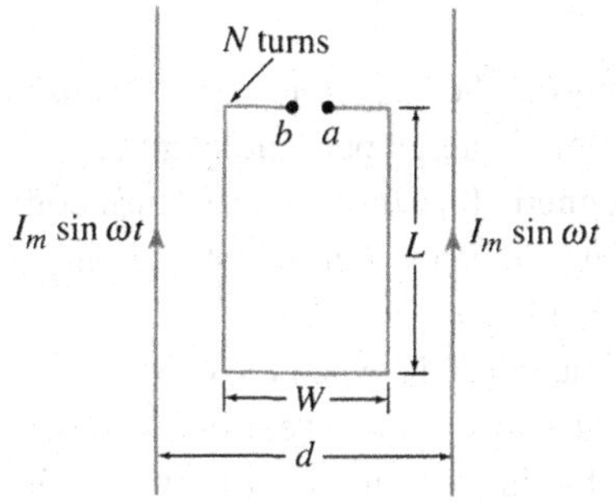

Figure 7.31 An N-turn coil situated midway between a parallel-wire transmission line

7.7 Compute the self-inductance of a toroid of square cross section if it is wound with 200 turns. The inner and the outer radii of the toroid are 20 cm and 25 cm, respectively. The relative permeability of the magnetic material is 500. If the current in the coil varies sinusoidally as $2 \sin 314t$ A, determine the induced emf in the coil.

7.8 Compute the self-inductance per unit length of a two-wire transmission line in free space if the radius of each conductor is a and their centers are length d apart. Assume the conductors are perfectly conducting and carry equal currents in opposite directions.

7.9 Calculate the mutual inductance between a very long, current-carrying conductor and a square loop of side a. The minimum separation between the conductor and the loop is b, as shown in Figure 7.32.

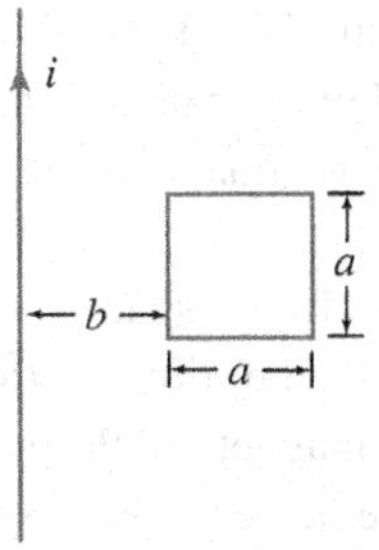

Figure 7.32

7.10 Two coils are coupled in such a way that the mutual inductance between them is 16 mH. If the inductance of one coil is 20 mH and that of the other coil is 80 mH, determine the coefficient of coupling between the coils.

7.11 Verify equation (7.48).

7.12 The self-inductances of two coils are 1.6 H and 4.9 H. Sketch the effective inductance of the two coils as a function of the coefficient of coupling when they are connected (a) in series aiding, (b) in series opposing, (c) in parallel aiding, and (d) in parallel opposing.

7.13 The volume current density inside a cylindrical wire of radius a and infinite length is given as

$$\begin{aligned}\vec{\mathbf{J}} &= J_0\vec{\mathbf{a}}_z \text{ A/m}^2 \quad \rho \le a \\ &= 0 \qquad\qquad\quad\;\; \rho \ge a\end{aligned}$$

Determine (a) the magnetic energy density at any point and (b) the internal stored energy per unit length.

7.14 If the magnetic flux linking an N-turn coil varies as $\Phi = a\sqrt{i}$, where a is a constant, show that the stored energy is $\frac{1}{3}N\Phi i$. Is the medium linear or nonlinear?

7.15 Repeat Example 7.12 when the coils are connected in series opposing.

7.16 Show that the displacement current density in a conductor such as copper is negligible in comparison with the conduction current density at all practical frequencies. Assume that the conduction current density in the conductor is $J_0 \cos \omega t \vec{\mathbf{a}}_z$ A/m^2, $\epsilon_{\text{Cu}} = \epsilon_0$, and $\sigma_{\text{Cu}} = 5.8 \times 10^7$ S/m.

7.17 In a source-free, dielectric medium the electric field intensity is given as $\vec{\mathbf{E}} = C \cos \alpha x \cos(\omega t - \beta z)\vec{\mathbf{a}}_y$ V/m, where C is the amplitude and α and β are constant quantities. Determine (a) the magnetic field intensity and (b) the electric flux density.

7.18 Using Maxwell's equations for time-varying fields, derive the boundary conditions as given in equations (7.84)–(7.89).

7.19 State the boundary conditions when (a) medium 1 is a perfect dielectric and medium 2 is a perfect conductor, (b) both media are perfect dielectrics, and (c) medium 1 is a perfect dielectric and medium 2 is a conductor.

7.20 Can the electric field given in Exercise 7.17 exist? If yes, what must be the condition? If not, why not?

7.21 A solid conductor of radius b and length L is carrying a uniformly distributed direct current I in the z direction. Show that the total power flowing into the conductor is equal to I^2R, where R is the conductor's resistance.

7.22 An electric field in free space is described by $\vec{\mathbf{E}} = 10\cos(\omega t + ky)\vec{\mathbf{a}}_x$ V/m. If the time period is 100 ns, determine (a) the constant k, (b) the magnetic field intensity, (c) the direction of power flow, (d) the average power density, (e) the energy density in the electric field, and (f) the energy density in the magnetic field.

7.23 Verify the results of Example 7.15 using phasors.

7.24 Repeat Exercise 7.22 using phasors.

7.25 The primary winding of a 10-kVA, two-winding transformer is rated at 500 V. If the a-ratio is 2 and the transformer is operating at 80% of its rated load (8 kVA) with 0.8 pf leading when the rated voltage is applied to the primary, determine (a) the primary winding current, (b) the secondary winding voltage and current.

7.26 A 100-kVA, two-winding transformer has $R_1 = 16\,\Omega$, $R_2 = 4\,\Omega$, $L_1 = 80$ mH, $N_1 = 500$ turns, and $f = 60$ Hz. The a-ratio is 2. When the transformer delivers a current of 40 A at 0.707 lagging pf, and the load voltage is 2500 V (rms), determine (a) N_2 and L_2, (b) the reluctance of the magnetic core, (c) the power output, (d) the power input, and (e) the efficiency of the transformer.

7.27 A 720-VA, 360/120-V, two-winding transformer is to be connected as an autotransformer. For each possible combination, determine (a) the primary winding voltage, (b) the secondary winding voltage, (c) the ratio of transformation, and (d) the nominal rating of the autotransformer.

7.28 The resistances of 360-V and 120-V windings of the two-winding transformer given in Exercise 7.27 are 4.5 Ω and 0.5 Ω, respectively. If it is connected as an autotransformer and delivers the rated load at the rated voltage and 0.8 pf lagging, determine the efficiency for each connection.

7.29 The time-harmonic current through the windings of an electromagnet produces a magnetic field $\vec{\mathbf{B}} = B_m \sin \omega t \vec{\mathbf{a}}_z$, where B_m is the maximum value at the stable orbit of the electron. Determine (a) the space average magnetic flux density, (b) the total flux enclosed when the radius of the orbit is a, (c) the induced electric field, and (d) the kinetic energy acquired by the electron in one trip around the orbit.

7.30 Repeat Exercise 7.29 when $B_m = 0.4$ T, $a = 84$ cm, and the frequency of the alternating current is 60 Hz. What is the average kinetic energy gained by the electron in one revolution? Express the energy in electron volts (eV).

7.18 Problems

7.1 Two conducting bars are sliding over two stationary conductors, as shown in Figure P7.1. What is the induced current in the closed loop thus formed when its resistance is 12 Ω?

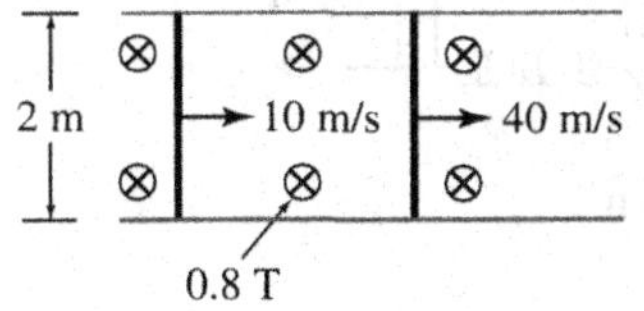

Figure P7.1

7.2 A square conducting loop of side 25 cm has a resistance of 12 Ω and lies in the yz plane. A uniform magnetic flux density of $0.8\vec{\mathbf{a}}_x$ T exists in a region bounded by $0 \le y \le 150$ cm and $0 \le z \le 12$ cm. The four corners of the loop at $t = 0$ are at (0, 0, 0) m, (0, 0.25, 0) m, (0, 0, 0.25) m, and (0, 0.25, 0.25) m. When the loop is moved through this region with a velocity of 100 m/s in the y direction, sketch the flux linking the loop and the current induced in it as a function of time.

7.3 A copper disc 20 cm in radius is rotating at a speed of 1200 revolutions per minute (rpm) about its axis in a uniform magnetic field of 250 mT. If the magnetic field makes an angle of 30° with the axis of the disc, determine the induced emf between the rim and the axis of the disc.

7.4 An aluminum disc 6 cm in diameter is located at the center of a very long solenoid with the disc axis coincident with the magnetic axis of the solenoid. The solenoid has 50,000 turns per unit length and carries a current of 12 A. If the disc rotates at a speed of 3600 rpm, what is the induced emf between the axis of the disc and its rim?

7.5 The magnetic flux density in a region varies as $\vec{\mathbf{B}} = 2.5 \sin 300t\, \vec{\mathbf{a}}_x + 1.75 \cos 300t\, \vec{\mathbf{a}}_y + 0.5 \cos 500t\, \vec{\mathbf{a}}_z$ mT. A closed conducting rectangular loop has its corners at (0, 0, 0), (3, 4, 0), (3, 4, 4), and (0, 0, 4), where all distances are in meters. Determine (a) the flux linking the loop and (b) the induced current in the loop if its resistance is 2 Ω.

7.6 A tightly and closely wound rectangular coil having 200 turns is rotating at 120 rad/s in a uniform magnetic field of 0.8 T. The axis of rotation of the coil is at right angles to the direction of the field, and its cross-sectional area is 40 cm^2. Calculate the induced emf in the coil.

7.7 Determine the induced emf in the coil of Problem 7.6 if the flux density has an amplitude of 0.8 T and it pulsates sinusoidally with an angular frequency of 120 rad/s.

7.8 A conductor of length ℓ moving with a velocity $\vec{\mathbf{u}} = u \cos \omega t\, \vec{\mathbf{a}}_y$ m/s is connected with flexible leads to a voltmeter, as shown in Figure P7.8. If the magnetic flux density in the region is $\vec{\mathbf{B}} = B \cos \omega t \vec{\mathbf{a}}_x$ T, determine the induced emf in the circuit using (a) the concepts of transformer and motional emfs and (b) Faraday's law of induction. Assume that the conductor is at $y = 0$ when $t = 0$.

Figure P7.8

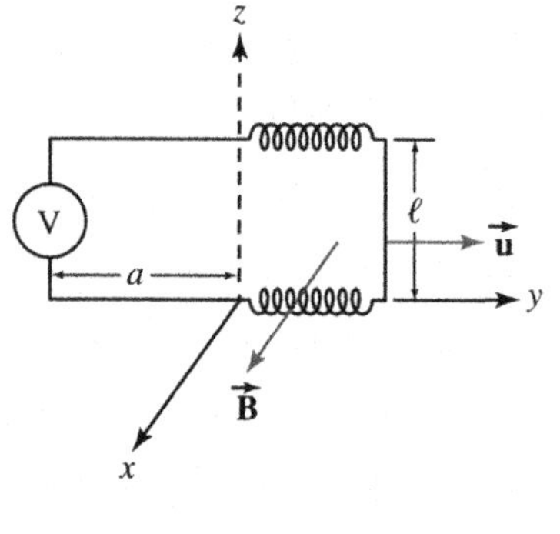

7.9 A rectangular metal strip of width 10 cm is moving parallel to the xy plane with a constant velocity of $\vec{\mathbf{u}} = -1000\vec{\mathbf{a}}_y$ m/s, as shown in Figure P7.9. If a magnetic flux density of $\vec{\mathbf{B}} = 0.2\vec{\mathbf{a}}_z$ T exists in the region, determine the reading on the voltmeter. Show the polarity of the induced voltage.

Figure P7.9

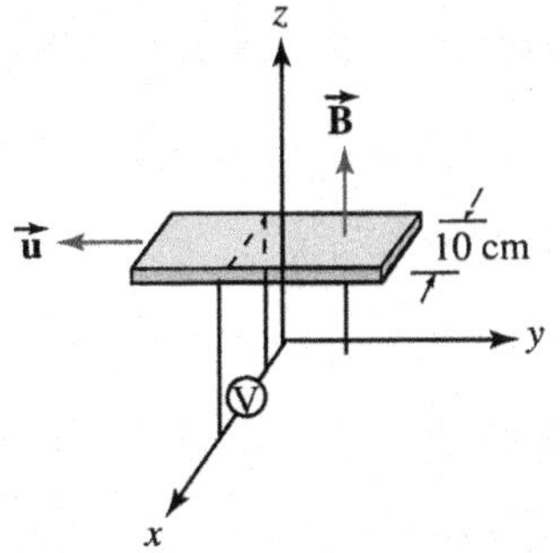

7.10 The inner and the outer radii of a magnetic core with a square cross section are 10 cm and 12 cm, respectively. The core is tightly wound with a 1200-turn coil. The magnetization characteristic of the magnetic material is given in Figure 5.37. When the coil carries a current of 0.75 A, determine the inductance of the system. Assume the flux density is uniformly distributed in the core.

7.11 A 1000-turn closely and tightly wound toroid has an inductance of 20 mH when the current in the coil is 2.5 A. What is the magnetic flux in the toroid?

7.12 The inner conductor of a coaxial cable is solid and has a radius of 2 mm. The outer conductor is very thin and has a radius of 4 mm. Determine the inductance per unit length of the cable if the current is uniformly distributed (a) over the surface of each conductor and (b) within the inner conductor and the inner surface of the outer conductor.

7.13 Three coils with 100, 150, and 200 turns are tightly and closely wound on a common magnetic circuit of 40 cm^2 in cross section and 80 cm in length. The 100-turn coil is connected to a current source of $10 \sin 800\pi t$ A. Determine (a) the self-inductance of each coil, (b) the mutual inductance between any two coils, and (c) the induced emf in each coil. Assume the relative permeability of the magnetic material as 500.

7.14 The effective inductances when two coupled coils are connected in series aiding and series opposing are 3.28 mH and 0.72 mH, respectively. If the self-inductance of one coil is four times the self-inductance of the other, determine (a) the self-inductance of each coil, (b) the mutual inductance, and (c) the coefficient of coupling.

7.15 Two coaxial loops of radii a and b, where $a \gg b$, are separated by a distance d. If the flux density is assumed to be the same at any point in the plane of the smaller loop, determine the mutual inductance.

7.16 Two concentric coplanar circular loops have radii a and b, where $a \gg b$. If the flux density is assumed to be the same in the plane bounded by the smaller loop, determine the mutual inductance.

7.17 Calculate the mutual inductance between the infinite straight conductor and the closed loop as shown in Figure P7.17.

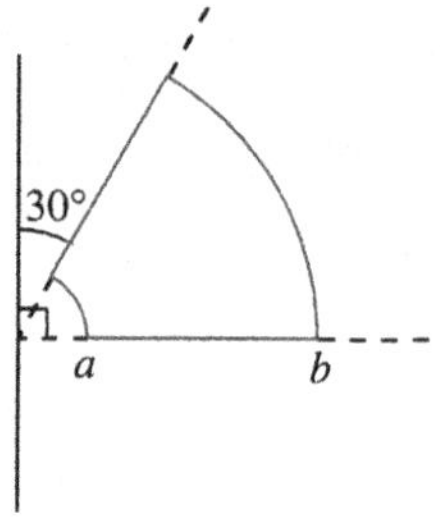

Figure P7.17

7.18 A very long, air-core solenoid 2 cm in diameter has two coils wound over each other. The inner coil has 400 turns/m, and the outer coil has 4000 turns/m. What is the mutual inductance of the two coils? When the current in the inner coil is 0.5 cos 200t A, what is the induced emf per unit length in the other coil?

7.19 The relationship between the current in an N-turn coil and the flux created by it is given as $\Phi = ai^n$, where a and n are constants. Determine the energy stored in the coil when the current varies from 0 to I.

7.20 The relationship between the current in an N-turn coil and the flux created by it is given as $\Phi = a \ln(bi)$, where a and b are constants. Determine the energy stored in the coil when the current varies from 0 to I.

7.21 The relationship between the current in an N-turn coil and the flux created by it is given as $\Phi = ai/(b + ci)$, where a, b, and c are constants. Determine the energy stored in the coil when the current varies from 0 to I.

7.22 A coil with an inductance of 30 mH and a resistance of 10 Ω is connected to a 200-V (dc) supply. What is the magnetic energy stored in the magnetic field under steady-state conditions? If the coil is wound over a nonmagnetic material of length 20 cm and diameter 5 cm, calculate the energy density and the magnetic flux density in the region. Assume uniform flux distribution.

7.23 Two hollow concentric cylinders having negligible wall thicknesses are used as a coaxial transmission line. The radius of the inner cylinder is 5 cm, and that of the outer cylinder is 10 cm. When the line carries a current of 1000 A, determine (a) the energy density and (b) the total energy stored in the system.

7.24 The magnetic flux density is typically 0.04 mT near the surface of the earth. What is the magnetic energy density? If the radius of the earth is approximated as 6400 km, and the magnetic flux density is assumed constant up to an altitude equal to the earth's radius, what is the total magnetic energy stored in the region above the earth's surface?

7.25 An inductive coil has a resistance of 0.5 Ω and an inductance of 2 H. It is required to store a magnetic energy of 6.4 kJ at all times. How much power is required to maintain such an energy storage?

7.26 A 500-turn toroid of square cross section has an inner radius of 10 cm and an outer radius of 15 cm. The relative permeability of the core is 1000. For a current of 10 A in the coil, calculate (a) the energy density, (b) the stored energy, and (c) the inductance of the toroid.

7.27 Starting with the equation of continuity, and assuming Ohm's law, show that the charge density in a conductor is given by the following first-order differential equation:

$$\frac{\partial \rho_v}{\partial t} + \frac{\sigma}{\epsilon}\rho_v = 0$$

where σ and ϵ are, respectively, the conductivity and the permittivity of the medium. Assume a linear, homogeneous, and isotropic medium.

7.28 Starting with Maxwell's equation from Faraday's law and the definition of vector magnetic potential $\vec{\mathbf{A}}$, show that the line integral of $(\vec{\mathbf{E}} + \partial\vec{\mathbf{A}}/\partial t)$ around a closed path is zero.

7.29 If the electric field intensity in a source-free, dielectric medium is given as $\vec{\mathbf{E}} = E_0[\sin(\alpha x - \omega t) + \sin(\alpha x + \omega t)]\,\vec{\mathbf{a}}_y$ V/m, determine the magnetic field intensity using Maxwell's equation from Faraday's law. What is the displacement current density in the medium?

7.30 If the magnetic field intensity in a source-free, dielectric medium is given as $\vec{\mathbf{H}} = H_0[\cos(\alpha x - \omega t) + \cos(\alpha x + \omega t)]\,\vec{\mathbf{a}}_z$ A/m, determine the electric field intensity using Maxwell's equation from Ampère's law. What is the displacement current density in the medium?

7.31 Determine the condition that must be satisfied for the existence of the electric field intensity as given in Problem 7.29.

7.32 Can the magnetic field intensity as given in Problem 7.30 exist? If yes, what must be the condition? If no, explain why the fields cannot exist.

7.33 If $\vec{\mathbf{E}} = E_0 \cos(\omega t - \beta z)\,\vec{\mathbf{a}}_x$ V/m in a dielectric medium, show that the electric energy density is equal to the magnetic energy density. Also compute (a) the Poynting vector, (b) the average power density, and (c) the time-average values of the energy densities.

7.34 Compute the time-average electric and magnetic energy densities for the fields given in Problem 7.29.

7.35 Determine the time-average electric and magnetic energy densities for the fields given in Problem 7.30.

7.36 The current through the leads of a parallel-plate capacitor is given by $i(t) = I_m \cos \omega t$ A. Show that the displacement current density in the capacitor is exactly equal to $i(t)$.

7.37 The conductivity of seawater is approximately 0.4 mS/m and its dielectric constant is 81. Determine the frequency at which the magnitude of the displacement current density is equal to the magnitude of the conduction current density. Comment on the electric behavior of seawater at very low and very high frequencies.

7.38 Two circular conducting plates, each of area 0.4 m^2, are separated by a lossy dielectric of thickness 5 mm. The dielectric constant and the conductivity of the medium are 4 and 0.02 S/m, respectively. If the potential difference across the plates is 141 sin $10^9 t$ V, determine (a) the conduction current, (b) the displacement current, and (c) the rms value of the total current in the lossy dielectric region.

7.39 The electric field intensity in a source-free dielectric medium is given as $\vec{\mathbf{E}} = E\cos(\omega t - ax - kz)\vec{\mathbf{a}}_y$ V/m. Find the corresponding $\vec{\mathbf{H}}$ field. What is the necessary condition for these fields to exist? Determine the time-average values of electric energy density, magnetic energy density, and the Poynting vector.

7.40 For electromagnetic fields to exist in a linear, homogeneous, isotropic, source-free conductive region, show that the $\vec{\mathbf{E}}$ field must satisfy the following equation:

$$\nabla^2\vec{\mathbf{E}} - \mu\epsilon\frac{\partial^2\vec{\mathbf{E}}}{\partial t^2} - \mu\sigma\frac{\partial\vec{\mathbf{E}}}{\partial t} = 0$$

7.41 For electromagnetic fields to exist in a linear, homogeneous, isotropic, source-free conductive region, show that the $\vec{\mathbf{H}}$ field must satisfy the following equation:

$$\nabla^2\vec{\mathbf{H}} - \mu\epsilon\frac{\partial^2\vec{\mathbf{H}}}{\partial t^2} - \mu\sigma\frac{\partial\vec{\mathbf{H}}}{\partial t} = 0$$

7.42 Repeat Problems 7.40 and 7.41 for a dielectric medium.

7.43 Derive the phasor form of Poynting's theorem for a dielectric region.

7.44 Derive the phasor form of Poynting's theorem for a source-free conductive region.

7.45 Repeat Problem 7.39 using phasors.

7.46 The fields inside an air-filled coaxial line having inner radius a and outer radius b are given as

$$\tilde{E}_\rho = \frac{V}{\rho\ln(b/a)}e^{-jkz}\ \text{V/m} \quad \text{and} \quad \tilde{H}_\phi = \frac{I}{2\pi\rho}e^{-jkz}\ \text{A/m}$$

where V and I are the peak values of the voltage and the current, which vary sinusoidally with an angular frequency of ω rad/s. Determine the condition for the fields to exist within the coaxial line. What is the direction of power flow? Compute the average power inside the coaxial line.

7.47 $\tilde{\mathbf{A}}$ and $\tilde{\mathbf{B}}$ are two complex vectors such that $\tilde{\mathbf{A}} = \vec{\mathbf{A}}_r + j\vec{\mathbf{A}}_i$ and $\tilde{\mathbf{B}} = \vec{\mathbf{B}}_r + j\vec{\mathbf{B}}_i$, where the subscripts r and i denote the real and imaginary vectors. Show that the time-average value of their scalar product is $\langle\vec{\mathbf{A}}(t)\cdot\vec{\mathbf{B}}(t)\rangle = \frac{1}{2}\text{Re}[\tilde{\mathbf{A}}\cdot\tilde{\mathbf{B}}^*]$.

7.48 Using the definitions for complex vectors $\tilde{\mathbf{A}}$ and $\tilde{\mathbf{B}}$ given in Problem 7.47, show that the time-average value of their cross product is given by $\langle\vec{\mathbf{A}}(t)\times\vec{\mathbf{B}}(t)\rangle = \frac{1}{2}\text{Re}[\tilde{\mathbf{A}}\times\tilde{\mathbf{B}}^*]$.

7.49 For electromagnetic fields to exist in a linear, homogeneous, isotropic, source-free conductive region, show that the $\tilde{\mathbf{E}}$ field must satisfy the following equation:

$$\nabla^2\vec{\mathbf{E}} + (\omega^2\mu\epsilon - j\omega\mu\sigma)\vec{\mathbf{E}} = 0$$

7.50 For electromagnetic fields to exist in a linear, homogeneous, isotropic, source-free conductive region, show that the $\vec{\mathbf{H}}$ field must satisfy the following equation:

$$\nabla^2\vec{\mathbf{H}} + (\omega^2\mu\epsilon - j\omega\mu\sigma)\vec{\mathbf{H}} = 0$$

7.51 In a dielectric medium ($\epsilon = 4\epsilon_0$ and $\mu = \mu_0$) the $\vec{\mathbf{E}}$ and $\vec{\mathbf{H}}$ fields are given as follows:

$$E_z = 1000 \cos\left(\omega t - \frac{\pi}{3}x\right) \text{ V/m}$$

and

$$H_y = -\frac{1000}{\eta} \cos\left(\omega t - \frac{\pi}{3}x\right) \text{ A/m}$$

Using phasor analysis, determine (a) ω and η, (b) the direction of power flow, and (c) the average power crossing the surface area bounded by the corners of a triangle at (2, 0, 0) m, (2, 4, 0) m, and (2, 4, 2) m.

7.52 A magnetically coupled equivalent circuit of a transformer is given in Figure P7.52. Under steady-state conditions determine the voltage drop across the capacitor when the source voltage is 120 cos(1000t) V and the coefficient of coupling is unity.

Figure P7.52

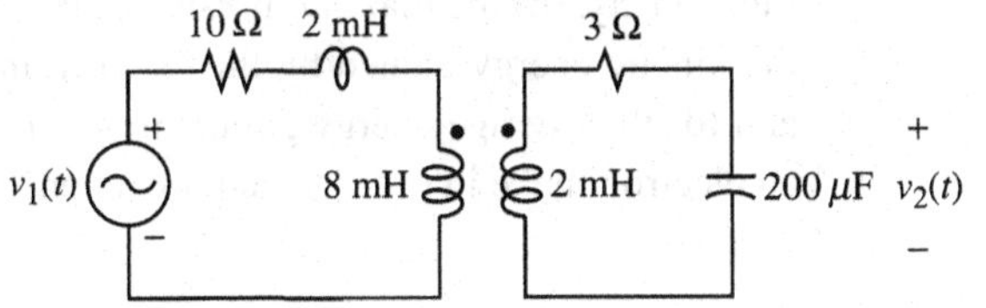

7.53 A capacitive load is connected to a source via two ideal transformers, as shown in Figure P7.53. Determine (a) the current supplied by the source, (b) the average power supplied by the source, (c) the power loss on the transmission line, (d) the load current, (e) the load voltage, (f) the power supplied to the load, and (g) the overall efficiency of the system.

Figure P7.53

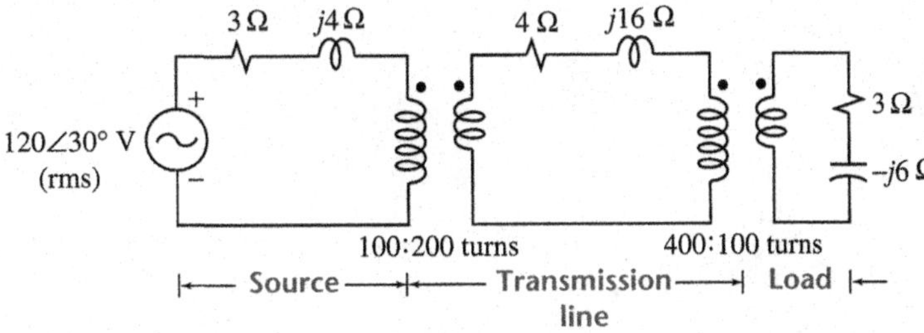

7.54 The primary winding of an ideal transformer has 30 turns and is connected to a 240-V (rms), 50-Hz source. The secondary winding has 750 turns and supplies a load of 4 A at 0.8 pf lagging. Determine (a) the ratio of transformation, (b) the current in the primary winding, and (c) the magnetic flux in the core.

7.55 A coil wound around a magnetic material sets up a flux of 1 mWb (rms value) when a 230-V, 60-Hz source is applied to the coil. Under ideal conditions, determine the number of turns in the coil.

7.56 A 1-kVA, 480/120-V, ideal transformer delivers the rated load at 0.6 power factor leading. Determine the load impedance.

7.57 A 4.8-kVA, 120/480-V, two-winding transformer is to be connected as an autotransformer. Determine the nominal rating for each possible connection.

7.58 If the primary and the secondary winding resistances of the transformer given in Problem 7.57 are 0.5 Ω and 12 Ω, respectively, determine the efficiency of the two-winding transformer when it delivers the rated load at 0.6 power factor lagging.

7.59 The transformer in Problem 7.58 is to be connected as an autotransformer. Determine the efficiency for each possible connection when the transformer delivers the rated load at 0.6 power factor lagging.

7.60 In a betatron, the space average flux passing through the surface enclosed by the circular path of an electron is given as $\Phi = 1.5 \sin(100\pi t)$ Wb. If the stable orbit is at a radius of 75 cm, determine (a) the space average flux density, (b) the maximum value of the flux density at the orbit, (c) the kinetic energy gained by the electron in one trip around the orbit, and (d) the average energy gained in one trip. How many trips should an electron make in order to acquire an energy of 90 MeV?

8

Plane wave propagation

8.1 Introduction

We have stated $\vec{\mathbf{a}}$ this in Chapter 7 and we state it again: Maxwell's equations contain all the information necessary to characterize the electromagnetic fields at any point in a medium. For the electromagnetic (EM) fields to exist they must satisfy the four Maxwell equations at the source where they are generated, at any point in a medium through which they propagate, and at the load where they are received or absorbed.

In this chapter, we concentrate mainly on the propagation of EM fields in a source-free medium. As the fields must satisfy the four coupled Maxwell equations involving four unknown variables, we first obtain an equation in terms of one unknown variable. Similar equations can then be obtained for the other variables. We refer to these equations as the general *wave equations.* We will show in Chapter 11 that the fields generated by time-varying sources propagate as *spherical waves.* However, in a small region far away from the radiating source, the spherical wave may be approximated as a *plane wave,* that is, one in which all the field quantities are in a plane normal to the direction of its propagation (the transverse plane). Consequently, a plane wave does not have any field component in its direction of propagation (the longitudinal direction).

We first seek the solution of a plane wave in an unbounded dielectric medium and show that the wave travels with the speed of light in free space. We then consider the general case of a finitely conducting medium. We show that the wave attenuates as a result of loss in energy as it travels in the conducting medium. Finally, we introduce the concept of reflection and transmission of a plane wave when it leaves one medium and enters another.

8.2 General wave equations

Let us consider a uniform but source-free medium having dielectric constant ϵ, magnetic permeability μ, and conductivity σ. The medium is considered to be source free as long as it does not contain the charges and currents necessary to generate the fields. However, the conduction current density as determined by Ohm's law ($\vec{\mathbf{J}} = \sigma\vec{\mathbf{E}}$) can exist in a finitely conducting medium. Under these conditions, Maxwell's equations are

$$\nabla \times \vec{\mathbf{E}} = -\mu \frac{\partial \vec{\mathbf{H}}}{\partial t} \tag{8.1}$$

$$\nabla \times \vec{\mathbf{H}} = \sigma\vec{\mathbf{E}} + \epsilon \frac{\partial \vec{\mathbf{E}}}{\partial t} \tag{8.2}$$

$$\nabla \cdot \vec{\mathbf{B}} = 0 \quad \Rightarrow \quad \nabla \cdot \vec{\mathbf{H}} = 0 \tag{8.3}$$

$$\nabla \cdot \vec{\mathbf{D}} = 0 \quad \Rightarrow \quad \nabla \cdot \vec{\mathbf{E}} = 0 \tag{8.4}$$

where $\vec{\mathbf{B}} = \mu\vec{\mathbf{H}}$ and $\vec{\mathbf{D}} = \epsilon\vec{\mathbf{E}}$.

For a *linear* ($\vec{\mathbf{D}}$ is parallel to $\vec{\mathbf{E}}$, and $\vec{\mathbf{B}}$ is parallel to $\vec{\mathbf{H}}$), *homogeneous* (medium properties are the same at all points), and *isotropic* (μ and ϵ are independent of direction) medium, both μ and ϵ are scalar constants. Such a medium is also referred to as a **uniform medium**. Unless stated otherwise, we shall always assume the medium to be linear, homogeneous, and isotropic.

Instead of four variables, the preceding coupled equations are in terms of two variables ($\vec{\mathbf{E}}$ and $\vec{\mathbf{H}}$). Let us now obtain an equation in terms of one variable, say the $\vec{\mathbf{E}}$ field only. To do this, we take the curl of (8.1) and obtain

$$\nabla \times \nabla \times \vec{\mathbf{E}} = -\mu \nabla \times \left(\frac{\partial \vec{\mathbf{H}}}{\partial t} \right) \tag{8.5}$$

Using the vector identity

$$\nabla \times \nabla \times \vec{\mathbf{E}} = \nabla(\nabla \cdot \vec{\mathbf{E}}) - \nabla^2 \vec{\mathbf{E}}$$

and substituting $\nabla \cdot \vec{\mathbf{E}} = 0$, we have

$$\nabla \times \nabla \times \vec{\mathbf{E}} = -\nabla^2 \vec{\mathbf{E}}$$

where the Laplacian of a vector quantity is defined in the rectangular coordinate system as

$$\nabla^2 \vec{\mathbf{E}} = \nabla^2 E_x \vec{\mathbf{a}}_x + \nabla^2 E_y \vec{\mathbf{a}}_y + \nabla^2 E_z \vec{\mathbf{a}}_z \tag{8.6}$$

and the Laplacian operator is

$$\nabla^2 = \frac{\partial^2}{\partial x^2} + \frac{\partial^2}{\partial y^2} + \frac{\partial^2}{\partial z^2} \tag{8.7}$$

Changing the order of differentiation with respect to space and time, we can rewrite (8.5) as

$$\nabla^2 \vec{\mathbf{E}} = \mu \frac{\partial}{\partial t} [\nabla \times \vec{\mathbf{H}}]$$

Substituting for $\nabla \times \vec{\mathbf{H}}$ from (8.2) in this equation, we get

$$\nabla^2\vec{\mathbf{E}} = \mu\sigma \frac{\partial \vec{\mathbf{E}}}{\partial t} + \mu\epsilon \frac{\partial^2 \vec{\mathbf{E}}}{\partial t^2} \tag{8.8}$$

which is a set of three scalar equations, one for each component of the $\vec{\mathbf{E}}$ field, in a conducting medium. We can also obtain a similar set of three equations in terms of the $\vec{\mathbf{H}}$ field as

$$\nabla^2\vec{\mathbf{H}} = \mu\sigma \frac{\partial \vec{\mathbf{H}}}{\partial t} + \mu\epsilon \frac{\partial^2 \vec{\mathbf{H}}}{\partial t^2} \tag{8.9}$$

The set of six independent equations given by (8.8) and (8.9) are known as the **general wave equations**. These equations govern the behavior of all electromagnetic fields in a uniform but source-free conducting medium. The presence of the first-order term in a second-order differential equation indicates that the fields decay (lose energy) as they propagate through the medium. For this reason, *a conducting medium is called a lossy medium.* We next solve these wave equations and show that each equation does, in fact, represent an electromagnetic wave.

8.3 Plane wave in a dielectric medium

Before we obtain the solution of the general wave equation, let us consider a dielectric medium in which the conduction current is almost nonexistent in comparison with the displacement current. Such a medium may be treated as a **perfect dielectric** or **lossless medium** ($\sigma = 0$). Thus, by setting $\sigma = 0$ in (8.8) and (8.9), we obtain the wave equations for a lossless medium as

$$\nabla^2\vec{\mathbf{E}} - \mu\epsilon \frac{\partial^2 \vec{\mathbf{E}}}{\partial t^2} = 0 \tag{8.10}$$

$$\nabla^2\vec{\mathbf{H}} - \mu\epsilon \frac{\partial^2 \vec{\mathbf{H}}}{\partial t^2} = 0 \tag{8.11}$$

These equations, called the time-dependent *Helmholtz equations*, still represent a set of six scalar equations. The absence of the first-order term signifies that the electromagnetic fields do not decay as they propagate in a lossless medium.

We now assume that the components of the field quantities $\vec{\mathbf{E}}$ and $\vec{\mathbf{H}}$ lie in a *transverse plane*, a plane perpendicular to the direction of propagation of the wave. We refer to such a wave as a **plane wave**. Let us consider that a plane wave propagates in the z direction. Then, the $\vec{\mathbf{E}}$ and $\vec{\mathbf{H}}$ fields have no components in the *longitudinal direction* (the direction of wave propagation). That is, $E_z = 0$ and $H_z = 0$. Such a wave is also called a **transverse electromagnetic wave (TEM wave)**.

In the family of plane waves, the *uniform plane wave* is one of the simplest to investigate and easiest to understand. The term *uniform* implies that, at any time, a field has the same magnitude and direction in a plane containing it. Thus, for a uniform plane wave propagating in the z direction, $\vec{\mathbf{E}}$ and $\vec{\mathbf{H}}$ are not functions of x and y. That is,

$$\frac{\partial \vec{\mathbf{E}}}{\partial x} = 0 \quad \frac{\partial \vec{\mathbf{E}}}{\partial y} = 0$$

$$\frac{\partial \vec{\mathbf{H}}}{\partial x} = 0 \quad \frac{\partial \vec{\mathbf{H}}}{\partial y} = 0$$

For a uniform plane wave propagating in the z direction, the Helmholtz equations can be expressed in scalar form as

$$\frac{\partial^2 E_x}{\partial z^2} - \mu\epsilon \frac{\partial^2 E_x}{\partial t^2} = 0 \tag{8.12a}$$

$$\frac{\partial^2 E_y}{\partial z^2} - \mu\epsilon \frac{\partial^2 E_y}{\partial t^2} = 0 \tag{8.12b}$$

$$\frac{\partial^2 H_x}{\partial z^2} - \mu\epsilon \frac{\partial^2 H_x}{\partial t^2} = 0 \tag{8.13a}$$

$$\frac{\partial^2 H_y}{\partial z^2} - \mu\epsilon \frac{\partial^2 H_y}{\partial t^2} = 0 \tag{8.13b}$$

where E_x, E_y, H_x, and H_y are the *transverse components* of the $\vec{\mathbf{E}}$ and $\vec{\mathbf{H}}$ fields. In addition, the field components are functions of z (the direction of propagation) and t (time) only.

Each of these four equations is a second-order differential equation with two possible solutions. Because these equations are similar, their solutions must also be similar. In other words, as soon as we know the solution of one of these equations, we immediately know the solutions for the others.

There are many possible functions that satisfy these wave equations. However, we are interested only in those functions that result in a traveling wave. A general function of the type $F(t \pm z/u)$, where u is the wave speed, is among the family of functions we are interested in. However, the class of the general function $F(t \pm z/u)$ and its attributes depend upon the nature of the sources creating the waves. Since most of the sources vary sinusoidally, the function $F(t \pm z/u)$ is also expected to follow sinusoidal variations. Instead of seeking a solution in terms of a general function, we focus our attention only on time-harmonic functions as possible solutions of the wave equations. This does not pose a considerable threat to the general solution of the wave equation because any periodic function may be represented in terms of infinite series of sinusoidal functions (Fourier series).

The preceding discussion suggests that, for time-harmonic fields, each wave equation can be expressed in its phasor equivalent form.

For example, we can write (8.12a) as

$$\frac{d^2\tilde{E}_x}{dz^2} + \omega^2\mu\epsilon\tilde{E}_x = 0 \tag{8.14}$$

where $\tilde{E}_x(z)$ is a phasor form of $E_x(z,t)$, $\omega = 2\pi f$ is the angular frequency of the wave in rad/s, and f is the frequency of oscillation in Hz. Note the use of the tilde ($\tilde{}$) notation to represent a phasor quantity. Similar wave equations can also be written for the field components $\tilde{E}_y(z)$, $\tilde{H}_x(z)$ and $\tilde{H}_y(z)$.

For a monochromatic wave propagating in a uniform medium, $\omega^2\mu\epsilon$ is a constant quantity. If we define a variable β as

$$\beta = \omega\sqrt{\mu\epsilon} \tag{8.15}$$

then we can rewrite the wave equation as

$$\frac{d^2\tilde{E}_x}{dz^2} + \beta^2\tilde{E}_x = 0 \tag{8.16}$$

We assume an exponential solution of the form

$$\tilde{E}_x(z) = \hat{C}e^{\hat{s}z}$$

where $\hat{C}$ and $\hat{s}$ may, in general, be complex quantities as indicated by the caret ($\hat{}$) notation.

Substituting the assumed solution in (8.16), we obtain

$$\hat{s} = \pm j\beta$$

As expected, there are two solutions for the x component of the $\vec{\mathbf{E}}$ field. One solution is obtained by considering the negative sign as

$$\tilde{E}_x(z) = \hat{E}_{xf}e^{-j\beta z} \tag{8.17a}$$

where $\hat{E}_{xf}$, in general, may be a complex constant. The other solution is obtained by choosing the positive sign as

$$\tilde{E}_x(z) = \hat{E}_{xb}e^{j\beta z} \tag{8.17b}$$

where $\hat{E}_{xb}$, in general, may be another complex constant. Since we are considering a solution of a second-order differential equation, the general solution is

$$\tilde{E}_x(z) = \hat{E}_{xf}e^{-j\beta z} + \hat{E}_{xb}e^{j\beta z} \tag{8.18}$$

If we express the two complex constants in equation (8.18) as

$$\hat{E}_{xf} = E_{xf}e^{j\theta_{xf}}$$

and

$$\hat{E}_{xb} = E_{xb}e^{j\theta_{xb}}$$

where E_{xf}, θ_{xf}, E_{xb}, and θ_{xb} are (real) constants, then we obtain

$$\tilde{E}_x(z) = E_{xf}e^{-j(\beta z - \theta_{xf})} + E_{xb}e^{j(\beta z + \theta_{xb})} \tag{8.19a}$$

as the general solution, in the phasor form, of the wave equation for the x component of the $\vec{\mathbf{E}}$ field. It can be written in the time domain as

$$E_x(z, t) = E_{xf}\cos(\omega t - \beta z + \theta_{xf}) + E_{xb}\cos(\omega t + \beta z + \theta_{xb}) \tag{8.19b}$$

8.3.1 The forward-traveling wave

Let us examine the first term, the function $F_x = E_{xf}\cos(\omega t - \beta z + \theta_{xf})$, on the right-hand side of (8.19b). Since βz refers to the part of the phase of the function, β is called the *phase constant.* In a dielectric medium, the phase constant β is given by (8.15) and is a linear function of frequency f or ω.

At any given point in a transverse plane (z = constant), the function F_x varies sinusoidally in time with the angular frequency ω and has an amplitude E_{xf}, as shown in Figure 8.1. The function reverts to its initial magnitude and phase when t increases by a time period T such that $\omega T = 2\pi$. The function F_x also varies with z. A plot of F_x as a function of z for several instants of time is given in Figure 8.2. Note that each point on the function moves to the right (forward direction) as time progresses. Hence, the first term of (8.19b), $E_{xf}\cos(\omega t - \beta z + \theta_{xf})$, represents a *forward-traveling wave.*

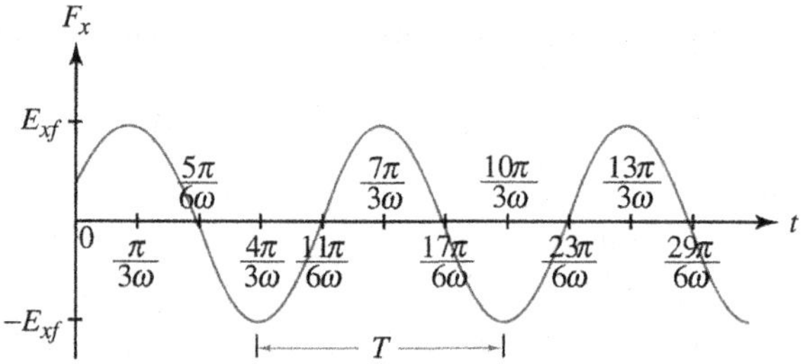

Figure 8.1 Plot of $F_x = E_{xf}\cos(\omega t - \beta z + \theta_{xf})$ in a plane where $\beta z - \theta_{xf} = \pi/3$

At any given time (t = constant) the wave returns to its original magnitude and phase when z increases by a wavelength λ such that $\beta\lambda = 2\pi$. Accordingly, by definition, the **wavelength** is the distance between two planes when the phase difference between them at any given time is 2π radians. Thus, the wavelength is

$$\lambda = \frac{2\pi}{\beta} \tag{8.20}$$

The time is not at a standstill, and the wave is not stationary. The only quantity that can be viewed as constant is the phase of the wave. That is,

$$\omega t - \beta z + \theta_{xf} = M \tag{8.21}$$

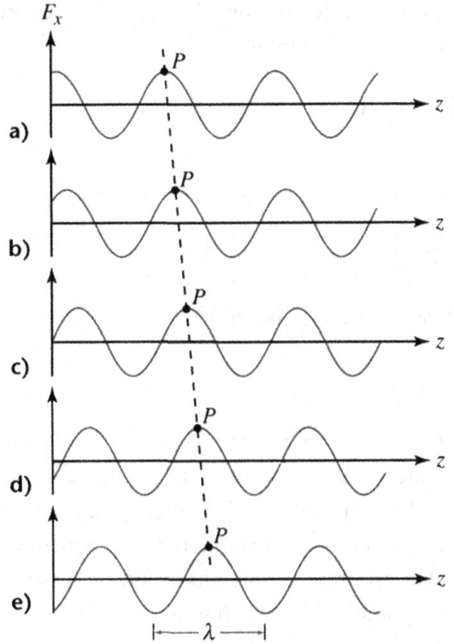

Figure 8.2 Plots of $F_x = E_{xf} \cos(\omega t - \beta z + \theta_{xf})$ when
(a) $\omega t = -\theta_{xf}$
(b) $\omega t = \frac{\pi}{4} - \theta_{xf}$
(c) $\omega t = \frac{\pi}{2} - \theta_{xf}$
(d) $\omega t = \frac{3\pi}{4} - \theta_{xf}$ and
(e) $\omega t = \pi - \theta_{xf}$

where M is a constant quantity. This equation stresses the point that z must increase as t increases to ensure that the phase of the function is constant. For example, *a constant phase must travel a distance of one wavelength in one time period.* At any time $t = t_a$, the constant phase dictates that

$$z_a = \frac{\omega t_a + \theta_{xf} - M}{\beta}$$

which represents a constant plane at $z = z_a$.

Differentiating (8.21) with respect to t, we obtain the *speed of a plane of constant phase (phase speed)* as

$$u_p = \frac{dz}{dt} = \frac{\omega}{\beta} \tag{8.22a}$$

Because ω and β are both positive quantities, the phase speed u_p is greater than zero. Hence, the wave propagates in the positive z direction. This equation also validates our previous contention that the function $E_{xf} \cos(\omega t - \beta z + \theta_{xf})$ represents a forward-traveling wave. Because θ_{xf} is a constant quantity, it is the minus sign in the argument $(\omega t - \beta z)$ that dictates the propagation of the wave in the forward direction.

The *phase velocity* of the forward-traveling wave is

$$\vec{\mathbf{u}}_p = \frac{\omega}{\beta} \vec{\mathbf{a}}_z \tag{8.22b}$$

Substituting for β from (8.15), we can express the phase velocity in a dielectric medium as

$$\vec{\mathbf{u}}_p = \frac{c}{n}\,\vec{\mathbf{a}}_z \tag{8.22c}$$

where

$$\frac{1}{\sqrt{\mu_0 \epsilon_0}} = c \approx 3 \times 10^8 \text{ m/s}$$

is the speed of light, and

$$n = \sqrt{\mu_r \epsilon_r} \tag{8.23}$$

is the *index of refraction.* From (8.22c) it is clear that the phase velocity is independent of frequency and depends only upon the permeability and the permittivity of the dielectric medium. A medium in which the phase velocity is not a function of frequency is said to be *nondispersive.* Stated differently, in a nondispersive medium the phase constant β is a linear function of the angular frequency ω.

8.3.2 The backward-traveling wave

Following the development of the preceding section, we can show that the second term on the right-hand side of (8.19b), $E_{xb}\cos(\omega t + \beta z + \theta_{xb})$, represents a *backward-traveling wave* because it moves in the negative z direction as time progresses. Thus, when the sign in the argument $(\omega t + \beta z)$ of a function is positive, the wave travels in the backward direction with a phase velocity of $-\omega/\beta\vec{\mathbf{a}}_z$. Equation (8.19a), therefore, represents the general solution of the wave equation for the x component of the $\vec{\mathbf{E}}$ field as it includes both the forward- and backward-traveling waves.

Following this development, we can also obtain a solution for the y component of the $\vec{\tilde{\mathbf{E}}}$ field, from (8.12b), as

$$\tilde{E}_y(z) = E_{yf}e^{-j(\beta z - \theta_{yf})} + E_{yb}e^{j(\beta z + \theta_{yb})} \tag{8.24a}$$

in the phasor form and

$$E_y(z, t) = E_{yf}\cos(\omega t - \beta z + \theta_{yf}) + E_{yb}\cos(\omega t + \beta z + \theta_{yb}) \tag{8.24b}$$

in the time domain. In these equations, E_{yf} and E_{yb} are the amplitudes of the forward- and backward-traveling waves for the y component of the $\vec{\tilde{\mathbf{E}}}$ field. Likewise, θ_{yf} and θ_{yb} are the corresponding phase shifts at $t = 0$ and $z = 0$.

Since $\vec{\tilde{\mathbf{E}}}$ and $\vec{\tilde{\mathbf{H}}}$ fields are coupled together by Maxwell's equations, and the $\vec{\tilde{\mathbf{E}}}$ field is known completely, we can obtain the components of the $\vec{\tilde{\mathbf{H}}}$ field from Maxwell's equation (8.1).

8.3.3 Boundless dielectric medium

Let us now assume that the dielectric medium is of infinite extent, and there is only one wave propagating along the z direction. For this discussion we will assume that it is the forward wave propagating in the positive z direction. Then, the x and y components of the $\tilde{\vec{\mathbf{E}}}$ field, in the phasor domain, are

$$\tilde{E}_x(z) = E_{xf}e^{-j(\beta z - \theta_{xf})} \tag{8.25a}$$

$$\tilde{E}_y(z) = E_{yf}e^{-j(\beta z - \theta_{yf})} \tag{8.25b}$$

Using Maxwell's equation (8.1), we obtain the x and y components of the $\tilde{\vec{\mathbf{H}}}$ field as

$$\tilde{H}_x(z) = -\sqrt{\frac{\epsilon}{\mu}}\,\tilde{E}_y(z) \tag{8.26a}$$

$$\tilde{H}_y(z) = -\sqrt{\frac{\epsilon}{\mu}}\,\tilde{E}_x(z) \tag{8.26b}$$

Equations (8.26a) and (8.26b) can also be written in concise form as

$$\vec{\mathbf{a}}_z \times \tilde{\vec{\mathbf{E}}} = \sqrt{\frac{\mu}{\epsilon}}\,\tilde{\vec{\mathbf{H}}}$$

or

$$\vec{\mathbf{a}}_z \times \tilde{\vec{\mathbf{E}}} = \eta\tilde{\vec{\mathbf{H}}} \tag{8.27}$$

where

$$\eta = \sqrt{\frac{\mu}{\epsilon}} \tag{8.28}$$

has the units of ohms because $\tilde{\vec{\mathbf{E}}}$ is in V/m and $\tilde{\vec{\mathbf{H}}}$ is in A/m. For this reason, η is called the **intrinsic (or wave) impedance**. For the wave propagating in a dielectric medium, η is a pure resistance. Thus, in a dielectric medium, the corresponding components of the $\tilde{\vec{\mathbf{E}}}$ and $\tilde{\vec{\mathbf{H}}}$ fields are in time phase with each other, as shown in Figure 8.3.

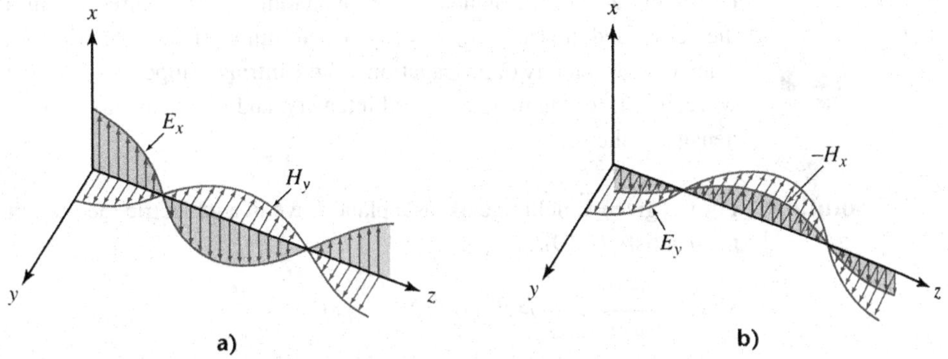

Figure 8.3 Components of $\tilde{\vec{\mathbf{E}}}$ and $\tilde{\vec{\mathbf{H}}}$ fields for a forward-traveling wave at t = constant: **(a)** E_x and H_y and **(b)** E_y and $-H_x$

In (8.27) the $\vec{\mathbf{a}}_z$ term must be perceived as the unit vector in the direction of wave propagation. For example, if the wave propagates in the y direction, then $\vec{\mathbf{a}}_z$ must be replaced by $\vec{\mathbf{a}}_y$ to compute one of the unknown fields in terms of the other.

We can now compute the average power density at any point in the medium as

$$\begin{aligned}\langle \hat{\mathbf{S}} \rangle &= \frac{1}{2}\,\mathrm{Re}[\vec{\tilde{\mathbf{E}}} \times \vec{\tilde{\mathbf{H}}}^*] \\ &= \frac{1}{2}\,\mathrm{Re}[\tilde{E}_x \tilde{H}_y^* - \tilde{E}_y \tilde{H}_x^*]\,\vec{\mathbf{a}}_z \\ &= \frac{1}{2\eta}\left[E_{xf}^2 + E_{yf}^2\right]\vec{\mathbf{a}}_z \\ &= \frac{1}{2\eta}\,E^2 \vec{\mathbf{a}}_z \end{aligned} \tag{8.29a}$$

where $E^2 = \vec{\tilde{\mathbf{E}}} \cdot \vec{\tilde{\mathbf{E}}}^*$ and $\vec{\tilde{\mathbf{E}}} = \tilde{E}_x \vec{\mathbf{a}}_x + \tilde{E}_y \vec{\mathbf{a}}_y$. Substituting for η from (8.28) in (8.29a) and expressing the result in terms of the phase velocity, we obtain the average power density as

$$\langle \mathbf{S} \rangle = \tfrac{1}{2}\epsilon E^2 \vec{\mathbf{u}}_p \tag{8.29b}$$

or

$$\langle \hat{\mathbf{S}} \rangle = \tfrac{1}{2}\mu H^2 \vec{\mathbf{u}}_p \tag{8.29c}$$

As expected, these equations confirm that, in a plane wave, *the velocity with which the electromagnetic energy propagates is equal to the phase velocity.* Note that $\frac{1}{2}\epsilon E^2$ (or $\frac{1}{2}\mu H^2$) represents the total average energy density in the medium. The average electric energy density is $\frac{1}{4}\epsilon E^2$ and the average magnetic energy density is $\frac{1}{4}\mu H^2$, so the preceding equations suggest that *the average electric energy density equals the average magnetic energy density when a plane wave propagates in a dielectric medium.*

EXAMPLE 8.1

WORKSHEET 7

Mathcad

If the electric field intensity as given by $\vec{\mathbf{E}} = 377\cos(10^9 t - 5y)\vec{\mathbf{a}}_z$ V/m represents a uniform plane wave propagating in the y direction in a dielectric medium ($\mu = \mu_0$, $\epsilon = \epsilon_r \epsilon_0$), determine (a) the dielectric constant, (b) the velocity of propagation, (c) the intrinsic impedance, (d) the wavelength, (e) the magnetic field intensity, and (f) the average power density of the wave.

Solution For the given $\vec{\mathbf{E}}$ field to exist as a plane wave in a dielectric medium, it must satisfy (8.10).

$$\nabla^2 \vec{\mathbf{E}} = \frac{\partial^2 E_z}{\partial y^2} = -9425\ \cos(10^9 t - 5y)$$

$$\frac{\partial^2 E_z}{\partial t^2} = -377 \times 10^{18}\ \cos(10^9 t - 5y)$$

Substituting in (8.10), we get

$$-9425\cos(10^9 t - 5y) + \mu\epsilon[377 \times 10^{18}]\cos(10^9 t - 5y) = 0$$

or

$$\mu\epsilon = 25 \times 10^{-18}$$

$$\text{Thus,}\quad \epsilon_r = \frac{25 \times 10^{-18}}{\mu_0\epsilon_0} = 25 \times 10^{-18} \times (3 \times 10^8)^2 = 2.25$$

a) The given field satisfies the wave equation as long as the dielectric constant of the medium is 2.25.

b) The phase of the $\vec{\mathbf{E}}$ field is constant when the argument of the cosine function is constant. That is,

$$10^9 t - 5y = M$$

where M is a constant. Differentiating with respect to t, we obtain

$$u_p = \frac{dy}{dt} = \frac{10^9}{5} = 2 \times 10^8 \text{ m/s}$$

since $dy/dt > 0$, the wave propagates in the y direction with a phase velocity $\vec{\mathbf{u}}_p = 2 \times 10^8 \vec{\mathbf{a}}_y$ m/s. We could have arrived at the same conclusion by examining the argument of the cosine function as follows:

1. The argument $(\omega t - \beta y)$ is a function of y, where $\omega = 10^9$ rad/s and $\beta = 5$ rad/m; therefore, the wave propagates in the *y direction.*
2. The minus sign in the argument $(\omega t - \beta y)$ suggests that the wave propagates in the *positive y direction.*
3. From (8.22a), the phase speed is

$$u_p = \frac{\omega}{\beta} = \frac{10^9}{5} = 2 \times 10^8 \text{ m/s}$$

c) The intrinsic impedance, from (8.28), is

$$\eta = \sqrt{\frac{\mu_0}{\epsilon_0\epsilon_r}} = \frac{120\pi}{\sqrt{2.25}} = 251.33\ \Omega$$

d) For the propagating wave, $\beta\lambda$ is always 2π. Thus, the wavelength is

$$\lambda = \frac{2\pi}{\beta} = \frac{2\pi}{5} = 1.257 \text{ m}$$

e) The electric field intensity in the phasor form is

$$\tilde{\vec{\mathbf{E}}} = 377 e^{-j5y}\, \vec{\mathbf{a}}_z \text{ V/m}$$

We can obtain the $\tilde{\vec{\mathbf{H}}}$ field from (8.27) as

$$\vec{\mathbf{a}}_y \times \tilde{\vec{\mathbf{E}}} = \eta\tilde{\vec{\mathbf{H}}}$$

Thus, $\tilde{\vec{\mathbf{H}}} = 1.5e^{-j5y}\,\vec{\mathbf{a}}_x$ A/m. We could also have used Maxwell's equation (8.1) to determine the $\vec{\mathbf{H}}$ field and avoided the memorization of (8.27).

The time domain expression for the $\vec{\mathbf{H}}$ field is

$$\vec{\mathbf{H}} = 1.5\ \cos(10^9 t - 5y)\vec{\mathbf{a}}_x\ \text{A/m}$$

f) The average power density in the medium is

$$\begin{aligned}\langle\hat{\mathbf{S}}\rangle &= \tfrac{1}{2}\text{Re}[\tilde{\vec{\mathbf{E}}} \times \tilde{\vec{\mathbf{H}}}^*] \\ &= \tfrac{1}{2} \times 377 \times 1.5[\vec{\mathbf{a}}_z \times \vec{\mathbf{a}}_x] \\ &= 282.75\vec{\mathbf{a}}_y\ \text{W/m}^2\end{aligned}$$

Note that we used the fundamental equation instead of (8.29) to determine the average power density in the medium. • • •

8.4 Plane wave in free space

Free space (or vacuum) is a special case of a dielectric medium in which $\epsilon = \epsilon_0$ and $\mu = \mu_0$. Although we can simply replace μ with μ_0 and ϵ with ϵ_0 in all the equations discussed in Section 8.3, it is worth discussing plane waves in free space because a great deal of wave propagation takes place in free space. Most of these waves are at the lower end of the electromagnetic spectrum and are called the *radio waves*. These waves include AM radio (535–1605 kHz), shortwave radio (2–26 MHz), VHF television and FM radio (54–216 MHz), and UHF television (470–806 MHz). Frequencies in the GHz range are especially used for radar and satellite communications.

Substituting $\epsilon = \epsilon_0$ and $\mu = \mu_0$ in (8.15), we obtain the *phase constant in free space* as

$$\beta_0 = \omega\sqrt{\mu_0\epsilon_0} = \frac{\omega}{c} \tag{8.30}$$

where $c = 1/\sqrt{\mu_0\epsilon_0} = 3 \times 10^8$ m/s is the speed of light. From (8.22a) we obtain the *wave speed in free space* as

$$u_p = \frac{\omega}{\beta_0} = c \tag{8.31}$$

This equation states that *an electromagnetic wave propagates in free space with the speed of light.* In fact, this result enabled Maxwell to suggest that light may be viewed as an electromagnetic phenomenon. This is probably one of the strongest arguments in favor of the modification of Ampère's law. Without the displacement current density term, it would have been impossible for Maxwell to predict the wave nature of electromagnetic fields.

The *wavelength in free space* is

$$\lambda_0 = \frac{2\pi}{\beta_0} = \frac{c}{f} \tag{8.32}$$

Finally, the *intrinsic impedance of free space* is

$$\eta_0 = \sqrt{\frac{\mu_0}{\epsilon_0}} = 120\pi \approx 377\ \Omega \tag{8.33}$$

EXAMPLE 8.2

The electric field intensity of a uniform plane wave in free space is given by $\vec{\mathbf{E}} = 94.25\cos(\omega t + 6z)\vec{\mathbf{a}}_x$ V/m. Determine (a) the velocity of propagation, (b) the wave frequency, (c) the wavelength, (d) the magnetic field intensity, and (e) the average power density in the medium.

Solution

a) The wave propagates in free space with the speed of light. Because the wave is traveling in the negative z direction, the phase velocity is

$$\vec{\mathbf{u}}_p = -3 \times 10^8\, \vec{\mathbf{a}}_z \text{ m/s}$$

b) $\beta_0 = 6$ rad/m, so the angular frequency of the wave is

$$\omega = \beta_0 u_p = 6 \times 3 \times 10^8 = 1.8 \times 10^9 \text{ rad/s}$$

c) The wavelength of the wave in free space is

$$\lambda_0 = \frac{2\pi}{\beta_0} = \frac{2\pi}{6} = 1.047 \text{ m}$$

d) The electric field intensity, in phasor form, is

$$\tilde{\vec{\mathbf{E}}} = 94.25 e^{j6z}\, \vec{\mathbf{a}}_x \text{ V/m}$$

The corresponding $\vec{\mathbf{H}}$ field for the backward-traveling wave, from (8.27), is

$$\tilde{\vec{\mathbf{H}}} = -\frac{94.25}{377} e^{j6z}\, \vec{\mathbf{a}}_y = -0.25 e^{j6z}\, \vec{\mathbf{a}}_y \text{ A/m}$$

or

$$\vec{\mathbf{H}}(z,t) = -0.25\cos(1.8 \times 10^9 t + 6z)\vec{\mathbf{a}}_y \text{ A/m}$$

e) The average power density in the medium is

$$\begin{aligned} \langle \hat{\mathbf{S}} \rangle &= \tfrac{1}{2}\mathrm{Re}[\tilde{\vec{\mathbf{E}}} \times \tilde{\vec{\mathbf{H}}}^*] \\ &= -\tfrac{1}{2} \times 94.25 \times 0.25\, \vec{\mathbf{a}}_z \\ &= -11.78\, \vec{\mathbf{a}}_z \text{ W/m}^2 \end{aligned}$$

• • •

8.5 Plane wave in a conducting medium

In the preceding sections, we obtained the steady-state solution of the wave equation in a dielectric medium and concluded that the plane wave propagates in the medium without any loss of energy. We now consider a general case of wave propagation in a medium having finite conductivity σ, permeability μ, and permittivity ϵ. Once again, we seek the solution of the wave equation with an understanding that the fields vary sinusoidally. To this end, we express the general wave equations, (8.8) and (8.9), in phasor form as

$$\nabla^2 \tilde{\vec{\mathbf{E}}} = (j\omega\mu\sigma - \omega^2\mu\epsilon)\tilde{\vec{\mathbf{E}}} \tag{8.34}$$

$$\nabla^2 \tilde{\vec{\mathbf{H}}} = (j\omega\mu\sigma - \omega^2\mu\epsilon)\tilde{\vec{\mathbf{H}}} \tag{8.35}$$

The complex coeffient in these equations may be expressed in somewhat compact form as

$$\begin{aligned} j\omega\mu\sigma - \omega^2\mu\epsilon &= j\omega\mu(\sigma + j\omega\epsilon) \\ &= -\omega^2\mu\epsilon\left[1 - j\frac{\sigma}{\omega\epsilon}\right] \\ &= -\omega^2\mu\hat{\epsilon} \end{aligned} \tag{8.36}$$

where

$$\hat{\epsilon} = \epsilon\left[1 - j\frac{\sigma}{\omega\epsilon}\right] \tag{8.37}$$

is called the *complex permittivity* of the medium. The complex permittivity is a function of frequency and is often given in the literature as

$$\hat{\epsilon} = \epsilon' - j\epsilon'' \tag{8.38}$$

where ϵ' is the permittivity ($\epsilon_r\epsilon_0$), and $\omega\epsilon''$ is the conductivity (σ) of the medium. The term $\sigma/\omega\epsilon$ in (8.37) is referred to as the *loss tangent* and will be discussed in the following paragraphs. The complex permittivity may also be given in the literature in terms of the loss tangent at a certain frequency.

In a conducting medium, the displacement current density $\tilde{\vec{\mathbf{J}}}_d$ and the conduction current density $\tilde{\vec{\mathbf{J}}}_c$ are

$$\tilde{\vec{\mathbf{J}}}_d = j\omega\epsilon\tilde{\vec{\mathbf{E}}}$$

$$\tilde{\vec{\mathbf{J}}}_c = \sigma\tilde{\vec{\mathbf{E}}}$$

The total current density is

$$\tilde{\vec{\mathbf{J}}}_t = \tilde{\vec{\mathbf{J}}}_c + \tilde{\vec{\mathbf{J}}}_d$$

Using $\tilde{\vec{\mathbf{E}}}$ as a reference phasor, we can sketch the phasor diagram in terms of the three current densities, as shown in Figure 8.4. It is apparent from this diagram that

$$\tan\delta = \frac{\sigma}{\omega\epsilon} \tag{8.39}$$

where $\tan\delta$ is the *loss tangent*, and δ is called the *loss tangent angle.* It is the angle between the displacement current density $\tilde{\vec{\mathbf{J}}}_d$ and the total current density $\tilde{\vec{\mathbf{J}}}_t$ in a conducting medium. The angle δ is zero for

Figure 8.4 Phasor diagram for $\tilde{\mathbf{J}}_c$, $\tilde{\mathbf{J}}_d$, and $\tilde{\mathbf{J}}_t$

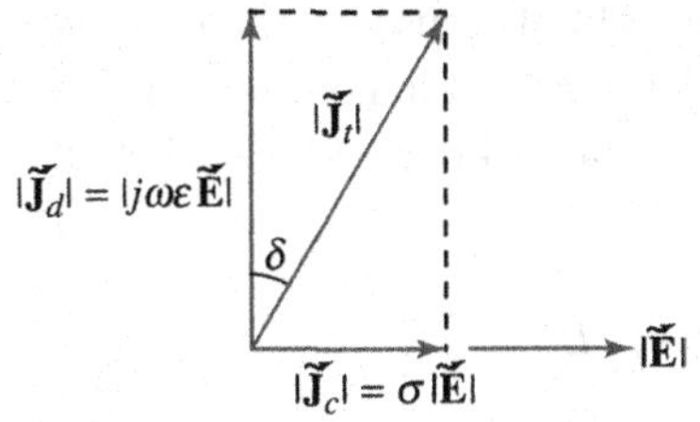

a perfectly dielectric medium and approaches 90° as the medium becomes perfectly conducting. Thus, the loss tangent provides an indirect measure of the conductivity of the medium.

The definition of the loss tangent enables us to express the complex permittivity, from (8.37), as

$$\hat{\epsilon} = \epsilon[1 - j \tan \delta] \tag{8.40}$$

Comparing (8.38) and (8.40), we find that

$$\epsilon'' = \epsilon \tan \delta \tag{8.41}$$

In terms of the complex permittivity, we can express the wave equations as

$$\nabla^2 \tilde{\mathbf{E}} = -\omega^2 \mu \hat{\epsilon} \tilde{\mathbf{E}} \tag{8.42}$$

$$\nabla^2 \tilde{\mathbf{H}} = -\omega^2 \mu \hat{\epsilon} \tilde{\mathbf{H}} \tag{8.43}$$

By replacing $\hat{\epsilon}$ with ϵ in these equations, we obtain the wave equations for a perfect dielectric medium. This means that an equation that is true for a perfect dielectric medium can be made valid for a conducting medium by replacing ϵ with $\hat{\epsilon}$. This is primarily why we chose to express the wave equations in terms of the complex permittivity of the medium. It is a common practice to write (8.42) and (8.43) as

$$\nabla^2 \tilde{\mathbf{E}} = \hat{\gamma}^2 \tilde{\mathbf{E}} \tag{8.44}$$

$$\nabla^2 \tilde{\mathbf{H}} = \hat{\gamma}^2 \tilde{\mathbf{H}} \tag{8.45}$$

where

$$\hat{\gamma}^2 = -\omega^2 \mu \hat{\epsilon} \tag{8.46}$$

and $\hat{\gamma}$ is called the *propagation constant,* which, in general, is a complex quantity.

Once again, we assume that (a) the wave propagates in the z direction, (b) the transverse components of the $\tilde{\mathbf{E}}$ and $\tilde{\mathbf{H}}$ fields are independent of variations with respect to x and y, and (c) the fields do not have longitudinal components. The time variations are implied implicitly; therefore, the partial derivatives of the $\tilde{\mathbf{E}}$ and $\tilde{\mathbf{H}}$ fields can now be treated as ordinary derivatives. Let us also imagine that the $\tilde{\mathbf{E}}$ field has only one component in the x direction. This assumption, however, constitutes no loss of generality because we can make use of the linearity theorem if the $\vec{\tilde{\mathbf{E}}}$ field also has a component in the y direction. These assumptions allow us to write (8.44) as a scalar equation in terms of the x component of the $\vec{\tilde{\mathbf{E}}}$ field as

$$\frac{d^2\tilde{E}_x(z)}{dz^2} = \hat{\gamma}^2\tilde{E}_x$$

This second-order differential equation has a solution of the form

$$\tilde{E}_x(z) = \hat{E}_f e^{-\hat{\gamma}z} + \hat{E}_b e^{\hat{\gamma}z} \tag{8.47}$$

where $\hat{E}_f$ and $\hat{E}_b$ are arbitrary constants of integration. In general, these constants may be complex quantities independent of t and z variations. We can express them as

$$\hat{E}_f = E_f e^{j\theta_f} \tag{8.48}$$

$$\hat{E}_b = E_b e^{j\theta_b} \tag{8.49}$$

As $\hat{\gamma}$ is a complex quantity, it can also be written in terms of its real and imaginary components as

$$\hat{\gamma} = j\omega\sqrt{\mu\hat{\epsilon}} = \sqrt{j\omega\mu(\sigma + j\omega\epsilon)} = \alpha + j\beta \tag{8.50}$$

Here α (the real part of $\hat{\gamma}$) is called the *attenuation constant* and is measured in nepers per meter (Np/m), and β (the imaginary part of $\hat{\gamma}$) is the *phase constant* and is measured in radians per meter (rad/m). As both radians and nepers are dimensionless quantities, the commonly used unit for the propagation constant is m^{-1}. When we separate the expression for $\hat{\gamma}$ into its components, we find that

$$\alpha = \omega\sqrt{\mu\epsilon \sec\,\delta}\,\sin\,(\delta/2) \tag{8.51a}$$

$$\beta = \omega\sqrt{\mu\epsilon \sec\,\delta}\,\cos\,(\delta/2) \tag{8.51b}$$

The preceding definitions allow us to write (8.47) as

$$\tilde{E}_x(z) = E_f e^{-\alpha z} e^{j(\theta_f - \beta z)} + E_b e^{\alpha z} e^{j(\theta_b + \beta z)} \tag{8.52a}$$

in the phasor form (frequency domain), and as

$$E_x(z,t) = E_f e^{-\alpha z}\cos(\omega t - \beta z + \theta_f) + E_b e^{\alpha z}\cos(\omega t + \beta z + \theta_b) \tag{8.52b}$$

in the time domain.

The first term on the right-hand side of (8.52) represents a time-harmonic uniform plane wave propagating in the positive z direction (forward-traveling wave). The factor $e^{-\alpha z}$ signifies that the wave attenuates as it proceeds in the z direction, as shown in Figure 8.5a. The second term represents a backward-traveling wave, which also attenuates as it makes its headway in the negative z direction, as shown in Figure 8.5b.

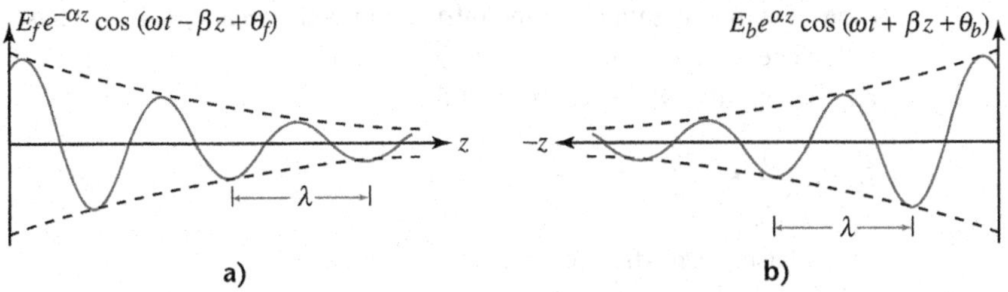

Figure 8.5 **(a)** Forward- and **(b)** backward-traveling waves in a conducting medium at time $t =$ constant

Equation (8.52) leads us to conclude that the solution of a wave equation in a conducting medium is an *attenuated (damped) wave.* The attenuation factor depends upon the conductivity of the medium. The higher the conductivity of the medium, the more pronounced the attenuation. Then the question arises: How far can the wave propagate in a conducting medium before its amplitude becomes insignificant? This question is usually answered in terms of skin depth. The **skin depth** is the distance traveled by the wave in a conducting medium at which its amplitude falls to $1/e$ of its value on the surface of that conducting medium. If we denote the skin depth by δ_c, the amplitude of the wave falls to $1/e$ when $\alpha\delta_c = 1$. Thus,

$$\delta_c = \frac{1}{\alpha} \tag{8.53}$$

The amplitude of the wave reduces to less than 1% after the wave has penetrated a distance equal to $5\delta_c$. Thereafter, the wave is assumed to be completely attenuated.

The wavelength, in a conducting medium, is

$$\lambda = \frac{2\pi}{\beta} \tag{8.54}$$

where β is the imaginary part of $\hat{\gamma}$. Note that, in a conducting medium, $\beta \neq \omega\sqrt{\mu\epsilon}$.

Once again, by setting the phase of the wave equal to a constant and differentiating with respect to t, we obtain the *phase speed* as

$$u_p = \frac{dz}{dt} = \frac{\omega}{\beta} \tag{8.55}$$

Using Maxwell's equation

$$\nabla \times \vec{\mathbf{E}} = -j\omega\mu\vec{\mathbf{H}}$$

we obtain the magnetic field intensity as

$$\vec{\mathbf{H}} = \frac{\hat{\gamma}}{j\omega\mu}[\hat{E}_f e^{-\hat{\gamma}z} - \hat{E}_b e^{-\hat{\gamma}z}]\vec{\mathbf{a}}_y \tag{8.56}$$

The intrinsic impedance $\hat{\eta}$ of the conducting medium, from (8.56), is

$$\hat{\eta} = \frac{j\omega\mu}{\hat{\gamma}} = \sqrt{\frac{\mu}{\hat{\epsilon}}} = \sqrt{\frac{j\omega\mu}{\sigma + j\omega\epsilon}} = \eta\underline{/\theta_\eta} \tag{8.57}$$

where η is the magnitude of the intrinsic impedance and θ_η is its phase angle. The intrinsic impedance, in general, is a complex quantity. We can now express (8.56) in terms of $\hat{\eta}$ as

$$\tilde{H}_y(z) = \frac{1}{\eta} E_f e^{-\alpha z} e^{j(\theta_f - \beta z - \theta_\eta)} - \frac{1}{\eta} E_b e^{\alpha z} e^{j(\theta_b + \beta z - \theta_\eta)} \tag{8.58a}$$

in the phasor form (frequency domain), and as

$$H_y(z,t) = \frac{1}{\eta} E_f e^{-\alpha z} \cos(\omega t - \beta z + \theta_f - \theta_\eta) - \frac{1}{\eta} E_b e^{\alpha z} \cos(\omega t + \beta z + \theta_b - \theta_\eta) \tag{8.58b}$$

in the time domain. Comparing (8.52) and (8.58) we find that the electric field of a traveling wave in a conducting medium leads the magnetic field by θ_η, as shown in Figure 8.6.

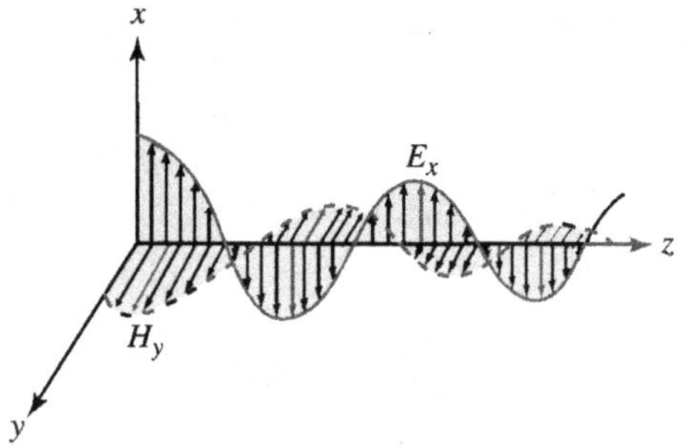

Figure 8.6 Forward-traveling wave in a conducting medium at t = constant

We can compute the average power density (power flow per unit area) from (8.52a) and (8.58a) as

$$\begin{aligned}\langle \hat{\mathbf{S}} \rangle &= \frac{1}{2}\mathrm{Re}[\tilde{\mathbf{E}} \times \tilde{\mathbf{H}}^*] \\ &= \frac{1}{2\eta} E_f^2 e^{-2\alpha z} \cos\theta_\eta \, \vec{\mathbf{a}}_z \\ &\quad - \frac{1}{2\eta} E_b^2 e^{2\alpha z} \cos\theta_\eta \, \vec{\mathbf{a}}_z \\ &\quad - \frac{1}{\eta} E_f E_b \sin(2\beta z - \theta_f + \theta_b) \sin\,\theta_\eta \, \vec{\mathbf{a}}_z \\ &= \langle \hat{\mathbf{S}}_f \rangle + \langle \hat{\mathbf{S}}_b \rangle + \langle \hat{\mathbf{S}}_{fb} \rangle \end{aligned} \tag{8.59}$$

where

$$\langle \hat{\mathbf{S}}_f \rangle = \frac{1}{2\eta} E_f^2 e^{-2\alpha z} \cos\theta_\eta \, \vec{\mathbf{a}}_z \tag{8.60a}$$

represents the average power density in the forward-travelling wave,

$$\langle \hat{\mathbf{S}}_b \rangle = -\frac{1}{2\eta} E_b^2 e^{2\alpha z} \cos\theta_\eta \vec{\mathbf{a}}_z \tag{8.60b}$$

yields the average power density in the backward-travelling wave, and

$$\langle \hat{\mathbf{S}}_{fb} \rangle = -\frac{1}{\eta} E_f E_b \sin(2\beta z - \theta_f + \theta_b) \sin\theta_\eta \, \vec{\mathbf{a}}_z \tag{8.60c}$$

is the average power density due to the cross-coupling between the forward and the backward waves. Note that the cross-coupling between the two waves varies as $\sin\theta_\eta$. Thus, the cross-coupling term disappears when $\theta_\eta = 0$—a condition that is true only when the medium is a perfect dielectric.

EXAMPLE 8.3

WORKSHEET 8

Mathcad

A 1.8-GHz wave propagates in a medium characterized by $\mu_r = 1.6$, $\epsilon_r = 25$, and $\sigma = 2.5$ S/m. The electric field intensity in the region is given by $\vec{\mathbf{E}} = 0.1e^{-\alpha z}\cos(2\pi f t - \beta z)\vec{\mathbf{a}}_x$ V/m. Determine the attenuation constant, the propagation constant, the intrinsic impedance, the phase velocity, the skin depth, and the wavelength of the wave. Obtain an expression for the $\vec{\mathbf{H}}$ field. Find the average power density in the medium.

Solution

$$\omega = 2\pi \times 1.8 \times 10^9 = 11.31 \times 10^9 \text{ rad/s}$$

$$\omega\epsilon = 11.31 \times 10^9 \times \frac{25 \times 10^{-9}}{36\pi} = 2.5$$

The complex permittivity is

$$\hat{\epsilon} = \epsilon\left[1 - j\frac{2.5}{2.5}\right] = \epsilon[1 - j1]$$

Thus, the propagation constant is

$$\begin{aligned}\hat{\gamma} &= j\omega\sqrt{\mu\hat{\epsilon}} = j\omega\sqrt{\mu_0\epsilon_0}\,\sqrt{\mu_r\epsilon_r}\,\sqrt{1 - j1} \\ &= j\frac{11.31 \times 10^9}{3 \times 10^8}\sqrt{1.6 \times 25}\,\sqrt{1 - j1} \\ &= 283.55\underline{/67.5^\circ} = 108.51 + j261.97 \text{ m}^{-1}\end{aligned}$$

The attenuation constant and the phase constant are

$$\alpha = 108.51 \text{ Np/m}$$

$$\beta = 261.97 \text{ rad/m}$$

The intrinsic impedance is

$$\hat{\eta} = \sqrt{\frac{\mu}{\hat{\epsilon}}} = \sqrt{\frac{\mu_0}{\epsilon_0}}\sqrt{\frac{\mu_r}{\epsilon_r}}\sqrt{\frac{1}{1.414\underline{/-45^\circ}}}$$

$$= (120\pi)\sqrt{\frac{1.6}{25}}\frac{1}{\sqrt{1.414}}\underline{/22.5^\circ} = 80.2\underline{/22.5^\circ}\ \Omega$$

The phase velocity: $\vec{\mathbf{u}}_p = \frac{\omega}{\beta}\vec{\mathbf{a}}_z = \frac{11.31 \times 10^9}{261.97}\vec{\mathbf{a}}_z = 4.32 \times 10^7\vec{\mathbf{a}}_z$ m/s

The wavelength: $\lambda = \frac{2\pi}{\beta} = \frac{2\pi}{261.97} = 0.024$ m or 24 mm

The skin depth: $\delta_c = \frac{1}{\alpha} = \frac{1}{108.51} = 9.22 \times 10^{-3}$ m or 9.22 mm

The electric field intensity phasor for the forward-traveling wave is

$$\tilde{E}_x(z) = 0.1e^{-108.51z}e^{-j261.97z} \text{ V/m}$$

The corresponding magnetic field intensity is

$$\tilde{H}_y(z) = \frac{1}{\hat{\eta}}\tilde{E}_x(z) = 1.25e^{-108.51z}e^{-j261.97z}e^{-j22.5^\circ} \text{ mA/m}$$

Thus, the average power density in the z direction is

$$\begin{aligned}\langle \hat{\mathbf{S}}_f \rangle &= \tfrac{1}{2}\text{Re}[\vec{\tilde{\mathbf{E}}} \times \vec{\tilde{\mathbf{H}}}^*] \\ &= \tfrac{1}{2} \times 0.1 \times 1.25 \times 10^{-3} \times e^{-217.02z}\cos(22.5^\circ)\vec{\mathbf{a}}_z \\ &= 57.7e^{-217.02z}\vec{\mathbf{a}}_z\ \mu\text{W/m}^2\end{aligned}$$

• • •

8.6 Plane wave in a good conductor

The total current in a conducting medium includes the conduction current and the displacement current, as was shown in Figure 8.4. Any increase in the conduction current is accompanied by an increase in the loss tangent angle δ and the loss tangent tan δ. Therefore, it is possible for the term tan $\delta(\sigma/\omega\epsilon)$ to dominate in (8.40). However, for this to happen either the conductivity σ of the medium is very high or the wave frequency is low. In either case, the conducting medium behaves as a *good conductor* (or a *high-loss material*) as long as $\sigma \gg \omega\epsilon$. In this book, we view a conducting medium as a good conductor when

$$\frac{\sigma}{\omega\epsilon} \geq 10 \tag{8.61}$$

Note that (8.61) is a very broad definition of a good conductor. For example, copper (5.8×10^7 S/m) is a good conductor even at very high frequencies (10^{16} Hz); however, seawater (4 S/m) behaves like a good conductor up to a frequency of 8 MHz. For a good conductor, we can approximate (8.40) as

$$\hat{\epsilon} \approx \frac{\sigma}{j\omega} \tag{8.62}$$

Substituting (8.62) in (8.50), we obtain the propagation constant as

$$\hat{\gamma} = j\omega\sqrt{\mu\hat{\epsilon}} \approx j\omega\sqrt{\frac{\mu\sigma}{j\omega}} = \sqrt{j\omega\mu\sigma} = \sqrt{\omega\mu\sigma}\,\underline{/45^\circ} \tag{8.63}$$

Thus, in a good conductor, the attenuation constant and the phase constant are

$$\alpha = \beta = \sqrt{\frac{\omega\mu\sigma}{2}} \tag{8.64}$$

We can also obtain the approximate equations for the intrinsic impedance, the phase speed, and the skin depth as

$$\hat{\eta} = \sqrt{\frac{\mu}{\hat{\epsilon}}} = \sqrt{\frac{\omega\mu}{\sigma}}\,\underline{/45^\circ} \tag{8.65}$$

$$u_p = \frac{\omega}{\beta} = \sqrt{\frac{2\omega}{\mu\sigma}} \tag{8.66}$$

$$\delta_c = \frac{1}{\alpha} = \sqrt{\frac{2}{\omega\mu\sigma}} = \sqrt{\frac{1}{f\pi\mu\sigma}} \tag{8.67}$$

From these equations it is obvious that α, β, $\hat{\eta}$, and u_p vary directly as $\sqrt{\omega}$. Thus, the shape of a wave comprising many different frequencies will keep on changing as it progresses; that is, the signal is distorted by the time it reaches its destination. A medium in which a signal becomes distorted is said to be a **dispersive medium**. A conducting medium is, in general, a dispersive medium.

For all practical purposes, the wave vanishes after traveling a distance of $5\delta_c$ in a conducting medium. The skin depth δ_c for copper at a frequency of 1 MHz is approximately 0.07 mm. The amplitude of the wave becomes insignificant after penetrating a distance of 0.35 mm. In good conductors, the wave attenuates very rapidly and the fields are confined to the region near the surface of the conductor. This phenomenon is called the **skin effect.**

8.6.1 Surface resistance

Let the electric field intensity in a good conductor be

$$\tilde{\vec{\mathbf{E}}} = Ee^{-\hat{\gamma}z}\vec{\mathbf{a}}_x$$

where $\hat{\gamma} = \alpha + j\beta = \sqrt{2}\,\alpha\underline{/45^\circ}$ and α is given by (8.64). Neglecting the displacement current density in a good conductor, the total current is

$$\tilde{\vec{\mathbf{J}}} = \sigma Ee^{-\gamma z}\vec{\mathbf{a}}_x$$

The average power dissipated (power loss) per unit volume is

$$\langle \hat{S}_d \rangle = \tfrac{1}{2}\tilde{\vec{\mathbf{E}}} \cdot \tilde{\vec{\mathbf{J}}}^* = \tfrac{1}{2}\sigma E^2 e^{-2\alpha z}$$

Let us concentrate on a region bounded by $0 \leq x \leq b, 0 \leq y \leq w$, and $0 \leq z \leq \infty$. Then, the total power dissipated within this region is

$$\langle P_d \rangle = \frac{1}{2}\sigma E^2 \int_0^\infty e^{-2\alpha z} dz \int_0^b dx \int_0^w dy$$
$$= \frac{1}{4\alpha} \sigma b w E^2 \tag{8.68}$$

The total current along the x direction is

$$\tilde{I} = \int_s \tilde{\mathbf{J}} \cdot \overrightarrow{d\mathbf{s}} = \sigma E \int_0^\infty e^{-\hat{\gamma} z} \, dz \int_0^w dy$$
$$= \frac{1}{\sqrt{2}\alpha} \sigma w E \underline{/-45^\circ}$$

If R is the resistance of the block, then the power that it dissipates is

$$\langle P_d \rangle = \frac{1}{2} I^2 R = \frac{1}{4}\left[\frac{\sigma w E}{\alpha}\right]^2 R \tag{8.69}$$

Comparing (8.68) and (8.69), we obtain the resistance of the block as

$$R = \frac{b\alpha}{\sigma w} = \frac{b}{\sigma w \delta_c} \tag{8.70}$$

where $\delta_c = 1/\alpha$ is the skin depth. As the current is along the x direction, b is the length of the block, and $w\delta_c$ is the equivalent cross-sectional area with w as its width and δ_c as its thickness (or depth).

The *skin resistance* or *surface resistivity* is defined as the resistance of a plane conductor of unit length ($b = 1$), unit width ($w = 1$), and thickness δ_c. Thus,

$$R_s = \frac{\alpha}{\sigma} = \frac{1}{\sigma \delta_c} \tag{8.71}$$

Note that the skin resistance is computed just like the dc resistance by assuming that the conductor is exactly one skin depth in thickness as long as its actual thickness is greater than the skin depth.

Although we have computed the skin resistance on the basis that a plane wave is propagating through a flat block, equation (8.71) can also be used to obtain an approximate value of the skin resistance for a cylindrical conductor. When the current is along the length of a cylindrical conductor of radius a such that $a > \delta_c$, the skin resistance per unit length is

$$R_s = \frac{1}{2\pi a \sigma \delta_c} \tag{8.72}$$

Note that this is also the resistance of a hollow conductor with outer radius a and thickness δ_c. Hence, by coating the surface of a dielectric material with a thin film of silver (thickness $= 1\delta_c$), we can make it behave like a good conductor.

EXAMPLE 8.4

Repeat Example 8.3 when the wave frequency is 1.8 kHz.

Solution

$$\omega = 2\pi \times 1800 = 11.31 \times 10^3 \text{ rad/s}$$

$$\omega\epsilon = 11.31 \times 10^3 \times \frac{25 \times 10^{-9}}{36\pi} = 2.5 \times 10^{-6}$$

$$\frac{\sigma}{\omega\epsilon} = \frac{2.5}{2.5 \times 10^{-6}} = 10^6$$

$\sigma/\omega\epsilon \gg 10$, so the medium acts like a good conductor. Thus, the propagation constant, from (8.63), is

$$\hat{\gamma} = \sqrt{11.31 \times 10^3 \times 1.6 \times 4\pi \times 10^{-7} \times 2.5 \underline{/45^\circ}}$$
$$= 0.2384\underline{/45^\circ} \text{ m}^{-1}$$

Hence,

$$\alpha = 0.1686 \text{ Np/m}$$
$$\beta = 0.1686 \text{ rad/m}$$

The intrinsic impedance, from (8.65), is

$$\hat{\eta} = \sqrt{\frac{11.31 \times 10^3 \times 1.6 \times 4\pi \times 10^{-7}}{2.5}\underline{/45^\circ}}$$
$$= 0.0954\underline{/45^\circ} \ \Omega$$

The phase speed: $\vec{u}_p = \frac{\omega}{\beta} = \frac{2\pi \times 1800}{0.1686} = 67.08 \times 10^3$ m/s

The wavelength: $\lambda = \frac{2\pi}{\beta} = \frac{2\pi}{0.1686} = 37.27$ m

The skin depth: $\delta_c = \frac{1}{\alpha} = \frac{1}{0.1686} = 5.93$ m

The electric field intensity, in phasor form, is

$$\tilde{E}_x(z) = 0.1e^{-0.1686z}e^{-j0.1686z} \text{ V/m}$$

The corresponding $\vec{\mathbf{H}}$ field is

$$\tilde{H}_y(z) = \frac{1}{\hat{\eta}}\tilde{E}_x(z)$$
$$= 1.048e^{-0.1686z}e^{-j0.1686z}e^{-j45^\circ}$$

Finally, the average power density is

$$\langle \hat{\mathbf{S}} \rangle = \tfrac{1}{2}\text{Re}[\vec{\mathbf{E}} \times \vec{\mathbf{H}}^*]$$
$$= \tfrac{1}{2} \times 0.1 \times 1.048 \times \cos(45^\circ)e^{-0.3372z}\,\vec{\mathbf{a}}_z$$
$$= 0.037e^{-0.3372z}\,\vec{\mathbf{a}}_z \text{ W/m}^2$$

• • •

8.7 Plane wave in a good dielectric

A good dielectric is a conducting medium in which the displacement current dominates the conduction current. In other words, a poorly conducting medium may be viewed as a good dielectric as long as $\sigma \ll \omega\epsilon$. In this book, we will classify a medium as being a good dielectric if

$$\frac{\sigma}{\omega\epsilon} \le 0.1 \tag{8.73}$$

Note that (8.73) is a very broad definition of a good dielectric. This condition is satisfied when the conductivity of the medium is low or the wave frequency is very high.

The first-order approximation for $\sqrt{\hat{\epsilon}}$ in a good dielectric medium, using the binomial expansion, is

$$\sqrt{\hat{\epsilon}} = \sqrt{\epsilon\left[1 - j\frac{\sigma}{\omega\epsilon}\right]} \approx \sqrt{\epsilon}\left[1 - j\frac{\sigma}{2\omega\epsilon}\right] \tag{8.74}$$

Using (8.74), we obtain the approximate expression for the propagation constant, from (8.50), as

$$\hat{\gamma} = \frac{\sigma}{2}\sqrt{\frac{\mu}{\epsilon}} + j\omega\sqrt{\mu\epsilon} \tag{8.75}$$

Thus, the attenuation constant and the phase constant β in a good dielectric are

$$\alpha = \frac{\sigma}{2}\sqrt{\frac{\mu}{\epsilon}} \tag{8.76}$$

$$\beta = \omega\sqrt{\mu\epsilon} \tag{8.77}$$

Equation (8.77) states that the phase constant in a good dielectric medium is essentially the same as that in a perfect dielectric. However, (8.76) signifies that the fields do attenuate as they travel in a good dielectric medium. The attenuation factor, however, is very small compared with that of a good conductor. Many books on this subject suggest that α may be considered as zero. We disagree; a good dielectric is not a perfect dielectric. In addition, a wave must attenuate as it propagates in a finitely conducting medium.

The intrinsic impedance for a good dielectric becomes

$$\hat{\eta} = \sqrt{\frac{\mu}{\epsilon}\left[1 + j\frac{\sigma}{2\omega\epsilon}\right]} \approx \sqrt{\frac{\mu}{\epsilon}} \tag{8.78}$$

EXAMPLE 8.5

A 180-MHz wave travels in a medium characterized by $\mu_r = 1$, $\epsilon_r = 25$, and $\sigma = 2.5$ mS/m. The electric field intensity is given by $\vec{\mathbf{E}} = 37.7e^{-\hat{\gamma}}\vec{\mathbf{a}}_x$ V/m. Determine the intrinsic impedance, the attenuation constant, the propagation constant, the phase speed, the skin depth, and the wavelength of the wave. Obtain an expression for the $\vec{\mathbf{H}}$ field. Find the average power density in the medium.

Solution

$$\omega = 2\pi \times 1.8 \times 10^6 = 1.131 \times 10^9 \text{ rad/s}$$

$$\omega\epsilon = 1.131 \times 10^9 \times \frac{25 \times 10^{-9}}{36\pi} = 0.25$$

$$\frac{\sigma}{\omega\epsilon} = \frac{2.5 \times 10^{-3}}{0.25} = 0.01$$

Thus, the medium behaves like a good dielectric. The intrinsic impedance, from (8.78), is

$$\eta = \sqrt{\frac{\mu_0}{\epsilon_0}}\sqrt{\frac{\mu_r}{\epsilon_r}} = 120\pi\frac{1}{\sqrt{25}} = 75.398\ \Omega$$

The attenuation constant, from (8.76), can also be expressed as

$$\begin{aligned}\alpha &= \tfrac{1}{2}\sigma\eta \\ &= \tfrac{1}{2} \times 2.5 \times 10^{-3} \times 75.398 = 0.094 \text{ Np/m}\end{aligned} \tag{8.79}$$

The phase constant is

$$\begin{aligned}\beta &= \omega\sqrt{\mu_0\epsilon_0}\sqrt{\mu_r\epsilon_r} \\ &= \frac{1.131 \times 10^9}{3 \times 10^8} \times 5 = 18.85 \text{ rad/m}\end{aligned}$$

Hence, the propagation constant is

$$\hat{\gamma} = \alpha + j\beta = 0.094 + j18.85 \text{ m}^{-1}$$

We can now compute the phase speed, skin depth, and wavelength as

$$u_p = \frac{\omega}{\beta} = \frac{1.131 \times 10^9}{18.85} = 6 \times 10^7 \text{ m/s}$$

$$\delta_c = \frac{1}{\alpha} = \frac{1}{0.094} = 10.64 \text{ m}$$

$$\lambda = \frac{2\pi}{\beta} = \frac{2\pi}{18.85} = 0.3333 \text{ m} \quad \text{or} \quad 33.33 \text{ cm}$$

The electric field intensity is given as

$$\tilde{\vec{\mathbf{E}}} = 37.7e^{-0.094z}e^{-j18.85z}\,\vec{\mathbf{a}}_x \text{ V/m}$$

The wave propagates in the z direction, thus the $\tilde{\vec{\mathbf{H}}}$ field, from (8.27), is

$$\tilde{\vec{\mathbf{H}}} = 0.5e^{-0.094z}e^{-j18.85z}\,\vec{\mathbf{a}}_y \text{ A/m}$$

Finally, the average power density in the medium is

$$\begin{aligned}\langle\hat{\mathbf{S}}\rangle &= \tfrac{1}{2}\text{Re}[\tilde{\vec{\mathbf{E}}} \times \tilde{\vec{\mathbf{H}}}^*] \\ &= 9.425e^{-0.188z}\,\vec{\mathbf{a}}_z \text{ W/m}^2\end{aligned}$$

• • •

8.8 Polarization of a wave

It is a common practice to describe an electromagnetic wave by its polarization. By definition, the **polarization** of a wave is the locus of the tip of the electric field at a given point as a function of time. When two or more waves of the same frequency propagate in the same direction, the polarization is then defined by the composite wave that is obtained by superimposing all the waves. A wave is said to be *linearly polarized* when at some point in the medium the electric field oscillates along a straight line as a function of time. If the tip of the electric field traces a circle, the wave is said to be *circularly polarized.* The wave is *elliptically polarized* when the electric field follows an elliptical path. An unpolarized wave, such as a light wave, is usually referred to as a *randomly polarized* wave.

The polarization of a wave depends upon the transmitting source (such as an antenna). In the standard broadcast frequency band, the vertical antenna is designed to transmit a *ground wave*, which is vertically polarized because the $\vec{\mathbf{E}}$ field from the antenna to ground is vertical. In other applications the antennas are placed in a horizontal plane to transmit a horizontally polarized wave. Both vertically and horizontally polarized waves are examples of linearly polarized waves.

8.8.1 A linearly polarized wave

The electric field intensity of a uniform plane wave in a conducting medium can be written in the time domain as

$$\vec{\mathbf{E}}(z,t) = E_0 e^{-\alpha z}\cos(\omega t - \beta z)\,\vec{\mathbf{a}}_x$$

This represents a linearly polarized wave because the $\vec{\mathbf{E}}$ field is always in the x direction in a z = constant plane. For example, the electric field intensity when $z = 0$ is

$$\vec{\mathbf{E}}(0,t) = E_0 \cos\omega t\,\vec{\mathbf{a}}_x$$

Its plot as a function of time is given in Figure 8.7.

Let us now consider a uniform plane wave having the following components of electric field intensity:

$$E_x(z,t) = E_{0x}e^{-\alpha z}\cos(\omega t - \beta z + \theta_x) \tag{8.80}$$

$$E_y(z,t) = E_{0y}e^{-\alpha z}\cos(\omega t - \beta z + \theta_y) \tag{8.81}$$

At any point in a $z = 0$ plane, these field components become

$$E_x(0,t) = E_{0x}\cos(\omega t + \theta_x) \tag{8.82}$$

$$E_y(0,t) = E_{0y}\cos(\omega t + \theta_y) \tag{8.83}$$

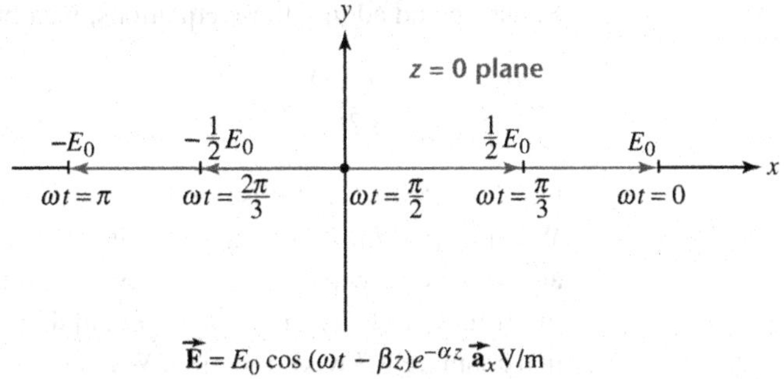

Figure 8.7 Linear polarization

When the two components are in phase; i.e., $\theta_x = \theta_y = \theta$, then the preceding equations yield

$$E_x(0, t) = \frac{E_{0x}}{E_{0y}} E_y(0, t) \tag{8.84}$$

This equation describes a linear relationship between the two components, as sketched in Figure 8.8. Hence, the wave consisting of two components of the electric field as given by (8.80) and (8.81) when $\theta_x = \theta_y$ is also a linearly polarized wave.

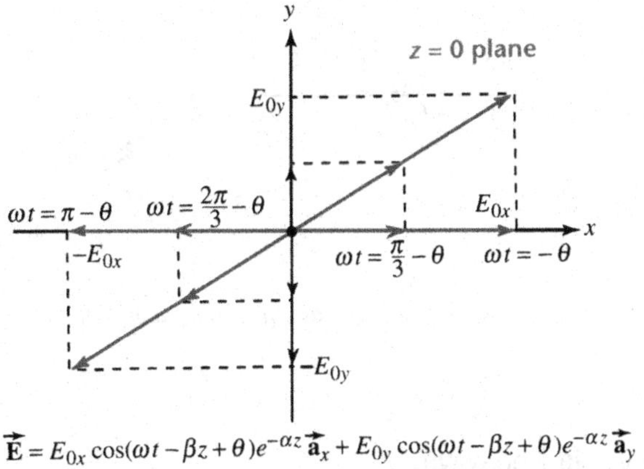

Figure 8.8 Another example of linear polarization

8.8.2 An elliptically polarized wave

When the field components in (8.80) and (8.81) are in quadrature, say $\theta_y = \theta_x - \pi/2$, then the two equations become

$$E_x(z, t) = E_{0x} e^{-\alpha z} \cos(\omega t - \beta z + \theta_x) \tag{8.85}$$

$$E_y(z, t) = E_{0y} e^{-\alpha z} \sin(\omega t - \beta z + \theta_x) \tag{8.86}$$

In a $z = 0$ plane, equations (8.85) and (8.86) yield

$$\frac{E_x(0, t)}{E_{0x}} = \cos(\omega t + \theta_x) \tag{8.87}$$

$$\frac{E_y(0, t)}{E_{0y}} = \sin(\omega t + \theta_x) \tag{8.88}$$

Squaring and adding these equations, we obtain

$$\frac{E_x^2(0,t)}{E_{0x}^2} + \frac{E_y^2(0,t)}{E_{0y}^2} = 1 \tag{8.89}$$

which describes an ellipse in the $z = 0$ plane, as shown in Figure 8.9. When $E_{0x} > E_{0y}$, the major axis is $2E_{0x}$ and the minor axis is $2E_{0y}$, and vice versa when $E_{0y} > E_{0x}$. By substituting the increasing values of t (time) in (8.87) and (8.88), we find that the tip of the $\vec{\mathbf{E}}$ field rotates in a counterclockwise direction. When we extend our right-hand thumb in the direction of propagation of the wave (z direction), we find that the fingers of the right hand curl in the direction of rotation of the $\vec{\mathbf{E}}$ field. Thus, (8.85) and (8.86) represent a *right-handed elliptically polarized wave.* However, if we had set $\theta_y = \theta_x + \pi/2$, we would have obtained a *left-handed elliptically polarized wave.*

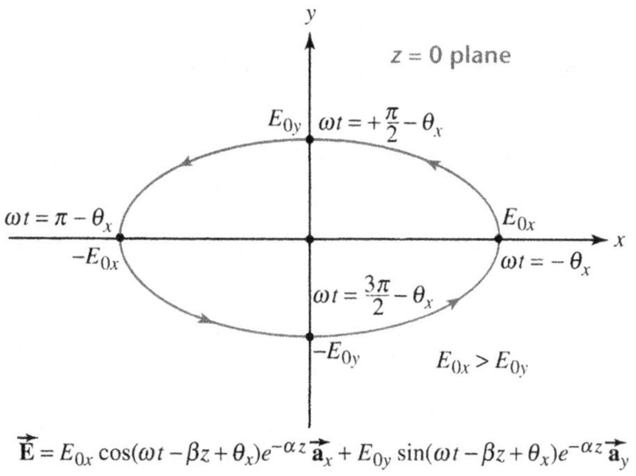

Figure 8.9 A right-handed elliptically polarized wave

8.8.3 A circularly polarized wave

When $E_{0x} = E_{0y} = E_0$, and $\theta_y = \theta_x - \pi/2$ in (8.80) and (8.81), the two electric field components become

$$E_x(z,t) = E_0 e^{-\alpha z}\cos(\omega t - \beta z + \theta_x) \tag{8.90}$$

$$E_y(z,t) = E_0 e^{-\alpha z}\sin(\omega t - \beta z + \theta_x) \tag{8.91}$$

In a $z = 0$ plane, the two components become

$$E_x(0,t) = E_0\cos(\omega t + \theta_x) \tag{8.92}$$

$$E_y(0,t) = E_0\sin(\omega t + \theta_x) \tag{8.93}$$

By squaring and adding these equations, we get

$$E_x^2(0,t) + E_y^2(0,t) = E_0^2 \tag{8.94}$$

which is an equation of a circle. By evaluating (8.92) and (8.93) for various values of t we find that the wave is *right-handed circularly polarized,* as shown in Figure 8.10. We would have obtained a *left-handed circularly polarized wave* if we had set $\theta_y = \theta_x + \pi/2$.

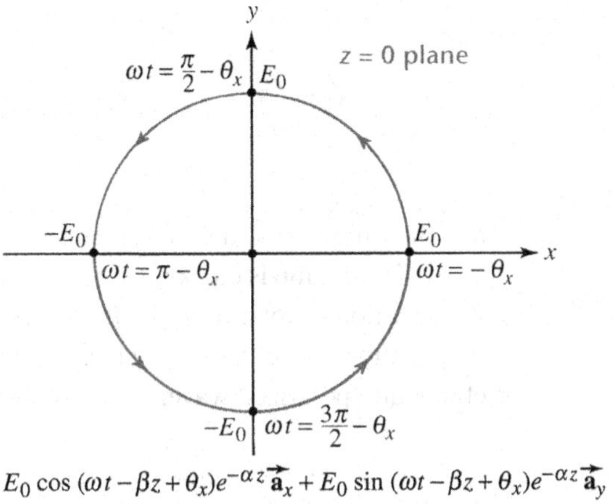

Figure 8.10 A right-handed circularly polarized wave

EXAMPLE 8.6

Determine the polarization of the wave if the electric field intensity in a region is given by $\tilde{\mathbf{E}} = (3\vec{\mathbf{a}}_x + j4\vec{\mathbf{a}}_y)e^{-0.2z}e^{-j0.5z}$ V/m.

Solution We can express the $\vec{\mathbf{E}}$ field in the time domain as

$$E_x(z,t) = 3e^{-0.2z}\cos(\omega t - 0.5z)$$
$$E_y(z,t) = -4e^{-0.2z}\sin(\omega t - 0.5z)$$

In a $z = 0$ plane, these components become

$$\tfrac{1}{3}E_x(0,t) = \cos\omega t \tag{8.95}$$
$$\tfrac{1}{4}E_y(0,t) = -\sin\omega t \tag{8.96}$$

Squaring and adding, we get

$$\tfrac{1}{9}E_x^2(0,t) + \tfrac{1}{16}E_y^2(0,t) = 1$$

which is an equation of an ellipse. Hence the wave is elliptically polarized. The major axis is along the y axis and the minor axis is along the x axis, as shown in Figure 8.11.

To determine the direction of rotation, let us substitute some values for t or ωt in (8.95) and (8.96). When $\omega t = 0$ $(t = 0)$, we have

$$E_x(0,0) = 3 \quad \text{and} \quad E_y(0,0) = 0$$

and the tip of the $\vec{\mathbf{E}}$ field is on the positive x axis, as shown in the figure. By letting $\omega t = \pi/2 (t = \pi/2\omega)$, we obtain

$$E_x(0,\pi/2\omega) = 0 \quad \text{and} \quad E_y(0,\pi/2\omega) = -4$$

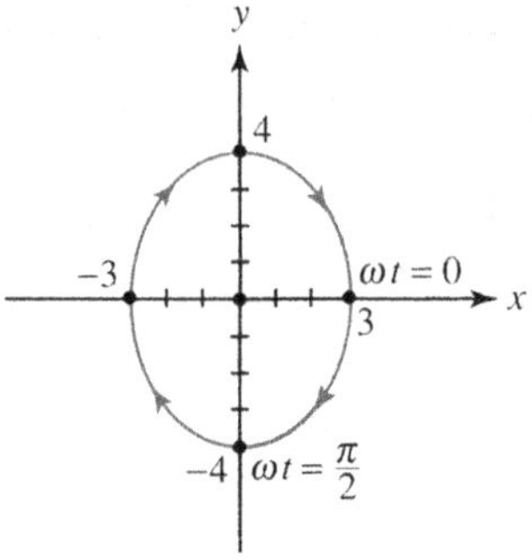

Figure 8.11 Left-handed elliptically polarized wave

and the tip of the $\vec{\mathbf{E}}$ field has moved to a point on the negative y axis. Thus, the rotation is clockwise. Since the fingers of the left-hand curl in the direction of rotation when the thumb is extended in the direction of propagation (z direction), the given electric field represents a left-handed elliptically polarized wave. • • •

8.9 Normal incidence of uniform plane waves

We have, to this point, focused our attention on the propagation of a uniform plane wave in an unbounded medium. We now consider a monochromatic uniform plane wave that travels through one medium and then enters another medium of infinite extent. At this stage we assume that the interface between the two media is normal to the direction of propagation of the incoming wave. We further presume that (a) the incoming wave, called the **incident wave**, is propagating in the z direction; (b) the interface is an infinite plane at $z = 0$; (c) the region to the left of the interface is medium 1 ($z \leq 0$); and (d) the region to the right of the interface is medium 2 ($z \geq 0$). At the interface, we expect a part of the wave to penetrate the boundary and continue its propagation in medium 2. This wave is referred to as the **transmitted wave**. The remainder of the wave is reflected at the interface and then propagates in the negative z direction. This wave is called the **reflected wave**. Thus, both the incident and transmitted waves propagate in the positive z direction, whereas the reflected wave propagates in the negative z direction. The incident and reflected waves are in medium 1, and the transmitted wave is in medium 2. If we treat the incident wave as the forward-traveling wave, the reflected wave is then the backward-traveling wave.

8.9.1 Conductor-conductor interface

Let us first assume that the interface is between two finitely conducting media, as shown in Figure 8.12. To simplify the discussion without any loss of generality, let us consider that the $\tilde{\vec{\mathbf{E}}}$ field of the incident wave

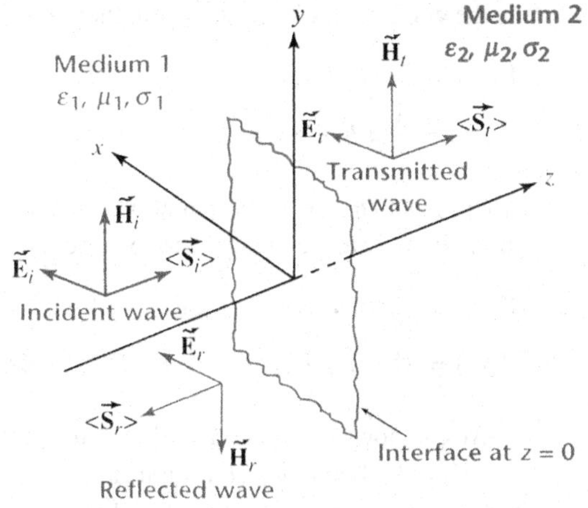

Figure 8.12 An interface normal to the direction of propagation of a uniform plane wave

is polarized in the x direction and has an amplitude of $\hat{E}_0$ at the interface. If $\hat{\gamma}_1$ is the propagation constant and $\hat{\eta}_1$ is the intrinsic impedance in medium 1, then we can express the electric and the magnetic field intensities of the incident wave as

$$\tilde{\mathbf{E}}_i(z) = \hat{E}_0 e^{-\hat{\gamma}_1 z}\, \vec{\mathbf{a}}_x \tag{8.97a}$$

$$\tilde{\mathbf{H}}_i(z) = \frac{1}{\hat{\eta}_1} \hat{E}_0 e^{-\hat{\gamma}_1 z}\, \vec{\mathbf{a}}_y \tag{8.97b}$$

where the subscript i stands for the incident wave. Subscripts r and t are used for the reflected and transmitted waves, respectively. By defining, at the interface, a complex quantity known as the *reflection coefficient*,

$$\hat{\rho} = \frac{\tilde{\mathbf{E}}_r(0)}{\tilde{\mathbf{E}}_i(0)} \tag{8.98}$$

we can write the reflected fields as

$$\tilde{\mathbf{E}}_r(z) = \hat{\rho} \hat{E}_0 e^{\hat{\gamma}_1 z}\, \vec{\mathbf{a}}_x \tag{8.99a}$$

$$\tilde{\mathbf{H}}_r(z) = -\frac{1}{\hat{\eta}_1} \hat{\rho} \hat{E}_0 e^{\hat{\gamma}_1 z}\, \vec{\mathbf{a}}_y \tag{8.99b}$$

The negative sign for the $\tilde{\mathbf{H}}$ field in (8.99b) is in accordance with the flow of energy in the negative z direction because in (8.99a) we have tacitly assumed that the reflected $\tilde{\mathbf{E}}$ field is in the same direction as its incident counterpart. Note that we could have assumed the incident and the reflected $\tilde{\mathbf{H}}$ fields to be in the same direction and changed the direction of the reflected $\tilde{\mathbf{E}}$ field to ascertain the energy propagation in the negative z direction.

If we define another complex quantity

$$\hat{\tau} = \frac{\tilde{\mathbf{E}}_t(0)}{\tilde{\mathbf{E}}_i(0)} \tag{8.100}$$

as the coefficient of transmission, then the transmitted fields are

$$\vec{\tilde{\mathbf{E}}}_t(z) = \hat{\tau}\hat{E}_0 e^{-\hat{\gamma}_2 z}\,\vec{\mathbf{a}}_x \tag{8.101a}$$

$$\vec{\tilde{\mathbf{H}}}_t(z) = \frac{1}{\hat{\eta}_2}\,\hat{\tau}\hat{E}_0 e^{-\hat{\gamma}_2 z}\,\vec{\mathbf{a}}_y \tag{8.101b}$$

where $\hat{\gamma}_2$ and $\hat{\eta}_2$ are the propagation constant and the intrinsic impedance in medium 2, respectively. The total fields in medium 1 are

$$\vec{\tilde{\mathbf{E}}}_1(z) = \vec{\tilde{\mathbf{E}}}_i(z) + \vec{\tilde{\mathbf{E}}}_r(z) = \hat{E}_0[e^{-\hat{\gamma}_1 z} + \hat{\rho}e^{\hat{\gamma}_1 z}]\,\vec{\mathbf{a}}_x \tag{8.102a}$$

$$\vec{\tilde{\mathbf{H}}}_1(z) = \vec{\tilde{\mathbf{H}}}_i(z) + \vec{\tilde{\mathbf{H}}}_r(z) = \frac{1}{\hat{\eta}_1}\hat{E}_0[e^{-\hat{\gamma}_1 z} - \hat{\rho}e^{\hat{\gamma}_1 z}]\,\vec{\mathbf{a}}_y \tag{8.102b}$$

We can now determine the reflection and transmission coefficients by applying the boundary conditions at $z = 0$. From the continuity of the tangential components of the $\vec{\tilde{\mathbf{E}}}$ field at the boundary, we obtain

$$1 + \hat{\rho} = \hat{\tau} \tag{8.103}$$

As both the media are finitely conducting, we do not expect any surface current density at the boundary. Hence the tangential components of the $\vec{\tilde{\mathbf{H}}}$ field are also continuous at the interface. Applying this boundary condition, we obtain

$$\frac{1}{\hat{\eta}_1} - \frac{1}{\hat{\eta}_1}\hat{\rho} = \frac{1}{\hat{\eta}_2}\hat{\tau}$$

or

$$1 - \hat{\rho} = \frac{\hat{\eta}_1}{\hat{\eta}_2}\hat{\tau} \tag{8.104}$$

Manipulating (8.103) and (8.104), we get

$$\hat{\rho} = \frac{\hat{\eta}_2 - \hat{\eta}_1}{\hat{\eta}_2 + \hat{\eta}_1} \tag{8.105}$$

and

$$\hat{\tau} = \frac{2\hat{\eta}_2}{\hat{\eta}_2 + \hat{\eta}_1} \tag{8.106}$$

as the reflection and transmission coefficients, respectively.

The average power density of the transmitted wave in medium 2 is

$$\begin{aligned}\langle \hat{\mathbf{S}}_t \rangle &= \frac{1}{2}\,\mathrm{Re}[\vec{\tilde{\mathbf{E}}}_t(z) \times \vec{\tilde{\mathbf{H}}}_t^*(z)] \\ &= \frac{1}{2\eta_2}\,\tau^2 E_0^2 e^{-2\alpha_2 z}\cos\theta_{\eta_2}\,\vec{\mathbf{a}}_z \end{aligned}\tag{8.107}$$

where

$$\hat{\gamma}_2 = \alpha_2 + j\beta_2$$
$$\hat{\eta}_2 = \eta_2 e^{j\theta_{\eta_2}}$$
$$\tau^2 = \hat{\tau}\hat{\tau}^*$$

The total power density in medium 1 is

$$\begin{aligned}\langle \hat{\mathbf{S}}_1 \rangle &= \frac{1}{2}\,\mathrm{Re}\,[\vec{\mathbf{E}}_1(z) \times \vec{\mathbf{H}}_1^*(z)] \\ &= \frac{1}{2\eta_1}\, E_0^2\, \mathrm{Re}[(e^{-2\alpha_1 z} - \rho^2 e^{2\alpha_1 z} + \hat{\rho} e^{j2\beta_1 z} - \hat{\rho}^* e^{-j2\beta_1 z}) e^{j\theta_{\eta 1}}]\, \vec{\mathbf{a}}_z\end{aligned}$$

where

$$\begin{aligned}\hat{\gamma}_1 &= \alpha_1 + j\beta_1 \\ \hat{\eta}_1 &= \eta_1 e^{j\theta_{\eta 1}} \\ \rho^2 &= \hat{\rho}\hat{\rho}^*\end{aligned}$$

If we write the reflection coefficient in exponential form as

$$\hat{\rho} = \rho e^{j\theta_\rho}$$

then the average power density in medium 1 becomes

$$\begin{aligned}\langle \hat{\mathbf{S}}_1 \rangle &= \frac{1}{2\eta_1}\, E_0^2 e^{-2\alpha_1 z} \cos\theta_{\eta_1}\, \vec{\mathbf{a}}_z \\ &\quad - \frac{1}{2\eta_1}\, \rho^2 E_0^2 e^{2\alpha_1 z} \cos\theta_{\eta_1}\, \vec{\mathbf{a}}_z \\ &\quad - \frac{1}{\eta_1}\, \rho E_0^2 \sin(2\beta_1 z + \theta_\rho) \sin\theta_{\eta_1} \vec{\mathbf{a}}_z \qquad (8.108a) \\ &= \langle \hat{\mathbf{S}}_i \rangle + \langle \hat{\mathbf{S}}_r \rangle + \langle \hat{\mathbf{S}}_{ir} \rangle \qquad (8.108b)\end{aligned}$$

where

$$\begin{aligned}\langle \hat{\mathbf{S}}_i \rangle &= \frac{1}{2}\,\mathrm{Re}\,[\vec{\mathbf{E}}_i(z) \times \vec{\mathbf{H}}_i^*(z)] \\ &= \frac{1}{2\eta_1}\, E_0^2 e^{-2\alpha_1 z} \cos\theta_{\eta_1}\, \vec{\mathbf{a}}_z \qquad (8.108c)\end{aligned}$$

is the average power density of the incident wave,

$$\begin{aligned}\langle \hat{\mathbf{S}}_r \rangle &= \frac{1}{2}\,\mathrm{Re}\,[\vec{\mathbf{E}}_r(z) \times \vec{\mathbf{H}}_r^*(z)] \\ &= -\frac{1}{2\eta_1}\, \rho^2 E_0^2 e^{2\alpha_1 z} \cos\theta_{\eta_1}\, \vec{\mathbf{a}}_z \qquad (8.108d)\end{aligned}$$

is the average power density of the reflected wave, and

$$\begin{aligned}\langle \hat{\mathbf{S}}_{ir} \rangle &= \frac{1}{2}\,\mathrm{Re}\,[\vec{\mathbf{E}}_i(z) \times \vec{\mathbf{H}}_r^*(z) + \vec{\mathbf{E}}_r(z) \times \vec{\mathbf{H}}_i^*(z)] \\ &= -\frac{1}{\eta_1}\, \rho E_0^2 \sin(2\beta_1 z + \theta_\rho) \sin\theta_{\eta_1} \vec{\mathbf{a}}_z \qquad (8.108e)\end{aligned}$$

is the average power density due to the cross-coupling of the incident and the reflected waves. The cross-coupling term varies directly as $\sin\theta_{\eta_1}$ and exists as long as the medium is conducting.

EXAMPLE 8.7

WORKSHEET 9

Mathcad

A 50-MHz uniform plane wave traveling in a medium ($\epsilon_r = 16$, $\mu_r = 1$, and $\sigma = 0.02$ S/m) strikes normally to the surface of another medium ($\epsilon_r = 25$, $\mu_r = 1$, and $\sigma = 0.2$ S/m). If the amplitude of the incident electric field intensity at the interface is 10 V/m, determine the average power density of the transmitted wave.

Solution For medium 1 with $\epsilon_{r1} = 16$, $\mu_{r1} = 1$, and $\sigma_1 = 0.02$ S/m, we have

$$\omega = 2\pi \times 50 \times 10^6 = 3.142 \times 10^8 \text{ rad/s}$$

$$\frac{\sigma_1}{\omega \epsilon_{r1} \epsilon_0} = \frac{0.02 \times 36\pi}{3.142 \times 10^8 \times 16 \times 10^{-9}} = 0.45$$

$$\begin{aligned}\hat{\epsilon}_1 &= 16 \times \frac{10^{-9}}{36\pi} \times (1 - j0.45) \\ &= (14.15 - j6.366)10^{-11}\end{aligned}$$

$$\begin{aligned}\hat{\gamma}_1 &= j3.142 \times 10^8 \sqrt{4\pi \times 10^{-7}(14.15 - j6.366)10^{-11}} \\ &= 0.92 + j4.29 \text{ m}^{-1}\end{aligned}$$

$$\hat{\eta}_1 = \sqrt{\frac{4\pi \times 10^{-7} \times 10^{11}}{14.15 - j6.366}} = 87.997 + j18.887 = 90\angle 12.11^\circ\ \Omega$$

For medium 2 with $\epsilon_{r2} = 25$, $\mu_{r2} = 1$, and $\sigma_2 = 0.2$ S/m, we have

$$\frac{\sigma_2}{\omega \epsilon_{r2} \epsilon_0} = \frac{0.2 \times 36\pi}{3.142 \times 10^8 \times 25 \times 10^{-9}} = 2.88$$

$$\begin{aligned}\hat{\epsilon}_2 &= 25 \times \frac{10^{-9}}{36\pi} \times (1 - j2.88) \\ &= (2.21 - j6.366)10^{-10}\end{aligned}$$

$$\begin{aligned}\hat{\gamma}_2 &= j3.142 \times 10^8 \sqrt{4\pi \times 10^{-7}(2.21 - j6.366)10^{-10}} \\ &= 5.30 + j7.45 \text{ m}^{-1}\end{aligned}$$

$$\begin{aligned}\hat{\eta}_2 &= \sqrt{\frac{4\pi \times 10^{-7} \times 10^{10}}{2.21 - j6.366}} = 35.188 + j25.031 \\ &= 43.182\angle 35.43^\circ\ \Omega\end{aligned}$$

The reflection coefficient, from (8.105), is

$$\begin{aligned}\hat{\rho} &= \frac{(35.188 + j25.031) - (87.997 + j18.887)}{35.188 + j25.031 + 87.997 + j18.887} \\ &= -0.365 + j0.18 = 0.407\angle 153.74^\circ\end{aligned}$$

The transmission coefficient, from (8.106), is

$$\begin{aligned}\hat{\tau} &= \frac{2(35.188 + j25.031)}{35.188 + j25.031 + 87.997 + j18.887} \\ &= 0.635 + j0.18 = 0.66\angle 15.81^\circ\end{aligned}$$

From (8.101), the transmitted fields are

$$\begin{aligned}\tilde{\vec{\mathbf{E}}}_t &= 6.6e^{-5.30z} e^{-j7.45z} e^{j15.81^\circ}\ \vec{\mathbf{a}}_x \text{ V/m} \\ \tilde{\vec{\mathbf{H}}}_t &= 0.153e^{-5.30z} e^{-j7.45z} e^{-j19.62^\circ}\ \vec{\mathbf{a}}_y \text{ A/m}\end{aligned}$$

Thus, the average power of the transmitted wave is

$$\begin{aligned}\langle \hat{\mathbf{S}}_t \rangle &= \tfrac{1}{2}\text{Re}[\vec{\mathbf{E}}_t \times \vec{\mathbf{H}}_t^*] \\ &= 0.41e^{-10.60z}\,\vec{\mathbf{a}}_z\ \text{W/m}^2\end{aligned}$$

• • •

8.9.2 Dielectric–dielectric interface

When both media are lossless ($\sigma_1 = 0$, and $\sigma_2 = 0$), the intrinsic impedance of each medium is a real quantity. Accordingly, the transmission and the reflection coefficients are also real quantities. That is,

$$\rho = \frac{\eta_2 - \eta_1}{\eta_2 + \eta_1} \tag{8.109}$$

$$\tau = \frac{2\eta_2}{\eta_2 + \eta_1} \tag{8.110}$$

where

$$\eta_1 = \sqrt{\frac{\mu_1}{\epsilon_1}} \quad \text{and} \quad \eta_2 = \sqrt{\frac{\mu_2}{\epsilon_2}}$$

In terms of $\hat{\gamma}_1 = j\omega\sqrt{\mu_1\epsilon_1} = j\beta_1$ and $\hat{\gamma}_2 = j\omega\sqrt{\mu_2\epsilon_2} = j\beta_2$, we can express the incident, reflected, and transmitted fields as

$$\vec{\mathbf{E}}_i(z) = E_0 e^{-j\beta_1 z}\,\vec{\mathbf{a}}_x \tag{8.111a}$$

$$\vec{\mathbf{H}}_i(z) = \frac{1}{\eta_1} E_0 e^{-j\beta_1 z}\,\vec{\mathbf{a}}_y \tag{8.111b}$$

$$\vec{\mathbf{E}}_r(z) = \rho E_0 e^{j\beta_1 z}\,\vec{\mathbf{a}}_x \tag{8.111c}$$

$$\vec{\mathbf{H}}_r(z) = -\frac{1}{\eta_1} \rho E_0 e^{j\beta_1 z}\,\vec{\mathbf{a}}_y \tag{8.111d}$$

$$\vec{\mathbf{E}}_t(z) = \tau E_0 e^{-j\beta_2 z}\,\vec{\mathbf{a}}_x \tag{8.111e}$$

$$\vec{\mathbf{H}}_t(z) = \frac{1}{\eta_2} \tau E_0 e^{-j\beta_2 z}\,\vec{\mathbf{a}}_y \tag{8.111f}$$

where we have assumed that E_0 is the maximum value of the incident $\vec{\mathbf{E}}$ field at the interface.

The average power densities of the incident, reflected, and transmitted waves are

$$\langle \hat{\mathbf{S}}_i \rangle = \frac{1}{2\eta_1} E_0^2\,\vec{\mathbf{a}}_z \tag{8.112a}$$

$$\langle \hat{\mathbf{S}}_r \rangle = -\frac{1}{2\eta_1} \rho^2 E_0^2\,\vec{\mathbf{a}}_z \tag{8.112b}$$

$$\langle \hat{\mathbf{S}}_t \rangle = \frac{1}{2\eta_1} \tau^2 E_0^2\,\vec{\mathbf{a}}_z \tag{8.112c}$$

The total fields in medium 1 are

$$\vec{\mathbf{E}}_1(z) = \vec{\mathbf{E}}_i(z) + \vec{\mathbf{E}}_r(z) = E_0 e^{-j\beta_1 z}[1 + \rho e^{j2\beta_1 z}]\,\vec{\mathbf{a}}_x \tag{8.113a}$$

$$\vec{\mathbf{H}}_1(z) = \vec{\mathbf{H}}_i(z) + \vec{\mathbf{H}}_r(z) = \frac{1}{\eta_1} E_0 e^{-j\beta_1 z}[1 - \rho e^{j2\beta_1 z}]\,\vec{\mathbf{a}}_y \tag{8.113b}$$

EXAMPLE 8.8

An electromagnetic wave propagates in a dielectric medium with $\epsilon = 9\epsilon_0$ along the z direction. It strikes another dielectric medium with $\epsilon = 4\epsilon_0$ at $z = 0$. If the incoming wave has a maximum value of 0.1 V/m at the interface, and its angular frequency is 300 Mrad/s, determine (a) the reflection coefficient, (b) the transmission coefficient, and (c) the power densities of the incident, reflected, and transmitted waves.

Solution Medium 1:

$$\hat{\gamma}_1 = j\beta_1 = j\omega\sqrt{\mu_0\epsilon_1} = j\frac{300 \times 10^6}{3 \times 10^8}\sqrt{9} = j3 \text{ m}^{-1}$$

$$\eta_1 = \sqrt{\frac{\mu_0}{\epsilon_1}} = \frac{120\pi}{\sqrt{9}} = 125.664\ \Omega$$

Medium 2:

$$\hat{\gamma}_2 = j\beta_2 = j\omega\sqrt{\mu_0\epsilon_2} = j\frac{300 \times 10^6}{3 \times 10^8}\sqrt{4} = j2 \text{ m}^{-1}$$

$$\eta_2 = \sqrt{\frac{\mu_0}{\epsilon_2}} = \frac{120\pi}{\sqrt{4}} = 188.496\ \Omega$$

Thus, the reflection and transmission coefficients are

$$\rho = \frac{\eta_2 - \eta_1}{\eta_2 + \eta_1} = \frac{188.486 - 125.664}{188.496 + 125.664} = 0.2$$

$$\tau = \frac{2\eta_2}{\eta_2 + \eta_1} = \frac{2 \times 188.496}{188.496 + 125.664} = 1.2$$

The expressions for the incident, reflected, and transmitted fields as well as the corresponding power densities are

$$\tilde{\vec{\mathbf{E}}}_i = 0.1e^{-j3z}\,\vec{\mathbf{a}}_x \text{ V/m}$$

$$\tilde{\vec{\mathbf{H}}}_i = \frac{0.1}{125.664}e^{-j3z}\,\vec{\mathbf{a}}_y \text{ A/m}$$

$$\langle\hat{\mathbf{S}}_i\rangle = \frac{1}{2}\left[\frac{0.1^2}{125.664}\right]\vec{\mathbf{a}}_z = 39.79 \times 10^{-6}\,\vec{\mathbf{a}}_z \text{ W/m}^2$$

$$\tilde{\vec{\mathbf{E}}}_r = 0.02e^{j3z}\,\vec{\mathbf{a}}_x \text{ V/m}$$

$$\tilde{\vec{\mathbf{H}}}_r = \frac{0.02}{125.664}e^{j3z}\vec{\mathbf{a}}_y \text{ A/m}$$

$$\langle\hat{\mathbf{S}}_r\rangle = -\frac{1}{2}\left[\frac{0.02^2}{125.664}\right]\vec{\mathbf{a}}_z = -1.59 \times 10^{-6}\,\vec{\mathbf{a}}_z \text{ W/m}^2$$

$$\tilde{\vec{\mathbf{E}}}_t = 0.12e^{-j2z}\,\vec{\mathbf{a}}_x \text{ V/m}$$

$$\tilde{\vec{\mathbf{H}}}_t = \frac{0.12}{188.496}e^{-j2z}\,\vec{\mathbf{a}}_y \text{ A/m}$$

$$\langle\hat{\mathbf{S}}_t\rangle = \frac{1}{2}\left[\frac{0.12^2}{188.496}\right]\vec{\mathbf{a}}_z = 38.2 \times 10^{-6}\,\vec{\mathbf{a}}_z \text{ W/m}^2$$

• • •

8.9.3 Dielectric–perfect conductor interface

We now consider the case when a wave traveling in a dielectric medium (medium 1) impinges normally upon a perfectly conducting medium (medium 2), as depicted in Figure 8.13. As the electromagnetic fields cannot exist inside a perfect conductor ($\sigma_2 = \infty$), $\hat{\eta}_2 = 0$. Thus, from (8.109) and (8.110), we obtain $\rho = -1$ and $\tau = 0$. In other words, the incident wave is totally reflected from the boundary.

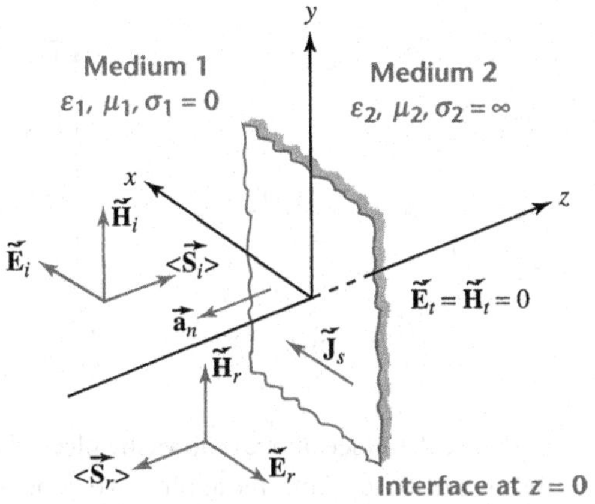

Figure 8.13 A perfectly conducting interface normal to the direction of propagation of a uniform plane wave

The incident, reflected, and total fields in the dielectric medium are

$$\tilde{\vec{E}}_i(z) = \hat{E}_0 e^{-j\beta_1 z}\,\vec{a}_x \tag{8.114a}$$

$$\tilde{\vec{H}}_i(z) = \frac{1}{\eta_1}\,\hat{E}_0 e^{-j\beta_1 z}\,\vec{a}_y \tag{8.114b}$$

$$\tilde{\vec{E}}_r(z) = -\hat{E}_0 e^{j\beta_1 z}\,\vec{a}_x \tag{8.114c}$$

$$\tilde{\vec{H}}_r(z) = \frac{1}{\eta_1}\,\hat{E}_0 e^{j\beta_1 z}\,\vec{a}_y \tag{8.114d}$$

$$\tilde{\vec{E}}_1(z) = -j2\hat{E}_0\,\sin(\beta_1 z)\,\vec{a}_x \tag{8.114e}$$

$$\tilde{\vec{H}}_1(z) = \frac{2}{\eta_1}\,\hat{E}_0\cos(\beta_1 z)\,\vec{a}_y \tag{8.114f}$$

where

$$\beta_1 = \omega\sqrt{\mu_1\epsilon_1}$$

and

$$\eta_1 = \sqrt{\frac{\mu_1}{\epsilon_1}}$$

If we consider that the incident electric field is maximum at the interface; that is, $\hat{E}_0 = E_0$, then the total fields in the dielectric medium can be written in the time domain as

$$\vec{E}_1(z) = 2E_0\sin(\beta_1 z)\sin\omega t\,\vec{a}_x \tag{8.115a}$$

$$\vec{H}_1(z) = \frac{2}{\eta_1}\,E_0\cos(\beta_1 z)\cos\omega t\,\vec{a}_y \tag{8.115b}$$

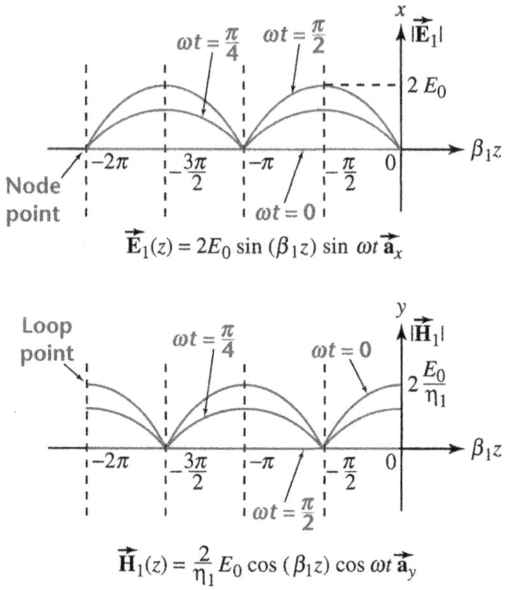

Figure 8.14 Magnitude plots of electric and magnetic field intensities in medium 1

Figure 8.14 (see above) shows the plots of these equations for various instants of time. From these plots we conclude that although the fields pulsate with time, they do not represent a propagating wave. These $\vec{\mathbf{E}}$ and $\vec{\mathbf{H}}$ fields are *pure standing waves* that are mutually orthogonal to each other. At any time, the magnitude of the $\vec{\mathbf{E}}$ field is maximum when

$$\sin(\beta_1 z) = \pm 1 \tag{8.116}$$

and zero when

$$\sin(\beta_1 z) = 0 \tag{8.117}$$

The point at which the field is maximum is called a *loop*; the point at which the field is zero is referred to as a *node*. Note that where there are loops of $\vec{\mathbf{E}}$ field there are nodes of $\vec{\mathbf{H}}$ field. In addition, when the $\vec{\mathbf{E}}$ field is maximum at any point in space, the $\vec{\mathbf{E}}$ field is zero, and vice versa. Thus, *the standing waves are 90° out of time and space phase.*

The absence of a normal component of the $\vec{\mathbf{E}}$ field in the dielectric medium implies that there are no induced charges on the surface of the perfect conductor. However, the presence of the tangential component of the $\vec{\mathbf{H}}$ field next to the boundary in the dielectric medium ensures the existence of a surface current on the surface of the perfect conductor. Applying the boundary condition for the tangential component of the $\tilde{\vec{\mathbf{E}}}$ field yields

$$\tilde{\vec{\mathbf{J}}}_s = -\vec{\mathbf{a}}_z \times \tilde{\vec{\mathbf{H}}}_1(0) \tag{8.118a}$$

$$= \frac{2}{\eta_1} E_0 \vec{\mathbf{a}}_x \tag{8.118b}$$

where $\tilde{\vec{\mathbf{J}}}_s$ is the surface current density on the surface of the perfect conductor. It can be expressed in the time domain as

$$\vec{\mathbf{J}}_s(t) = \frac{2}{\eta_1} E_0 \cos \omega t \, \vec{\mathbf{a}}_x \tag{8.118c}$$

We can now determine the *radiation pressure* (force per unit area) at the interface, from the Lorentz force equation, as

$$\vec{\mathbf{P}} = \frac{d\vec{\mathbf{F}}}{ds} = \vec{\mathbf{J}}_s \times \vec{\mathbf{B}} \tag{8.119a}$$

For sinusoidally varying fields, the average radiation pressure is

$$\langle \tilde{\vec{\mathbf{P}}} \rangle = \tfrac{1}{2}\mathrm{Re}[\tilde{\vec{\mathbf{J}}}_s \times \tilde{\vec{\mathbf{B}}}^*] \tag{8.119b}$$

where the magnetic flux density $\tilde{\vec{\mathbf{B}}}$ at the interface is from other sources elsewhere in space. As the incident fields are, in fact, produced by sources elsewhere in space, the magnetic flux density at the interface is

$$\tilde{\vec{\mathbf{B}}} = \tilde{\vec{\mathbf{B}}}_i(0) = \mu_1 \tilde{\vec{\mathbf{H}}}_i(0) = \frac{1}{\eta_1} \mu_1 E_0 \, \vec{\mathbf{a}}_y$$

Thus, the average radiation pressure is

$$\langle \mathbf{P} \rangle = \frac{\mu_1}{\eta_1^2} E_0^2 \, \vec{\mathbf{a}}_z = \epsilon_1 E_0^2 \, \vec{\mathbf{a}}_z \tag{8.120}$$

This equation yields the average force per unit area being experienced at the interface by the perfect conductor due to the normal incidence of an electromagnetic wave.

Fabry–Perot resonator

From the magnitude plot of the electric field intensity in Figure 8.14 it is evident that the $\vec{\mathbf{E}}$ field in the dielectric medium is zero (node) when $\beta_1 z = n\tau$, where n is an integer (0, 1, 2, etc.). This simply means that we can insert a perfectly conducting plane at any node location without perturbing the standing wave pattern. This is illustrated in Figure 8.15 for $n = 3$. The design of the Fabry–Perot resonator is based upon this principle. The resonator is used to measure frequencies in millimeter and submillimeter wavelength ranges. The separation between the two perfectly conducting plates of a Fabry–Perot resonator can be expressed as

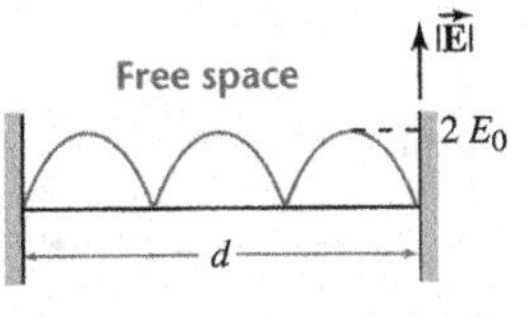

Figure 8.15

$$d = \frac{n\pi}{\beta_1} \tag{8.121}$$

If the medium between the two plates is free space, then

$$\beta_1 = \omega\sqrt{\mu_0 \epsilon_0} = \frac{\omega}{c} = \frac{2\pi f}{c}$$

Hence, the frequency f of the electromagnetic wave in terms of d is

$$f = \frac{nc}{2d} \tag{8.122}$$

EXAMPLE 8.9

A uniform plane wave propagating in a dielectric medium ($\mu_r = 1$, $\epsilon_r = 16$) strikes normally upon a perfect conductor. If the angular frequency of the wave is 96 Grad/s and the amplitude of the incident $\vec{\mathbf{E}}$ field at the interface is 100 V/m, obtain expressions for the incident, reflected, and total fields in the dielectric medium. If the frequency of the wave is to be measured using a Fabry–Perot resonator, what is the minimum distance at which the other perfectly conducting plate must be placed?

Solution Since the fields exist in the dielectric medium only, there is no need for the subscript to identify the medium. Let us also assume that the incident $\vec{\mathbf{E}}$ field is polarized in the x direction and the interface is at $z = 0$.

$$\beta = \omega\sqrt{\mu\epsilon} = \frac{96 \times 10^9}{3 \times 10^8}\sqrt{16} = 1280 \text{ rad/m}$$

$$\eta = \frac{120\pi}{\sqrt{16}} = 94.248\ \Omega$$

$$u_p = \frac{\omega}{\beta} = \frac{96 \times 10^9}{1280} = 7.5 \times 10^7 \text{ m/s}$$

The incident, reflected, and total fields in the dielectric medium are

$$\tilde{\mathbf{E}}_i = 100e^{-j1280z}\,\vec{\mathbf{a}}_x$$

$$\tilde{\mathbf{H}}_i = 1.061e^{-j1280z}\,\vec{\mathbf{a}}_y$$

$$\tilde{\mathbf{E}}_r = -100e^{j1280z}\,\vec{\mathbf{a}}_x$$

$$\tilde{\mathbf{H}}_r = 1.061e^{j1280z}\,\vec{\mathbf{a}}_y$$

$$\tilde{\vec{\mathbf{E}}} = -j200\sin(1280z)\,\vec{\mathbf{a}}_x$$

$$\tilde{\vec{\mathbf{H}}} = 2.122\cos(1280z)\,\vec{\mathbf{a}}_y$$

For the separation between the two plates to be minimum, $n = 1$. Hence, the distance between the plates must be

$$d = \frac{n\pi}{\beta} = \frac{\pi}{1280} = 2.45 \times 10^{-3} \text{ m} \quad \text{or} \quad 2.45 \text{ mm}$$

• • •

8.9.4 Dielectric–conductor interface

The last case for normal incidence of a uniform plane wave involves the interface between a dielectric and a finitely conducting medium. The method of attack is basically the same as that for the general case, thus we illustrate this case with the following example.

EXAMPLE 8.10

WORKSHEET 10

Mathcad

A uniform plane wave propagating in free space strikes normally upon a lossy medium having a dielectric constant of 18 and a conductivity of 0.6 mS/m. The frequency of the wave polarized in the x direction is 300 kHz. If the electric field intensity has an amplitude of 10 V/m at the interface at $t = 0$, determine the average power density in each region.

Solution

$$f = 300\text{ kHz},\ \omega = 2\pi f = 1.885 \times 10^6\text{ rad/s}$$

Medium 1: Because medium 1 is free space,

$$\hat{\gamma}_1 = j\omega\sqrt{\mu_0\epsilon_0} = j\frac{\omega}{c} = j\frac{1.885 \times 10^6}{3 \times 10^8} = j6.283 \times 10^{-3}\text{ m}^{-1}$$

Hence,

$$\beta_1 = 6.283 \times 10^{-3}\text{ rad/m}$$

$$\eta_1 = \sqrt{\frac{\mu_0}{\epsilon_0}} = 120\pi \approx 377\ \Omega$$

$$u_{p1} = \frac{\omega}{\beta_1} = \frac{1.885 \times 10^6}{6.283 \times 10^{-3}} = 3 \times 10^8\text{ m/s}$$

The propagation constant in free space, being imaginary, is the same as the phase constant, and there is no loss in energy. The intrinsic impedance is real, and the wave propagates with the velocity of light.

Medium 2:

$$\omega\epsilon_2 = 1.885 \times 10^6 \times 18 \times \frac{10^{-9}}{36\pi} = 300 \times 10^{-6}$$

Loss tangent: $\tan\delta_2 = \dfrac{\sigma_2}{\omega\epsilon_2} = \dfrac{0.6 \times 10^{-3}}{300 \times 10^{-6}} = 2$

Complex permittivity: $\hat{\epsilon}_2 = \epsilon_2[1 - j\tan\delta_2]$

$$= 18(1 - j2)\epsilon_0 = 40.25\,\epsilon_0\angle{-63.44^\circ}$$

Thus,

$$\hat{\gamma}_2 = j\omega\sqrt{\mu_2\hat{\epsilon}_2} = j\frac{\omega}{c}\sqrt{40.25\angle{-63.44^\circ}}$$

$$= 0.021 + j0.034 = 0.04\angle{58.28^\circ}\text{ m}^{-1}$$

$$\alpha_2 = 0.021\text{ Np/m} \quad \text{and} \quad \beta_2 = 0.034\text{ rad/m}$$

$$\hat{\eta}_2 = \sqrt{\frac{\mu_0}{\hat{\epsilon}_2}} = \frac{120\pi}{\sqrt{40.25\angle{-63.44^\circ}}} = 59.42\angle{-31.72^\circ}\ \Omega$$

$$= 50.55 + j31.24\ \Omega$$

$$u_{p2} = \frac{1.885 \times 10^6}{0.034} = 55.44 \times 10^6\text{ m/s}$$

The finite conductivity of medium 2 means that the propagation constant and the intrinsic impedance are both complex quantities. The wave front propagates slowly in the conducting medium. As medium 2 is finitely conducting, there will be no surface charge density at the interface. Thus, equations (8.105) and (8.106) are also valid in this case. The reflection coefficient is

$$\hat{\rho} = \frac{\hat{\eta}_2 - \eta_1}{\hat{\eta}_2 + \eta_1} = \frac{50.55 + j31.24 - 377}{50.55 + j31.24 + 377} = 0.765\underline{/170.35^\circ}$$

and the transmission coefficient is

$$\hat{\tau} = \frac{2\hat{\eta}_2}{\eta_1 + \hat{\eta}_2} = \frac{2(50.55 + j31.24)}{50.55 + j31.24 + 377} = 0.277\underline{/27.54^\circ}$$

We can now write the expressions for the incident, reflected, and transmitted fields:

$$\tilde{\vec{\mathbf{E}}}_i = 10e^{-j6.283\times10^{-3}z}\,\vec{\mathbf{a}}_x \text{ V/m}$$
$$\tilde{\vec{\mathbf{H}}}_i = \frac{10}{120\pi}\,e^{-j6.283\times10^{-3}z}\,\vec{\mathbf{a}}_y \text{ A/m}$$
$$\tilde{\vec{\mathbf{E}}}_r = 7.65e^{j170.35^\circ}e^{j6.283\times10^{-3}z}\,\vec{\mathbf{a}}_x \text{ V/m}$$
$$\tilde{\vec{\mathbf{H}}}_r = -\frac{7.65}{377}\,e^{j170.35^\circ}\,e^{j6.283\times10^{-3}z}\,\vec{\mathbf{a}}_y \text{ A/m}$$
$$\tilde{\vec{\mathbf{E}}}_t = 2.77e^{j27.54^\circ}e^{-(0.021+j0.034)z}\,\vec{\mathbf{a}}_x \text{ V/m}$$
$$\tilde{\vec{\mathbf{H}}}_t = \frac{2.77}{59.42}\,e^{-j4.18^\circ}\,e^{-(0.021+j0.034)z}\vec{\mathbf{a}}_y \text{ A/m}$$

The average power density of the incident wave is

$$\langle\hat{\mathbf{S}}_i\rangle = \tfrac{1}{2}\text{Re}\,[\tilde{\vec{\mathbf{E}}}_i \times \tilde{\vec{\mathbf{H}}}_i^*] = 0.133\,\vec{\mathbf{a}}_z \text{ W/m}^2$$

The average power density of the reflected wave is

$$\langle\hat{\mathbf{S}}_r\rangle = \tfrac{1}{2}\text{Re}\,[\tilde{\vec{\mathbf{E}}}_r \times \tilde{\vec{\mathbf{H}}}_r^*] = 0.078\,\vec{\mathbf{a}}_z \text{ W/m}^2$$

The average power density of the transmitted wave is

$$\langle\hat{\mathbf{S}}_t\rangle = \tfrac{1}{2}\text{Re}[\tilde{\vec{\mathbf{E}}}_t \times \tilde{\vec{\mathbf{H}}}_t^*] = 0.065e^{-0.042z}\,\vec{\mathbf{a}}_z \text{ W/m}^2$$

The skin depth, or depth of penetration, of the wave in medium 2 is

$$\delta_c = \frac{1}{\alpha_2} = \frac{1}{0.021} = 47.62 \text{ m}$$

Hence, the electromagnetic wave will practically vanish in medium 2 after traveling a distance of 5 skin depths ($\approx$ 238 m). • • •

8.10 Oblique incidence on a plane boundary

When an electromagnetic wave strikes a plane boundary any arbitrary angle, we refer to it as **oblique incidence**. In fact, normal incidence is a special case of oblique incidence. We discuss oblique incidence because it leads to three well-known laws in optics: Snell's law of reflection, Snell's law of refraction, and Brewster's law directing polarization by reflection.

Once again, we consider a wave that is linearly polarized and a boundary that constitutes a plane at $z = 0$. The plane that includes the unit normal to the boundary ($\vec{\mathbf{a}}_n$) and the propagation constant of the incident wave ($\hat{\gamma}_1$) is called the **plane of incidence**, as shown in Figure 8.16. In general, the $\vec{\mathbf{E}}$ field of the incident wave can make any angle with the plane of incidence; however, we will limit our discussion to two special cases. First, we will assume that the $\vec{\mathbf{E}}$ field is normal to the plane of incidence and refer to it as a *perpendicularly polarized wave* as shown in Figure 8.17 where the $\vec{\mathbf{E}}$ field is in the $\vec{\mathbf{a}}_x$ direction and yz is the plane of incidence. In this case, the $\vec{\mathbf{E}}$ field is parallel to the interface. In our second case, the $\vec{\mathbf{E}}$ field lies in the plane of incidence and is called a *parallel polarized wave.* In this case, the $\vec{\mathbf{H}}$ field lies in the plane parallel to the interface. The superposition theorem allows us to obtain all the necessary information for an incident wave that makes an arbitrary angle with the plane of incidence.

When we envision the interface between free space and the earth, we usually refer to the perpendicular polarization as *horizontal polarization* because the $\vec{\mathbf{E}}$ field is in a horizontal plane with respect to the earth.

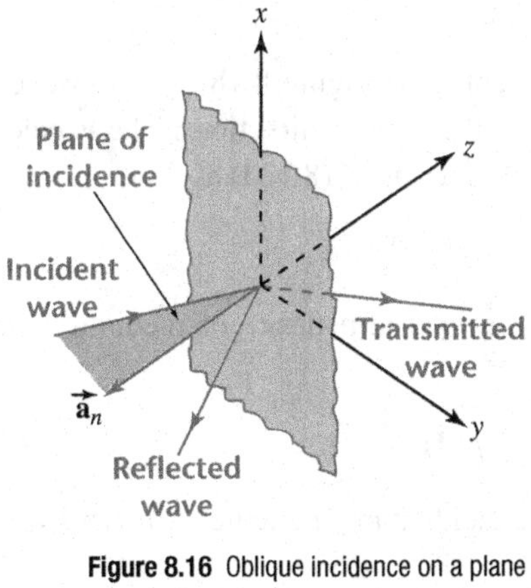

Figure 8.16 Oblique incidence on a plane boundary

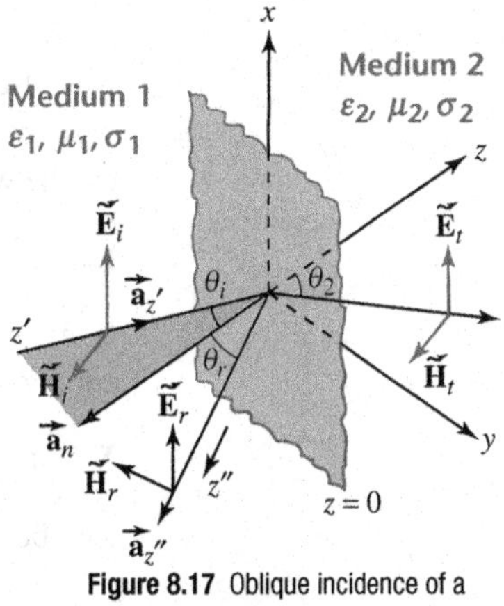

Figure 8.17 Oblique incidence of a perpendicularly polarized wave

8.10.1 Perpendicular polarization

Let us consider a general case in which the interface separates two linear, isotropic, and homogeneous but finitely conducting media, as shown in Figure 8.17. The incident wave is propagating in the z' direction and makes an angle θ_i with the unit normal $\vec{\mathbf{a}}_n$. If $\hat{E}_0$ is the amplitude of the incident $\vec{\mathbf{E}}$ field at $t = 0$ and $z = 0$, we can express the incident electric field intensity at any point in medium 1 as

$$\tilde{\mathbf{E}}_i = \hat{E}_0 e^{-\hat{\gamma}_1 z'} \vec{\mathbf{a}}_x$$

Note that the exponent $\hat{\gamma}_1 z'$ represents the propagation of the wave front of the incident wave. That is, the wave front has propagated from a to b in the z' direction in time t with a propagation constant of $\hat{\gamma}_1$, as shown in Figure 8.18. We can also express the exponent as

$$\begin{aligned}\hat{\gamma}_1 z' &= [\hat{\gamma}_1 \vec{\mathbf{a}}'_z] \cdot [z' \vec{\mathbf{a}}'_z] \\ &= [\hat{\gamma}_1 \cos\theta_i \vec{\mathbf{a}}_z + \hat{\gamma}_1 \sin\theta_i \vec{\mathbf{a}}_y] \cdot [z\vec{\mathbf{a}}_z + y\vec{\mathbf{a}}_y] \\ &= \hat{\gamma}_1 [z\cos\theta_i + y\sin\theta_i] \end{aligned} \tag{8.123}$$

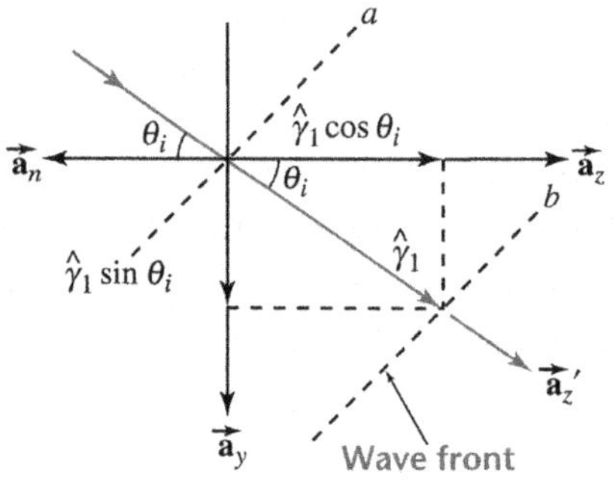

Figure 8.18 Wave front propagation of incident wave from a to b in $\vec{\mathbf{a}}_{z'}$ direction

It is also evident from Figure 8.18 that the wave front of the incident wave travels in the y and z directions. The $\vec{\mathbf{E}}$ field of the incident wave can be written in terms of (8.123) as

$$\tilde{\mathbf{E}}_i = \hat{E}_0 e^{-\hat{\gamma}_1 (z\cos\theta_i + y\sin\theta_i)} \vec{\mathbf{a}}_x \tag{8.124a}$$

If $\hat{\eta}_1$ is the intrinsic impedance of medium 1, then from Maxwell's equation

$$\nabla \times \tilde{\mathbf{E}} = -j\omega\mu_1 \tilde{\mathbf{H}}$$

we obtain the incident magnetic field intensity as

$$\tilde{\mathbf{H}}_i = -\frac{1}{\hat{\eta}_1} \hat{E}_0 e^{-\hat{\gamma}_1 (z\cos\theta_i + y\sin\theta_i)} [-\cos\theta_i\, \vec{\mathbf{a}}_y + \sin\theta_i\, \vec{\mathbf{a}}_z] \tag{8.124b}$$

Let us assume that the reflected wave propagates in the z'' direction and makes an angle θ_r with the unit normal $\vec{\mathbf{a}}_n$, as shown in Figure 8.17. The propagation of the wave front for the reflected wave is shown in Figure 8.19. The wave front propagates in the positive y direction and the negative z direction. Taking that into account and assuming that the reflected $\tilde{\mathbf{E}}$ field is still polarized in the x direction with $\hat{\rho}_n$ as the coefficient of reflection, we can express the reflected fields in

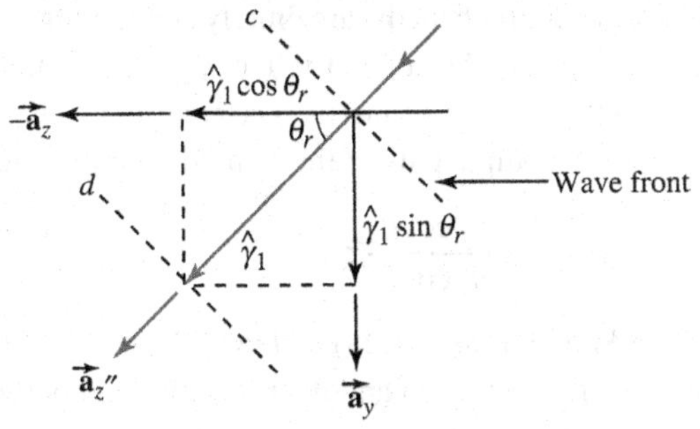

Figure 8.19 Wave front propagation of reflected wave from c to d along $\vec{\mathbf{a}}_z$

medium 1 as

$$\tilde{\vec{\mathbf{E}}}_r = \hat{E}_0\hat{\rho}_n e^{\hat{\gamma}_1(z\cos\theta_r - y\sin\theta_r)}\,\vec{\mathbf{a}}_x \tag{8.125a}$$

$$\tilde{\vec{\mathbf{H}}}_r = -\frac{1}{\hat{\eta}_1}\hat{E}_0\hat{\rho}_n e^{\hat{\gamma}_1(z\cos\theta_r - y\sin\theta_r)}[-\cos\theta_r\,\vec{\mathbf{a}}_y + \sin\theta_r\,\vec{\mathbf{a}}_z] \tag{8.125b}$$

If $\hat{\tau}_n$ is the coefficient of transmission of the perpendicularly polarized wave in medium 2, θ_2 is the angle with respect to the z direction as shown in Figure 8.17, $\hat{\gamma}_2$ is the propagation constant, and $\hat{\eta}_2$ is the intrinsic impedance, then the $\tilde{\vec{\mathbf{E}}}$ and $\tilde{\vec{\mathbf{H}}}$ fields in medium 2 are

$$\tilde{\vec{\mathbf{E}}}_t = \hat{E}_0\hat{\tau}_n e^{-\hat{\gamma}_2(z\cos\theta_2 + y\sin\theta_2)}\,\vec{\mathbf{a}}_x \tag{8.126a}$$

$$\tilde{\vec{\mathbf{H}}}_t = -\frac{1}{\hat{\eta}_2}\hat{E}_0\hat{\tau}_n e^{-\hat{\gamma}_2(z\cos\theta_2 + y\sin\theta_2)}[-\cos\theta_2\vec{\mathbf{a}}_y + \sin\theta_2\vec{\mathbf{a}}_z] \tag{8.126b}$$

The continuity of the tangential components at the boundary, $z = 0$, yields

$$e^{-\hat{\gamma}_1 y\sin\theta_i} + \hat{\rho}_n e^{-\hat{\gamma}_1 y\sin\theta_r} = \hat{\tau}_n e^{-\hat{\gamma}_2 y\sin\theta_2} \tag{8.127}$$

This equation must be true for all values of y; however, at $y = 0$, it becomes

$$1 + \hat{\rho}_n = \hat{\tau}_n \tag{8.128}$$

For (8.127) and (8.128) to hold for all values of y, we must have

$$\hat{\gamma}_1\sin\theta_i = \hat{\gamma}_1\sin\theta_r = \hat{\gamma}_2\sin\theta_2 \tag{8.129}$$

From the first equality in this equation, we obtain

$$\theta_i = \theta_r = \theta_1 \tag{8.130}$$

which states that *the angle of incidence is equal to the angle of reflection.* This is a well-known relation in optics and is called *Snell's law of reflection.*

From the other equality in (8.129), we obtain

$$\hat{\gamma}_1\sin\theta_1 = \hat{\gamma}_2\sin\theta_2 \tag{8.131}$$

which is *Snell's law of refraction* for finitely conducting media. For this reason, the transmitted wave is usually referred to as the *refracted wave.*

Since both media are finitely conducting, we do not expect any surface currents at the boundary. Thus, the tangential components of the $\vec{\mathbf{H}}$ field at the boundary must also be continuous. By setting $z = 0$ and equating the y components of the $\vec{\mathbf{H}}$ field on either side, we obtain

$$1 - \hat{\rho}_n = \frac{\hat{\eta}_1 \cos\theta_2}{\hat{\eta}_2 \cos\theta_1} \hat{\tau}_n \tag{8.132}$$

Manipulating (8.128) and (8.132), we obtain the following expressions for the reflection and the transmission coefficients:

$$\hat{\rho}_n = \frac{\hat{\eta}_2 \cos\theta_1 - \hat{\eta}_1 \cos\theta_2}{\hat{\eta}_2 \cos\theta_1 + \hat{\eta}_1 \cos\theta_2} \tag{8.133}$$

$$\hat{\tau}_n = \frac{2\hat{\eta}_2 \cos\theta_1}{\hat{\eta}_2 \cos\theta_1 + \hat{\eta}_1 \cos\theta_2} \tag{8.134}$$

Equations (8.133) and (8.134) enable us to express the reflected and transmitted fields in terms of the incident fields. We can now compute the total fields in medium 1 and the power densities in both regions.

Let us now consider some of the special cases of oblique incidence for a perpendicularly polarized wave.

Dielectric–dielectric interface

In this case, $\sigma_1 = 0$ and $\sigma_2 = 0$. Since the permeability of any dielectric material is nearly the same as that of free space, we will also assume that $\mu_1 = \mu_2 = \mu_0$.

We can express the ratio of the propagation constants as

$$\frac{\hat{\gamma}_1}{\hat{\gamma}_2} = \frac{j\beta_1}{j\beta_2} = \sqrt{\frac{\epsilon_{r1}}{\epsilon_{r2}}} \tag{8.135}$$

and the ratio of the intrinsic impedances as

$$\frac{\eta_1}{\eta_2} = \sqrt{\frac{\epsilon_{r2}}{\epsilon_{r1}}} \tag{8.136}$$

Snell's law of refraction, from (8.131), becomes

$$\sin\theta_2 = \sqrt{\frac{\epsilon_{r1}}{\epsilon_{r2}}} \sin\theta_1 \tag{8.137}$$

Also

$$\cos\theta_2 = \sqrt{1 - \sin^2\theta_2} = \sqrt{1 - \frac{\epsilon_{r1}}{\epsilon_{r2}} \sin^2\theta_1} \tag{8.138}$$

These simplifications allow us to write (8.133) as

$$\hat{\rho}_n = \frac{\cos\theta_1 - \sqrt{\dfrac{\epsilon_{r2}}{\epsilon_{r1}} - \sin^2\theta_1}}{\cos\theta_1 + \sqrt{\dfrac{\epsilon_{r2}}{\epsilon_{r1}} - \sin^2\theta_1}} \tag{8.139}$$

The discussion of the following three cases explains why the reflection coefficient may still be a complex quantity.

Case I If medium 2 is denser than medium 1, $\epsilon_{r2} > \epsilon_{r1}$, the reflection coefficient, from (8.139), is a real quantity. Consequently, from (8.132), the transmission coefficient is also a real quantity.

Case II If medium 1 is denser than medium 2, $\epsilon_{r1} > \epsilon_{r2}$, the reflection coefficient is real as long as

$$\sin\theta_1 \le \sqrt{\frac{\epsilon_{r2}}{\epsilon_{r1}}} \tag{8.140}$$

If the left-hand side of (8.140) is equal to the right-hand side when $\theta_1 = \theta_c$, then (a) θ_c is called the *critical angle*, (b) $\rho = 1$ and $\tilde{\vec{\mathbf{E}}}_r = \tilde{\vec{\mathbf{E}}}_i$, (c) from (8.136), $\theta_2 = 90°$, (d) the transmitted wave will propagate entirely parallel to the interface, as shown in Figure 8.20, and (e) the y component of the transmitted $\tilde{\vec{\mathbf{H}}}$ field will be zero as $\cos\theta_2 = 0$. Simply stated, there will be no power propagating along the z direction in medium 2, and the reflected power density will be equal to the incident power density. This is called *total reflection.* For this reason, the critical angle is also referred to as the *angle of total reflection.*

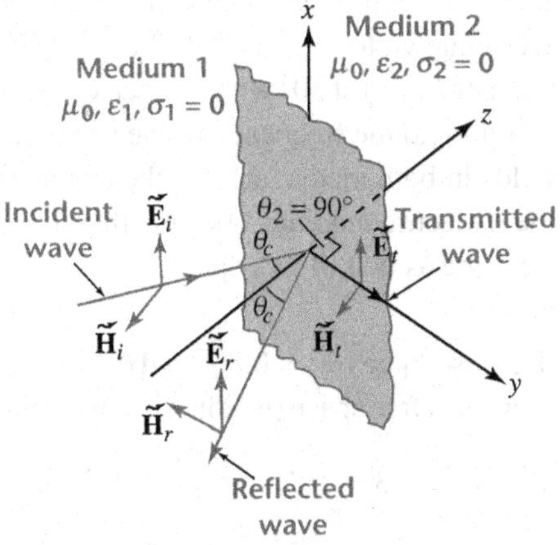

Figure 8.20 A uniform plane wave striking an interface between two dielectrics at a critical angle

Case III If the angle of incidence is greater than the critical angle, $\theta_i > \theta_c$, the reflection coefficient will be a complex number with a magnitude of unity. Also, from (8.137), we have

$$\sin\theta_2 > 1$$

and

$$\cos\theta_2 = \pm j\sqrt{\frac{\epsilon_{r1}}{\epsilon_{r2}}\sin^2\theta_1 - 1} \tag{8.141}$$

neither of which is a possibility for the uniform plane wave to exist in medium 2. However, the fields in medium 2 cannot be zero, because a perfect dielectric cannot suddenly behave like a perfect conductor. To determine the nature of the wave, let us substitute (8.141) in (8.126a) with $\hat{\gamma}_2 = j\beta_2$. We obtain

$$\tilde{\mathbf{E}}_t = \hat{E}_0 \hat{\tau}_n e^{-j\beta_2 y \sin\theta_2} e^{-\beta_2 z \sqrt{\frac{\epsilon_{r1}}{\epsilon_{r2}} \sin^2\theta_2 - 1}} \vec{\mathbf{a}}_x \tag{8.142a}$$

where we have only retained the negative sign for the second exponential term because the wave cannot grow as a function of z. This equation shows that the wave is propagating along the y direction (parallel to the interface) and is attenuating in the z direction with an attenuation constant of

$$\alpha_2 = \beta_2 \sqrt{\frac{\epsilon_{r1}}{\epsilon_{r2}} \sin^2\theta_1 - 1} \tag{8.142b}$$

These are the traits of a *nonuniform plane wave.* As it is decaying in the z direction and is propagating parallel to the interface, it is also called the *surface wave.* Once again, there is no real power flow in the z direction. Thus, based upon experimental observations in optics, we expect total reflection for all angles of incidence greater than the critical angle.

EXAMPLE 8.11

WORKSHEET 11 Mathcad

The electric field intensity of a uniform plane wave propagating in free space is known to be $377 e^{-j0.866z} e^{-j0.5y} \vec{\mathbf{a}}_x$ V/m. It strikes a dielectric medium ($\epsilon_r = 9$) at 30° with respect to the normal to the plane interface. Determine (a) the frequency of the wave, (b) expressions for the $\tilde{\mathbf{E}}$ and $\tilde{\mathbf{H}}$ fields in both media, and (c) the average power density of the wave in the dielectric medium. Assume that the permeability of the medium is the same as that of free space.

Solution

a) Because $\beta_1 \cos\theta_1 = 0.866$ and $\theta_1 = 30°$, $\beta_1 = 1$ rad/m. As the wave travels in free space (medium 1), the angular frequency of the wave is

$$\omega = c\beta_1 = 3 \times 10^8 \times 1 = 3 \times 10^8 \text{ rad/s} \quad \text{or} \quad 300 \text{ Mrad/s}$$

The propagation constant in the dielectric medium (medium 2) is

$$\beta_2 = \frac{\omega}{c}\sqrt{\epsilon_{r2}} = 1\sqrt{9} = 3 \text{ rad/m}$$

The intrinsic impedances are

$$\eta_1 = 120\pi \ \Omega$$

and

$$\eta_2 = \frac{120\pi}{\sqrt{9}} = 40\pi \ \Omega$$

From Snell's law of refraction, we get

$$\sin\theta_2 = \frac{j1}{j3}\sin 30^\circ = 0.167 \Rightarrow \theta_2 = 9.594^\circ$$

and

$$\cos\theta_2 = 0.986$$

The reflection coefficient, from (8.133), is

$$\rho_n = \frac{40\pi \times 0.866 - 120\pi \times 0.986}{40\pi \times 0.866 + 120\pi \times 0.986} = -0.547$$

and the transmission coefficient, from (8.128), is

$$\tau_n = 1 + \rho_n = 0.453$$

b) The incident fields, from (8.124a) and (8.124b), are

$$\tilde{\vec{\mathbf{E}}}_i = 377e^{-j0.866z}e^{-j0.5y}\,\vec{\mathbf{a}}_x \text{ V/m}$$
$$\tilde{\vec{\mathbf{H}}}_i = -[-0.866\,\vec{\mathbf{a}}_y + 0.5\,\vec{\mathbf{a}}_z]e^{-j0.866z}e^{-j0.5y} \text{ A/m}$$

The reflected fields, from (8.125a) and (8.125b), are

$$\tilde{\vec{\mathbf{E}}}_r = -206.22e^{-j0.5y}e^{j0.866z}\,\vec{\mathbf{a}}_x \text{ V/m}$$
$$\tilde{\vec{\mathbf{H}}}_r = [0.474\,\vec{\mathbf{a}}_y + 0.274\,\vec{\mathbf{a}}_z]e^{-j0.5y}e^{j0.866z} \text{ A/m}$$

Finally, the transmitted fields, from (8.126a) and (8.126b), are

$$\tilde{\vec{\mathbf{E}}}_t = 170.78e^{-j2.958z}e^{-j0.5y}\,\vec{\mathbf{a}}_x \text{ V/m}$$
$$\tilde{\vec{\mathbf{H}}}_t = -[-1.34\,\vec{\mathbf{a}}_y + 0.227\,\vec{\mathbf{a}}_z]e^{-j2.958z}e^{-j0.5y} \text{ A/m}$$

c) The average power density in medium 2 is

$$\langle\hat{\mathbf{S}}_t\rangle = \tfrac{1}{2}\mathrm{Re}[\tilde{\vec{\mathbf{E}}}_t \times \tilde{\vec{\mathbf{H}}}_t^*] = 114.4\,\vec{\mathbf{a}}_z + 19.4\,\vec{\mathbf{a}}_y \text{ W/m}^2$$

Thus, the average power flow per unit area in medium 2 in the z direction is 114.4 W/m^2. • • •

Dielectric–perfect conductor interface

A perpendicularly polarized uniform plane wave propagating through a dielectric medium and striking obliquely a perfectly conducting medium is shown in Figure 8.21 (page 400). The absence of electromagnetic fields inside a perfect conductor suggests that the transmission coefficient must be zero. The reflection coefficient, from (8.128), is

$$\rho_n = -1$$

Let the electric field intensity of the incident wave be maximum at the interface at $t = 0$. The total electric field at any point in the dielectric

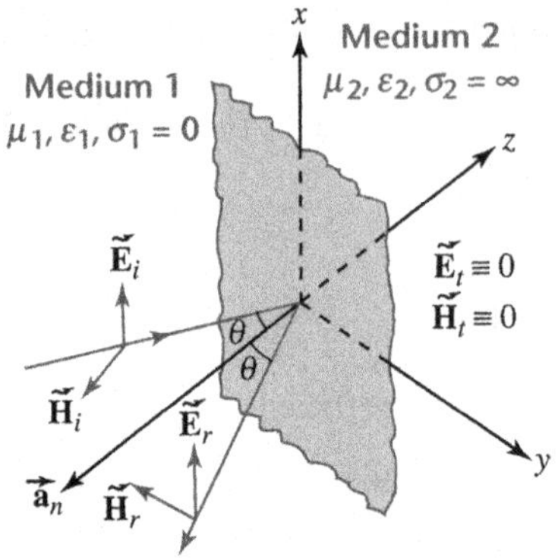

Figure 8.21 An interface between a dielectric and a perfectly conducting medium

medium, from (8.124a) and (8.125a), is

$$\tilde{\vec{\mathbf{E}}} = \tilde{\vec{\mathbf{E}}}_i + \tilde{\vec{\mathbf{E}}}_r$$
$$= j2E_0 \sin(\beta z \cos\theta) e^{-j\beta y \sin\theta}\, \vec{\mathbf{a}}_x \tag{8.143a}$$

where E_0 is the maximum value of the incident electric field, θ is the angle of incidence, and β is the phase constant. Note that we have dropped the subscript 1 because the fields exist only in the dielectric medium. The total magnetic field intensity in the dielectric medium, from (8.124b) and (8.125b), is

$$\tilde{\vec{\mathbf{H}}} = \tilde{\vec{\mathbf{H}}}_i + \tilde{\vec{\mathbf{H}}}_r$$
$$= \frac{2}{\eta} E_0 [\cos\theta \cos(\beta z \cos\theta)\, \vec{\mathbf{a}}_y + j \sin\theta \sin(\beta z \cos\theta)\, \vec{\mathbf{a}}_z] e^{-j\beta y \sin\theta} \tag{8.143b}$$

where η is the intrinsic impedance of the dielectric medium. The average power density in the dielectric medium is

$$\langle \hat{\mathbf{S}} \rangle = \frac{1}{2} \text{Re}[\tilde{\vec{\mathbf{E}}} \times \tilde{\vec{\mathbf{H}}}^*]$$
$$= \frac{2}{\eta} E_0^2 \sin\theta \sin^2(\beta z \cos\theta)\, \vec{\mathbf{a}}_y \tag{8.144}$$

This equation shows that there is no power flow in the z direction. This is, however, obvious as the x component of $\tilde{\vec{\mathbf{E}}}$ and the y component of $\tilde{\vec{\mathbf{H}}}$ are in space quadrature. However, there is a power flow, but it is in the y direction (parallel to the interface). As we will discuss later, this observation led to the design of the parallel-plate waveguide.

At the interface ($z = 0$), the magnetic field intensity in the dielectric medium

$$\tilde{\vec{\mathbf{H}}}(0) = \frac{2}{\eta} E_0 \cos\theta e^{-j\beta y \sin\theta}\, \vec{\mathbf{a}}_y$$

is tangential to the interface. From the boundary condition for the tangential components of the $\vec{\mathbf{H}}$ field, the surface current density at $z = 0$ is

$$\vec{\mathbf{J}}_s(0) = \frac{2}{\eta} E_0 \cos\theta e^{-j\beta y \sin\theta} \vec{\mathbf{a}}_x \tag{8.145}$$

By expressing the $\vec{\mathbf{E}}$ and $\vec{\mathbf{H}}$ fields in the time domain as

$$E_x(x, y, z, t) = 2E_0 \sin(\omega t - \beta y \sin\theta) \sin(\beta z \cos\theta) \tag{8.146a}$$

$$H_y(x, y, z, t) = \frac{2}{\eta} E_0 \cos\theta \cos(\beta z \cos\theta) \cos(\omega t - \beta y \sin\theta) \tag{8.146b}$$

$$H_z(x, y, z, t) = -\frac{2}{\eta} E_0 \sin\theta \sin(\beta z \cos\theta) \sin(\omega t - \beta y \sin\theta) \tag{8.146c}$$

we realize that the fields propagate in the y direction and form a standing wave pattern along the z direction. The nodal points for the electric field intensity and the y component of the magnetic field intensity are at

$$\beta z \cos\theta = -m\pi \quad \text{for} \quad m = 0, 1, 2, \ldots$$

or at

$$z = -\frac{m\pi}{\beta \cos\theta} \tag{8.147}$$

$\beta \cos\,\theta$ is the component of the phase constant in the z direction, thus we can define the wavelength in that direction as

$$\lambda_z = \frac{2\pi}{\beta \cos\theta} \tag{8.148}$$

In terms of the wavelength in the z direction, the nodal points are at

$$z = -\frac{m}{2}\lambda_z \quad \text{for} \quad m = 1, 2, 3, \ldots \tag{8.149}$$

Thus, the $\vec{\mathbf{E}}$ field has nodes at $z = 0$ and at multiples of half-wavelengths from the interface in the negative z direction. Similarly, we can show that the y component of the $\vec{\mathbf{H}}$ field has nodal points at odd multiples of quarter-wavelengths in the negative z direction.

From (8.146a), (8.146b), and (8.146c) we also realize that the phase velocity of the wave in the y direction is

$$u_{py} = \frac{\omega}{\beta \sin\theta} = \frac{1}{\sin\theta} u_p \tag{8.150}$$

where u_p is the phase velocity of the wave in the unbounded dielectric medium and is given as

$$u_p = \frac{\omega}{\beta} = \frac{1}{\sqrt{\mu\epsilon}}$$

Since $\sin\theta \leq 1$, we expect that $u_{py} \geq u_p$. In other words, if the medium is free space, we can expect the phase velocity in the y direction to be

Figure 8.22 Motion of a point of constant phase as viewed from different directions

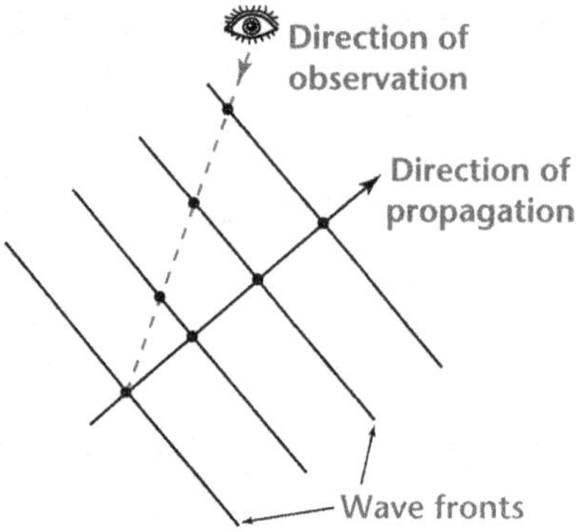

greater than the speed of light. Does this mean that the energy propagation in the y direction can occur with a velocity greater than the speed of light? The answer, of course, is no. Note that the phase velocity is the velocity of a point of constant phase on the wave. A point of constant phase on the wave always appears to move faster when viewed along directions other than its direction of propagation, as depicted in Figure 8.22. The energy, however, propagates with a velocity known as the *group velocity*. To determine the group velocity, we will calculate the average power flow using the Poynting vector and then the energy density.

a) **Average power flow using Poynting vector:** The average power per unit area, from (8.144), can be expressed as

$$\langle \hat{\mathbf{S}} \rangle = 2\epsilon E_0^2 \sin^2(\beta z \cos\theta) u_p \sin\theta \, \vec{\mathbf{a}}_y$$

As the power density is independent of variations along the x direction, we can calculate the total average power per unit length along the x direction and between any two nodal points in the z direction as

$$\langle P \rangle = 2\epsilon E_0^2 u_p \sin\theta \int_{z_1}^{z_2} [\sin(\beta z \cos\theta)]^2 dz$$
$$= \frac{\epsilon E_0^2 u_p \sin\theta}{\beta\cos\theta}(m-n)\pi \tag{8.151}$$

where

$$z_1 = -\frac{m\pi}{\beta\cos\theta} \quad \text{and} \quad z_2 = -\frac{n\pi}{\beta\cos\theta}$$

are two distinct node points for the $\vec{\mathbf{E}}$ field such that $m \neq n$.

b) **Average power from energy density:** The average energy density in the medium is

$$\langle w \rangle = \tfrac{1}{4}\,\mathrm{Re}[\tilde{\vec{\mathbf{D}}} \cdot \tilde{\vec{\mathbf{E}}}^* + \tilde{\vec{\mathbf{B}}} \cdot \tilde{\vec{\mathbf{H}}}^*]$$
$$= \tfrac{1}{4}[\epsilon E^2 + \mu H^2]$$

for a linear, isotropic, and homogeneous medium with $\vec{\mathbf{D}} = \epsilon\vec{\mathbf{E}}$ and $\vec{\mathbf{B}} = \mu\vec{\mathbf{H}}$. After substituting for the fields and making some simplifications, we get

$$\langle w \rangle = 2\epsilon E_0^2 \sin^2(\beta z \cos\theta) \tag{8.152}$$

The average power flow in the y direction between the same two nodes z_1 and z_2 in the z direction and per unit length in the x direction is

$$\langle P \rangle = \int_{z_1}^{z_2} \langle w \rangle u_{gy}\, dz$$

where u_{gy} is the group velocity of the wave in the y direction. Integrating this equation, we obtain

$$\langle P \rangle = \frac{\epsilon E_0^2 u_{gy}}{\beta \cos\theta}[m - n]\pi \tag{8.153}$$

Since both points of view should result in the same average power, comparing (8.151) and (8.153) we conclude that

$$u_{gy} = u_p \sin\theta \tag{8.154}$$

Thus, the group velocity can never be greater than the phase velocity of the wave in an unbounded medium. From (8.150) and (8.154), we have

$$u_{py} u_{gy} = u_p^2 \tag{8.155}$$

This equation states that as the phase velocity of the wave in the y direction increases, the group velocity with which the energy is propagating in that direction decreases.

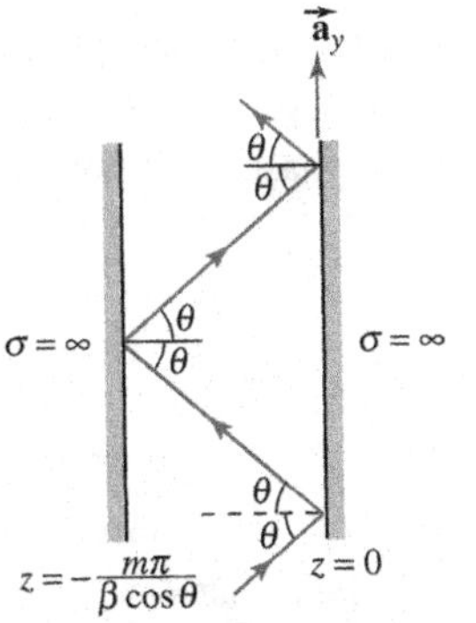

Figure 8.23 A parallel-plate waveguide

As explained earlier, the fields in the dielectric medium represent a pure standing wave in the z direction and a traveling wave in the y direction. We can place another conducting plate at any of the nodal points, Figure 8.23, without disturbing the field patterns. In this case, it appears as if the fields are being guided by the two perfectly conducting planes. These perfectly conducting planes are said to form a *parallel-plate waveguide*, and the preceding equations are the solutions of Maxwell's equations for this waveguide.

Instead of memorizing the preceding equations for the $\vec{\mathbf{E}}$ and $\vec{\mathbf{H}}$ fields and trying to correlate the geometry of a problem so that these results can be used, it is always easier to write the wave equations and from them obtain the necessary results. The following example illustrates this procedure.

EXAMPLE 8.12

A uniform plane wave propagating in free space strikes the surface of a perfect conductor at 60° with respect to the normal to the plane of

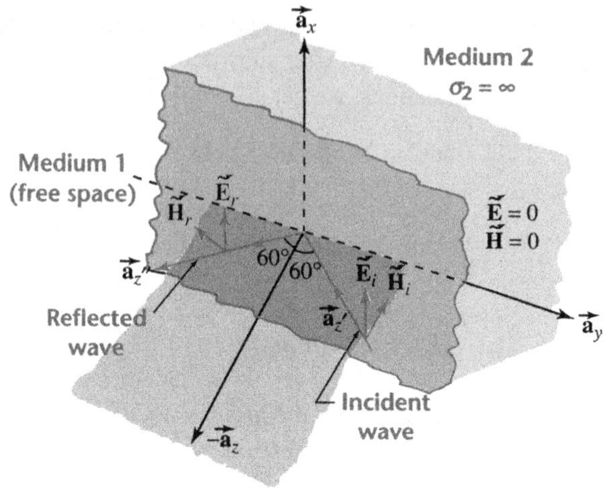

Figure 8.24 Plane wave striking a perfect conductor for Example 8.12

the conductor (Figure 8.24). The incident electric field is perpendicular to the plane of incidence and has an amplitude of 37.7 V/m at $z = 0$ and $t = 0$. If the angular frequency of the wave is 3 Mrad/s, write the expressions for the incident, reflected, and total fields in free space. Calculate the average power density in the medium, the direction of its flow, and the velocity of its propagation.

Solution The propagation constant in free space is

$$\beta = \frac{\omega}{c} = \frac{3 \times 10^9}{3 \times 10^8} = 10 \text{ rad/m}$$

Let us assume that the incident wave is traveling in the z' direction, as shown in Figure 8.24. The propagation of the incident wave is shown in Figure 8.25a. From this figure, it is clear that the wave propagates in the positive z direction with a phase constant of $\beta \cos(60°) = 5$ rad/m, and in the negative y direction with a phase constant of

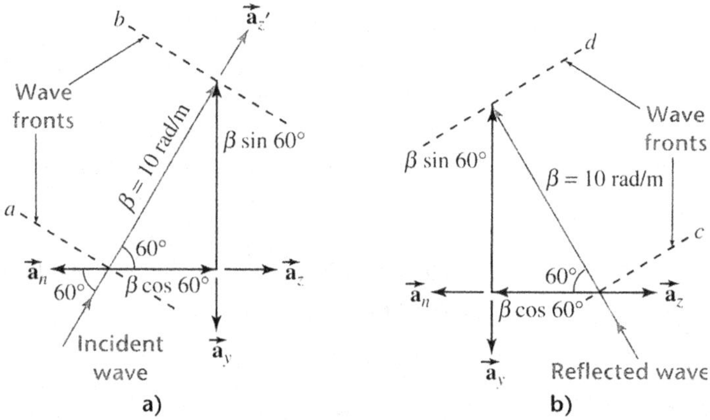

Figure 8.25 Constant wave fronts of **(a)** incident wave and **(b)** reflected wave

$\beta \sin(60°) = 8.66$ rad/m. With this information, we can express the incident electric field intensity in free space as

$$\tilde{\vec{\mathbf{E}}}_i = 37.7e^{-j\beta z'}\,\vec{\mathbf{a}}_x = 37.7e^{-j5z}e^{j8.66y}\,\vec{\mathbf{a}}_x \text{ V/m}$$

Using Maxwell's equation, $\nabla \times \tilde{\vec{\mathbf{E}}} = -j\omega\mu_0\tilde{\vec{\mathbf{H}}}$, we obtain the magnetic field intensity of the incident wave as

$$\tilde{\vec{\mathbf{H}}}_i = [0.05\,\vec{\mathbf{a}}_y + 0.087\,\vec{\mathbf{a}}_z]e^{-j5z}e^{j8.66y} \text{ A/m}$$

The tangential component of the total $\vec{\mathbf{E}}$ field on the surface of a perfect conductor must be zero, thus the reflection coefficient is $\rho_n = -1$, and the angle of reflection is 60°. If we assume that the reflected wave is propagating in the z'' direction, as shown in Figure 8.25b, then the phase constant in the negative z direction is $\beta \cos(60°) = 5$ rad/m, and that in the negative y direction is $\beta \sin(60°) = 8.66$ rad/m. Hence, the electric field intensity of the reflected wave is

$$\tilde{\vec{\mathbf{E}}}_r = -37.7e^{-j\beta z''}\,\vec{\mathbf{a}}_x = -37.7e^{j5z}e^{j8.66y}\,\vec{\mathbf{a}}_x \text{ V/m}$$

Once again, using the same Maxwell equation, we obtain the reflected component of the magnetic field intensity as

$$\tilde{\vec{\mathbf{H}}}_r = [0.05\,\vec{\mathbf{a}}_y - 0.087\,\vec{\mathbf{a}}_z]e^{j5z}e^{j8.66y} \text{ A/m}$$

Summing the incident and the reflected $\vec{\mathbf{E}}$ fields in free space, we obtain the total $\vec{\mathbf{E}}$ field as

$$\begin{aligned}\tilde{\vec{\mathbf{E}}} &= -37.7(e^{j5z} - e^{-j5z})e^{j8.66y}\,\vec{\mathbf{a}}_x \\ &= -j75.4\sin(5z)e^{j8.66y}\,\vec{\mathbf{a}}_x\end{aligned}$$

Similarly, we obtain the total $\tilde{\vec{\mathbf{H}}}$ field as

$$\tilde{\vec{\mathbf{H}}} = 0.1\cos(5z)e^{j8.66y}\,\vec{\mathbf{a}}_y - j0.174\sin(5z)e^{j8.66y}\,\vec{\mathbf{a}}_z$$

We can express these fields in the time domain as

$$\begin{aligned}E_x(x, y, z, t) &= 75.4\sin(5z)\sin(3\times 10^9 t + 8.66y) \text{ V/m}\\ H_y(x, y, z, t) &= 0.1\cos(5z)\cos(3\times 10^9 t + 8.66y) \text{ A/m}\\ H_z(x, y, z, t) &= 0.174\sin(5z)\sin(3\times 10^9 t + 8.66y) \text{ A/m}\end{aligned}$$

The phase velocity can be obtained by setting

$$3\times 10^9 t + 8.66y = k$$

where k is a constant, and differentiating with respect to t. Thus,

$$u_{py} = \frac{dy}{dt} = -3.46\times 10^8 \text{ m/s}$$

The negative sign shows that the wave propagates in the negative y direction. The phase velocity of the wave in the unbounded medium (free space in our case) is 3×10^8 m/s; thus, we obtain the group velocity of the wave, from (8.155), as

$$u_{gy} = -\frac{[3 \times 10^8]^2}{3.46 \times 10^8} = -2.6 \times 10^8 \text{ m/s}$$

Finally, the average power flow per unit area is

$$\begin{aligned}\langle \hat{\mathbf{S}} \rangle &= \tfrac{1}{2}\text{Re}[\tilde{\vec{\mathbf{E}}} \times \tilde{\vec{\mathbf{H}}}^*] \\ &= -6.56 \sin^2(5z)\,\vec{\mathbf{a}}_y \text{ W/m}^2\end{aligned}$$

• • •

8.10.2 Parallel polarization

We now consider the case when the electric field intensity is parallel to the plane of incidence, as depicted in Figure 8.26. As the $\tilde{\vec{\mathbf{H}}}$ field is parallel to the interface, let us assume that it is polarized along the x direction. If the magnetic field intensity of the incident wave is $\hat{H}_0$ at the interface, then we can express the incident $\tilde{\vec{\mathbf{H}}}$ field as

$$\tilde{\vec{\mathbf{H}}}_i = \hat{H}_0 e^{-\hat{\gamma}_1 z'}\,\vec{\mathbf{a}}_x$$

where $\vec{\mathbf{a}}_{z'}$ is the direction of propagation of the incident wave, and $\hat{\gamma}_1$ is its propagation constant. Expressing the propagation in terms of y and z coordinates, we obtain

$$\tilde{\vec{\mathbf{H}}}_i = \hat{H}_0 e^{-\hat{\gamma}_1(z\cos\theta_i + y\sin\theta_i)}\,\vec{\mathbf{a}}_x \tag{8.156a}$$

The corresponding $\tilde{\vec{\mathbf{E}}}$ field from Maxwell's equation, $\nabla \times \tilde{\vec{\mathbf{H}}} = j\omega\hat{\epsilon}\tilde{\vec{\mathbf{E}}}$, is

$$\tilde{\vec{\mathbf{E}}}_i = -[\cos\theta_i\,\vec{\mathbf{a}}_y - \sin\theta_i\,\vec{\mathbf{a}}_z]\hat{\eta}_1\hat{H}_0 e^{-\hat{\gamma}_1(z\cos\theta_i + y\sin\theta_i)} \tag{8.156b}$$

where $\hat{\eta}_1$ is the intrinsic impedance of medium 1.

If $\hat{\rho}_p$ and $\hat{\tau}_p$ are, respectively, the reflection and transmission coefficients, the fields associated with the reflected and transmitted

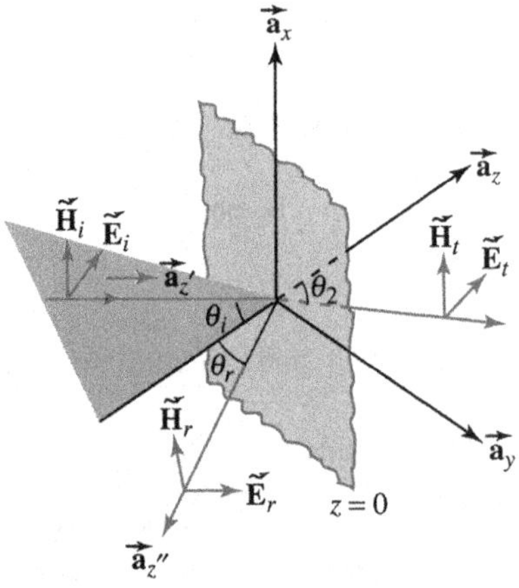

Figure 8.26 Oblique incidence–parallel polarization

waves are

$$\vec{\mathbf{H}}_r = \hat{\rho}_p \hat{H}_0 e^{-\hat{\gamma}_1(y \sin\theta_r - z\cos\theta_r)} \vec{\mathbf{a}}_x \tag{8.157a}$$

$$\vec{\mathbf{E}}_r = [\cos\theta_r \vec{\mathbf{a}}_y + \sin\theta_r \vec{\mathbf{a}}_z]\hat{\rho}_p \hat{\eta}_1 \hat{H}_0 e^{-\hat{\gamma}_1(y\sin\theta_r - z\cos\theta_r)} \tag{8.157b}$$

$$\vec{\mathbf{H}}_t = \frac{\hat{\eta}_1}{\hat{\eta}_2} \hat{\tau}_p \hat{H}_0 e^{-\hat{\gamma}_2(z\cos\theta_2 + y\sin\theta_2)} \vec{\mathbf{a}}_x \tag{8.158a}$$

$$\vec{\mathbf{E}}_t = -[\cos\theta_2 \vec{\mathbf{a}}_y - \sin\theta_2 \vec{\mathbf{a}}_z]\hat{\tau}_p \hat{\eta}_1 \hat{H}_0 e^{-\hat{\gamma}_2(z\cos\theta_2 + y\sin\theta_2)} \tag{8.158b}$$

where $\hat{\gamma}_2$ is the propagation constant and $\hat{\eta}_2$ is the intrinsic impedance of medium 2. Each medium is considered to be finitely conducting, so we rule out the possibility of a surface current at the interface. Thus, the continuity of the tangential components of the $\vec{\mathbf{H}}$ field at the interface yields

$$e^{-\hat{\gamma}_1 y \sin\theta_i} + \hat{\rho}_p e^{-\hat{\gamma}_1 y \sin\theta_r} = \frac{\hat{\eta}_1}{\hat{\eta}_2} \hat{\tau}_p e^{-\hat{\gamma}_2 y \sin\theta_2} \tag{8.159a}$$

This equation must be valid for all values of y; therefore, by setting $y = 0$ we obtain

$$1 + \hat{\rho}_p = \frac{\hat{\eta}_1}{\hat{\eta}_2} \hat{\tau}_p \tag{8.159b}$$

In addition, for (8.159a) to be true in general, we must have

$$\hat{\gamma}_1 \sin\theta_i = \hat{\gamma}_1 \sin\theta_r = \hat{\gamma}_2 \sin\theta_2 \tag{8.159c}$$

Therefore, the angle of incidence must be equal to the angle of reflection:

$$\theta_i = \theta_r = \theta_1 \tag{8.160a}$$

Equation (8.160a) is simply a mathematical statement of Snell's law of reflection. In addition, the equality in (8.159c) yields

$$\hat{\gamma}_1 \sin\theta_1 = \hat{\gamma}_2 \sin\theta_2 \tag{8.160b}$$

which is Snell's law of refraction.

Equating the tangential component of the $\vec{\mathbf{E}}$ field at the interface, we obtain

$$(1 - \hat{\rho}_p) = \frac{\cos\theta_2}{\cos\theta_1} \hat{\tau}_p \tag{8.161}$$

Manipulating (8.159b) and (8.161), we get

$$\hat{\tau}_p = \frac{2\hat{\eta}_2 \cos\theta_1}{\hat{\eta}_1 \cos\theta_1 + \hat{\eta}_2 \cos\theta_2} \tag{8.162}$$

as the coefficient of transmission and

$$\hat{\rho}_p = \frac{\hat{\eta}_1 \cos\theta_1 - \hat{\eta}_2 \cos\theta_2}{\hat{\eta}_1 \cos\theta_1 + \hat{\eta}_2 \cos\theta_2} \tag{8.163}$$

as the coefficient of reflection for the parallel polarized wave.

Dielectric-dielectric interface

As a special case of parallel polarization, let us consider that the interface is between two dielectric media and the permeability of each medium is the same as that of free space. With these assumptions, we can express the coefficient of reflection as

$$\hat{\rho}_p = \frac{\cos\theta_1 - \sqrt{\dfrac{\epsilon_1}{\epsilon_2}}\cos\theta_2}{\cos\theta_1 + \sqrt{\dfrac{\epsilon_1}{\epsilon_2}}\cos\theta_2} \tag{8.164}$$

From (8.160b), we obtain

$$\sin^2\theta_2 = \frac{\epsilon_1}{\epsilon_2}\sin^2\theta_1 \tag{8.165}$$

Expressing $\cos\theta_2$ in terms of $\sin\theta_2$ and substituting for $\sin\theta_2$ from (8.165) into (8.164), we get

$$\hat{\rho}_p = \frac{\dfrac{\epsilon_2}{\epsilon_1}\cos\theta_1 - \sqrt{\dfrac{\epsilon_2}{\epsilon_1} - \sin^2\theta_1}}{\dfrac{\epsilon_2}{\epsilon_1}\cos\theta_1 + \sqrt{\dfrac{\epsilon_2}{\epsilon_1} - \sin^2\theta_1}} \tag{8.166}$$

Let us assume that when $\theta_1 = \theta_p$, the reflection coefficient becomes zero. For this to happen, the numerator must go to zero when the angle of incidence is θ_p. That is,

$$\left(\frac{\epsilon_2}{\epsilon_1}\right)^2 \cos^2\theta_p = \frac{\epsilon_2}{\epsilon_1} - \sin^2\theta_p$$

which yields

$$\sin^2\theta_p = \frac{\epsilon_2}{\epsilon_1 + \epsilon_2} \tag{8.167a}$$

$$\cos^2\theta_p = \frac{\epsilon_1}{\epsilon_1 + \epsilon_2} \tag{8.167b}$$

$$\tan\theta_p = \sqrt{\frac{\epsilon_2}{\epsilon_1}} \tag{8.167c}$$

These equations state that when an incident wave traveling in a dielectric medium impinges upon another dielectric medium at an angle θ_p, there is no reflected wave. The equations of (8.167) are known as *Brewster's law,* and the angle θ_p is called the *Brewster angle.* When an incident wave having both parallel and perpendicular components impinges upon a plane boundary between two dielectric media at the Brewster angle, the reflected wave will have only a perpendicular component. That is, an elliptically or circularly polarized wave will become a linearly polarized wave upon reflection. For this reason, the Brewster angle is also called the *polarizing angle.*

Further analysis of (8.166) reveals that when the wave propagates from a denser to a lighter medium, $\epsilon_1 \geq \epsilon_2$, and the angle of incidence is such that

$$\sin^2\theta_1 = \frac{\epsilon_2}{\epsilon_1} \tag{8.168}$$

the wave will experience total reflection. That is,

$$\rho_p = 1 \tag{8.169}$$

Equation (8.168), in fact, defines the critical angle for total reflection as

$$\theta_c = \sin^{-1}\sqrt{\frac{\epsilon_2}{\epsilon_1}} \tag{8.170}$$

Just as in perpendicular polarization, a parallel polarized wave will experience a total reflection as long as the angle of incidence is greater than or equal to the critical angle as given by (8.170).

Dielectric–perfect conductor interface

This case is very similar to the one discussed under perpendicular polarization. For this reason, we use the following example to explain it.

EXAMPLE 8.13

A parallel polarized uniform plane wave propagating in free space impinges at the surface of a perfect conductor at 45°, as shown in Figure 8.27. The magnetic field intensity of the incident wave is in the x direction and has an amplitude of 0.1 A/m at the interface. If the angular frequency of the wave is 600 Mrad/s, write expressions for the incident, reflected, and total fields in free space. Calculate the average power density in the region.

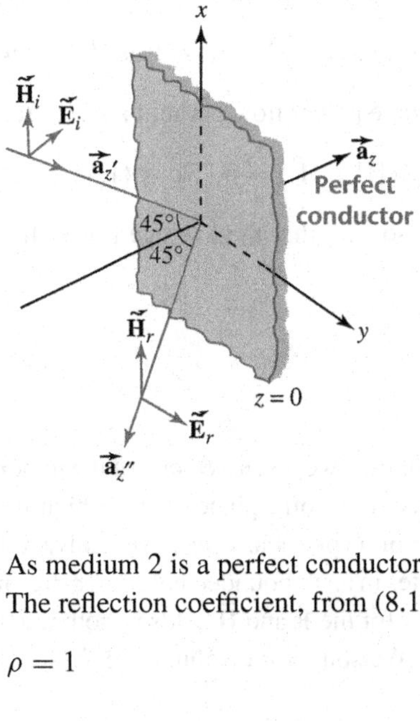

Figure 8.27 Oblique incidence of a parallel polarized wave on a perfect conductor

Solution As medium 2 is a perfect conductor, the fields exist only in free space. The reflection coefficient, from (8.163), is

$$\rho = 1$$

The propagation constant for free space is

$$\hat{\gamma} = j\omega\sqrt{\mu_0\epsilon_0} = j\frac{\omega}{c} = j\frac{6 \times 10^8}{3 \times 10^8} = j2\ \text{m}^{-1}$$

Thus, the phase constant is $\beta = 2$ rad/m and the intrinsic impedance for free space is $\eta = 377\ \Omega$. For an incident angle of 45° and $\rho = 1$, the incident and the reflected fields are

$$\tilde{\vec{\mathbf{H}}}_i = 0.1e^{-j1.414z}e^{-j1.414y}\,\vec{\mathbf{a}}_x\ \text{A/m}$$
$$\tilde{\vec{\mathbf{E}}}_i = -26.658(\vec{\mathbf{a}}_y - \vec{\mathbf{a}}_z)e^{-j1.414z}e^{-j1.414y}\ \text{V/m}$$
$$\tilde{\vec{\mathbf{H}}}_r = 0.1e^{j1.414z}e^{-j1.414y}\,\vec{\mathbf{a}}_x\ \text{A/m}$$
$$\tilde{\vec{\mathbf{E}}}_r = 26.658(\vec{\mathbf{a}}_y + \vec{\mathbf{a}}_z)e^{j1.414z}e^{-j1.414y}\ \text{V/m}$$

The total fields in free space are

$$\tilde{\vec{\mathbf{H}}} = \tilde{\vec{\mathbf{H}}}_i + \tilde{\vec{\mathbf{H}}}_r = 0.2\cos(1.414z)e^{-j1.414y}\,\vec{\mathbf{a}}_x\ \text{A/m}$$
$$\tilde{\vec{\mathbf{E}}} = \tilde{\vec{\mathbf{E}}}_i + \tilde{\vec{\mathbf{E}}}_r = 53.32[\cos(1.414z)\,\vec{\mathbf{a}}_z + j\sin(1.414z)\,\vec{\mathbf{a}}_y]e^{-j1.414y}\ \text{V/m}$$

The surface current density at the interface due to the tangential component of the $\tilde{\vec{\mathbf{H}}}$ field at $z = 0$ is

$$\tilde{\mathbf{J}}_s(0) = -0.2e^{-j1.414y}\,\vec{\mathbf{a}}_y\ \text{A/m}$$

The tangential component of the electric field intensity in free space at $z = 0$ is zero. However, the normal component of the $\tilde{\vec{\mathbf{E}}}$ field is

$$\tilde{\vec{\mathbf{E}}}(0) = 53.32e^{-j1.414y}\,\vec{\mathbf{a}}_z\ \text{V/m}$$

This gives rise to the surface charge density at the interface as

$$\tilde{\rho}_s = -53.32\epsilon_0 e^{-j1.414y} = 471.5e^{-j1.414y}\,\text{pC/m}^2$$

The average power flow per unit area in free space is

$$\langle\hat{\mathbf{S}}\rangle = \tfrac{1}{2}\text{Re}[\tilde{\vec{\mathbf{E}}} \times \tilde{\vec{\mathbf{H}}}^*] = 5.33\cos^2(1.414z)\,\vec{\mathbf{a}}_y\ \text{W/m}^2$$

clearly showing that the average power flow is along the y direction.

• • •

8.11 Summary

In this chapter, we discussed one of the major applications of Maxwell's equations: a uniform plane wave, which may be considered as a special case of a spherical wave. We analyzed how a uniform plane wave propagates in an unbounded but source-free medium. We obtained wave equations for the $\vec{\mathbf{E}}$ and $\vec{\mathbf{H}}$ fields in both the time domain and the phasor domain. We sought the solution of these waves in the phasor domain and in the process defined the propagation constant:

$$\hat{\gamma} = j\omega\sqrt{\mu\hat{\epsilon}} = \alpha + j\beta$$

where ω is the angular frequency of the wave (rad/s), μ is the permeability of the medium, $\hat{\epsilon}$ is the complex permittivity, α is the attenuation constant, and β is the phase constant.

As the wave progresses in a conductive medium it attenuates and disappears after penetrating a depth equal to $5\delta_c$, where $\delta_c = 1/\alpha$ is the skin depth. The phase velocity in the direction of propagation is u_p and is given as

$$u_p = \frac{\omega}{\beta}$$

The complex permittivity was defined as

$$\hat{\epsilon} = \epsilon\left[1 - j\frac{\sigma}{\omega\epsilon}\right] = \epsilon' - j\epsilon'' = \epsilon[1 - j\tan\delta]$$

where σ is the conductivity of the medium and ϵ is its permittivity. We defined the wavelength as the distance between two planes when the phase difference between them at any give time is 2π. That is,

$$\beta\lambda = 2\pi$$

As the wave travels through the medium, it appears that the medium has some impedance, called the intrinsic impedance. It is defined mathematically as

$$\hat{\eta} = \sqrt{\frac{\mu}{\hat{\epsilon}}}$$

For a uniform plane, the relationship between $\tilde{\vec{\mathbf{E}}}$ and $\tilde{\vec{\mathbf{H}}}$ fields is expressed in terms of $\hat{\eta}$ as

$$\vec{\mathbf{a}}_z \times \tilde{\vec{\mathbf{E}}} = \hat{\eta}\tilde{\vec{\mathbf{H}}}$$

where $\vec{\mathbf{a}}_z$ is the direction of propagation of the wave. Since $\hat{\eta}$ is, in general, a complex quantity, the $\tilde{\vec{\mathbf{E}}}$ and $\tilde{\vec{\mathbf{H}}}$ fields are not in phase. The $\tilde{\vec{\mathbf{E}}}$ and $\tilde{\vec{\mathbf{H}}}$ fields are in phase only when the medium is dielectric (free space is a special case of a dielectric medium).

Using the $\vec{\mathbf{E}}$ field as a reference, we explained how to determine the polarization of a wave. A wave may be linearly, circularly, or elliptically polarized. An unpolarized wave is usually said to be randomly polarized. It is, however, possible to represent a circularly polarized wave as a set of two linearly polarized waves. Similarly, an elliptically polarized wave can be represented by two circularly polarized waves.

When a wave enters one medium from another, it can strike the boundary either normally (normal incidence) or at an angle (oblique incidence). Normal incidence may be treated as a special case of oblique incidence. Part of the wave, called the reflected wave, is reflected back. The remainder of the wave penetrates into the second medium and continues its propagation in that medium. The interface can be between two conducting media, two dielectric media, or between a dielectric and a conducting medium. The general equations derived for the two conducting media also apply to the other cases. In these cases, the tangential components of both the $\vec{\mathbf{E}}$ and $\vec{\mathbf{H}}$ fields are continuous. These boundary conditions enable us to determine the reflection coefficient $\hat{\rho}$ and the transmission coefficient $\hat{\tau}$.

When the incident wave strikes a perfectly conducting medium, the tangential components of the $\vec{\mathbf{H}}$ field result in a surface current density on the surface of a conductor. Only when the $\vec{\mathbf{E}}$ field is in the plane of incidence and the wave strikes the interface at some angle of incidence, does there exist a surface charge density on the surface of a perfect conductor. For all other cases, the surface charge density is zero.

8.12 Review questions

8.1 Explain the differences between a plane wave and a uniform plane wave.

8.2 Can a uniform plane wave exist in "real life"?

8.3 What is a uniform medium?

8.4 A plane wave in a dielectric medium has an $\vec{\mathbf{E}}$ component of the form $\vec{\mathbf{E}} = E_0 \sin(\beta x - \omega t)\vec{\mathbf{a}}_z$ V/m. Derive expressions for the associated $\vec{\mathbf{H}}$ field and the instantaneous Poynting vector. What is the average power density in the medium?

8.5 What are the wavelength and frequency of a wave propagating in free space when $\beta = 2$?

8.6 The electric field intensity in a dielectric medium is given as $\vec{\mathbf{E}} = 100\cos(3 \times 10^6 t - 0.1z)\vec{\mathbf{a}}_x$ V/m. Plot this as a function of time from 0 to $4\pi/T$ on a common scale for $z = 50$ m and $z = 100$ m.

8.7 Plot the $\vec{\mathbf{E}}$ field in Question 8.6 as a function of z from 0 to $4\pi/\lambda$ on a common scale for $t = \pi/\omega$ and $t = 2\pi/\omega$.

8.8 A uniform plane wave propagates in free space in the x direction. If the $\vec{\mathbf{E}}$ field is in the z direction and has an amplitude of 100 V/m, what must be the associated $\vec{\mathbf{H}}$ field? Compute the time-average Poynting vector.

8.9 Justify that, in a plane wave propagating in a dielectric medium, the average energy stored is equally divided between the electric and magnetic fields.

8.10 Show that the intrinsic impedance has the dimensions of resistance.

8.11 Explain the following terms: the propagation constant, the attenuation constant, the phase constant, the wavelength, the intrinsic impedance, and the skin depth.

8.12 State the condition for a "high-loss" medium. Calculate the phase speed, the wavelength, and the skin depth in copper at (a) 60 Hz, (b) 60 kHz, (c) 60 MHz, and (d) 2.4 GHz. Assume $\sigma = 5.8 \times 10^7$ S/m.

8.13 Compute $\hat{\gamma}$, α, β, $\hat{\eta}$, λ, δ_c, and $\tan\delta$ for moist soil ($\epsilon_r = 16$, and $\sigma = 5$ mS/m) at a frequency of 100 MHz.

8.14 Compute $\hat{\gamma}$, α, β, $\hat{\eta}$, λ, δ_c, and $\tan\delta$ for rocky soil ($\epsilon_r = 12$, and $\sigma = 1.25$ mS/m) at a frequency of 100 MHz.

8.15 Explain why the resistance of a wire increases with frequency.

8.16 Define skin depth and explain surface resistance.

8.17 Verify (8.72). Can you still use this equation when the diameter of a wire is less than $2\delta_c$?

8.18 What is a good dielectric? What is a good conductor? How does a good conductor differ from a perfect conductor?

8.19 Explain the polarization of a wave. What is the major difference between an elliptically and circularly polarized wave?

8.20 What is the significance of knowing whether a circularly polarized wave is right-handed or left-handed?

8.21 Define the reflection and transmission coefficients. Can these be defined at any point in the direction of propagation? Why did we define them at the interface?

8.22 The standing wave ratio (SWR) was defined in Exercise 8.21. Using that definition, determine the SWR in medium 1 for Example 8.8.

8.23 Will it make sense to define the SWR in a conductive medium? Cite reasons to justify your answers.

8.24 How do you differentiate between a standing wave and a propagating wave?

8.25 Can a standing wave, in general, be represented by two waves propagating in opposite directions?

8.26 Explain radiation pressure and justify its definition as given by (8.119).

8.27 Explain the principle of operation of a Fabry-Perot resonator.

8.28 What do we mean by the plane of incidence?

8.29 What is the difference between a perpendicularly polarized wave and a parallel polarized wave?

8.30 What is Snell's law of reflection?

8.31 Explain Snell's law of reflection.

8.32 What is meant by a critical angle of reflection? What is its significance? What happens to a wave that hits the interface at an angle less than the critical angle? What happens if the angle of incidence is greater than the critical angle?

8.33 What is a surface wave? What is the condition under which a uniform plane wave, after passing through a dielectric boundary, becomes a surface wave?

8.34 Explain the difference between phase velocity and group velocity. Cite some examples to justify your answer.

8.35 Explain the propagation within a parallel-plate waveguide. Can the wave between the two parallel plates be referred to as a guided wave? Do you know of any other examples of a guided wave?

8.36 Is it justifiable to refer to the voltage and current waveforms on a transmission line as a guided wave?

8.37 Explain the Brewster angle. What is the difference between the critical angle and the Brewster angle? Why is the Brewster angle also called a polarizing angle?

8.38 Is the Brewster angle possible for a perpendicularly polarized wave? Justify your answer mathematically.

8.13 Exercises

8.1 Derive (8.9).

8.2 Write (8.8) and (8.9) as a set of six scalar equations.

8.3 Verify equations (8.26a), (8.26b), and (8.27).

8.4 Show that the electric field intensity as given by $\vec{\mathbf{E}} = 100\sin(10^8 t + x/\sqrt{3})\vec{\mathbf{a}}_z$ V/m is a valid solution of the wave equation in a dielectric medium. Also determine (a) the dielectric constant of the medium assuming that $\mu_r = 1$, (b) the velocity of propagation, (c) the magnetic field intensity, (d) the intrinsic impedance, (e) the wavelength, and (f) the average power density.

8.5 Using (8.27), show that the $\vec{\mathbf{E}}$ and $\vec{\mathbf{H}}$ fields are normal to each other.

8.6 The electric field intensity of a uniform plane wave in free space is given by $\vec{\mathbf{E}} = 120\cos(2\pi \times 10^9 t - \beta y)\vec{\mathbf{a}}_z$ V/m. Determine (a) the phase constant, (b) the magnetic field intensity, (c) the wavelength, (d) the average power density in the medium, (e) the average energy density in the electric field, and (f) the average energy density in the magnetic field.

8.7 The magnetic field intensity of a plane wave in free space is given as $\vec{\mathbf{H}} = 0.1\cos(200\pi \times 10^6 t + \beta z)\vec{\mathbf{a}}_x$ A/m. Determine (a) the phase constant, (b) the velocity of propagation, (c) the wavelength, (d) the electric field intensity, (e) the displacement current density, and (f) the average power flow per unit area.

8.8 Using Maxwell's equations in phasor form and the definitions of $\hat{\epsilon}$ and $\hat{\gamma}$, derive (8.44) and (8.45).

8.9 Verify the expressions for α (8.51a) and β (8.51b).

8.10 A uniform plane wave at a frequency of 100 MHz propagates in the z direction in a conducting medium characterized by $\mu_r = 1$, $\epsilon_r = 2.25$ and $\sigma = 9.375$ mS/m. Determine $\hat{\gamma}, \alpha, \beta, \lambda, \delta_c, \vec{\mathbf{u}}_p$, and $\tan\delta$. If the maximum value of the electric field intensity in the y direction is 125 V/m and occurs at $z = 0$ and $t = 0$, write expressions for the $\vec{\mathbf{E}}$ and $\vec{\mathbf{H}}$ fields. Obtain a general expression for the power density in the medium. What is the total current density in the medium?

8.11 A glass material is coated with silver ($\sigma = 6.1 \times 10^7$ S/m) to make it a good conductor at 2.4 GHz. What is the surface resistance of the coating?

8.12 Calculate the skin depth of aluminum at 60 Hz and at 60 MHz. Assume that $\sigma = 3.5 \times 10^7$ S/m, $\mu = \mu_0$, and $\epsilon = \epsilon_0$. If the outer diameter of a 100-m-long hollow conductor is 2.54 cm and the thickness is 5 mm, determine the surface resistance at each frequency.

8.13 Verify equations (8.75) and (8.78).

8.14 An antenna just beneath the surface of a moist soil ($\epsilon_r = 16, \mu_r = 1, \sigma = 5$ mS/m) is transmitting a signal at 60 MHz. Assuming that the signal propagates as a uniform plane wave, a receiver buried in the soil receives the signal with a strength of $\frac{1}{10}$ of its original value. Compute the distance of the receiver.

8.15 The electric field intensity of a wave in a region is given by $\vec{\mathbf{E}} = 3\cos(\omega t - \beta x - 45°)\vec{\mathbf{a}}_y + 4\sin(\omega t - \beta x + 45°)\vec{\mathbf{a}}_z$ V/m. Determine the polarization of the wave.

8.16 Find the polarization of a wave if the electric field intensity is given by $\tilde{\vec{\mathbf{E}}} = (-j25\,\vec{\mathbf{a}}_x + 25\,\vec{\mathbf{a}}_z)e^{-(0.01+j120y)}$ V/m.

8.17 The electric field intensity of a wave in a region is given by $\vec{\mathbf{E}} = 30\cos(\omega t - \beta x - 30°)\vec{\mathbf{a}}_y + 40\cos(\omega t - \beta x + 60°)\vec{\mathbf{a}}_z$ V/m. Determine the polarization of the wave.

8.18 Write expressions for the incident, reflected, and total fields in medium 1 of Example 8.7. Also compute the incident, reflected, and total power densities in the region. Is the power density in medium 1 equal to the power density in medium 2 at the interface?

8.19 Repeat Example 8.7 when the wave frequency is 500 MHz. What sort of conclusions can you draw by comparing the results?

8.20 For the uniform plane wave discussed in Example 8.8, determine (a) the phase velocity and (b) the wavelength in each dielectric region. Show that the total power density in region $z \leq 0$ is the sum of the incident and the reflected power densities.

8.21 If we define the **standing wave ratio (SWR)** as the ratio of the maximum to the minimum values of the total $\vec{\mathbf{E}}$ field in medium 1, show that

$$SWR = \frac{|E|_{\max}}{|E|_{\min}} = \frac{1 + |\rho|}{1 - |\rho|}$$

where $|\rho|$ is the magnitude of the reflection coefficient.

8.22 An electromagnetic wave propagates in a dielectric medium with $\epsilon = 2.25\epsilon_0$ along the x direction. It strikes another dielectric medium with $\epsilon = 9\epsilon_0$ at $x = 0$. If the incoming wave is polarized in the z direction, has a maximum value of 250 mV/m at the interface, and has an angular frequency of 300 Mrad/s, determine (a) the reflection coefficient, (b) the transmission coefficient, and (c) the power densities of the incident, reflected, and transmitted waves.

8.23 Repeat Example 8.9 when the angular frequency of the incident wave is 600 Mrad/s. Compute the surface current density and the radiation pressure on the surface of the perfect conductor. Is it practically possible to measure the wave frequency using a Fabry–Perot resonator?

8.24 Write the expressions for the incident, reflected, and total fields in the time domain for Exercise 8.23. Compute the average power density of the incident and reflected waves. What is the average power density of the total wave?

8.25 A uniform plane wave propagating in air strikes normally upon seawater ($\epsilon_r = 81, \sigma = 4$ S/m). If the angular frequency of the wave is 96 Mrad/s, and the amplitude of the incident $\vec{\mathbf{E}}$ field at the interface is 100 V/m, obtain expressions for the incident, reflected, and transmitted power densities. What is the skin depth of the wave in the conducting medium?

8.26 Repeat Exercise 8.25 if the wave frequency is 9.6 Grad/s. What kind of conclusions can you draw about the medium by comparing the skin depths?

8.27 A uniform plane wave propagating in free space strikes the surface of a conducting medium at an angle of 30° with respect to the normal ($\sigma = 0.4$ S/m, $\epsilon = \epsilon_0$, $\mu = \mu_0$). If the frequency of the incident wave is 96 Mrad/s, and the amplitude of the wave at the interface is 100 V/m, obtain the expressions for the incident, reflected, and transmitted fields. Compute the average power density in each medium.

8.28 What is the critical angle of the wave discussed in Example 8.11? Write the expressions for the fields and average power densities in each region when the angle of incidence is equal to the critical angle. Can you now substantiate the statements we made for this case?

8.29 Verify equation (8.145).

8.30 Redo Example 8.12 if the incident wave was propagating in a medium with $\epsilon = 9\epsilon_0$ and $\mu = \mu_0$.

8.31 Verify (8.158a).

8.32 Show that $\rho = \dfrac{\tan(\theta_1 - \theta_2)}{\tan(\theta_1 + \theta_2)}$ for a dielectric–dielectric interface, where θ_1 and θ_2 are, respectively, the angles of the incident and transmitted waves. In order to derive this result, is it necessary to assume that the permeabilities of the two dielectric media are the same?

8.33 A parallel polarized uniform plane wave propagating in free space impinges at the surface of a conducting medium ($\sigma = 0.4$ S/m) at 30°. The magnetic field intensity of the incident wave is in the x direction and has an amplitude of 10 A/m at the interface. If the angular frequency of the wave is 30 Mrad/s, write expressions for the incident, reflected, and transmitted fields. Calculate the average power density in the conducting medium.

8.14 Problems

8.1 Derive the Helmholtz equations in the time domain for a source-free, linear, homogeneous, and isotropic dielectric medium.

8.2 Derive the general wave equations in the time domain for a linear, homogeneous, and isotropic but finitely conducting medium containing volume current and charge densities. Show that the wave equations for the $\vec{\mathbf{E}}$ and $\vec{\mathbf{H}}$ fields are not identical.

8.3 Derive the wave equations for a highly conductive medium in which the displacement current density can be neglected. Are the wave equations for the $\vec{\mathbf{E}}$ and $\vec{\mathbf{H}}$ fields similar?

8.4 Show that $\vec{\mathbf{E}} = E_0 \cos \omega t \cos \beta z\, \vec{\mathbf{a}}_x$ V/m satisfies the wave equation in a source-free dielectric medium when $\beta = \pm\omega\sqrt{\mu\epsilon}$. What is the corresponding $\vec{\mathbf{H}}$ field? Express these fields in terms of forward- and backward-traveling waves.

8.5 Show that $E_x(z, t) = F_x(t + \sqrt{\mu\epsilon} z)$ satisfies the wave equation in a dielectric medium. Also obtain expressions for the $\vec{\mathbf{H}}$ field and the Poynting vector.

8.6 Show that a general function $F(t - z/u)$ satisfies the wave equation in a dielectric medium as long as $u = \pm c/n$, where c is the speed of light and n is the index of refraction.

8.7 If $\vec{\mathbf{H}} = 100 \cos(30,000t + \beta z)\, \vec{\mathbf{a}}_x$ A/m is the magnetic field intensity in free space of a uniform plane wave, determine (a) the phase constant, (b) the wavelength, (c) the velocity of propagation, (d) the $\vec{\mathbf{E}}$ field, and (e) the time average power flow per unit area.

8.8 A 100-MHz uniform plane wave is traveling in the y direction in a lossless unbounded medium ($\epsilon_r = 4$ and $\mu_r = 1.0$). If the $\vec{\mathbf{E}}$ field has only an x component and its amplitude is 500 V/m when $t = 0$ and $y = 0$, determine (a) the phase velocity, (b) the phase constant, (c) the $\vec{\mathbf{H}}$ field, (d) the wavelength, and (e) the average power flow through a cross-sectional area of 16 cm^2.

8.9 A uniform plane wave is propagating through a dielectric medium having a dielectric constant of 2.5. If the breakdown strength of the medium is 70.7 kV/m (rms), what is the maximum power flow per unit area that can take place without a breakdown? What is the corresponding magnetic field intensity?

8.10 The $\vec{\mathbf{E}}$ field of a plane monochromatic wave traveling in a uniform medium is given as

$$\vec{\mathbf{E}} = E_m \cos(\beta z - \omega t)\, \vec{\mathbf{a}}_x + E_m \sin(\beta z - \omega t)\, \vec{\mathbf{a}}_y$$

where $\beta = \omega\sqrt{\mu\epsilon}$ and E_m is a constant. Find the corresponding $\vec{\mathbf{H}}$ field and the Poynting vector.

8.11 Start with Maxwell's equations in phasor form for a source-free, uniform conducting medium and obtain the wave equations in terms of the $\tilde{\mathbf{E}}$ and $\tilde{\mathbf{H}}$ fields.

8.12 Start with Maxwell's equations in phasor form for a source-free, uniform, highly conducting medium ($\sigma \gg \omega\epsilon$) and obtain the wave equations in terms of the $\tilde{\mathbf{E}}$ and $\tilde{\mathbf{H}}$ fields. Solve these equations and verify the expressions for α, β, δ, and u_p for a good conductor.

8.13 A uniform plane wave at a frequency of 50 MHz is transmitted in moist soil ($\epsilon_r = 16$, $\mu_r = 1$, and $\sigma = 0.02$ S/m). If the amplitude of the tangential component of the $\vec{\mathbf{E}}$ field just beneath the surface is 120 V/m, determine (a) the propagation, attenuation, and phase constants, (b) the phase velocity, (c) the wavelength, (d) the intrinsic impedance, (e) the skin depth, and (f) the average power density. If the range is to be taken as a distance at which 90% of the wave amplitude is attenuated, find the range of the signal being transmitted.

8.14 A uniform plane wave at a frequency of 10 kHz propagates in ferrite ($\epsilon_r = 9$, $\mu_r = 4$, and $\sigma = 0.01$ S/m). If the amplitude of the tangential component of the $\vec{\mathbf{E}}$ field just beneath the surface is 100 V/m, determine (a) the propagation, attenuation, and phase constants, (b) the phase velocity, (c) the wavelength, (d) the intrinsic impedance, (e) the skin depth, and (f) the average power density.

8.15 The magnetic field intensity in a medium is given by

$$\vec{\mathbf{H}} = 0.1e^{-77.485y}\cos(2\pi \times 10^9 t - 203.8y)\,\vec{\mathbf{a}}_x \text{ A/m}$$

If the medium is characterized by the free space permeability, determine the dielectric constant and the conductivity of the medium. Obtain the associated component of the $\vec{\mathbf{E}}$ field. Compute the average power density.

8.16 The amplitude of the tangential electric field intensity just below the surface of copper ($\epsilon_r = 1$, $\mu_r = 1$, and $\sigma = 5.8 \times 10^7$ S/m) is 100 V/m. If the frequency of the uniform plane wave is 10 kHz, write time domain expressions for $\vec{\mathbf{E}}$, $\vec{\mathbf{J}}$, and $\vec{\mathbf{H}}$ on a plane at $z = 0.2\delta_c$, where δ_c is the skin depth.

8.17 A uniform plane wave is propagating in a good conductor. If the magnetic field intensity is given by

$$\vec{\mathbf{H}} = 0.1e^{-15z}\cos(2\pi \times 10^8 t - 15z)\,\vec{\mathbf{a}}_x \text{ A/m}$$

determine the conductivity and the corresponding component of the $\vec{\mathbf{E}}$ field. Calculate the average power loss in a block of unit area and δ thickness.

8.18 Show that $\theta_n = \delta/2$ where θ_n is the angle by which the magnetic field lags the electric field and δ is the loss tangent angle.

8.19 A uniform plane wave at 20 MHz is propagating through a nonmagnetic lossy medium. The amplitude of the wave reduces by 20% per meter. The magnetic field lags the electric field by 20°. Compute α, β, σ, $\hat{\eta}$, and δ_c.

8.20 If the intrinsic impedance of a nonmagnetic medium is $60\pi\,\underline{/30^\circ}\ \Omega$, and the phase constant is 1.2 rad/m, determine (a) the dielectric constant, (b) the wave frequency, (c) the attenuation constant, and (d) the skin depth. Express the attenuation in dB/m.

8.21 A 500-kHz uniform plane wave propagates beneath the earth's surface. The amplitude of the $\vec{\mathbf{E}}$ field just beneath the surface is 120 V/m. If the relative permeability, the dielectric constant, and the conductivity of the earth are taken as 1, 16, and 0.02 S/m, respectively, determine (a) the propagation, attenuation, and phase constants, (b) the phase velocity, (c) the wavelength, (d) the intrinsic impedance, (e) the skin depth, and (f) the average power density. If the range is considered to be the distance at which 90% of the wave is attenuated, find the range of the signal being transmitted.

8.22 Find the polarization of the following waves:

a) $\vec{\mathbf{E}} = 100e^{-j300x}\,\vec{\mathbf{a}}_y + 100e^{-j300x}\,\vec{\mathbf{a}}_z$ V/m

b) $\vec{\mathbf{E}} = 16e^{j\pi/4}e^{-j100z}\,\vec{\mathbf{a}}_x - 9e^{-j\pi/4}e^{-j100z}\,\vec{\mathbf{a}}_y$ V/m

c) $\vec{\mathbf{E}} = 3\cos(t - 0.5y)\,\vec{\mathbf{a}}_x - 4\sin(t - 0.5y)\,\vec{\mathbf{a}}_z$ V/m

8.23 A uniform plane wave with an $\vec{\mathbf{E}}$ of $12\,\cos(\omega t - \beta z)\vec{\mathbf{a}}_x - 5\sin(\omega t - \beta z)\vec{\mathbf{a}}_y$ V/m is propagating in a lossless medium ($\epsilon_r = 2.5$, $\mu_r = 1$) at 200 Mrad/s. Determine the corresponding $\vec{\mathbf{H}}$ field, the phase constant β, the wavelength λ, the intrinsic impedance η, the phase velocity $\vec{\mathbf{u}}_p$, and the polarization of the wave.

8.24 Show that a linearly polarized plane wave, $E_0\cos(\omega t - \beta z)\vec{\mathbf{a}}_x + E_0\cos(\omega t - \beta z)\vec{\mathbf{a}}_y$, can be expressed as a sum of left- and right-handed circularly polarized waves of equal amplitudes.

8.25 Show that an elliptically polarized wave, $3\;E_0\cos(\omega t - \beta z)\vec{\mathbf{a}}_x + 4E_0\sin\,(\omega t - \beta z)\vec{\mathbf{a}}_y$, can be expressed as a sum of left- and right-handed circularly polarized waves of unequal amplitudes.

8.26 A 100-MHz uniform plane wave traveling in a conducting medium ($\epsilon_r = 2.25$, $\mu_r = 1$, $\sigma = 2$ mS/m) strikes normally on the surface of another conducting medium $\epsilon_r = 1$, $\mu_r = 1$, $\sigma = 20$ mS/m. If the amplitude of the incident wave is maximum at the interface and has a value of 10 V/m, determine the average power densities of the incident, reflected, and transmitted waves. How far will the transmitted wave travel before its amplitude becomes vanishingly small?

8.27 A 200-MHz uniform wave in a conducting medium ($\sigma = 0.04$ S/m, $\epsilon_r = 1$, $\mu_r = 1$) impinges normally on a conducting medium ($\sigma = 4$ S/m, $\epsilon_r = 1$, $\mu_r = 1$). The incident electric field has a maximum value of 50 V/m at the interface. Determine the average power densities of the incident, reflected, and transmitted waves. What is the depth of penetration of the transmitted wave?

8.28 The electric field intensity of a uniform plane wave in air is given as $\vec{\mathbf{E}} = 100\sin(\omega t - \beta z)\vec{\mathbf{a}}_x + 200\cos(\omega t - \beta z)\vec{\mathbf{a}}_y$ V/m, where $\omega =$ 90 Mrad/s. If it strikes normally at a conducting medium ($\sigma =$ 0.4 S/m, $\epsilon_r = 81$, $\mu_r = 1$), write expressions for the incident, reflected, and transmitted fields. Show that the average power density in each medium is the same at the interface ($z = 0$).

8.29 A 60-Mrad/s uniform plane wave is propagating in a dielectric medium ($\epsilon_r = 9$, $\mu_r = 1$) and has an electric field intensity of $150\sin(\omega t - \beta z)\vec{\mathbf{a}}_x$ V/m. If it enters free space normal to the z axis at $z = 0$, obtain the expressions for the power densities in both media.

8.30 A 120-Mrad/s uniform plane wave is propagating in a dielectric medium ($\epsilon_r = 9$, $\mu_r = 1$) and has an electric field intensity of $50\,\sin(\omega t +$

$\beta z)\vec{\mathbf{a}}_x$ V/m. If it enters free space normal to the z axis at $z = 0$, obtain the expressions for the power densities in both media.

8.31 A uniform plane wave is traveling from one dielectric medium into another dielectric medium. The dielectric constant of one medium is 1.25. If the permeability of each medium is the same as that of free space, and the reflection coefficient is 0.25, determine the dielectric constant of the other medium.

8.32 A uniform plane wave is traveling from one dielectric medium into another dielectric medium. The dielectric constant of one medium is 2.25. If the permeability of each medium is the same as that of free space, and the reflection coefficient is −0.75, determine the dielectric constant of the other medium.

8.33 A uniform plane wave propagating in free space strikes the plane surface of a dielectric ($\epsilon_r = 2.25$, $\mu_r = 1$) at an angle θ with respect to the normal. If the interface is at $z = 0$, and the electric field intensity of the incident wave is $50\cos(3 \times 10^6 t - 0.766z + 0.643y)\vec{\mathbf{a}}_x$ V/m, determine θ, the phase velocity, and the group velocity. Write expressions for the incident, reflected, and transmitted waves. Compute the power density in each region.

8.34 If the incident wave in Problem 8.33 strikes at an angle θ on the surface of a perfect conductor at $z = 0$, write expressions for the incident, reflected, and total fields. What is the surface current density on the surface of the conductor?

9

Transmission lines

9.1 Introduction

In Chapter 8, we discussed the propagation of a plane wave in an unbounded medium. Because the wave had neither electric nor magnetic field components in the longitudinal direction, we referred to it as a *transverse electromagnetic wave*. We now consider those waves that can exist in regions bounded by conductors. We refer to these waves as **guided waves**. There are, in fact, three types of guided waves: the transverse electromagnetic wave, the transverse electric wave, and the transverse magnetic wave. Guided waves require conductors for their existence and propagate along the length of the conductors.

When the magnetic field of a guided wave has a component in the direction of its propagation in addition to its other components in the transverse direction, and the electric field is entirely in the transverse direction, the guided wave is called a **transverse electric (TE) wave**. The guided wave is referred to as a **transverse magnetic (TM) wave** when the magnetic field is entirely in the transverse direction and the electric field has a component in its direction of propagation. TE and TM waves can exist within a single hollow conductor; we discuss them in Chapter 10, Waveguides and cavity resonators.

The transverse electromagnetic (TEM) wave, also known as the **principal wave** (or **principal mode**), requires two or more conductors for its existence. The wave propagates along the length of the conductors with its electric and magnetic fields entirely transverse to its direction of propagation. Neither the electric field nor the magnetic field has a component in the direction of propagation of the wave. Such a wave, very similar to a plane wave, can be excited at any frequency and is used to transmit signals along transmission lines. We devote this chapter to the study of these waves.

Among the endless varieties of transmission lines are parallel-wire lines, coaxial lines, and microstrip as shown in Figure 9.1 (page 422). The two conductors forming a transmission line are separated by a

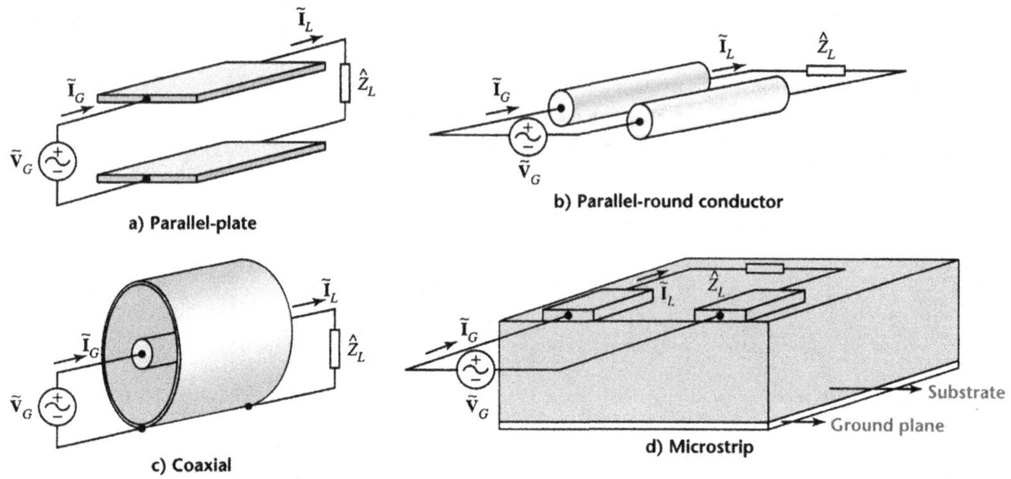

Figure 9.1 Different kinds of transmission lines

dielectric medium. The properties of the dielectric medium are assumed to be the same throughout the length of the transmission line. Also, each conductor of the transmission line is made of the same material, has the same dimensions, and has the same cross-sectional area along the length of the line. A transmission line satisfying these criteria is usually referred to as a **uniform transmission line**. These conditions enable

Figure 9.2 Overhead power lines (courtesy of Detroit Edison Company)

Figure 9.3 Underground power lines (courtesy of Detroit Edison Company)

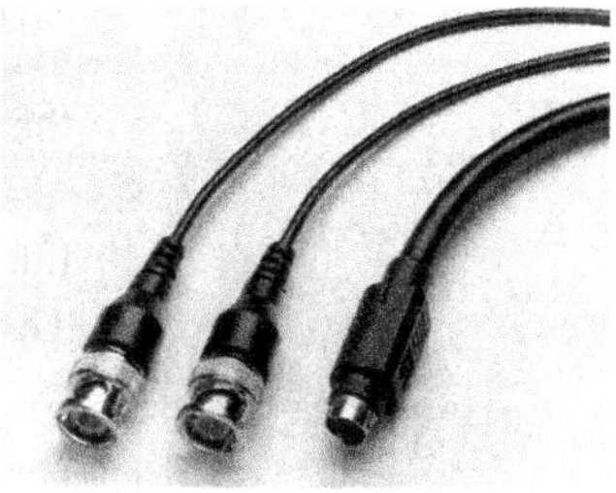

Figure 9.4 Coaxial cables (courtesy of Keithley Instruments, Inc.)

us to model a transmission line on a per-unit-length basis. Our reference to a transmission line will always imply a uniform transmission line.

Transmission lines range in length from a few millimeters in microwave circuits (microstrips) to hundreds of kilometers for low-power signal (telephone) lines and high-power transmissions. They can be strung over poles as parallel transmission lines or buried underground as coaxial cables (see Figures 9.2–9.4). A transmission line may also be used as a resonant circuit, as a filter, and as part of a wave-shaping network.

A useful graphical technique to analyze a transmission line is referred to as a *Smith chart*. Its development and applications are given in Appendix A.

9.2 A parallel-plate transmission line

We will develop the theory for transmission lines based on the discussion of uniform plane wave propagation in Chapter 8. The electric and magnetic fields of a uniform plane wave are orthogonal to each other, and the wave propagates along the direction perpendicular to the plane containing the electromagnetic fields. In this chapter, we will confine our study to only uniform and homogeneous transmission lines in order to reduce the mathematical intricacies. A uniform transmission line offers the same cross-sectional view throughout its entire length. A transmission line is said to be homogeneous if the medium surrounding the conductors is homogeneous everywhere.

In contrast to our earlier discussions on uniform plane waves, let us consider two isolated, infinitely long metal plates parallel to the direction of propagation acting as a guide for the plane wave to propagate in the medium (see Figure 9.5). Because the energy is transmitted from one point to another as guided by the two plates, this arrangement is known as a parallel-plate transmission line. Since we have assumed plane wave propagation in this transmission line, the electric and magnetic fields are transverse to the direction of propagation, resulting in only the **TEM** mode of propagation. Strictly speaking, this assumption applies only if the transmission line is lossless. However, for transmission lines with

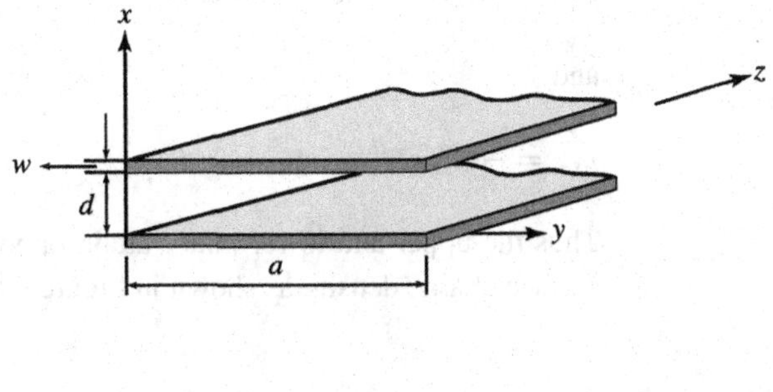

Figure 9.5 Parallel-plate transmission line, $a \gg d$

finite conductivity, it is still fairly valid that the electric and magnetic fields are transverse up to about 100 MHz. The most important simplification of considering only the TEM mode of propagation is that the transmission line can be represented by an electric circuit model with distributed parameters. Later in this section we will develop such a model using Maxwell's equations.

Assume that we have two parallel plates made of perfect conductors ($\sigma = \infty$) and separated by a perfect dielectric ($\sigma_d = 0$). We will refer to such a line as a **lossless transmission line**. When the distance between the two parallel plates is very small in comparison with the width of the plates, then the distribution of electric and magnetic fields will be almost uniform between the plates along the direction of propagation at any particular instant of time.

Let us consider a parallel-plate transmission line, as shown in Figure 9.6 (page 425). As the plates are of infinite extent, we expect the electromagnetic wave to propagate along only one direction with no backward-traveling wave within the dielectric medium. Therefore, the electric and magnetic fields in the medium can be expressed in phasor form as

$$\tilde{\mathbf{E}}(z) = \hat{E}_x e^{-j\beta z}\, \vec{\mathbf{a}}_x$$
$$\tilde{\mathbf{H}}(z) = \hat{H}_y e^{-j\beta z}\, \vec{\mathbf{a}}_y \tag{9.1}$$

or

$$\tilde{\mathbf{H}}(z) = \frac{\hat{E}_x}{\eta} e^{-j\beta z}\, \vec{\mathbf{a}}_y \tag{9.2}$$

Here, $\hat{E}_x = E_x e^{j\theta}$ and $\hat{H}_y = H_y e^{j\theta}$ are arbitrary constants defining the strength of the fields (see Chapter 8), and η and β are the intrinsic impedance and the phase constant of the lossless medium, respectively. The pattern of the fields is given in Figure 9.6a.

Because the fields exist in a dielectric medium bounded by two perfectly conducting plates, we have to satisfy the boundary conditions to determine the charges and the currents on the inner surfaces of the perfect conductors. From the normal components of the electric flux density, we can determine the surface charge densities on the plates as

$$\tilde{\rho}_{s+} = \epsilon \hat{E}_x e^{-j\beta z} \quad \text{at} \quad x = 0 \tag{9.3}$$

and

$$\tilde{\rho}_{s-} = -\epsilon \hat{E}_x e^{-j\beta z} \quad \text{at} \quad x = d \tag{9.4}$$

Thus the upper and lower plates attain opposite but equal amounts of surface charge density as shown in Figure 9.6b.

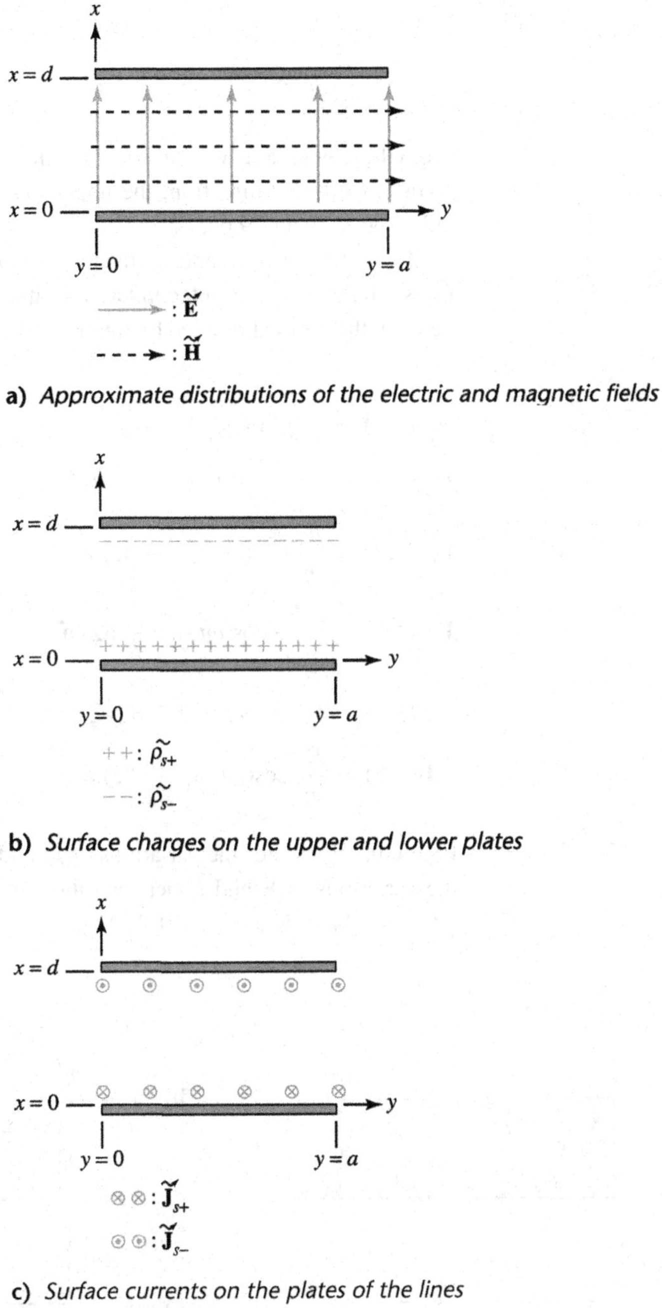

Figure 9.6 A parallel-plate transmission line

a) *Approximate distributions of the electric and magnetic fields*

b) *Surface charges on the upper and lower plates*

c) *Surface currents on the plates of the lines*

Since the magnetic field in the medium is parallel to the two perfectly conducting plates, the boundary conditions pertaining to the tangential components of the magnetic field yield the surface current densities as

$$\tilde{\mathbf{J}}_{s+} = \frac{\hat{E}_x}{\eta} e^{-j\beta z} \vec{\mathbf{a}}_z \quad \text{at} \quad x = 0 \tag{9.5}$$

and

$$\vec{\mathbf{J}}_{s-} = -\frac{\hat{E}_x}{\eta} e^{-j\beta z}\, \vec{\mathbf{a}}_z \quad \text{at} \quad x = d \tag{9.6}$$

From these equations we can conclude that when the lower plate carries a current in the z direction, the upper plate provides a return path as indicated in Figure 9.6c.

The surface charge and surface current densities on the inner surfaces of the two perfect conductors, and the electric and magnetic fields in the region bounded by them, can be written in the time domain as

$$\rho_{s+}(z,t) = \epsilon E_x \cos(\omega t + \theta - \beta z) \tag{9.7a}$$

$$\rho_{s-}(z,t) = -\epsilon E_x \cos(\omega t + \theta - \beta z) \tag{9.7b}$$

$$\vec{\mathbf{J}}_{s+}(z,t) = \frac{E_x}{\eta} \cos(\omega t + \theta - \beta z)\, \vec{\mathbf{a}}_z \tag{9.8a}$$

$$\vec{\mathbf{J}}_{s-}(z,t) = -\frac{E_x}{\eta} \cos(\omega t + \theta - \beta z)\, \vec{\mathbf{a}}_z \tag{9.8b}$$

$$\vec{\mathbf{E}}(z,t) = E_x \cos(\omega t + \theta - \beta z)\, \vec{\mathbf{a}}_x \tag{9.9}$$

$$\vec{\mathbf{H}}(z,t) = \frac{E_x}{\eta} \cos(\omega t + \theta - \beta z)\, \vec{\mathbf{a}}_y \tag{9.10}$$

Figure 9.7 illustrates the variations of ρ_s and $\vec{\mathbf{J}}_s$ on the lower plate, and the variations of $\vec{\mathbf{E}}$ and $\vec{\mathbf{H}}$ between the plates as a function of z when $\omega t + \theta = 2n\pi$, where $n = 0, 1, 2, 3, \ldots$.

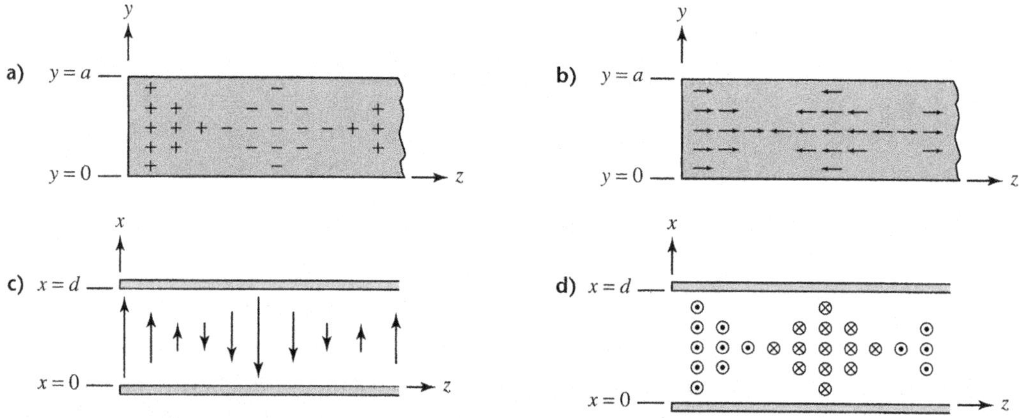

Figure 9.7 (a) Distribution of the surface charge density, ρ_s, on the lower plate; (b) distribution of the surface current density, $\vec{\mathbf{J}}_s$, on lower plate; (c) distribution of the electric field intensity, $\vec{\mathbf{E}}$, between the parallel plates; (d) distribution of the magnetic field intensity, $\vec{\mathbf{H}}$, between the parallel plates

EXAMPLE 9.1

The applied voltage to a 100-m-long, parallel-plate, lossless transmission line is given by $v(t) = 100 \cos 10^5 t$ V. The separation between the plates is 2 mm, and the width of each plate is 10 mm. Determine the variation of the surface charge and surface current densities on each plate if the relative permittivity and relative permeability of the dielectric medium are 4 and 1, respectively.

Solution Let us assume that at any time t the charge distribution on the lower plate of the parallel-plate transmission line is positive with respect to the upper plate. Then the electric field intensity between the two plates at $z = 0$ is

$$E_x(t) = \frac{v}{d} = \frac{100 \cos 10^6 t}{2 \times 10^{-3}} = 50 \times 10^3 \cos 10^5 t \text{ V/m}$$

The phase constant and the intrinsic impedance of the medium are

$$\beta = \omega\sqrt{\mu\epsilon} = 10^5\sqrt{4\pi \times 10^{-7} \times 4 \times 8.85 \times 10^{-12}}$$
$$= 6.67 \times 10^{-4} \text{ rad/m}$$

and

$$\eta = \sqrt{\frac{4\pi \times 10^{-7}}{4 \times 8.85 \times 10^{-12}}} = 188.41\ \Omega$$

The electric field at any point on the line becomes

$$\vec{\mathbf{E}}(z, t) = 50 \times 10^3 \cos(10^5 t - 6.67 \times 10^{-4} z)\, \vec{\mathbf{a}}_x \text{ V/m}$$

Using (9.7a), the surface charge density on the lower plate can be determined as

$$\rho_s = 4 \times 8.85 \times 10^{-12} \times 50 \times 10^3 \cos(10^5 t - 6.67 \times 10^{-4} z) \text{ C/m}^2$$
$$= 1.77 \cos(10^5 t - 6.67 \times 10^{-4} z)\ \mu\text{C/m}^2$$

From (9.8a), the surface current density on the lower plate is

$$\vec{\mathbf{J}}_s(z, t) = \frac{50 \times 10^3}{188.41} \cos(10^5 t - 6.67 \times 10^{-4} z)$$
$$= 265.38 \cos(10^5 t - 6.67 \times 10^{-4} z)\, \vec{\mathbf{a}}_z \text{ A/m}$$

• • •

9.2.1 Parameters of a parallel-plate transmission line

Before we continue our discussion of a transmission line, we need to digress and determine its equivalent circuit parameters on a unit-length basis. Later these parameters will enable us to represent a transmission line by its equivalent circuit.

Capacitance of a parallel-plate transmission line

Let ρ_ℓ be the electric charge per unit length on the plate at $x = 0$ and let V be the potential of that plate with respect to the plate at $x = d$. Then, the capacitance per unit length of this transmission line is

$$C_\ell = \frac{\rho_\ell}{V} \tag{9.11}$$

The charge per unit length of the transmission line can be determined as

$$\begin{aligned}\tilde{\rho}_\ell &= \int_0^a \hat{\rho}_{s+}\, dy = \int_0^a \epsilon \hat{E}_x\, e^{-j\beta z}\, dy \\ &= \epsilon \hat{E}_x a\, e^{-j\beta z}\end{aligned} \tag{9.12}$$

The potential difference between the two parallel plates can be obtained from Faraday's law

$$\nabla \times \tilde{\mathbf{E}} = -j\omega \tilde{\mathbf{B}} \tag{9.13}$$

which can also be expressed in terms of the vector potential $\tilde{\mathbf{A}}$ as

$$\nabla \times \tilde{\mathbf{E}} = -j\omega \nabla \times \tilde{\mathbf{A}} \tag{9.14a}$$

or

$$\nabla \times (\tilde{\mathbf{E}} + j\omega \tilde{\mathbf{A}}) = 0 \tag{9.14b}$$

where $\tilde{\mathbf{A}}$ is due to the current density $\vec{\mathbf{J}}_s$. Using the vector identity ($\nabla \times \nabla \tilde{V} = 0$) in (9.14b), we can express the gradient of the electric potential as

$$\nabla \tilde{V} = -(\tilde{\mathbf{E}} + j\omega \tilde{\mathbf{A}}) \tag{9.15}$$

The negative sign in (9.15) is in compliance with the definition of electric potential as introduced in Chapter 3.

The potential of the lower plate with respect to the upper plate for the parallel-plate transmission line is

$$\int_{\tilde{V}}^0 d\tilde{V} = \int_0^d \nabla \tilde{V} \cdot dx\, \vec{\mathbf{a}}_x = -\int_0^d \tilde{\mathbf{E}} \cdot dx\, \vec{\mathbf{a}}_x - j\omega \int_0^d \tilde{\mathbf{A}} \cdot dx\, \vec{\mathbf{a}}_x \tag{9.16}$$

Because $\tilde{\mathbf{A}}$ is in the same direction as $\tilde{\mathbf{J}}_s$, $\int \tilde{\mathbf{A}} \cdot dx\, \vec{\mathbf{a}}_x = 0$, and

$$\tilde{V} = \int_0^d \tilde{\mathbf{E}} \cdot dx\, \vec{\mathbf{a}}_x = \hat{E}_x e^{-j\beta z} d \tag{9.17}$$

Therefore, the potential difference between the two parallel plates is the product of the separation and the electric field intensity in the dielectric medium. It is also clear that the vector potential in (9.16) does not contribute to the potential difference because it is perpendicular to the path of integration. Consequently, the distribution of the potential difference for the TEM mode is the same as for electrostatic fields.

Using equations (9.11), (9.12), and (9.17), we can determine the capacitance per unit length of the parallel-plate transmission lines as

$$C_\ell = \frac{\epsilon a}{d} \text{ F/m} \tag{9.18}$$

Inductance of a parallel-plate transmission line

Using a similar approach, we can also determine the inductance per unit length of the same parallel-plate transmission line. The inductance of a linear magnetic system was given in Chapter 5 as

$$L = \frac{\lambda}{I} \tag{9.19}$$

where λ is the total flux linkages and I is the total current in the system. Our interest is the inductance per unit length of the transmission line, thus it will be sufficient to calculate the flux linkages per unit length as

$$\begin{aligned} \tilde{\lambda}_\ell &= \int_0^d \mu \frac{\hat{E}_x}{\eta} e^{-j\beta z}\, dx \\ &= \mu d \frac{\hat{E}_x}{\eta} e^{-j\beta z} \text{ Wb/m} \end{aligned} \tag{9.20}$$

The total current on the transmission line is

$$\begin{aligned} \tilde{I} &= \int_0^a \tilde{J}_s\, dy \\ &= \int_0^a \frac{\hat{E}_x}{\eta} e^{-j\beta z}\, dy = \frac{\hat{E}_x}{\eta} e^{-j\beta z} a = \tilde{H}_y a \end{aligned} \tag{9.21}$$

Substitution of (9.20) and (9.21) in (9.19) yields the inductance per unit length of the parallel-plate transmission line

$$L_\ell = \frac{\mu d}{a} \text{ H/m} \tag{9.22}$$

The parameters of other types of transmission lines can be similarly determined and are given in Table 9.1 for brevity. Note that σ is the conductivity of each conductor and σ_d is the conductivity of the region between the two conductors.

9.2.2 Equivalent circuit of a parallel-plate transmission line

We can now represent a unit section of the transmission line by means of an equivalent electric circuit in terms of L_ℓ and C_ℓ. In order to do this we need to develop relationships for the voltage and current along the line. These relationships can be obtained from Maxwell's equations. If we apply Maxwell's equation $\nabla \times \vec{\mathbf{E}} = -j\omega\mu\vec{\mathbf{H}}$ to the parallel-plate transmission line configuration with $\vec{\mathbf{E}} = \tilde{E}_x\, \vec{\mathbf{a}}_x$, and $\vec{\mathbf{H}} = \tilde{H}_y\, \vec{\mathbf{a}}_y$,

Table 9.1.

Parameter	Parallel Plate 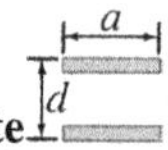	Two Parallel-Wire 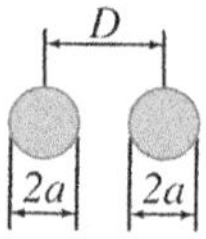	Coaxial Cable 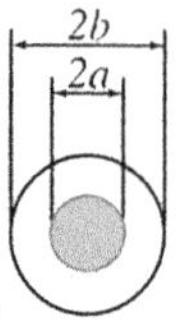	Unit
Resistance (R_ℓ)	$\frac{2}{a}\sqrt{\frac{\pi f \mu}{\sigma}}$	$\frac{1}{a}\sqrt{\frac{f\mu}{\pi\sigma}}$	$\frac{1}{2}\sqrt{\frac{f\mu}{\pi\sigma}}\left(\frac{1}{a}+\frac{1}{b}\right)$	$\frac{\Omega}{\text{m}}$
Inductance (L_ℓ)	$\frac{\mu d}{a}$	$\frac{\mu}{\pi}\cosh^{-1}\left(\frac{D}{2a}\right)$	$\frac{\mu}{2\pi}\ln\left(\frac{b}{a}\right)$	$\frac{\text{H}}{\text{m}}$
Capacitance (C_ℓ)	$\frac{\epsilon a}{d}$	$\frac{\pi\epsilon}{\cosh^{-1}(D/2a)}$	$\frac{2\pi\epsilon}{\ln(b/a)}$	$\frac{\text{F}}{\text{m}}$
Conductance (G_ℓ)	$\frac{\sigma_d a}{d}$	$\frac{\pi\sigma_d}{\cosh^{-1}(D/2a)}$	$\frac{2\pi\sigma_d}{\ln(b/a)}$	$\frac{\text{S}}{\text{m}}$

we obtain

$$\frac{\partial \tilde{E}_x}{\partial z} = -j\omega\mu\tilde{H}_y \tag{9.23}$$

Integration of (9.23) from $x = 0$ to $x = d$ yields

$$\frac{\partial}{\partial z}\int_0^d \tilde{E}_x\,dx = -j\omega\mu\int_0^d \tilde{H}_y\,dx \tag{9.24}$$

Both $\tilde{E}_x$ and $\tilde{H}_y$ are constant with respect to x, so we can write

$$\frac{\partial \tilde{V}(z)}{\partial z} = -j\omega\mu\tilde{H}_y d$$

from equations (9.24) and (9.17). Using (9.21) and (9.22) the preceding equation can be written as

$$\frac{\partial \tilde{V}(z)}{\partial z} = -j\omega\,\frac{\mu d}{a}\,\tilde{I}(z)$$

or

$$\frac{\partial \tilde{V}(z)}{\partial z} = -j\omega L_\ell \tilde{I}(z) \tag{9.25}$$

Similarly, from Maxwell's equation $\nabla\times\tilde{\mathbf{H}} = j\omega\epsilon\tilde{\mathbf{E}}$ for a dielectric medium, we can obtain

$$\frac{\partial \tilde{H}_y}{\partial z} = -j\omega\epsilon\tilde{E}_x \tag{9.26}$$

The integration of both sides from $x = 0$ to $x = d$ yields

$$\frac{\partial}{\partial z}\int_0^d \tilde{H}_y\,dx = -j\omega\epsilon\int_0^d \tilde{E}_x\,dx$$

or

$$\frac{\partial}{\partial z}(\tilde{H}_y d) = -j\omega\epsilon\tilde{V}(z) \tag{9.27}$$

A direct comparison of (9.26) and (9.27) lets us express the rate of change of the current with respect to the length of the line as

$$\frac{\partial\tilde{I}(z)}{\partial z} = -j\omega C_\ell\tilde{V}(z) \tag{9.28}$$

Equations (9.25) and (9.28) govern the variations of voltage and current in a lossless transmission line. These equations are known as the *transmission line equations*, or *telegraphist's equations*.

Equations (9.25) and (9.28) can be rearranged as

$$d\tilde{V}(z) = -j\omega L_\ell\, dz\, \tilde{I}(z) \tag{9.29a}$$

and

$$d\tilde{I}(z) = -j\omega C_\ell\, dz\, \tilde{V}(z) \tag{9.29b}$$

where $d\tilde{V}$ is the potential drop across the inductive reactance, and $d\tilde{I}$ is the current through the capacitive reactance of the differential length dz of the transmission line.

Equations (9.29a) and (9.29b) can be approximated by

$$\Delta\tilde{V} = \tilde{V}(z+\Delta z) - \tilde{V}(z) \tag{9.30a}$$

and

$$\Delta\tilde{I} = \tilde{I}(z+\Delta z) - \tilde{I}(z) \tag{9.30b}$$

for a very short section of a transmission line of length Δz, such that $\Delta z \to 0$. An equivalent electric circuit model can then be given, as shown in Figure 9.8 (page 432), using (9.30a) and (9.30b) for a Δz section of the transmission line.

In order to determine the voltage and/or current distribution along the transmission line we have to first obtain a differential equation in terms of one variable. To do so, let us differentiate equation (9.25) with respect to z to obtain

$$\frac{d^2\tilde{V}(z)}{dz^2} = -j\omega L_\ell\,\frac{d\tilde{I}(z)}{dz} \tag{9.31}$$

which, with the help of (9.28), becomes

$$\frac{d^2\tilde{V}(z)}{dz^2} = -\omega^2 L_\ell C_\ell\tilde{V}(z) = \hat{\gamma}^2\tilde{V}(z) \tag{9.32}$$

where

$$\hat{\gamma} = j\omega\sqrt{L_\ell C_\ell} = j\beta \tag{9.33}$$

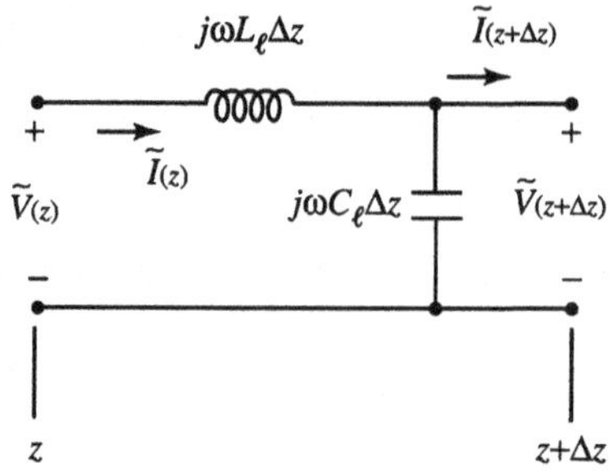

Figure 9.8 Electric circuit model for a very short (Δz) section of a transmission line

is the propagation constant and β is the phase constant of the lossless line.

Similarly, we can obtain a wave equation in terms of I as

$$\frac{d^2\tilde{I}(z)}{dz^2} = -\omega^2 C_\ell L_\ell \tilde{I}(z) = \hat{\gamma}^2 \tilde{I}(z) \tag{9.34}$$

Equations (9.32) and (9.34) have the same mathematical form. Hence, their general solutions are similar. The solutions are given by

$$\tilde{V}(z) = \hat{V}^+ e^{-\hat{\gamma} z} + \hat{V}^- e^{\hat{\gamma} z} \tag{9.35}$$

and

$$\tilde{I}(z) = \hat{I}^+ e^{-\hat{\gamma} z} + \hat{I}^- e^{\hat{\gamma} z} \tag{9.36}$$

where $\hat{V}^+$ and $\hat{I}^+$ are arbitrary constants for the forward voltage and current waves traveling along the positive z direction, and $\hat{V}^-$ and $\hat{I}^-$ are the arbitrary constants for the backward voltage and current waves traveling along the negative z direction. The negative sign preceding $\hat{I}^-$ indicates that the backward wave carries power in the negative z direction.

As there is no backward-traveling wave for an infinite transmission line, therefore (9.35) and (9.36) become

$$\tilde{V}(z) = \hat{V}^+ e^{-\hat{\gamma} z} = \hat{V}^+ e^{-j\beta z} \tag{9.37}$$

and

$$\tilde{I}(z) = \hat{I}^+ e^{-\hat{\gamma} z} = \hat{I}^+ e^{-j\beta z} \tag{9.38}$$

The velocity of propagation of the wave in a lossless transmission line is

$$u_p = \frac{\omega}{\beta} = \frac{1}{\sqrt{L_\ell C_\ell}} = \frac{1}{\sqrt{\mu\epsilon}} \tag{9.39}$$

We can define the **transit time** t_t as the time elapsed for the wave to travel from one end of the line to the other. Thus, the transit time is

$$t_t = \frac{\ell}{u_p} \tag{9.40}$$

where ℓ is the length of the transmission line.

The characteristic impedance $\hat{Z}_c$ of the transmission line is defined as the ratio of voltage to current for the wave propagating in the positive z direction. To determine $\hat{Z}_c$, we substitute (9.35) and (9.36) into (9.27), assuming that the backward-traveling waves are zero, and obtain

$$\frac{d}{dz}(\hat{V}^+e^{-\hat{\gamma}z}) = -j\omega L_\ell \hat{I}^+e^{-\hat{\gamma}z}$$

or

$$-\hat{\gamma}\hat{V}^+e^{-\hat{\gamma}z} = -j\omega L_\ell \hat{I}^+e^{-\hat{\gamma}z}$$

Thus,

$$\hat{Z}_c = \frac{\hat{V}^+e^{-\hat{\gamma}z}}{\hat{I}^+e^{-\hat{\gamma}z}} = \frac{j\omega L_\ell}{\hat{\gamma}} = \sqrt{\frac{L_\ell}{C_\ell}} \tag{9.41}$$

Substituting for L_ℓ and C_ℓ from (9.22) and (9.18) in (9.41), we obtain the characteristic impedance of a parallel-plate transmission line as

$$\hat{Z}_c = \eta\,\frac{d}{a} \tag{9.42}$$

in terms of the intrinsic impedance of the unbounded medium. Note that the characteristic impedance of a lossless transmission line is a real quantity. However, we will see later that the characteristic impedance is, in general, a complex quantity for transmission lines with losses.

A relationship for the voltage and the current travelling in the negative z direction can also be obtained using the same approach we employed in determining (9.41):

$$\hat{Z}_c = \frac{\hat{V}^-e^{\hat{\gamma}z}}{-\hat{I}^-e^{\hat{\gamma}z}} = \frac{j\omega L_\ell}{\hat{\gamma}} = \sqrt{\frac{L_\ell}{C_\ell}} \tag{9.43}$$

From (9.41) and (9.43), we can express $\hat{I}^+$ and $\hat{I}^-$ as

$$\hat{I}^+ = \frac{\hat{V}^+}{\hat{Z}_c} \tag{9.44}$$

and

$$\hat{I}^- = -\frac{\hat{V}^-}{\hat{Z}_c} \tag{9.45}$$

If we substitute (9.44) and (9.45) into (9.36), we can obtain the current in terms of the voltage and the characteristic impedance of the line:

$$\tilde{I}(z) = \frac{\hat{V}^+}{\hat{Z}_c}\,e^{-\hat{\gamma}z} - \frac{\hat{V}^-}{\hat{Z}_c}\,e^{\hat{\gamma}z} \tag{9.46}$$

Equations (9.35) and (9.46) are the general solutions for the voltage and current along the line, and can be written as

$$\tilde{V}(z) = \hat{V}^+e^{-\hat{\gamma}z}[1 + \hat{\rho}(z)] \tag{9.47a}$$

and

$$\tilde{I}(z) = \frac{\hat{V}^+ e^{-\hat{\gamma} z}}{\hat{Z}_c}[1 - \hat{\rho}(z)] \tag{9.47b}$$

where

$$\hat{\rho}(z) = \frac{\hat{V}^- e^{\hat{\gamma} z}}{\hat{V}^+ e^{-\hat{\gamma} z}} = \frac{\hat{V}^-}{\hat{V}^+} e^{2\hat{\gamma} z} \tag{9.48}$$

is defined as the reflection coefficient at a distance z on the transmission line.

EXAMPLE 9.2

A 100-m-long lossless transmission line has a total inductance and capacitance of 27.72 μH and 18 nF, respectively. Determine (a) the velocity of propagation and the phase constant for an operating frequency of 100 kHz and (b) the characteristic impedance of the transmission line.

Solution

a) The inductance and capacitance per unit length of the transmission line are

$$L_\ell = \frac{27.72 \times 10^{-6}}{100} = 0.2772 \ \mu\text{H/m}$$

$$C_\ell = \frac{18 \times 10^{-9}}{100} = 0.18 \ \text{nF/m}$$

The velocity of propagation is

$$u_p = \frac{1}{\sqrt{L_\ell C_\ell}} = \frac{1}{\sqrt{0.2772 \times 10^{-6} \times 0.18 \times 10^{-9}}} = 1.416 \times 10^8 \ \text{m/s}$$

and the propagation constant is

$$\beta = \frac{\omega}{u_p} = \frac{2\pi \times 100 \times 10^3}{1.416 \times 10^8} = 4.439 \times 10^{-3} \ \text{rad/m}$$

b) The characteristic impedance of the transmission line is

$$\hat{Z}_c = \sqrt{\frac{L_\ell}{C_\ell}} = \sqrt{\frac{0.2772 \times 10^{-6}}{0.18 \times 10^{-9}}} = 39.243 \ \Omega$$

• • •

EXAMPLE 9.3

The characteristic impedance of a 10-m-long lossless coaxial cable is 50 Ω. The dielectric material between the inner and outer conductors of the cable has $\epsilon_r = 3.5$ and $\mu_r = 1$. If the radius of the inner conductor is 1 mm, what should be the outer radius of this cable?

Solution

The per unit length inductance and capacitance of a coaxial cable are given by

$$L_\ell = \frac{\mu}{2\pi} \ln\left(\frac{b}{a}\right) \quad \text{and} \quad C_\ell = \frac{2\pi\epsilon}{\ln\left(\frac{b}{a}\right)}$$

where a and b are, respectively, the inner and outer radii of the cable.

$$\hat{Z}_c = \sqrt{\frac{L_\ell}{C_\ell}}$$

$$= \frac{\ln\left(\frac{b}{a}\right)}{2\pi}\sqrt{\frac{\mu}{\epsilon}}$$

$$50 = \frac{\ln\left(\frac{b}{10^{-3}}\right)}{2\pi}\sqrt{\frac{4\pi \times 10^{-7}}{3.5 \times 8.85 \times 10^{-12}}}$$

$$b = 4.75 \text{ mm}$$

• • •

9.2.3 Lossless transmission line of finite length

As long as we know the distributed parameters of a lossless line, we can use (9.47a) and (9.47b) to determine the voltage and current at any point on the line.

Let us now consider a lossless transmission line of finite length ℓ that is terminated into arbitrary load impedance $\hat{Z}_L$ as shown in Figure 9.9. If we represent the load voltage and load current at $z = \ell$ as $\tilde{V}_R = \tilde{V}(z = \ell)$ and $\tilde{I}_R = \tilde{I}(z = \ell)$, then from (9.47a) and (9.47b) we obtain

$$\tilde{V}_R = \hat{V}^+ e^{-\hat{\gamma}\ell}\,[1 + \hat{\rho}_R] \tag{9.49a}$$

$$\tilde{I}_R = \frac{\hat{V}^+ e^{-\hat{\gamma}\ell}}{\hat{Z}_c}\,[1 - \hat{\rho}_R] \tag{9.49b}$$

where $\hat{\rho}_R = \hat{\rho}(z = \ell)$ is the reflection coefficient at $z = \ell$ such that

$$\hat{\rho}_R = \frac{\hat{V}^-}{\hat{V}^+}\,e^{2\hat{\gamma}\ell} \tag{9.49c}$$

This equation can also be written as

$$\frac{\hat{V}^-}{\hat{V}^+} = \hat{\rho}_R\, e^{-2\hat{\gamma}\ell} \tag{9.50}$$

If we denote the input voltage and current at $z = 0$ as $\tilde{V}_s = \tilde{V}(z = 0)$ and $\tilde{I}_s = \tilde{I}(z = 0)$ respectively, then we can express $\tilde{V}_s$ and $\tilde{I}_s$ from (9.47a) and (9.47b) as

$$\tilde{V}_s = \hat{V}^+\,[1 + \hat{\rho}_S] \tag{9.51a}$$

$$\tilde{I}_s = \frac{\hat{V}^+}{\hat{Z}_c}\,[1 - \hat{\rho}_S] \tag{9.51b}$$

where $\hat{\rho}_S = \hat{\rho}(z = 0)$ is the reflection coefficient at the input (source) side of the line and is given, from (9.48) and using (9.50), as

$$\hat{\rho}_S = \frac{\hat{V}^-}{\hat{V}^+} = \hat{\rho}_R\, e^{-2\hat{\gamma}\ell} \tag{9.51c}$$

Equation (9.51c) lets us express the reflection coefficient at $z = 0$ in terms of the reflection coefficient at $z = \ell$. In fact, we can express the reflection coefficient at any point on the line as given by (9.48) in terms of the reflection coefficient at $z = \ell$ as

$$\hat{\rho}(z) = \hat{\rho}_R \, e^{-2\hat{\gamma}(\ell - z)} \tag{9.52}$$

Since we know the voltage and current at any point on the line, their ratio yields the equivalent impedance at that point as

$$\hat{Z}(z) = \frac{\tilde{V}(z)}{\tilde{I}(z)} = \hat{Z}_c \left[\frac{1 + \hat{\rho}(z)}{1 - \hat{\rho}(z)}\right] \tag{9.53}$$

Equation (9.53) helps us replace the load impedance as well as the portion of the line on the load side of the point by an equivalent impedance $\hat{Z}(z)$. However, when we set $z = 0$ in (9.53), we obtain an expression for what is usually termed as the **input impedance** $\hat{Z}_{\text{in}}$ of a length ℓ of the transmission line. That is,

$$\hat{Z}_{\text{in}} = \frac{\tilde{V}(z = 0)}{\tilde{I}(z = 0)} = \hat{Z}_c \left[\frac{1 + \hat{\rho}(z = 0)}{1 - \hat{\rho}(z = 0)}\right]$$

In terms of the terminology we have defined earlier, the above expression for the input impedance can be written concisely as

$$\hat{Z}_{\text{in}} = \frac{\tilde{V}_S}{\tilde{I}_S} = \hat{Z}_c \left[\frac{1 + \hat{\rho}_S}{1 - \hat{\rho}_S}\right] \tag{9.54}$$

The input impedance is a concept that has been extensively exploited in the design of a transmission line. Therefore, we will have more to say about it in a later section.

The equivalent impedance in (9.53) at $z = \ell$ must be nothing but the load impedance. Setting $z = \ell$ in (9.53), we obtain

$$\hat{Z}_L = \hat{Z}_c \left[\frac{1 + \hat{\rho}_R}{1 - \hat{\rho}_R}\right] \tag{9.55}$$

This equation can also be expressed as

$$\hat{\rho}_R = \frac{\hat{Z}_L - \hat{Z}_c}{\hat{Z}_L + \hat{Z}_c} \tag{9.56}$$

This is a very powerful equation in the sense that it allows us to compute the reflection coefficient at the load, $\hat{\rho}_R$, especially when the load impedance is known. Once $\hat{\rho}_R$ is known, we can then compute the reflection coefficient at the source $\hat{\rho}_S$ using (9.51c), the reflection coefficient at any point on the line using (9.52), the equivalent impedance at any point on the line using (9.53), and the input impedance using (9.54).

Note that the reflection coefficient in (9.56) is zero when $\hat{Z}_L = \hat{Z}_c$. In this case, there is no reflected wave on the line. The voltage and current expressions for the forward wave are as given in (9.37) and

(9.38), respectively. A line terminated into an impedance equal to its characteristic impedance is called a **matched line**. For a matched line the amplitude of the voltage is the same at any point on the line. The same is true for the current. It is clear from (9.53) that the equivalent impedance at any point on the matched line is exactly equal to its characteristic impedance. The same is true for the input impedance. Likewise we can say that when a transmission line is terminated in some arbitrary impedance such that $\hat{Z}_L \neq \hat{Z}_c$, there is always a reflected wave on the line. The input impedance at any point on the line differs from its characteristic impedance. The amplitudes of the voltage and the current vary as we traverse along the length of the line.

The input impedance enables us to represent the entire line and its load impedance at the source side. The voltage drop across the input impedance yields the applied voltage at the input (source) side, $\tilde{V}_S = \tilde{V}(z = 0)$. Likewise, the current through the input impedance yields the applied current at the source side, $\tilde{I}_S = \tilde{I}(z = 0)$. Once we know $\tilde{V}_S$ and $\tilde{I}_S$, we can then compute the voltage and current at any point on the transmission line. The following example highlights the technique to determine the voltages and currents at various points on the transmission line.

EXAMPLE 9.4

A 500-m long lossless transmission line is used to transmit power at a frequency of 10 kHz. Its distributed parameters at the operating frequency are: $L_\ell = 2.6\ \mu$H/m and $C_\ell = 28.7$ pF/m. The line is terminated into an inductive impedance of $(75 + j150)\Omega$. The rms value of the applied voltage source is 120 V and its internal impedance is $(1 + j9)\ \Omega$. Determine (a) the voltage and current at the input side, (b) voltage and current at the load, (c) the power input to the transmission line, and (d) power delivered to the load. Obtain expressions for the forward and reflected waves.

Solution $\omega = 2\pi f = 2\pi \times 10 \times 10^3 = 62.83 \times 10^3$ rad/m

Using the distributed parameters of the line, we can compute its characteristic impedance, the propagation constant, and the phase constant as

$$\hat{Z}_c = \sqrt{\frac{L_\ell}{C_\ell}} - \sqrt{\frac{2.6 \times 10^{-6}}{28.7 \times 10^{-12}}} = 300.986\ \Omega$$

$$\hat{\gamma} = j\omega\sqrt{L_\ell C_\ell} = j62.83 \times 10^3\sqrt{2.6 \times 10^{-6} \times 28.7 \times 10^{-12}}$$
$$= j5.428 \times 10^{-4}$$
$$\beta = 5.428 \times 10^{-4}\ \text{rad/m}$$

As expected, the characteristic impedance of the lossless transmission line is purely resistive and the propagation constant is purely imaginary.

The phase velocity and the wavelength can be determined as

$$u_p = \frac{\omega}{\beta} = \frac{62.83 \times 10^3}{5.428 \times 10^{-4}} = 1.158 \times 10^8 \text{ m/s}$$

$$\lambda = \frac{2\pi}{\beta} = \frac{2\pi}{5.428 \times 10^{-4}} = 11.576 \times 10^3 \text{ m} \quad \text{or} \quad 11.576 \text{ km}$$

Quite often, the phase velocity of the wave on the line is also expressed as a percentage of the velocity of light. In our case, the phase velocity is 36.8% $\lfloor (u_p/c)100 \rfloor$ of the velocity of light.

We can also express the length of the line in terms of its wavelength. In our case, the length of the line is 0.043 $\lambda[\ell/\lambda]$.

We can now calculate the reflection coefficient at the load using (9.56) as

$$\hat{\rho}_R = \frac{\hat{Z}_L - \hat{Z}_c}{\hat{Z}_L + \hat{Z}_c} = \frac{-225.986 + j150}{375.986 + j150} = 0.67\underline{/124.68^\circ}$$

The factor $\beta\ell$ will be needed later. So let us compute it as

$$\beta\ell = 5.428 \times 10^{-4} \times 500 = 0.2714 \text{ rad or } 15.55^\circ$$

Note that $\beta\ell$ is the phase difference of the forward and the reflected waves over the 500-m length of the line.

Let us now compute the reflection coefficient at the input side using (9.51c) as

$$\hat{\rho}_S = \hat{\rho}_R\, e^{-2\gamma\ell} = \hat{\rho}_R\, e^{-j2\beta\ell} = 0.67\, e^{j124.68^\circ}\, e^{-2\times 15.55^\circ} = 0.67\underline{/93.58^\circ}$$

Equation (9.54) now yields the input impedance at $z = 0$ as

$$\hat{Z}_{\text{in}} = \hat{Z}_c\left[\frac{1+\hat{\rho}_S}{1-\hat{\rho}_S}\right] = 300.986\left[\frac{1+0.67\underline{/93.58^\circ}}{1-0.67\underline{/93.58^\circ}}\right] = 284.087\underline{/67.61^\circ}\ \Omega$$

Using the voltage-division rule, we obtain the input voltage on the line as

$$\tilde{V}_S = \tilde{V}_G\, \frac{\hat{Z}_{\text{in}}}{\hat{Z}_G + \hat{Z}_{\text{in}}} = 120\left[\frac{284.087\underline{/67.61^\circ}}{1 + j9 + 284.087\underline{/67.61^\circ}}\right]$$
$$= 116.429\underline{/-0.49^\circ}\text{ V}$$

The corresponding current on the transmission line on the input side is

$$\tilde{I}_S = \frac{\tilde{V}_S}{\hat{Z}_{\text{in}}} = \frac{116.429\underline{/-0.49^\circ}}{284.087\underline{/67.61^\circ}} = 0.41\underline{/-68.10^\circ}\text{ A}$$

The power input to the transmission line can be calculated as

$$P_{\text{in}} = \text{Re}\,[\tilde{V}_S \tilde{I}_S^*] = \text{Re}\,[(116.429\underline{/-0.49^\circ})(0.41\underline{/68.10^\circ})] = 18.18\text{ W}$$

In order to compute the voltage and current at the receiving side, let us first compute $\hat{V}^+$ from (9.51a) as

$$\hat{V}^+ = \frac{\tilde{V}_S}{1+\hat{\rho}_S} = \frac{116.429\underline{/-0.49^\circ}}{1+0.67\underline{/93.58^\circ}} = 99.642\underline{/-35.40^\circ}\text{ V}$$

Now from (9.49a), the voltage at the load side is

$$\begin{aligned}\tilde{V}_R &= \hat{V}^+ e^{-j\beta\ell}[1+\hat{\rho}_R] \\ &= 99.642\, e^{-j35.40^\circ} e^{-j15.55^\circ}[1+0.67\,\underline{/124.68^\circ}] = 82.561\,\underline{/-9.27^\circ}\text{ V}\end{aligned}$$

The corresponding load current is

$$\tilde{I}_R = \frac{\tilde{V}_R}{\hat{Z}_L} = \frac{82.561\underline{/-9.27^\circ}}{75+j150} = 0.492\underline{/-72.7^\circ}\text{ A}$$

Finally, the power delivered to the load is

$$P_O = \text{Re}\left[\tilde{V}_R\,\tilde{I}_R^*\right] = \text{Re}\,[(82.561\underline{/-9.27^\circ})(0.492\underline{/72.7^\circ})] = 18.18\text{ W}$$

Since the transmission line is lossless, there is no power loss on the line. All the power applied at the input side is delivered to the load on the receiving side. Since $\hat{V}^+$ and $\hat{\rho}_S$ are now known, we can determine $\hat{V}^-$ from (9.51c) as

$$\hat{V}^- = \hat{\rho}_S\hat{V}^+ = (0.67\underline{/93.58^\circ})(99.642\underline{/-35.40^\circ}) = 66.765\underline{/58.18^\circ}\text{ V}$$

The corresponding values of $\hat{I}^+$ and $\hat{I}^-$ from (9.44) and (9.45), respectively, are

$$\hat{I}^+ = \frac{\hat{V}^+}{\hat{Z}_c} = \frac{99.642\underline{/-35.40^\circ}}{300.986} = 0.331\underline{/-35.40^\circ}\text{ A}$$

$$\hat{I}^- = \frac{\hat{V}^-}{\hat{Z}_c} = \frac{66.765\underline{/58.18^\circ}}{300.986} = 0.222\underline{/58.18^\circ}\text{ A}$$

Thus, the voltage and current expressions for the forward wave are

$$\begin{aligned}\tilde{V}_f(z) &= \hat{V}^+ e^{-\hat{\gamma}z} = 99.642\ e^{-j35.40^\circ}\ e^{-j5.428\times 10^{-4}z}\text{ V} \\ \tilde{I}_f(z) &= \hat{I}^+ e^{-\hat{\gamma}z} = 0.331\ e^{-j35.40^\circ}\ e^{-j5.428\times 10^{-4}z}\text{ A}\end{aligned}$$

in the phasor domain in terms of their rms values, and

$$\begin{aligned}v_f(z,t) &= 140.915\ \cos(6.283\times 10^4 t - 5.428\times 10^{-4}z - 35.40^\circ)\text{ V} \\ i_f(z,t) &= 0.331\ \cos(6.283\times 10^4 t - 5.428\times 10^{-4}z - 35.40^\circ)\text{ A}\end{aligned}$$

in the time domain. Note that an expression in the time domain is always in terms of its amplitude.

Similarly, the voltage and current expressions for the backward wave are

$$\tilde{V}_b(z) = \hat{V}^- e^{\hat{\gamma} z} = 66.765\ e^{j58.18^\circ}\ e^{j5.428\times 10^4 z}\ \text{V}$$
$$\tilde{I}_b(z) = -\hat{I}^- e^{\hat{\gamma} z} = -0.222\ e^{j58.18^\circ}\ e^{j5.428\times 10^4 z}\ \text{A}$$

in the phasor domain in terms of their rms values, and

$$v_b(z,t) = 94.42\ \cos(6.283\times 10^4 t + 5.428\times 10^{-4} z - 58.18^\circ)\ \text{V}$$
$$i_f(z,t) = -0.314\ \cos(6.283\times 10^4 t + 5.428\times 10^{-4} z - 58.18^\circ)\ \text{A}$$

in the time domain. • • •

9.3 Voltage and current in terms of sending-end and receiving-end variables

Let us consider a lossless transmission line, as shown in Figure 9.9. One end of the line is connected to the source, and the other end is terminated into a load. The source side is usually referred to as the *sending end*, and the load side is called the *receiving end*. If the voltage and the current at the sending end are specified as $\tilde{V}(0) = \tilde{V}_S$ and $\tilde{I}(0) = \tilde{I}_S$, we can uniquely compute the voltage and current at any point along the transmission line.

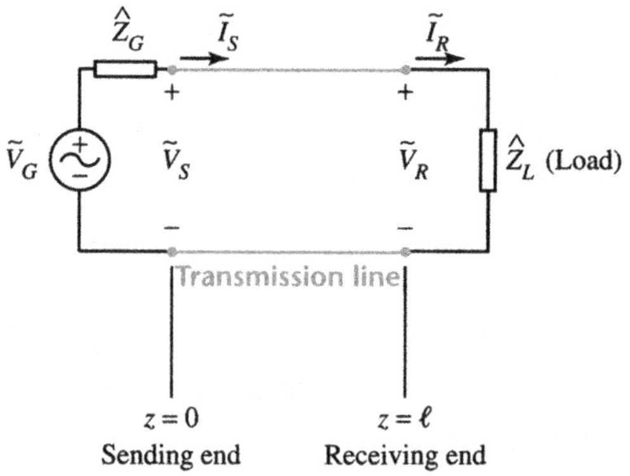

Figure 9.9 A tranmission line of length ℓ

The voltage and current at $z = 0$, from (9.35) and (9.46), are

$$\tilde{V}_S = \hat{V}^+ + \hat{V}^-$$

and

$$\tilde{I}_S = \frac{\hat{V}^+}{\hat{Z}_c} - \frac{\hat{V}^-}{\hat{Z}_c}$$

The solution of the above equations yields

$$\hat{V}^+ = \frac{\tilde{V}_S + \hat{Z}_c \tilde{I}_S}{2}$$

and

$$\hat{V}^- = \frac{\tilde{V}_S - \hat{Z}_c\tilde{I}_S}{2}$$

By substituting for $\hat{V}^+$ and $\hat{V}^-$ from the preceding equations in (9.35) and (9.46), we get

$$\tilde{V}(z) = \left(\frac{\tilde{V}_S + \hat{Z}_c\tilde{I}_S}{2}\right) e^{-j\beta z} + \left(\frac{\tilde{V}_S - \hat{Z}_c\tilde{I}_S}{2}\right) e^{j\beta z}$$

and

$$\tilde{I}(z) = \left(\frac{\tilde{V}_S + \hat{Z}_c\tilde{I}_S}{2\hat{Z}_c}\right) e^{-j\beta z} - \left(\frac{\tilde{V}_S - \hat{Z}_c\tilde{I}_S}{2\hat{Z}_c}\right) e^{j\beta z}$$

When we group $\tilde{V}_S$ and $\tilde{I}_S$ terms, we get

$$\tilde{V}(z) = \tilde{V}_S \cos\beta z - j\hat{Z}_c\tilde{I}_S \sin\beta z$$

$$\tilde{I}(z) = -j\frac{\tilde{V}_S}{\hat{Z}_c}\sin\beta z + \tilde{I}_S\cos\beta z$$

These equations can be expressed in matrix form as

$$\begin{bmatrix} \tilde{V}(z) \\ \tilde{I}(z) \end{bmatrix} = \begin{bmatrix} \cos\beta z & -j\hat{Z}_c\sin\beta z \\ -j\dfrac{1}{\hat{Z}_c}\sin\beta z & \cos\beta z \end{bmatrix} \begin{bmatrix} \tilde{V}_S \\ \tilde{I}_S \end{bmatrix} \tag{9.57}$$

This matrix equation enables us to compute the voltage and current at any point z along the transmission line when their counterparts at the sending end are known.

By letting $z = \ell$, we can determine the receiving-end voltage and current as

$$\begin{bmatrix} \tilde{V}_R \\ \tilde{I}_R \end{bmatrix} = \begin{bmatrix} \cos\beta\ell & -j\hat{Z}_c\sin\beta\ell \\ -j\dfrac{1}{\hat{Z}_c}\sin\beta\ell & \cos\beta\ell \end{bmatrix} \begin{bmatrix} \tilde{V}_S \\ \tilde{I}_S \end{bmatrix} \tag{9.58}$$

where $\tilde{V}_R = \tilde{V}(\ell)$, and $\tilde{I}_R = \tilde{I}(\ell)$.

We can also express $\tilde{V}_S$ and $\tilde{I}_S$ in terms of $\tilde{V}_R$ and $\tilde{I}_R$ as

$$\begin{bmatrix} \tilde{V}_S \\ \tilde{I}_S \end{bmatrix} = \begin{bmatrix} \cos\beta\ell & j\hat{Z}_c\sin\beta\ell \\ j\dfrac{1}{\hat{Z}_c}\sin\beta\ell & \cos\beta\ell \end{bmatrix} \begin{bmatrix} \tilde{V}_R \\ \tilde{I}_R \end{bmatrix} \tag{9.59}$$

If we substitute (9.59) into (9.57), we can predict the voltage and the current along the transmission line in terms of the receiving-end variables $\tilde{V}_R$ and $\tilde{I}_R$ as

$$\begin{bmatrix} \tilde{V}(z) \\ \tilde{I}(z) \end{bmatrix} = \begin{bmatrix} \cos\beta(\ell - z) & j\hat{Z}_c\sin\beta(\ell - z) \\ j\dfrac{1}{\hat{Z}_c}\sin\beta(\ell - z) & \cos\beta(\ell - z) \end{bmatrix} \begin{bmatrix} \tilde{V}_R \\ \tilde{I}_R \end{bmatrix} \tag{9.60}$$

Thus, we can use (9.60) to determine voltage $\tilde{V}(z)$ and current $\tilde{I}(z)$ at a distance z from the source when the load voltage $\tilde{V}_R$ and current $\tilde{I}_R$ are known.

EXAMPLE 9.5

The inductance and capacitance of a 50-m-long coaxial cable are 0.25 μH/m and 50 pF/m, respectively, and its frequency of operation is 100 kHz. (a) Calculate the characteristic impedance and the phase constant of the line. (b) If the permeability of the medium is the same as that of free space, what must be the dielectric constant of the medium? (c) Determine the delay generated by the transmission line.

Solution a) The characteristic impedance of the cable is

$$\hat{Z}_c = \sqrt{\frac{L_\ell}{C_\ell}} = \sqrt{\frac{0.25 \times 10^{-6}}{50 \times 10^{-12}}} = 70.71\ \Omega$$

and its phase constant is

$$\begin{aligned}\beta &= \omega\sqrt{L_\ell C_\ell} \\ &= 2\pi \times 100 \times 10^3 \times \sqrt{0.25 \times 10^{-6} \times 50 \times 10^{-12}} \\ &= 2.22 \times 10^{-3}\ \text{rad/m}\end{aligned}$$

b) The velocity of propagation is

$$u_p = \frac{1}{\sqrt{L_\ell C_\ell}} = \frac{1}{\sqrt{0.25 \times 10^{-6} \times 50 \times 10^{-12}}} = 2.83 \times 10^8\ \text{m/s}$$

Also, from equation (9.39), $u_p = \dfrac{1}{\sqrt{\mu\epsilon}}$. Therefore,

$$\begin{aligned}\epsilon_r &= \frac{1}{\epsilon_0 u_p^2 \mu_0} \\ &= \frac{1}{8.85 \times 10^{-12} \times (2.83 \times 10^8)^2 \times 4\pi \times 10^{-7}} = 1.12\end{aligned}$$

c) The delay on the transmission line is

$$t_d = \frac{\ell}{u_p} = \frac{50}{2.83 \times 10^8} = 176.68 \times 10^{-9}\ \text{s} \quad \text{or} \quad 176.68\ \text{ns}$$

• • •

EXAMPLE 9.6

A 100-m-long, lossless transmission line with a distributed inductance of 296 nH/m and a distributed capacitance of 46.2 pF/m operates at no load. The power to the line is delivered by a voltage source connected to the input terminals of the transmission line. The open circuit terminal voltage of the voltage source is $v_s(t) = 100 \cos 10^6 t$ V, and its internal impedance is negligible. Calculate (a) the characteristic impedance and the phase constant of the line, (b) the voltage across the receiving end terminals and the current supplied by the source, and (c) the power delivered by the source.

Solution a) The characteristic impedance:

$$\hat{Z}_c = \sqrt{\frac{L_\ell}{C_\ell}} = \sqrt{\frac{296 \times 10^{-9}}{46.2 \times 10^{-12}}} \approx 80\ \Omega$$

The phase constant:

$$\begin{aligned}\beta &= \omega\sqrt{L_\ell C_\ell} \\ &= 10^6\sqrt{296 \times 10^{-9} \times 46.2 \times 10^{-12}} = 3.698 \times 10^{-3}\ \text{rad/m}\end{aligned}$$

b) Using (9.58), we have

$$\begin{bmatrix}\tilde{V}_R \\ 0\end{bmatrix} = \begin{bmatrix}\cos(0.3698) & -j80\ \sin(0.3698) \\ -j\frac{1}{80}\ \sin(0.3698) & \cos(0.3698)\end{bmatrix}\begin{bmatrix}100\angle 0^\circ \\ \tilde{I}_S\end{bmatrix}$$

From this matrix equation we first compute the current through the source as

$$I_S = \frac{\left[j\frac{1}{80}\ \sin(0.3698)\right] 100\angle 0^\circ}{\cos(0.3698)} = j0.458\ \text{A}$$

The voltage across the receiving end terminals is

$$\tilde{V}_R = 100\ \cos(0.3698) - j80 \times (j0.485)\ \sin(0.3698) = 107.26\ \text{V}$$

Note that the magnitude of the receiving-end voltage is higher than that of the voltage at the sending end. This is due to the capacitance of the transmission line when it operates at no load.

c) The complex power at the sending end is

$$\begin{aligned}\hat{S}_s &= \tfrac{1}{2}\tilde{V}_S\tilde{I}_S^* = P_s + jQ_s \\ &= \tfrac{1}{2}(100\angle 0^\circ)(-j0.485) \\ &= -j24.25\ \text{VA}\end{aligned}$$

Thus, the average power at the sending end, P_s, is zero. The reactive power is $Q_s = -24.25$ VAR. • • •

9.4 The input impedance

The input impedance $\hat{Z}_{in}(z)$ at any point z along the lossless transmission line, shown in Figure 9.10, is calculated as the ratio of the voltage to current at that point. That is,

$$\hat{Z}_{in}(z) = \frac{\tilde{V}(z)}{\tilde{I}(z)} = \frac{\tilde{V}_R\ \cos\beta(\ell - z) + j\hat{Z}_c\tilde{I}_R\sin\beta(\ell - z)}{j\frac{\tilde{V}_R}{\hat{Z}_c}\sin\beta(\ell - z) + \tilde{I}_R\cos\beta(\ell - z)} \tag{9.61}$$

Also, we know that

$$\hat{Z}_L = \frac{\tilde{V}_R}{\tilde{I}_R} \tag{9.62}$$

where $\hat{Z}_L$ is the load impedance.

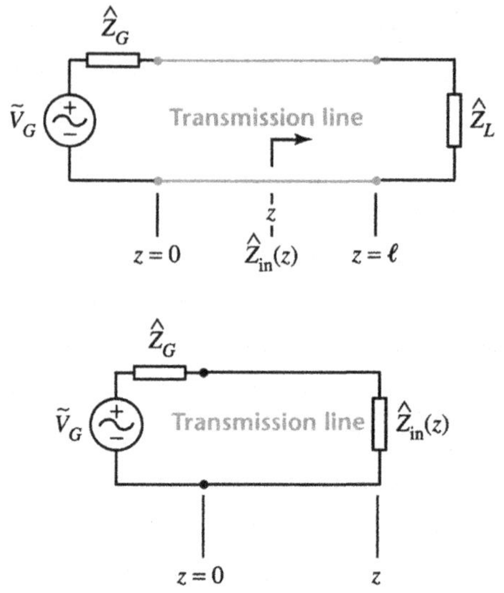

Figure 9.10 Representation of a transmission line in terms of its input impedance

Substituting (9.62) into (9.61), we obtain

$$\hat{Z}_{in}(z) = \hat{Z}_c \frac{\hat{Z}_L + j\hat{Z}_c \tan \beta(\ell - z)}{\hat{Z}_c + j\hat{Z}_L \tan \beta(\ell - z)} \tag{9.63}$$

As is apparent from (9.63), the input impedance of the line depends on the load impedance, the characteristic impedance, the length of the line, the phase constant, and also the location along the line. By letting $z = 0$ we can obtain the input impedance of the transmission line at the sending end as

$$\hat{Z}_{in}(0) = \hat{Z}_c \frac{\hat{Z}_L + j\hat{Z}_c \tan \beta\ell}{\hat{Z}_c + j\hat{Z}_L \tan \beta\ell} \tag{9.64}$$

When a short circuit exists at the receiving end of a line, the input impedance at $z = 0$ becomes

$$\hat{Z}_{in}(0)|_{sc} = j\hat{Z}_c \tan \beta\ell \tag{9.65}$$

However, an open circuit at the receiving end yields an input impedance of

$$\hat{Z}_{in}(0)|_{oc} = -j\hat{Z}_c \cot \beta\ell \tag{9.66}$$

at $z = 0$. In either case, the input impedance for a given transmission line depends upon the length of the line. The dependence of $\hat{Z}_{in}$ on the length of a short-circuited and an open-circuited transmission line is shown in Figures 9.11 and 9.12, respectively. It is also evident from equations (9.65) and (9.66) that

$$\hat{Z}_c = \sqrt{\hat{Z}_{in}(0)|_{oc}\hat{Z}_{(0)_{in}}|_{sc}}$$

Therefore, by performing open-circuit and short-circuit tests on a line, we can determine its characteristic impedance.

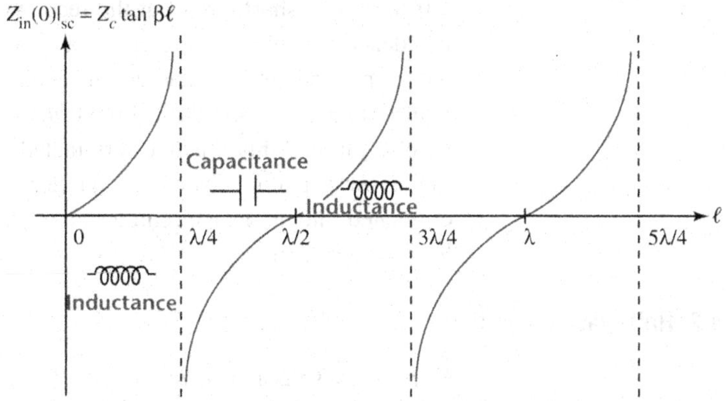

Figure 9.11 Input impedance of a short-circuited transmission line as a function of its length

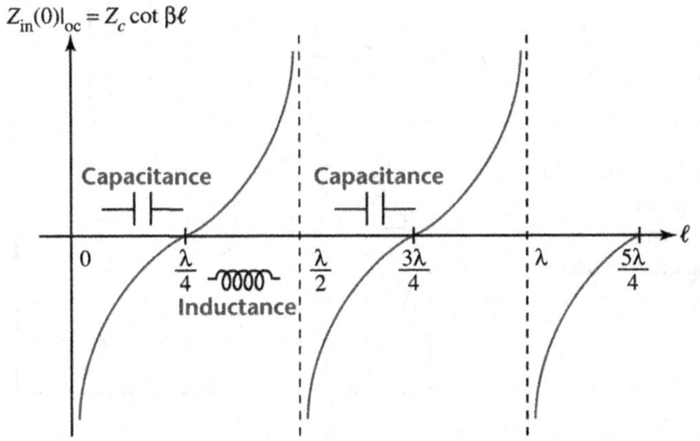

Figure 9.12 Input impedance of an open-circuited transmission line as a function of its length

In the following subsections we investigate the effect of the length of transmission line on the input impedance at the sending end.

9.4.1 Quarter-wavelength line

The input impedance of a quarter-wavelength ($\ell = \lambda/4$) transmission line at the sending end is determined from (9.64) as

$$\hat{Z}_{\text{in}}(0) = \hat{Z}_c \frac{\hat{Z}_L + j\hat{Z}_c \tan\left(\dfrac{2\pi}{\lambda}\dfrac{\lambda}{4}\right)}{\hat{Z}_c + j\hat{Z}_L \tan\left(\dfrac{2\pi}{\lambda}\dfrac{\lambda}{4}\right)} \tag{9.67}$$

$$= \frac{\hat{Z}_c^2}{\hat{Z}_L}$$

Therefore, a quarter-wavelength transmission line transforms an impedance into an admittance. This fact becomes even more evident when we define impedances in terms of their normalized values with respect to the characteristic impedance $\hat{Z}_c$. When we write the normalized input impedance as $\hat{z}_{\text{in}} = \hat{Z}_{\text{in}}/\hat{Z}_c$, and the normalized load impedance as $\hat{z}_L = \hat{Z}_L/\hat{Z}_c$, then from (9.67) we get $\hat{z}_{\text{in}} = 1/\hat{z}_L$.

If there is a short circuit at the receiving end of the line, the input impedance becomes infinite. Thus, a quarter-wavelength transmission line terminated into a short circuit behaves as an open circuit at its input terminals, as shown in Figure 9.13. For this reason, a quarter-wavelength transmission line terminated into a short circuit may be used as an insulator. Likewise, a quarter-wavelength transmission line terminated into an open circuit acts like a short circuit as viewed from the input terminals.

9.4.2 Half-wavelength line

When a transmission line is one-half wavelength long ($\ell = \lambda/2$), it presents an input impedance of

$$\hat{Z}_{\text{in}}(0) = \hat{Z}_c \frac{\hat{Z}_L + j\hat{Z}_c \tan\left(\dfrac{2\pi}{\lambda}\dfrac{\lambda}{2}\right)}{\hat{Z}_c + j\hat{Z}_L \tan\left(\dfrac{2\pi}{\lambda}\dfrac{\lambda}{2}\right)} = \hat{Z}_L \tag{9.68}$$

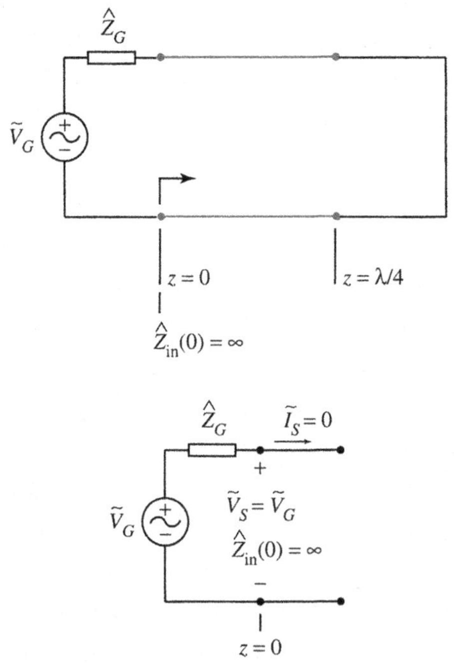

Figure 9.13 Equivalent circuit of a short-circuited quarter-wavelength transmission line at its sending end

Thus the load impedance is replicated at every one-half wavelength in a transmission line, as illustrated in Figure 9.14. Except for the determination of the transit time of the wave, we can make all the calculations by assuming the transmission line to be $n\lambda/2$ less than the original length, where n is an integer. If $\tilde{V}(z)$ and $\tilde{I}(z)$ are the voltage and current at any point z on the transmission line, then $\tilde{V}(z + \lambda/2) = -\tilde{V}(z)$, and $\tilde{I}(z + \lambda/2) = -\tilde{I}(z)$. In other words, the voltage and current reverse their directions every one-half wavelength along the line.

Figure 9.14 Equivalent circuit of a short-circuited half-wavelength transmission line at its sending end

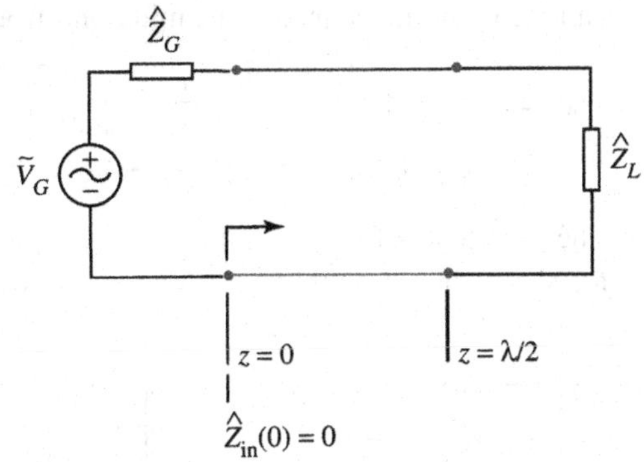

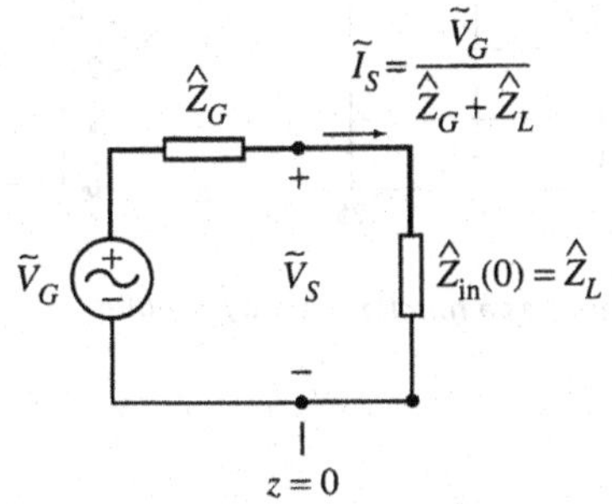

EXAMPLE 9.7

WORKSHEET 12

Mathcad

A 25-m-long lossless transmission line is terminated with a load having an equivalent impedance of $40 + j30\ \Omega$ at 10 MHz. The per-unit length inductance and capacitance of the line are 310.4 nH/m and 38.28 pF/m, respectively. Calculate the input impedance at the sending end and the midpoint of the line. Also, plot the variations of the magnitude of the input impedance and its angle as a function of location on the transmission line.

Solution The characteristic impedance and the phase constant are, respectively,

$$\hat{Z}_c = \sqrt{\frac{310.4 \times 10^{-9}}{38.28 \times 10^{-12}}} \approx 90\ \Omega$$

and

$$\beta = 2\pi \times 10 \times 10^6 \sqrt{310.4 \times 10^{-9} \times 38.28 \times 10^{-12}} = 0.217$$

The input impedance at the sending end, from (9.64), is

$$\hat{Z}_{in}(0) = 90\left[\frac{40 + j30 + j90\tan(0.217 \times 25)}{90 + j(40 + j30)\tan(0.217 \times 25)}\right]$$
$$= 57\underline{/-41.29^\circ}\ \Omega = 42.83 - j37.62\ \Omega$$

and the input impedance at the midpoint, from (9.63), is

$$\hat{Z}_{\text{in}}(12.5) = 90\left[\frac{40 + j30 + j90\tan(0.217 \times 12.5)}{90 + j(40 + j30)\tan(0.217 \times 12.5)}\right]$$
$$= 35.49\angle{-5.61^\circ}\ \Omega = 35.32 - j3.47\ \Omega$$

The variations of the magnitude of the input impedance and its angle are shown in Figure 9.15 as a function of location on the transmission line.

• • •

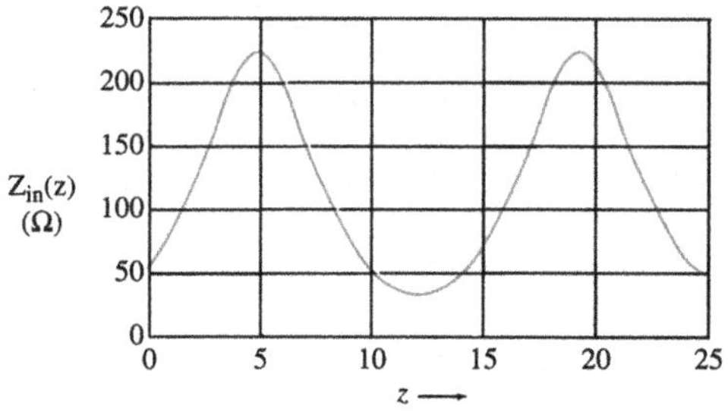

a) *Magnitude of the input impedance as a function of location*

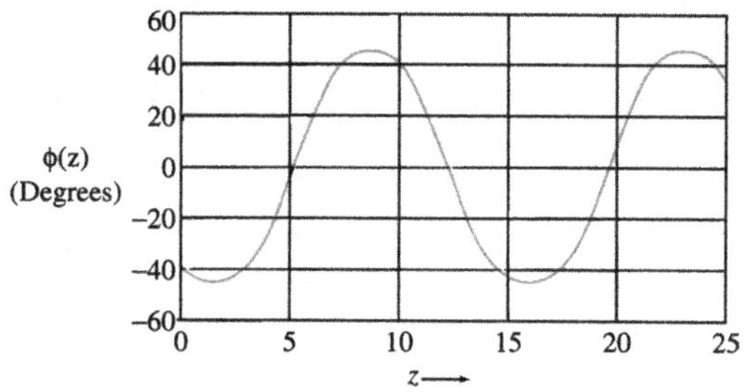

b) *Input impedance angle as a function of location*

Figure 9.15

EXAMPLE 9.8

A quarter-wavelength transmission line with a characteristic impedance of 400 Ω is short-circuited at the receiving end. Across the sending-end terminals of the line there is a generator with an open-circuit voltage of $v(t) = 50\cos 10^6 t$ V and an internal resistance of 100 Ω. Determine the voltage across the receiving end and at the midpoint of the transmission line.

Solution Figure 9.16 illustrates the transmission line and its equivalent circuit. From equation (9.67) the input impedance at the sending end for a short-circuited quarter-wavelength line is $\hat{Z}_{\text{in}}(0) = \infty$. The voltage and the current at the sending end are

$$v_s(t) = v(t) = 50\cos 10^6 t\ \Omega, \quad i_S(t) = 0$$

or, in the phasor form

$$\tilde{V}_S = 50\angle 0^\circ V,\ \tilde{I}_S = 0$$

The voltage across the load can be obtained from (9.58) as

$$\tilde{V}_R = (50\angle 0^\circ)\cos\left(\frac{2\pi}{\lambda}\frac{\lambda}{4}\right) = 0$$

At the midpoint, $z = \frac{\lambda}{8}$, however, the voltage, from (9.57), is

$$\tilde{V}\left(\frac{\lambda}{8}\right) = (50\angle 0^\circ)\cos\left(\frac{2\pi}{\lambda}\frac{\lambda}{8}\right)$$
$$= \frac{50}{\sqrt{2}}\angle 0^\circ\ \text{V}$$

or, in the time domain,

$$v\left(\frac{\pi}{8}, t\right) = \frac{50}{\sqrt{2}} \cos 10^6 t \text{ V}$$

• • •

Figure 9.16

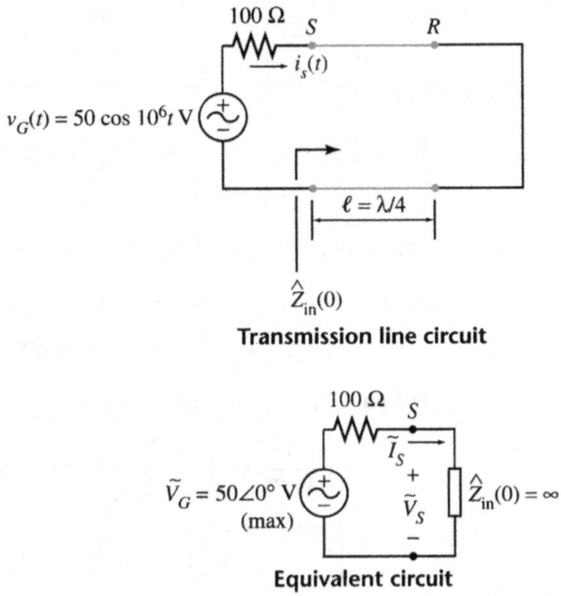

EXAMPLE 9.9

A generator having an internal resistance of 50 Ω applies a signal with a wavelength of 100 cm to a lossless transmission line. The transmission line is 50 cm long and terminated with a load having an impedance of $50 + j20\,\Omega$. The peak value of the open-circuit terminal voltage at the source is 10 V. The per-unit-length inductance and capacitance of the line are 0.17 μH/m and 70 pF/m, respectively. Calculate (a) the frequency of the source and (b) the voltage, current, and power at the sending and receiving ends of the transmission line.

Solution The velocity of propagation and the characteristic impedance of the line are

$$u_p = \frac{1}{\sqrt{L_\ell C_\ell}} = \frac{1}{\sqrt{0.17 \times 10^{-6} \times 70 \times 10^{-12}}} = 2.899 \times 10^8 \text{ m/s}$$

and

$$\hat{Z}_c = \sqrt{\frac{L_\ell}{C_\ell}} = \sqrt{\frac{0.17 \times 10^{-6}}{70 \times 10^{-12}}} = 49.28\ \Omega$$

a) The frequency of the source is

$$f = \frac{u_p}{\lambda} = \frac{2.899 \times 10^8}{1} = 2.899 \times 10^8 \text{ Hz}$$

b) Because the length of the line $\ell = 50$ cm $= \lambda/2$ (a half-wavelength transmission line), the input impedance at the sending end is

$$\hat{Z}_{\text{in}}(0) = \hat{Z}_L = 50 + j20\ \Omega$$

Consequently, the sending-end current and voltage are

$$\tilde{I}_S = \frac{\tilde{V}_G}{R_G + \hat{Z}_{\text{in}}(0)} = \frac{10\underline{/10^\circ}}{50 + 50 + j20} = 98 \times 10^{-3}\underline{/-11.31^\circ}\ \text{A}$$

and

$$\tilde{V}_S = \tilde{I}_S \hat{Z}_{\text{in}}(0) = (98 \times 10^{-3}\underline{/-11.31^\circ})(50 + j20)$$
$$= 5.28\underline{/10.49^\circ}\ \text{V}$$

The complex power at the sending end is

$$\hat{S}_S = \tfrac{1}{2}\tilde{V}_S \tilde{I}_S^* = \tfrac{1}{2}(5.28\underline{/10.49^\circ})(98 \times 10^{-3}\underline{/11.31^\circ})$$
$$= 0.259\underline{/21.8^\circ}\ \text{VA} = 0.24 + j0.096\ \text{VA}$$

The current at the receiving end is

$$\tilde{I}_R = \left[-j\frac{1}{49.28}\sin\left(\frac{2\pi}{\lambda}\frac{\lambda}{2}\right)\right][5.28\underline{/10.49^\circ}]$$
$$+\left[\cos\left(\frac{2\pi}{\lambda}\frac{\lambda}{2}\right)\right][98 \times 10^{-3}\underline{/-11.31^\circ}]$$

$$\tilde{I}_R = -98 \times 10^{-3}\underline{/-11.31^\circ}\ \text{A} \quad \text{or} \quad \tilde{I}_R = 98 \times 10^{-3}\underline{/168.69^\circ}\ \text{A}$$

and the voltage at the receiving end is

$$\tilde{V}_R = \left[\cos\left(\frac{2\pi}{\lambda}\frac{\lambda}{2}\right)\right][5.28\underline{/10.49^\circ}]$$
$$-j49.28[98 \times 10^{-3}\underline{/-11.31^\circ}]\left[\sin\left(\frac{2\pi}{\lambda}\frac{\lambda}{2}\right)\right]$$

$$\tilde{V}_R = -5.28\underline{/10.49^\circ}\ \text{V} \quad \text{or} \quad \tilde{V}_R = 5.28\underline{/190.49^\circ}\ \text{V}$$

The complex power at the receiving end is

$$\hat{S}_R = \tfrac{1}{2}\tilde{V}_R \tilde{I}_R^* = \tfrac{1}{2}(5.28\underline{/190.49^\circ})(98 \times 10^{-3}\underline{/-168.69^\circ})$$
$$= 0.259\underline{/21.8^\circ}\ \text{VA} = 0.24 + j0.096\ \text{VA}$$

As expected, for a lossless transmission line, the power delivered to the load is exactly equal to the power input at the sending end. However, the power supplied by the generator is

$$\hat{S}_G = \tfrac{1}{2} \times 10 \times \tilde{I}_S^* = 0.49\underline{/11.31^\circ} = 0.48 + j0.096\ \text{VA}$$

Hence, the overall efficiency of the system is

$$\eta = \frac{0.24}{0.48} = 0.5 \quad \text{or} \quad 50\%$$

• • •

9.5 Reflections at discontinuity points along transmission lines

A discontinuity point along a transmission line is a point at which the characteristic impedance of the line changes. For example, when two transmission lines with different characteristic impedances are connected in tandem, the connection point is a discontinuity point. The sending and receiving ends of a transmission line may also be considered as points of discontinuity because the internal impedance of the source on one hand and the impedance of the load on the other may not be the same as the characteristic impedance of the line.

When a wave is incident at a discontinuity point, a part of it is transmitted beyond the point of discontinuity, and the remainder is reflected back as a backward-traveling wave, as illustrated in Figure 9.17. By using (9.47a) and (9.47b), we can express $\hat{Z}_L$ as

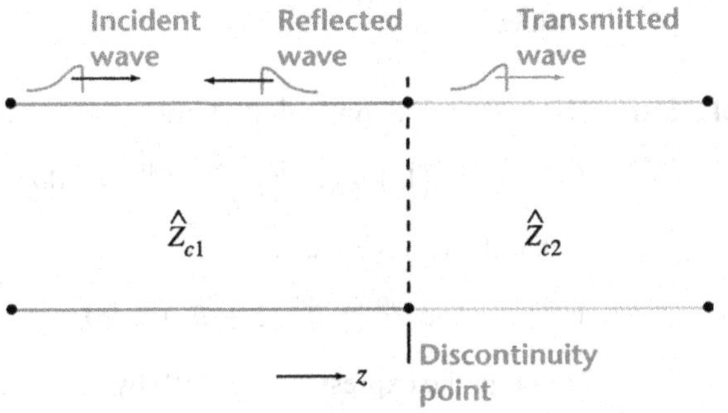

Figure 9.17 Incident, reflected, and transmitted waves at a discontinuity point on a transmission line

$$\hat{Z}_L = \frac{\tilde{V}(\ell)}{\tilde{I}(\ell)} = \hat{Z}_c \frac{1+\hat{\rho}_R}{1-\hat{\rho}_R} \tag{9.69}$$

where $\hat{\rho}_R$ is the reflection coefficient at $z = \ell$. Thus, from (9.48),

$$\hat{\rho}_R = \frac{\hat{V}-}{\hat{V}+} e^{2\hat{\gamma}\ell}$$

Rearranging (9.69), we obtain the reflection coefficient at $z = \ell$ as

$$\hat{\rho}_R = \frac{\hat{Z}_L - \hat{Z}_c}{\hat{Z}_L + \hat{Z}_c} \tag{9.70}$$

The reflection coefficient at any point z can also be expressed, in terms of $\hat{\rho}_R$, from (9.48), as

$$\hat{\rho}(z) = \hat{\rho}_R e^{-2\hat{\gamma}(\ell-z)} \tag{9.71}$$

The voltage at the receiving end, $z = \ell$, from (9.47a) is

$$\tilde{V}(\ell) = \hat{V}^+ e^{-\hat{\gamma}\ell}(1+\hat{\rho}_R) \tag{9.72}$$

thus we can also define the **transmission coefficient** as

$$\hat{\tau}_R = 1 + \hat{\rho}_R \tag{9.73a}$$

or

$$\hat{\tau}_R = \frac{2\hat{Z}_L}{\hat{Z}_L + \hat{Z}_c} \tag{9.73b}$$

in terms of $\hat{Z}_L$ and $\hat{Z}_c$. The transmission coefficient helps us determine the transmitted wave in terms of the incident wave.

EXAMPLE 9.10

The phase constant and the characteristic impedance of a 100-m-long lossless transmission line are 3×10^{-3} rad/m and 60 Ω, respectively. The rms voltages at the sending and receiving ends are measured as 100 V and 90 V, respectively, and the receiving-end voltage lags the sending-end voltage by an angle of 2°. Determine the transmission coefficient and the load impedance at the receiving end. Also, write the general expressions for the forward and backward voltage and current waves at any point along the line. Compute the average power associated with each wave.

Solution Using (9.71) along with (9.47a), at the sending end, $z = 0$,

$$\tilde{V}(0) = \hat{V}^+\left(1 + \hat{\rho}_R e^{-j2\times 0.003\times 100}\right) = 100\underline{/0^\circ}$$

and at the receiving end, $z = \ell$,

$$\tilde{V}(\ell) = \hat{V}^+ e^{-j0.003\times 100}(1 + \hat{\rho}_R) = 90\underline{/-2^\circ}$$

Dividing the expression for $\tilde{V}(0)$ by the expression for $\tilde{V}(\ell)$, we obtain

$$\frac{1 + \hat{\rho}_R e^{-j0.6}}{(1 + \hat{\rho}_R)e^{-j0.3}} = \frac{10}{9}\underline{/2^\circ}$$

and we can calculate the reflection coefficient as

$$\hat{\rho}_R = 0.814\underline{/56.03^\circ}$$

The transmission coefficient is

$$\hat{\tau}_R = 1 + \hat{\rho}_R = 1 + 0.814\underline{/56.03^\circ} = 1.604\underline{/24.89^\circ}$$

The load impedance can be determined from (9.69) as

$$\hat{Z}_L = 60\left[\frac{1 + 0.814\underline{/56.03^\circ}}{1 - 0.814\underline{/56.03^\circ}}\right] = 110.9\underline{/75.97^\circ}\ \Omega$$

or

$$\hat{Z}_L = 26.89 + j107.59\ \Omega$$

To obtain the general $\tilde{V}(z)$ and $\tilde{I}(z)$ at any point we have to first determine $\hat{V}^+$ and $\hat{I}^+$:

$$\hat{V}^+ = \frac{90\underline{/-2^\circ}}{e^{-j0.3} \times 1.604\underline{/24.89^\circ}} = 56.12\underline{/-9.7^\circ}\ \text{V}$$

$$\hat{I} = \frac{\hat{V}^+}{\hat{Z}_c} = \frac{56.12\underline{/-9.7^\circ}}{60\underline{/0^\circ}} = 0.94\underline{/-9.7^\circ}\ \text{A}$$

The voltage and current at any point on the line, in terms of their rms values, are

$$\tilde{V}(z) = 56.12e^{-j(0.003z+9.7°)}\left[1 + 0.814e^{-j[0.006(\ell-z)-56.03°]}\right]$$
$$\tilde{I}(z) = 0.94e^{-j(0.003z+9.7°)}\left[1 - 0.814e^{-j[0.006(\ell-z)-56.03°]}\right]$$

The forward and backward voltage and current waves, in terms of their rms values are,

$$\tilde{V}_f(z) = 56.12e^{-j0.003z}e^{-j9.7°} \text{ V}$$
$$\tilde{I}_f(z) = 0.94e^{-j0.003z}e^{-j9.7°} \text{ A}$$
$$\tilde{V}_b(z) = 45.682e^{j0.003z}e^{j12.05°} \text{ V}$$
$$\tilde{I}_b(z) = -0.765e^{j0.003z}e^{j12.05°} \text{ V}$$

in the phasor domain where the subscripts f and b are for the forward and backward fields, and

$$v_f(z,t) = 79.366 \cos(\omega t - 0.003z - 9.7°) \text{ V}$$
$$i_f(z,t) = 1.329 \cos(\omega t - 0.003z - 9.7°) \text{ A}$$
$$v_b(z,t) = 64.604 \cos(\omega t + 0.003z + 12.05°) \text{ V}$$
$$i_b(z,t) = -1.082 \cos(\omega t + 0.003z + 12.05°) \text{ A}$$

in the time domain.

The power associated with each wave is

$$P_f(z) = \text{Re}[\tilde{V}_f(z)\tilde{I}_f^*(z)] = 52.75 \text{ W}$$
$$P_b(z) = \text{Re}[\tilde{V}_b(z)\tilde{I}_b^*(z)] = -34.95 \text{ W}$$

The net power flow in the z direction along the line is

$$P(z) = 52.75 - 34.95 = 17.8 \text{ W}$$

Note that the average power associated with the cross-coupling of the fields is zero for a lossless transmission line. • • •

9.6 Standing waves in transmission lines

The voltage and current at any point z along a transmission line are, in general,

$$\tilde{V}(z) = \hat{V}^+e^{-j\beta z}\left[1 + \rho_R e^{j\phi}e^{-j2\beta(\ell-z)}\right] \tag{9.74}$$

and

$$\tilde{I}(z) = \frac{\hat{V}^+e^{-j\beta z}}{\hat{Z}_c}\left[1 - \rho_R e^{j\phi}e^{-j2\beta(\ell-z)}\right] \tag{9.75}$$

where $\hat{\rho}_R = \rho_R e^{j\phi}$ is the reflection coefficient at $z = \ell$. Each equation is the result of a forward- and backward-traveling wave. As explained in Chapter 8, the combination of forward- and backward-traveling waves forms a standing wave. Thus, (9.74) and (9.75) are the expressions for the voltage and current standing waves along a transmission line.

The magnitudes of the voltage and current in (9.74) and (9.75), for an arbitrary load, when $\hat{V}^+ = V^+\underline{/0^\circ}$, are

$$V(z) = V^+\sqrt{1 + \rho_R^2 + 2\rho_R \cos[2\beta(\ell - z) - \phi]} \tag{9.76}$$

and

$$I(z) = \frac{V^+}{Z_c}\sqrt{1 + \rho_R^2 - 2\rho_R \cos[2\beta(\ell - z) - \phi]} \tag{9.77}$$

The variations of $V(z)$ and $I(z)$ for $\hat{Z}_c = 32\ \Omega$, $V^+ = 100$ V, $\beta =$ 8.38 rad/m, $\ell = 1$m, and $\hat{\rho}_R = 0.531\underline{/7.77^\circ}$ are sketched in Figure 9.18. Note that when the voltage is maximum ($z = \lambda/8 = 0.125\lambda$), the current is minimum and vice versa ($z = 3\lambda/8 = 0.375\lambda$). The minima and maxima are always a quarter-wavelength apart, and the distance between two successive maxima or minima is always a half-wavelength.

When a lossless transmission line is terminated into a resistive load, the reflection coefficient at the receiving end is a real quantity. In this case, (9.76) and (9.77) become

$$V(z) = V^+\sqrt{1 + \rho_R^2 + 2\rho_R \cos[2\beta(\ell - z)]} \tag{9.78}$$

and

$$I(z) = \frac{V^+}{Z_c}\sqrt{1 + \rho_R^2 - 2\rho_R \cos[2\beta(\ell - z)]} \tag{9.79}$$

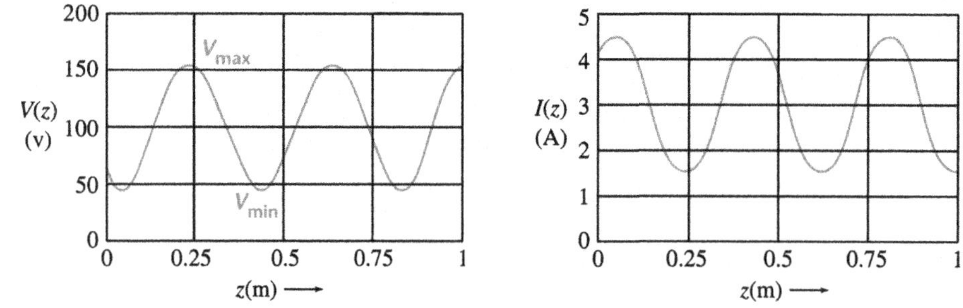

Figure 9.18 Voltage and current standing waves on a transmission line (see text)

Let us use these equations to investigate the variations of $V(z)$ and $I(z)$ with respect to z when (a) $R_L > Z_c$ and (b) $R_L < Z_c$.

a) $R_L > Z_c \Rightarrow \rho_R > 0$: Plots of $V(z)$ and $I(z)$ using (9.78) and (9.79) when $\hat{Z}_c = 32\ \Omega$, $V^+ = 100$ V, $\beta = 8.38$ rad/m, and $R_L = 288\ \Omega$, are shown in Figure 9.19a. From this figure, we can see that the standing wave in a lossless transmission line causes a voltage maximum and a current minimum at the receiving end, $z = \ell$. The same trend follows at every half-wavelength on the transmission line.

b) $R_L < Z_c \Rightarrow \rho_R < 0$: When $\hat{Z}_c = 32\ \Omega$, $V^+ = 100$ V, $\beta = 8.38$ rad/m, $\ell = 1$m, and $R_L = 3.55\Omega$, a voltage minimum and current maximum occur at the receiving end, as shown in Figure 9.19b. In fact, where there is a minimum for $R_L > Z_c$, there is a voltage maximum for $R_L < Z_c$, and vice versa.

In the following sections we consider three special cases, where the transmission line terminates into a short circuit, an open circuit, and a matched load. We also discuss a measure known as the voltage standing wave ratio.

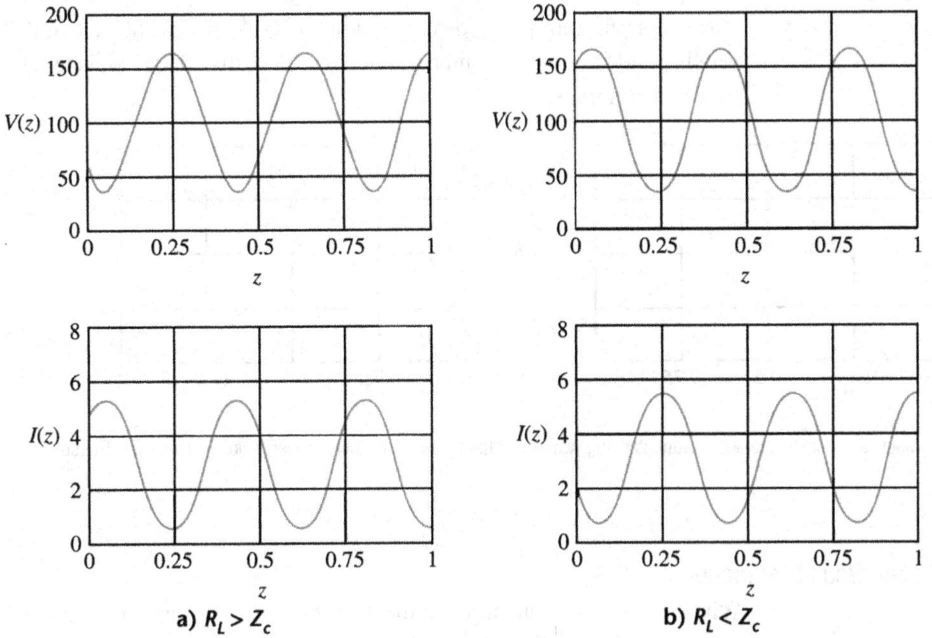

Figure 9.19 Magnitudes of voltage and current standing waves on a transmission line terminated into resistance

Case I: Short-circuit termination

When a transmission line is terminated into a short circuit, $\rho_R = -1$, and the voltage and current magnitudes at any point z along the transmission line are

$$V(z) = V^+\sqrt{2 - 2\cos[2\beta(\ell - z)]} \tag{9.80a}$$

$$= |2V^+ \sin\beta(\ell - z)| \tag{9.80b}$$

and

$$I(z) = \left(\frac{V^+}{Z_c}\right)\sqrt{2 + 2\cos[2\beta(\ell - z)]} \tag{9.81a}$$

$$= \left|\frac{2V^+}{Z_c}\cos\beta(\ell - z)\right| \tag{9.81b}$$

The variations of $V(z)$ and $I(z)$ for $\hat{Z}_c = 32\ \Omega$, $V^+ = 100$ V, $\beta =$ 8.38 rad/m, and $\ell = 1$m are shown in Figure 9.20, and the standing wave patterns are evident for both the voltage and current. The standing wave pattern for a transmission line with a short-circuit termination is similar to that of a plane wave incident on a perfect conductor (see Figure 8.12). As you can see from Figure 9.20, the voltage magnitude is maximum while the current is zero at locations that are odd multiples of a quarter-wavelength from the receiving end. However, the current is maximum and the voltage is zero at points on the line that are odd multiples of one-half of a wavelength. These observations are indeed consistent with our previous discussions of the input impedance at the sending end for short-circuited quarter-wavelength and half-wavelength transmission lines.

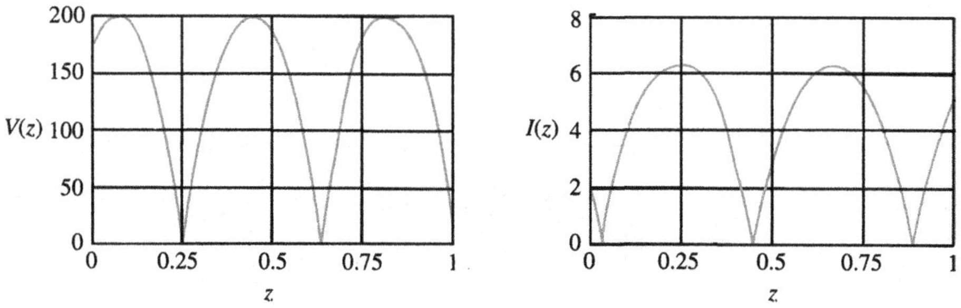

Figure 9.20 Magnitudes of voltage and current standing waves at different locations along transmission line with short-circuit termination

Case II: Open-circuit termination

When there is an open circuit at the receiving end of a transmission line, $\rho_R = 1$, and the voltage and current magnitudes at any point z along the transmission line are

$$V(z) = V^+\sqrt{2 + 2\cos[2\beta(\ell - z)]} \tag{9.82a}$$

$$= |2V^+ \cos\beta(\ell - z)| \tag{9.82b}$$

and

$$I(z) = \left(\frac{V^+}{Z_c}\right)\sqrt{2 - 2\cos[2\beta(\ell - z)]} \tag{9.83a}$$

$$= \left|\frac{2V^+}{Z_c}\sin\beta(\ell - z)\right| \tag{9.83b}$$

Figure 9.21 illustrates the changes in the magnitude of the voltage and current wave-forms along the transmission line when $\hat{Z}_c = 32\ \Omega$, $\beta =$ 8.38 rad/m, $\ell = 1$ m, and $V^+ = 100$ V. As expected, the current is zero and the voltage is maximum at the receiving end. At a distance of a quarter-wavelength from the receiving end, the voltage becomes zero while the current is maximum. Thus, a quarter-wavelength transmission line terminated into an open circuit behaves as a short circuit when viewed from the sending end. However, at a distance of a half-wavelength away from the receiving end, the transmission line behaves like an open circuit.

Case III: Termination with $R_L = Z_c$

If a lossless transmission line is terminated by a load having a resistance identical to the characteristic impedance of the line; i.e., $R_L = Z_c$, the reflection coefficient ρ_R at the receiving end will be zero, and the incident wave will be fully absorbed by the load. As a result, no standing wave develops along the transmission line. Also, from (9.76) and (9.77), we can conclude that the voltage and current magnitudes remain unchanged along the transmission line. Plots of $V(z)$ and $I(z)$ when $R_L = \hat{Z}_c = 32\ \Omega$, $\beta = 8.38$ rad/m, $V^+ = 100$ V, and $\ell = 1$ m are shown in Figure 9.22. Because there are no reflections at the receiving end, the transmission line is said to be *perfectly matched* or simply *matched* at its termination.

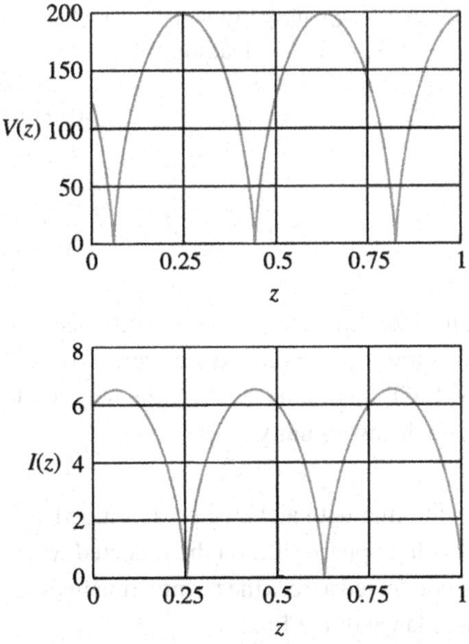

Figure 9.21 Changes in magnitude of voltage and current standing waves at different locations in open-circuited transmission line

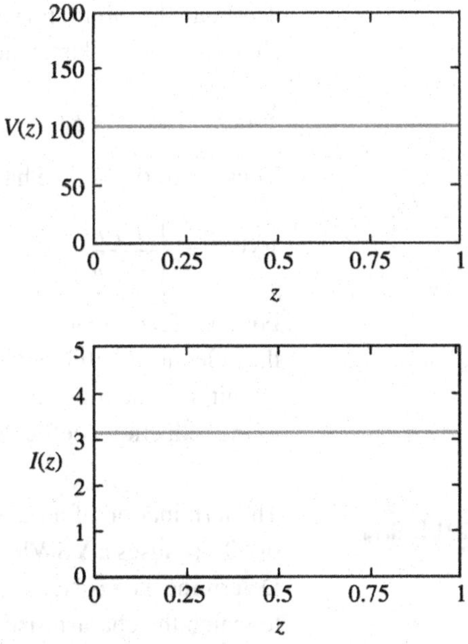

Figure 9.22 Voltage and current magnitudes along a perfectly matched transmission line

9.6.1 Voltage standing wave ratio

We can define a measure, known as the **voltage standing wave ratio, VSWR**, to evaluate the degree of mismatch of a load connected to a lossless transmission line. The VSWR is defined as the ratio of the maximum to the minimum value of the voltage standing wave on the transmission line:

$$VSWR = \frac{V_{\max}}{V_{\min}} \tag{9.84}$$

where $V_{\max}$ and $V_{\min}$ are as indicated in Figure 9.18. For a matched transmission line, $V_{\max} = V_{\min}$ and the VSWR will be unity.

The voltage standing wave ratio can also be conveniently expressed in terms of the reflection coefficient at the receiving end. The voltage along a transmission line can be given by

$$\tilde{V}(z) = \hat{V}^{+} e^{-j\beta z}[1 + \rho_R e^{j\phi} e^{-j2\beta(\ell - z)}] \tag{9.85}$$

where $\hat{\rho}_R = \rho_R e^{j\phi}$ and ρ_R is the magnitude of the reflection coefficient at the receiving end. In order to make the magnitude of $V(z)$ maximum, $[\phi - 2\beta(\ell - z)]$ should be $2n\pi$, where $n = 0, 1, 2, 3, \ldots$. Hence,

$$V_{\max} = V^{+}(1 + \rho_R) \tag{9.86a}$$

To obtain the minimum value of the voltage magnitude, $[\phi - 2\beta(\ell - z)]$ should be $(2n - 1)\pi$, where $n = 0, 1, 2, 3, \ldots$, leading to

$$V_{\min} = V^{+}(1 - \rho_R) \tag{9.86b}$$

Thus, from (9.84), we have

$$VSWR = \frac{1 + \rho_R}{1 - \rho_R} \tag{9.87}$$

For a perfectly matched transmission line, the VSWR is unity because the reflection coefficient is zero. However, for the short-circuit and open-circuit terminations, the VSWR will be infinity because the magnitude of the reflection coefficient in each case is unity.

EXAMPLE 9.11

The termination of a transmission line with a characteristic impedance of 50 Ω causes a VSWR of 2 with no phase shift on the reflected wave. Determine the resistance to be connected across the termination in order to match the characteristic impedance of the line.

Solution

$$2 = \frac{1 + \rho_R}{1 - \rho_R}$$

$$\rho_R = \frac{1}{3}$$

Because the reflected wave is in phase with the incident wave,

$$\rho_R = \tfrac{1}{3}\underline{/0^\circ}$$

Now, we can determine the existing load impedance that causes the reflection at the termination as

$$\tfrac{1}{3}\underline{/0^\circ} = \frac{\hat{Z}_L - 50}{\hat{Z}_L + 50}$$

$$\hat{Z}_L = 100\ \Omega$$

In order to obtain a unity VSWR for a matched transmission line, ρ_R must be zero. This condition can be achieved if $\hat{Z}_L = \hat{Z}_c = 50\ \Omega$. The additional resistance R_m to be connected in parallel to the load is

$$Z_m = \frac{R_m 100}{R_m + 100} = 50\ \Omega$$

$$R_m = 100\ \Omega$$

• • •

EXAMPLE 9.12

WORKSHEET 13

Mathcad

A 75-Ω, 100-m-long lossless transmission line feeds a resistive load of 45 Ω. The voltage across the load is 30 V (max) and the operating frequency is 1 MHz. If the transit time of the line is 0.357 μs, determine the voltage at the sending end of the line.

Solution The receiving-end reflection coefficient is

$$\hat{\rho}_R = \frac{45 - 75}{45 + 75} = -0.25$$

The phase constant is

$$\beta = \frac{\omega}{u_p} = \frac{\omega t_t}{\ell} = \frac{2\pi \times 10^6 \times 0.357 \times 10^{-6}}{100} = 2.24 \times 10^{-2}\ \text{rad/m}$$

From equation (9.47a) we can determine the arbitrary constant as

$$\hat{V}^+ = \frac{\tilde{V}_R}{(1 + \hat{\rho}_R)e^{-j\beta\ell}}$$

$$= \frac{30\underline{/0^\circ}}{(1 - 0.25)e^{-j0.0224\times 100}} = 40\underline{/128.34^\circ}\ \text{V}$$

The reflection coefficient at the sending end is

$$\hat{\rho}_S = \hat{\rho}(0) = \rho_R e^{-j2\beta\ell}$$

$$= -0.25e^{-j2\times 0.0224\times 100}$$

$$= 0.25\underline{/-76.69^\circ}$$

Employing (9.47) with $z = 0$ yields the sending-end voltage as

$$\tilde{V}_S = \hat{V}^+(1 + \hat{\rho}_S)$$

$$= 40\underline{/128.34^\circ}(1 + 0.25\underline{/-76.69^\circ})$$

$$= 43.41\underline{/115.39^\circ}\ \text{V}$$

$\tilde{V}_S$ can also be calculated using (9.59):

$$\tilde{V}_S = \tilde{V}_R \cos\beta\ell + j\hat{Z}_c\tilde{I}_R \sin\beta\ell$$

The receiving-end current $\tilde{I}_R$ is

$$\tilde{I}_R = \frac{30\underline{/0^\circ}}{45} = 0.667\underline{/0^\circ}\ \text{A}$$

Thus,

$$\begin{aligned}\tilde{V}_S &= (30\underline{/0^\circ})\cos(0.0224 \times 100) + j75(0.667\underline{/0^\circ})\sin(0.0224 \times 100)\\ &= 43.41\underline{/115.39^\circ}\ \text{V}\end{aligned}$$

• • •

9.7 Impedance matching with shunt stub lines

In Section 9.6 we observed that standing waves occur along a transmission line when it is connected to a load having an impedance other than the characteristic impedance of the line. We also learned that the standing waves disappear when the load is perfectly matched to the transmission line. A transmission line is said to be perfectly matched when its equivalent load impedance is exactly equal to its characteristic impedance. As the characteristic impedance of a lossless line is purely resistive; i.e., $Z_c = R_c$, the equivalent load impedance must also be equal to R_c. From (9.63), it is evident that the input impedance of a transmission line varies with z for a given load impedance. In addition, the input impedance at a half-wavelength from the load is the same as that of the load.

At some point D, as close to the load as possible, along the transmission line, the real part of the input impedance is equal to R_c. Let that point D be d meters away from the load, where $0 < d < \lambda/2$. At that location the reactive component of the input impedance may be positive (inductive) or negative (capacitive). We also know that the input impedence of a short-circuited line is always reactive; i.e., purely inductive or capacitive. Therefore, we can connect a short-circuited line to the transmission line at point D in order to cancel the reactive component of its input impedance, as shown in Figure 9.23. When a short-circuited line is used in this fashion, it is referred to as a *stub line*. Because the input impedances of the stub line and the transmission line are parallel, we can add their admittances to obtain an equivalent admittance of $1/R_c$, where R_c is the desired input impedance at point D on the transmission line.

Let us assume that the input admittance of the transmission line at point D is

$$\hat{Y}_{\text{line}} = \frac{1}{R_c} + jB \tag{9.88}$$

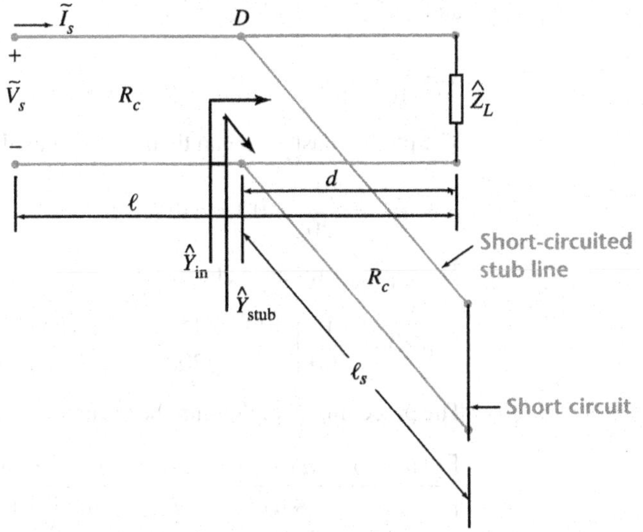

Figure 9.23 Short-circuited stub line connected to transmission line

where $1/R_c$ is the desired conductance and B is the susceptance. Note that B can be either positive (capacitive) or negative (inductive). The input admittance of a short-circuited stub line of length ℓ_s, from (9.65), is

$$\hat{Y}_{stub} = -j\frac{1}{R_c \tan \beta\ell_s} \tag{9.89}$$

where the per-unit-length parameters of the stub line are the same as those of the transmission line.

The equivalent input admittance at point D is then,

$$\begin{aligned}\hat{Y}_{in} &= \hat{Y}_{line} + \hat{Y}_{stub}\\ &= \frac{1}{R_c} + jB - j\frac{1}{R_c \tan \beta\ell_s}\end{aligned} \tag{9.90}$$

For a perfect match at point D, Y_{in} is equal to $1/R_c$; therefore, the susceptance B of the transmission line at $z = \ell - d$ must be

$$B = \frac{1}{R_c \tan \beta\ell_s} \tag{9.91}$$

EXAMPLE 9.13

WORKSHEET 14

Mathcad

A 100-Ω, 200-m-long lossless transmission line operates at 10 MHz and is terminated into an impedance of $50 - j200\ \Omega$. The transit time of the line is 1 μs. Determine the length and the location of a short-circuited stub line.

Solution The velocity of propagation and the wavelength on this line are, respectively,

$$u_p = \frac{200}{10^{-6}} = 2 \times 10^8 \text{ m/s}$$

and

$$\lambda = \frac{2 \times 10^8}{10 \times 10^6} = 20 \text{ m}$$

The phase constant β can then be calculated as

$$\beta = \frac{2\pi}{\lambda} = \frac{2\pi}{20} = 0.1\pi \text{ rad/m}$$

The input admittance at a distance $d = \ell - z$ from the load is

$$\hat{Y}_{\text{line}}(d) = \frac{1}{100}\left[\frac{100 + j(50 - j200)\tan(0.1\pi d)}{(50 - j200) + j100\tan(0.1\pi d)}\right] = \frac{1}{100} + jB$$

The preceding complex-number equation can be arranged as

$$\left[\frac{[2\tan^2(0.1\pi d) + 2] + j[-8\tan^2(0.1\pi d) + 13\tan(0.1\pi d) + 8]}{4\tan^2(0.1\pi d) - 16\tan(0.1\pi d) + 17}\right]\frac{1}{100}$$
$$= G + jB$$

where $G = \dfrac{1}{100}$ and $B = \dfrac{1}{R_c \tan \beta \ell_s}$ from the stub line.

From the real part of the preceding complex equation

$$\left[\frac{[2\tan^2(0.1\pi d) + 2]}{4\tan^2(0.1\pi d) - 16\tan(0.1\pi d) + 17}\right]\frac{1}{100} = \frac{1}{100}$$

we have two solutions for d. Selecting the shorter distance, we obtain $d = 2.63$ m. This is the distance from the load to the connection point of the stub line on the transmission line. From the imaginary part of the same equation

$$\left[\frac{[-8\tan^2(0.1 \times 2.63\pi) + 13\tan(0.1 \times 2.63\pi) + 8]}{4\tan^2(0.1 \times 2.63\pi) - 16\tan(0.1 \times 2.63\pi) + 17}\right]\frac{1}{100}$$
$$= \frac{1}{100\tan(0.1\pi \ell_s)}$$

we also get two solutions for the length of the stub line. Selection of the shorter length yields $\ell_s = 1.05$ m. This is the minimum length of the stub line to match the load with the transmission line. • • •

9.8 Transmission lines with imperfect materials

In the preceding sections we have studied the fundamentals of transmission lines having perfect conductors and dielectrics. In these cases, the transmission line is lossless, and the electromagnetic fields are in a plane perpendicular to the direction of the wave propagation (TEM wave). As imperfect materials introduce losses along a transmission line, the line cannot sustain a TEM wave in the strictest sense. Nevertheless, for all practical purposes, TEM wave approximation is still a fairly accurate method for the analysis of such transmission lines. The imperfections in a transmission line are basically due to (a) the finite resistance of the two conductors and (b) the finite conductance of the dielectric material separating the conductors.

The resistance of the line and the conductance of the dielectric material result in power loss as the wave propagates from one end of the transmission line to the other. Thus, the design of transmission lines usually emphasizes the minimization of these losses. As the existence of these losses cannot be ruled out, we must find ways to include them in the analysis of a transmission line. For this reason, we now discuss the analysis of transmission lines with imperfect materials.

9.8.1 Wave equations

When current flows through the conductors, the finite conductivity of each conductor causes the transmission line to experience a power loss. This power loss can be conveniently represented by a resistance determined from $R = P/I^2$, where P is the total power loss and I is the current in the transmission line. In accordance with our earlier discussion on transmission lines, we can express the per-unit-length resistance of the line as $R_\ell = R/\ell$, where ℓ is the length of the line. Since the current responsible for the power loss is the same current that produces the magnetic field in the transmission line, we can consider the per-unit-length resistance in series with the per-unit-length inductance of the line and express an impedance

$$\hat{Z}_\ell = R_\ell + j\omega L_\ell \tag{9.92}$$

where R_ℓ is the per-unit-length resistance of the two conductors, L_ℓ is the per-unit-length inductance, and $\hat{Z}_\ell$ is the per-unit-length *series impedance* of the transmission line.

Equation (9.25) can then be modified as

$$\frac{\partial \tilde{V}(z)}{\partial z} = -(R_\ell + j\omega L_\ell)\tilde{I} \tag{9.93}$$

For the parallel-plate transmission line shown in Figures 9.5 and 9.6, the per-unit-length resistance R_ℓ is $R_\ell = 2/\sigma\delta_c a$ as long as $\delta_c < w$, where w is the thickness of each plate, σ is the conductivity of each conductor, and δ_c is the skin depth. The per-unit-length capacitance of the parallel-plate line was determined in (9.18) as

$$C_\ell = \epsilon \frac{a}{d} \tag{9.94a}$$

If σ_d is the conductivity of the medium between the two conductors, the complex permittivity of the medium is $\hat{\epsilon} = \epsilon\left[1 - j\dfrac{\sigma_d}{\omega\epsilon}\right]$. By substituting $\hat{\epsilon}$ for ϵ in (9.94a), we obtain

$$\hat{C}_\ell = \epsilon\left(1 - j\,\frac{\sigma_d}{\omega\epsilon}\right)\frac{a}{d} \tag{9.94b}$$
$$= \frac{\epsilon a}{d} - j\,\frac{\sigma_d a}{\omega d}$$

Replacing C_ℓ with $\hat{C}_\ell$ in (9.28), we obtain

$$\frac{d\tilde{I}(z)}{dz} = -j\omega\left(\frac{\epsilon a}{d} - j\,\frac{\sigma_d a}{\omega d}\right)\tilde{V}$$
$$= -\left(\frac{\sigma_d a}{d} + j\omega\,\frac{\epsilon a}{d}\right)\tilde{V} \tag{9.95}$$
$$= -(G_\ell + j\omega C_\ell)\tilde{V}$$

where $G_\ell = \dfrac{\sigma_d a}{d}$ is the per-unit-length conductance of the parallel-plate transmission line. Differentiating (9.93) and (9.94) with respect to z, we get

$$\frac{d^2\tilde{V}}{dz^2} = \hat{Z}_\ell \hat{Y}_\ell \tilde{V} \tag{9.96}$$

and

$$\frac{d^2\tilde{I}}{dz^2} = \hat{Z}_\ell \hat{Y}_\ell \tilde{V} \tag{9.97}$$

where $\hat{Y}_\ell = G_\ell + j\omega C_\ell$ is the *shunt admittance* per unit length of the transmission line.

The equivalent electric circuit given in Figure 9.8 can now be modified for a transmission line with imperfect materials, as shown in Figure 9.24. The solutions of (9.96) and (9.97) are

$$\tilde{V}(z) = \hat{V}^+ e^{-\hat{\gamma} z} + \hat{V}^- e^{\hat{\gamma} z} \tag{9.98}$$

and

$$\tilde{I}(z) = \frac{\hat{V}^+}{\hat{Z}_c}\,e^{-\hat{\gamma} z} - \frac{\hat{V}^-}{\hat{Z}_c}\,e^{\hat{\gamma} z} \tag{9.99}$$

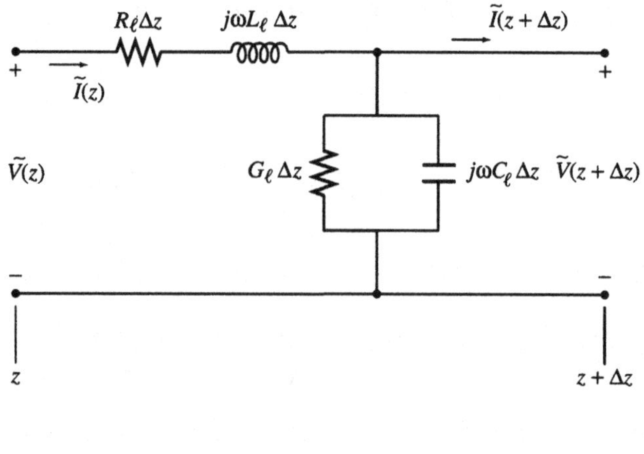

Figure 9.24 Electric circuit model for very short section of transmission line with imperfect materials

where

$$\hat{Z}_c = \sqrt{\frac{\hat{Z}_\ell}{\hat{Y}_\ell}} \tag{9.100}$$

is the characteristic impedance, and

$$\hat{\gamma} = \sqrt{\hat{Z}_\ell \hat{Y}_\ell} = \alpha + j\beta \tag{9.101}$$

is the equation for the propagation constant, where α is the attenuation constant along the line, and β is the phase constant. The attenuation constant is measured in nepers/meter (Np/m) in the SI unit system. However, decibels/meter (dB/m) is another commonly used unit for the attenuation, where the number of decibels/meter due to 1-Np attenuation is $20 \log_e = 8.69$ dB/m for voltage, i.e., $\alpha_{\text{dB}} = 8.69\alpha_{\text{Np}}$, and $10 \log_e = 4.34$ dB/m for power, i.e., $\alpha_{\text{dB}} = 4.34\alpha_{\text{Np}}$.

EXAMPLE 9.14

A transmission line operating at a frequency of 1.5 MHz has the following parameters: $R_\ell = 2.6\ \Omega/\text{m}$, $L_\ell = 0.82\ \mu\text{H/m}$, $G_\ell = 0$, $C_\ell = 22$ pF/m. Calculate the characteristic impedance, propagation constant, attenuation and phase constants, and the velocity of propagation.

Solution The per-unit-length series impedance of the line is calculated from equation (9.92) as

$$\hat{Z}_\ell = 2.6 + j2\pi \times 1.5 \times 10^6 \times 0.82 \times 10^{-6} = 2.6 + j7.73$$
$$= 8.16\underline{/71.41^\circ}\ \Omega$$

and the per-unit-length shunt admittance is determined as

$$\hat{Y}_\ell = j2\pi \times 1.5 \times 10^6 \times 22 \times 10^{-12} = j20.73 \times 10^{-5}\ \text{S/m}$$

Thus, the characteristic impedance and the propagation constant are obtained from (9.100) and (9.101), respectively, as

$$\hat{Z}_c = \sqrt{\frac{8.16\underline{/71.41^\circ}}{20.73 \times 10^{-5}\underline{/90^\circ}}} = 198.40\underline{/-9.3^\circ}\ \Omega$$

and

$$\hat{\gamma} = \sqrt{\hat{Z}_\ell \hat{Y}_\ell} = \sqrt{(8.16\underline{/71.41^\circ})(20.73 \times 10^{-5}\underline{/90^\circ})}$$
$$= 41.13 \times 10^{-3}\underline{/80.71^\circ} = 6.64 \times 10^{-3} + j40.59 \times 10^{-3}\ \text{m}^{-1}$$

Therefore, the attenuation constant is

$$\alpha = 6.64 \times 10^{-3}\ \text{Np/m} \quad \text{or} \quad \alpha = 0.0577\ \text{dB/m}$$

and the phase constant is

$$\beta = 40.59 \times 10^{-3}\ \text{rad/m}$$

The velocity of propagation can be calculated as

$$u_p = \frac{\omega}{\beta} = \frac{2\pi \times 1.5 \times 10^6}{40.59 \times 10^{-3}} = 2.322 \times 10^8 \text{ m/s}$$

• • •

9.8.2 Voltage and current relationships

As was done in (9.53) and (9.54), the voltage and current at any point along a transmission line can be expressed in terms of the sending-end voltage $\tilde{V}_S$ and the sending-end current $\tilde{I}_S$ as

$$\tilde{V}(z) = \frac{\tilde{V}_S + \hat{Z}_c\tilde{I}_S}{2} e^{-\hat{\gamma}z} + \frac{\tilde{V}_S - \hat{Z}_c\tilde{I}_S}{2} e^{\hat{\gamma}z} \tag{9.102}$$

and

$$\tilde{I}(z) = \frac{\tilde{V}_S + \hat{Z}_c\tilde{I}_S}{2\hat{Z}_c} e^{-\hat{\gamma}z} - \frac{\tilde{V}_S - \hat{Z}_c\tilde{I}_S}{2\hat{Z}_c} e^{\hat{\gamma}z} \tag{9.103}$$

After performing the necessary mathematical substitutions and simplifications in (9.102) and (9.103), we obtain

$$\tilde{V}(z) = \tilde{V}_S \cosh \hat{\gamma}z - \tilde{I}_S\hat{Z}_c \sinh \hat{\gamma}z \tag{9.104}$$

and

$$\tilde{I}(z) = -\frac{\tilde{V}_S}{\hat{Z}_c} \sinh \hat{\gamma}z + \tilde{I}_S \cosh \hat{\gamma}z \tag{9.105}$$

or in matrix form,

$$\begin{bmatrix} \tilde{V}(z) \\ \tilde{I}(z) \end{bmatrix} = \begin{bmatrix} \cosh \hat{\gamma}z & -\hat{Z}_c \sinh \hat{\gamma}z \\ -\dfrac{1}{\hat{Z}_c} \sinh \hat{\gamma}z & \cosh \hat{\gamma}z \end{bmatrix} \begin{bmatrix} \tilde{V}_S \\ \tilde{I}_S \end{bmatrix} \tag{9.106}$$

This equation helps us to determine the voltage and current at any location z along the transmission line when the voltage and current at the sending (input) end are known.

$\tilde{V}(z)$ and $\tilde{I}(z)$ can also be expressed in terms of $\tilde{V}_R$ and $\tilde{I}_R$ as

$$\begin{bmatrix} \tilde{V}(z) \\ \tilde{I}(z) \end{bmatrix} = \begin{bmatrix} \cosh \hat{\gamma}(\ell - z) & \hat{Z}_c \sinh \hat{\gamma}(\ell - z) \\ \dfrac{1}{\hat{Z}_c} \sinh \hat{\gamma}(\ell - z) & \cosh \hat{\gamma}(\ell - z) \end{bmatrix} \begin{bmatrix} \tilde{V}_R \\ \tilde{I}_R \end{bmatrix} \tag{9.107}$$

Using this equation, we can determine the voltage and current at any point along the transmission line when the voltage and current at the receiving end (load) are known.

The input impedance at any point in the transmission line, which is the ratio of $\tilde{V}(z)$ to $\tilde{I}(z)$, becomes

$$\hat{Z}_{\text{in}}(z) = \hat{Z}_c \left[\frac{\hat{Z}_L + \hat{Z}_c \tanh \hat{\gamma}(\ell - z)}{\hat{Z}_c + \hat{Z}_L \tanh \hat{\gamma}(\ell - z)} \right] \tag{9.108}$$

Strictly speaking, the concept of voltage standing wave ratio (VSWR) cannot be used for transmission lines with imperfect materials because the voltage maximum and minimum are not clearly defined. However, in transmission lines with a small loss, the VSWR still serves as a useful tool to evaluate the degree of mismatching.

EXAMPLE 9.15

A 1000-m-long communication line has the following per-unit-length parameters: $R_\ell = 22$ mΩ/m, $L_\ell = 0.63$ μH/m, $G_\ell = 0.1$ μS/m, $C_\ell =$ 31 pF/m. The resistive load at the receiving end of this line absorbs 10 W at 50 V (rms). Determine the sending-end voltage, current, and power for an operating frequency of 10 kHz.

Solution Let us first determine the current at the receiving end of the transmission line:

$$I_R = \frac{10}{50} = 0.2 \text{ A}$$

In phasor form, $\tilde{I}_R = 0.2\underline{/0^\circ}$ A and $\tilde{V}_R = 50\underline{/0^\circ}$ V. The series impedance and the shunt admittance of the line are

$$\begin{aligned}\hat{Z}_\ell &= R_\ell + j\omega L_\ell \\ &= 22 \times 10^{-3} + j2\pi \times 10 \times 10^3 \times 0.63 \times 10^{-6} \\ &= 4.53 \times 10^{-2}\underline{/60.95^\circ} \ \Omega/\text{m}\end{aligned}$$

and

$$\begin{aligned}\hat{Y}_\ell &= G_\ell + j\omega C_\ell \\ &= 10^{-7} + j2\pi \times 10 \times 10^3 \times 31 \times 10^{-12} \\ &= 10^{-7} + j1.948 \times 10^{-6} \text{ S/m} \\ &= 1.95 \times 10^{-6}\underline{/87.06^\circ} \text{ S/m}\end{aligned}$$

Then, the characteristic impedance and the propagation constant of this line can be calculated as

$$\hat{Z}_c = \sqrt{\frac{4.53 \times 10^{-2}\underline{/60.95^\circ}}{1.95 \times 10^{-6}\underline{/87.06^\circ}}} = 152.42\underline{/-13.06^\circ} \ \Omega$$

and

$$\begin{aligned}\hat{\gamma} &= \sqrt{(4.53 \times 10^{-2}\underline{/60.95^\circ})(1.95 \times 10^{-6}\underline{/87.06^\circ})} \\ &= 81.89 \times 10^{-6} + j285.69 \times 10^{-6} \text{ m}^{-1}\end{aligned}$$

Using (9.107) with $z = 0$, we obtain the sending-end voltage as

$$\begin{aligned}\tilde{V}_S &= (50\underline{/0^\circ}) \cosh(81.89 \times 10^{-6} \times 10^3 + j285.69 \times 10^{-6} \times 10^3) \\ &\quad + (152.42\underline{/-13.06^\circ})(0.2\underline{/0^\circ}) \sinh(81.89 \times 10^{-6} \times 10^3 \\ &\quad + j285.69 \times 10^{-6} \times 10^3) \\ &= 53.19\underline{/9.75^\circ} \text{ V}\end{aligned}$$

and the sending-end current as

$$\tilde{I}_S = \frac{50\angle 0^\circ}{152.42\angle -13.06^\circ} \sinh(81.89 \times 10^{-6} \times 10^3 + j285.69 \times 10^{-6} \times 10^3) + (0.2\angle 0^\circ) \cosh(81.89 \times 10^{-6} \times 10^3 + j285.69 \times 10^{-6} \times 10^3) = 0.221\angle 27.14^\circ \text{ A}$$

The sending-end power can be calculated as

$$P_S = \text{Re}\,[\tilde{V}_S \tilde{I}_S^*] = \text{Re}\,[(53.19\angle 9.75^\circ)(0.221\angle -27.14^\circ)] = 11.22 \text{ W}$$

Note that 1.22 W is the power dissipated by the transmission line.

• • •

9.9 Transients in transmission lines

Thus far we have examined the behavior of a transmission line at steady state with a sinusoidal excitation. However, there are cases when transmission lines are subjected to sudden changes in the input voltage and/or current waveforms. We will refer to such impulses as *transient waveforms*. Some of these waveforms are due to natural causes; others are man-made. For example, lightning, as simulated by an impulse of very short duration in Figure 9.25, may strike an overhead power transmission line, causing an overvoltage as it propagates along this line. Such sudden changes in the voltages and currents on the transmission line can adversely affect not only the transmission line but also other pertinent equipment unless protective measures are built into the system. In order to be able to predict and then reduce the adverse effects of lightning, we must evaluate the behavior of a transmission line when subjected to such impulses.

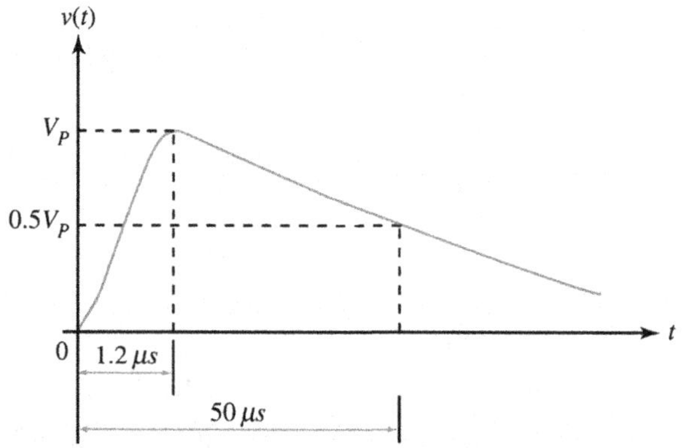

Figure 9.25 Standard lightning impulse voltage waveform as specified by International Electrotechnical Commission (IEC)

As another example, the transmission lines used in communication networks are routinely subjected to sudden changes in voltage: the "on" and "off" pulses that are necessary for the transmission of a signal. In this case, the design engineer must evaluate whether or not the pulse code, which results in transients along the transmission line, is successfully transmitted from one end to the other. Transient waveforms are also a natural outcome when a transmission line is connected to or disconnected from a source. For all these reasons, we devote this section to the study of transients on transmission lines. We will examine the behavior of a lossless transmission line when subjected to short and long pulses.

9.9.1 Transmission line equations in the time domain

The general equations for a lossless transmission line can be written in the time domain as

$$\frac{\partial V(z,t)}{\partial z} = -L_\ell \frac{\partial I(z,t)}{\partial t} \tag{9.109}$$

and

$$\frac{\partial I(z,t)}{\partial z} = -C_\ell \frac{\partial V(z,t)}{\partial t} \tag{9.110}$$

After some mathematical manipulations, we obtain the following wave equations:

$$\frac{\partial^2 V(z,t)}{\partial z^2} = L_\ell C_\ell \frac{\partial^2 V(z,t)}{\partial t^2} \tag{9.111}$$

and

$$\frac{\partial^2 I(z,t)}{\partial z^2} = L_\ell C_\ell \frac{\partial^2 I(z,t)}{\partial t^2} \tag{9.112}$$

We can write the general solution of the voltage wave equation by following a procedure similar to that in Chapter 8 to obtain

$$V(z,t) = V^+(z - u_p t) + V^-(z + u_p t) \tag{9.113}$$

and

$$I(z,t) = I^+(z - u_p t) - I^-(z + u_p t) \tag{9.114}$$

where V^+ and V^- are arbitrary functions for the forward- and backward-traveling voltage waves, respectively, and $u_p = 1/\sqrt{L_\ell C_\ell}$ is the phase velocity of the wave.

Substituting (9.113) in (9.110), we obtain the solution of (9.112) as

$$I(z,t) = \frac{V^+(z - u_p t)}{R_c} - \frac{V^-(z + u_p t)}{R_c} \tag{9.115}$$

where $R_c = \sqrt{L_\ell / C_\ell}$ is simply the characteristic resistance of the lossless transmission line.

9.9.2 Transient response of a lossless transmission line

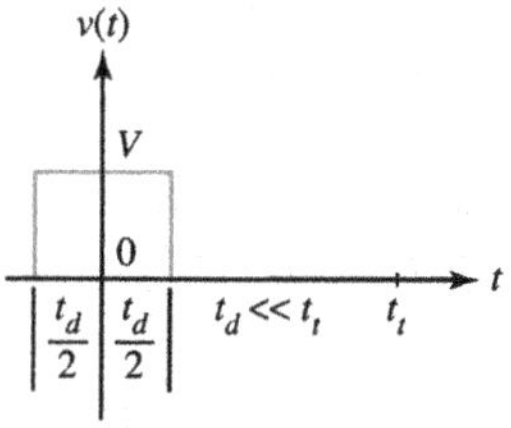

Figure 9.26 Impulse function

Before determining the transient response of a lossless transmission line, let us first clarify what we mean by a short pulse and a long pulse. We define a **short pulse** as one having a pulse duration much shorter than the transit time of the wave propagating along the transmission line (see Figure 9.26). Such a pulse will be treated as an impulse function. A **long pulse** has a pulse duration much longer than the transit time of the traveling wave (see Figure 9.27). This pulse will be considered as a step function.

Impulse response

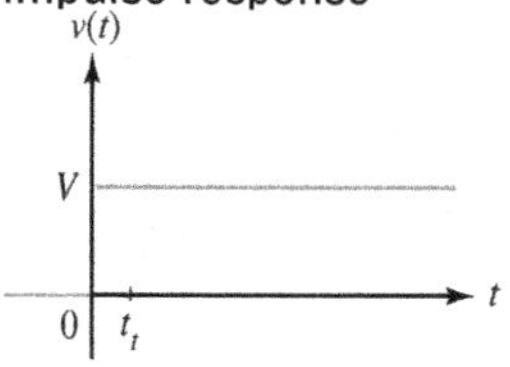

Figure 9.27 Step function

Let us consider a lossless transmission line of length ℓ, energized from a voltage source having an internal resistance R_G, and terminated into a resistive load R_L. (See Figure 9.28.) Let us also assume that there is no voltage or current on the transmission line for $t < 0$. At $t = 0$, a voltage pulse of very short duration is applied to the transmission line. Because of the finite length of the line and the velocity of propagation of the wave, the pulse will arrive at the other end in one transit time, t_t. A part of the incoming wave at the receiving end is reflected due to the mismatch of the load to the transmission line. It will take one more transit time for the reflected traveling wave to arrive at the sending end. Thus, for $0 < t < 2t_t$, the voltage and current will be zero at the sending end. However, at $t = 0$ (see Figure 9.29a) the voltage, current, and power at the sending end are, respectively,

$$V_S(0) = V^+ = \frac{R_c}{R_G + R_c} V_G = V \tag{9.116}$$

$$I_S(0) = I^+ = \frac{V_G}{R_G + R_c} = I \tag{9.117}$$

$$P_S(0) = P^+ = V^+ I^+ = VI = P \tag{9.118}$$

After one-half transit time the wave reaches the midpoint of the line with

Figure 9.28 Lossless transmission line subjected to an impulse voltage

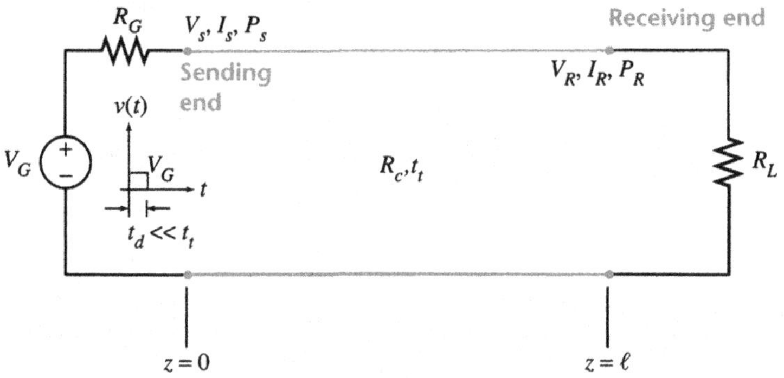

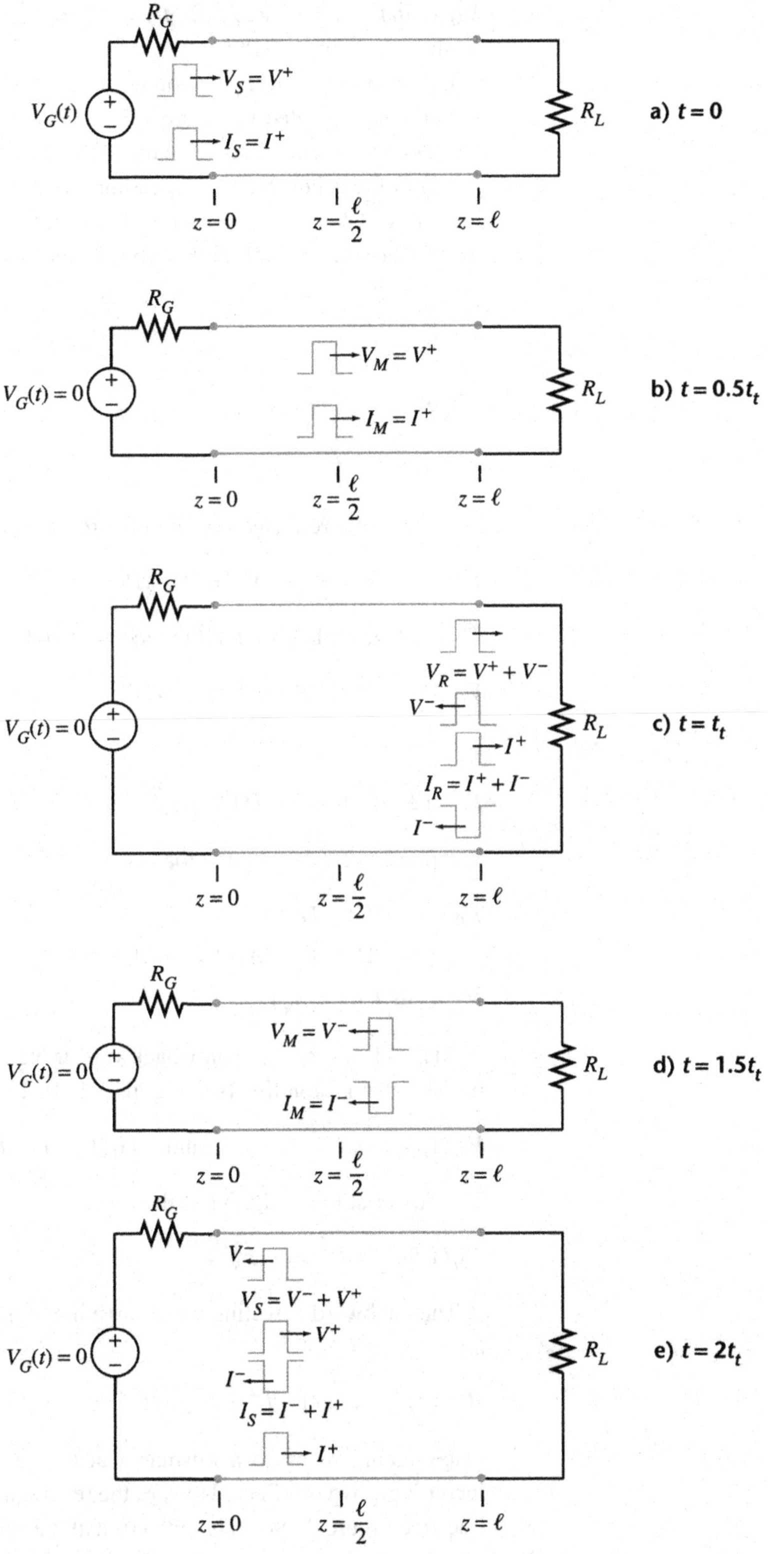

Figure 9.29 Impulse waveforms travelling along a transmission line

$V_M(0.5t_t) = V^+ = V$, $I_M(0.5t_t) = I^+ = I$, and $P_M(0.5t_t) = P^+ = P$, as shown in Figure 9.29b.

One transit time later the voltage and current waves reach the receiving end, as indicated in Figure 9.29c. Because the transmission line is lossless, the incoming power at the receiving end is the same as the input power at the sending end. Depending upon the load resistance R_L, the incoming voltage and current at the receiving end are partly reflected toward the source. Such a reflection produces backward-traveling waves as

$$V^- = \rho_R V^+ = \rho_R V \quad \text{and} \quad I^- = -\rho_R I^+ = -\rho_R I$$

where

$$\rho_R = \frac{R_L - R_c}{R_L + R_c}$$

Therefore, the reflected power at the receiving end becomes

$$P^- = V^- I^- = -\rho_R^2 VI = -\rho_R^2 P$$

The voltage and current at the receiving end (load) at $t = t_t$ are

$$V_R(t_t) = V^+ + V^- = V(1 + \rho_R)$$

and

$$I_R(t_t) = I^+ + I^- = I(1 - \rho_R)$$

The power absorbed by the load is

$$\begin{aligned} P_R(t_t) &= V_R(t_t) I_R(t_t) \\ &= [V(1 + \rho_R)][I(1 - \rho_R)] \\ &= P\left(1 - \rho_R^2\right) \end{aligned}$$

At $t = 1.5t_t$, there are only backward-traveling waves at the midpoint of the transmission line (see Figure 9.29d). They are

$$V_M(1.5t_t) = V^- = \rho_R V \quad \text{and} \quad I_M(1.5t_t) = I^- = -\rho_R I$$

The power at the midpoint of the line at $t = 1.5t_t$ is

$$P_M(1.5t_t) = P^- = -\rho_R^2 VI$$

The backward-traveling waves arriving at the sending end at $t = 2t_t$ are

$$V^- = \rho_R V \quad \text{and} \quad I^- = -\rho_R I$$

Once again, owing to a mismatch at the sending end, a part of the incoming wave is reflected toward the receiving end of the transmission line (see Figure 9.29e). The reflection at the sending end produces new forward waves as

$$V^+ = \rho_S V^- = \rho_S \rho_R V \quad \text{and} \quad I^+ = -\rho_S I^- = \rho_S \rho_R I$$

where

$$\rho_S = \frac{R_G - R_c}{R_G + R_c}$$

Thus the power in the forward-traveling wave at the sending end is

$$P^+ = \rho_S^2 \rho_R^2 P$$

The voltage and current at the sending end at $t = 2t_t$ are

$$V_S(2t_t) = V^+ + V^- = \rho_S \rho_R V + \rho_R V = \rho_R(\rho_S + 1)V$$

and

$$I_S(2t_t) = I^+ + I^- = \rho_S \rho_R I - \rho_R I = \rho_R(\rho_S - 1)I$$

The power dissipation in the internal resistance of the source is

$$P_S(2t_t) = VI(\rho_S^2 \rho_R^2 - \rho_R^2) = P(\rho_S^2 \rho_R^2 - \rho_R^2) = \rho_R^2(\rho_S^2 - 1)P$$

Because this is a lossless line, the reflections at the sending and receiving ends keep on occurring until the input power is completely dissipated by the two resistances R_L and R_G.

From the preceding discussion we can write general equations for the voltage, current, and power for a short pulse as follows:

At $z = 0$, for $t > 0$

$$V_S(2nt_t) = \rho_S^{n-1} \rho_R^n (1 + \rho_S)V \tag{9.119a}$$

$$I_S(2nt_t) = \rho_S^{n-1} \rho_R^n (\rho_S - 1)I \tag{9.119b}$$

$$P_S(2nt_t) = \left(\rho_S^{n-1} \rho_R^n\right)^2 \left(\rho_S^2 - 1\right)P \tag{9.119c}$$

where $n = 1, 2, 3, \ldots$.

At $z = \ell$, for $t = 0$, $V_R(0) = 0$, $I_R(0) = 0$, $P_R(0) = 0$.

At $z = \ell$, for $t > 0$

$$V_R[(2n-1)t_t] = \rho_S^{n-1} \rho_R^{n-1} (1 + \rho_R)V \tag{9.120a}$$

$$I_R[(2n-1)t_t] = \rho_S^{n-1} \rho_R^{n-1} (1 - \rho_R)I \tag{9.120b}$$

$$P_R[(2n-1)t_t] = \left(\rho_S^{n-1} \rho_R^{n-1}\right)^2 \left(1 - \rho_R^2\right)P \tag{9.120c}$$

where $n = 1, 2, 3, \ldots$.

Step response

To examine the behavior of a transmission line resulting from a long pulse, let us consider the line shown in Figure 9.30. In this case, the source voltage is

$$v_G(t) = \begin{cases} 0 & t < 0 \\ V_G & t > 0 \end{cases}$$

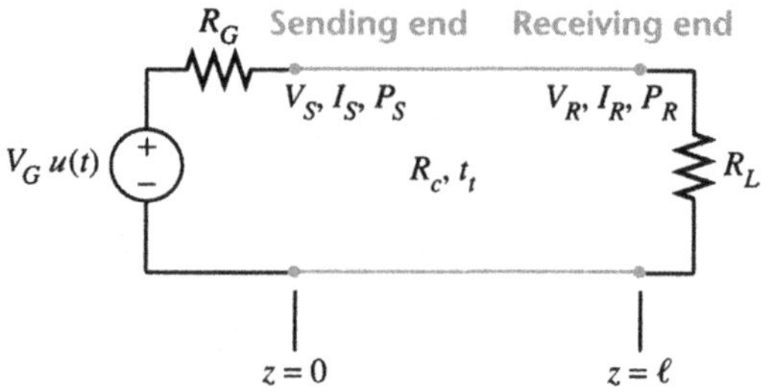

Figure 9.30 Transmission line with load R_L subjected to a step function

The voltage and the current at the sending end are

$$V_S(t) = V^+ = \frac{R_c}{R_G + R_c} V_G = V \tag{9.121}$$

$$I_S(t) = I^+ = \frac{V_G}{R_G + R_c} = I \tag{9.122}$$

at $t = 0$.

At $t = 0$ the forward voltage and current waves begin propagating along the transmission line and reach the receiving end (load) at $t = t_t$, as shown in Figure 9.31. For $t < t_t$, the voltage and current at the load are zero. At $t = t_t$, the mismatch of the load to the transmission line causes reflections of the incoming voltage and current waves.

$$V^- = \rho_R V \quad \text{and} \quad I^- = -\rho_R I$$

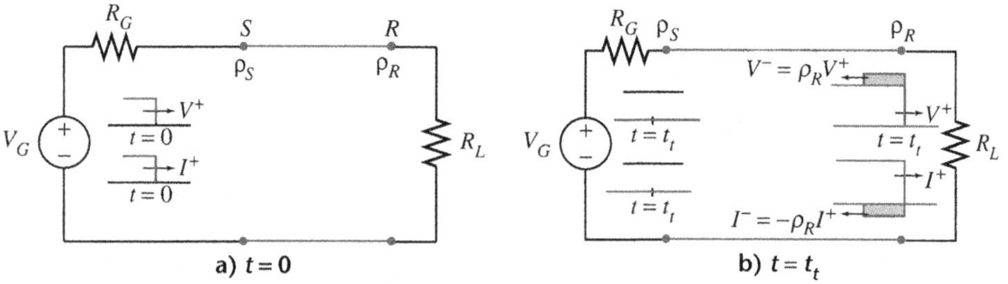

Figure 9.31 Voltages and currents on the line at $t = 0$ and $t = t_t$

where ρ_R is the reflection coefficient at the receiving end. Thus, for $t_t < t < 2t_t$, the voltage and the current at any distance z along the transmission line are expressed as

$$V(z) = V + \rho_R V \quad \text{and} \quad I(z) = I - \rho_R I \tag{9.123a}$$

as shown in Figure 9.31b.

From now on let us concentrate on the voltage waveform; we can use a similar process to account for the current waveform. At $t = 2t_t$, the reflected wave reaches the sending end, where it may be reflected back toward the load if there is a mismatch at the sending end. Due to the mismatch, if there were any, the strength of the voltage wave would be

$\rho_S\rho_R V$. Thus, for $2t_t < t < 3t_t$, the voltage at any distance z along the line is

$$V(z) = V + \rho_R V + \rho_S\rho_R V \tag{9.123b}$$

At $t = 3t_t$, the reflected wave at the sending end reaches the load, where it undergoes another reflection. This wave travels toward the source end with a strength of $\rho_S\rho_R^2 V$. The total voltage at a distance z along the line for $3t_t < t < 4t_t$ builds up to

$$V(z) = V + \rho_R V + \rho_S\rho_R V + \rho_S\rho_R^2 V \tag{9.123c}$$

At $t = 4t_t$, the reflected wave appears at the sending end and encounters another reflection. This wave has a strength of $\rho_S^2\rho_R^2 V$ and as it reaches the load it is reflected back toward the sending end with a strength of $\rho_S^2\rho_R^3 V$. This process continues until the steady state is reached. During this process the voltage along the transmission line at a distance of z from the sending end is

$$\begin{aligned} V(z) &= V + \rho_R V + \rho_S\rho_R V + \rho_S\rho_R^2 V + \rho_S^2\rho_R^3 V + \cdots \\ &= V(1+\rho_R)\left[1 + \rho_S\rho_R V + \rho_S^2\rho_R^2 V + \cdots\right] \end{aligned} \tag{9.123d}$$

As long as $\rho_S\rho_R < 1$, the infinite series in equation (9.123d) becomes

$$1 + \rho_S\rho_R + \rho_S^2\rho_R^2 + \cdots = \frac{1}{1-\rho_S\rho_R}$$

Thus, when $t \to \infty$,

$$V(z) = V\left(\frac{1+\rho_R}{1-\rho_S\rho_R}\right) \tag{9.124a}$$

We can obtain a similar expression for the current as

$$I(z) = I\left(\frac{1-\rho_R}{1-\rho_S\rho_R}\right) \tag{9.124b}$$

Substituting

$$\rho_R = \frac{R_L - R_c}{R_L + R_c} \quad \text{and} \quad \rho_S = \frac{R_G - R_c}{R_G + R_c}$$

in (9.124a) and (9.124b), we obtain

$$V(z) = \left[\frac{R_L(R_c + R_L)}{R_c(R_L + R_G)}\right] V$$

and

$$I(z) = \left(\frac{R_c + R_G}{R_L + R_G}\right) I$$

Finally, when V and I are expressed in terms of V_G, we get

$$V(z) = \left(\frac{R_L}{R_L + R_G}\right) V_G \tag{9.125}$$

and

$$I(z) = \frac{V_G}{R_L + R_G} \tag{9.126}$$

These are the expressions that we obtain without a transmission line between the load and the source. Thus, a transmission line connected to a dc source appears as a short circuit when the transients subside and the steady-state condition is reached.

9.9.3 Lattice diagrams

The determination of voltage and current along a transmission line in the time domain is somewhat involved because we have to keep track of the forward- and backward-traveling waves at different points and times along the transmission line. A very useful technique for tracking the forward and reflected waves is the *lattice diagram or bounce diagram.* A **lattice diagram** is a time–distance diagram; the horizontal axis is for the distance along the line, and the vertical axis is for the time.

In order to explain the development of a lattice diagram, let us assume that a dc voltage source of strength V_G with an internal resistance of R_G is connected to a lossless transmission line at $t = 0$. The characteristic resistance of the lossless line is R_c, and the other end of the line is terminated into a load resistance R_L, as shown in Figure 9.32a.

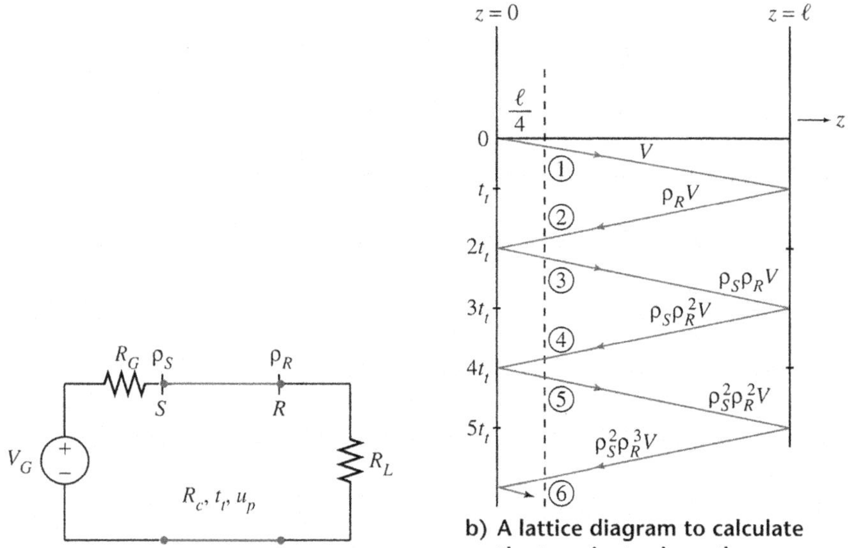

a) A transmission line connects a source to a load

b) A lattice diagram to calculate the transients along the transmission line shown in (a)

Figure 9.32 Transmission line and its lattice diagram

At $t = 0$, the applied voltage across the line at $z = 0$ is

$$V = \frac{R_c}{R_G + R_c} V_G$$

As time progresses, the voltage wave of strength V travels along the transmission line of length ℓ with a velocity of u_p and reaches the load end at $t = t_t$. This wave is shown by arrow 1 in Figure 9.32b. If ρ_R is the reflection coefficient at the load end, a part of the incoming wave at the load is reflected back toward the source. This wave is shown by arrow 2 in the figure and has a magnitude of $\rho_R V$. At $t = 2t_t$, the reflected wave arrives at the source end. If ρ_S is the reflection coefficient at the source end, then a part of the incoming wave of magnitude $\rho_R V$ is reflected toward the load. The strength of this reflected wave is $\rho_S \rho_R V$, and it is indicated by arrow 3 in the lattice diagram. At $t = 3t_t$, this wave reaches the load and creates another wave of strength $\rho_S \rho_R^2 V$, which then travels toward the source, as indicated by arrow 4. We have continued this technique to draw two more reflections in Figure 9.32b. The process can, however, be continued forever.

Let us examine the voltage buildup at some point on the transmission line, say at $z = \ell/4$. The voltage at this point is zero until $t = 0.25t_t$, when the forward voltage wave of strength V arrives at the observation point. At that time the voltage becomes V and it stays at that level until $t = 1.75t_t$, when the reflected wave as given by arrow 2 reaches the point and the voltage becomes $V + \rho_R V$. The voltage stays at this level until the reflected wave as given by arrow 3 reaches the observation point at $z = \ell/4$ at $t = 2.25t_t$. The voltage now rises to $V + \rho_R V + \rho_S \rho_R V$ and remains the same until $t = 3.75t_t$, when the reflected wave given by arrow 4 reaches the point. The total voltage is now $V + \rho_R V + \rho_S \rho_R V + \rho_S \rho_R^2 V$. Such a voltage buildup at the observation point at $z = \ell/4$ is illustrated in Figure 9.33.

When we continue the process for a long time, the voltage buildup at the point of observation on the line would be

$$V(z) = V + \rho_R V + \rho_S \rho_R V + \rho_S \rho_R^2 V + \rho_S^2 \rho_R^3 V + \cdots$$

This equation is the same as (9.123d), but its development has been made much easier by the lattice diagram. We can also draw a lattice

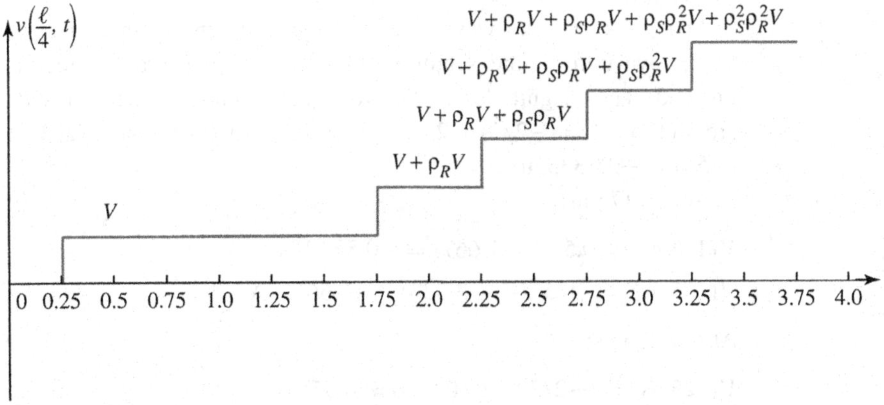

Figure 9.33 Voltage buildup as a function of time at $z = \frac{\ell}{4}$

diagram for the current on the transmission line in a similar fashion. We use the following example to illustrate the buildup of the voltage and current waveforms using lattice diagrams.

EXAMPLE 9.16

A 50-Ω transmission line having a transit time of 50 ns connects a pulse generator to a 150-Ω load resistance. The internal resistance of the pulse generator is 10 Ω, and its voltage magnitude is 10 V at no load. Determine the variation of the voltage at the midpoint of the line for a duration of $5t_t$ if the pulse width is 1 ns.

Solution The reflection coefficients at the receiving and sending ends are, respectively,

$$\rho_R = \frac{R_L - R_c}{R_L + R_c} = \frac{150 - 50}{150 + 50} = 0.5$$

and

$$\rho_S = \frac{R_S - R_c}{R_S + R_c} = \frac{10 - 50}{10 + 50} = -\frac{2}{3}$$

Because the pulse duration (1 ns) is substantially shorter than the transit time of the transmission line ($t_t = 50$ ns), we will treat the applied pulse as an impulse.

From equations (9.116) and (9.117), the voltage and the current at the sending end at $t = 0$ are

$$V_S(0) = \left(\frac{50}{10 + 50}\right) 6 = 5 \text{ V}$$

$$I_S(0) = \frac{6}{10 + 50} = 0.1 \text{ A} \quad \text{or} \quad 100 \text{ mA}$$

The impulse voltage and current waves traveling along the line reach the midpoint at $t = 25$ ns. The strength of each wave is the same as that at $t = 0$. Before and after $t = 25$ ns, both the voltage and current are zero at the midpoint. At $t = 75$ ns, the reflected wave from the receiving end arrives at the midpoint with $V(75 \text{ ns}) = 0.5 \times 5 = 2.5$ V and $I(75 \text{ ns}) = -0.5 \times 100 = -50$ mA.

At $t = 125$ ns, the wave occurs at the midpoint again as the reflected wave from the sending end propagates toward the receiving end. The magnitudes of the voltage and current at this instant are $V(125 \text{ ns}) = (-2/3) \times 2.5 = -1.667$ V and $I(125 \text{ ns}) = (2/3) \times (-50) = -33.333$ mA.

At $t = 175$ ns:

$$V(175 \text{ ns}) = 0.5 \times (-1.667) = -0.833 \text{ V}$$

$$I(175 \text{ ns}) = -0.5 \times (-33.333) = 16.667 \text{ mA}$$

At $t = 225$ ns:

$$V(225 \text{ ns}) = (-2/3) \times (-0.833) = 0.555 \text{ V}$$

$$I(225 \text{ ns}) = (2/3) \times (16.667) = 11.111 \text{ mA}$$

At $t = 275$ ns:

$$V(275 \text{ ns}) = 0.5 \times 0.555 = 0.278 \text{ V}$$

$$I(275 \text{ ns}) = -0.5 \times 11.111 = -5.556$$

Figure 9.34 shows the voltage and current impulses in time at the midpoint of the transmission line. As can be noted from the graph, after $5t_t$ the voltage and current decayed significantly from their initial values.

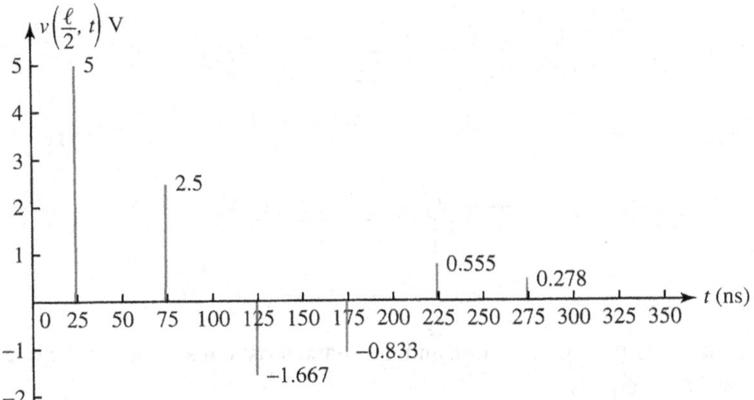

a) Variation of the voltage at $z = \frac{\ell}{2}$

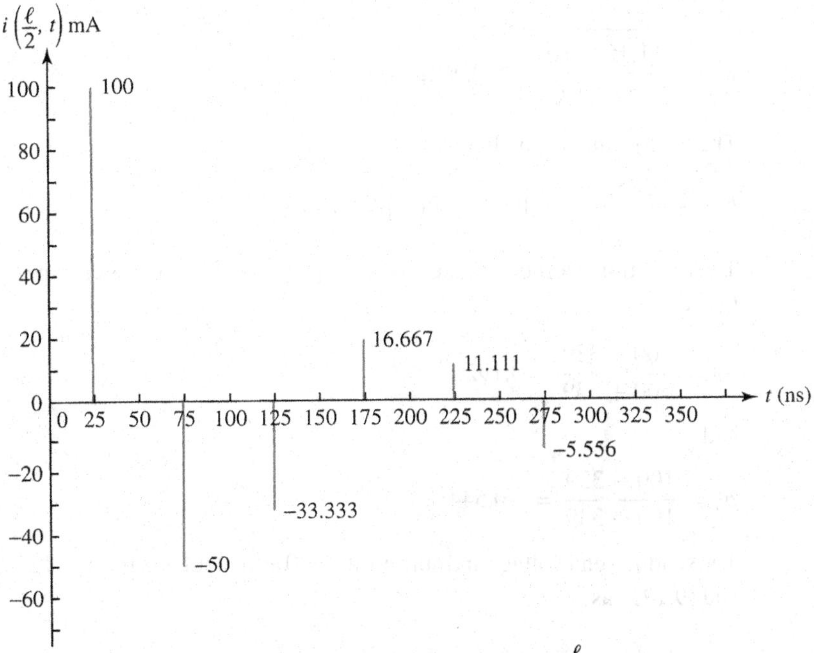

b) Variation of the current at $z = \frac{\ell}{2}$

Figure 9.34 Voltage and current impulses in time

Ultimately, the load and source resistances dissipate the energy from the waves, leading to zero voltage and current on the line. • • •

EXAMPLE 9.17

A 2950-m-long, lossless telephone line is subjected to a 24-V $u(t)$ source having an internal resistance of 100 Ω, as shown in Figure 9.35. The per-unit-length inductance and capacitance of the line are 1.15 μH/m and 10 pF/m, respectively. Sketch the voltage and current waveforms as a function of time at the midpoint when the transmission line is terminated into a load resistance of 500 Ω.

Figure 9.35

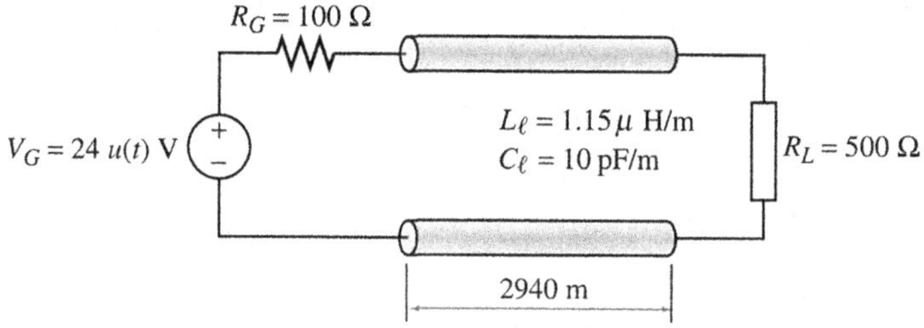

Solution The velocity of propagation and the characteristic resistance of the line are, respectively,

$$u_p = \sqrt{\frac{1}{1.15 \times 10^{-6} \times 10 \times 10^{-12}}} = 2.95 \times 10^8 \text{ m/s}$$

and

$$R_c = \sqrt{\frac{1.15 \times 10^{-6}}{10 \times 10^{-12}}} \approx 339\ \Omega$$

The transit time of the line is

$$t_t = \frac{2950}{2.95 \times 10^8} = 10^{-5} \text{ s} \quad \text{or} \quad t_t = 10\ \mu\text{s}$$

The reflection coefficients at the receiving and sending ends are, respectively,

$$\rho_R = \frac{500 - 339}{500 + 339} = 0.192$$

and

$$\rho_S = \frac{100 - 339}{100 + 339} = -0.544$$

The sending-end voltage and current at $t = 0$ are calculated from (9.121) and (9.122) as

$$V_S(0) = \left(\frac{24}{100 + 339}\right) 339 = 18.533 \text{ V}$$

and

$$I_S(0) = \frac{24}{100 + 339} = 54.67 \times 10^{-3}\ \text{A} \quad \text{or} \quad 54.67\ \text{mA}$$

The corresponding lattice diagram for the voltage is constructed in Figure 9.36. From the lattice diagram the voltage waveform at the midpoint can be plotted as follows. Until 5 μs no wave has arrived at the midpoint of the line, as shown in Figure 9.36. At $t = 5$ μs, a forward-traveling wave with a magnitude of 18.533 V reaches the midpoint of the line. The voltage at the midpoint remains at 18.533 V until $t = 15$ μs. At $t = 15$ μs, a reflected wave from the receiving end reaches the midpoint of the line with a magnitude of 3.558 V. Now, the voltage at $z = \ell/2$ becomes 22.091 V. From the lattice diagram, there are no more changes in voltage at the midpoint of the line until $t = 25$ μs. At $t = 25$ μs, a voltage wave of -1.936 V reaches the midpoint from the sending end of the line, and causes the voltage to become 20.155 V. At $t = 35$ μs, the voltage at the midpoint becomes 19.783 V. The reflections continue until the steady state is achieved, as can be seen from Figure 9.37. At that time the expected value of the voltage is

$$V(\infty) = \left(\frac{24}{100 + 500}\right) 500 = 20\ \text{V}$$

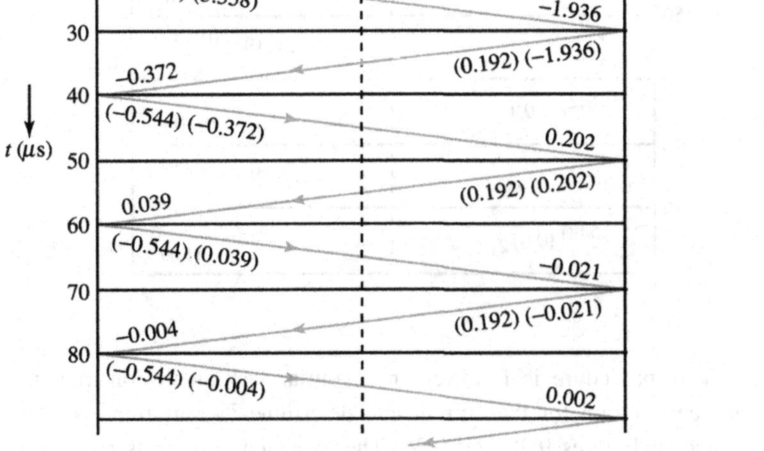

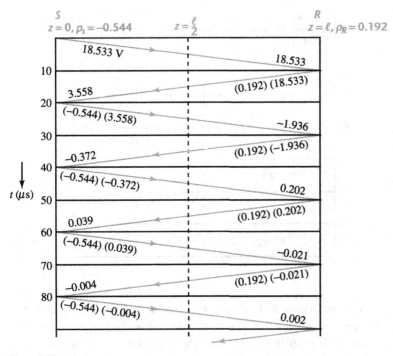

Figure 9.36

Figure 9.37

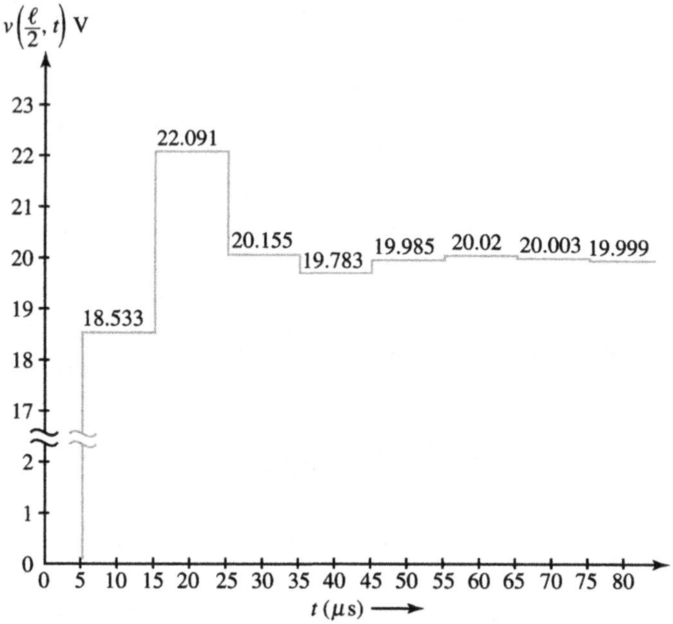

Figure 9.38 Lattice diagram for current wave for Example 9.17

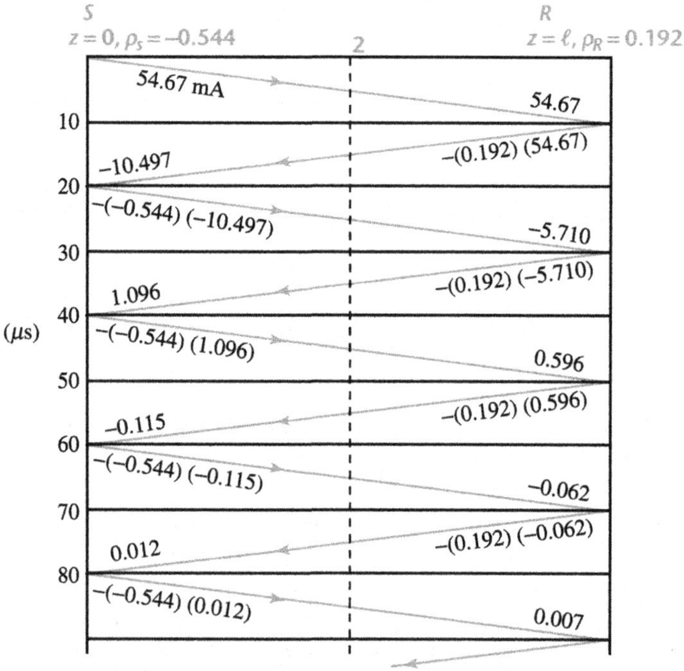

This procedure is followed in a similar fashion to construct the lattice diagram for the current and determine its variation, as illustrated in Figures 9.38 and 9.39. The expected current as $t \to \infty$ is 40 mA.

• • •

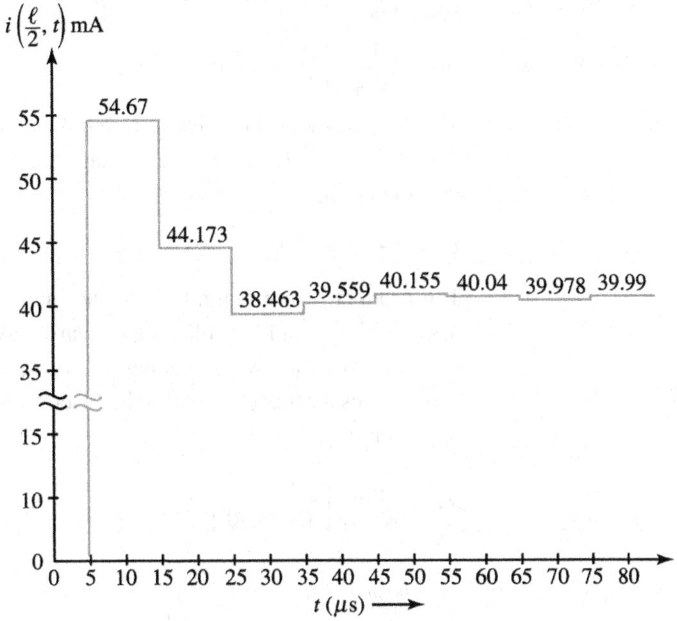

Figure 9.39 Variation of current for Example 9.17

EXAMPLE 9.18

An 800-m-long, overhead transmission line having a transit time of 4 μs is connected to a 400-m-long cable with a transit time of 2 μs at point A, as shown in Figure 9.40. The characteristic resistances of the overhead line and cable are 200 Ω and 50 Ω, respectively. The receiving end of the cable is open-circuited, and the internal resistance of the source is assumed to be zero. Determine the variation of the voltage for a duration of 30 μs at the receiving end of the cable when the overhead line is subjected to a unit step voltage.

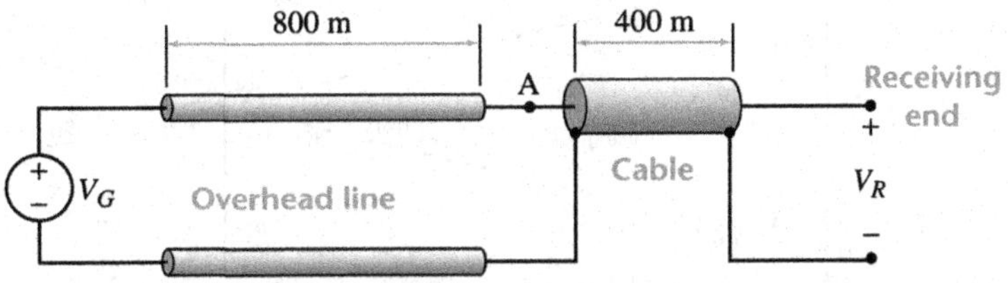

Figure 9.40 The transmission lines for Example 9.18

Solution The reflection coefficient at the sending end toward the source:

$$\rho_S = \frac{0-200}{0+200} = -1$$

As the forward wave arrives at point A, it experiences a reflection toward the source due to the mismatch between the two transmission lines. Thus, the reflection coefficient of the wave that is reflected toward the source is

$$\rho_1 = \frac{50 - 200}{50 + 200} = -0.6$$

The remainder of the wave at point A becomes an incidence wave in the 50-Ω cable. To determine its magnitude, we can compute the transmission coefficient at A as

$$\tau_1 = 1 - 0.6 = 0.4$$

The product of the magnitude of the incoming wave on the overhead line at point A and τ_1 yields the forward wave in the 50-Ω cable.

As the forward wave reaches the open-circuit end of the cable, it experiences a reflection due to the discontinuity. Thus, the reflection at the receiving end is

$$\rho_R = \lim_{R_L \to \infty} \left[\frac{R_L - 50}{R_L + 50} \right] = 1$$

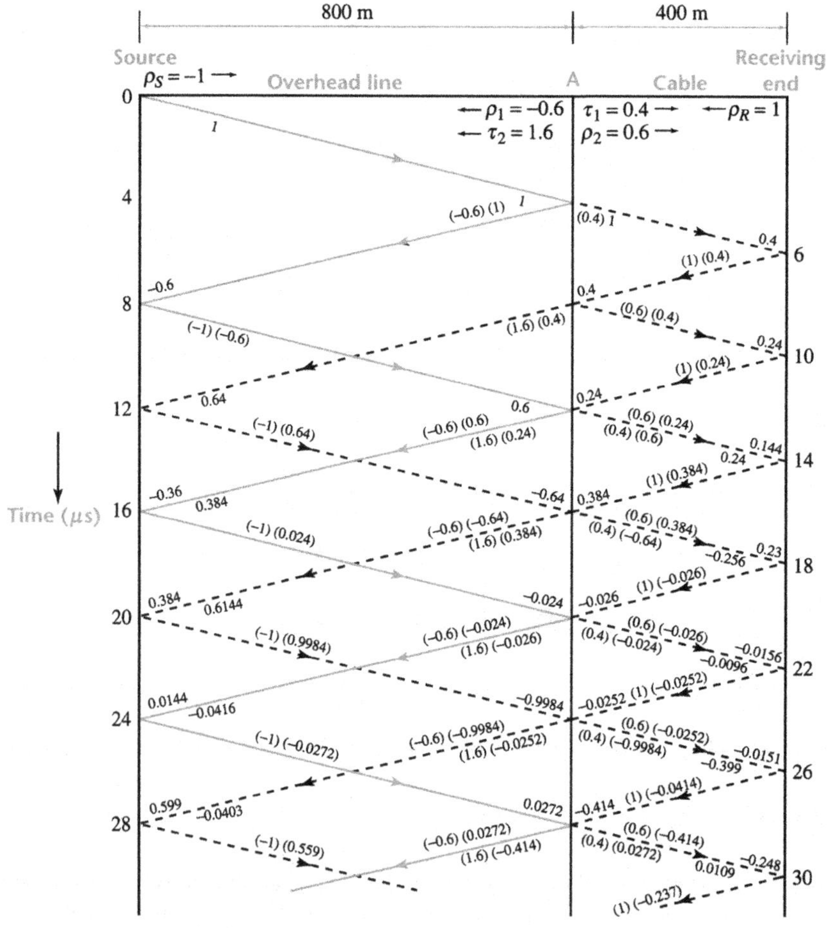

Figure 9.41 (Example 9.18) Lattice diagram for the reflections and transmissions along the overhead line and the cable. Numbers on the lines are for the reflected waves. Numbers under the lines are for transmitted waves

As the reflected wave in the 50-Ω cable arrives at point A, it also experiences a reflection and heads toward the receiving end. The reflection coefficient of this wave is

$$\rho_2 = \frac{200 - 50}{200 + 50} = 0.6$$

The remainder of the wave at point A becomes an incident wave for the overhead transmission line. To determine its magnitude we can calculate the transmission coefficient at A as

$$\tau_2 = 1 + 0.6 = 1.6$$

The details of the lattice diagram for the analysis are given in Figure 9.41. From this lattice diagram the variation of the voltage at the receiving end is obtained, as shown in Figure 9.42.

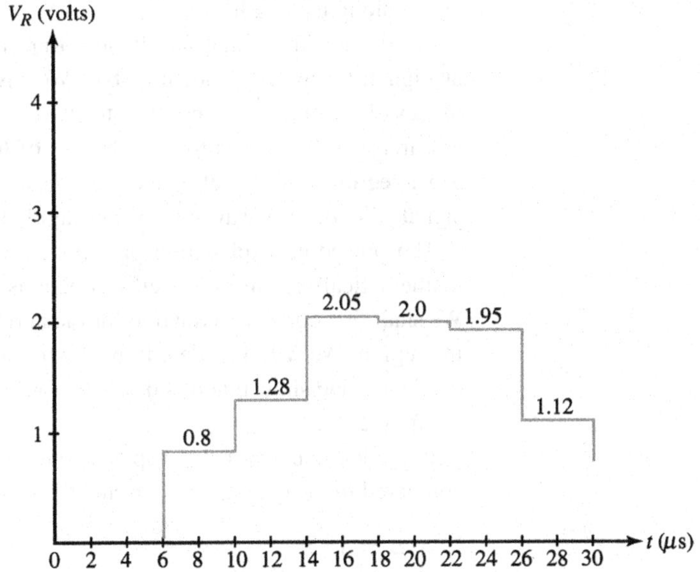

Figure 9.42 Variation of receiving-end voltage as a function of time (Example 9.18)

• • •

9.10 Skin effect and resistance

In calculating the resistance and inductance of a conductor, we may assume that the current distributes uniformly over its cross section. Strictly speaking, this assumption is valid only when the rate of change of the current (di/dt) in the conductor is zero. Stated differently, a conductor can support a uniform current density when it carries a direct current (dc). The current distribution can still be taken as uniform as long as the rate of change of the current is small. This premise becomes especially convincing for thin conductors operating at low frequencies.

However, the condition of nonuniformity is highly accentuated when the rate of change of the current becomes very large, as in the case of a high-frequency circuit. It is the magnetic field created by a high-frequency current in a conductor that induces the largest emf in the central region of the conductor. Since the induced emf produces an induced current in a closed circuit, the induced current is the largest at the center of the conductor. Because the induced current is always in a direction to reduce the original current, it forces the current to confine itself near the outer surface of the conductor. Thus, the inner region becomes practically devoid of any current. The concentration of the current in a thin layer next to the outer surface of the conductor results in an increase in its resistance. As the resistance of a conductor increases, so does the power loss in that conductor. This phenomenon is called the *skin effect*. It is the skin effect that makes a wire-type transmission line very inefficient at high (microwave) frequencies because of the high attenuation of a signal along its length.

We do not imply that an efficient transmission of a guided signal at high frequencies is not possible. We just have to search for new modes of transmission other than the wire-type transmission line. Such a search has steered us toward the use of hollow conductors, known as **waveguides**, for the effective transmission of high-frequency (GHz) signals. We discuss waveguides thoroughly in Chapter 10.

The preceding explanation on the skin effect can be substantiated mathematically using Maxwell's equations in a conducting medium. We shall first derive an equation for the current density in the conducting region. We will then determine the resistance of a current-carrying conductor and show that it is directly proportional to the square root of the frequency.

In a good conductor the displacement current is usually very small compared to the conduction current. Thus, we can write

$$\nabla \times \tilde{\mathbf{H}} = \tilde{\mathbf{J}} \tag{9.127}$$

and from Faraday's law

$$\nabla \times \tilde{\mathbf{E}} = -j\omega\mu\tilde{\mathbf{H}} \tag{9.128}$$

The curl of (9.128) yields

$$\nabla \times \nabla \times \tilde{\mathbf{E}} = -j\omega\mu\nabla \times \tilde{\mathbf{H}} \tag{9.129}$$

or

$$\nabla(\nabla \cdot \tilde{\mathbf{E}}) - \nabla^2\tilde{\mathbf{E}} = -j\omega\mu\nabla \times \tilde{\mathbf{H}} \tag{9.130}$$

from the vector identity. As the medium does not contain any free charges, $\nabla \cdot \tilde{\mathbf{E}} = 0$ and (9.130) can be rewritten as

$$\nabla^2\tilde{\mathbf{E}} = j\omega\mu\tilde{\mathbf{J}} \tag{9.131}$$

In source-free conductive medium, $\tilde{\mathbf{J}} = \sigma\tilde{\mathbf{E}}$. Therefore, equation (9.131) can now be expressed in terms of $\tilde{\mathbf{J}}$ as

$$\nabla^2\tilde{\mathbf{J}} = j\omega\mu\sigma\tilde{\mathbf{J}} \tag{9.132}$$

This is a general wave equation for the current density $\tilde{\mathbf{J}}$ in a conductive medium. In fact, this is the equation that governs eddy currents in conductors. For a round conductor, shown in Figure 9.43, we can express equation (9.132) as

$$\frac{\partial^2 \tilde{J}_z}{\partial r^2} + \frac{1}{r}\frac{\partial \tilde{J}_z}{\partial r} + \frac{\partial^2 \tilde{J}_z}{\partial z^2} = j\omega\sigma\mu\tilde{J}_z \tag{9.133}$$

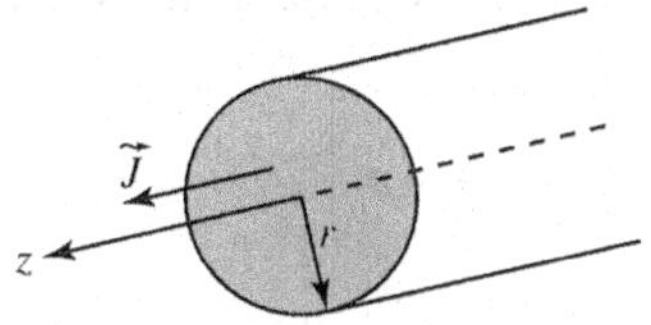

Figure 9.43 A solid, round conductor

When the current density is in the z direction (i.e., along the length of the conductor), for all practical purposes, $\tilde{J}_z$ does not vary with respect to z, and (9.133) can be rewritten as

$$r^2\frac{\partial^2 \tilde{J}_z}{\partial r^2} + r\frac{\partial \tilde{J}_z}{\partial r} - j\omega\mu\sigma r^2\tilde{J}_z = 0 \tag{9.134}$$

This is a form of *Bessel's equation*, and its solution requires the knowledge of Bessel functions. We will not attempt to solve this equation here. Instead, we want to make the reader understand how to formulate the skin effect problem. To do so, we will simplify the problem and calculate the internal impedance of a round wire in order to determine the resistance and the internal inductance of that wire.

Let us first consider an idealized problem in which a current-carrying conductor fills the region $y \geq 0$. The conductor is assumed to be of finite length ℓ in the x direction, as shown in Figure 9.44. The total current $\tilde{I}$ in the conductor is distributed in the form of current density $\tilde{\mathbf{J}}$ in the z direction such that $\tilde{\mathbf{J}} = \tilde{J}_0\mathbf{a}_z$ at $y = 0$ and $\tilde{\mathbf{J}} = 0$ for $y < 0$ (dielectric region). To maintain the finite current $\tilde{I}$ in the conductor, $\tilde{\mathbf{J}} \to 0$ as $y \to \infty$. The current density $\tilde{\mathbf{J}}$ must only be a function of y because of

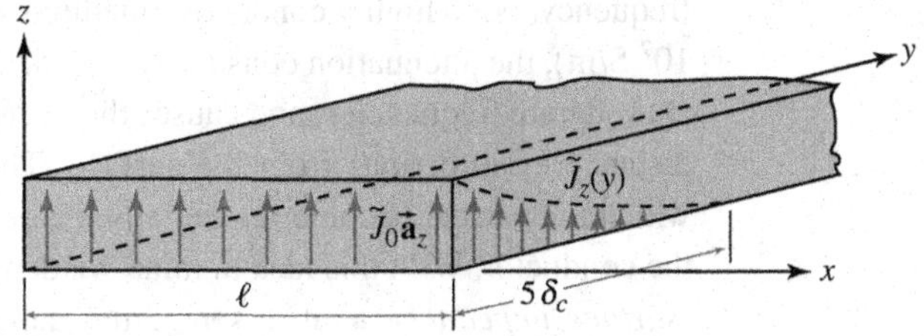

Figure 9.44 A current-carrying conducting slab with indicated current density distribution

its uniform distribution in the x direction. The wave equation (9.132) in $\tilde{J}_z$ can now be written in the rectangular coordinate system as

$$\nabla^2 \tilde{J}_z(y) = j\omega\mu\sigma \tilde{J}_z(y)$$

or

$$\frac{\partial^2 \tilde{J}_z}{\partial y^2} - j\omega\mu\sigma \tilde{J}_z = 0 \tag{9.135}$$

The general solution of the wave equation is

$$\tilde{J}_z(y) = \hat{A}\, e^{-\sqrt{j\omega\mu\sigma}\,y} + \hat{B}\, e^{\sqrt{j\omega\mu\sigma}\,y} \tag{9.136}$$

We expect $\tilde{\mathbf{J}}$ to decrease from $\tilde{J}_0$ at $y = 0$ to 0 at $y = \infty$; therefore, $\hat{A} = \tilde{J}_0$ and $\hat{B} = 0$. Thus, the current distribution within the conducting slab is

$$\tilde{J}_z(y) = \tilde{J}_0 e^{-\sqrt{j\omega\mu\sigma}\,y} = J_0 e^{-\alpha y} e^{-j\beta y} e^{j\phi} \tag{9.137}$$

where

$$\alpha = \beta = \sqrt{\frac{\omega\mu\sigma}{2}} \quad \text{and} \quad \tilde{J}_0 = J_0 e^{j\phi}$$

The exponential decay in the magnitude of $J_z(y) = J_0 e^{-\alpha y}$ is shown in Figure 9.44.

The total current $\tilde{I}$ in the conductor is

$$\begin{aligned} \tilde{I} &= \int_0^{\ell} \int_0^{\infty} \tilde{J}_0 e^{-\sqrt{j\omega\mu\sigma}\,y}\, dy\, dx \\ &= \frac{\ell \tilde{J}_0}{\sqrt{j\omega\mu\sigma}} = \frac{\ell \tilde{J}_0}{\alpha + j\beta} \end{aligned} \tag{9.138}$$

Because $\tilde{\mathbf{J}} = \sigma\tilde{\mathbf{E}}$, the electric field intensity inside the conductor is

$$\tilde{E}_z = \frac{\tilde{J}_0 e^{-\sqrt{j\omega\mu\sigma}\,y}}{\sigma}$$

or

$$\tilde{E}_z = \frac{\tilde{I}\sqrt{j\omega\mu\sigma}\, e^{-\sqrt{j\omega\mu\sigma}\,y}}{\ell\sigma} = \frac{\tilde{I}}{\ell\sigma}(\alpha + j\beta) e^{-\alpha y} e^{-j\beta y} \tag{9.139}$$

in terms of the total current $\tilde{I}$ in the conductor.

Note that the attenuation constant α is directly proportional to the square root of the frequency; thus, it increases with the increase in frequency. For a highly conductive medium such as copper ($\sigma = 5.8 \times 10^7$ S/m), the attenuation constant ($\alpha = 15.13\sqrt{f}$) becomes large even at moderate frequencies, and causes the decay of fields with increasing distance (y direction) from the surface. The extreme case is that in which the current becomes essentially a current sheet on the surface of the conductor. With this idea in mind we define the *internal impedance* (*surface impedance* as it is sometimes called) per unit length in the

z direction as the ratio of the electric field at $y = 0$ to the current. That is,

$$\hat{Z}_i = \frac{\tilde{E}_z(0)}{\tilde{I}} = \frac{1}{\ell\sigma}(\alpha + j\beta) \tag{9.140}$$

or

$$\hat{Z}_i = \frac{1}{\ell\sigma\delta_c} + j\frac{1}{\ell\sigma\delta_c} \tag{9.141}$$

where

$$\delta_c = \frac{1}{\alpha} = \sqrt{\frac{2}{\omega\mu\sigma}}$$

is the skin depth or depth of penetration of fields in the y direction. After penetrating a distance of $5\delta_c$, the field strength becomes less than 1% of its initial value at the surface, and the power reduces to less than 0.01%. The skin depth for copper at a frequency of 10 kHz is 0.66 mm, and that at 10 MHz is 0.02 mm. Thus, the fields essentially vanish inside copper after traveling a distance of 0.1 mm ($5\delta_c$) at 10 MHz. These calculations enable us to justify the concept of surface current at the boundary of a good conductor. In turn, the concept of surface current led us to the previous definition of internal impedance (surface impedance).

The internal impedance consists of an internal resistance

$$R_{\ell i} = \frac{1}{\ell\sigma\delta_c} \tag{9.142}$$

and an internal inductance

$$L_{\ell i} = \frac{1}{\omega\ell\sigma\delta_c} \tag{9.143}$$

If a and b are the inner and outer radii of a conducting shell and its thickness $(b - a)$ is greater than the skin depth δ_c, then we can use the preceding equations to define the internal impedance (per unit length) of a cylindrical conductor as

$$\hat{Z}_{\ell i} = \frac{1}{2\pi b\sigma\delta_c} + j\frac{1}{2\pi b\sigma\delta_c} \tag{9.144}$$

where $2\pi b$ is the circumference of the conductor at the outer radius b. From this equation it is clear that $2\pi b\delta_c$ is the cross-sectional area of a shell with b as its outer radius and δ_c as its thickness, as shown in Figure 9.45. We can therefore conclude that the current distribution can be considered uniform within this region.

EXAMPLE 9.19

Determine the resistances and the internal inductances per unit length of a round copper conductor having a diameter of 10 mm for frequencies of 1 kHz and 1 MHz. The conductivity of copper is 5.8×10^7 S/m.

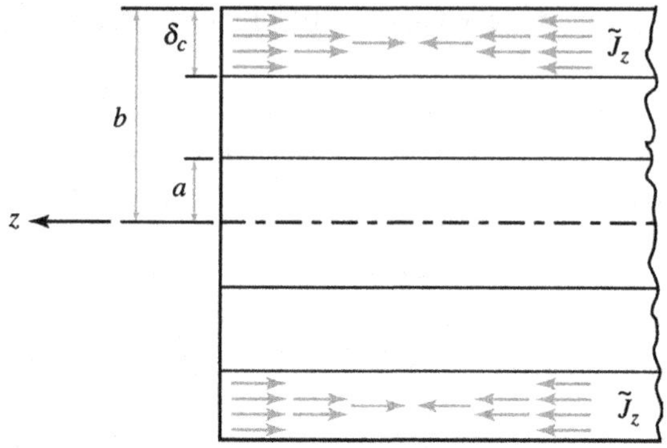

Figure 9.45 Approximate current distribution in a cylindrical shell at high frequencies

Solution The skin depth of the conductor at 1 kHz:

$$\delta_c = \frac{1}{\sqrt{\pi f \sigma \mu}} = \frac{1}{\sqrt{\pi \times 10^3 \times 5.8 \times 10^7 \times 4\pi \times 10^{-7}}}$$

$$= 2.09 \times 10^{-3} \text{ m or } 2.09 \text{ mm}$$

The internal resistance from (9.144) is

$$R_{\ell i} = \frac{1}{2\pi \times 5 \times 10^{-3} \times 5.8 \times 10^7 \times 2.09 \times 10^{-3}}$$

$$= 262.59 \times 10^{-6}\ \Omega/\text{m or } 262.59\ \mu\Omega/\text{m}$$

The inductive reactance of the internal impedance is the same as its internal resistance. Thus, the internal inductance is

$$L_{\ell i} = \frac{262.59 \times 10^{-6}}{2\pi \times 10^3} = 41.79 \times 10^{-9} \text{ H/m or } 41.79 \text{ nH/m}$$

$$\hat{Z}_{\ell i} = 262.59 + j262.59\, \mu\Omega/\text{m} = 371.36\underline{/45^\circ}\ \mu\Omega/\text{m}$$

At 1 MHz:

$$\delta_c = \frac{1}{\sqrt{\pi \times 10^6 \times 5.8 \times 10^7 \times 4\pi \times 10^{-7}}}$$

$$= 66.09 \times 10^{-6} \text{ m or } 66.09\ \mu\text{m}$$

$$R_{\ell i} = \frac{1}{2\pi \times 5 \times 10^{-3} \times 5.8 \times 10^7 \times 66.09 \times 10^{-6}}$$

$$= 8.30 \times 10^{-3}\ \Omega/\text{m or } 8.3 \text{ m}\Omega/\text{m}$$

$$L_{\ell i} = \frac{8.30 \times 10^{-3}}{2\pi \times 10^6} = 1.32 \times 10^{-9} \text{ H/M or } 1.32 \text{ nH/m}$$

$$\hat{Z}_{\ell i} = 8.3(1 + j) \text{ m}\Omega/\text{m} = 11.74\underline{/45^\circ} \text{ m}\Omega/\text{m}$$

As expected, the internal impedance has increased with the increase in the frequency. • • •

9.11 Summary

Transmission lines are used to transmit electrical energy from one location to another. The transmitted energy can be very low in the case of a communication or a measuring signal, and very high for transmission lines connecting a generating plant to a consumer location. Transmission lines may have different configurations depending on their applications. The most popular ones are parallel-conductor lines, coaxial cables, and microstrips. Microstrips are usually employed on printed circuit boards.

A lossless transmission line does not present any power loss along the line, and the voltage and current waves along such a line are given as

$$\tilde{V}(z) = \hat{V}^{+} e^{-\hat{\gamma} z} + \hat{V}^{-} e^{\hat{\gamma} z}$$
$$\tilde{I}(z) = \frac{\hat{V}^{+}}{\hat{Z}_c} e^{-\hat{\gamma} z} - \frac{\hat{V}^{-}}{\hat{Z}_c} e^{\hat{\gamma} z}$$

where $\hat{Z}_c = \sqrt{\frac{L_\ell}{C_\ell}}$ is the characteristic impedance and $\hat{\gamma} = j\sqrt{L_\ell C_\ell}$ is the propagation constant of the line. $\hat{V}^{+}$ and $\hat{V}^{-}$ are arbitrary constants. L_ℓ and C_ℓ are the per-unit-length inductance and capacitance of the transmission line.

The input impedance at any point along a lossless transmission line of length ℓ is

$$\hat{Z}_{\text{in}}(z) = \hat{Z}_c \left[\frac{\hat{Z}_L + j\hat{Z}_c \tan\beta(\ell - z)}{\hat{Z}_c + j\hat{Z}_L \tan\beta(\ell - z)} \right]$$

where $\hat{Z}_L$ is the load impedance. The input impedance repeats itself at every half-wavelength along the line.

A discontinuity point on a transmission line is the location where the characteristic impedance of the line changes. A discontinuity causes reflection of an incoming wave.

At the receiving end, we defined a reflection coefficient as

$$\hat{\rho}_R = \frac{\hat{Z}_L - \hat{Z}_c}{\hat{Z}_L + \hat{Z}_c}$$

to determine the amount of reflected voltage wave. The transmission coefficient, $\hat{\tau}_R = 1 + \hat{\rho}_R$, however, helps us determine the amount of voltage transmitted into the load. When the reflection coefficient is zero, the line is said to be perfectly matched, implying that no reflection will take place along the line.

The voltage standing wave ratio (VSWR) is a measure to evaluate the degree of mismatch along a transmission line. Mathematically, it is the ratio of the maximum value to the minimum value of the voltage on a lossless transmission line. That is,

$$VSWR = \frac{V_{\max}}{V_{\min}}$$

The "real-world" transmission lines have some losses because of the imperfect conductors and dielectrics. But the use of good conductors with low losses (such as copper and aluminum) and good dielectrics with very little leakage (air, polyethylene) help keep these losses to a minimum.

The characteristic impedance, propagation constant, and input impedance at any point along a transmission line with imperfect materials are given as follows:

Characteristic impedance: $$\hat{Z}_c = \sqrt{\frac{\hat{Z}_\ell}{\hat{Y}_\ell}}$$

Propagation constant: $$\hat{\gamma} = \sqrt{\hat{Z}_\ell \hat{Y}_\ell} = \alpha + j\beta$$

Input impedance: $$\hat{Z}_{\text{in}}(z) = \hat{Z}_c \left[\frac{\hat{Z}_L + \hat{Z}_c \tanh \hat{\gamma}(\ell - z)}{\hat{Z}_c + \hat{Z}_L \tanh \hat{\gamma}(\ell - z)} \right]$$

Here $\hat{Z}_\ell = R_\ell + j\omega L_\ell$ and $\hat{Y}_\ell = G_\ell + j\omega C_\ell$ are the per-unit-length series impedance and shunt admittance of the line, respectively. $\hat{Z}_L$ is the load impedance, and α and β are the attenuation and phase constants, respectively.

To study the transient response of a transmission line, we considered short and long pulses as the two different kinds of inputs to the line. We can use lattice diagrams to determine the response of a transmission line excited by a long pulse.

When a conductor carries a current at high frequency, the skin effect makes the current distribution in the conductor become more pronounced in a region near its outer surface. This, in turn, causes an increase in the resistance of a conductor as the frequency increases. The per-unit-length resistance and inductance of a round conductor are approximately

$$R_{\ell i} = \frac{1}{2\pi b \sigma \delta_c}$$

and

$$L_{\ell i} = \frac{1}{2\pi b \sigma \delta_c \omega}$$

where δ_c is the skin depth, b is the outer radius of the conductor, and ω is the operating angular frequency.

9.12 Review questions

9.1 What are the typical applications of transmission lines?

9.2 What is TEM wave propagation?

9.3 Do parallel-plane transmission lines satisfy TEM wave propagation?

9.4 What are the transmission line equations?

9.5 Express the voltage and current waves in a transmission line in terms of the forward- and backward-traveling waves.

9.6 Why do we need the boundary conditions for the voltage and/or current in a transmission line?

9.7 Express the propagation constant and the characteristic impedance of a transmission line in terms of its parameters.

9.8 How do we express the delay in a transmission line?

9.9 Define the input impedance of a transmission line.

9.10 What is the input impedance of a quarter-wavelength transmission line?

9.11 What is the input impedance of a half-wavelength transmission line?

9.12 Express the voltage reflection coefficient in a transmission line in terms of its characteristic and load impedances.

9.13 Define the transmission coefficient in terms of the reflection coefficient.

9.14 What is a standing wave in a transmission line?

9.15 What are the voltage and current magnitudes in a transmission line when it is terminated on a short circuit?

9.16 Repeat Question 9.15 when the transmission line is open-circuited.

9.17 How do we define a perfectly matched transmission line?

9.18 What is the voltage standing wave ratio?

9.19 What is the difference between a perfect conductor and an imperfect conductor?

9.20 What is the propagation constant in an imperfect transmission line?

9.21 What are the units for the attenuation constant and phase constant for an imperfect transmission line?

9.22 What are the possible reasons for transients along transmission lines?

9.23 How does a short pulse differ from a long pulse?

9.24 What is a lattice diagram? What is its usefulness?

9.25 Explain the phenomenon called the *skin effect.*

9.26 Is it efficient to use solid conductors at microwave frequencies?

9.27 A lossless transmission line of length less than $\lambda/4$ is terminated in a short circuit. Is the input impedance inductive or capacitive?

9.28 A lossless transmission line is terminated in a short circuit. What is the minimum length of the line so that the input impedance is capacitive?

9.29 A lossless transmission line is terminated in an open circuit. What is the minimum length of the line so that the input impedance is inductive?

9.30 What is the minimum length of the transmission line that is needed to transform normalized impedance into normalized admittance?

9.13 Exercises

9.1 Determine the capacitance and inductance per unit length of a coaxial transmission line with inner and outer radii of 3 mm and 6 mm, respectively. The insulating material that separates the conductors is polyethylene with a permittivity of $2.5\epsilon_0$.

9.2 A 600-m-long lossless transmission line has an inductance and a capacitance of 0.4 μH/m and 85 pF/m, respectively. Determine (a) the velocity of propagation and the phase constant for an operating frequency of 100 kHz and (b) the characteristic impedance of the transmission line.

9.3 A lossless coaxial cable is used to delay a pulse by 100 ns. The inductance and the capacitance per meter length of the cable are 0.20 μH/m and 60 pF/m, respectively. Determine the length of the cable.

9.4 A 20-m-long, lossless transmission line with a per-unit-length inductance and capacitance of 0.35 μH/m and 45 pF/m is terminated into a resistive load. If the load dissipates 20 W at 50 V (rms) when the operating frequency is 1 MHz, determine (a) the characteristic impedance and the phase constant of the line, (b) the voltages and currents at both the receiving and sending ends of the line, and (c) the power delivered by the source.

9.5 Calculate the sending-end input impedance of a 50-m-long transmission line with 0.3 μH/m and 40 pF/m when it delivers $10 + j2$ VA to a load at 20 V and 100 kHz.

9.6 A 2-m-long coaxial cable operates at 10 MHz. The distributed inductance and capacitance of the cable are specified as 0.215 μH/m and 48.28 pF/m, respectively. The input impedance of the cable at the sending end is measured as $50 + j25\ \Omega$. Determine the impedance of the load. The cable is assumed to be lossless.

9.7 A 3-m-long transmission line is connected to a source of $20\ \cos(3.14 \times 10^8 t)$ V to feed a load with an impedance of $100 + j20\ \Omega$. Determine the current and the average power delivered by the source. The distributed inductance and capacitance of the line are 0.4 μH/m and 40 pF/m, respectively.

9.8 A 10-m-long lossless transmission line feeds a load having an impedance of $35 + j10\ \Omega$. The load voltage is $\sqrt{2} \times 50\ \cos 10^8 t$ V. The voltage applied to the line is $\sqrt{2} \times 66\ \cos(10^7 t + 31^\circ)$ V. Calculate the distributed inductance and capacitance of the line.

9.9 The transmission coefficient at the end of a 50-m-long lossless transmission line with a capacitance of 75 pF/m is $0.75\underline{/9^\circ}$. The average power delivered to an inductive load at a 0.9 power factor is 10 W at 20 V (rms). Determine (a) the reflection coefficient at the receiving end, (b) the characteristic impedance, and (c) the phase constant of the line at 1 kHz. Also, show that the sending-end average power is the same as that at the receiving end. Write expressions for the voltage and current waves in phasor and time domains. What is the average power associated with each wave?

9.10 A resistive load connected to a 75-Ω cable causes a VSWR of 1.3 with no phase shift on the reflected wave. Determine the shunt resistance to be connected across the load in order not to have standing waves along the line.

9.11 A 50-Ω cable transmits a 1-MHz signal to a load having a 100-Ω resistance in series with a 10-μH inductance. Calculate the VSWR along the cable.

9.12 A 75-Ω, 10-m-long lossless transmission line operates at 150 MHz, and is terminated into an impedance of $150 + j225$ Ω. If the velocity of propagation on the line is 2.95×10^8 m/s, what should be the length and the location of a short-circuited stub line?

9.13 A 50-Ω, 2-m-long lossless transmission line operating at 60 MHz is matched by a 50-cm-long, short-circuited stub line at 60 cm away from the load point. If the delay of the signal from one end of the transmission line to the other is 7 ns, determine the impedance of the load.

9.14 A power supply injects $2\sin(314 \times 10^3 t)$ A to a 10-m-long coaxial cable at a voltage of 50 $\cos(314 \times 10^3 t)$ V. The parameters of the total length of the cable are: $R = 0.25$ Ω, $L = 6.5$ μH, $G = 0$, $C = 320$ pF. Determine the impedance of the load, the power input, the power output, and the efficiency of the transmission line.

9.15 A 40-m-long transmission line has a characteristic impedance of $75\underline{/-4^\circ}$ Ω and an attenuation constant of 0.001 dB/m at a frequency of 2 MHz. The velocity of propagation of the wave in the line is 250,000 km/s at the given frequency. The applied voltage to the line is $60\underline{/0^\circ}$ V, and the line operates at no load. Determine (a) the receiving-end voltage and (b) the sending-end current. What is the power input to the line?

9.16 A short pulse of magnitude 5 V is applied to a 10-m-long, lossless transmission line whose characteristic impedance is 50 Ω. Determine the receiving-end voltage, current, and power after five transit times if $u_p = 2.85 \times 10^8$ m/s. The load resistance is 100 Ω, and the source resistance is negligible.

9.17 A 10-V step voltage is applied to a 20-m-long, 75-Ω, lossless transmission line from a voltage source having an internal resistance of 75 Ω. The line is terminated with a load resistance of 100 Ω. Determine the variation of the voltage and current waveforms 5 m away from the sending end of the source. The velocity of propagation in the transmission line is assumed to be the speed of light.

9.18 Two cables with characteristic impedances of 50 Ω and 75 Ω are connected to each other. The 50-Ω cable feeds a 30-Ω load, and a 10-V dc supply with an internal resistance of 25 Ω delivers power to the 75-Ω cable. The transit times of the 50-Ω and 75-Ω cables are 2 μs and 3 μs, respectively. Determine the variation of the voltage across the load for a duration of 60 μs after the supply is connected to the line.

9.19 What is the skin depth of a 2-mm radius aluminum, round conductor operating at (a) 60 Hz, (b) 1000 Hz, and (c) 1 MHz? The conductivity of aluminum is 3.55×10^7 S/m.

9.20 Redo Example 9.19 with the copper conductor being replaced with an aluminum conductor.

9.14 Problems

9.1 Show that the capacitance per-unit-length of a parallel two-wire transmission line having conductors of circular cross section is

$$C_\ell = \frac{\pi \epsilon}{\cosh^{-1} \dfrac{d}{2a}} \text{ F/m}$$

where d is the center-to-center separation of the conductors and a is the radius of each conductor.

9.2 A lossless tranmission line 20-km long has a characteristic impedance of 150 Ω. Calculate the total inductance and capacitance of this line if the velocity of propagation is 90% of the speed of light.

9.3 Determine the inductance and the capacitance per unit length of a coaxial cable having an inner and outer radii of a and b, respectively.

9.4 A coaxial cable is to be designed for a specific application to give a delay of 14 ns at 1 MHz. The length of the cable is limited to 2 m. Determine the relative permittivity of the dielectric medium suitable for this delay cable. The permeability of the medium is that of free space.

9.5 A 300-m-long transmission line has a characteristic impedance of 75 Ω and a phase velocity of 220,000 km/s at the operating frequency of 3 MHz. The line is terminated in an impedance of $150 + j400$ Ω. If the voltage across the receiving end is $50\underline{/0^\circ}$ V (rms), determine the current through the load and the voltage input to the transmission line.

9.6 A 2-m-long transmission line operating at a frequency of 15 MHz has a phase constant of 369.6×10^{-3} rad/m. The complex power absorbed by the load is $3.5 - j1.5$ VA at an rms voltage of $50\underline{/0^\circ}$ V. Calculate the characteristic impedance of the line if the voltage input to the line is 34 V.

9.7 A coaxial power transmission cable transmits $100 + j30$ MVA over a distance of 100 km at a frequency of 60 Hz. The rms voltage at the load is 110 kV, and the inductance and capacitance of the cable are 0.372 μH/m and 76 pF/m, respectively. Calculate (a) the characteristic impedance, (b) the phase constant and phase velocity, (c) the sending-end voltage, and (d) the voltage drop along the cable. Neglect the resistance of the cable.

9.8 Two different kinds of transmission lines of length ℓ_1 and ℓ_2 are connected to each other to transmit signals over a distance of $(\ell_1 + \ell_2)$. For these lines the characteristic impedances are $\hat{Z}_{c1}$, $\hat{Z}_{c2}$ and the phase constants are β_1 and β_2. Find a transformation $\hat{A}$ such that

$$\begin{bmatrix} \tilde{V}(0) \\ \tilde{I}(0) \end{bmatrix} = \hat{A} \begin{bmatrix} \tilde{V}(\ell_1 + \ell_2) \\ \tilde{I}(\ell_1 + \ell_2) \end{bmatrix}$$

9.9 A lossless transmission line of length 20 m is excited from a signal generator operating at 10 MHz and the load connected to that line has an impedance of $100 + j60\ \Omega$. Calculate the input impedance at the midpoint of the line. The inductance and capacitance of the line are 3×10^{-7} H/m and 40×10^{-12} F/m, respectively.

9.10 A 90-m-long, 50-Ω transmission line operates at 500 kHz. The phase velocity of the wave is 2.8×10^8 m/s, and the input impedance measured at the sending-end terminals is $60 - j20\ \Omega$. Calculate the impedance of the load.

9.11 A 2-m-long, lossless transmission line with $\hat{Z}_c = 75\ \Omega$ and $u_p = 2.6 \times 10^8$ m/s is terminated in a load of $\hat{Z}_L = 120 + j90\ \Omega$. If the operating voltage across the load, in time domain, is $v_R(t) = 150\ \cos(1.26 \times 10^8 t)$ V, calculate (a) the reflection coefficient $\hat{\rho}(z)$, (b) the forward and backward voltage and current waves in time domain at the sending end, (c) the *VSWR*, (d) the voltage drop, (e) the average power at any point associated with the forward and backward waves, and (f) the efficiency of the line.

9.12 A 50-m-long lossless transmission line with $L_\ell = 0.5\ \mu$H/m and $C_\ell =$ 50 pF/m is connected to a voltage of $v_S(t) = 280\ \cos(6.28 \times 10^7 t)$ V when the load at the receiving end is 250 Ω. (a) Determine the reflection coefficient at the receiving end. (b) Calculate the forward and backward voltage and current waves at any point of the line. (c) Find the average power carried by each wave.

9.13 A quarter-wavelength transmission line is 1.5 m long and terminated in $20 - j10\ \Omega$. If the total capacitance of this line is 166.66 pF, what should be the rms voltage across the sending end in order to maintain a current of $i(t) = \sqrt{2}\ \cos(6 \times 10^8 t)$ A at the sending end?

9.14 A 7-m-long lossless transmission line absorbs the maximum average power from the supply with $\hat{Z}_G = 28\angle{-20^\circ}\ \Omega$ when the voltage and current are $50\angle{0^\circ}$ V (rms), and $2\angle{0^\circ}$ A (rms) at the receiving end. Calculate the characteristic impedance and the phase constant of the transmission line.

9.15 A 75-Ω coaxial cable is terminated in an impedance of $10 - j40\ \Omega$. Calculate the voltage and current reflection coefficients at the termination.

9.16 A 50-Ω coaxial cable transmits a 1-MHz measuring signal to an oscilloscope. The input resistance of the oscilloscope is 1 MΩ. Calculate the resistance needed across the oscilloscope in order not to produce reflections along the line.

9.17 The input impedance at the sending end of a lossless, 1.2-m-long transmission line is $120 - j80\ \Omega$. The wavelength and the operating frequency of the transmission line are 2 m and 50 MHz, respectively. Calculate the characteristic impedance of the line.

9.18 A 50-Ω, 10-m coaxial cable is fed from a signal generator of 200 kHz, $50\underline{/0^\circ}$ V (rms). A 75-Ω, 2-m cable is connected to the 50-Ω cable to transmit signals to a device with an input impedance of $120 - j200$ Ω. The transit times of the 50-Ω and 75-Ω cables are 36 ns and 8 ns, respectively. Calculate (a) the reflection and transmission coefficients and (b) the voltage and current at the junction of the two cables.

9.19 The load connected to a 50-Ω transmission line causes a VSWR of 1.5 with no phase shift on the reflected wave. Calculate the resistance to be connected in parallel to the load in order to achieve no standing waves along the transmission line.

9.20 A 10-m long transmission line operating at a frequency of 50 MHz has a characteristic impedance of 80 Ω and a phase constant of 1.18 rad/m. The line has to provide a voltage of 100 V (rms) across a resistive load of 1500 Ω. Calculate the necessary voltage across the sending end of the line.

9.21 Determine the peak values of the voltage and current and their locations along the transmission line described in Problem 9.20. Also, calculate the voltage standing wave ratio of the transmission line.

9.22 The VSWR of a lossless, 75-Ω transmission line is 2. Calculate the impedance at the voltage maxima and minima in the standing wave pattern along the line.

9.23 A coaxial cable operating at a frequency of 10 MHz has a characteristic impedance of 55 Ω. It is terminated into a 20-Ω resistor parallel to a capacitance of 100 pF. If the phase constant is 0.22 rad/m, calculate the VSWR along the line and determine the location of the nearest voltage maximum in the standing wave pattern from the receiving end.

9.24 A 50-Ω, 12-m-long, lossless transmission line with a phase velocity of 2.7×10^8 m/s is connected to a load having an impedance of $\hat{Z}_L = 150$ Ω. The open-circuit voltage of the supply that feeds the line is $v_G(t) = 25 \cos(8 \times 10^5 t)$ V and internal impedance of $\hat{Z}_G = 10 - j5$ Ω. Calculate the input impedance, voltage, current, and power at (a) the sending end, (b) the receiving end, (c) 3 m from the supply, (d) 3 m from the load, and (e) determine the voltage drop along the line.

9.25 A 50-Ω, 100-m-long, lossless transmission line is terminated into a load having an impedance of $40 - j100$ Ω. The transit time of the line is 0.5 μs. In order to match the transmission line to the load, determine the length and the location of a short-circuited stub line if the operating frequency is 20 MHz.

9.26 A 15-m-long polythylene ($\epsilon_r = 2.5$) coaxial cable operating at 125 MHz is terminated into a $150 + j225$ Ω. The diameters of the inner and outer conductors are 2.5 mm and 6 mm, respectively. In order to perfectly match the termination, what should be the length and location of a short-circuited stub line?

9.27 A 100-m-long, two-parallel-wire copper transmission line has the following parameters: conductor radius = 5 mm, $L = 150\ \mu\text{H}$, $G = 0$, and $C = 2000$ pF. Calculate the characteristic impedance, propagation constant, attenuation and phase constants, and the phase velocity at 10 kHz, 100 kHz, 1 MHz, 10 MHz, 100 MHz, and 1 GHz. Plot the variation of α and β with frequency on a logarithmic graph.

9.28 A 100-km-long cable operating at 60 Hz has the following per-unit-length parameters: $R_\ell = 34.63 \times 10^{-6}\ \Omega/\text{m}$, $L_\ell = 1.5 \times 10^{-6}$ H/m, $G_\ell = 0$, $C_\ell = 55 \times 10^{-12}$ F/m. The inductive load at the receiving end requires 100 MW at a power factor of 0.9 and a voltage of 100 kV. Determine (a) the voltage, current, and power at the sending end of the cable, and (b) the efficiency and the voltage regulation along the cable.

9.29 A 200-m-long coaxial cable transmits communication signals to a dipole antenna whose input impedance is $74 + j42.5\ \Omega$ at a carrier frequency of 90 MHz. The per-unit-length parameters of the cable are: $R_\ell = 1.4 \times 10^{-3}\ \Omega/\text{m}$, $L_\ell = 220 \times 10^{-9}$ H/m, $C_\ell = 177 \times 10^{-12}$ F/m, $G_\ell = 0.1\ \mu\text{S/m}$. (a) Calculate the input impedance of the cable at the terminals of the signal source, and (b) determine the average power delivered to the cable, and the transmitted power by the antenna, if the applied current to the cable is $10\underline{/0^\circ}$ A (rms).

9.30 A 500-m-long transmission line has a characteristic impedance of $50\underline{/-5^\circ}\ \Omega$ and an attenuation constant of 50×10^{-3} dB/m at the operating frequency of 2.5 MHz. The velocity of propagation along the line is 230,000 km/s at the operating frequency. The line is terminated in an impedance of $200 - j300\ \Omega$. Calculate the current through the load impedance if the sending-end voltage is $20\underline{/0^\circ}$ V.

9.31 Verify equations (9.106), (9.107), and (9.108).

9.32 A 50-m-long transmission line having a characteristic impedance of $\hat{Z}_c = 40\underline{/-5^\circ}\ \Omega$ is terminated in a load $\hat{Z}_L = 280\ \Omega$. The power supply at the sending end has an open-circuit voltage $v_G(t) = 20\cos(6 \times 10^6 t)$ V and an internal impedance $\hat{Z}_G = 30 + j40\ \Omega$. If the sending-end voltage is $v_S(t) = 18\cos(6 \times 10^6 t - 12^\circ)$ V, determine (a) the input impedance $\hat{Z}_{\text{in}}(0)$, (b) the reflection coefficient $\hat{\rho}(z)$, (c) the propagation constant, (d) the voltage and current waveforms in time domain at the receiving end, (e) the average powers at any point related to the forward and backward waves, and (f) the overall efficiency of the line.

9.33 A 25-m-long transmission line with $L_\ell = 0.4\ \mu\text{H/m}$, $C_\ell = 45$ pF/m, $R_\ell = 8\ \Omega/\text{m}$, and $G_\ell = 0$ is connected to a voltage $v_S(t) = 60\cos(7 \times 10^6 t)$ V when the load at the receiving end is 160 Ω. Calculate (a) the reflection coefficient $\hat{\rho}(z)$, (b) the forward and backward voltage and current waves at any point, (c) the average power associated with each wave, (d) the efficiency, and (e) the voltage drop along the line.

9.34 The substrate of a 10-cm-long parallel-plate stripline is epoxy resin with a thickness of 0.2 mm. The plates are made of copper with a width of 5 mm and a thickness of 0.01 mm. If the operating frequency of this stripline is 100 MHz, determine (a) the characteristic impedance, (b) the propagation constant, and (c) the velocity of propagation. Neglect the fringing effects and conduction through the substrate. Epoxy resin: $\epsilon = 3.5\epsilon_0$, $\mu = \mu_0$; copper: $\sigma = 5.8 \times 10^7$ S/m.

9.35 The stripline given in Problem 9.34 is connected to $v_S = 5$, $\cos(6.28 \times 10^8 t)$ V. Determine the power at the sending end when the current through the load is $5\underline{/0^\circ}$ mA (peak).

9.36 A 2-m-long coaxial cable has copper conductors and polyethylene dielectric. The radii of the inner and outer conductors are 1 mm and 5 mm, respectively, and the cable is connected to a voltage of $v_S(t) = 10\cos(5 \times 10^{10} t)$ V. Determine the power loss along the cable if the load current is $i_R(t) = 0.5\cos(5 \times 10^{10} t - 10^\circ)$ mA. Discuss your results. Copper: $\sigma = 5.8 \times 10^7$ S/m; polyethylene: $\epsilon = 2.5\epsilon_0$, $\mu = \mu_0$.

9.37 A 350-km-long, single-phase power transmission line is loaded by 150 MW at 300 kV, 60 Hz with a unity power factor. Determine (a) the characteristic impedance, (b) the propagation constant, (c) the velocity of propagation, (d) the wavelength, (e) the forward and backward voltage waves at any point, (f) the sending-end voltage from the forward and backward voltages, (g) the applied power at the sending end, (h) the efficiency, and (i) the voltage drop along the line. The parameters of the line are given as: $R_\ell = 0.1$ Ω/km, $L_\ell = 1.5$ mH/km, $C_\ell = 7.9$ nF/km, and $G_\ell = 0$.

9.38 A transmission line having a property of $R_\ell C_\ell = L_\ell G_\ell$ is known as a *distortionless line* because its characteristic impedance is purely resistive and its velocity of propagation is frequency independent. Show that $\hat{Z}_c = \sqrt{L_\ell / C_\ell}$, $\alpha = \sqrt{R_\ell G_\ell}$, and $\beta = \omega\sqrt{L_\ell C_\ell}$ in a distortionless line.

9.39 A distortionless transmission line as described in Problem 9.38 has $L_\ell = 0.4$ μH/m, $C_\ell = 86$ pF/m, and $R_\ell = 11$ mΩ/m. Determine its characteristic impedance, propagation constant, and the velocity of propagation if the operating frequency is 95 MHz.

9.40 A 20-m-long distortionless cable with the properties as described in Problem 9.38 has a characteristic impedance of 75 Ω. A signal applied to this cable is delayed by 90 ns before it is measured at the receiving end. Also, at the receiving end an attenuation of 0.1 dB in magnitude is observed. Determine the parameters R_ℓ, G_ℓ, L_ℓ, and C_ℓ of this cable.

9.41 A distortionless coaxial cable as described in Problem 9.38 operates at 100 kHz. The radii of the inner conductor and outer sheath of this cable are 1.5 mm and 3 mm, respectively. Determine the characteristic impedance, attenuation constant, and velocity of propagation in this cable. Conductivity of the conductors is 5.8×10^7 S/m and the permittivity of the dielectric medium is $2.2\epsilon_0$.

9.42 A pulse of magnitude 10 V and a duration of 1 ns is injected into a lossless 10-m-long coaxial cable with an inductance of 2 μH and a capacitance of 2000 pF. The internal resistance of the power supply is 10 Ω and the resistance of the load at the receiving end is 100 Ω. (a)

Calculate the transit time of the line and (b) plot the variation of the voltage, current, and power at the sending and receiving ends of the transmission line for a period of 0.4 μs.

9.43 A step-function voltage of magnitude 100 V is applied to a 100-m-long, lossless transmission line of characteristic impedance 90 Ω. The velocity of propagation along the transmission line is 250,000 km/s. The source and load resistances are 50 Ω and 250 Ω, respectively. (a) Plot the variation of voltage, current, and power for a duration of 4 μs, and (b) calculate the voltage, current, and power at both ends of the line when a steady state is achieved.

9.44 A lossless transmission line delivers 1 kW to a resistive load at a voltage of 120 V (dc). The length of the line is 500 m and its characteristic impedance is 90 Ω. The circuit breaker at the sending end is suddenly closed, and the line is energized with 130 V from a dc supply having a 50-Ω internal resistance. Determine the variation of the voltage at the sending end, the receiving end, and 100 m away from the sending end for a period of 25 μs after the switch is closed. The velocity of propagation of the line is 210,000 km/s. Use a lattice diagram.

9.45 A 50-Ω cable perfectly matched at its termination is connected to a transmission line having a characteristic resistance of 100 Ω. The dc power supply that feeds the 100-Ω transmission line has an internal resistance of 50 Ω and applies 24 V to the line suddenly at $t = 0$. The transit times of the 50-Ω and 100-Ω lines are 2 μs and 4 μs, respectively. Determine the variation of voltage both across the load and at the junction point of the two lines for a duration of 50 μs.

9.46 Calculate the inductance and resistance per unit length of a round copper conductor of diameter 4 mm for frequencies of 100 Hz, 1 kHz, 10 kHz, 100 kHz, 1 MHz, 10 MHz, 100 MHz, and 1 GHz. Plot the variation of both the resistance and inductance with frequency on a logarithmic graph.

10

Waveguides and cavity resonators

10.1 Introduction

In our discussion of transmission lines, we pointed out that the resistance of a conductor increases with an increase in the signal frequency, leading to an increase in power loss along the line. This power loss becomes intolerable at microwave frequencies (in the GHz range) and makes the transmission line almost impractical. At such high frequencies hollow conductors, known as waveguides, are employed to guide electrical signals efficiently. Figure 10.1 shows a typical waveguide assembly.

In the study of transmission lines with at least two conductors, we found that the propagating wave has field components in the transverse direction and is referred to as the transverse electromagnetic (TEM) wave. However, as a waveguide consists of only one hollow conductor, we do not expect it to support the TEM wave. In this chapter, we show that a waveguide can support the other two types of waves, the *transverse magnetic (TM)* and the *transverse electric (TE)* waves. These waves can exist inside a hollow conductor under certain conditions. TM and TE waves can also propagate in a region bounded by a parallel-plate transmission line, in which case the two conducting plates are said to form a *parallel-plate waveguide*.

The propagation of an electromagnetic wave inside a waveguide is quite different than the propagation of a TEM wave. When a wave is introduced at one end of the waveguide, it is reflected from the wall of the waveguide whenever it strikes it. Because the wave is entirely enclosed by conducting walls, we expect the wave to experience multiple reflections as it progresses along the waveguide. These various reflected waves interact with each other to produce an infinite number of discrete characteristic patterns called **modes**. The existence of a discrete mode depends upon (a) the shape and size of the waveguide, (b) the medium within the waveguide, and (c) the operating frequency.

Unlike the TEM mode, which can be excited at any frequency, the TM or TE modes can only propagate when the wave frequency is higher than a certain frequency, called the *cutoff frequency*. The cutoff frequency is different for each mode. When the operating frequency of the wave is

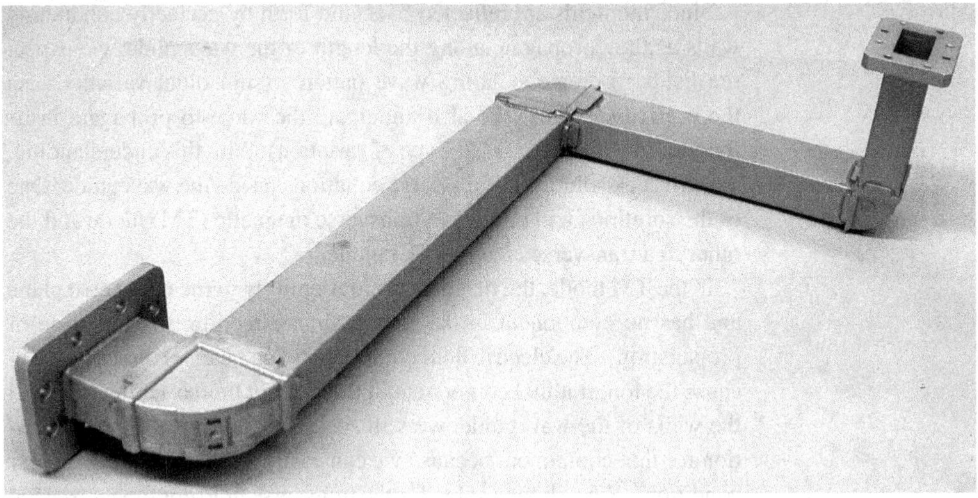

Figure 10.1 A rectangular waveguide assembly (courtesy of Space-machine and Engineering Corp.)

lower than the cutoff frequency of the lowest mode, the wave experiences attenuation and disappears after traveling a very short distance. On the other hand, all those modes with cutoff frequencies lower than the operating frequency can exist simultaneously inside the waveguide. In order to avoid the existence of multiple modes, the waveguide is operated at a frequency that lies between the cutoff frequencies of the lowest and the next lowest modes. Thus, the lowest-order mode can propagate, and all other modes are attenuated.

Two types of waveguides are commonly used for guiding signals along their lengths at microwave frequencies. One has a rectangular cross section and is aptly called a *rectangular waveguide*. The other, of circular cross section, is appropriately referred to as a *cylindrical waveguide*. The rectangular waveguide is the one most commonly used and is also comparatively easy to analyze. For this reason, we will discuss only rectangular waveguides in this chapter.

In our discussion of waveguides, we will presume that (a) the four sides of the waveguide are bounded by perfectly conducting walls, and (b) the medium enclosed by these perfectly conducting walls is a perfect dielectric. These assumptions will help us develop wave equations in terms of the $\breve{\mathbf{E}}$ and $\breve{\mathbf{H}}$ fields within a perfect dielectric medium. The solution of each of these wave equations must also satisfy the boundary conditions at each of the four walls of the waveguide. The boundary conditions are immensely simplified with the assumption of perfectly conducting walls. This assumption is not really that farfetched because either the waveguide is made of a highly conducting material such as copper or brass, or the inside walls of the waveguide are coated with a thin layer of silver.

Since the fields are reflected back and forth by perfectly conducting walls as they propagate along the length of the waveguide, we expect the fields to create standing wave patterns (sinusoidal variations) in the transverse plane. We also anticipate the wave to propagate in its longitudinal direction ($e^{-j\beta z}$ type of variation). With this understanding, we will seek solutions of the wave equations inside the waveguide. One of the solutions will apply to a transverse magnetic (TM) mode and the other to a transverse electric (TE) mode.

In the TM mode, the magnetic field is entirely in the transverse plane and has no component in the longitudinal direction (the direction of propagation). The electric field components can be in all directions. Because the longitudinal component of the electric field is tangential to all the walls of the waveguide, we will seek a solution of the wave equation for that component because we can easily apply the four boundary conditions. We will then solve for the transverse field components using Maxwell's equations.

The electric field is entirely in the transverse plane when the waveguide supports a TE mode. The components of the magnetic field can now be in all directions. Here, we will seek a solution of the wave equation for the longitudinal component of the magnetic field. The unknown constants will then be determined by applying the boundary conditions for the tangential components of the electric field. To do this, we have to express the transverse components of the electric field in terms of the longitudinal component of the magnetic field. Once these constants are known, the other components can then be obtained using Maxwell's equations.

When the two openings along the length of a waveguide are closed by two perfectly conducting planes to form a completely enclosed structure, the device is called a *cavity resonator*. In this case, we expect standing waves in all directions. A cavity resonator at microwave frequencies is like a resonant circuit at low frequencies. A frequency meter designed to measure microwave frequencies is, in fact, a calibrated cavity resonator. We will explore the conditions that the fields must satisfy in order to exist inside a cavity resonator.

Our discussion of waveguides will also show that the phase velocity of the wave within the waveguide may be greater than the speed of light. Is this a contradiction of the principle of relativity? The answer, of course, is no, as we will explain later.

In previous chapters, we studied signals that were propagating in the same direction as the phase velocity. Thus, the velocity of signal propagation was the same as the phase velocity—which we mathematically showed while discussing plane waves. However, the waves inside a waveguide are reflected back and forth by its walls as they propagate along its length. Thus, the direction of the phase velocity within the waveguide is not the same as that of the signal propagation. Our theoretical development in this chapter will reveal that the velocity with which the signal propagates (called the *group velocity*) along the waveguide is always less than the speed of light. We will also prove that the phase velocity is a function of frequency. As the operating frequency approaches the cutoff frequency, the phase velocity becomes infinite, while the group velocity decreases to zero. In fact, we will derive a relationship that associates the phase velocity of the wave inside the waveguide, the group velocity with which the signal propagates in the waveguide, and the phase velocity of the wave in an unbounded medium.

10.2 Wave equations in Cartesian coordinates

During our discussion of plane waves in Chapter 8, we derived general wave equations in terms of the $\vec{\tilde{\mathbf{E}}}$ and $\vec{\tilde{\mathbf{H}}}$ fields using the concept of complex permittivity. We also stated that, for a perfect dielectric medium, the complex permittivity becomes simply the permittivity of the medium. The medium under consideration in a waveguide is also a perfect dielectric; therefore, the same wave equations, (8.42) and (8.43), must apply for the $\vec{\tilde{\mathbf{E}}}$ and $\vec{\tilde{\mathbf{H}}}$ fields within the waveguide. Replacing $\hat{\epsilon}$ with ϵ, we can write a set of wave equations in scalar form as

$$\frac{\partial^2 \tilde{E}_x}{\partial x^2} + \frac{\partial^2 \tilde{E}_x}{\partial y^2} + \frac{\partial^2 \tilde{E}_x}{\partial z^2} = -\omega^2 \mu\epsilon \tilde{E}_x \tag{10.1a}$$

$$\frac{\partial^2 \tilde{E}_y}{\partial x^2} + \frac{\partial^2 \tilde{E}_y}{\partial y^2} + \frac{\partial^2 \tilde{E}_y}{\partial z^2} = -\omega^2 \mu\epsilon \tilde{E}_y \tag{10.1b}$$

$$\frac{\partial^2 \tilde{E}_z}{\partial x^2} + \frac{\partial^2 \tilde{E}_z}{\partial y^2} + \frac{\partial^2 \tilde{E}_z}{\partial z^2} = -\omega^2 \mu\epsilon \tilde{E}_z \tag{10.1c}$$

$$\frac{\partial^2 \tilde{H}_x}{\partial x^2} + \frac{\partial^2 \tilde{H}_x}{\partial y^2} + \frac{\partial^2 \tilde{H}_x}{\partial z^2} = -\omega^2 \mu\epsilon \tilde{H}_x \tag{10.1d}$$

$$\frac{\partial^2 \tilde{H}_y}{\partial x^2} + \frac{\partial^2 \tilde{H}_y}{\partial y^2} + \frac{\partial^2 \tilde{H}_y}{\partial z^2} = -\omega^2 \mu\epsilon \tilde{H}_y \tag{10.1e}$$

$$\frac{\partial^2 \tilde{H}_z}{\partial x^2} + \frac{\partial^2 \tilde{H}_z}{\partial y^2} + \frac{\partial^2 \tilde{H}_z}{\partial z^2} = -\omega^2 \mu\epsilon \tilde{H}_z \tag{10.1f}$$

where $\tilde{E}_x$, $\tilde{E}_y$, $\tilde{E}_z$, $\tilde{H}_x$, $\tilde{H}_y$, and $\tilde{H}_z$ are the components of the electric and magnetic fields.

These wave equations satisfy the fields inside the waveguide shown in Figure 10.2. The rectangular waveguide with perfectly conducting walls is intended to carry the signal in the longitudinal direction (z direction), preferably with the least amount of attenuation. Thus, right at the outset, we can assume that each field within the waveguide varies as $e^{-\hat{\gamma}z}$, where $\hat{\gamma} = \alpha + j\beta$ is the propagation constant, α is the attenuation constant, and β is the phase constant.

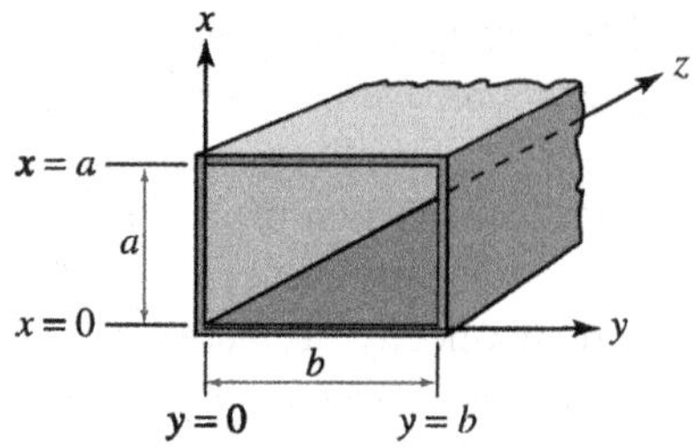

Figure 10.2 A rectangular waveguide

When we consider only the forward wave, we can write the $\tilde{\vec{\mathbf{E}}}$ and $\tilde{\vec{\mathbf{H}}}$ fields as

$$\tilde{\vec{\mathbf{E}}}(x, y, z) = \tilde{E}_x(x, y)e^{-\hat{\gamma} z}\vec{\mathbf{a}}_x + \tilde{E}_y(x, y)e^{-\hat{\gamma} z}\vec{\mathbf{a}}_y + \tilde{E}_z(x, y)e^{-\hat{\gamma} z}\vec{\mathbf{a}}_z$$

and

$$\tilde{\vec{\mathbf{H}}}(x, y, z) = \tilde{H}_x(x, y)e^{-\hat{\gamma} z}\vec{\mathbf{a}}_x + \tilde{H}_y(x, y)e^{-\hat{\gamma} z}\vec{\mathbf{a}}_y + \tilde{H}_z(x, y)e^{-\hat{\gamma} z}\vec{\mathbf{a}}_z$$

Thus we can rewrite (10.1) as

$$\frac{\partial^2 \tilde{E}_x}{\partial x^2} + \frac{\partial^2 \tilde{E}_x}{\partial y^2} = -(\omega^2 \mu\epsilon + \hat{\gamma}^2)\tilde{E}_x \tag{10.2a}$$

$$\frac{\partial^2 \tilde{E}_y}{\partial x^2} + \frac{\partial^2 \tilde{E}_y}{\partial y^2} = -(\omega^2 \mu\epsilon + \hat{\gamma}^2)\tilde{E}_y \tag{10.2b}$$

$$\frac{\partial^2 \tilde{E}_z}{\partial x^2} + \frac{\partial^2 \tilde{E}_z}{\partial y^2} = -(\omega^2 \mu\epsilon + \hat{\gamma}^2)\tilde{E}_z \tag{10.2c}$$

$$\frac{\partial^2 \tilde{H}_x}{\partial x^2} + \frac{\partial^2 \tilde{H}_x}{\partial y^2} = -(\omega^2 \mu\epsilon + \hat{\gamma}^2)\tilde{H}_x \tag{10.2d}$$

$$\frac{\partial^2 \tilde{H}_y}{\partial x^2} + \frac{\partial^2 \tilde{H}_y}{\partial y^2} = -(\omega^2 \mu\epsilon + \hat{\gamma}^2)\tilde{H}_y \tag{10.2e}$$

$$\frac{\partial^2 \tilde{H}_z}{\partial x^2} + \frac{\partial^2 \tilde{H}_z}{\partial y^2} = -(\omega^2 \mu\epsilon + \hat{\gamma}^2)\tilde{H}_z \tag{10.2f}$$

Because these equations are similar to one another, their solutions must also be similar. As the field components also satisfy Maxwell's equations, the solution of one equation in terms of one field component will enable us to determine the remaining field components. This is the approach that is commonly used in determining the field components inside a waveguide. Therefore, before solving these equations, let us first determine how the field components are related to each other.

For the forward-traveling wave, we can express Maxwell's equation $\nabla \times \tilde{\vec{\mathbf{E}}} = -j\omega\mu\tilde{\vec{\mathbf{H}}}$ in scalar form as

$$\frac{\partial \tilde{E}_z}{\partial y} + \hat{\gamma}\tilde{E}_y = -j\omega\mu\tilde{H}_x \tag{10.3a}$$

$$\frac{\partial \tilde{E}_z}{\partial x} + \hat{\gamma}\tilde{E}_x = j\omega\mu\tilde{H}_y \tag{10.3b}$$

$$\frac{\partial \tilde{E}_y}{\partial x} - \frac{\partial \tilde{E}_x}{\partial y} = -j\omega\mu\tilde{H}_z \tag{10.3c}$$

Likewise, we obtain the scalar components of $\nabla \times \tilde{\mathbf{H}} = j\omega\tilde{\mathbf{E}}$ as

$$\frac{\partial \tilde{H}_z}{\partial y} + \hat{\gamma}\tilde{H}_y = j\omega\epsilon\tilde{E}_x \tag{10.4a}$$

$$\frac{\partial \tilde{H}_z}{\partial x} + \hat{\gamma}\tilde{H}_x = -j\omega\epsilon\tilde{E}_y \tag{10.4b}$$

$$\frac{\partial \tilde{H}_y}{\partial x} - \frac{\partial \tilde{H}_x}{\partial y} = j\omega\epsilon\tilde{E}_z \tag{10.4c}$$

From equations (10.3) and (10.4), we can express the x and y components of the electric and magnetic fields in terms of their z components as

$$\tilde{E}_x = -\frac{1}{\hat{\gamma}^2 + \omega^2\mu\epsilon}\left(\hat{\gamma}\frac{\partial \tilde{E}_z}{\partial x} + j\omega\mu\frac{\partial \tilde{H}_z}{\partial y}\right) \tag{10.5a}$$

$$\tilde{E}_y = \frac{1}{\hat{\gamma}^2 + \omega^2\mu\epsilon}\left(-\hat{\gamma}\frac{\partial \tilde{E}_z}{\partial y} + j\omega\mu\frac{\partial \tilde{H}_z}{\partial x}\right) \tag{10.5b}$$

$$\tilde{H}_x = \frac{1}{\hat{\gamma}^2 + \omega^2\mu\epsilon}\left(-\hat{\gamma}\frac{\partial \tilde{H}_z}{\partial x} + j\omega\epsilon\frac{\partial \tilde{E}_z}{\partial y}\right) \tag{10.5c}$$

$$\tilde{H}_y = -\frac{1}{\hat{\gamma}^2 + \omega^2\mu\epsilon}\left(\hat{\gamma}\frac{\partial \tilde{H}_z}{\partial y} + j\omega\epsilon\frac{\partial \tilde{E}_z}{\partial x}\right) \tag{10.5d}$$

The set of equations (10.5) suggests that we need to solve (10.2c) and (10.2f) to obtain solutions of $\tilde{E}_z$ and $\tilde{H}_z$. But that poses no problem—as soon as we obtain a solution for, say, $\tilde{E}_z$ from (10.2c) we can write a similar solution for $\tilde{H}_z$. In order to solve the two-dimensional second-order differential equation we begin with a product solution of (10.2c) as

$$\tilde{E}_z(x, y) = \tilde{X}(x)\tilde{Y}(y) \tag{10.6}$$

and evaluate $\tilde{X}(x)$ and $\tilde{Y}(y)$ using the technique of separation of variables. By substituting (10.6) in (10.2c), we obtain

$$\frac{\partial^2}{\partial x^2}[\tilde{X}(x)\tilde{Y}(y)] + \frac{\partial^2}{\partial y^2}[\tilde{X}(x)\tilde{Y}(y)] = -(\omega^2\mu\epsilon + \hat{\gamma}^2)\tilde{X}(x)\tilde{Y}(y)$$

or

$$\frac{1}{\tilde{X}(x)}\frac{\partial^2\tilde{X}(x)}{\partial x^2} + \frac{1}{\tilde{Y}(y)}\frac{\partial^2\tilde{Y}(y)}{\partial y^2} = -(\omega^2\mu\epsilon + \hat{\gamma}^2) \tag{10.7}$$

The left-hand side of (10.7) is the algebraic sum of two terms, each a function of one variable only, and their sum is equal to a constant. Therefore, each term must then be a constant as follows:

$$\frac{1}{\tilde{X}(x)}\frac{\partial^2\tilde{X}(x)}{\partial x^2} = -M^2 \tag{10.8a}$$

and

$$\frac{1}{\tilde{Y}(y)}\frac{\partial^2\tilde{Y}(y)}{\partial y^2} = -N^2 \tag{10.8b}$$

where M and N are two arbitrary constants.

Substituting (10.8) into (10.7), we obtain

$$M^2 + N^2 = \omega^2\mu\epsilon + \hat{\gamma}^2$$

or

$$\hat{\gamma} = \sqrt{\omega^2\mu\epsilon - (M^2 + N^2)} \tag{10.9}$$

The solutions to (10.8) are

$$\tilde{X}(x) = \hat{C}_1 \sin(Mx) + \hat{C}_2 \cos(Mx) \tag{10.10a}$$

and

$$\tilde{Y}(y) = \hat{C}_3 \sin(Ny) + \hat{C}_4 \cos(Ny) \tag{10.10b}$$

respectively. Here, $\hat{C}_1$, $\hat{C}_2$, $\hat{C}_3$, and $\hat{C}_4$ are arbitrary complex constants and will be determined from the boundary conditions. Thus, the general solution of (10.2c) yields $\tilde{E}_z$ for the forward-traveling wave as

$$\begin{aligned}\tilde{E}_z(x, y, z) = {}& [\hat{C}_1 \sin(Mx) + \hat{C}_2 \cos(Mx)] \\ & \times [\hat{C}_3 \sin(Ny) + \hat{C}_4 \cos(Ny)]e^{-\hat{\gamma}z}\end{aligned} \tag{10.11a}$$

Similarly, the solution of (10.2f) for the z component of the magnetic field, $\tilde{H}_z$, can be written as

$$\begin{aligned}\tilde{H}_z(x, y, z) = {}& [\hat{K}_1 \sin(Mx) + \hat{K}_2 \cos(Mx)] \\ & \times [\hat{K}_3 \sin(Ny) + \hat{K}_4 \cos(Ny)]e^{-\hat{\gamma}z}\end{aligned} \tag{10.11b}$$

where $\hat{K}_1$, $\hat{K}_2$, $\hat{K}_3$, and $\hat{K}_4$ are unknown complex constants. We can now determine the general solutions for other field components using the set of equations in (10.5).

In the following sections, instead of continuing with general solutions, we will tailor our solutions for the two types of modes we have mentioned earlier: the transverse magnetic (TM) mode and the transverse electric (TE) mode.

10.3 Transverse magnetic (TM) mode

In the TM mode, the magnetic field has components only in the transverse (xy) plane. In other words, the longitudinal (z-direction) component of the magnetic field $\tilde{H}_z$ is zero, and $\tilde{E}_x$, $\tilde{E}_y$, and $\tilde{E}_z$ can all exist.

The general solution of $\tilde{E}_z$ as given in (10.11a) must satisfy the boundary conditions at each of the four perfectly conducting walls of the waveguide. Note that $\tilde{E}_z$ represents a tangential field at each boundary. The tangential components of the $\vec{\mathbf{E}}$ field must be continuous, and the $\vec{\mathbf{E}}$ field inside a perfectly conducting wall is zero; therefore, $\tilde{E}_z$ must be zero at each boundary. Thus, the four boundary conditions are

$$\text{At} \quad x = 0 \qquad \tilde{E}_z(0, y, z) = 0 \tag{10.12a}$$

$$\text{At} \quad y = 0 \qquad \tilde{E}_z(x, 0, z) = 0 \tag{10.12b}$$

$$\text{At} \quad x = a \qquad \tilde{E}_z(a, y, z) = 0 \tag{10.12c}$$

$$\text{At} \quad y = b \qquad \tilde{E}_z(x, b, z) = 0 \tag{10.12d}$$

Applying the boundary conditions at $x = 0$ and $y = 0$, equation (10.11a), we obtain $\hat{C}_2 = 0$ and $\hat{C}_4 = 0$. Therefore,

$$\tilde{E}_z(x, y, z) = \hat{C}_1\hat{C}_3 \sin(Mx)\sin(Ny)e^{-\hat{\gamma}z}$$

or

$$\tilde{E}_z(x, y, z) = \hat{E}_{zm} \sin(Mx)\sin(Ny)e^{-\hat{\gamma}z} \tag{10.13}$$

where $\hat{E}_{zm} = \hat{C}_1\hat{C}_3 = E_{zm}\underline{/\phi}$, and E_{zm} is the amplitude of $\hat{E}_{zm}$.

When we apply the boundary conditions at $x = a$, we obtain

$$\sin(Ma) = 0$$

or

$$M = \frac{m\pi}{a} \quad \text{for } m = 0, 1, 2, \ldots \tag{10.14}$$

Finally, the application of boundary conditions at $y = b$ yields

$$\sin(Nb) = 0$$

or

$$N = \frac{n\pi}{b} \quad \text{for } n = 0, 1, 2, \ldots \tag{10.15}$$

The z component of the electric field inside the waveguide can now be written as

$$\tilde{E}_z = \hat{E}_{zm} \sin\left(\frac{m\pi}{a}x\right) \sin\left(\frac{n\pi}{b}y\right) e^{-\hat{\gamma}z} \tag{10.16}$$

With $\tilde{E}_z$ as given in equation (10.16) and $\tilde{H}_z = 0$, we can determine the other field components for the TM mode using (10.5) as

$$\tilde{E}_x = -\frac{\hat{\gamma}}{\hat{\gamma}^2 + \omega^2\mu\epsilon} M\hat{E}_{zm} \cos(Mx)\sin(Ny)e^{-\hat{\gamma}z} \tag{10.17a}$$

$$\tilde{E}_y = -\frac{\hat{\gamma}}{\hat{\gamma}^2 + \omega^2\mu\epsilon} N\hat{E}_{zm} \sin(Mx)\cos(Ny)e^{-\hat{\gamma}z} \tag{10.17b}$$

$$\tilde{H}_x = \frac{j\omega\epsilon}{\hat{\gamma}^2 + \omega^2\mu\epsilon} N\hat{E}_{zm} \sin(Mx)\cos(Ny)e^{-\hat{\gamma}z} \tag{10.17c}$$

$$\tilde{H}_y = -\frac{j\omega\epsilon}{\hat{\gamma}^2 + \omega^2\mu\epsilon} M\hat{E}_{zm} \cos(Mx)\sin(Ny)e^{-\hat{\gamma}z} \tag{10.17d}$$

where $\hat{\gamma}^2 + \omega^2\mu\epsilon = M^2 + N^2$

As can be noted from these equations, a rectangular waveguide can support an infinite number of TM modes for each integer value of m and n, and they are denoted by TM_{mn}. However, if either m or n is zero, no field will exist in the waveguide as is apparent from (10.16). Therefore, the lowest possible TM mode is the TM_{11} mode, whose field plots and excitation schemes are given in Figures 10.3 and 10.4 (see below). The excitation of a particular mode in a waveguide can be accomplished either by coupling the electric field by means of a probe (antenna), as shown in Figure 10.4a, or by coupling the magnetic field by means of a loop (loop antenna), as shown in Figure 10.4b, at locations where the corresponding fields are the highest in the waveguide.

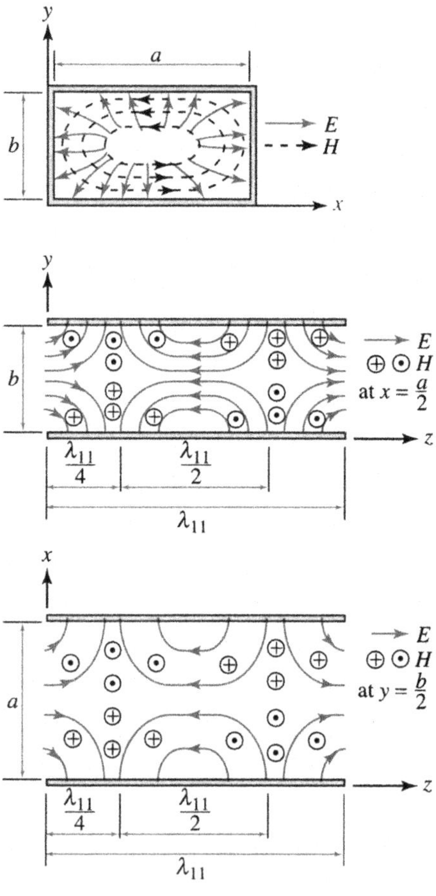

Figure 10.3 Field configurations of TM_{11} mode

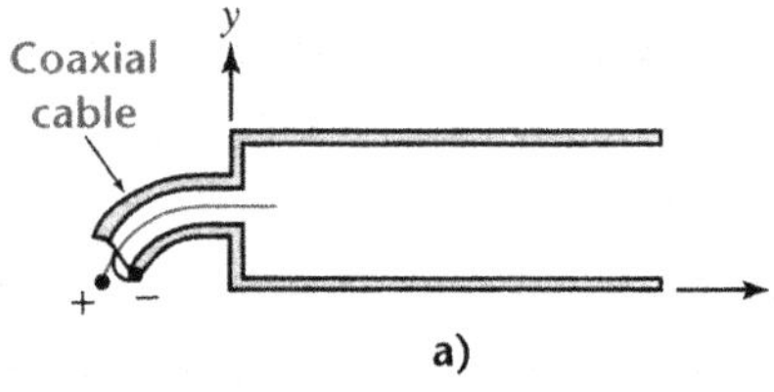

Figure 10.4(a) Excitation of TM_{11} by a probe

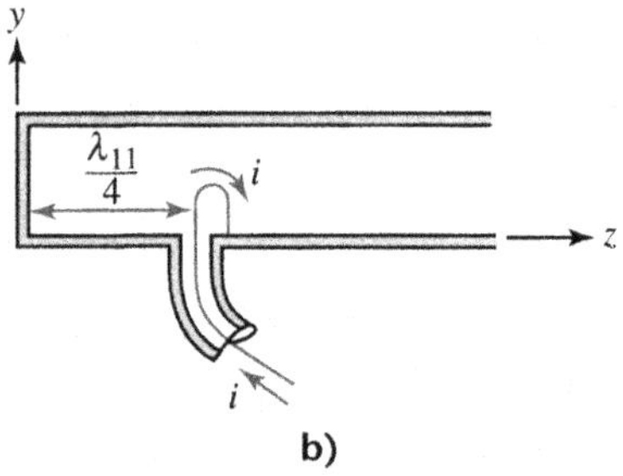

Figure 10.4(b) Excitation of TM_{11} by a loop

From (10.9), the general expression for the propagation constant is

$$\hat{\gamma}_{mn} = \sqrt{\left(\frac{m\pi}{a}\right)^2 + \left(\frac{n\pi}{b}\right)^2 - \omega^2\mu\epsilon} = \alpha_{mn} + j\beta_{mn} \tag{10.18}$$

where α_{mn} and β_{mn} are the attenuation and phase constants, respectively, for the TM_{mn} mode. From (10.18), if $\hat{\gamma}_{mn}$ is real, i.e., $\hat{\gamma}_{mn} = \alpha_{mn}$, no wave propagation will take place because the electric and magnetic fields inside the waveguide will be sharply attenuated. On the other hand, if $\hat{\gamma}_{mn}$ is imaginary, that is, $\hat{\gamma}_{mn} = j\beta_{mn}$, electromagnetic waves will travel along the waveguide with no attenuation.

In order for $\hat{\gamma}_{mn}$ to be real or imaginary, the argument of the square-root function in (10.18) must be positive or negative, respectively. Therefore at one frequency, $\omega = \omega_{cmn}$, the argument must be zero; that is,

$$\left(\frac{m\pi}{a}\right)^2 + \left(\frac{n\pi}{b}\right)^2 - \omega_{cmn}^2\mu\epsilon = 0 \tag{10.19}$$

which yields

$$\omega_{cmn} = u_p\sqrt{\left(\frac{m\pi}{a}\right)^2 + \left(\frac{n\pi}{b}\right)^2} \tag{10.20}$$

where $\omega_{cmn} = 2\pi f_{cmn}$ is known as the *angular cutoff frequency* of the TM_{mn} mode, and $u_p = \dfrac{1}{\sqrt{\mu\epsilon}}$ is the phase velocity in the unbounded medium.

Equation (10.20) can also be expressed as

$$f_{cmn} = \frac{u_p}{2}\sqrt{\left(\frac{m}{a}\right)^2 + \left(\frac{n}{b}\right)^2} \tag{10.21}$$

where f_{cmn} is the *cutoff frequency* of the TM_{mn} mode. We call f_{cmn} the cutoff frequency because it is the frequency above which electromagnetic wave propagation occurs in the waveguide. Let us now discuss the two distinct operating regions of a rectangular waveguide: namely excitations with frequencies less than ($f < f_{cmn}$), and greater than the cutoff frequency ($f > f_{cmn}$).

10.3.1 Operation below cutoff frequency

If the operating frequency f is less than the cutoff frequency f_{cmn} for a given rectangular waveguide, then

$$\left(\frac{m\pi}{a}\right)^2 + \left(\frac{n\pi}{b}\right)^2 - \omega^2\mu\epsilon > 0 \tag{10.22}$$

The inequality criterion in (10.22) leads to

$$\hat{\gamma}_{mn} = \alpha_{mn} = \sqrt{\left(\frac{m\pi}{a}\right)^2 + \left(\frac{n\pi}{b}\right)^2 - \omega^2\mu\epsilon} \tag{10.23}$$

which may also be expressed in conjunction with (10.20) as

$$\alpha_{mn} = \frac{1}{u_p}\sqrt{\omega_{cmn}^2 - \omega^2} \tag{10.24}$$

or

$$\alpha_{mn} = \frac{1}{u_p}\sqrt{\left(\frac{f_{cmn}}{f}\right)^2 - 1} \tag{10.25}$$

Substituting α_{mn} for $\hat{\gamma}_{mn}$ in (10.16), we get

$$\tilde{E}_z = \hat{E}_{zm} \sin\left(\frac{m\pi}{a}x\right) \sin\left(\frac{n\pi}{b}y\right) e^{-\alpha_{mn} z} \tag{10.26}$$

As is evident from (10.26), $\tilde{E}_z$ attenuates along the direction of propagation. As all field components are dependent on $\tilde{E}_z$ they will also attenuate inside the waveguide. Therefore, no wave propagation will take place for a given mode if the operating frequency of the waveguide is less than the cutoff frequency of that mode. Such a wave is said to be *evanescent*.

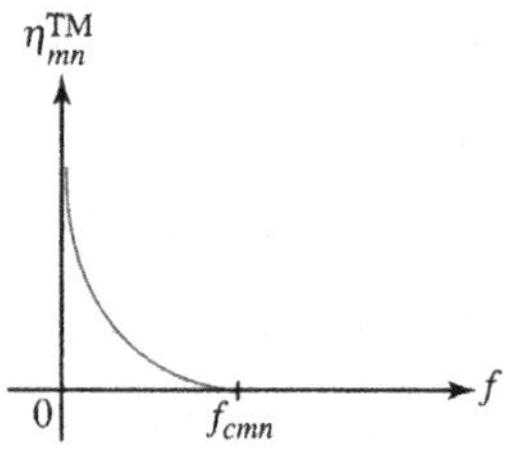

Figure 10.5 Magnitude of wave impedance for $f < f_{cmn}$

The ratio of $\tilde{E}_x$ to $\tilde{H}_y$ yields the wave impedance as

$$\hat{\eta}^{TM}_{mn} = -j\frac{\alpha_{mn}}{\omega\epsilon} \tag{10.27}$$

when the operating frequency is less than the cutoff frequency of a particular TM_{mn} mode. Equation (10.27) points out, once again, that the average power flow will not exist in the waveguide because the wave impedance is purely capacitive, causing only a reactive power, or energy storage, in the waveguide. Figure 10.5 shows the variation of the wave impedance as a function of frequency for $0 \leq f \leq f_{cmn}$.

10.3.2 Operation above cutoff frequency

With an operating frequency f greater than the cutoff frequency f_{cmn}, we have

$$\left(\frac{m\pi}{a}\right)^2 + \left(\frac{n\pi}{b}\right)^2 - \omega^2\mu\epsilon < 0 \tag{10.28}$$

which gives rise to

$$\hat{\gamma}_{mn} = j\beta_{mn} = j\sqrt{\omega^2\mu\epsilon - \left[\left(\frac{m\pi}{a}\right)^2 + \left(\frac{n\pi}{b}\right)^2\right]} \tag{10.29}$$

The phase constant β_{mn} can also be expressed in terms of the cutoff frequency as

$$\beta_{mn} = \frac{1}{u_p}\sqrt{\omega^2 - \omega^2_{cmn}} \tag{10.30}$$

or

$$\beta_{mn} = \frac{\omega}{u_p}\sqrt{1 - \left(\frac{f_{cmn}}{f}\right)^2}$$

$$= \beta\sqrt{1 - \left(\frac{f_{cmn}}{f}\right)^2} \tag{10.31}$$

where $\beta = \omega/u_p$ is the phase constant in the unbounded medium.

The wavelength of this wave within the waveguide can be calculated from $\lambda_{mn} = \dfrac{2\pi}{\beta_{mn}}$ as

$$\lambda_{mn} = \frac{2\pi u_p}{\omega\sqrt{1 - \left(\dfrac{f_{cmn}}{f}\right)^2}}$$

$$= \frac{\lambda}{\sqrt{1 - \left(\dfrac{f_{cmn}}{f}\right)^2}} \tag{10.32}$$

where $\lambda = 2\pi/\beta$ is the wavelength in an unbounded medium. Equation (10.32) can also be expressed as

$$\frac{1}{\lambda_{mn}^2} = \frac{1}{\lambda^2} - \frac{1}{\lambda_{cmn}^2} \tag{10.33}$$

where $\lambda_{cmn} = \dfrac{u_p}{f_{cmn}}$ is the wavelength at the cutoff frequency and is referred to as the cutoff wavelength.

The phase velocity of the wave traveling along the waveguide is

$$u_{pmn} = \frac{\omega}{\beta_{mn}} \tag{10.34}$$

or

$$u_{pmn} = \frac{u_p}{\sqrt{1 - \left(\dfrac{f_{cmn}}{f}\right)^2}} \tag{10.35}$$

where u_p is the phase velocity in the unbounded medium. As can be noted from (10.35), the phase velocity in a rectangular waveguide is greater than the speed of the wave in the unbounded medium. In fact, it can even approach infinity as the operating frequency approaches the cutoff frequency of the mode. Therefore, a waveguide acts as a dispersive medium.

The phase velocity is actually the speed of the constant phase points on the wave as the wave travels. In other words, it is not the velocity of the propagating energy within the waveguide. The energy propagates at a speed equal to the group velocity, u_{gmn}, of the wave. The group velocity of a wave propagating in any medium was defined in Chapter 8 as $u_g = \left(\dfrac{d\beta}{d\omega}\right)^{-1}$. Thus, the group velocity of the TM_{mn} mode is

$$u_{gmn} = \left(\frac{d\beta_{mn}}{d\omega}\right)^{-1} = u_p\sqrt{1 - \left(\frac{f_{cmn}}{f}\right)^2} \tag{10.36}$$

Figure 10.6 shows the variations of the phase and group velocities with the operating frequency. Figure 10.7 illustrates α_{mn} and β_{mn} as functions of frequency.

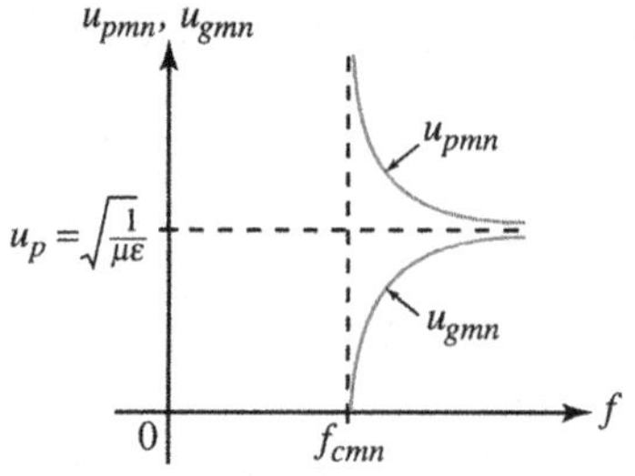

Figure 10.6 Variation of u_{pmn} and u_{gmn} with the frequency

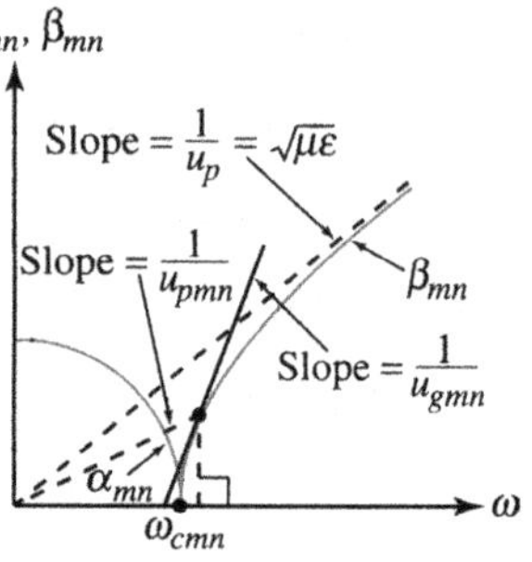

Figure 10.7 α_{mn} and β_{mn} as functions of ω

Using equations (10.34), (10.35), and (10.36), we can also obtain a relationship between the three velocities u_p, u_{pmn}, and u_{gmn}, as

$$\sqrt{u_{pmn}u_{gmn}} = u_p \tag{10.37}$$

The wave impedance for the propagating modes can be obtained from (10.17a) and (10.17d) as

$$\hat{\eta}_{mn}^{\text{TM}} = \frac{\hat{\gamma}_{mn}}{j\omega\epsilon}$$

or

$$\eta_{mn}^{\text{TM}} = \eta\sqrt{1 - \left(\frac{f_{cmn}}{f}\right)^2} \tag{10.38}$$

where $\eta = \sqrt{\mu/\epsilon}$ is the intrinsic impedance of the unbounded medium. Also, it should be noted that the wave impedance in (10.38) is purely resistive and always smaller than that of an unbounded medium. The variation of the wave impedance as a function of frequency is given in Figure 10.8.

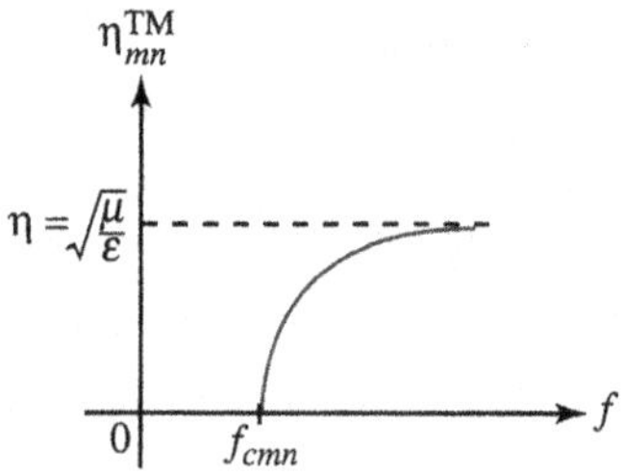

Figure 10.8 Wave impedance as a function of frequency ($f > f_{cmn}$)

10.3.3 Power flow in TM mode

In order to determine the average power flow in a waveguide, let us first compute the average power density $\langle \hat{S} \rangle$ inside the waveguide from

$$\langle \hat{S} \rangle = \tfrac{1}{2}\,\text{Re}[\vec{\mathbf{E}} \times \vec{\mathbf{H}}^*] \tag{10.39}$$

For the TM_{mn} mode the average power density in the z direction becomes

$$\langle \hat{S}_{mn} \rangle_z = \tfrac{1}{2}\mathrm{Re}\left[\tilde{E}_x \tilde{H}_y^* - \tilde{E}_y \tilde{H}_x^*\right] \tag{10.40}$$

or

$$\langle \hat{S}_{mn} \rangle_z = \frac{1}{2\eta_{mn}^{\mathrm{TM}}}\left(E_x^2 + E_y^2\right) \tag{10.41}$$

Substituting for E_x and E_y and replacing the $\hat{\gamma}$'s with $j\beta_{mn}$, we obtain the average power density as

$$\langle \vec{\mathbf{S}}_{mn} \rangle = \left[\frac{\beta_{mn}^2}{2\left(\omega^2\mu\epsilon - \beta_{mn}^2\right)^2} M^2 \frac{E_{zm}^2}{\eta_{mn}^{\mathrm{TM}}} \cos^2(Mx)\sin^2(Ny) + \frac{\beta_{mn}^2}{2\left(\omega^2\mu\epsilon - \beta_{mn}^2\right)^2} N^2 \frac{E_{zm}^2}{\eta_{mn}^{\mathrm{TM}}} \sin^2(Mx)\cos^2(Ny)\right] \vec{\mathbf{a}}_z \tag{10.42}$$

The average power flow within the waveguide can now be computed as

$$\begin{aligned}\langle P_{mn} \rangle &= \int_s \langle \vec{\mathbf{S}}_{mn} \rangle \cdot \vec{d\mathbf{s}} \\ &= \frac{\beta_{mn}^2 E_{zm}^2}{2\left(\omega^2\mu\epsilon - \beta_{mn}^2\right)^2 \eta}\left[\int_0^b \int_0^a M^2 \cos^2(Mx)\sin^2(Ny)\,dx\,dy + \int_0^b \int_0^a N^2 \sin^2(Mx)\cos^2(Ny)\,dx\,dy\right]\end{aligned} \tag{10.43}$$

as

$$\langle P_{mn} \rangle = \frac{\beta_{mn}^2 ab E_{zm}^2}{8\eta_{mn}^{\mathrm{TM}}(M^2+N^2)^2}(M^2+N^2) \tag{10.44}$$

After substituting $M = m\pi/a$ and $N = n\pi/b$ in this equation, we obtain the average power within the waveguide as

$$\langle P_{mn} \rangle = \frac{\beta_{mn}^2 a^3 b^3}{8\pi^2 \eta_{mn}^{\mathrm{TM}}(n^2a^2 + m^2b^2)} E_{zm}^2 \tag{10.45}$$

The following examples are intended to show how to (a) write the field equations inside a waveguide, (b) determine whether a given mode can propagate or not, and (c) calculate the phase velocity, the group velocity, the wavelength, etc., associated with a mode that can exist in a waveguide.

EXAMPLE 10.1

Determine the cutoff frequency, electric and magnetic field distributions, phase constant, characteristic impedance, and average power flow for the lowest-order TM_{11} mode in a waveguide.

Solution From (10.21) the cutoff frequency is

$$f_{c11} = \frac{u_p}{2}\sqrt{\frac{1}{a^2} + \frac{1}{b^2}} = \frac{u_p}{2a}\sqrt{1 + \left(\frac{a}{b}\right)^2} \tag{10.46}$$

For a propagating TM_{11} mode, the field components in a rectangular waveguide, from (10.16) and (10.17), are

$$\tilde{E}_x = -\frac{j\beta_{11}\pi \hat{E}_{zm}}{(\omega^2\mu\epsilon - \beta_{11}^2)a}\cos\left(\frac{\pi}{a}x\right)\sin\left(\frac{\pi}{b}y\right)e^{-j\beta_{11}z} \tag{10.47a}$$

$$\tilde{E}_y = -\frac{j\beta_{11}\pi \hat{E}_{zm}}{(\omega^2\mu\epsilon - \beta_{11}^2)b}\sin\left(\frac{\pi}{a}x\right)\cos\left(\frac{\pi}{b}y\right)e^{-j\beta_{11}z} \tag{10.47b}$$

$$\tilde{E}_z = \hat{E}_{zm}\sin\left(\frac{\pi}{a}x\right)\sin\left(\frac{\pi}{b}y\right)e^{-j\beta_{11}z} \tag{10.47c}$$

$$\tilde{H}_x = j\frac{\omega\epsilon\pi \hat{E}_{zm}}{(\omega^2\mu\epsilon - \beta_{11}^2)b}\sin\left(\frac{\pi}{a}x\right)\cos\left(\frac{\pi}{b}y\right)e^{-j\beta_{11}z} \tag{10.47d}$$

$$\tilde{H}_y = -j\frac{\omega\epsilon\pi \hat{E}_{zm}}{(\omega^2\mu\epsilon - \beta_{11}^2)a}\cos\left(\frac{\pi}{a}x\right)\sin\left(\frac{\pi}{b}y\right)e^{-j\beta_{11}z} \tag{10.47e}$$

$$\tilde{H}_z = 0 \tag{10.47f}$$

where

$$\beta_{11} = \frac{\omega}{u_p}\sqrt{1 - \left(\frac{f_{c11}}{f}\right)^2}$$

or in terms of the waveguide dimensions

$$\beta_{11} = \frac{\pi}{u_p}\sqrt{4f^2 - u_p^2\left(\frac{1}{a^2} + \frac{1}{b^2}\right)} \tag{10.48}$$

For the TM_{11} mode, the wave impedance can be determined from (10.38) as

$$\eta_{11}^{\mathrm{TM}} = \eta\sqrt{1 - \left(\frac{f_{c11}}{f}\right)^2}$$

or

$$\eta_{11}^{\mathrm{TM}} = \frac{\eta}{2f}\sqrt{4f^2 - u_p^2\left(\frac{1}{a^2} + \frac{1}{b^2}\right)} \tag{10.49}$$

and the average power flow can be obtained from (10.45) as

$$\langle P_{11}\rangle = \frac{\beta_{11}^2 a^3 b^3}{8\pi^2\eta_{11}^{\mathrm{TM}}(a^2 + b^2)}E_{zm}^2 \tag{10.50}$$

• • •

EXAMPLE 10.2

WORKSHEET 15

Mathcad

An air-filled rectangular waveguide operates at a frequency of 1 GHz and has $a = 5$ cm and $b = 2$ cm. (a) Show that the TM_{21} mode cannot propagate at this frequency (evanescent mode). (b) Determine the distance from the source at which the z-component of the electric field reduces to 0.5% of its amplitude at $z = 0$. The amplitude of the z-component of the electric field at $z = 0$ is 1 kV/m.

Solution a) The propagation constant for TM_{21} mode is obtained from (10.18) as

$$\hat{\gamma}_{21} = \sqrt{\left(\frac{2\pi}{5 \times 10^{-2}}\right)^2 + \left(\frac{\pi}{2 \times 10^{-2}}\right)^2 - (2\pi \times 10^9)^2 \times 4\pi \times 10^{-7} \times 8.85 \times 10^{-12}}$$

$$\hat{\gamma}_{21} = 200 + j0 \quad \alpha_{21} = 200 \text{ Np/m}$$

The imaginary part of the propagation constant $\hat{\gamma}_{21}$ is zero, thus the TM_{21} mode cannot propagate at 1 GHz and, for all practical purposes, decays after traveling a distance of $5/\alpha_{21}$ within the waveguide.

b) Let the amplitude of the electric field at $z = d$ be 0.5% of that at $z = 0$. Then,

$$E_z(d) = 0.005 \times 1000 = 5 \text{ V/m}$$

As $E_z(d) = 1000e^{-\alpha_{21}d}$ and $E_z(d) = 5$ V,

$$d = \frac{1}{200} \ln\left(\frac{1000}{5}\right) = 26.5 \times 10^{-3} \text{ m or } 26.5 \text{ mm}$$

• • •

EXAMPLE 10.3

Excite the waveguide given in Example 10.2 at a frequency of 20 GHz.

a) Calculate the cutoff frequency for TM_{21} and determine if the mode represents an evanescent wave. If not, determine the phase constant, phase velocity, group velocity, and wavelength of propagation.
b) Calculate the wave impedance of TM_{21}.
c) Find the amount of average power transmitted by the waveguide if the amplitude of the applied electric field is 500 V/m.

Solution a) The cutoff frequency for TM_{21} can be calculated from (10.21) as

$$f_{c21} = \frac{3 \times 10^8}{2}\sqrt{\left(\frac{2}{0.05}\right)^2 + \left(\frac{1}{0.02}\right)^2}$$
$$= 9.6 \times 10^9 \text{ Hz}$$

The cutoff frequency of the waveguide for TM_{21} is smaller than the operating frequency of 20 GHz. Hence, wave propagation takes place inside the waveguide.

The phase constant is calculated from (10.31):

$$\beta_{21} = \frac{2\pi \times 20 \times 10^9}{3 \times 10^8}\sqrt{1 - \left(\frac{9.6 \times 10^9}{20 \times 10^9}\right)^2}$$
$$= 367.47 \text{ rad/m}$$

The phase velocity is

$$u_{p21} = \frac{2\pi \times 20 \times 10^9}{367.47} = 3.42 \times 10^8 \text{ m/s}$$

and the group velocity is obtained from (10.37) as

$$u_{g21} = \frac{(3 \times 10^8)^2}{3.42 \times 10^8} = 2.63 \times 10^8 \text{ m/s}$$

Finally, the wavelength for TM_{21} is calculated using (10.32) as

$$\lambda_{21} = \frac{2\pi}{\beta_{21}} = \frac{2\pi}{367.47} = 0.017 \text{ m or } 17 \text{ mm}$$

b) The characteristic impedance of TM_{21} is

$$\eta_{21}^{TM} = -j\frac{\hat{\gamma}_{21}}{\omega\epsilon} = -j\frac{j367.47}{2\pi \times 20 \times 10^9 \times 8.85 \times 10^{-12}}$$
$$= 330.42\ \Omega$$

c) The average power flow in this lossless waveguide can be evaluated using (10.45) as

$$\langle P_{21} \rangle = \frac{(367.47)^2 \times (0.05)^3(0.02)^3}{8\pi^2 \times 330.42 \times [(0.05)^2 + 2^2 \times (0.02)^2]} 500^2$$
$$= 0.3156 \text{ W or } 315.6 \text{ mW}$$

• • •

10.4 Transverse electric (TE) mode

The TE mode is another mode that can be excited in a rectangular waveguide. In this case, the electric field is always transverse to the direction of propagation; that is, the $\tilde{E}_z$ component is zero when the wave propagates in the z direction. However, the components of the magnetic field exist in all directions.

Let us once again consider the rectangular waveguide shown in Figure 10.2. The general solution for $\tilde{H}_z$ is given in (10.11b). From (10.5a), the x component of the electric field, in terms of $\tilde{H}_z$ is

$$\tilde{E}_x = -\frac{j\omega\mu}{\hat{\gamma}^2 + \omega^2\mu\epsilon}\left(\frac{\partial \tilde{H}_z}{\partial y}\right) \tag{10.51}$$

For E_x to be zero at the boundaries, this equation suggests that

$$\frac{\partial \tilde{H}_z}{\partial y} = 0 \tag{10.52}$$

at $y = 0$ and $y = b$. Differentiating $\tilde{H}_z$ with respect to y, we get

$$\frac{\partial \tilde{H}_z}{\partial y} = [\hat{K}_1 \sin(Mx) + \hat{K}_2 \cos(Mx)]$$
$$\times [\hat{K}_3 N \cos(Ny) - \hat{K}_4 N \sin(Ny)]e^{-\hat{\gamma}z}$$

Applying the boundary condition at $y = 0$, we obtain

$$\hat{K}_3 = 0 \tag{10.53}$$

Because N can have nonzero values, the application of boundary conditions at $y = b$ yields

$$\sin(Nb) = 0$$

or

$$N = \frac{n\pi}{b} \quad n = 0, 1, 2, \ldots \tag{10.54}$$

$\tilde{H}_z$ can now be expressed as

$$\tilde{H}_z(x, y) = [\hat{K}_1 \sin(Mx) + \hat{K}_2 \cos(Mx)]\left[\hat{K}_3 \cos\left(\frac{n\pi}{b} y\right)\right] \tag{10.55}$$

Equation (10.5b) lets us express $\tilde{E}_y$ in terms of $\tilde{H}_z$ as

$$\tilde{E}_y = \frac{j\omega\mu}{\hat{\gamma}^2 + \omega^2\mu\epsilon}\left(\frac{\partial \tilde{H}_z}{\partial x}\right) \tag{10.56}$$

However, $\tilde{E}_y$ must be zero at $x = 0$, and $x = a$. This requirement implies that

$$\frac{\partial \tilde{H}_z}{\partial x} = 0 \tag{10.57}$$

at $x = 0$ and $x = a$. Differentiating (10.55) with respect to x, we obtain

$$\frac{\partial \tilde{H}_z}{\partial x} = [\hat{K}_1 M \cos(Mx) - \hat{K}_2 M \sin(Mx)]\left[\hat{K}_4 \cos\left(\frac{n\pi}{b} y\right)\right]e^{-\hat{\gamma} z}$$

Applying the boundary condition at $x = 0$, we obtain

$$\hat{K}_1 = 0 \tag{10.58}$$

Because M, in general, can be nonzero, the other boundary condition at $x = a$ yields

$$\sin(Ma) = 0$$

or

$$M = \frac{m\pi}{a} \quad m = 0, 1, 2, \ldots \tag{10.59}$$

Finally, the complete expression for the z component of the magnetic field for the TE_{mn} mode is

$$\tilde{H}_z(x, y, z) = \hat{H}_{zm} \cos\left(\frac{m\pi}{a} x\right) \cos\left(\frac{n\pi}{b} y\right) e^{-\hat{\gamma} z} \tag{10.60}$$

where $\hat{H}_{zm} = \hat{K}_2 \hat{K}_4$.

We can now obtain all the other field components for a TE mode from (10.5) as

$$\tilde{H}_x = \frac{\hat{\gamma}}{\hat{\gamma}^2 + \omega^2\mu\epsilon} M\hat{H}_{zm} \sin(Mx)\cos(Ny)e^{-\hat{\gamma}z} \tag{10.61a}$$

$$\tilde{H}_y = \frac{\hat{\gamma}}{\hat{\gamma}^2 + \omega^2\mu\epsilon} N\hat{H}_{zm} \cos(Mx)\sin(Ny)e^{-\hat{\gamma}z} \tag{10.61b}$$

$$\tilde{E}_x = \frac{j\omega\mu}{\hat{\gamma}^2 + \omega^2\mu\epsilon} N\hat{H}_{zm} \cos(Mx)\sin(Ny)e^{-\hat{\gamma}z} \tag{10.61c}$$

$$\tilde{E}_y = \frac{j\omega\mu}{\hat{\gamma}^2 + \omega^2\mu\epsilon} M\hat{H}_{zm} \sin(Mx)\cos(Ny)e^{-\hat{\gamma}z} \tag{10.61d}$$

$$\tilde{E}_z = 0 \tag{10.61e}$$

As can be noted from (10.60), $\tilde{H}_z(x, y, z)$ exists even when both m and n are zero, but in such a case all other field components will disappear, as is evident from (10.61). Thus, the TE_{00} mode cannot exist in a waveguide. However, the field components for the TE_{01} and TE_{10} modes are as follows.

a) TE_{01} mode:

$$\tilde{H}_x = 0, \quad \tilde{E}_y = 0, \quad \tilde{E}_z = 0, \tag{10.62a}$$

$$\tilde{H}_y = j\frac{\beta_{01}b}{\pi}\hat{H}_{zm} \sin\left(\frac{\pi}{b}y\right)e^{-j\beta_{01}z} \tag{10.62b}$$

$$\tilde{H}_z = \hat{H}_{zm} \cos\left(\frac{\pi}{b}y\right)e^{-j\beta_{01}z} \tag{10.62c}$$

$$\tilde{E}_x = j\frac{\omega\mu b}{\pi}\hat{H}_{zm} \sin\left(\frac{\pi}{b}y\right)e^{-j\beta_{01}z} \tag{10.62d}$$

b) TE_{10} mode:

$$\tilde{H}_y = 0, \quad \tilde{E}_x = 0, \quad \tilde{E}_z = 0, \tag{10.63a}$$

$$\tilde{H}_x = j\frac{\beta_{10}a}{\pi}\hat{H}_{zm} \sin\left(\frac{\pi}{a}x\right)e^{-j\beta_{10}z} \tag{10.63b}$$

$$\tilde{H}_z = \hat{H}_{zm} \cos\left(\frac{\pi}{a}x\right)e^{-j\beta_{10}z} \tag{10.63c}$$

$$\tilde{E}_y = -j\frac{\omega\mu a}{\pi}\hat{H}_{zm} \sin\left(\frac{\pi}{a}x\right)e^{-j\beta_{10}z} \tag{10.63d}$$

The presence of $\tilde{E}_x$ and $\tilde{H}_y$ field components for the TE_{01} mode indicates that the power flow can occur in the z direction; thus, this mode can exist in the waveguide. Similarly, the field components $\tilde{E}_y$ and $\tilde{H}_x$ for the TE_{10} mode result in the power flow in the z direction, suggesting that this is a viable mode. Thus, for the TE_{mn} mode, either m or n can be zero, but not both. If $a > b$, then TE_{10} is the lowest-order mode and TE_{01} is the next lowest. The field patterns and the excitation schemes of TE_{10} are illustrated in Figures 10.9 and 10.10.

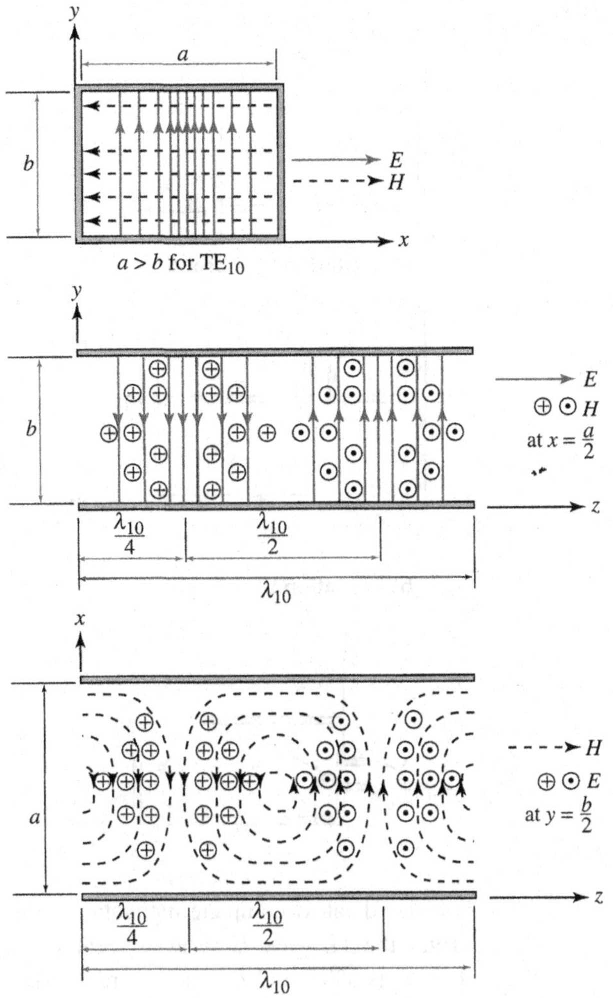

Figure 10.9 Field patterns of TE_{10} mode

As in a TM wave, two distinct operating frequency regions exist for a TE mode separated by the cutoff frequency

$$f_{cmn} = \frac{u_p}{2}\sqrt{\left(\frac{m}{a}\right)^2 + \left(\frac{n}{b}\right)^2} \tag{10.64}$$

Let us now investigate the various characteristics of TE modes for the two distinct operating frequency regions.

10.4.1 Operation below cutoff frequency

When the operating frequency is less than the cutoff frequency of a TE_{mn} wave, we can write the propagation constant from equation (10.22) as

$$\hat{\gamma}_{mn} = \alpha_{mn} = \frac{1}{u_p}\sqrt{\omega_{cmn}^2 - \omega^2} \tag{10.65}$$

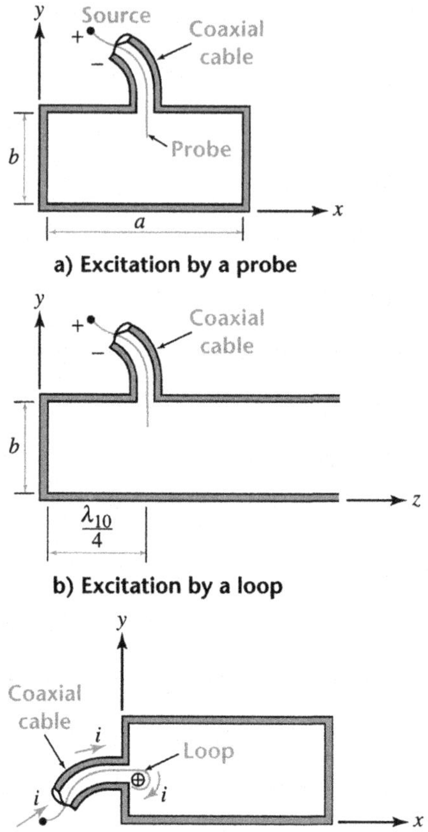

Figure 10.10 Excitation schemes of TE_{10} mode

which indicates a complete attenuation of the wave inside the waveguide. Thus, the TE wave is an evanescent wave. Although the attenuation constants of TE and TM modes are the same, the wave impedance of a TE mode is different than that of a TM mode. It is given by

$$\hat{\eta}_{mn}^{TE} = \frac{\tilde{E}_x}{\tilde{H}_y} = -\frac{\tilde{E}_y}{\tilde{H}_x} = j\frac{\omega\mu}{\alpha_{mn}} \tag{10.66}$$

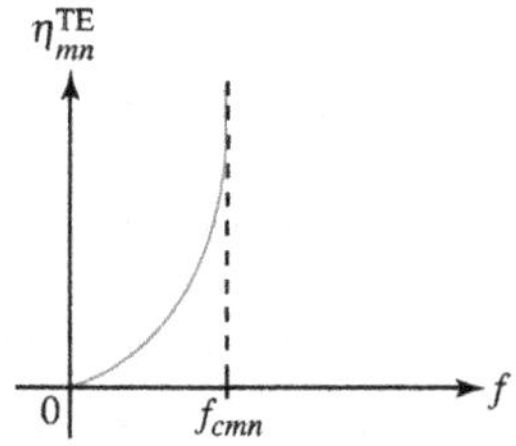

Figure 10.11 $\hat{\eta}_{TE}$ as a function of frequency ($f < f_{cmn}$)

From (10.66) it is evident that no average power flow will exist in the waveguide because the wave impedance is purely inductive. Figure 10.11 illustrates the variation of $\hat{\eta}_{TE}$ as a function of frequency when the operating frequency is less than the cutoff frequency.

10.4.2 Operation above cutoff frequency

When the operating frequency is greater than the cutoff frequency of a TE_{mn} wave, the propagation constant $\hat{\gamma}$ will be an imaginary quantity, as given in (10.29) for a TE_{mn} wave. Since $\hat{\gamma}$ is the same for TE and TM modes, the expressions for the TE phase constant β_{mn}, wavelength

λ_{mn}, phase velocity u_{pmn}, and group velocity u_{gmn} are also exactly the same as those for the TE_{mn} mode.

The wave impedance, on the other hand, is given by

$$\hat{\eta}_{mn}^{TE} = \frac{\tilde{E}_x}{\tilde{H}_y} = -\frac{\tilde{E}_y}{\tilde{H}_x} = \frac{\omega\mu}{\beta_{mn}} \tag{10.67}$$

or

$$\eta_{mn}^{TE} = \frac{\eta}{\sqrt{1-\left(\frac{f_{cmn}}{f}\right)^2}} \tag{10.68}$$

which is different than the wave impedance of a TM wave. The frequency dependence of η_{mn}^{TM} is shown in Figure 10.12. In (10.68) η is the intrinsic impedance of the unbounded medium. Because the wave impedance in (10.68) is resistive, average power flow occurs in the waveguide.

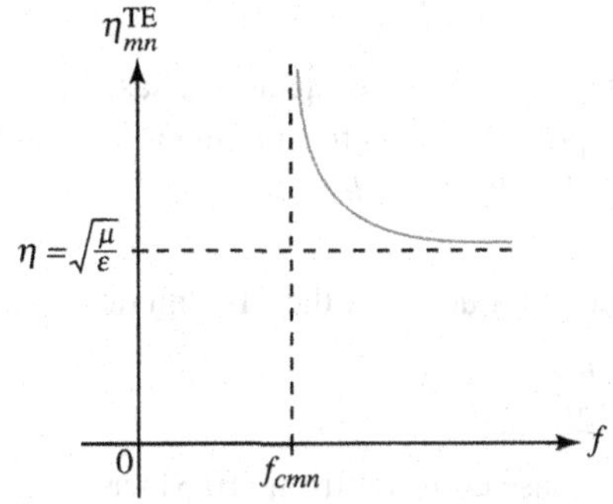

Figure 10.12 Variation of $\hat{\eta}_{mn}^{TE}$ with frequency ($f > f_{cmn}$)

It is also interesting to note, from (10.38) and (10.68), that

$$\sqrt{\eta_{mn}^{TM}\eta_{mn}^{TE}} = \eta = \sqrt{\frac{\mu}{\epsilon}} \tag{10.69}$$

10.4.3 Power flow in TE mode

The average power density for the TE_{mn} mode is

$$\langle S_{mn}\rangle_z = \tfrac{1}{2}\mathrm{Re}(\tilde{E}_x\tilde{H}_y^* - \tilde{E}_y\tilde{H}_x^*)$$

or

$$\langle S_{mn}\rangle_z = \frac{1}{2\eta_{mn}^{TE}}\left(E_x^2 + E_y^2\right) \tag{10.70}$$

or

$$\langle \vec{\mathbf{S}}_{mn}\rangle_z = \left[\eta_{mn}^{TE}\frac{n^2\pi^2\beta_{mn}^2}{2(\omega^2\mu\epsilon - \beta_{mn}^2)^2 b^2}H_{zm}^2\cos^2(Mx)\sin^2(Ny) + \eta_{mn}^{TE}\frac{m^2\pi^2\beta_{mn}^2}{2(\omega^2\mu\epsilon - \beta_{mn}^2)^2 a^2}H_{zm}^2\sin^2(Mx)\cos^2(Ny)\right]\vec{\mathbf{a}}_z \tag{10.71}$$

Therefore, the average power in the waveguide is

$$\langle P_{mn}\rangle = \int_s \langle \vec{\mathbf{S}}_{mn}\rangle \cdot \overrightarrow{d\mathbf{s}}$$

$$= \eta_{mn}^{\mathrm{TE}} \frac{\pi^2 \beta_{mn}^2 H_{zm}^2}{2(\omega^2\mu\epsilon - \beta_{mn}^2)} \left[\int_0^b \int_0^a \frac{n^2}{b^2} \cos^2(Mx) \sin^2(Ny)\, dx\, dy \right.$$
$$\left. + \int_0^b \int_0^a \frac{m^2}{a^2} \sin^2(Mx) \cos^2(Ny)\, dx\, dy \right]$$

or

$$\langle P_{mn}\rangle = \eta_{mn}^{\mathrm{TE}} \left[\frac{\beta_{mn}^2 ab H_{zm}^2}{8(\omega^2\mu\epsilon - \beta_{mn}^2)} \right] \left[\left(\frac{m\pi}{a}\right)^2 + \left(\frac{n\pi}{b}\right)^2 \right] \tag{10.72a}$$

or

$$\langle P_{mn}\rangle = \eta_{mn}^{\mathrm{TE}} \left[\frac{\beta_{mn}^2 a^3 b^3 H_{zm}^2}{8\pi^2 (b^2 m^2 + a^2 n^2)} \right] \tag{10.72b}$$

EXAMPLE 10.4 Determine the cutoff frequency, phase constant, wave impedance, and average power flow for the dominant TE_{10} mode in a lossless rectangular waveguide when $a > b$.

Solution The cutoff frequency of the TE_{10} mode can be obtained from (10.64) as

$$f_{c10} = \frac{u_p}{2a} \tag{10.73}$$

and the phase constant from (10.31) as

$$\beta_{10} = \frac{\omega}{u_p} \sqrt{1 - \left(\frac{f_{c10}}{f}\right)^2}$$

The wave impedance of TE_{10} is

$$\eta_{10}^{\mathrm{TE}} = \frac{\eta}{\sqrt{1 - \left(\dfrac{u_p}{2af}\right)^2}} \tag{10.74}$$

and the average power flow can be calculated for this mode from (10.71) as

$$\langle P_{10}\rangle = \eta_{10}^{\mathrm{TE}} \left[\frac{\beta_{10}^2 \pi^2 H_{zm}^2 b}{4(\omega^2\mu\epsilon - \beta_{10}^2)^2 a} \right] \tag{10.75a}$$

or

$$\langle P_{10}\rangle = \eta_{10}^{\mathrm{TE}} \left[\frac{\beta_{10}^2 a^3 b H_{zm}^2}{4\pi^2} \right] \tag{10.75b}$$

For practical applications, for a given frequency, the dimensions of a waveguide are selected in such a way that only the energy associated with the dominant mode is supported inside the waveguide. • • •

EXAMPLE 10.5

An air-filled waveguide operates at a frequency of 10 GHz. If the dimensions of the waveguide are $a = 2$ cm and $b = 1$ cm, determine the mode of propagation in the waveguide.

Solution Let us first consider the two lowest-order modes, TE_{01} and TE_{10}, and calculate the cutoff frequencies for both of them.

$$\text{TE}_{01}: \quad f_{c01} = \frac{u_p}{2b} = \frac{3 \times 10^8}{0.02} = 1.5 \times 10^{10} \text{ Hz} \quad \text{or} \quad 15 \text{ GHz}$$

Because the operating frequency ($f = 10$ GHz) is smaller than $f_{c01} = 15$ GHz, TE_{01} cannot be a possible propagating mode.

$$\text{TE}_{10}: \quad f_{c10} = \frac{u_p}{2a} = \frac{3 \times 10^8}{2 \times 2 \times 10^{-2}} = 0.75 \times 10^{10} \text{ Hz} \quad \text{or} \quad 7.55 \text{ GHz}$$

Therefore, TE_{10} is the only mode that can propagate in the waveguide.

• • •

EXAMPLE 10.6

WORKSHEET 16

Mathcad

An air-filled waveguide operates at 7 GHz. The dimensions of the waveguide are $a = 3$ cm and $b = 2$ cm. Calculate the maximum average power transmitted along the waveguide without causing any breakdown inside the waveguide at the TE_{10} mode. The dielectric strength of air is 30 kV/cm. Use a safety factor of 10.

Solution The cutoff frequency of TE_{10} is

$$f_{c10} = \frac{u_p}{2a} = \frac{3 \times 10^8}{2 \times 0.03} = 5 \times 10^9 \text{ Hz}$$

The existing component of the electric field for TE_{10}, from (10.61d), is

$$\begin{aligned} \tilde{E}_y &= -j\frac{\omega\mu a}{\pi}\hat{H}_{zm} \sin\left(\frac{\pi}{a}x\right)e^{-j\beta_{10}z} \\ &= \hat{E}_{ym} \sin\left(\frac{\pi}{a}x\right)e^{-j\beta_{10}z} \end{aligned}$$

The dielectric strength of air is given as 30 kV/cm or 3 MV/m, so the maximum value of E_y is 0.3 MV/m with a safety factor of 10. Thus,

$$\tilde{E}_y = 3 \times 10^5 \sin\left(\frac{\pi}{a}x\right)e^{-j\beta_{10}z}$$

The intrinsic impedance of the TE_{10} mode is

$$\eta_{10}^{\text{TE}} = \frac{\eta}{\sqrt{1 - \left(\dfrac{f_{c10}}{f}\right)^2}} = \frac{377}{\sqrt{1 - \left(\dfrac{5}{7}\right)^2}} = 538.68\ \Omega$$

and the maximum average power density, from (10.70), is

$$\langle \vec{\mathbf{S}}_{10} \rangle = \frac{1}{2}\left[\frac{(3 \times 10^5)^2 \sin^2\left(\frac{\pi}{a}x\right)}{538.68}\right]\vec{\mathbf{a}}_z = 83.54 \sin^2\left(\frac{\pi}{a}x\right)\vec{\mathbf{a}}_z \text{ MW/m}^2$$

The maximum power that can be safely transmitted along the waveguide is

$$\langle P_{10}\rangle = \int_0^b \int_0^a 83.54 \times 10^6 \sin^2\left(\frac{\pi}{a}x\right) dx\, dy = 25.06\ \text{kW}$$

• • •

EXAMPLE 10.7

The phase constant of the TE_{10} mode of an air-filled waveguide with $b = 1$ cm is 102.65 rad/m. If the operating frequency of the waveguide is 12 GHz, and the only mode of propagation is TE_{10}, calculate the length a of the waveguide.

Solution The cutoff frequency of TE_{10} is obtained from

$$\beta_{10} = \beta\sqrt{1 - \left(\frac{f_{c10}}{f}\right)^2}$$

$$\beta = \omega\sqrt{\mu_0\epsilon_0} = 2\pi \times 12 \times 10^9\sqrt{4\pi \times 10^{-7} \times 8.85 \times 10^{-12}}$$
$$= 251.44\ \text{rad/m}$$

$$102.65 = 251.44\sqrt{1 - \left(\frac{f_{c10}}{12 \times 10^9}\right)^2}$$

Thus, $f_{c10} = 10.95$ GHz

$$f_{c10} = \frac{u_p}{2a}$$

Therefore,

$$a = \frac{u_p}{2f_{c10}} = \frac{3 \times 10^8}{2 \times 10.95 \times 10^9} = 0.0136\ \text{m}$$
$$a = 1.36\ \text{cm}$$

• • •

10.5 Losses in a waveguide

Our study of fields inside an ideal waveguide, a waveguide with perfectly conducting walls enclosing a perfect dielectric medium, has disclosed that a mode can propagate along the length of the waveguide without any power loss as long as the operating frequency is greater than the cutoff frequency of the mode. In a real waveguide, the dielectric medium is not a perfect dielectric, and the walls are not perfect conductors. Therefore, we expect some power loss in a real waveguide as the wave propagates. In other words, we will expect some attenuation of the fields as they propagate inside the waveguide.

10.5.1 Perfect dielectric medium with finitely conducting walls

Most waveguides are made of copper, brass, or some other type of material, with a thin coating of silver. The conductivity of a coated waveguide, although very high, is not infinite. With such a finite conductivity of the walls, the determination of fields becomes quite complicated because the fields can now exist inside the walls of the waveguide. In order to determine power loss and the attenuation constant inside the walls, we have to first obtain the fields inside the dielectric medium as well as the walls of the waveguide and then apply the boundary conditions at each of the four walls. As you may have guessed, the process, although exact, is quite involved. Because of the high conductivity of the walls, we anticipate the skin depth to be very small. In other words, we envision (a) the fields to concentrate in the region near the inner surface (surface next to the dielectric) of each wall and (b) the losses to be very small. That is, we expect the actual fields inside the waveguide to differ slightly from those when the walls are perfectly conducting. We can assume that the two sets of fields are almost the same. However, because of the finite conductivity of the walls, the fields inside the waveguide attenuate exponentially with an attenuation constant α_{cmn}.

When the fields decay as $e^{-\alpha_{cmn} z}$, the power at any point z in the longitudinal direction of the waveguide can be expressed as

$$\langle P_{mn}(z)\rangle = P_0 e^{-2\alpha_{cmn} z} \tag{10.76}$$

where P_0 is the power at the reference location ($z = 0$). The power loss per unit length in the walls of the waveguide can be computed from the rate of change of power in the z direction:

$$\langle P_d(z)\rangle = -\frac{\partial \langle P_{mn}(z)\rangle}{\partial z} \tag{10.77}$$

where the minus sign accounts for the decrease in power as it propagates in the z direction and $\langle P_d(z)\rangle$ is the average power loss per unit length (expressed in W/m).

Equation (10.76) allows us to express $\langle P_d(z)\rangle$ as

$$\langle P_d(z)\rangle = 2P_0\alpha_{cmn} e^{-2\alpha_{cmn} z} \tag{10.78}$$

Thus, the unknown attenuation constant α_{cmn} is

$$\alpha_{cmn} = \frac{\langle P_d(z)\rangle}{2\langle P_{mn}(z)\rangle} \tag{10.79}$$

We now explain how to find the attenuation constant by using the expressions for the fields for the TE_{10} mode inside a waveguide when the conductivity of each wall is σ_c. We can approximate the fields inside the waveguide for the TE_{10} mode, from (10.60) and (10.61), as

$$\tilde{H}_x = j\frac{\beta_{10} a}{\pi}\hat{H}_{zm} \sin\left(\frac{\pi}{a}x\right) e^{-j\beta_{10} z} e^{-\alpha_{c10} z} \tag{10.80a}$$

$$\tilde{H}_z = \hat{H}_{zm} \cos\left(\frac{\pi}{a}x\right) e^{-j\beta_{10} z} e^{-\alpha_{c10} z} \tag{10.80b}$$

$$\tilde{E}_y = -j\frac{\omega\mu a}{\pi}\hat{H}_{zm} \sin\left(\frac{\pi}{a}x\right) e^{-j\beta_{10} z} e^{-\alpha_{c10} z} \tag{10.80c}$$

where $\hat{\gamma}_{10} = \alpha_{c10} + j\beta_{10}$

The average power density in the z direction is

$$\langle S_{10}\rangle_z = \tfrac{1}{2}\mathrm{Re}[(\tilde{\vec{\mathbf{E}}} \times \tilde{\vec{\mathbf{H}}}^*] \cdot \vec{\mathbf{a}}_z = \tfrac{1}{2}\mathrm{Re}[-\tilde{E}_y \tilde{H}_x^*]$$
$$= \frac{\omega\mu a^2}{2\pi^2}\beta_{10}\, H_{zm}^2 \sin^2\left(\frac{\pi x}{a}\right) e^{-2\alpha_{c10}z} \tag{10.81}$$

The average power flow in the z direction is

$$\langle P_{10}\rangle_z = \int_0^a \int_0^b \langle S_{10}\rangle_z \, dx\, dy$$
$$= \frac{\omega\mu a^2}{2\pi^2}\beta_{10} H_{zm}^2 e^{-2\alpha_{c10}z}\left(\frac{ab}{2}\right)$$
$$= \frac{\omega\mu a^3 b}{4\pi^2}\beta_{10} H_{zm}^2 e^{-2\alpha_{c10}z} \tag{10.82}$$

In order to compute the power loss per unit length in each wall, we will first determine the surface current density by applying the boundary conditions at each surface. We will always assume that region 1 is the dielectric medium inside the waveguide and region 2 is the region inside the conducting wall. We will determine the surface current by assuming that the magnetic field in region 2 is zero as the skin depth in region 2 is negligibly small. That is, $\tilde{\vec{\mathbf{J}}}_s = \vec{\mathbf{a}}_n \times \tilde{\vec{\mathbf{H}}}_1$, where $\vec{\mathbf{a}}_n$ is the unit normal to the surface pointing into region 1. Assuming that the current is uniformly distributed inside each wall within a length of one skin depth δ_c, we can determine the power loss within the walls using the surface resistance, $R_s = 1/\sigma_c\delta_c$, as

$$P_d = \tfrac{1}{2}R_s I^2 \tag{10.83}$$

where I is the amplitude of the total current on the surface of the wall. This procedure is in accordance with our discussion of the skin effect in Chapter 9.

There are four walls and we have to determine the power loss in each of them. But the power loss in the wall at $x = 0$ must be equal to the power loss in the wall at $x = a$. Similarly, the power loss in the wall at $y = 0$ is also equal to that at $y = b$. Thus, the total power loss per unit length is

$$\langle P_d\rangle = 2[\langle P_d\rangle_{x=0} + \langle P_d\rangle_{y=0}] \tag{10.84}$$

To compute the power loss in the wall at $x = 0$, we can determine the surface current density on the wall from

$$\tilde{\vec{\mathbf{J}}}_s = \vec{\mathbf{a}}_n \times \tilde{\vec{\mathbf{H}}}_1 \tag{10.85}$$

In our case, from Figure 10.2, $\vec{\mathbf{a}}_n = \vec{\mathbf{a}}_x$; thus, the surface current density is

$$\tilde{\vec{\mathbf{J}}}_s = -\hat{H}_{zm} e^{-j\beta_{10}z} e^{-\alpha_{c10}z}\vec{\mathbf{a}}_y \tag{10.86}$$

and the power loss per unit length in the z direction is

$$\langle P_d \rangle_{x=0} = \frac{1}{2}\left(\frac{1}{\sigma_c \delta_c}\right) \int_0^b H_{zm}^2 e^{-2\alpha_{c10} z}\, dy = \frac{H_{zm}^2 b}{2\sigma_c \delta_c} e^{-2\alpha_{c10} z} \tag{10.87}$$

The current density at the surface of the wall at $y = 0$ is

$$\begin{aligned} \tilde{\mathbf{J}}_s &= \vec{\mathbf{a}}_y \times (\tilde{H}_x \vec{\mathbf{a}}_x + \tilde{H}_z \vec{\mathbf{a}}_z) \\ &= -j \frac{\beta_{10} a}{\pi} \hat{H}_{zm} \sin\left(\frac{\pi}{a} x\right) e^{-j\beta_{10} z} e^{-\alpha_{c10} z} \vec{\mathbf{a}}_z \\ &\quad + \hat{H}_{zm} \cos\left(\frac{\pi}{a} x\right) e^{-j\beta_{c10} z} e^{-\alpha_{c10} z} \vec{\mathbf{a}}_x \end{aligned} \tag{10.88}$$

and the corresponding power loss is

$$\begin{aligned} \langle P_d \rangle_{y=0} &= \frac{1}{2}\left(\frac{1}{\sigma_c \delta_c}\right) \int_0^a \left[\frac{(\beta_{10} a)^2}{\pi^2} H_{zm}^2 \sin^2\left(\frac{\pi}{a} x\right) e^{-2\alpha_{c10} z} \right. \\ &\quad \left. + H_{zm}^2 \cos^2\left(\frac{\pi}{a} x\right) e^{-2\alpha_{c10} z}\right] dx \\ &= \frac{H_{zm}^2 a}{4\sigma_c \delta_c}\left[1 + \left(\frac{\beta_{10} a}{\pi}\right)^2\right] e^{-2\alpha_{c10} z} \end{aligned} \tag{10.89}$$

The total power loss per unit length is

$$\langle P_d \rangle = \frac{H_{zm}^2}{\sigma_c \delta_c}\left(b + \frac{a}{2} + \frac{\beta_{10}^2 a^3}{2\pi^2}\right) e^{-2\alpha_{c10} z}$$

or

$$\langle P_d \rangle = \frac{H_{zm}^2}{\sigma_c \delta_c}\left[b + \frac{a}{2}\left(\frac{f}{f_{c10}}\right)^2\right] e^{-2\alpha_{c10} z} \tag{10.90}$$

which is obtained by using $\beta_{10}^2 = \omega^2 \mu\epsilon - (\pi/a)^2$ and $f_{c10} = \dfrac{1}{2a\sqrt{\mu\epsilon}}$.

From (10.79), we can now determine α_{c10} as

$$\begin{aligned} \alpha_{c10} &= \frac{1}{2}\left(\frac{\langle P_d \rangle}{\langle P_{10} \rangle_z}\right) = \frac{\dfrac{H_{zm}^2}{\sigma_c \delta_c}\left[b + \dfrac{a}{2}\left(\dfrac{f}{f_{c10}}\right)^2\right] e^{-2\alpha_{c10} z}}{\dfrac{\omega\mu a^3 b}{2\pi^2} \beta_{10} H_{zm}^2 e^{-2\alpha_{c10} z}} \\ &= \left(\frac{1}{\sigma_c \delta_c \eta b}\right) \frac{\left(1 + \dfrac{2b}{a}\left(\dfrac{f_{c10}}{f}\right)^2\right)}{\sqrt{1 - \left(\dfrac{f_{c10}}{f}\right)^2}} \end{aligned} \tag{10.91}$$

This equation shows that α_{c10} is a complicated function of frequency. Its dependency upon frequency is illustrated in Figure 10.13.

10.5.2 Imperfect dielectric medium with perfectly conducting walls

For this case, the conductivity σ_d, however small, of the dielectric medium in the waveguide can be taken into consideration by simply

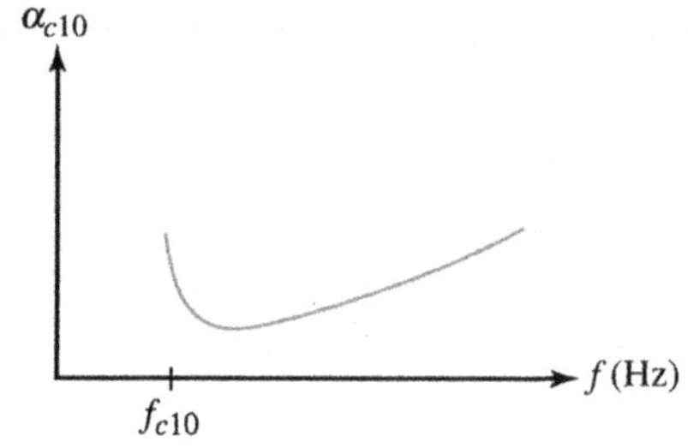

Figure 10.13 The variation of α_{c10} as a function of frequency

replacing ϵ with $\hat{\epsilon}$ where $\hat{\epsilon} = \epsilon - j\sigma_d/\omega$. In this case, the waveguide walls are assumed to be perfect conductors. We can then follow the preceding steps to determine the fields inside the waveguide as long as we assume the walls to be perfectly conducting. This assumption enables us to state that (a) there are no fields inside the walls, and (b) the tangential components of the electric field at the walls are zero. The propagation constant for TE_{10} mode becomes

$$\begin{aligned}\hat{\gamma} &= j\omega\sqrt{\mu\hat{\epsilon}}\sqrt{1-\left(\frac{f_{c10}}{f}\right)^2}\\ &= j\omega\sqrt{\mu\epsilon}\sqrt{1-j\frac{\sigma_d}{\omega\epsilon}}\sqrt{1-\left(\frac{f_{c10}}{f}\right)^2}\\ &= j\beta\sqrt{1-\left(\frac{f_{c10}}{f}\right)^2}\sqrt{1-j\frac{\sigma_d}{\omega\epsilon}}\end{aligned} \tag{10.92}$$

As long as $\sigma_d/\omega\epsilon \ll 1$ for a dielectric medium, the term

$$\sqrt{1-j\frac{\sigma_d}{\omega\epsilon}}$$

can be approximated, using the first-order approximation of the binomial series, as

$$\sqrt{1-j\frac{\sigma_d}{\omega\epsilon}} \cong 1 - j\frac{1}{2}\left(\frac{\sigma_d}{\omega\epsilon}\right) \tag{10.93}$$

Thus,

$$\begin{aligned}\hat{\gamma} &= j\beta\sqrt{1-\left(\frac{f_{c10}}{f}\right)^2}\left[1-j\frac{1}{2}\left(\frac{\sigma_d}{\omega\epsilon}\right)\right]\\ &= \sqrt{1-\left(\frac{f_{c10}}{f}\right)^2}\left[\frac{\beta}{2}\left(\frac{\sigma_d}{\omega\epsilon}\right)+j\beta\right] = \alpha_{d10} + j\beta_{10}\end{aligned} \tag{10.94}$$

where

$$\alpha_{d10} = \frac{\beta}{2}\left(\frac{\sigma_d}{\omega\epsilon}\right)\sqrt{1-\left(\frac{f_{c10}}{f}\right)^2} = \frac{\sigma_d}{2}\eta\sqrt{1-\left(\frac{f_{c10}}{f}\right)^2} \tag{10.95}$$

and

$$\beta_{10} = \beta\sqrt{1 - \left(\frac{f_{c10}}{f}\right)^2} \quad \text{with } \eta = \sqrt{\mu/\epsilon}$$

We must note that the phase constant in a good dielectric medium is the same as that in a perfect dielectric medium as long as $\sigma_d/\omega\epsilon \ll 1$. For good dielectrics at microwave frequencies, the loss tangent $(\sigma_d/\omega\epsilon)$ is usually in the range of 10^{-4} or so. When the walls are finitely conducting and the medium is a good dielectric, the fields can be treated as if they decay with an attenuation constant $\alpha = \alpha_c + \alpha_d$.

EXAMPLE 10.8

Calculate the attenuation constants in an air-filled rectangular waveguide with dimensions $a = 3$ cm and $b = 2$ cm when it is excited with TE_{10} mode at 6 GHz. The loss tangent in air is 0.001, and the conductivity of the copper walls is 5.76×10^7 S/m.

Solution The cutoff frequency:

$$f_{c10} = \frac{u_p}{2a} = \frac{3 \times 10^8}{2 \times 0.03} = 5 \times 10^9 \text{ Hz}$$

The phase constant:

$$\beta_{10} = \omega\sqrt{\mu\epsilon}\sqrt{1 - \left(\frac{f_{c10}}{f}\right)^2}$$

$$= 2\pi \times 6 \times 10^9 \frac{1}{3 \times 10^8}\sqrt{1 - \left(\frac{5 \times 10^9}{6 \times 10^9}\right)^2} = 69.5 \text{ rad/m}$$

The skin depth:

$$\delta_c = \frac{1}{\sqrt{\sigma_c \mu f \pi}} = \frac{1}{\sqrt{5.76 \times 10^7 \times 4\pi \times 10^{-7} \times 6 \times 10^9 \pi}}$$
$$= 8.56 \times 10^{-7} \text{ m} \quad \text{or} \quad 0.856 \text{ mm}$$

The conductivity of air:

$$\sigma_d = \omega\epsilon \tan\delta = 2\pi \times 6 \times 10^9 \times 0.001 \times 8.85 \times 10^{-12}$$
$$= 3.336 \times 10^{-4} \text{ S/m}$$

Thus, the attenuation constant due to the finite conductivity of the walls, from (10.91), is

$$\alpha_{c10} = \frac{\left[1 + \frac{2 \times 0.02}{0.03}\left(\frac{5 \times 10^9}{6 \times 10^9}\right)^2\right]}{5.76 \times 10^7 \times 8.56 \times 10^{-7} \times 377 \times 0.02\sqrt{1 - \left(\frac{5}{6}\right)^2}}$$
$$= 9.372 \times 10^{-3} \text{ Np/m}$$

From (10.95), the attenuation constant due to the dielectric medium is

$$\alpha_{d10} = \tfrac{1}{2} \times 3.336 \times 10^{-4} \times 377\sqrt{1 - (\tfrac{5}{6})^2} = 0.035 \text{ Np/m}$$

• • •

10.6 Cavity resonators

When we close both ends of a waveguide with conducting plates, as shown in Figure 10.14, the closed structure forms a cavity and is referred to as a *cavity resonator*. In a cavity resonator, we expect the wave to yield a standing wave because of its reflections at the closed ends. Only those standing wave patterns (modes) will exist that satisfy the boundary conditions at each of the six walls. When a mode can exist at a certain frequency, called the resonant frequency, it traps energy at that frequency. The entrapment of energy at some discrete frequencies allows us to use a cavity as a frequency meter to measure microwave frequencies. In this case, one wall, say at $z = 0$, is fixed, whereas the opposite wall can be moved back and forth. The movable wall changes the length of the cavity in the z direction, which, in turn, changes its resonant frequency.

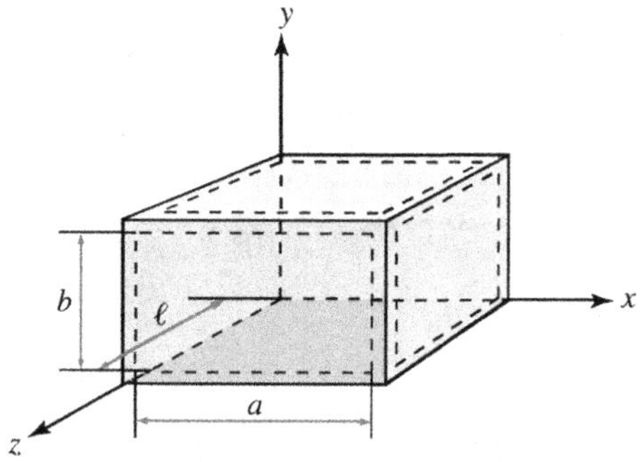

Figure 10.14 A rectangular cavity resonator

In this section we will investigate the field distributions inside the cavity for both TE and TM modes, determine the resonant frequency in each case, and compute the corresponding quality factor.

10.6.1 Transverse magnetic (TM) mode

In a cavity resonator, we expect both the forward and the backward waves to exist. With that understanding, we can express the electric field along the z axis as

$$\tilde{E}_z = \hat{E}^+_{zm} \sin(Mx)\sin(Ny)e^{-\hat{\gamma}z} + \hat{E}^-_{zm} \sin(Mx)\sin(Ny)e^{\hat{\gamma}z} \quad (10.96a)$$

where $\hat{E}^+_{zm}$ and $\hat{E}^-_{zm}$ are the amplitudes of the forward and backward electric fields,

$$M = \frac{m\pi}{a}, N = \frac{n\pi}{b}, m = 1, 2, 3, \ldots, \text{and } n = 1, 2, 3, \ldots.$$

Substituting (10.96a) into (10.5), and replacing $\hat{\gamma}$ with $(-\hat{\gamma})$ in these equations for the backward wave, we obtain

$$\tilde{E}_x = -\frac{\hat{\gamma} M}{\hat{\gamma}^2 + \omega^2 \mu\epsilon}[\hat{E}^+_{zm} \cos(Mx)\sin(Ny)e^{-\hat{\gamma}z} - \hat{E}^-_{zm} \cos(Mx)\sin(Ny)e^{\hat{\gamma}z}] \tag{10.96b}$$

$$\tilde{E}_y = -\frac{\hat{\gamma} N}{\hat{\gamma}^2 + \omega^2 \mu\epsilon}[\hat{E}^+_{zm} \sin(Mx)\cos(Ny)e^{-\hat{\gamma}z} - \hat{E}^-_{zm} \sin(Mx)\cos(Ny)e^{\hat{\gamma}z}] \tag{10.96c}$$

$$\tilde{H}_x = \frac{j\omega\epsilon N}{\hat{\gamma}^2 + \omega^2 \mu\epsilon}[\hat{E}^+_{zm} \sin(Mx)\cos(Ny)e^{-\hat{\gamma}z} + \hat{E}^-_{zm} \sin(Mx)\cos(Ny)e^{\hat{\gamma}z}] \tag{10.96d}$$

$$\tilde{H}_y = -\frac{j\omega\epsilon M}{\hat{\gamma}^2 + \omega^2 \mu\epsilon}[\hat{E}^+_{zm} \cos(Mx)\sin(Ny)e^{-\hat{\gamma}z} + \hat{E}^-_{zm} \cos(Mx)\sin(Ny)e^{\hat{\gamma}z}] \tag{10.96e}$$

$$\tilde{H}_z = 0 \tag{10.96f}$$

When we assume that the walls of the cavity resonator are perfect conductors, the boundary conditions for the tangential components of the electric field are

$$\tilde{E}_x = \tilde{E}_y = 0 \quad \text{at both } z = 0 \text{ and } z = \ell \tag{10.97}$$

The application of boundary conditions to (10.96b) at $z = 0$ yields

$$\hat{E}^+_{zm} = \hat{E}^-_{zm} = \hat{E}_{zm} \tag{10.98}$$

For the existence of a standing wave in a rectangular cavity, $\hat{\gamma}$ must be an imaginary quantity such that $\hat{\gamma} = jP$. Then, we can express (10.96b) as

$$\tilde{E}_x = \frac{jPM}{-P^2 + \omega^2 \mu\epsilon}\hat{E}_{zm} \cos(Mx)\sin(Ny)[j2\sin(Pz)] \tag{10.99}$$

When we apply the boundary condition given in (10.97) at $z = \ell$, we obtain

$$\sin P\ell = 0$$

or

$$P = \frac{p\pi}{\ell} \tag{10.100}$$

where $p = 0, 1, 2, \ldots$.

As a result, we can express the fields for a TM_{mnp} mode in a rectangular cavity resonator as

$$\tilde{E}_z = 2\hat{E}_{zm}\sin(Mx)\sin(Ny)\cos(Pz) \tag{10.101a}$$

$$\tilde{E}_x = -\frac{2MP}{M^2+N^2}\hat{E}_{zm}\cos(Mx)\sin(Ny)\sin(Pz) \tag{10.101b}$$

$$\tilde{E}_y = -\frac{2NP}{M^2+N^2}\hat{E}_{zm}\sin(Mx)\cos(Ny)\sin(Pz) \tag{10.101c}$$

$$\tilde{H}_x = \frac{j2\omega\epsilon N}{M^2+N^2}\hat{E}_{zm}\sin(Mx)\cos(Ny)\cos(Pz) \tag{10.101d}$$

$$\tilde{H}_y = \frac{-j2\omega\epsilon M}{M^2+N^2}\hat{E}_{zm}\cos(Mx)\sin(Ny)\cos(Pz) \tag{10.101e}$$

$$\tilde{H}_z = 0 \tag{10.101f}$$

From these equations we can draw two important conclusions.

1. The electric and magnetic fields do not propagate along the z axis, rather they oscillate in time at a given location.
2. The lowest-order TM mode for a rectangular cavity resonator is characterized by $m = 1$, $n = 1$, and $p = 0$.

The resonant frequency of a TM_{mnp} mode can be determined from (10.9) by replacing $\hat{\gamma}$ with jP as

$$f_{mnp} = \frac{1}{2\sqrt{\mu\epsilon}}\sqrt{\left(\frac{m}{a}\right)^2 + \left(\frac{n}{b}\right)^2 + \left(\frac{p}{\ell}\right)^2} \tag{10.102}$$

where $m = 1, 2, 3, \ldots, n = 1, 2, 3, \ldots,$ and $p = 0, 1, 2, 3, \ldots$.

The frequency for the lowest-order TM mode ($m = 1, n = 1, p = 0$) is

$$f_{110} = \frac{1}{2\sqrt{\mu\epsilon}}\sqrt{\frac{1}{a^2} + \frac{1}{b^2}} \tag{10.103}$$

10.6.2 Transverse electric (TE) mode

In a cavity resonator, for a transverse electric (TE) mode, we can write the z component of the magnetic field as

$$\tilde{H}_z = \hat{H}_{zm}^{+}\cos(Mx)\cos(Ny)e^{-\hat{\gamma}z} + \hat{H}_{zm}^{-}\cos(Mx)\cos(Ny)e^{\hat{\gamma}z} \tag{10.104}$$

where $\hat{H}_{zm}^{+}$ and $\hat{H}_{zm}^{-}$ are the amplitudes of the forward and reflected magnetic fields. The other field components, using (10.5) and (10.104), are given in (10.105). Also, $(-\hat{\gamma})$ has been replaced by $(\hat{\gamma})$ for the fields associated with the backward-travelling wave in (10.5).

$$\tilde{H}_x = \frac{\hat{\gamma} M}{\hat{\gamma}^2 + \omega^2 \mu \epsilon} \sin(Mx) \cos(Ny) \left(\hat{H}_{zm}^{+} e^{-\hat{\gamma} z} - \hat{H}_{zm}^{-} e^{\hat{\gamma} z} \right) \quad (10.105a)$$

$$\tilde{H}_y = \frac{\hat{\gamma} N}{\hat{\gamma}^2 + \omega^2 \mu \epsilon} \cos(Mx) \sin(Ny) \left(\hat{H}_{zm}^{+} e^{-\hat{\gamma} z} - \hat{H}_{zm}^{-} e^{\hat{\gamma} z} \right) \quad (10.105b)$$

$$\tilde{E}_x = \frac{j\omega\mu N}{\hat{\gamma}^2 + \omega^2 \mu \epsilon} \cos(Mx) \sin(Ny) \left(\hat{H}_{zm}^{+} e^{-\hat{\gamma} z} + \hat{H}_{zm}^{-} e^{\hat{\gamma} z} \right) \quad (10.105c)$$

$$\tilde{E}_y = -\frac{j\omega\mu M}{\hat{\gamma}^2 + \omega^2 \mu \epsilon} \sin(Mx) \cos(Ny) \left(\hat{H}_{zm}^{+} e^{-\hat{\gamma} z} - \hat{H}_{zm}^{-} e^{\hat{\gamma} z} \right) \quad (10.105d)$$

$$\tilde{E}_z = 0 \quad (10.105e)$$

By applying the boundary conditions given in (10.97) and defining $\hat{\gamma} = jP$, we obtain

$$\hat{H}_{zm}^{+} = -\hat{H}_{zm}^{-} = \hat{H}_{zm} \quad (10.106)$$

and

$$P = \frac{p\pi}{\ell} \qquad p = 0, 1, 2, 3, \ldots \quad (10.107)$$

Therefore, the fields for a TE_{mnp} mode can be expressed in a cavity resonator as

$$\tilde{H}_x = j\frac{2MP}{M^2 + N^2} \hat{H}_{zm} \sin(Mx) \cos(Ny) \cos(Pz) \quad (10.108a)$$

$$\tilde{H}_y = j\frac{2PN}{M^2 + N^2} \hat{H}_{zm} \cos(Mx) \sin(Ny) \cos(Pz) \quad (10.108b)$$

$$\tilde{H}_z = -j2\hat{H}_{zm} \cos(Mx) \cos(Ny) \sin(Pz) \quad (10.108c)$$

$$\tilde{E}_x = \frac{2\omega\mu N}{M^2 + N^2} \hat{H}_{zm} \cos(Mx) \sin(Nx) \sin(Pz) \quad (10.108d)$$

$$\tilde{E}_y = -\frac{2\omega\mu M}{M^2 + N^2} \hat{H}_{zm} \sin(Mx) \cos(Ny) \sin(Pz) \quad (10.108e)$$

$$\tilde{E}_z = 0 \quad (10.108f)$$

Just as in the TM mode, the fields of the TE mode do not travel in the cavity, but they oscillate in time. For the lowest-order TE mode, however, either m or n can be 0, but p must be at least 1 for both the electric and magnetic fields to exist in the resonator.

By following the same procedure as for the TM mode, we can determine the resonant frequency for the TE mode:

$$f_{mnp} = \frac{1}{2\sqrt{\mu\epsilon}} \sqrt{\left(\frac{m}{a}\right)^2 + \left(\frac{n}{b}\right)^2 + \left(\frac{p}{\ell}\right)^2} \quad (10.109)$$

Although (10.109) is the same as (10.102), the constraints on m, n, and p are different.

The resonant frequency of the lowest-order mode ($m = 1$, $n = 0$, $p = 1$), when $a > b$, is

$$f_{101} = \frac{1}{2\sqrt{\mu\epsilon}}\sqrt{\frac{1}{a^2} + \frac{1}{\ell^2}} \tag{10.110}$$

The electric and magnetic fields of TE_{101} are

$$\tilde{E}_x = 0 \tag{10.111a}$$

$$\tilde{E}_y = -\frac{2\omega\mu a}{\pi}\hat{H}_{zm}\sin\left(\frac{\pi x}{a}\right)\sin\left(\frac{\pi z}{\ell}\right) \tag{10.111b}$$

$$\tilde{E}_z = 0 \tag{10.111c}$$

$$\tilde{H}_x = j\frac{2a}{\ell}\hat{H}_{zm}\sin\left(\frac{\pi x}{a}\right)\cos\left(\frac{\pi z}{\ell}\right) \tag{10.111d}$$

$$\tilde{H}_y = 0 \tag{10.111e}$$

$$\tilde{H}_z = -j2\hat{H}_{zm}\cos\left(\frac{\pi x}{a}\right)\sin\left(\frac{\pi z}{\ell}\right) \tag{10.111f}$$

10.6.3 Quality factor

Along with its resonant frequency, the quality factor Q is another important feature of a cavity resonator. By definition, the quality factor of a resonating system is

$$Q = \omega_0\frac{W}{P_d} \tag{10.112}$$

where ω_0 is the angular resonant frequency in rad/s, W is the time-average stored energy in joules, and P_d is the power dissipation, in watts, in the system. Thus, the higher the quality factor, the better the cavity resonator. In fact, under ideal conditions $\sigma_d \to 0$ and $\sigma_c \to \infty$, $Q \to \infty$.

Let us consider the TE_{101} mode and develop an expression for the Q factor of a rectangular cavity resonator. The total energy storage in a cavity resonator is the sum of the energy stored in the electric (W_e) and the magnetic (W_m) fields. That is,

$$W = W_e + W_m \tag{10.113}$$

It can be shown that at the resonant frequency $W_e = W_m$. Thus,

$$W = 2W_e = 2W_m \tag{10.114}$$

The time-average stored energy in the electric field is

$$W_e = \frac{1}{4}\int_v \epsilon E^2\, dv \tag{10.115}$$

and the total energy storage in the resonator is

$$W = \frac{1}{2}\int_v \epsilon E^2\,dv$$
$$= \frac{\epsilon}{2}\int_0^{\ell}\int_0^b\int_0^a \left[-\frac{2\omega\mu a H_{zm}}{\pi}\right]^2 \left[\sin^2\left(\frac{\pi x}{a}\right)\sin^2\left(\frac{\pi z}{\ell}\right)dx\,dy\,dz\right] \tag{10.116}$$

or

$$W = \frac{\omega^2\mu^2 a^3 H_{zm}^2 \epsilon b\ell}{2\pi^2} \tag{10.117a}$$

or

$$W = 2\epsilon b\ell a^3 f^2\mu^2 H_{zm}^2 \tag{10.117b}$$

The cavity resonator under discussion is assumed to have a perfect dielectric. Therefore, the total power loss is simply the loss on each wall face of the resonator.

Let us consider the cavity resonator shown in Figure 10.14. The walls of a resonator are generally made of a good conductor, such as copper. Therefore, it will be a fairly good approximation to assume that the tangential component of the magnetic field just inside the inner surface of a cavity wall is zero because of negligible skin depth. Then, we can express the boundary condition on the surface of the wall as

$$\vec{\mathbf{J}}_s = \vec{\mathbf{a}}_n \times \vec{\mathbf{H}}_1 \tag{10.118}$$

where $\vec{\mathbf{H}}_1$ is the magnetic field next to the wall surface in the dielectric medium, and $\vec{\mathbf{J}}_s$ is the surface current density on the wall's surface. By using the tangential component of the magnetic field, we can determine the magnitude of the power density on the wall surface as

$$\langle\hat{S}\rangle = \tfrac{1}{2}\mathrm{Re}\left[\eta H_{t1}^2\right] \tag{10.119a}$$

or

$$\langle\hat{S}\rangle = \frac{1}{2}H_{t1}^2\,\mathrm{Re}\left[\sqrt{\frac{j\omega\mu}{\sigma_c + j\omega\epsilon}}\right] \tag{10.119b}$$

For a good conductor ($\sigma_c \gg \omega\epsilon$) the power density can be approximated as

$$\langle\hat{S}\rangle = \frac{1}{2}H_{t1}^2\,\mathrm{Re}\left(\sqrt{\frac{j\omega\mu}{\sigma_c}}\right)$$
$$= \frac{1}{2}H^2 R_s \tag{10.120}$$

where $R_s = \sqrt{\omega\mu/2\sigma_c}$ is the resistive part of the intrinsic impedance.

R_s can also be expressed in terms of the skin depth of the wall as

$$R_s = \frac{1}{\sigma_c \delta_c} \tag{10.121}$$

where $\delta_c = \sqrt{2/\omega\mu\sigma_c}$ is the skin depth of the conductor.

The total power dissipation now can be calculated by

$$P_d = \oint_s \langle \vec{\mathbf{S}} \rangle \cdot \vec{d\mathbf{s}} \tag{10.122}$$

where s is the closed inner surface of the cavity resonator.

The tangential components of the magnetic field, the average power density, and the power dissipated on the walls are as follows.

On $z = 0$ plane:

$$H_t|_{z=0} = |\vec{\mathbf{a}}_z \times \vec{\mathbf{H}}| = \left(\frac{2a}{\ell}\right) H_{zm} \sin\left(\frac{\pi}{a}x\right)$$

$$\langle S \rangle|_{z=0} = \frac{2}{\sigma_c \delta_c} \left(\frac{a}{\ell}\right)^2 H_{zm}^2 \sin^2\left(\frac{\pi}{a}x\right)$$

On $z = \ell$ plane:

$$H_t|_{z=\ell} = |\vec{\mathbf{a}}_z \times \vec{\mathbf{H}}| = \left(\frac{2a}{\ell}\right) H_{zm} \sin\left(\frac{\pi}{a}x\right)$$

$$\langle S \rangle|_{z=\ell} = \frac{2}{\sigma_c \delta_c} \left(\frac{a}{\ell}\right)^2 H_{zm}^2 \sin^2\left(\frac{\pi}{a}x\right)$$

On $y = 0$ plane:

$$H_t|_{y=0} = \vec{\mathbf{a}}_y \times \vec{\mathbf{H}}|$$

$$= \sqrt{4\left(\frac{a}{\ell}\right)^2 H_{zm}^2 \sin^2\left(\frac{\pi}{a}x\right)\cos^2\left(\frac{\pi}{\ell}z\right) + 4H_{zm}^2 \cos^2\left(\frac{\pi}{a}x\right)\sin^2\left(\frac{\pi}{\ell}z\right)}$$

$$\langle S \rangle|_{y=0} = \frac{1}{\sigma_c \delta_c}\left[2\left(\frac{a}{\ell}\right)^2 H_{zm}^2 \sin^2\left(\frac{\pi}{a}x\right)\cos^2\left(\frac{\pi}{\ell}z\right) + 2H_{zm}^2 \cos^2\left(\frac{\pi}{a}x\right)\sin^2\left(\frac{\pi}{\ell}z\right)\right]$$

On $y = b$ plane:

$$H_t|_{y=b} = |\vec{\mathbf{a}}_y \times \vec{\mathbf{H}}|$$

$$= \sqrt{4\left(\frac{a}{\ell}\right)^2 H_{zm}^2 \sin^2\left(\frac{\pi}{a}x\right)\cos^2\left(\frac{\pi}{\ell}z\right) + 4H_{zm}^2 \cos^2\left(\frac{\pi}{a}x\right)\sin^2\left(\frac{\pi}{\ell}z\right)}$$

$$\langle S \rangle|_{y=b} = \frac{1}{\sigma_c \delta_c}\left[2\left(\frac{a}{\ell}\right)^2 H_{zm}^2 \sin^2\left(\frac{\pi}{a}x\right)\cos^2\left(\frac{\pi}{\ell}z\right) + 2H_{zm}^2 \cos^2\left(\frac{\pi}{a}x\right)\sin^2\left(\frac{\pi}{\ell}z\right)\right]$$

On $x = 0$ plane:

$$H_t|_{x=0} = |\vec{\mathbf{a}}_x \times \vec{\mathbf{H}}| = 2H_{zm} \sin\left(\frac{\pi}{\ell}z\right)$$

$$\langle S\rangle|_{x=0} = \frac{2}{\sigma_c\delta_c} H_{zm}^2 \sin^2\left(\frac{\pi}{\ell}z\right)$$

On $x = a$ plane:

$$H_t|_{x=a} = |\vec{\mathbf{a}}_x \times \vec{\mathbf{H}}| = 2H_{zm} \sin\left(\frac{\pi}{\ell}z\right)$$

$$\langle S\rangle|_{x=a} = \frac{2}{\sigma_c\delta_c} H_{zm}^2 \sin^2\left(\frac{\pi}{\ell}z\right)$$

The total power dissipation on the walls is

$$\begin{aligned} P_d &= \int_0^b\int_0^a \langle S\rangle|_{z=0}\,dx\,dy + \int_0^\ell\int_0^a \langle S\rangle|_{y=0}\,dx\,dz + \int_0^\ell\int_0^b \langle S\rangle|_{x=0}\,dy\,dz \\ &\quad + \int_0^b\int_0^a \langle S\rangle|_{z=\ell}\,dx\,dy + \int_0^\ell\int_0^a \langle S\rangle|_{y=b}\,dx\,dz + \int_0^\ell\int_0^b \langle S\rangle|_{x=a}\,dy\,dz \\ &= \frac{H_{zm}^2}{\sigma_c\delta_c\ell^2}(2a^3b + a^3\ell + a\ell^3 + 2b\ell^3) \end{aligned} \quad (10.123)$$

Finally, the quality factor Q can be determined from (10.112) by direct substitution of (10.117b) and (10.123) as

$$Q = \frac{4\pi f^3 a^3\ell^3\mu^2\epsilon b\sigma_c\delta_c}{2a^3b + a^3\ell + a\ell^3 + 2b\ell^3} \quad (10.124)$$

EXAMPLE 10.9

A rectangular cavity resonator made of copper has dimensions $a = 3$ cm, $b = 1$ cm, and $\ell = 4$ cm, and operates at the dominant mode. Determine the resonant frequency and the quality factor of this resonator. The conductivity of copper is 5.76×10^7 S/m.

Solution The TE_{101} mode is the dominant mode of a rectangular cavity resonator, and the corresponding resonant frequency is

$$\begin{aligned} f_{101} &= \frac{u_p}{2}\sqrt{\left(\frac{1}{a}\right)^2 + \left(\frac{1}{\ell}\right)^2} \\ &= \frac{3\times 10^8}{2}\sqrt{\left(\frac{1}{0.03}\right)^2 + \left(\frac{1}{0.04}\right)^2} \\ &= 6.25 \times 10^9 \text{ Hz} \quad \text{or} \quad 6.25 \text{ GHz} \end{aligned}$$

The skin depth δ_c is

$$\begin{aligned} \delta_c &= \frac{1}{\sqrt{\pi f\sigma_c\mu}} = \frac{1}{\sqrt{\pi \times 6.25\times 10^9 \times 5.76\times 10^7 \times 4\pi\times 10^{-7}}} \\ &= 8.39 \times 10^{-7} \text{ m} \end{aligned}$$

The quality factor, from (10.124), is:

$Q \approx 7427$

• • •

10.7 Summary

A waveguide is a hollow conductor that is used to transmit electromagnetic signals at microwave frequencies where conventional transmission lines fail to operate efficiently because of their high transmission losses. Waveguides may have different geometries depending on their applications. Rectangular, circular, and ridged waveguides are the ones that are commonly used. In our study, we concentrated on rectangular waveguides because they are relatively easy to analyze.

A rectangular waveguide can support either a transverse magnetic field (TM) mode or a transverse electric field (TE) mode. Dictated by its dimensions, a rectangular waveguide has a definite cutoff frequency for each mode of operation. In general, the cutoff frequency for the TM_{mn} or TE_{mn} mode is

$$f_{cmn} = \frac{u_p}{2}\sqrt{\left(\frac{m}{a}\right)^2 + \left(\frac{n}{b}\right)^2}$$

where a is the dimension of the waveguide along the x direction, and b is the width along the y direction. In order to transmit a signal along the length of the waveguide, the frequency of the signal must be greater than the cutoff frequency. Otherwise, no transmission takes place along the waveguide.

The energy travels inside the waveguide with a group velocity given as

$$u_{gmn} = u_p\sqrt{1 - \left(\frac{f_{cmn}}{f}\right)^2}$$

The phase velocity determines the speed of the constant phase points on the wave as it travels. In a rectangular waveguide the phase velocity is greater than the speed of the wave in an unbounded medium and is expressed by

$$u_{pmn} = \frac{u_p}{\sqrt{1 - \left(\frac{f_{cmn}}{f}\right)^2}}$$

The phase constant of a TM_{mn} or TE_{mn} mode is

$$\beta_{mn} = \beta\sqrt{1 - \left(\frac{f_{cmn}}{f}\right)^2} \qquad \text{where } \beta = \omega\sqrt{\mu\epsilon}$$

The wave impedance is

$$\hat{\eta}_{mn}^{TM} = \eta\sqrt{1 - \left(\frac{f_{cmn}}{f}\right)^2}$$

for the TM_{mn} mode, and

$$\hat{\eta}_{mn}^{TE} = \frac{\eta}{\sqrt{1 - \left(\frac{f_{cmn}}{f}\right)^2}}$$

for the TE_{mn} mode.

The average power flow for a TM_{mn} mode is

$$\langle P_{mn} \rangle = \frac{\beta_{mn}^2 a^3 b^3}{8\pi^2 \eta_{mn}^{TM}(n^2a^2 + m^2b^2)} E_{zm}^2$$

For a TE_{mn} mode it is

$$\begin{aligned} \langle P_{mn} \rangle &= \eta_{mn}^{TM} \left[\frac{\beta_{mn}^2 a^3 b^3 H_{zm}^2}{8\pi^2(b^2m^2 + a^2n^2)} \right] && \text{for } m, n \neq 0 \\ &= \eta_{mn}^{TM} \left[\frac{\beta_{mn}^2 a^3 b^3 H_{zm}^2}{4\pi^2(m^2b^2 + n^2a^2)} \right] && \text{for either } m = 0 \text{ or } n = 0 \end{aligned}$$

In a waveguide, neither the conductor walls nor the dielectric medium is a perfect material. Thus, attenuation of the transmitted signal in a waveguide is inevitable. We have explained how to account for these losses.

A cavity resonator is a metal box that is used to tune circuits in microwave frequencies. In a cavity resonator electromagnetic fields do not propagate along the z axis, but instead oscillate in time at a given location. The lowest-order TM_{mnp} mode is TM_{110} and the lowest-order TE_{mnp} modes are TE_{101} or TE_{011} in a cavity resonator. The resonant frequency for both the TM_{mnp} and TE_{mnp} modes is given by

$$f_{mnp} = \frac{1}{2\sqrt{\mu\epsilon}} \sqrt{\left(\frac{m}{a}\right)^2 + \left(\frac{n}{b}\right)^2 + \left(\frac{p}{\ell}\right)^2}$$

The quality factor for the TE_{101} mode is given by

$$Q = \frac{4\pi f^3 a^3 \ell^3 \mu^2 \epsilon b \sigma_c \delta_c}{2a^3 b + a^3 \ell + a\ell^3 + 2b\ell^3}$$

The quality factor is the measure of merit of a cavity resonator. The higher the quality factor, the less the power loss in the walls.

10.8 Review questions

10.1 Why do we use waveguides?

10.2 Can a TEM wave be supported in a waveguide?

10.3 What is the TM mode of propagation?

10.4 What is the TE mode of propagation?

10.5 How many possible TM modes can occur in a rectangular waveguide?

10.6 What is the lowest possible mode of a TM wave?

10.7 How do we define the cutoff frequency in a waveguide?

10.8 What is an evanescent wave?

10.9 If the operating frequency is below the cutoff frequency, will there be average power flow in the waveguide?

10.10 What is the nature of the wave impedance for a TM mode when a rectangular waveguide is excited at a frequency less than its cutoff frequency?

10.11 How do we calculate the group velocity for a TM wave?

10.12 What is the power flow expression for a TM mode?

10.13 What is the lowest-order mode of a TE mode?

10.14 Calculate the cutoff frequency for the dominant mode in a rectangular waveguide.

10.15 A waveguide excited by a maximum field of 2 A/m in the z direction at 10 GHz has a square cross section of 4 cm^2. Will there be a wave propagation in this waveguide? If so, which mode will exist, and how much power will be transmitted?

10.16 Why does signal attenuation occur in a waveguide although it is excited above the cutoff frequency of the dominant mode?

10.17 Why can we not use an ordinary LC circuit for tuning in at microwave frequencies?

10.18 What is a cavity resonator?

10.19 How does the resonance occur in a cavity resonator? Explain.

10.20 What is the resonant frequency in a cavity resonator?

10.21 How do we define the quality factor for a cavity resonator?

10.9 Exercises

10.1 A 2-m-long, air-filled rectangular waveguide of $a = 2$ cm and $b = 1$ cm supports the TM$_{11}$ mode with a propagation constant of $\hat{\gamma}_{11} = j200$. What is the operating frequency of the waveguide?

10.2 Write expressions for the fields both in the phasor and time domains inside the waveguide given in Exercise 10.1 when the amplitude of the electric field intensity in the z-direction is 2 kV/m at $z = 0$.

10.3 A polyethylene-filled ($\epsilon_r = 2.5$) rectangular waveguide of $a = 1$ cm and $b = 0.5$ cm supports the TM$_{11}$ mode when the operating frequency is 9 GHz. The amplitude of the z-component of the electric field is 1.5 kV/m at $z = 0$. Calculate (a) the cutoff frequency, (b) the propagation constant, (c) the phase and group velocities, (d) the intrinsic impedance, and (e) the average power flow.

10.4 An air-filled, rectangular waveguide of $b = 1$ cm is supposed to operate at 12 GHz with $\beta_{10} = 150$ rad/m. What should a be in order to support the TE$_{10}$ mode?

10.5 An air-filled, rectangular waveguide of $a = 1$ cm and $b = 1.5$ cm supports the lowest-order TE mode with a propagation constant of 100 rad/m. What is the operating frequency of the waveguide? (a) Write the field expressions both in the time and phasor domains, if the amplitude of the electric field intensity is 500 V/m, and (b) determine the average power in the waveguide.

10.6 An air-filled rectangular waveguide of $a = 2$ cm and $b = 1$ cm supports the TE_{10} mode while operating at 9 GHz. The waveguide is excited by 20 V/cm at $z = 0$. Calculate (a) the cutoff frequency, (b) the propagation constant, (c) the phase and group velocities, (d) the intrinsic impedance, and (e) the average power flow in the waveguide.

10.7 Repeat Exercise 10.6 when the dielectric material in the waveguide is replaced with polyethylene ($\epsilon_r = 2.5$, $\mu_r = 1$).

10.8 Determine the attenuation constants for the waveguide given in Example 10.8 if the dielectric material in the waveguide is replaced with polyethylene and the operating frequency is 4 GHz. The loss tangent for polyethylene is 10^{-13}, relative permittivity is 2.5 and $\mu = \mu_0$.

10.9 Redo Example 10.9 with aluminum as the cavity material. Compare the results. The conductivity of aluminum is 3.55×10^7 S/m.

10.10 A copper, rectangular cavity resonator supports the TM_{101} mode. Calculate the resonant frequency, quality factor, and power dissipation on the cavity walls for the given mode. The dimensions of the resonator are $a = 2$ cm, $b = 1$ cm, and $c = 4$ cm.

10.10 Problems

10.1 A parallel-plate waveguide (transmission line) is formed by placing two perfect-conductor sheets at $x = 0$ and $x = a$. The sheets may be treated as of infinite extent in the y direction. If they are separated by a dielectric medium, show that the electric field

$$\tilde{\mathbf{E}} = \hat{E}_o \sin\left(\frac{n\pi}{a}x\right) e^{-\hat{\gamma}z}\vec{\mathbf{a}}_y \quad \text{where } n = 1, 2, 3, \ldots$$

defines a set of n possible solutions for the forward-traveling wave in the z direction with $\hat{\gamma} = \sqrt{(n\pi/a)^2 - \omega^2\mu\epsilon}$. These solutions are classified as a set of n transverse electric (TE_n) modes and only exist above a cutoff frequency f_c, whereby $\hat{\gamma} = 0$ for $f = f_c$. Determine the other field components and the cutoff frequency.

10.2 For the parallel-plate waveguide (transmission line) of Problem 10.1, obtain an expression for the power carried per unit width (in the y direction) by the wave. What are the surface charge and surface current densities?

10.3 Show that the magnetic field

$$\tilde{\mathbf{H}} = \hat{H}_o \cos\left(\frac{n\pi}{a}x\right) e^{-\hat{\gamma}z}\vec{\mathbf{a}}_y \quad \text{where } n = 0, 1, 2, 3, \ldots$$

in the parallel-plate waveguide (transmission line) of Problem 10.1 defines a set of n possible solutions for the forward-traveling wave in the

z direction with $\hat{\gamma} = \sqrt{(n\pi/b)^2 - \omega^2\mu\epsilon}$. These solutions are classified as a set of n transverse magnetic (TM_n) modes. Determine the other field components, the cutoff frequency, the surface current densities, and surface charge densities.

10.4 Show that the TM_0 mode of the parallel-plate waveguide in Problem 10.3 is actually a TEM mode. Determine the power carried per-unit width (in the y direction) by the wave.

10.5 A rectangular waveguide with cross-sectional dimensions of $a = 4$ cm, and $b = 3$ cm operates at a frequency of 20 GHz. If the amplitude of the z component of the electric field is 600 V/m, (a) determine the phase constant, and (b) calculate the electric and magnetic field components at $x = 1$ cm, $y = 1.5$ cm, and $z = 50$ cm for the dominant TM mode.

10.6 A 2-cm-square waveguide operates at 12 GHz with the TM_{11} mode. Determine the (a) cutoff frequency, (b) cutoff wavelength, (c) wavelength inside the waveguide, (d) phase velocity, and (e) group velocity. Write general expressions for the fields, surface current densities, and surface charge densities.

10.7 The group velocity of the wave in a square waveguide operating at 3 GHz with the dominant mode is determined as 2×10^8 m/s. Calculate the size of the square cross section if the dielectric inside the guide is characterized by a permittivity of $\epsilon = 2\epsilon_0$.

10.8 The x and y components of the magnetic and electric fields for the TE_{10} mode are given in a rectangular waveguide with $b = 1$ cm as follows:

$$\tilde{E}_y = -j100 \sin\left(\frac{\pi x}{a}\right) e^{-j\beta_{10}z}$$

$$\tilde{H}_x = j0.1 \sin\left(\frac{\pi x}{a}\right) e^{-j\beta_{10}z}$$

If the frequency of operation of this waveguide is 10 GHz, and the only possible mode of propagation is TE_{10}, determine the critical dimension a of the waveguide. What are the other field components? Express the fields in the time domain. What is the average power transmitted by the waveguide?

10.9 The phase constant in an air-filled rectangular waveguide is 165 rad/m when it is excited with the TM_{21} mode. Calculate the wavelength of the wave if the excitation frequency is 10% higher than the cutoff frequency of the operating mode.

10.10 The magnitude of the applied electric field in an air-filled rectangular waveguide is 500 V/m when it operates with TE_{10} at 10 GHz. Determine the average power that is transmitted along the waveguide. The length of the waveguide is 2 m, $a = 3$ cm, and $b = 2$ cm. Write expressions for the field components, the surface current, and the surface charge densities in the time domain.

10.11 An air-filled waveguide has dimensions $a = 2$ cm and $b = 1$ cm. Determine the frequency range for this waveguide so that it will operate only at its dominant mode.

10.12 A rectangular waveguide with $a = 2$ cm and $b = 3$ cm is filled with a dielectric of $\epsilon_r = 3$, and is operated at 50 GHz. Calculate the cutoff frequency, wavelength, phase constant, phase velocity, group velocity, and wave impedance for the TE_{22} mode. Write general expressions for the fields, the surface current, and the surface charge densities both in time domain and phasor from.

10.13 An air-filled lossless rectangular waveguide with $a = 2$ cm and $b = 1$ cm operates in the TE_{10} mode at 15 GHz. Calculate the magnitudes of the applied electric and magnetic fields if the average power transmitted by the waveguide is 1 kW. Write expressions for the fields, the surface current, and the surface charge densities both in time domain and phasor form.

10.14 An air-filled, 3 cm × 1 cm, 1-m-long waveguide operates at 12 GHz in the TE_{10} mode. Calculate the attenuation constant in the waveguide due to the imperfect dielectric and conductor. Consider a loss tangent of 10^{-4} in the dielectric and a conductivity of 5.8×10^7 S/m in the conductor.

10.15 Calculate the power dissipation along the waveguide given in Problem 10.14 if the magnitude of the applied electric field is 800 V/m, its operating frequency is 12 GHz, and the TE_{10} mode of excitation is used. Write approximate expressions for the surface charge and surface current densities.

10.16 An air-filled, 10-m-long rectagular waveguide with dimensions $a = 4$ cm and $b = 3$ cm is excited in the TE_{10} mode at 4 GHz. The magnitude of the applied electric field is 1000 V/m. The loss tangent in air is 0.0001, and the conductivity of the copper walls is 5.8×10^7 S/m. Calculate the average power delivered to the load.

10.17 Calculate the power dissipation along the waveguide given in Problem 10.16 if the magnitude of the applied electric field is 800 V/m, and the waveguide is excited by the TE_{10} mode and is operating at 4 GHz.

10.18 A rectangular cavity resonator has dimensions $a = 5$ cm, $b = 2$ cm, and $\ell = 7$ cm. Calculate the resonant frequency if the resonator is coupled with the TM_{110} mode.

10.19 The resonant frequency of a cubic cavity resonator is 9 GHz when it is excited in the TE_{101} mode. Calculate the dimensions of the resonator.

10.20 A rectangular cavity resonator operates with the dominant mode. Calculate (a) the resonant frequency, (b) the quality factor, and (c) the amount of energy storage if $H_{zm} = 2$ A/m. The dimensions of the resonator are $a = 3$ cm, $b = 1$ cm, and $\ell = 5$ cm, and it is made of copper ($\sigma_{Cu} = 5.8 \times 10^7$ S/m).

10.21 Design a lumped parallel *RLC* resonant circuit that performs the same as the resonator given in Problem 10.20. Consider ideal elements in the analysis with $C = 1$ pF.

10.22 A cubic cavity resonator is to operate in its dominant mode at a resonant frequency of 10 GHz. It is filled with air, and its walls are copper. Determine the size of the resonator in order to minimize the losses.

10.23 Compare the resonant frequency, quality factor, and amount of energy storage in a rectangular cavity resonator with copper and aluminum walls. The dimensions are $a = 2$ cm, $b = 3$ cm, and $\ell = 5$ cm, and the waveguide operates in its dominant mode. $\sigma_{\mathrm{Cu}} = 5.8 \times 10^7$ S/m and $\sigma_{\mathrm{Al}} = 3.5 \times 10^7$ S/m.

10.24 A cubic cavity resonator is to operate in its dominant mode at a resonant frequency of 20 GHz. It is filled with polyethylene of $\epsilon_r = 2.5$, and its walls are copper. Determine the size of the resonator in order to minimize the losses.

11

Antennas

11.1 Introduction

We now understand how fields propagate in an unbounded medium as plane waves and carry energy from one point to the other, how energy is transmitted along a transmission line, and how energy is channeled through a waveguide. We will now discuss systems that not only generate electromagnetic fields but also radiate them effectively.

Maxwell's equations dictate that in order to create electromagnetic fields we require time-varying sources such as charges and currents. When the fields created by these sources are confined to propagate as a wave along a transmission line or inside a waveguide, the wave is usually referred to as a *guided wave*. When these sources, finite in size, create waves that propagate away from them in an unbounded medium, they collectively form a **radiating system**, and the process is called the *radiation of electromagnetic waves*. The device at the end of a radiating system is referred to as a **transmitting antenna** (see Figure 11.1). A transmitting antenna, when fed by a transmission line as shown in Figure 11.2, is called a *dipole antenna*. Figure 11.3 shows a *horn antenna* fed by a waveguide. Among other types of antennas are a *slot antenna* (a slot in a large conducting plane fed by a waveguide), Figure 11.4, and a *microstrip antenna* (a thin patch of metal on a grounded dielectric substrate), Figure 11.5. When an antenna is used to capture the radiated energy, it is called a **receiving antenna**.

The power radiated by a transmission line at 60 Hz is so small that it cannot be considered a radiating system. Because the function of a transmission line is to guide the fields along its length, it is not designed to serve as an antenna. For this reason, we use coaxial cables as transmission lines at high frequencies. Note that a coaxial cable does not radiate at any frequency. A plane wave falls within the broad scope of a radiating wave. However, the generation of a plane wave requires an infinitely large planar radiating system, which in "real life" is not possible.

A finite-size antenna produces time-varying electromagnetic fields that propagate in space as **spherical waves**. These are the waves that

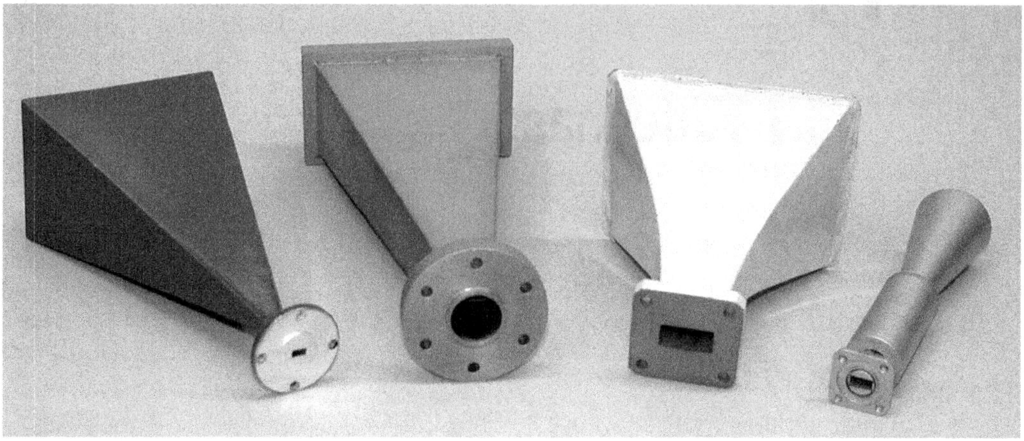

Figure 11.1 Various types of horn antennas (courtesy of Microstar Inc.)

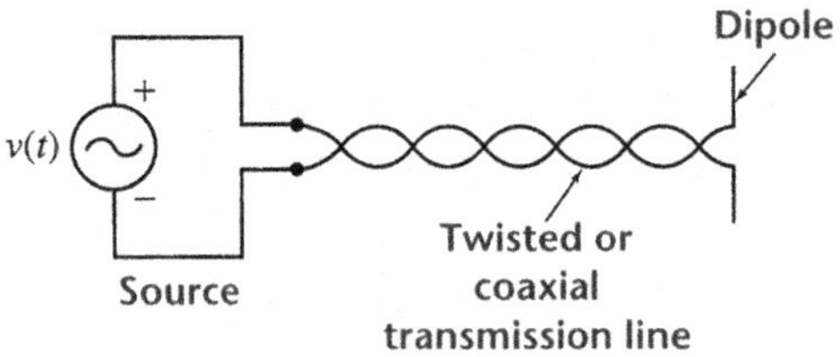

Figure 11.2 A dipole antenna

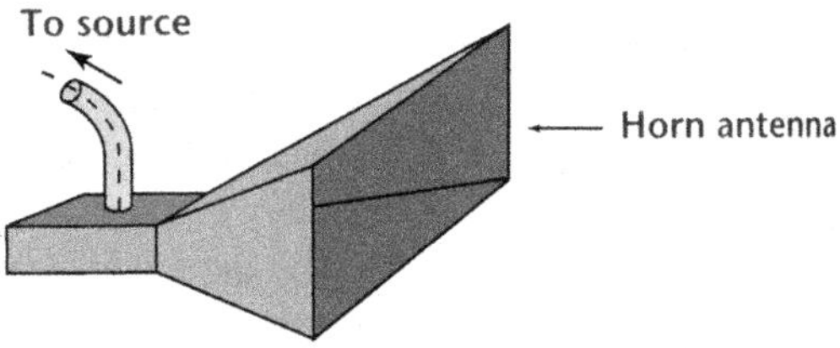

Figure 11.3 A horn antenna

are of interest to us in this chapter. The fields must satisfy the four Maxwell equations in a region containing time-varying charge and current sources. In other words, from Maxwell's equations, we must obtain a wave equation for the $\vec{\mathbf{E}}$ or $\vec{\mathbf{H}}$ field in terms of the time-varying sources, express it in the spherical coordinate system, and then solve it. As you might guess, the problem is quite complex because of the time-varying sources. However, the problem becomes more manageable when we seek a solution of the wave equation in terms of scalar and vector potentials. We will still be able to analyze only a few simple types

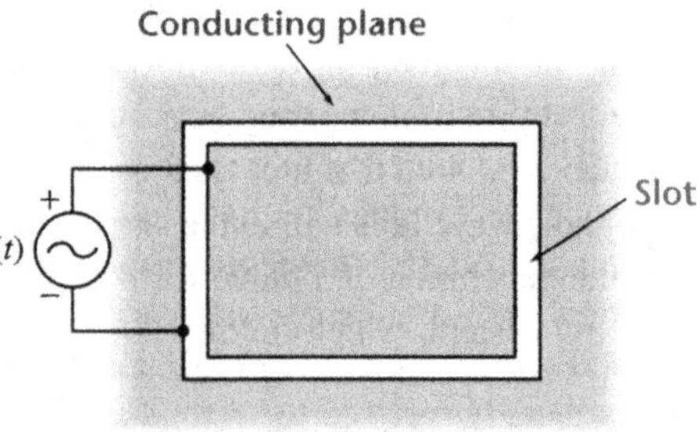

Figure 11.4 A slot antenna

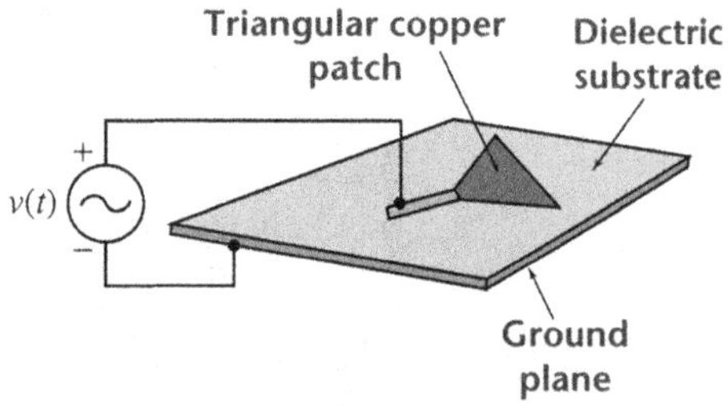

Figure 11.5 A triangular-patch microstrip antenna

of antennas, but the analysis will help us understand the radiating fields. We begin our discussion by deriving the wave equations in terms of the electric scalar potential V and the magnetic vector potential $\vec{\mathbf{A}}$.

11.2 Wave equations in terms of potential functions

We begin our discussion by stating Maxwell's equations in a lossless (dielectric), linear, isotropic, and homogeneous medium in their time-dependent form:

$$\nabla \times \vec{\mathbf{E}} = -\frac{\partial \vec{\mathbf{B}}}{\partial t} \tag{11.1}$$

$$\nabla \times \vec{\mathbf{H}} = \vec{\mathbf{J}} + \frac{\partial \vec{\mathbf{D}}}{\partial t} \tag{11.2}$$

$$\nabla \cdot \vec{\mathbf{B}} = 0 \tag{11.3}$$

$$\nabla \cdot \vec{\mathbf{D}} = \rho \tag{11.4}$$

where ρ and $\vec{\mathbf{J}}$ are, respectively, the volume charge and volume current densities that exist in the medium as time-varying sources. The constitutive relationships between the field quantities are

$$\vec{\mathbf{D}} = \epsilon \vec{\mathbf{E}} \tag{11.5}$$

$$\vec{\mathbf{B}} = \mu \vec{\mathbf{H}} \tag{11.6}$$

where ϵ and μ are, respectively, the permittivity and permeability of the dielectric medium.

As $\vec{\mathbf{B}}$ is a continuous (solenoidal, or divergence free) field, it can be defined in terms of another vector field $\vec{\mathbf{A}}$ as

$$\vec{\mathbf{B}} = \nabla \times \vec{\mathbf{A}} \tag{11.7}$$

where $\vec{\mathbf{A}}$ is the *magnetic vector potential* (or simply *vector potential*). This is the same definition for the vector potential that we used in the study of magnetostatics. When we substitute (11.7) in (11.1), we obtain

$$\nabla \times \vec{\mathbf{E}} = -\frac{\partial}{\partial t}[\nabla \times \vec{\mathbf{A}}] = -\nabla \times \left[\frac{\partial \vec{\mathbf{A}}}{\partial t}\right]$$

or

$$\nabla \times \left[\vec{\mathbf{E}} + \frac{\partial \vec{\mathbf{A}}}{\partial t}\right] = 0 \tag{11.8}$$

Because (11.8) is true, in general, at all points for all time, we can define a *scalar potential* V such that

$$\vec{\mathbf{E}} = -\nabla V - \frac{\partial \vec{\mathbf{A}}}{\partial t} \tag{11.9}$$

The vector potential $\vec{\mathbf{A}}$ in (11.7) and the scalar potential V in (11.9) are both functions of time. However, (11.9) reduces to the familiar equation $\vec{\mathbf{E}} = -\nabla V$ for static fields. Once we have obtained solutions for the potential functions $\vec{\mathbf{A}}$ and V, we can use (11.7) and (11.9) to determine the time-varying magnetic and electric fields. The knowledge of these fields is essential to calculate the power density in the dielectric region.

We now proceed to determine the wave equations in terms of these potential functions. Multiplying (11.2) by μ and substituting (11.7) for $\vec{\mathbf{B}}(\mu\vec{\mathbf{H}})$ and (11.9) for $\vec{\mathbf{D}}(\epsilon\vec{\mathbf{E}})$, we obtain

$$\nabla \times \nabla \times \vec{\mathbf{A}} = \mu\vec{\mathbf{J}} + \mu\epsilon\frac{\partial}{\partial t}\left[-\nabla V - \frac{\partial\vec{\mathbf{A}}}{\partial t}\right] \tag{11.10}$$

Realizing that

$$\nabla \times \nabla \times \vec{\mathbf{A}} = \nabla(\nabla \cdot \vec{\mathbf{A}}) - \nabla^2\vec{\mathbf{A}}$$

we can rewrite (11.10), after some simplifications, as

$$\nabla^2\vec{\mathbf{A}} - \mu\epsilon\frac{\partial^2\vec{\mathbf{A}}}{\partial t^2} = -\mu\vec{\mathbf{J}} + \nabla\left[\nabla \cdot \vec{\mathbf{A}} + \mu\epsilon\frac{\partial V}{\partial t}\right] \tag{11.11}$$

The uniqueness theorem states that to specify a vector field uniquely we must define the divergence and the curl of that field. We have already specified the curl of $\vec{\mathbf{A}}$. By defining the divergence of $\vec{\mathbf{A}}$ as

$$\nabla \cdot \vec{\mathbf{A}} + \mu\epsilon\frac{\partial V}{\partial t} = 0 \tag{11.12}$$

we simplify (11.11) to

$$\nabla^2\vec{\mathbf{A}} - \mu\epsilon\frac{\partial^2\vec{\mathbf{A}}}{\partial t^2} = -\mu\vec{\mathbf{J}} \tag{11.13}$$

which is the wave equation for the vector potential $\vec{\mathbf{A}}$. Equation (11.12) is called the *Lorentz condition* and it yields $\nabla \cdot \vec{\mathbf{A}} = 0$ for static fields, a condition we have used earlier. Similarly, by substituting for $\vec{\mathbf{D}}$ $(\epsilon\vec{\mathbf{E}})$ in (11.4) and applying the Lorentz condition, we obtain a wave equation in terms of the scalar potential as

$$\nabla^2 V - \mu\epsilon\frac{\partial^2 V}{\partial t^2} = -\frac{1}{\epsilon}\rho \tag{11.14}$$

Equations (11.13) and (11.14) are known as the *inhomogeneous Helmholtz equations* in terms of potential functions. They are, in fact, a set of four similar scalar equations. If we can obtain a solution for one of them, we can then write similar solutions for the others. Instead of seeking a general function as the solution for these wave equations, let us assume that the sources vary sinusoidally with time. This assumption allows us to write the wave equations and the Lorentz condition in phasor form as

$$\nabla^2\tilde{\mathbf{A}} + \beta^2\tilde{\mathbf{A}} = -\mu\tilde{\mathbf{J}} \tag{11.15}$$

$$\nabla^2\tilde{V} + \beta^2\tilde{V} = -\frac{1}{\epsilon}\tilde{\rho} \tag{11.16}$$

$$\nabla\cdot\tilde{\mathbf{A}} + j\omega\mu\epsilon\tilde{V} = 0 \tag{11.17}$$

where $\tilde{\mathbf{A}}$, $\tilde{V}$, $\tilde{\mathbf{J}}$, and $\tilde{\rho}$ are all phasor quantities, ω is the angular frequency of the wave in rad/s, and

$$\beta = \omega\sqrt{\mu\epsilon} \tag{11.18}$$

is the wave number of the unbounded medium.

Let us consider a point charge varying sinusoidally with time. Since the time variation is implied in the phasor domain, we expect the solution at a point to be a function only of the distance r from the charge. If we consider the potential function at a point away from the charge, then we can express (11.16) in the spherical coordinate system as

$$\frac{1}{r^2}\frac{\partial}{\partial r}\left[r^2\frac{\partial\tilde{V}}{\partial r}\right] + \beta^2\tilde{V} = 0 \tag{11.19}$$

By making the substitution

$$\tilde{V} = \frac{1}{r}\tilde{G} \tag{11.20}$$

where $\tilde{G}(r)$ is a function of r, in equation (11.19), we obtain

$$\frac{d^2\tilde{G}}{dr^2} + \beta^2\tilde{G} = 0 \tag{11.21}$$

which is a well-known wave equation for a simple harmonic motion. Its solutions are $Me^{-j\beta r}$ and $Ne^{j\beta r}$, where M and N are constants. We expect an outgoing wave, so the only solution of interest to us is $Me^{-j\beta r}$. Thus, the potential function, from (11.20), is

$$\tilde{V}(r) = \frac{M}{r}e^{-j\beta r} \tag{11.22}$$

in the phasor domain, and

$$V(r,t) = \frac{M}{r}\cos(\omega t - \beta r) = \frac{M}{r}\cos\omega(t - r/u) \tag{11.23}$$

where

$$u = \frac{\omega}{\beta} \tag{11.24}$$

is the phase velocity of the outgoing wave in the medium. Note that for free space, $u = c$, where c is the velocity of light. In (11.23), r/u represents a time delay between the response function (the potential V at a distance r from the charge) and the source (the time-varying charge). In other words, any change in the time-varying source at time $t = t_0$

will be reflected in the potential function at time $t = t_0 + r/u$. For this reason, we refer to the potential function as the *retarded scalar potential.*

When the observation point is very close to the time-varying charge $Q(t) = Q\cos(\omega t + \theta)$, we expect the delay time to be so small that it can be neglected. Then, the potential function $V(r, t)$ at a distance r will be approximately

$$V(r,t) = \frac{Q\cos(\omega t + \theta)}{4\pi\epsilon r} \tag{11.25}$$

in the time domain and

$$\tilde{V}(r) = \frac{\tilde{Q}}{4\pi\epsilon r} \tag{11.26}$$

in the phasor domain, where $\tilde{Q} = Qe^{-j\theta}$. By comparing (11.22) and (11.26), we find that the retarded potential at a distance r from the point source must be

$$\tilde{V}(r) = \frac{\tilde{Q}}{4\pi\epsilon r}e^{-j\beta r} \tag{11.27}$$

Equation (11.27) can now be generalized for a sinusoidally varying volume charge distribution as

$$\tilde{V} = \frac{1}{4\pi\epsilon}\int_v \frac{\tilde{\rho}}{r}e^{-j\beta r}dv \tag{11.28}$$

Similarly, we can obtain a phasor expression for the *retarded magnetic vector potential* as

$$\tilde{\vec{\mathbf{A}}} = \frac{\mu}{4\pi}\int_v \frac{\tilde{\vec{\mathbf{J}}}}{r}e^{-j\beta r}dv \tag{11.29}$$

When the charge and current distributions are known, we can use (11.28) and (11.29) to determine the scalar potential $\tilde{V}$ and the magnetic vector potential $\tilde{\vec{\mathbf{A}}}$. We can then use (11.7) and (11.9) to calculate the magnetic and electric fields, respectively.

11.3 Hertzian dipole

In our quest to understand the operation of an antenna, let us consider the fields produced at any point in a dielectric medium by a very small filament of current, as shown in Figure 11.6. In order to justify the existence of such a short current-carrying element, we can imagine it as two small, fixed, spherical conductors a distance ℓ apart connected by a thin straight wire. Each conductor is a seat for a time-varying charge. When the charge on one conductor is $q(t)$, at that time the charge on the other conductor is $-q(t)$ and the current between them is $i(t) = dq/dt$. This is the reason why a small current element is usually referred to as an **electric dipole**.

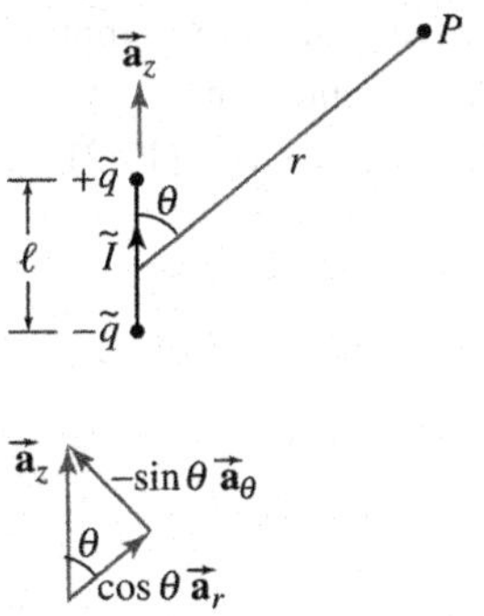

Figure 11.6 Hertzian dipole

By expressing the current in phasor form as $\tilde{I}$, where $\tilde{I} = j\omega\tilde{q}$, and replacing $\vec{\mathbf{J}}\,dv$ in (11.29) by $\tilde{I}\,dz\,\vec{\mathbf{a}}_z$, we can express the magnetic vector potential as

$$\tilde{A}_z = \frac{\mu}{4\pi}\int_c \frac{\tilde{I}}{r}e^{-j\beta r}\,dz$$

where μ is the permeability of the dielectric medium and r is the radial distance of point P from the center of the dipole as shown in Figure 11.6. If we assume that over the short length ℓ of the dipole the current is the same and the point of observation is far away, then the preceding integral can be approximated as

$$\tilde{A}_z = \frac{\mu}{4\pi r}\tilde{I}\ell e^{-j\beta r} \tag{11.30}$$

When we write this equation in the time domain as

$$A_z(r,t) = \frac{\mu}{4\pi r}I_0\ell\,\cos(\omega t - \beta r) \tag{11.31}$$

where $i(t) = I_0\cos(\omega t)$ has been assumed. We realize that (11.31) represents a wave propagating away from the dipole in the r direction with a phase constant β. The amplitude of the wave is decreasing inversely with the distance, and its phase velocity is

$$u_p = \frac{\omega}{\beta} \tag{11.32}$$

For free space, the phase constant is β_0, and the phase velocity is the speed of light.

The wavelength of the wave in the medium is

$$\lambda = \frac{2\pi}{\beta} = \frac{u_p}{f} \tag{11.33}$$

Since $\tilde{\mathbf{A}}$ is now known, we can use (11.7) to determine $\tilde{\mathbf{B}}$. To ease the calculations, let us first express $\tilde{\mathbf{A}}$ in spherical coordinates as

$$\tilde{\mathbf{A}} = \frac{\mu\tilde{I}\ell}{4\pi r}e^{-j\beta r}(\cos\theta\,\vec{\mathbf{a}}_r - \sin\theta\,\vec{\mathbf{a}}_\theta) \tag{11.34}$$

by using the following transformation:

$$\vec{\mathbf{a}}_z = \cos\theta\,\vec{\mathbf{a}}_r - \sin\theta\,\vec{\mathbf{a}}_\theta$$

Using (11.7), we obtain

$$\begin{aligned}\tilde{\mathbf{H}} &= \frac{1}{\mu}[\nabla \times \tilde{\mathbf{A}}]\\ &= \frac{j\beta\tilde{I}\ell}{4\pi r}\left(1 + \frac{1}{j\beta r}\right)\sin\theta\, e^{-j\beta r}\vec{\mathbf{a}}_\phi\end{aligned} \tag{11.35}$$

In order to compute $\tilde{\mathbf{E}}$ using (11.9) we have first to determine scalar potential $\tilde{V}$ either directly (see Problem 11.3) or from (11.12) in terms of $\tilde{\vec{\mathbf{A}}}$. We already know $\tilde{\vec{\mathbf{H}}}$, so it appears relatively easy to compute $\tilde{\vec{\mathbf{E}}}$ directly from Maxwell's equation (11.2). At a point away from the dipole in a dielectric medium, $\tilde{\vec{\mathbf{J}}} = 0$. Hence, we can write (11.2) in phasor form, as

$$\tilde{\vec{\mathbf{E}}} = \frac{1}{j\omega\epsilon}[\nabla \times \tilde{\vec{\mathbf{H}}}]$$

which yields

$$\begin{aligned}\tilde{\vec{\mathbf{E}}} &= \frac{\eta\tilde{I}\ell}{2\pi r^2}\left(1 + \frac{1}{j\beta r}\right)\cos\theta e^{-j\beta r}\vec{\mathbf{a}}_r \\ &+ \frac{j\tilde{I}\ell\eta\beta}{4\pi r}\left(1 + \frac{1}{j\beta r} - \frac{1}{\beta^2 r^2}\right)\sin\theta e^{-j\beta r}\vec{\mathbf{a}}_\theta \end{aligned} \tag{11.36}$$

where $\eta = \sqrt{\mu/\epsilon}$ is the intrinsic impedance of the dielectric medium.

11.3.1 Near-zone fields

Equations (11.35) and (11.36) have terms that vary as $1/r$, $1/r^2$, and $1/r^3$. The terms varying as $1/r^2$ and $1/r^3$ are predominant as long as $\beta r \ll 1$ and make up the **near-zone fields**. In this case, the exponential term $e^{-j\beta r}$ can be approximated as unity. With this approximation, we can express the near-zone fields as

$$\tilde{\vec{\mathbf{H}}} = \frac{\tilde{I}\ell\sin\theta}{4\pi r^2}\vec{\mathbf{a}}_\phi \tag{11.37}$$

$$\tilde{\vec{\mathbf{E}}} = \frac{\tilde{I}\ell\eta}{4\pi r^2}\left[\frac{1}{j\beta r} + 1\right](2\cos\theta\vec{\mathbf{a}}_r + \sin\theta\vec{\mathbf{a}}_\theta) \tag{11.38}$$

The term $1/j\beta r + 1$ can be approximated as $1/j\beta r$ as $\beta r \ll 1$. Replacing $\tilde{I}$ with $j\omega\tilde{q}$ and η/β with $1/\omega\epsilon$, the $\tilde{\vec{\mathbf{E}}}$ field can also be written as

$$\tilde{\vec{\mathbf{E}}} = \frac{\tilde{q}\ell}{4\pi\epsilon}\left[\frac{2\cos\theta}{r^3}\vec{\mathbf{a}}_r + \frac{\sin\theta}{r^3}\vec{\mathbf{a}}_\theta\right] \tag{11.39}$$

We have expressed the near-zone fields in such a way that we can easily recognize them. The expression for the electric field intensity as given in (11.39) is the same as that produced by a static electric dipole. For this reason, the terms varying as $1/r^3$ are called the *electrostatic field terms*. Equation (11.37) gives the static magnetic field intensity created by a short filament of current. As it varies as $1/r^2$, it is referred to as an *induction term*.

From (11.37) and (11.39), we can compute the power density in the near-zone fields as

$$\begin{aligned}\tilde{\vec{\mathbf{S}}} &= \tfrac{1}{2}[\tilde{\vec{\mathbf{E}}} \times \tilde{\vec{\mathbf{H}}}^*] \\ &= -j\frac{I^2\ell^2}{32\pi^2 r^5\omega\epsilon}\sin^2\theta\vec{\mathbf{a}}_r\end{aligned} \tag{11.40}$$

which is purely reactive. In other words, the average power in the near-zone fields is zero. The presence of $-j$ in (11.40) indicates that the near-zone region behaves like a capacitor. Note that $I^2 = \tilde{I}\tilde{I}^*$.

11.3.2 Radiation fields

Let us now consider that the observation point P is far away from the dipole so that $\beta r \gg 1$. In this case, the term $1/r$ dominates and the other terms become vanishingly small. With this understanding, we can express the far-field components as

$$\tilde{\vec{\mathbf{H}}} = \frac{j\beta\tilde{I}\ell}{4\pi r}\sin\theta e^{-j\beta r}\vec{\mathbf{a}}_\phi \tag{11.41}$$

$$\tilde{\vec{\mathbf{E}}} = \frac{j\beta\eta\tilde{I}\ell}{4\pi r}\sin\theta e^{-j\beta r}\vec{\mathbf{a}}_\theta \tag{11.42}$$

These equations show that (a) the far fields propagate in the radial direction, (b) the far fields have only transverse components, and (c) the electric and magnetic fields are perpendicular to each other. These are simply the traits of a TEM (transverse electromagnetic) wave. In short, the far fields represent a *spherical wave*. The ratio of $\tilde{E}_\theta$ to $\tilde{H}_\phi$ is the intrinsic impedance η of the medium. For free space, $\eta_0 = 120\pi \approx 377\,\Omega$.

A linear, current-carrying element along the z direction creates a magnetic vector potential also in the z direction, as given in (11.30). Comparing (11.30) and (11.41), we find that the magnetic field intensity in the far zone is related to the magnetic vector potential as

$$\tilde{H}_\phi = \frac{j\beta}{\mu}\sin\theta\tilde{A}_z \tag{11.43}$$

We will use this equation whenever we are interested in determining the fields produced by linear, current-carrying elements.

We also observe that each field component varies as $\sin\theta$ and its magnitude is zero when $\theta = 0°$ or $\theta = 180°$. The magnitude of each field is maximum when $\theta = 90°$. Thus, the fields are zero along the axis of the dipole and are maximum in a plane perpendicular to its axis (broadside to the dipole). The phase of each field is constant over the surfaces $r =$ constant. We can plot the field patterns in the far zone by normalizing the field components. A *normalized field component* is the ratio of its magnitude at a point to its maximum value. Thus, by definition, the normalized electric field component is

$$E_\theta(\theta, \phi)|_n = \frac{E_\theta(\theta, \phi)}{E_\theta(\theta, \phi)|_{\max}} \tag{11.44a}$$

Applying this definition, we obtain the normalized electric field component for the Hertzian dipole as

$$E_\theta(\theta, \phi)|_n = \sin\theta \tag{11.44b}$$

Its plot is shown in Figure 11.7. Note that the normalized component of the magnetic field also varies as $\sin\theta$.

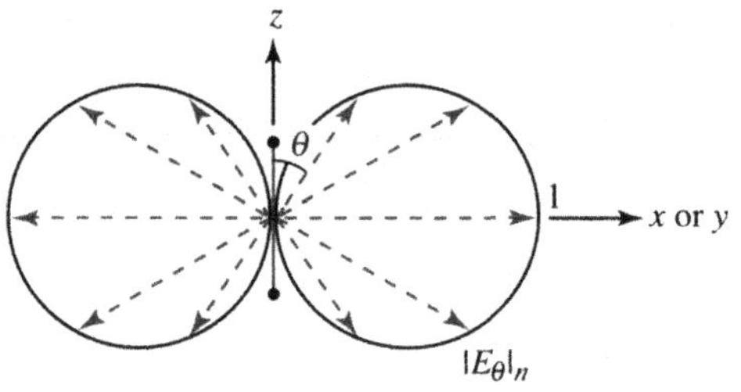

Figure 11.7 Normalized radiation field pattern of an electric dipole

The complex power density in the far region, from (11.41) and (11.42), is

$$\begin{aligned}\vec{\mathbf{S}} &= \frac{1}{2}[\tilde{\vec{\mathbf{E}}} \times \tilde{\vec{\mathbf{H}}}^*] = \frac{1}{2\eta}E_\theta^2\vec{\mathbf{a}}_r \\ &= \frac{I^2\ell^2}{32\pi^2 r^2}\beta^2\eta\sin^2\theta\vec{\mathbf{a}}_r\end{aligned} \tag{11.45}$$

Since the complex power density is purely real and is directed radially outward, it represents the average power per unit area $\langle\hat{S}\rangle$ being consumed by the medium. The question arises: How can a perfect dielectric medium consume power? The only reasonable answer is that the wave carries it away radially. Therefore, this is the *radiated power per unit area*, and the fields associated with it are called **radiated fields**. The **normalized power** (ratio of power density to its maximum value) radiated by the dipole may be expressed as

$$f(\theta, \phi) = \sin^2\theta \tag{11.46}$$

The term $f(\theta, \phi)$ is referred to as the *power density pattern function*, and its plot, known as the *power pattern*, is given in Figure 11.8.

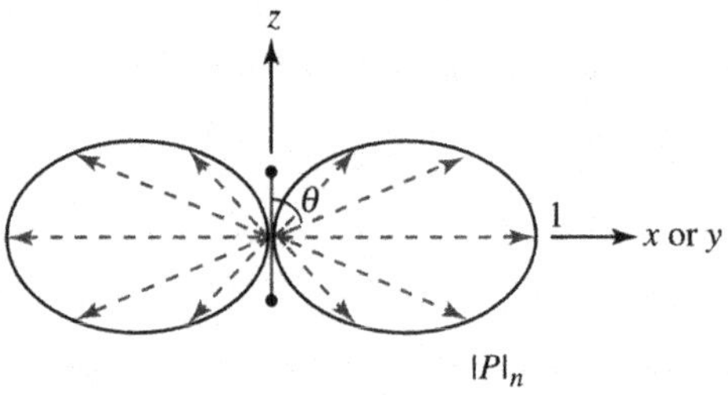

Figure 11.8 Normalized radiated power pattern of an electric dipole

The total power passing through a closed spherical surface at r, P_{rad}, is

$$P_{\text{rad}} = \oint_s \langle \vec{\mathbf{S}} \rangle \cdot \overrightarrow{ds}$$

$$= \frac{I^2 \ell^2 \beta^2 \eta}{32\pi^2} \int_0^{\pi} \sin^3\theta \, d\theta \int_0^{2\pi} d\phi$$

$$= \frac{\eta}{12\pi} \beta^2 \ell^2 I^2 \tag{11.47a}$$

or

$$P_{\text{rad}} = \frac{\pi}{3} \eta \left(\frac{\ell}{\lambda}\right)^2 I^2 \tag{11.47b}$$

This is the total power radiated by the electric dipole. In this equation, ℓ/λ denotes the length of the current filament as a fraction of the wavelength in the dielectric medium, η is the intrinsic impedance of the medium, and I is the maximum value of the current.

Using $\beta_0 = 2\pi/\lambda_0$ and $\eta_0 = 120\pi$ for free space, the power radiated by the electric dipole in free space becomes

$$P_{\text{rad}} = 40\pi^2 \left(\frac{\ell}{\lambda_0}\right)^2 I^2 \tag{11.47c}$$

11.3.3 Radiation resistance

As the electric dipole radiates power, the power must be supplied by the source connected to the dipole; that is, the power supplied by the source is dissipated by the current filament. Because the complex power is purely real, the far region can be simulated by a resistance called the *radiation resistance*. When calculating the power density we have assumed that the fields are in terms of their maximum values. Therefore, the current $\tilde{I}$ is also in terms of its maximum value. Thus, the power dissipated by the radiation resistance R_{rad} is

$$P_{\text{rad}} = \tfrac{1}{2} I^2 R_{\text{rad}} \tag{11.48}$$

Comparing (11.47a,b) and (11.48), we obtain

$$R_{\text{rad}} = \frac{2\pi}{3} \eta \left(\frac{\ell}{\lambda}\right)^2 \tag{11.49}$$

as the radiation resistance in a dielectric medium and

$$R_{\text{rad}} = 80\pi^2 \left(\frac{\ell}{\lambda_0}\right)^2 \tag{11.50}$$

in free space.

11.3.4 Directive gain and directivity

It is evident from (11.45) and Figure 11.8 that the average radiated power density varies as $\sin^2\theta$. Thus, the radiated power is zero along the axis of the electric dipole ($\theta = 0°$) and is maximum in a plane perpendicular to its axis ($\theta = 90°$). In other words, the power radiated by an electric dipole is directional, and the parameter called **directive gain**, G, is a measure of it. We define directive gain as the ratio of the power density radiated by the dipole to the average power density ($P_{rad}/4\pi r^2$). Mathematically, we can express it as

$$G = \frac{4\pi r^2 \langle \vec{\mathbf{S}} \rangle}{P_{rad}} \tag{11.51a}$$

Substituting for the average power density from (11.45) and the radiated power from (11.47a), we obtain the directive gain of the electric dipole as

$$G = 1.5 \sin^2\theta \tag{11.51b}$$

When G is maximum, it is called the *directivity*, D, of the dipole. Thus, the directivity of a current filament is

$$D = 1.5 \tag{11.51c}$$

We can make a formal statement for the directivity of the electric dipole as the ratio of maximum radiated power density to average power density. The directivity gain is generally expressed in decibels as

$$D = 10\,\log_{10}(1.5) = 1.76\text{ dB} \tag{11.51d}$$

EXAMPLE 11.1

An electric dipole of length 50 cm is situated in free space. If the maximum value of the current is 25 A and its frequency is 10 MHz, determine (a) the electric and magnetic fields in the far zone, (b) the average power density, and (c) the radiation resistance.

Solution Since the dipole is radiating in free space, the fields propagate with the speed of light, $c = 3 \times 10^8$ m/s.

$$\omega = 2\pi f = 6.283 \times 10^7 \text{ rad/s}$$

The phase constant: $\beta = \dfrac{\omega}{c} = 0.209$ rad/m

From the given data, we have $\tilde{\mathbf{I}} = 25\,\underline{/0°}$ A and $L = 0.5$ m. Substituting in (11.41) and (11.42), we obtain

$$\tilde{\mathbf{H}} = \frac{j0.208}{r}\sin\theta e^{-j0.209r}\vec{\mathbf{a}}_\phi \text{ A/m}$$

$$\tilde{\mathbf{E}} = \frac{j78.416}{r}\sin\theta e^{-j0.209r}\vec{\mathbf{a}}_\theta \text{ V/m}$$

Thus, the average power density in the radial direction, from (11.40), is

$$\langle S_r \rangle = \frac{8.15}{r^2} \sin^2\theta \ \text{W/m}^2$$

and the total power crossing a spherical surface at r, from (11.47a), is

$$P_{\text{rad}} = 68.25 \ \text{W/m}^2$$

Finally, the radiation resistance, from (11.48), is

$$R_{\text{rad}} = \frac{2}{25^2} \times 68.25 = 0.22 \ \Omega$$

• • •

11.4 A magnetic dipole

A small circular loop of wire carrying a current $I = I_0 \cos \omega t$ is shown in Figure 11.9 (see below). It is usually referred to as a **magnetic dipole** because the radiated fields that it generates are the magnetic analogue of electric dipole fields. We have tacitly assumed that the current distribution does not vary in the ϕ direction and $\beta a \ll 1$, where a is the radius of the loop as shown in the figure. The distance vector $\vec{\mathbf{R}}$ of point $P(r, \theta, \phi)$ from the current element $\tilde{I} d\phi'$ at $Q(a, \phi', 0)$ is

$$\vec{\mathbf{R}} = r\vec{\mathbf{a}}_r - a\vec{\mathbf{a}}_{\rho'} \tag{11.52}$$

from which we get

$$R^2 = r^2 + a^2 - 2ar \sin\theta \cos(\phi - \phi')$$

Under the assumption that $r \gg a$, the distance R can be approximated as

$$R = r - a \sin\theta \cos(\phi - \phi') \tag{11.53}$$

Thus, the magnetic vector potential at P is

$$\tilde{\vec{\mathbf{A}}} = \frac{\mu \tilde{I} a}{4\pi r} e^{-j\beta r} \int_0^{2\pi} e^{j\beta a \sin\theta \cos(\phi - \phi')} d\phi' \vec{\mathbf{a}}_{\phi'}$$

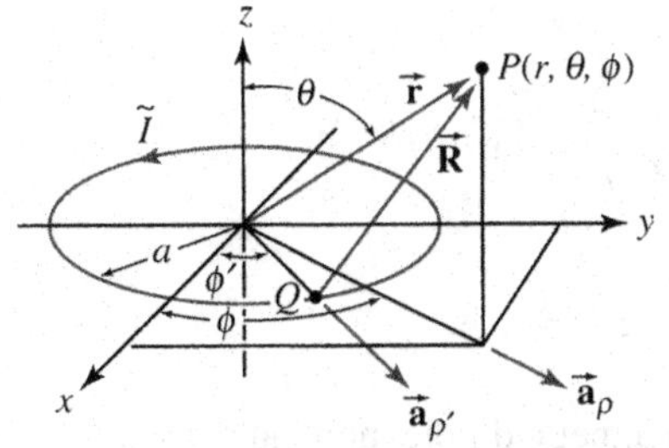

Figure 11.9 A magnetic dipole

where $\tilde{I} = I_0$ is the phasor equivalent of $i(t) = I_0 \cos \omega t$. Because the unit vector $\vec{\mathbf{a}}_{\phi'}$ is a function of ϕ', it can be expressed in terms of the unit vectors at $P(r, \theta, \phi)$ as

$$\vec{\mathbf{a}}_{\phi'} = \cos(\phi - \phi')\vec{\mathbf{a}}_{\phi} + \sin(\phi - \phi')\vec{\mathbf{a}}_r \tag{11.54}$$

It appears as if $\tilde{\vec{\mathbf{A}}}$ has a component in the radial direction. This should not be true because a current in the ϕ direction can only create a component in the ϕ direction. We leave it as an exercise for you to show that the radial component is zero.

The ϕ component of the magnetic vector potential is

$$\tilde{A}_{\phi} = \frac{\mu \tilde{I} a}{4\pi r} e^{-j\beta r} \int_0^{2\pi} e^{j\beta a \sin\theta \cos(\phi-\phi')} d\phi' \tag{11.55}$$

The integral in (11.55) is quite difficult to evaluate. In fact, it involves a knowledge of Bessel functions. However, the exponential term can be approximated as

$$e^{j\beta a \sin\theta \cos(\phi-\phi')} \approx 1 + j\beta a \sin\theta \cos(\phi - \phi') \tag{11.56}$$

as long as $\beta a \ll 1$. This approximation enables us to evaluate (11.55) as

$$\int_0^{2\pi} [1 + j\beta a \sin\theta \cos(\phi - \phi')] \cos(\phi - \phi')\, d\phi' = j\beta a\pi \sin\theta \tag{11.57}$$

Substituting this result in (11.55), we obtain the magnetic vector potential as

$$\tilde{A}_{\phi} = \frac{j\mu \tilde{I} a^2 \beta}{4r} \sin\theta e^{-j\beta r} \tag{11.58}$$

Note that $\tilde{A}_{\phi}$ does not vary with ϕ. Therefore,

$$\nabla \cdot \tilde{\vec{\mathbf{A}}} = 0$$

and, from (11.17), we obtain $\tilde{V} = 0$. We could have just stated this fact, as the net charge is zero. Thus, the far-field electric field intensity, from (11.9), is

$$\tilde{\vec{\mathbf{E}}} = -j\omega \tilde{\vec{\mathbf{A}}}$$

or

$$\tilde{E}_{\phi} = \frac{\omega\mu\beta}{4\pi r} \tilde{M} \sin\theta e^{-j\beta r} \tag{11.59}$$

where

$$\tilde{M} = \pi a^2 \tilde{I} \tag{11.60}$$

is the magnetic dipole moment.

From Maxwell's equation

$$\nabla \times \tilde{\vec{\mathbf{E}}} = -j\omega\mu \tilde{\vec{\mathbf{H}}}$$

we obtain the far-field magnetic field intensity as

$$\tilde{H}_\theta = -\frac{\omega\mu\beta}{4\pi r\eta}\tilde{M}\sin\theta e^{-j\beta r} \tag{11.61}$$

where η is the intrinsic impedance of the dielectric medium. Once again, equations (11.59) and (11.61) represent a TEM wave that lies in a plane normal to the radial direction and the magnitude of each field decays inversely with the distance. Since the fields vary as $\sin\theta$, the field patterns for the magnetic dipole are similar to those of a Hertzian dipole antenna (Figure 11.6). Although we have computed the magnetic vector potential, the electric field, and the magnetic fields only for a circular loop, equation (11.60) helps us to compute these fields for any current-carrying loop as long as we replace πa^2 by the area of the loop.

The complex power density associated with the far fields is

$$\begin{aligned}\langle\vec{\mathbf{S}}\rangle &= \tfrac{1}{2}[\tilde{\vec{\mathbf{E}}} \times \tilde{\vec{\mathbf{H}}}^*] \\ &= \frac{1}{2\eta}M^2\left(\frac{\omega\mu\beta}{4\pi r}\right)^2 \sin^2\theta\vec{\mathbf{a}}_r \end{aligned}\tag{11.62}$$

where $M^2 = \tilde{M}\tilde{M}^* = (\pi a^2 I_0)^2$. Since the complex power density is a real quantity, it represents the time-average power density in the medium. Once again, the power flow is in the radial direction, and the power pattern is similar to that given in Figure 11.7. The total power radiated by the current-carrying loop is found by integrating (11.62) over a large spherical surface of radius r as

$$\begin{aligned}P_{\text{rad}} &= \frac{1}{2\eta}M^2\left(\frac{\omega\mu\beta}{4\pi}\right)^2\int_0^\pi \sin^3\theta\, d\theta\int_0^{2\pi} d\phi \\ &= \frac{4}{3}\pi^3\eta\left(\frac{M}{\lambda^2}\right)^2\end{aligned}\tag{11.63a}$$

or

$$P_{\text{rad}} = \frac{\pi}{12}\eta I^2(\beta a)^4 \tag{11.63b}$$

where $\beta = \omega\sqrt{\mu\epsilon}$ is the phase constant (rad/m), $\lambda = 2\pi/\beta$ is the wavelength (m), and $\eta = \sqrt{\mu/\epsilon}$ is the intrinsic impedance (Ω).

Using (11.49), as we did for the Hertzian dipole, we can determine the radiation resistance for the loop antenna of a magnetic dipole as

$$R_{\text{rad}} = \frac{8}{3}\pi^3\eta\left(\frac{\pi a^2}{\lambda^2}\right)^2 \tag{11.64a}$$

or

$$R_{\text{rad}} = \frac{\pi}{6}\eta(\beta a)^4 \tag{11.64b}$$

Once again the radiation resistance of the loop antenna is low because βa is usually small.

Using the definition of directive gain as given by (11.51), we obtain the directive gain from (11.62) and (11.63a) as

$$G = 1.5\sin^2\theta \tag{11.65}$$

and the directivity as

$$D = 1.5 \quad \text{or} \quad D = 1.76\ \text{dB} \tag{11.66}$$

These are the same results that we obtained for the short current filament.

EXAMPLE 11.2 The current in a small loop of radius 10 cm is $100\cos(\omega t - 30°)$ A, where ω is 300 Mrad/s. If the medium is free space, write the expressions for the far fields in the time domain. Compute the power radiated by the loop and its radiation resistance.

Solution

$$\tilde{I} = 100e^{-j\pi/6}\ \text{A}$$
$$M = \pi a^2 \tilde{I} = 3.142e^{-j\pi/6}$$

The phase constant, the intrinsic impedance, and the wavelength in free space are

$$\beta_0 = \frac{\omega}{c} = \frac{300 \times 10^6}{3 \times 10^8} = 1\ \text{rad/m}$$
$$\eta_0 = 120\pi \approx 377\ \Omega$$
$$\lambda_0 = \frac{2\pi}{\beta_0} = 2\pi = 6.283\ \text{m}$$

Let us compute the following factor:

$$\frac{\omega\mu\beta}{4\pi}\tilde{M} = \frac{300 \times 10^6 \times 4\pi \times 10^{-7} \times 1}{4\pi}3.142e^{-j\pi/6} = 94.26e^{-j\pi/6}$$

Hence, from (11.59), the electric field intensity in the far region is

$$\tilde{E}_\phi = \frac{94.26}{r}\sin\theta e^{-j(r+\pi/6)}$$

which can be written in the time domain as

$$E_\phi(r,\theta,\phi,t) = \frac{94.26}{r}\sin\theta\cos(\omega t - r - \pi/6)\ \text{V/m}$$

The magnetic field intensity, from (11.61), is

$$\tilde{H}_\theta = -\frac{0.25}{r}\sin\theta e^{-j(r+\pi/6)}$$

This equation can be written in the time domain as

$$H_\theta(r,\theta,\phi,t) = \frac{0.25}{r}\sin\theta\cos(\omega t - r + 5\pi/6)\ \text{A/m}$$

The radiation resistance, from (11.64b), is

$$R_{\text{rad}} = \frac{\pi}{6} \times 377 \times (1 \times 0.1)^4 = 19.74 \text{ m}\Omega$$

Finally, the total power radiated by the loop is

$$P_{\text{rad}} = \tfrac{1}{2} I^2 R_{\text{rad}} = 0.5 \times 100^2 \times 19.74 \times 10^{-3} = 98.7 \text{ W}$$

• • •

11.5 A short dipole antenna

The discussion of a Hertzian dipole was necessary to understand how a current-carrying element radiates power in a dielectric medium, but such a radiating element of infinitesimal length is not practically realizable. For this reason, we now consider a center-fed dipole antenna of short length ℓ such that $\beta\ell \ll 1$, as shown in Figure 11.10. Since the current at either end of the antenna must be zero, we assume that the current distribution on the antenna is such that it decreases uniformly from its maximum value at the center to zero at the ends as shown. Let the current at the center be exactly the same as that of a Hertzian dipole. Then we can express the current distribution as

$$\tilde{I}(z) = \begin{cases} \tilde{I}(1 - 2z/\ell) & 0 \le z \le \ell/2 \\ \tilde{I}(1 + 2z/\ell) & -\ell/2 \le z \le 0 \end{cases}$$

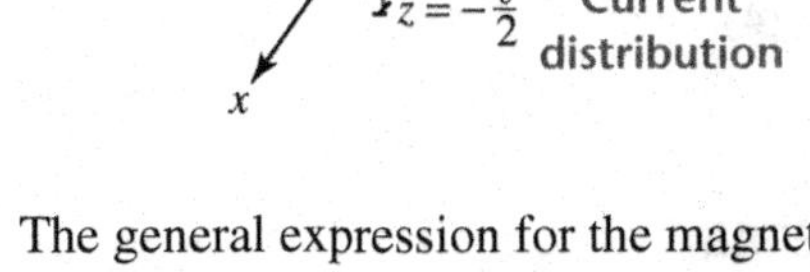

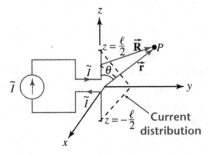

Figure 11.10 A short, center-fed antenna

The general expression for the magnetic vector potential becomes

$$\tilde{A}_z = \frac{\mu}{4\pi} \int_c \frac{1}{R} \tilde{I} \left(1 - \frac{2z}{\ell}\right) e^{-j\beta R}\, dz \qquad (11.67)$$

Under the assumption that $\beta\ell \ll 1$, we can approximate $R = r$ in the denominator and

$$e^{-j\beta R} = e^{-j\beta r} e^{j\beta z \cos\theta} \approx (1 + j\beta z \cos\theta) e^{-j\beta r} \qquad (11.68)$$

Substituting (11.68) in (11.67) and carrying out the integration, we obtain

$$\tilde{A}_z = \frac{\mu}{8\pi r}\tilde{I}\ell e^{-j\beta r}\left(1 + \frac{j\beta\ell}{6}\cos\theta\right) \tag{11.69}$$

The corresponding radiation fields, from (11.69) using (11.43), are

$$\vec{\tilde{\mathbf{H}}} = \frac{j\beta\tilde{I}\ell}{8\pi r}e^{-j\beta r}\sin\theta\left(1 + \frac{j\beta\ell}{6}\cos\theta\right)\vec{\mathbf{a}}_\phi \tag{11.70}$$

$$\vec{\tilde{\mathbf{E}}} = \frac{j\beta\tilde{I}\ell}{8\pi r}\eta e^{-j\beta r}\sin\theta\left(1 + \frac{j\beta\ell}{6}\cos\theta\right)\vec{\mathbf{a}}_\theta \tag{11.71}$$

If we assume that $(j\beta/6)\cos\theta \ll 1$ and compare these fields with their counterparts in (11.41) and (11.42), we find that the field intensities at every point in space due to the short antenna are one-half of those due to a Hertzian dipole of the same length and carrying the same magnitude of current.

The average power density of the short antenna is

$$\langle\vec{\mathbf{S}}\rangle = \frac{I^2\ell^2\beta^2}{128\pi^2 r^2}\eta\sin^2\theta\left(1 + \frac{\beta^2\ell^2}{36}\cos^2\theta\right)\vec{\mathbf{a}}_r \tag{11.72}$$

A direct comparison with (11.45) shows that the power radiated by the short antenna is one-quarter of that for the Hertzian dipole of the same length as long as $(\beta\ell/6)\cos\theta \ll 1$. The same assumption yields the radiation resistance of the short antenna as

$$R_{\text{rad}} = \frac{2\pi}{12}\eta\left(\frac{\ell}{\lambda}\right)^2 \tag{11.73}$$

This is one-quarter of that of a Hertzian dipole when the term $(\beta\ell/6)\cos\theta$ is ignored.

Although we have developed the equations for the short dipole antenna under the assumption that $\beta\ell \ll 1$, they are good approximations for center-fed antennas of length up to one-quarter wavelength ($\ell \leq \lambda/4$).

11.6 A half-wave dipole antenna

As the power radiated by an antenna is directly proportional to its radiation resistance, and the radiation resistance varies as ℓ^2, we need longer antennas to radiate reasonable amounts of power. For this reason, we employ dipole antennas having lengths equal to half-wavelength and full-wavelength. In order to calculate the fields radiated by a long antenna we must know the current distribution along its length. Except at the ends of the dipole antenna, where the current must be zero, we really have no way of knowing how it is distributed elsewhere. However, if we view a center-fed antenna as an open-circuit transmission line, we can speculate that the current distribution may be sinusoidal. Assuming such a current distribution, we can then calculate the radiated power and verify it by making measurements. A good correlation between the predicted and observed readings will justify the assumed distribution. For thin antennas, an assumed sinusoidal current distribution has already been verified. Therefore, we begin our discussion of a linear, half-wavelength, dipole antenna (see Figure 11.11) by assuming that the current distribution on the antenna is

$$\tilde{I} = I_0 \cos \beta z \tag{11.74}$$

where I_0 is the maximum value of the current.

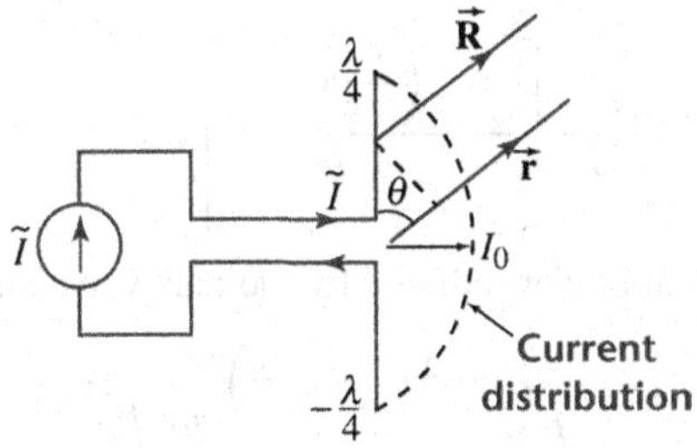

Figure 11.11 A linear, half-wave, dipole antenna

At a point $P(r, \theta, \phi)$ from the antenna, the magnetic vector potential is

$$\tilde{A}_z = \frac{\mu}{4\pi} I_0 \int_{-\lambda/4}^{\lambda/4} \frac{\cos \beta z}{R} e^{-j\beta R}\, dz$$

Once again, when $P(r, \theta, \phi)$ is far away from the antenna, we can approximate $R = r - z\cos\theta$ to account for the phase shift and $R \approx r$ for the distance in the denominator. Expressing $\cos \beta z$ as

$$\cos \beta z = \frac{e^{j\beta z} + e^{-j\beta z}}{2}$$

we can rewrite the approximate expression for $\tilde{A}_z$ as

$$\tilde{A}_z = \frac{\mu}{8\pi r} I_0 e^{-j\beta r} \int_{-\lambda/4}^{\lambda/4} (e^{j\beta z} + e^{-j\beta z}) e^{j\beta z \cos\theta}\, dz$$

Carrying out the integration, we obtain

$$\tilde{A}_z = \frac{\mu}{2\pi \beta r} I_0 e^{-j\beta r} \left[\frac{\cos\left(\frac{\pi}{2}\cos\theta\right)}{\sin^2\theta} \right] \tag{11.75}$$

The magnetic field intensity in the radiation zone for a z-directed antenna, from (11.43), is

$$\tilde{\vec{\mathbf{H}}} = \frac{j\beta}{\mu} \sin\theta \tilde{A}_z \vec{\mathbf{a}}_\phi \tag{11.76a}$$

$$= \frac{j}{2\pi r} I_0 e^{-j\beta r} \left[\frac{\cos\left(\frac{\pi}{2}\cos\theta\right)}{\sin\theta} \right] \vec{\mathbf{a}}_\phi \tag{11.76b}$$

The corresponding electric field intensity is

$$\tilde{\vec{\mathbf{E}}} = \frac{j}{2\pi r} \eta I_0 e^{-j\beta r} \left[\frac{\cos\left(\frac{\pi}{2}\cos\theta\right)}{\sin\theta} \right] \vec{\mathbf{a}}_\theta \tag{11.77}$$

Thus, the average power per unit area radiated by the half-wave antenna is

$$\langle \vec{\mathbf{S}} \rangle = \frac{\eta I_0^2}{8\pi^2 r^2} \left[\frac{\cos^2\left(\frac{\pi}{2}\cos\theta\right)}{\sin^2\theta} \right] \vec{\mathbf{a}}_r \tag{11.78}$$

The total power radiated by the half-wave antenna is

$$P_{\text{rad}} = \frac{\eta I_0^2}{8\pi^2} \int_0^\pi \frac{\cos^2\left(\frac{\pi}{2}\cos\theta\right)}{\sin\theta} d\theta \int_0^{2\pi} d\phi$$

The integral with respect to θ can be evaluated using numerical integration. We used MathCAD® and obtained its value as 1.21882. We will approximate it as 1.219. The power radiated by the half-wave antenna becomes

$$P_{\text{rad}} = \frac{1.219}{4\pi} \eta I_0^2 \tag{11.79}$$

Finally, the radiation resistance of the half-wave antenna is

$$R_{\text{rad}} = \frac{1.219}{2\pi} \eta \tag{11.80}$$

which is 73.14 Ω when the medium is free space. The high value of the radiation resistance makes a half-wave dipole antenna very effective for radiating considerable amounts of power. In this case, the radiation pattern is slightly more directive than the electric dipole. The input impedance of a half-wave antenna does have an inductive component of about 43 Ω. This can, however, be eliminated by reducing the length of the antenna by about 5%.

EXAMPLE 11.3

WORKSHEET 17

Mathcad

The amplitude of the electric field intensity broadside to a half-wave dipole antenna at a distance of 15 km is 0.1 V/m in free space. If the operating frequency is 100 MHz, determine the length of the antenna and the total power that it radiates. Also, write the general expressions for the electric and magnetic field intensities in the time domain.

Solution

$$f = 100 \text{ MHz}$$

$$\omega = 2\pi f = 628.319 \text{ Mrad/s}$$

$$\beta = \frac{\omega}{c} = \frac{628.319 \times 10^6}{3 \times 10^8} = 2.094 \text{ rad/m}$$

$$\lambda = \frac{2\pi}{\beta} = \frac{2\pi}{2.094} = 3 \text{ m}$$

Hence, the length of the half-wave dipole antenna is 1.5 m. From (11.77), when $\theta = 90^\circ$ and $r = 15$ km, the maximum amplitude of the current is

$$I_0 = \frac{2\pi r}{\eta_0}|E| = \frac{2\pi \times 15 \times 10^3}{120\pi}(0.1) = 25 \text{ A}$$

The radiation resistance of a half-wave dipole antenna is 73.14 Ω, thus the power radiated by the antenna is

$$P_{\text{rad}} = \tfrac{1}{2} I_0^2 R_{\text{rad}} = 0.5 \times 25^2 \times 73.14 = 22.86 \text{ kW}$$

The general expressions for the $\vec{\mathbf{E}}$ and $\vec{\mathbf{H}}$ fields, from (11.77) and (11.76b), in the time domain are

$$E_\theta(r, \theta, \phi, t) = -\frac{1500}{r}\left[\frac{\cos\left(\frac{\pi}{2}\cos\theta\right)}{\sin\theta}\right]\sin(6.283 \times 10^8 t - 2.094r) \text{ V/m}$$

$$H_\phi(r, \theta, \phi, t) = -\frac{3.98}{r}\left[\frac{\cos\left(\frac{\pi}{2}\cos\theta\right)}{\sin\theta}\right]\sin(6.283 \times 10^8 t - 2.094r) \text{ A/m}$$

• • •

11.7 Antenna arrays

You may realize by now that a linear antenna transmits power equally well in any plane perpendicular to its axis. The reason, of course, is that the power pattern is independent of variations in the ϕ direction. Stated differently, the directive gain of a linear antenna is the same in a $\theta =$ constant plane. A radiating system with high directive gain, and thereby high directivity, in a certain direction can be constructed by combining simple antenna elements into *arrays*. An antenna array consists of similar antennas pointing in the same direction. A proper arrangement of these antennas can create a radiation pattern that is in phase at some points in space and exactly 180° out of phase at other points in space. Our ability to modify the power pattern enables us to design an antenna array that transmits all the energy in the desired direction, with almost no radiation in any other direction. The design of such a radiation pattern is possible because we can control (a) the number of elements in the array, (b) the separation between each element of the array, and (c) the magnitude and the phase of current feeding each element.

We begin our discussion with a two-element array, as shown in Figure 11.12 (see below), with a separation d between the two elements. If we assume the current in element (0) as a reference such that

$$\tilde{I} = I_0 \underline{/0^\circ}$$

where I_0 is its maximum value, then we can define the current in element (1) as

$$\tilde{I}_1 = kI_0 \underline{/\alpha}$$

where k is the ratio of the magnitude of current in (1) to that in (0), and α is the phase angle by which the current in (1) leads the current in (0).

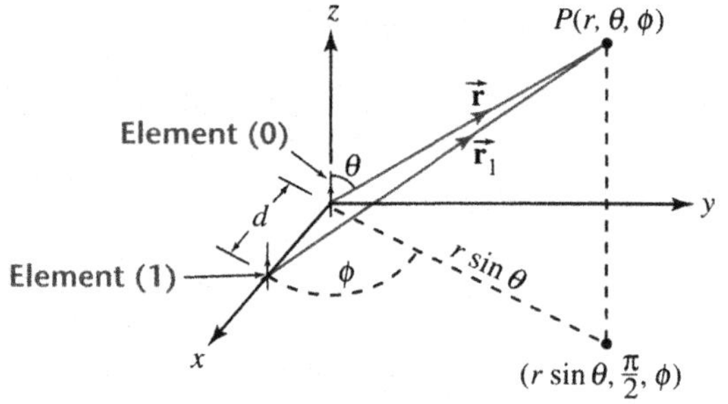

Figure 11.12 A linear, two-element array

Let us express the radiated electric field intensity produced by each antenna in the far zone as

$$\tilde{E}_\theta = E_m F(\theta, \phi) \frac{1}{r} e^{-j\beta r} \tag{11.81}$$

where E_m is the maximum value of the $\tilde{\mathbf{E}}$ field, and $F(\theta, \phi)$ is the associated field pattern. For example, the maximum value of the radiated electric field due to a Hertzian dipole, from (11.42), is

$$E_m = \frac{\beta \ell}{4\pi} \eta I$$

and

$$F(\theta, \phi) = \sin\theta$$

Because the two antennas are similar and are oriented in the same direction, the total electric field intensity at a point $P(r, \theta, \phi)$ far away from the array is

$$\tilde{E}_\theta = E_m F(\theta, \phi) \left[\frac{1}{r} e^{-j\beta r} + \frac{k}{r_1} e^{-j\beta r_1} e^{j\alpha} \right] \tag{11.82}$$

As long as the point of observation is far away from the array, it is possible to write

$$\frac{1}{r_1} \approx \frac{1}{r}$$

and

$$r_1 = r - d \sin\theta \cos\phi$$

With these approximations, we can express (11.82) as

$$\tilde{E}_\theta = E_m F(\theta, \phi) \frac{1}{r} e^{-j\beta r}[1 + k e^{j\psi}] \tag{11.83}$$

where $\psi = \beta d \sin\theta \cos\phi + \alpha$. The magnitude of the total electric field intensity, from (11.83), is

$$E_\theta = \frac{1}{r} E_m F(\theta, \phi)[(1 + k \cos\psi)^2 + (k \sin\psi)^2]^{1/2} \tag{11.84}$$

If we define $F(\psi)$ as

$$F(\psi) = [(1 + k \cos\psi)^2 + (k \sin\psi)^2]^{1/2} \tag{11.85}$$

then $F(\psi)$ is called the *normalized array pattern*. In terms of $F(\psi)$, we can express (11.84) as

$$E_\theta = \frac{1}{r} E_m F(\theta, \phi) F(\psi) \tag{11.86}$$

Thus, the total field pattern of an array of similar elements is the product of the element pattern $F(\theta, \phi)$ and the array pattern $F(\psi)$. This is known as the *principle of pattern multiplication*.

EXAMPLE 11.4

WORKSHEET 18

Mathcad

Sketch the field patterns of two Hertzian dipole antennas in the plane perpendicular to their axes when (a) $d = \lambda/2, k = 1$, and $\alpha = 0°$; (b) $d = \lambda/2, k = 1$, and $\alpha = \pi$; and (c) $d = \lambda, k = 1$, and $\alpha = -\pi/2$.

Solution Assuming that the axes of the two Hertzian dipole antennas are along the z direction, then $\theta = 90°$ is the plane perpendicular to their axes. Thus, the field pattern of each element, from (11.42), is

$$F(\theta, \phi) = \sin(\pi/2) = 1$$

a) In this case, $\beta d = \pi, k = 1, \alpha = 0$, and $\psi = \pi \cos\phi$. Substituting for ψ and k in (11.85), we obtain an equation as a function of ϕ. By varying ϕ from 0 to 2π, we can sketch the field plot. We have used MathCAD® to do so and the field pattern is given in Figure 11.13.
b) This plot is obtained from (a) by setting $\alpha = \pi$ and is given in Figure 11.14.
c) In this case, $\beta d = 2\pi$ and $\alpha = -\pi/2$. With these changes, we obtain a field pattern as illustrated in Figure 11.15.

Figure 11.13 Field pattern of an array consisting of two Hertzian dipoles when $k = 1$, $d = \lambda/2$ and $\alpha = 0$

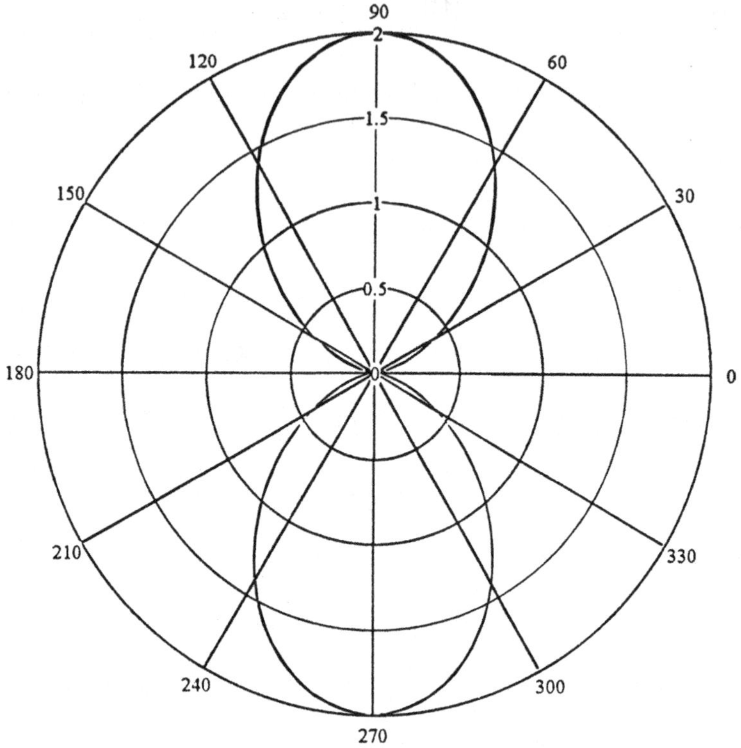

Figure 11.14 Field pattern of an array consisting of two Hertzian dipoles when $k = 1$, $d = \lambda/2$, and $\alpha = \pi$

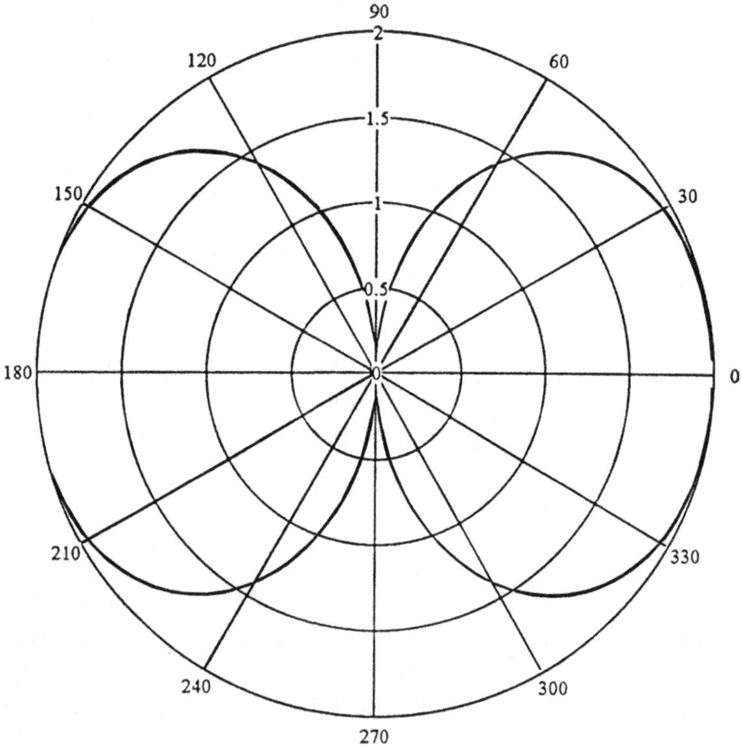

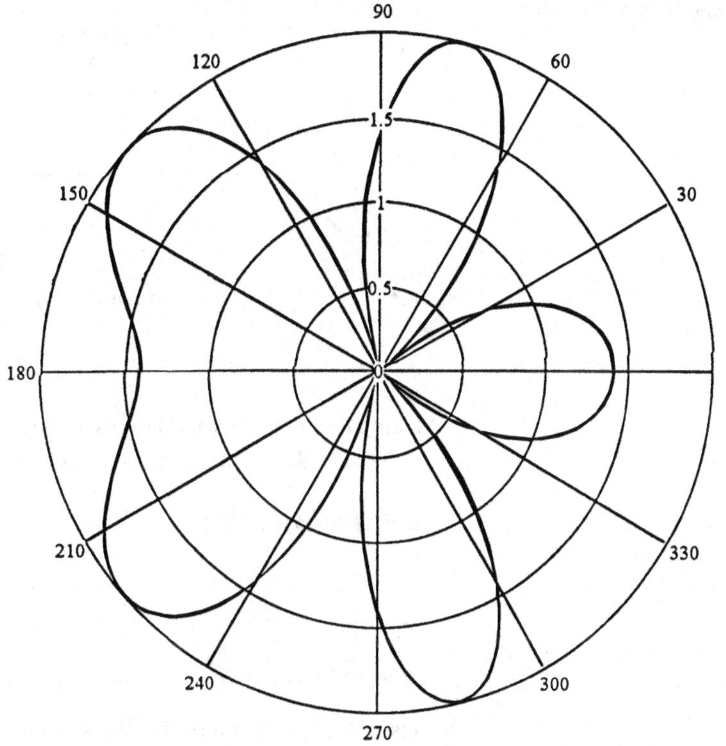

Figure 11.15 Field pattern of an array consisting of two Hertzian dipoles when $k = 1$, $d = \lambda$, and $\alpha = -\pi/2$

• • •

11.8 Linear arrays

From the discussion of a two-element array it must be apparent that we can control the field pattern only to a limited extent. In order to achieve better directivity for point-to-point communication, we need an array with more than two elements. In this section, we consider an n-element *uniform linear array*. In this context, the word *linear* implies that all elements of the array are spaced equally along a straight line, as shown in Figure 11.16. The word *uniform* is used to indicate that the magnitude of the current in each element is the same, and the phase shift is progressive. Thus, if the current in the element (k) is

$$\tilde{I}_k = I_0 e^{jk\alpha} \tag{11.87}$$

then the current in the element ($k + 1$) is

$$\tilde{I}_{k+1} = I_0 e^{j(k+1)\alpha} \tag{11.88}$$

We have assumed that element (0) carries the reference current and has a maximum value of I_0. In the preceding equations, α is the progressive phase shift from one element to the next.

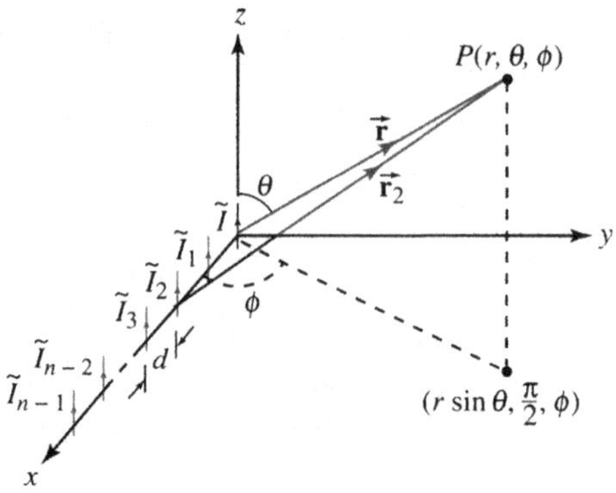

Figure 11.16 An n-element linear array

Making the same approximations as we made in Section 11.7, we can generalize (11.83) for an n-element array as

$$\tilde{E}_\theta = E_m F(\theta, \phi)\frac{1}{r}e^{-j\beta r}[1 + e^{j\psi} + e^{j2\psi} + e^{j3\psi} + \cdots + e^{j(n-1)\psi}] \tag{11.89}$$

where

$$\psi = \beta d \sin\theta \cos\phi + \alpha \tag{11.90}$$

Because the term within the brackets of (11.89) represents a geometric series, we can multiply (11.89) by $e^{j\psi}$ and then subtract the product from (11.89) to obtain

$$\tilde{E}_\theta = \frac{1}{r}F(\theta, \phi)e^{-j\beta r}\left[\frac{1 - e^{jn\psi}}{1 - e^{j\psi}}\right] \tag{11.91}$$

Equation (11.91) can also be rewritten as

$$\tilde{E}_\theta = \frac{1}{r}F(\theta, \phi)e^{-j\beta r}\frac{e^{jn\psi/2}}{e^{j\psi/2}}\left[\frac{e^{jn\psi/2} - e^{-jn\psi/2}}{e^{j\psi/2} - e^{-j\psi/2}}\right]$$

$$= \frac{1}{r}F(\theta, \phi)e^{-j\beta r}e^{j(n-1)\psi/2}\left[\frac{\sin(n\psi/2)}{\sin(\psi/2)}\right] \tag{11.92}$$

Thus, the normalized array pattern is

$$F(\psi) = \frac{\sin(n\psi/2)}{\sin(\psi/2)} \tag{11.93}$$

In the xy plane (the plane perpendicular to the axis of the array), $\theta = 90°$, and $F(\theta, \phi) = 1$ for both the Hertzian dipole and the half-wave antennas. Thus, the field pattern depends only upon $F(\psi)$ at $\theta = 90°$.

The maximum value of (11.93) is n and it occurs at $\psi = 0°$. This is referred to as the *principal maximum* of the array. For a fixed point of observation $P(r, \pi/2, \phi)$, ϕ is fixed. Thus, from (11.90), the progressive phase-shift when $\psi = 0°$ is

$$\alpha = -\beta d \cos\phi \tag{11.94}$$

By setting (11.93) to zero we obtain values of ψ for which the field intensity is zero. Each of these points is called the *null of the pattern.* The null points occur when

$$\psi = \pm\frac{2p\pi}{n} \qquad p = 1, 2, 3, \ldots \tag{11.95}$$

Between any two consecutive null points, the field pattern exhibits a secondary maximum point. We can obtain these points by setting $\sin(n\psi/2) = 1$. That is,

$$\psi = \pm\frac{(2q+1)\pi}{n} \qquad q = 1, 2, 3, \ldots \tag{11.96}$$

The first secondary maximum, from (11.96), occurs when

$$\psi = \frac{3\pi}{n}$$

and the amplitude of the first secondary maximum (**lobe**), from (11.93), is

$$\frac{1}{\sin(1.5\pi/n)} \approx \frac{n}{1.5\pi} \tag{11.97}$$

when n is very large. Thus, the ratio of the first secondary maximum to the principal maximum is 21.22% $(100/1.5\pi)$. In other words, the magnitude of the first secondary maximum is 13.56 dB below the principal maximum.

EXAMPLE 11.5

Sketch the field patterns in the xy and xz planes of a 20-element, Hertzian dipole, linear array with a spacing of $\lambda/8$ and a phase shift of 0°.

Solution From the given values, we have

$$\alpha = 0^\circ, \qquad \beta d = \pi/4, \quad \text{and} \quad n = 20$$

a) Field pattern in the xy plane:

$$\theta = 90^\circ \Rightarrow F(\theta, \phi) = 1$$

$$\psi = \frac{\pi}{4}\cos(\phi) \quad \text{and} \quad F(\psi) = \frac{\sin(10\psi)}{\sin(0.5\psi)}$$

Setting $\psi = 0$, we find that the principal maxima are along $\phi = 90^\circ$ and $\phi = 270^\circ$. The field pattern in the xy plane is shown in Figure 11.17. Note that when the currents are in phase and the antennas are arranged along the x axis, the principal lobe is in the y direction ($\phi = 90^\circ$ or 270°).

When the field pattern is maximum in a direction perpendicular to the array, it is called a *broadside array.* To obtain a null in the x direction, the spacing between the element must be $\lambda/2$. Do you know why? By setting $d = \lambda/2$ and $\alpha = \pi$, we can obtain the field

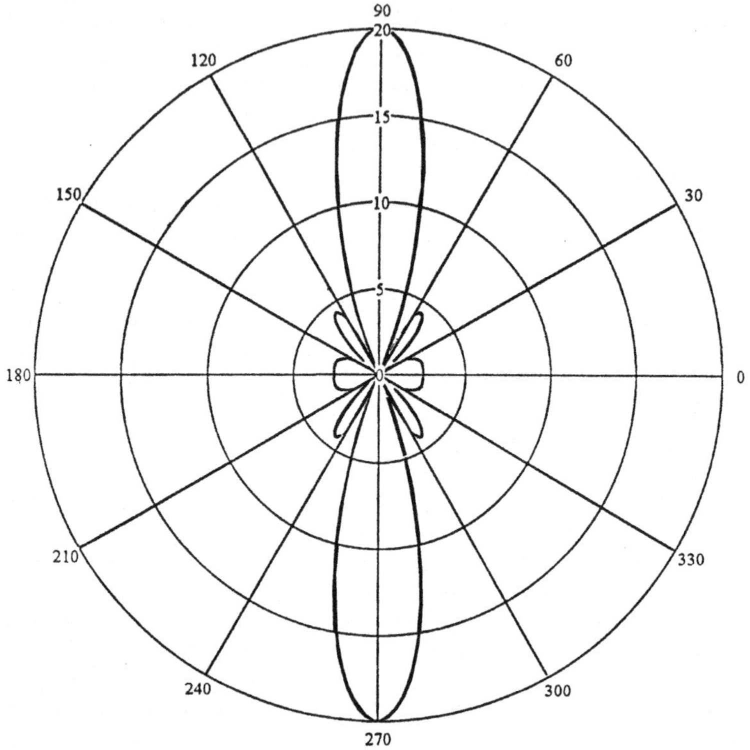

Figure 11.17 Field pattern of a 20-element uniform array when $d = \lambda/8$, $\theta = 90^\circ$, and $\alpha = 0$

pattern along the x axis (see Exercise 11.15). An array that directs power along its direction is referred to as an *end-fire array*.

b) To obtain a field pattern in the xz plane, we set $\phi = 0^\circ$. For a Hertzian dipole, the field pattern, from (11.44b), is

$$F(\theta, \phi) = \sin\theta$$

$$\psi = \frac{\pi}{4}\sin(\theta) \quad \text{and} \quad F(\psi) = \frac{\sin(10\psi)}{\sin(0.5\psi)}$$

The field pattern, $F(\theta, \phi)F(\psi)$, in the xz plane is given in Figure 11.18 (see page 575). • • •

11.9 Efficiency of an antenna

We now have a clear picture about the dielectric region surrounding an antenna. It appears as a resistance, which we call the radiation resistance R_{rad}. The antenna itself is made of a conducting material such as copper and has its own resistance R_c. Thus, the total resistance of the antenna as seen by the source connected at its terminals is

$$R_a = R_{\text{rad}} + R_c \tag{11.98}$$

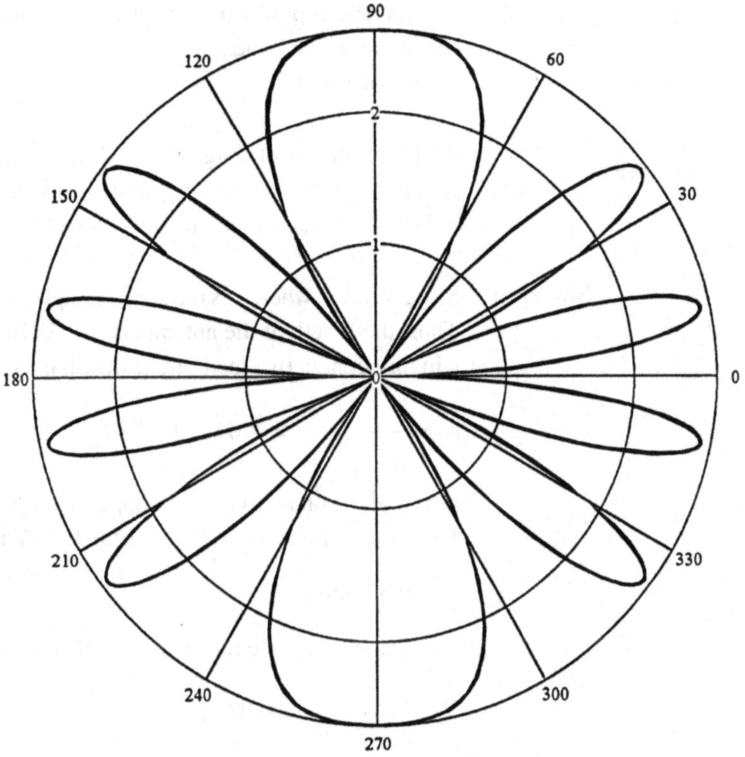

Figure 11.18 Field pattern of a 20-element linear array in the xz plane when $d = \lambda/8$, $\phi = 0^\circ$, and $\alpha = 0$

If the phasor current $\tilde{I}$ on the surface of a dipole antenna is given in terms of its maximum value, then the power delivered to a transmitting antenna by a source connected at its terminals is

$$P_{\text{in}} = \frac{1}{2} I^2 R_a \tag{11.99}$$

The power radiated by the antenna in terms of its radiation resistance is

$$P_{\text{rad}} = \frac{1}{2} I^2 R_{\text{rad}} \tag{11.100}$$

Hence, the efficiency η_e of an antenna is

$$\begin{aligned} \eta_e &= \frac{P_{\text{rad}}}{P_{\text{in}}} = \frac{R_{\text{rad}}}{R_a} \\ &= \frac{R_{\text{rad}}}{R_c + R_{\text{rad}}} \end{aligned} \tag{11.101}$$

From equation (11.101), it is obvious that the efficiency approaches 100% as long as $R_{\text{rad}} \gg R_c$. This condition is satisfied by a half-wave dipole antenna. In order to account for the power loss in the antenna, we can define the *power gain* as the product of directive gain and the efficiency. That is, the power gain in a given direction from the antenna is $\eta_e G$. The input impedance of a half-wave dipole antenna, as mentioned in Section 11.6, is $\hat{Z}_a = R_a + jX_a$, where $X_a \approx 43\,\Omega$. According to the maximum power transfer theorem, a transmitting antenna radiates maximum power when the transmission line feeding the antenna has an impedance of $\hat{Z}_a^*$.

EXAMPLE 11.6

A short antenna of length $\lambda/10$ is radiating power in free space at a frequency of 1 GHz. If the diameter of the copper wire is 1.02 mm (AWG# 18), determine the efficiency of the antenna.

Solution Since $f\lambda = c$ where c is the speed of light, the wavelength is $\lambda = 0.3$ m. Thus, the length of the antenna is $\ell = \lambda/10 = 3$ cm. The skin depth in a highly conductive medium, from Chapter 8, is

$$\delta_c = \frac{1}{\sqrt{\mu_0 \pi f \sigma}} = 2.09 \times 10^{-6} \text{ m}$$

where the conductivity of copper is $\sigma = 5.8 \times 10^7$ S/m, $f = 1 \times 10^9$ Hz, and $\mu_0 = 4\pi \times 10^{-7}$ H/m. The radius of the copper wire is

$$a = 0.51 \text{ mm}$$

Because $a \gg \delta_c$, the resistance of the antenna is

$$R_c = \frac{\ell}{2\pi a \delta_c \sigma} = 0.077\ \Omega$$

The radiation resistance of a short antenna, from (11.73), is

$$R_{\text{rad}} = \frac{2\pi}{12}\eta_0 \left[\frac{\ell}{\lambda}\right]^2 = 1.974\ \Omega$$

where $\eta_0 = 120\pi$ is the intrinsic impedance of free space. Hence, the antenna efficiency is

$$\eta_e = \frac{R_{\text{rad}}}{R_c + R_{\text{rad}}} = 0.962 \quad \text{or} \quad 96.2\%$$

• • •

11.10 Receiving antenna and Friis equation

When an antenna is placed in a medium such as free space to intercept some form of electromagnetic energy, it is referred to as a *receiving antenna.* The power radiated by a transmitting antenna decreases inversely with the square of the distance, and spreads as it propagates in the medium, therefore a receiving antenna captures only a very small portion of the total power. For this reason, a receiving antenna must not only be very efficient in capturing the available power, but it must also be matched to the load in order to deliver maximum power to the load. What this means is that the load impedance must be the conjugate of the antenna impedance.

The power-capturing capability of a receiving antenna is defined in terms of its *effective area* or *effective aperture*. The effective area is the ratio of the average power received by the receiving antenna to the average power density of the incident wave. That is,

$$A_{\text{er}} = \frac{P_r}{\langle S \rangle} \tag{11.102}$$

where P_r denotes the average power received by the antenna, $\langle S \rangle$ is the average power density at the location of the receiving antenna, and A_{er} is the effective area of the receiving antenna.

To derive an expression for the effective area A_{er} in terms of the wavelength λ and the directive gain of the receiving antenna G_r, let us consider that the transmitting and receiving antennas are Hertzian dipoles with their axes in the z direction, as illustrated in Figure 11.19. If the distance between the two antennas is R, then we express the electric field at the location of the receiving antenna, from (11.42), as

$$\tilde{E}_\theta = E_0 e^{-j\psi} \tag{11.103}$$

where

$$\psi = \beta R - \pi/2 - \alpha$$

when the current distribution on the antenna is given by

$$\tilde{I} = I e^{j\alpha}$$

In addition, from (11.42), the magnitude E_0 is

$$E_0 = \frac{\beta \eta I \ell}{4\pi R} \sin\theta \tag{11.104}$$

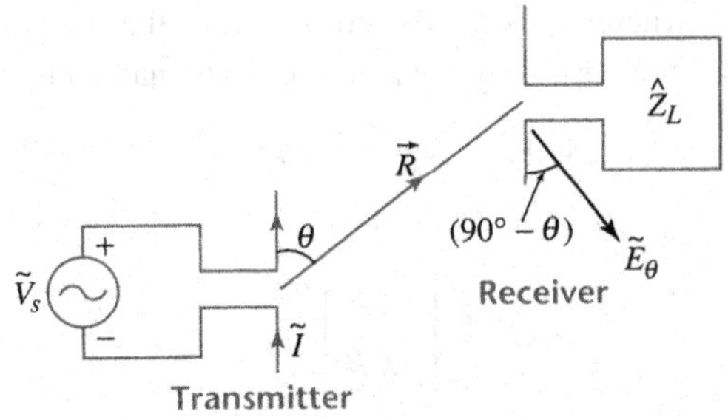

Figure 11.19 A setup for transmitting and receiving antennas

Since the electric field makes an angle of $90° - \theta$ with the axis of the receiving antenna as shown, it is the tangential component of the electric field, $E_0 \sin\theta$, that is responsible for the induced voltage. Hence, the induced voltage in the receiving antenna of length ℓ is

$$V_0 = E_0 \ell \sin\theta \tag{11.105}$$

When the load is matched to the antenna, the load impedance is $\langle \hat{Z}_L \rangle = R_a - jX_a$. Furthermore, for a lossless antenna, $R_a = R_{\text{rad}}$. Thus, the total impedance of the antenna and the load is $2R_{\text{rad}}$. The power delivered to the load is

$$P_r = \frac{1}{2}\left[\frac{V_0}{2R_{\text{rad}}}\right]^2 R_{\text{rad}}$$
$$= \frac{1}{8R_{\text{rad}}} E_0^2 \ell^2 \sin^2\theta \tag{11.106}$$

The available average power density at the location of the antenna is

$$\langle S \rangle = \frac{1}{2\eta} E_0^2 \tag{11.107}$$

From (11.102), (11.106), and (11.107), we obtain the effective area of the antenna as

$$A_{\text{er}} = \frac{\eta}{4R_{\text{rad}}} \ell^2 \sin^2\theta \tag{11.108}$$

Substituting for R_{rad}, from (11.50), in this equation, we obtain

$$A_{\text{er}} = \frac{\lambda^2}{4\pi}(1.5\sin^2\theta) = \frac{\lambda^2}{4\pi} G_r \tag{11.109}$$

where $G_r = 1.5\sin^2\theta$ is the directive gain of the Hertzian dipole, and λ is the wavelength of the fields in the medium. Note that the effective area is independent of the length of the antenna. Therefore, equation (11.109), though derived for a Hertzian dipole, is true in general.

From (11.45) and (11.47a), we can rewrite the average power density at a distance R from the transmitting antenna as

$$\langle S \rangle = \frac{P_{\text{rad}} G_t}{4\pi R^2} \tag{11.110}$$

where G_t is the directive gain of the transmitting antenna. The power intercepted by the receiving antenna, from (11.102), is

$$P_r = \langle S \rangle A_{\text{er}} \tag{11.111}$$

or

$$P_r = P_{\text{rad}} G_t G_r \left[\frac{\lambda}{4\pi R}\right]^2 \tag{11.112}$$

This equation is usually referred to as the *Friis transmission formula*. It provides a relationship between the power received by a receiving antenna and the power radiated by a transmitting antenna. From (11.109), we realize that the ratio of the effective area of an antenna and its directive gain is always constant. Thus, we can write a similar equation for the transmitting antenna in terms of its directive gain G_t and effective area A_{et} as

$$A_{et} = \frac{\lambda^2}{4\pi} G_t \tag{11.113}$$

The Friis equation (11.112) can also be expressed in terms of the effective areas of the two antennas as

$$P_r = \left[\frac{1}{\lambda R}\right]^2 A_{et} A_{er} P_{rad} \tag{11.114}$$

Once again, we remind you that these equations are valid as long as $R \gg \lambda$.

EXAMPLE 11.7

A half-wave dipole antenna radiates 10 kW at a frequency of 100 MHz. A short dipole antenna situated at a distance of 25 km is used as a receiving antenna. If both antennas are symmetrically placed in the xy plane and the medium is free space, determine the effective area of each antenna and the power absorbed by the receiving antenna.

Solution As both antennas are in the xy plane, $\theta = 90°$. Thus, the directive gains for the transmitting and the receiving antennas are $G_t = 1.64$ and $G_r = 1.5$, respectively. The wavelength in free space at a frequency of 100 MHz is 3 m. Hence, the effective areas are

$$A_{et} = \frac{3^2}{4\pi} \times 1.64 = 1.175$$

$$A_{er} = \frac{3^2}{4\pi} \times 1.5 = 1.074$$

The power absorbed by the receiving antenna at a distance of 25 km, from (11.114), is

$$P_r = 10 \times 10^3 \times 1.175 \times 1.074 \left[\frac{1}{3 \times 25,000}\right]^2$$
$$= 2.243\ \mu\text{W}$$

• • •

11.11 The radar system

The *radar,* an acronym for radio detection and ranging, is an electromagnetic system capable of transmitting and receiving high-frequency signals. The signal is usually in the form of a time-harmonic pulse of very short duration. In this case, the transmitting part of the device (transmitter) directs a signal toward an object in space, the reflection from the object causes a part of the signal to be reflected back (scattered) toward the radar, and a fraction of the total signal reflected by the object is received and analyzed by the receiving part of the device (receiver).

Most often, the radar system uses the same antenna for transmitting and receiving signals. This is done with the help of a *send–receive (SR) switch.*

If R is the distance of the object from the radar, and t is the elapsed time between the transmitted and the received signals, then

$$R = \frac{ct}{2} \tag{11.115}$$

where c is the speed of light. The transmitted power varies as $1/R^2$, so we expect the received power to vary as $1/R^4$. In addition, the receiver requires a minimum amount of detectable power to separate the incoming signal from the noise. Therefore, there is a maximum distance R beyond which the radar cannot detect the object. This is called the **maximum range** of the radar, and we now proceed to determine it.

Let us consider a general system that employs two different antennas for transmitting and receiving, as shown in Figure 11.20. If G_t is the directive gain of the transmitting antenna, and R_1 is the distance from the antenna to the object, the average incident power density $\langle \hat{S} \rangle_{\text{inc}}$ at the location of the object, from (11.110), is

$$\langle S \rangle_{\text{inc}} = \frac{P_{\text{rad}} G_t}{4\pi R_1^2} \tag{11.116}$$

If A_{eo} is the effective area of the object, usually referred to as the *scattering cross section*, responsible for the isotropic power reflected back (scattered), then the total power reflected by the object, P_{ref}, is

$$P_{\text{ref}} = A_{\text{eo}} \langle S \rangle_{\text{inc}} \tag{11.117}$$

Thus, the average power density received by the receiving antenna $\langle S \rangle_r$

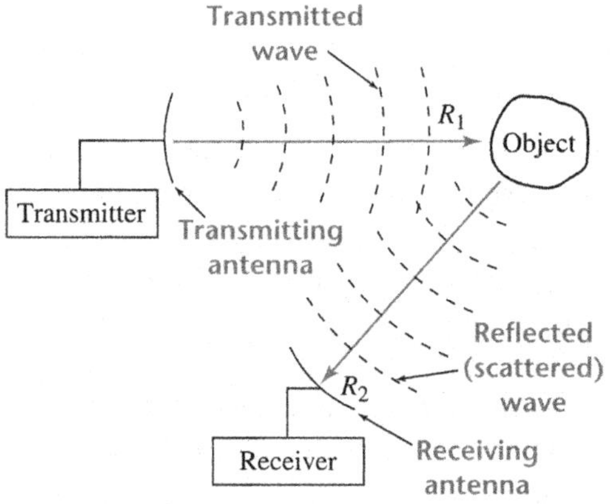

Figure 11.20 A basic radar system employing dedicated transmitting and receiving antennas

at a distance R_2 from the object is

$$\langle S \rangle_r = \frac{P_{\text{ref}}}{4\pi R_2^2} = \frac{P_{\text{rad}} G_t A_{\text{eo}}}{(4\pi)^2 R_1^2 R_2^2} \tag{11.118}$$

If A_{er} is the effective area of the receiving antenna, as defined in (11.109), then the power received by the receiving antenna, P_r, is

$$P_r = \frac{P_{\text{rad}} G_t A_{\text{eo}} A_{\text{er}}}{(4\pi)^2 R_1^2 R_2^2}$$

Substituting for A_{er} from (11.109) in this equation, we obtain

$$P_r = \frac{1}{4\pi} G_t G_r A_{\text{eo}} \left[\frac{\lambda}{4\pi R_1 R_2} \right]^2 P_{\text{rad}} \tag{11.119}$$

where G_r is the directive gain of the receiving antenna, and λ is the wavelength. Equation (11.119) is called the *radar equation for a bistatic radar*, a radar having separate transmitting and receiving antennas.

For a *monostatic radar*, a radar system that employs the same antenna for transmitting and receiving signals, $R_1 = R_2 = R$, and $G_t = G_r = G$. With these simplifications, (11.119) becomes

$$P_r = \frac{1}{4\pi} \left[\frac{G\lambda}{4\pi R^2} \right]^2 A_{\text{eo}} P_{\text{rad}} \tag{11.120}$$

From (11.120), we obtain an equation in terms of R as

$$R = \left[\frac{\lambda^2 G^2 P_{\text{rad}}}{(4\pi)^3 P_r} A_{\text{eo}} \right]^{1/4} \tag{11.121}$$

11.11.1 Doppler effect

For a moving target, the frequency of the received signal is different than that of the transmitted signal. This is known as the **Doppler effect**. This difference in frequency is exploited in the design of a traffic-control radar to determine the speed of the target in the radial direction. If f is the frequency of the transmitted signal, and u is the speed of the target, then the frequency of the received signal from an approaching target is

$$f_r = \left[1 + \frac{2u}{c} \right] f \tag{11.122}$$

We change the sign from positive to negative in (11.122) when the target is receding.

EXAMPLE 11.8

A radar system is capable of transmitting 100 kW at a frequency of 3 GHz. If the antenna gain is 20 dB, the cross-sectional area of the target is 4 m^2, and the minimum detectable signal power is 2 pW, determine the maximum range of the radar system.

Solution The antenna gain is 20 dB, therefore $G = 100$. At a frequency of 10 GHz, the wavelength λ_0 is 0.1 m. Substituting in (11.121), we obtain

$$R = \left[\frac{0.1^2 \times 100^2 \times 100 \times 10^3}{(4\pi)^3 \times 2 \times 10^{-12}} \times 4\right]^{1/4}$$
$$\approx 10 \text{ km}$$

• • •

11.12 Summary

An antenna is designed to radiate energy in its surrounding medium. Various types of antennas have been designed, including the commonly used half-wave dipole antenna, the loop antenna, horn antenna, slot antenna, and dish antenna. Each antenna is designed to satisfy a specific purpose. A dipole antenna is a close approximation of an isotropic antenna. Its radiation pattern is symmetric about its axis. However, an antenna array enables us to channel the energy in a given direction.

We began our study of antennas by formulating wave equations in terms of the magnetic vector potential and electric potential. We then sought the solutions of these wave equations. We examined only a few types of antennas because a general solution of these equations is very difficult. Among the antennas we analyzed were the Hertzian dipole, loop antenna, short dipole antenna, and half-wave antenna. In each case, we concentrated on the far-zone (radiation) fields. The average power density associated with these fields varies inversely with the square of the distance.

We also sketched the normalized field patterns and the power patterns. The field pattern is usually a plot of the magnitude of the normalized electric field intensity. The power pattern is a plot of the normalized magnitude of average power density.

We defined the directive gain, the directivity, and the radiation resistance for each antenna. To direct a signal in the desired direction, we explained the use of arrays.

We developed the Friis transmission formula to compute the total power intercepted by a receiving antenna when the total power radiated by a transmitting antenna is known. In this case, we also know either the effective areas of both antennas or we have an idea of their directivities.

A system that sends a signal toward an object in space and then receives a part of the power reflected by that object is called a *radar system*. We developed an equation that permits us to determine the distance between the radar and the object. The frequency of the reflected signal received by the radar is different than the transmitted signal for a moving object. This is called the *Doppler effect*. This concept enables us to compute the speed with which an object is moving closer to or farther away from the radar.

11.13 Review questions

11.1 Does a current-carrying conductor act as an antenna?

11.2 Do you think that every electric circuit has the capability to radiate electromagnetic energy?

11.3 State the differences between a poor and an efficient radiating system.

11.4 What is the role played by the length of the radiating system?

11.5 Define an isotropic antenna.

11.6 What is an omnidirectional antenna?

11.7 Why is the radiation from a power transmission line ignored?

11.8 What are the similarities and differences between spherical and plane waves?

11.9 What is meant by a retarded field?

11.10 Why is a Hertzian dipole referred to as an electric dipole?

11.11 If a term in the expression for the $\vec{\mathbf{E}}$ field varies as $1/r^3$, the term represents an ———— field.

11.12 If a term in the expression for the $\vec{\mathbf{H}}$ field varies as $1/r^2$, the term represents an ———— field.

11.13 Those terms in the $\vec{\mathbf{E}}$ or $\vec{\mathbf{H}}$ field constitute radiation terms only when they vary as ————.

11.14 What is the difference between directive gain and directivity?

11.15 What is the significance of radiation resistance? If the resistance of an antenna can be neglected, what must be the characteristic impedance of the transmission line feeding the antenna so that the antenna radiates maximum power?

11.16 What do you conclude if the directivity of an antenna is unity?

11.17 What is a magnetic dipole? How does it differ from an electric dipole?

11.18 Define a monopole antenna. Can you cite examples of such antennas that are used in our daily lives?

11.19 Can radiation fields exist if the electric scalar potential $\tilde{V}$ is zero? Cite appropriate equations for justification.

11.20 Can radiation fields exist if the magnetic vector potential $\vec{\mathbf{A}}$ is zero? Justify your answer by citing appropriate equations.

11.21 What is the significance of a half-wave dipole antenna? Can we use a quarter-wave monopole above a conducting surface to replace a half-wave dipole? What will the radiation resistance of a quarter-wave monopole be?

11.22 Why is it essential to use arrays for radiation purposes? What are the traits of a uniform linear array?

11.23 State the principle of pattern multiplication. What is its significance?

11.24 What is a uniform linear array? What is the difference between a broadside and an end-fired array?

11.25 What is meant by the effective area of an antenna? What is its importance? How is it related to the wavelength and the directive gain?

11.26 What is the significance of the Friis transmission formula?
11.27 Explain the difference between a monostatic and a bistatic radar system.
11.28 Can the radar equation be derived from the Friis transmission formula? Cite appropriate reasons for your answer.
11.29 Explain the Doppler effect. How does it change the frequency of a signal?
11.30 What is meant by the maximum range of a radar?
11.31 An antenna radiates at a frequency of 100 MHz. What is the wavelength of the wave? How much time does the wave take to travel a distance of 10,000 km?
11.32 A dipole antenna has a length of $\lambda/8$ m. What is its radiation resistance?

11.14 Exercises

11.1 Using Maxwell's equations and the Lorentz condition, derive (11.14).
11.2 Express (11.15) as a set of three scalar equations.
11.3 Show all the necessary steps to arrive at (11.21) from (11.19) using (11.20).
11.4 Show that the induction and the radiation fields have equal amplitude when $r = \lambda/2\pi$.
11.5 Verify equations (11.35) and (11.36).
11.6 Show that the approximate expressions for the radiated fields given in (11.41) and (11.42) do not satisfy Maxwell's equations.
11.7 Verify equations (11.53), (11.57), and (11.61).
11.8 Repeat Example 11.2 when (a) $\omega = 3$ Mrad/s and (b) $\omega = 30$ Mrad/s. What kind of conclusions can be drawn about the effectiveness of the loop as an antenna?
11.9 Verify equations (11.69), (11.70), and (11.71).
11.10 A short antenna of height $h = \ell/2$ mounted on a conducting plane as shown in Figure 11.21 is called a *monopole*. Show that its radiation resistance is one-half that of a short dipole antenna of length ℓ and carrying the same current.

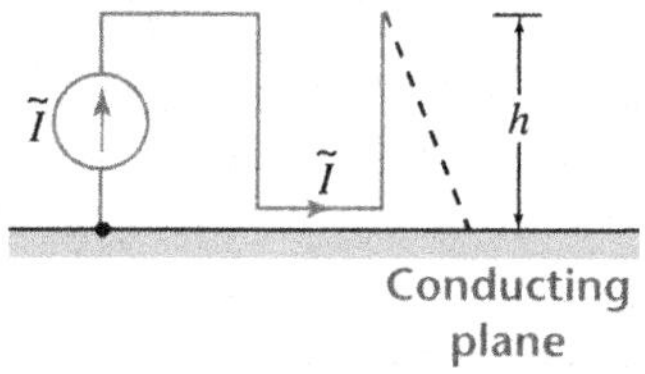

Figure 11.21 A short monopole mounted above a conducting plane

11.11 Find the directive gain and the directivity of a linear half-wave antenna. Sketch its fields and power patterns.
11.12 The amplitude of the electric field intensity of a half-wave dipole antenna at $r = 5$ km and $\theta = \pi/6$ rad is 0.01 V/m in free space. If the operating frequency is 30 MHz, determine its length and the total power that

it radiates. Also, write expressions for the electric and magnetic field intensities in the time domain.

11.13 Sketch the field patterns of two Hertzian dipole antennas in the plane perpendicular to their axes when (a) $d = \lambda/4$, $k = 1$, and $\alpha = -\pi/2$, and (b) $d = \lambda$, $k = 1$, and $\alpha = 0$.

11.14 Sketch the field patterns of two half-wave dipole antennas in the plane perpendicular to their axes when (a) $d = \lambda/4$, $k = 1$, and $\alpha = -\pi/2$, and (b) $d = \lambda$, $k = 1$, and $\alpha = 0$.

11.15 Sketch the field pattern in the xy plane of a 20-element, Hertzian dipole, linear array when the spacing between elements is $\lambda/2$ and the phase shift is π. Cite reasons why the field pattern is along the x axis.

11.16 Plot the field pattern in the xy plane for a 10-element, half-wave dipole, linear array with a spacing of $\lambda/2$ and phase shift of $-90°$.

11.17 A short antenna of length $\lambda/10$ is radiating power in free space at a frequency of 300 MHz. If the diameter of the copper wire is 0.813 mm (AWG# 22), determine the efficiency of the antenna.

11.18 A half-wave dipole antenna is radiating power in free space at a frequency of 600 MHz. If the diameter of the copper wire is 0.813 mm (AWG# 22), determine the efficiency of the antenna.

11.19 Repeat Example 11.7 if the receiving antenna is also a half-wave dipole antenna.

11.20 A receiving antenna with a directivity of 12 dB is placed at a distance of 100λ from a transmitting antenna having a directivity of 20 dB. If both antennas are placed symmetrically in the xy plane in free space and the power intercepted by the receiving antenna is 10 μW, what is the power radiated by the transmitting antenna?

11.21 Express equation (11.120) in terms of the effective area of the antenna A_e.

11.22 If the object in Example 11.8 is 2 km away from the radar, determine the power absorbed by the antenna from the scattered wave.

11.15 Problems

11.1 Express Maxwell's equations in spherical coordinates in a source-free dielectric medium away from the antenna.

11.2 Obtain an expression for the scalar electric potential of an electric (Hertzian) dipole antenna.

11.3 Show that electric field intensity can be expressed entirely in terms of magnetic vector potential as

$$\tilde{\mathbf{E}} = -j\omega\left[\tilde{\mathbf{A}} + \frac{\nabla(\nabla \cdot \tilde{\mathbf{A}})}{\beta^2}\right]$$

where $\beta = \omega\sqrt{\mu\epsilon}$ is the phase constant in the unbounded medium.

11.4 Show that the magnetic vector potential $\vec{\mathbf{A}} = \sin \beta y \cos \omega t \vec{\mathbf{a}}_x$ in a source-free medium is the solution of the wave equation (11.13) when $\beta = \omega\sqrt{\mu\epsilon}$. Determine the electric and magnetic fields associated with $\vec{\mathbf{A}}$.

11.5 When the magnetic vector potential $\tilde{A}_z$ due to a current-carrying element along the z direction is known, show that the magnetic field intensity of the radiated wave is given by

$$\tilde{H}_\phi = \frac{j\beta}{\mu} \sin\theta \tilde{A}_z$$

11.6 Determine the radiation resistance of a short antenna if its length is $0.1\lambda_0$, where λ_0 is the wavelength in free space. If the antenna is designed to radiate 500 W, calculate the maximum value of the antenna current.

11.7 A short antenna causes a maximum field intensity of 6 mV/m at a distance of 10 km. Write the expressions for the fields and calculate the total power radiated by the antenna.

11.8 A center-fed, short dipole antenna of length 0.1λ has a current of 7.07 A (rms) at its terminals. If the operating frequency is 300 Mrad/s, and the medium is free space, what are the field intensities at a distance of 3 km in a direction that is 30° from the dipole axis?

11.9 Verify the equation for the magnetic vector potential for a half-wave dipole antenna as given in equation (11.75).

11.10 Using expressions for the radiation fields produced by a Hertzian dipole, verify the expressions for the radiation fields of a half-wave dipole antenna.

11.11 A quarter-wave monopole antenna is mounted on a reflecting plane. Write expressions for the fields, average power density, total radiated power, and radiation resistance. What is the radiation resistance in free space?

11.12 The maximum value of the current at the input terminals of a center-fed dipole antenna is 5 A, and its frequency is 50 MHz. Find the length of the antenna in free space. Write general expressions for the fields and compute the total power radiated by the antenna.

11.13 Consider a half-wave dipole antenna of radius b. If the current $\tilde{I}_0$ on the antenna is taken to be uniformly distributed, and $b \gg \delta_c$, where δ_c is the skin depth, determine the surface current density and the resistance of the antenna.

11.14 Repeat Problem 11.13 if the current distribution is assumed to be $\tilde{I}_0 \cos \beta z$. Why is the resistance one-half of the value in the original Problem 11.13?

11.15 The range of a broadcasting radio station is defined in terms of the minimum electric field intensity of 25 mV/m broadside to the antenna. In order to maintain the minimum electric field at a distance of 100 km, determine the maximum current on the half-wave antenna. What is the total power radiated by the antenna?

11.16 Repeat Problem 11.15 if the transmitting antenna is a quarter-wave monopole over the ground.

11.17 What must be the current at the center of a short antenna of length $\lambda/10$ in order to radiate 100 W? What is the electric field intensity broadside to the antenna at a distance of 10 km? The operating frequency of the antenna is 100 MHz.

11.18 What must be the current at the center of a half-wave dipole antenna in order to radiate 100 W? What is the electric field intensity broadside to the antenna at a distance of 10 km? The operating frequency of the antenna is 100 MHz.

11.19 What must be the current at the center of a quarter-wave monopole antenna on the ground in order to radiate 100 W? What is the electric field intensity broadside to the antenna at a distance of 10 km? The operating frequency of the antenna is 100 MHz.

11.20 What must be the magnetic dipole moment of a magnetic dipole in order to radiate 100 W? What is the electric field intensity broadside to the dipole at a distance of 10 km? The operating frequency of the antenna is 100 MHz.

11.21 A center-fed dipole antenna of length ℓ has a current distribution of the form

$$\tilde{I}(z) = I_0 \sin \beta(\ell/2 - z) \quad z \geq 0$$
$$\tilde{I}(z) = I_0 \sin \beta(\ell/2 + z) \quad z \leq 0$$

Show that the electric field intensity in the far-zone area is

$$\tilde{E}_\theta = j\frac{e^{-j\beta r}}{2\pi r \sin\theta}\eta I_0 \left[\cos\left(\frac{\beta\ell}{2}\cos\theta\right) - \cos\left(\frac{\beta\ell}{2}\right)\right] \text{ V/m}$$

Determine the corresponding magnetic field intensity. What is the average power density in the radial direction from the antenna?

11.22 Show that the normalized radiated field pattern of a full-wave antenna is given as

$$E = \frac{\cos(\pi \cos\theta) + 1}{\sin\theta}$$

and plot it.

11.23 Show that the normalized radiated field pattern of a one-and-a-half-wave antenna is given as

$$E = \frac{\cos(1.5\pi \cos\theta)}{\sin\theta}$$

and plot it.

11.24 Show that the field pattern of a 4-element half-wave dipole antenna array in the xy plane is a broadside field pattern when the currents are in phase and the spacing between the elements is one-half wavelength.

11.25 Show that the field pattern of a 4-element half-wave dipole antenna array in the xy plane is an endfire field pattern when the currents are -180° out of phase and the spacing between the elements is one-half wavelength.

11.26 Sketch the end-fire array pattern of an 8-element half-wave dipole antenna array when the spacing between the elements is one-quarter wavelength and the currents are -180° out of phase.

11.27 Sketch the end-fire array pattern of an 8-element half-wave dipole antenna array when the spacing between the elements is one-quarter wavelength and the currents are -90° out of phase.

11.28 The electric field intensity in the far zone from an antenna is given in terms of its maximum input current I_0 as

$$\tilde{E}_\theta = \frac{15}{r} I_0 \text{ V/m}$$

Obtain the corresponding expression for the magnetic field. What is the total power radiated by the antenna? What is the radiation resistance? Can this antenna be called an isotropic antenna? What must I_0 be to radiate a total power of 75 kW?

11.29 The electric field intensity in the far zone from an antenna is given in terms of its maximum input current I_0 as

$$\tilde{E}_\theta = \frac{15}{r} I_0 \sin\theta \text{V/m}$$

Obtain the corresponding expression for the magnetic field. What is the total power radiated by the antenna? What is the radiation resistance? Can this antenna be called an isotropic antenna? What must I_0 be to radiate a power of 75 kW?

11.30 A copper wire of 5-mm radius is used as a loop antenna. The antenna radiates at a frequency of 3 MHz. If the loop radius is 0.5 m, and the maximum current in the loop is 100 A, determine (a) the power radiated by the loop, (b) the radiation resistance of the loop, and (c) the radiation efficiency.

11.31 Using numerical integration, compute the radiation resistance of a dipole antenna of length (a) $\ell = \lambda$, (b) $\ell = 1.5\lambda$, and (c) $\ell = 2\lambda$.

11.32 Using numerical integration, plot the radiation resistance of a dipole antenna as a function of its length. What kind of inference can you draw by examining the plot?

11.33 Two identical antennas are used for transmitting and receiving purposes and they are separated by a distance of 300 m. The directive gain of each antenna is 20 dB. If the power being received by the receiving antenna is 10 mW at a frequency of 100 MHz, what is the power transmitted by the transmitting antenna?

11.34 Repeat Problem 11.33 if the two antennas are half-wave dipoles.

11.35 An omnidirectional antenna fitted inside a hot-air balloon is in direct touch with the base station. The base station also has an omnidirectional antenna. The base station receives a power of 10 mW when the balloon is 500 m from it. If the minimum power for detection for the base unit is 10 μW, how far must the balloon travel before it loses contact with the base station?

11.36 A monostatic radar system can transmit a power of 10 kW at a frequency of 5 GHz and detect a signal of 3 pW. The direct gain of the antenna is 30 dB. What is the maximum range for detecting a target of 1.5-m^2 cross section?

12

Computer-aided analysis of electromagnetic fields

12.1 Introduction

The evaluation of the electric and magnetic fields in an electromagnetic system is of utmost importance for its efficient design. For example, in an insulating material, to isolate conductors from each other in a system, we want to keep the intensity of the electric fields below the breakdown strength of the insulating medium. In a magnetic switch, the magnetic field intensity should produce a sufficient force to activate the switch. And for the efficient design of a transmitting system, such as an antenna, knowledge of the electromagnetic field distribution in the medium surrounding the antenna is obviously essential.

To analyze electromagnetic fields, we start with the mathematical formulation of the problem. Depending on the nature of the electromagnetic system, Laplace's or Poisson's equation may be suitable to model the system for static and quasistatic (low-frequency) operating conditions. However, in high-frequency applications we must solve the wave equation in either the time domain or the frequency domain to accurately predict the electric and magnetic fields. In any case, the solution of one or more partial differential equations subject to boundary conditions is needed in order to determine the electric and magnetic fields inside and around an electromagnetic system.

Analytical solutions are available only for problems of regular geometry (rectangular, circular, etc.) with the most simple boundary conditions. In the preceding chapters solutions were given for several such configurations using analytical methods.

In this chapter, we will study three numerical techniques to compute electric and magnetic fields: the finite-difference method (FDM), the finite-element method (FEM), and the method of moments (MOM). In principle, each method discretizes a continuous domain into a finite number of sections, and then requires a solution of a set of algebraic equations instead of differential or integral equations. We have developed computer programs for the three numerical methods, and a listing of these programs is given in Appendix B.

12.2 Finite-difference method

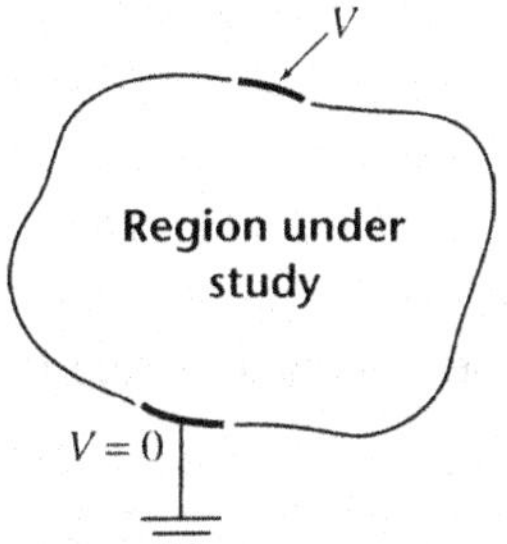

Figure 12.1 An arbitrary region for the study of electric potential

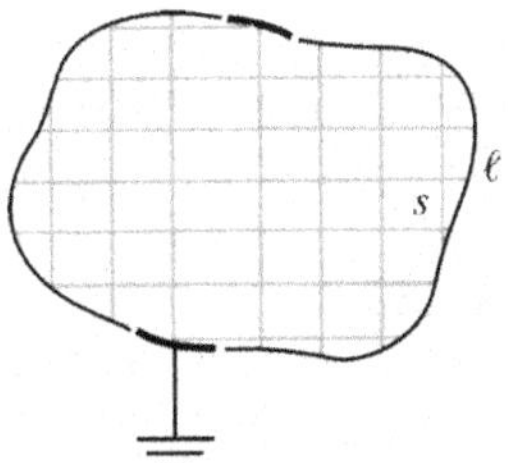

Figure 12.2 Mesh distribution for the solution region

The finite-difference method (FDM) is one of the most powerful numerical techniques for solving partial differential equations of any kind. Because all electromagnetic field problems are represented by scalar or vector partial differential equations, the FDM can be utilized to solve the spatial as well as the temporal distributions of electric and magnetic fields in various media. The **finite-difference method** is a technique that divides the solution domain into finite discrete points and replaces the partial differential equation with a set of difference equations. Hence, the solution is not exact, but approximate. However, the error in the solution can be minimized to an acceptable level if the discrete points are selected close to one another.

Although the determination of the electromagnetic fields may produce three-dimensional variations, within the scope of this book we will confine our discussion to variations in two dimensions only. Let us consider the two-dimensional Poisson equation

$$\nabla^2 V(x, y) = \frac{\partial^2 V(x, y)}{\partial x^2} + \frac{\partial^2 V(x, y)}{\partial y^2} = -\frac{\rho_v}{\epsilon} \tag{12.1}$$

where $V(x, y)$ is the unknown spatial distribution of the electrostatic potential, ρ_v is the volume charge density, and ϵ is the permittivity of the medium.

Our objective is to determine $V(x, y)$ in the region, shown in Figure 12.1, subject to the boundary conditions. First, we divide the region into a finite number of meshes, as illustrated in Figure 12.2. These meshes can be square, rectangular, triangular, etc., but here we will discuss rectangular and square meshes only. Let us consider a mesh with dimensions a, b, c, and d, and potentials $V_1 = V(x, y + a)$, $V_2 = V(x - b, y)$, $V_3 = V(x, y - c)$, $V_4 = V(x + d, y)$, and $V_0 = V(x, y)$ at the nodes, as shown in Figure 12.3, in order to find an approximate finite-difference equation to replace Poisson's equation.

Figure 12.3 Mesh configuration with unequal arms

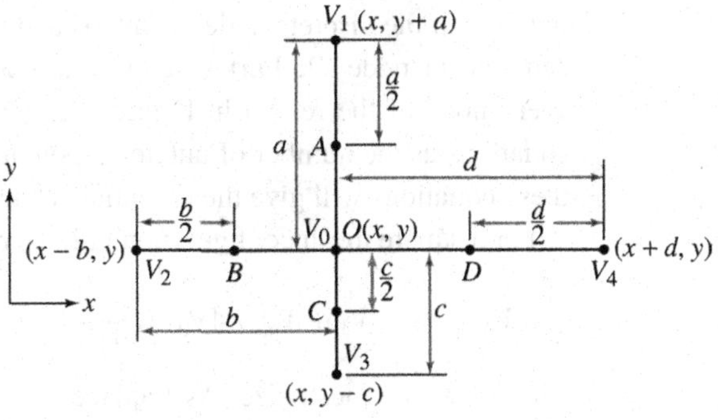

The first derivatives of $V(x, y)$ with respect to x at B and D are, approximately,

$$\left.\frac{\partial V}{\partial x}\right|_B = \left.\frac{\Delta V}{\Delta x}\right|_B = \frac{V_0 - V_2}{b} \tag{12.2}$$

$$\left.\frac{\partial V}{\partial x}\right|_D = \left.\frac{\Delta V}{\Delta x}\right|_D = \frac{V_4 - V_0}{d} \tag{12.3}$$

Similarly, the first derivatives can be approximated at points A and C as

$$\left.\frac{\partial V}{\partial y}\right|_A = \left.\frac{\Delta V}{\Delta y}\right|_A = \frac{V_1 - V_0}{a} \tag{12.4}$$

$$\left.\frac{\partial V}{\partial y}\right|_C = \left.\frac{\Delta V}{\Delta y}\right|_C = \frac{V_0 - V_3}{c} \tag{12.5}$$

The second-order partial derivatives of $V(x, y)$ can be approximated at point O as

$$\left.\frac{\partial^2 V}{\partial x^2}\right|_O = \frac{\left.\frac{\Delta V}{\Delta x}\right|_D - \left.\frac{\Delta V}{\Delta x}\right|_B}{\Delta x} = \frac{\frac{V_4 - V_0}{d} - \frac{V_0 - V_2}{b}}{\frac{d}{2} + \frac{b}{2}} \tag{12.6}$$

$$\left.\frac{\partial^2 V}{\partial x^2}\right|_O = 2\frac{(V_4 - V_0)b - (V_0 - V_2)d}{bd(d + b)} \tag{12.7}$$

$$\left.\frac{\partial^2 V}{\partial y^2}\right|_O = \frac{\left.\frac{\Delta V}{\Delta y}\right|_A - \left.\frac{\Delta V}{\Delta y}\right|_C}{\Delta x} = \frac{\frac{V_1 - V_0}{a} - \frac{V_0 - V_3}{c}}{\frac{a}{2} + \frac{c}{2}} \tag{12.8}$$

$$\left.\frac{\partial^2 V}{\partial y^2}\right|_O = 2\frac{(V_1 - V_0)c - (V_0 - V_3)a}{ac(a + c)} \tag{12.9}$$

By using these approximations, equation (12.1) becomes

$$\frac{1}{a(a+c)}V_1 + \frac{1}{b(b+d)}V_2 + \frac{1}{c(a+c)}V_3 + \frac{1}{d(d+b)}V_4 - \left(\frac{1}{bd} + \frac{1}{ac}\right)V_0 = -\frac{\rho}{2\epsilon} \tag{12.10}$$

in terms of the discrete node potentials and the dimensions of the mesh centered at node O. Further application of these approximations to every node of the region in Figure 12.2 will yield as many algebraic equations as the number of unknown potential nodes. The solutions of these equations will give the potential at each node.

For a square mesh configuration (12.10) reduces to

$$\frac{1}{h^2}(V_1 + V_2 + V_3 + V_4 - 4V_0) = -\frac{\rho_v}{\epsilon} \tag{12.11}$$

where h is the mesh size. As Laplace's equation is essentially a special case of Poisson's equation with a zero on the right-hand side, the finite-difference equation for Laplace's equation can be expressed as

$$\frac{1}{a(a+c)}V_1 + \frac{1}{b(b+d)}V_2 + \frac{1}{c(a+c)}V_3 + \frac{1}{d(b+d)}V_4 - \left(\frac{1}{bd}+\frac{1}{ac}\right)V_0 = 0 \tag{12.12}$$

which for a square mesh becomes

$$V_1 + V_2 + V_3 + V_4 - 4V_0 = 0 \tag{12.13}$$

12.2.1 Boundary conditions

As the partial differential equations describing electromagnetic fields are functions of space coordinates, a unique solution can be obtained only with a specified set of boundary conditions. Most electromagnetic field problems deal with three kinds of boundary conditions: the Dirichlet type of boundary, the Neumann type of boundary, and mixed boundary conditions.

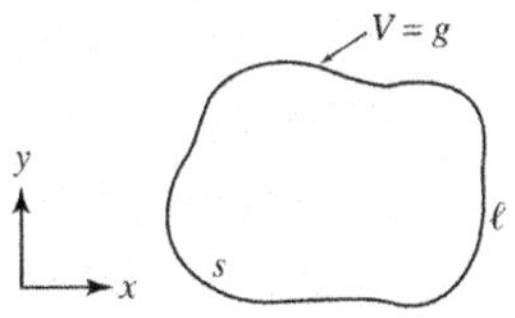

Figure 12.4 Dirichlet-type boundary

Let us consider a region s bounded by a curve ℓ, as shown in Figure 12.4. If we want to determine the potential distribution V in region s such that the potential along ℓ is $V = g$, where g is a prespecified continuous potential function, then the condition along the boundary ℓ is known as the *Dirichlet boundary* condition.

Some electromagnetic field problems may involve conditions along the boundary such that the normal derivative of the potential function at the boundary is specified as a continuous function (Figure 12.5). This boundary condition can be represented mathematically as

$$\frac{dV}{dn} = f \tag{12.14}$$

and is called the *Neumann boundary* condition.

Finally, there are problems having the Dirichlet condition and the Neumann condition along ℓ_1 and ℓ_2 portions of ℓ, respectively, as illustrated in Figure 12.6. This is defined as a *mixed boundary* condition.

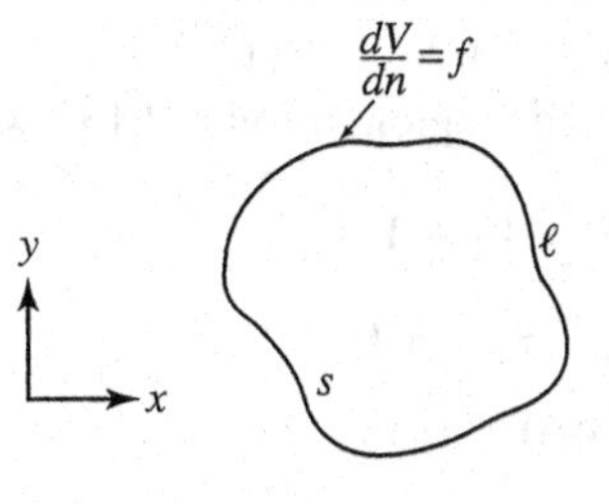

Figure 12.5 Neumann-type boundary

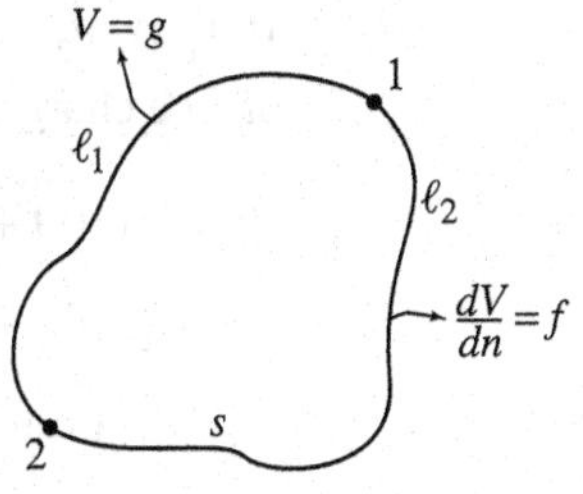

Figure 12.6 Mixed boundaries

The following example illustrates the concepts of the FDM that we have discussed so far.

EXAMPLE 12.1

Determine the electrostatic potential distribution inside the region, given in Figure 12.7, with boundary conditions as specified.

Solution As can be noted in Figure 12.7, boundaries ranging between $0 < y < 3$ at $x = 0$, between $0 < x < 3$ at $y = 0$, and between $0 < y < 3$ at $x = 3$ have zero potentials ($V = 0$). In other words, the potential along these boundaries is constant and, therefore, satisfies the Dirichlet condition. The boundary along $0 < x < 3$ at $y = 3$ carries a constant potential of 100 V and, hence, is another Dirichlet boundary condition. Note that, at $y = 3$, x is not strictly equal to 0 or 3, but can be very close to these boundaries. This is because of the small separation δ between the upper horizontal boundary and the vertical boundaries. These small separations are desirable as the upper horizontal boundary is maintained at a different potential than the others.

Figure 12.7

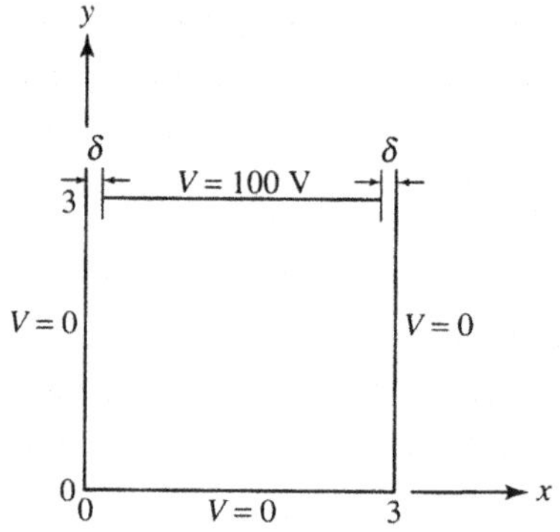

In order to use the FDM to determine the potential distribution, we divide the region into square meshes with $h = 1$ as shown in Figure 12.8. With the given number of meshes the problem reduces to determining the potentials at nodes (1, 2), (2, 2), (1, 1), and (2, 1). The potentials at nodes (1, 3) and (2, 3) are given as 100 V, and the potentials at (0, 3), (0, 2), (0, 1), (0, 0), (1, 0), (2, 0), (3, 0), (3, 1), (3, 2), and (3, 3) are all specified as zero. Let us rename the unknown potentials as $V_1 = V(1,2)$, $V_2 = V(2,2)$, $V_3 = V(1,1)$, and $V_4 = V(2,1)$. In the absence of free charge in the region, using (12.13), we can write

$$
\begin{aligned}
V_1 &= \tfrac{1}{4}(100 + 0 + V_3 + V_2) \\
V_2 &= \tfrac{1}{4}(100 + V_1 + V_4 + 0) \\
V_3 &= \tfrac{1}{4}(V_1 + 0 + 0 + V_4) \\
V_4 &= \tfrac{1}{4}(V_2 + V_3 + 0 + 0)
\end{aligned}
\tag{12.15}
$$

Figure 12.8

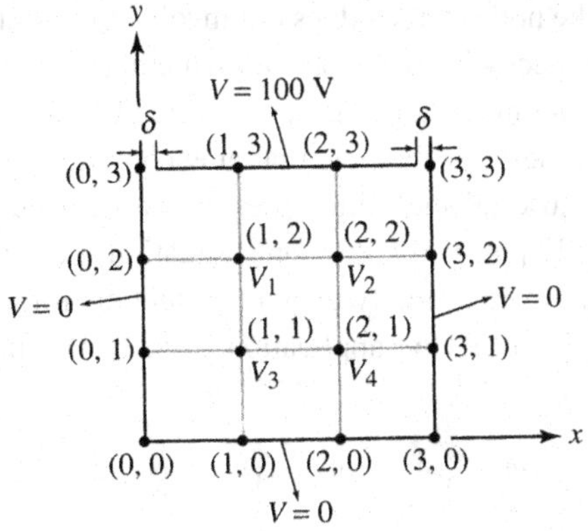

By arranging the set of linear algebraic equations in (12.15), we obtain

$$\begin{aligned} 4V_1 - V_2 - V_3 &= 100 \\ -V_1 + 4V_2 - V_4 &= 100 \\ -V_1 + 4V_3 - V_4 &= 0 \\ -V_2 - V_3 + 4V_4 &= 0 \end{aligned} \tag{12.16}$$

or in matrix form,

$$\begin{bmatrix} 4 & -1 & -1 & 0 \\ -1 & 4 & 0 & -1 \\ -1 & 0 & 4 & -1 \\ 0 & -1 & -1 & 4 \end{bmatrix} \begin{bmatrix} V_1 \\ V_2 \\ V_3 \\ V_4 \end{bmatrix} = \begin{bmatrix} 100 \\ 100 \\ 0 \\ 0 \end{bmatrix} \tag{12.17}$$

Equation (12.17) is a standard form for a system of linear equations. It is expressed in compact form as

$$\mathbf{AV} = \mathbf{b} \tag{12.18}$$

where $\mathbf{A}$ is a square matrix, $\mathbf{V}$ is the unknown potential vector, and $\mathbf{b}$ is the input vector. Thus the potentials are obtained from

$$\mathbf{V} = \mathbf{A}^{-1}\mathbf{b}$$

which gives $V_1 = 37.5$ V, $V_2 = 37.5$ V, $V_3 = 12.5$ V, and $V_4 = 12.5$ V.

• • •

12.2.2 Iterative solution of finite-difference equations

In Example 12.1, we tacitly selected the mesh size so that we had only four unknown potentials to be solved. However, for higher accuracy further subdivision of the region becomes necessary, which causes matrix $\mathbf{A}$ to become large. An efficient method to determine the potentials at the nodes which does not involve the solution of (12.18) is known as the successive over-relaxation method. The **successive over-relaxation (SOR)** method is basically an iterative algorithm that requires an initial guess for the potential at each node to begin the iterative process. Because all the initial guesses may not be correct, they will not satisfy Laplace's or Poisson's equation. For example if V_1^0, V_2^0, V_3^0, V_4^0, and V_0^0 are initially guessed potentials for a square mesh, as shown in Figure 12.9, the application of (12.13) will result in a residual R as follows:

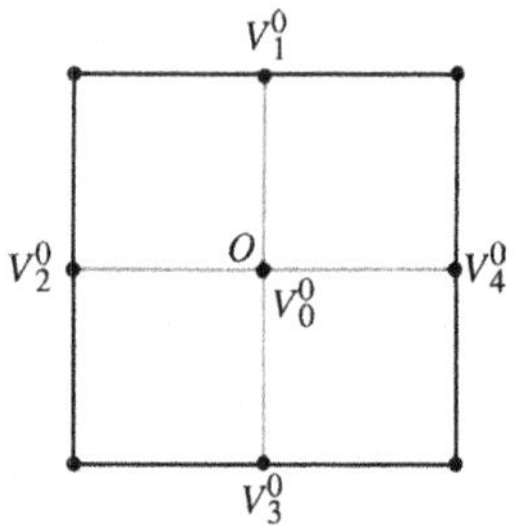

Figure 12.9 Initial guesses for a square mesh

$$V_1^0 + V_2^0 + V_3^0 + V_4^0 - 4V_0^0 = R \tag{12.19}$$

In order to obtain accurate potentials, R has to be minimized through the SOR iterative process.

Let V_0^n be the potential at node O after the nth iteration. Then the modified potential for the $(n+1)$th iteration, according to SOR, becomes

$$V_0^{n+1} = V_0^n + \frac{\alpha}{4} R^n \tag{12.20}$$

Here α is known as the acceleration factor, and for successful convergence $1 \leq \alpha < 2$. As is evident from (12.20), if the correct solution is achieved, then in the next iteration no further improvement on the potential will occur because R will be zero. The computational process for achieving a zero residual at every node is very time consuming. Hence, an error criterion, $|V_0^{n+1} - V_0^n| \ll 1$, is set in the beginning of the iterations. When this error criterion is satisfied at every node potential, the execution of the iterative process stops.

The residual R^n in (12.20), after the nth iteration at node O, is calculated from

$$R^n = V_1^{n+1} + V_2^{n+1} + V_3^n + V_4^n - 4V_0^n \tag{12.21}$$

and its substitution into (12.20) gives

$$V_0^{n+1} = V_0^n + \frac{\alpha}{4}(V_1^{n+1} + V_2^{n+1} + V_3^n + V_4^n - 4V_0^n) \tag{12.22}$$

which enables us to determine V_0 in terms of the potentials of the neighboring nodes.

EXAMPLE 12.2

Solve the node potentials in the geometry given in Example 12.1 using the SOR method.

Solution Let us consider an initial guess of 50 V at nodes 1, 2, 3, and 4, and an error criterion of 0.1. Also, let us select the acceleration factor 1.

Iteration 1:

$$V_1^{(1)} = 50 + 0.25(100 + 0 + 50 + 50 - 200) = 50$$
$$\left|V_1^{(1)} - V_1^{(0)}\right| = |50 - 50| = 0$$
$$V_2^{(1)} = 50 + 0.25(100 + 50 + 50 + 0 - 200) = 50$$
$$\left|V_2^{(1)} - V_2^{(0)}\right| = |50 - 50| = 0$$
$$V_3^{(1)} = 50 + 0.25(50 + 0 + 0 + 50 - 200) = 25$$
$$\left|V_3^{(1)} - V_3^{(0)}\right| = |25 - 50| = 25$$
$$V_4^{(1)} = 50 + 0.25(50 + 25 + 0 + 0 - 200) = 18.75$$
$$\left|V_4^{(1)} - V_4^{(0)}\right| = |18.75 - 50| = 31.25$$

Iteration 2:

$$V_1^{(2)} = 50 + 0.25(100 + 0 + 25 + 50 - 200) = 43.75$$
$$\left|V_1^{(2)} - V_1^{(1)}\right| = |43.75 - 50| = 6.25$$
$$V_2^{(2)} = 50 + 0.25(100 + 43.75 + 18.75 + 0 - 200) = 40.63$$
$$\left|V_2^{(2)} - V_2^{(1)}\right| = |40.63 - 50| = 9.37$$
$$V_3^{(2)} = 25 + 0.25(43.75 + 0 + 0 + 18.75 - 100) = 15.63$$
$$\left|V_3^{(2)} - V_3^{(1)}\right| = |15.63 - 25| = 9.37$$
$$V_4^{(2)} = 18.75 + 0.25(40.63 + 15.63 + 0 + 0 - 75) = 14.07$$
$$\left|V_4^{(2)} - V_4^{(1)}\right| = |14.07 - 18.75| = 4.68$$

After six iterations the solution converges to $V_1 = 37.5$ V, $V_2 = 37.5$ V, $V_3 = 12.5$ V, and $V_4 = 12.5$ V. The values of the voltages after each iteration are indicated in Figure 12.10.

When we solve Example 12.2 using square meshes with size of 5 mm, the potentials at (1 cm, 1 cm), (1 cm, 2 cm), (2 cm, 1 cm), and (2 cm, 2 cm) become $V(1, 1) = 38.1$ V,

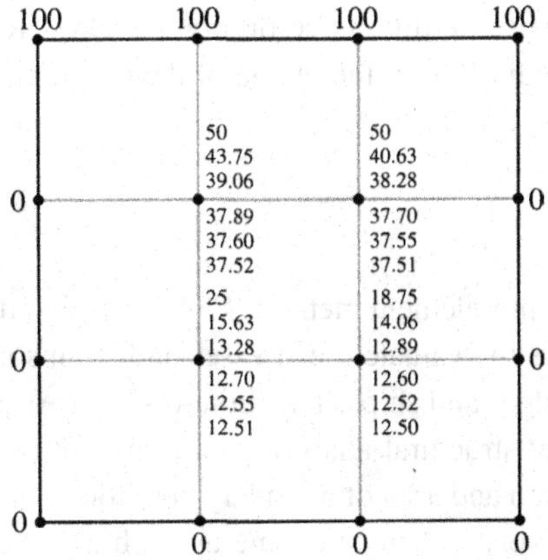

Figure 12.10 Voltages after each iteration (Example 12.2)

$V(1, 2) = 38.1$ V, $V(2, 1) = 12.3$ V, and $V(2, 2) = 12.3$ V. Note that the node potentials are slightly different than those obtained with the 1-cm mesh size. This discrepancy is essentially due to the discretization error in the FDM. • • •

EXAMPLE 12.3

The high-voltage and low-voltage windings of a 60-Hz transformer are placed as shown in Figure 12.11a. Determine the voltage distribution between the high- and low-voltage windings when the high-voltage winding is at 100 V, and the low-voltage winding is at 0 V. Assume that the current in the insulation is negligibly small at 60 Hz. Consider a square mesh configuration with a size of 0.5 cm.

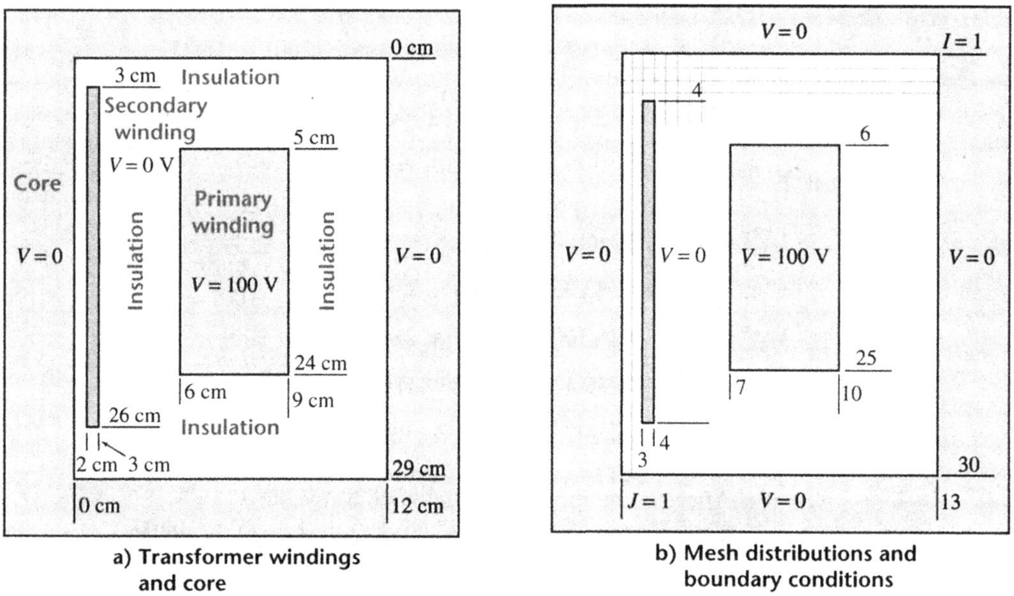

Figure 12.11

Solution In Figure 12.11b the partial model of the problem is shown in order to use the finite-difference program (FDM.TR). The results are given in Appendix B after the listing of the program. • • •

12.3 Finite-element method

The finite-element method (FEM) was first developed by structural engineers to evaluate the stresses and strains in such complex structures as bridges and ships. The finite-difference method is suitable for carrying out structural analysis, but it always requires a partial differential equation and a set of boundary conditions as a starting point. It is sometimes very difficult to come up with a partial differential equation for a complex structural problem. Hence, structural engineers developed the FEM, based on engineering insight and fundamental principles of physics. When the method was finally generalized, it became apparent that it is an approximate minimization of a functional such as the potential energy of a system. It was also discovered that the minimization of the functional was essentially nothing but the application of the variational principle. With those recognitions, the FEM was also expanded to problems whose partial differential equations could be appropriately replaced by a functional through a variational principle.

During the past twenty-five years, the FEM has been used extensively in solving electromagnetic field problems. In this section, we will study the fundamentals of the FEM as applied to electromagnetic field problems.

In electrostatic field analysis the functional to be minimized is the electrostatic energy

$$W = \frac{1}{2}\int_v \epsilon E^2 \, dv \tag{12.23}$$

inside a bounded volume. Equation (12.23) can also be expressed in terms of the electrostatic potential V as

$$W = \frac{1}{2}\epsilon \int_v \left[\left(\frac{\partial V}{\partial x}\right)^2 + \left(\frac{\partial V}{\partial y}\right)^2 + \left(\frac{\partial V}{\partial z}\right)^2\right] dv \tag{12.24}$$

Let us consider the energy functional for a two-dimensional case

$$W = \frac{1}{2}\epsilon \int_s \left[\left(\frac{\partial V}{\partial x}\right)^2 + \left(\frac{\partial V}{\partial y}\right)^2\right] ds \tag{12.25}$$

Figure 12.12 Region s is bounded by ℓ

for a bounded surface s, as shown in Figure 12.12. In the FEM, we minimize the energy functional in (12.25) because the variation of energy in the system is insignificant with respect to small changes in dV within the bounded region s. Hence, we can determine the potential distribution inside the region s by setting the differential change in energy equal to zero:

$$dW = 0 \tag{12.26}$$

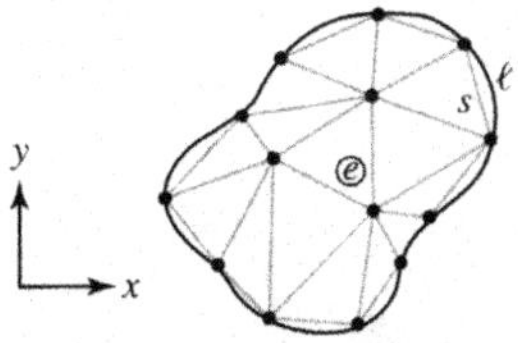

Figure 12.13 Triangular mesh configuration for the solution region s

In finite-element analysis, we divide the region under study into a finite number of n triangular meshes*, known as elements, as shown in Figure 12.13. If there are m number of nodes at which the potentials are unknown, (12.26) can be rewritten as

$$dW = \frac{\partial W}{\partial V_1} dV_1 + \frac{\partial W}{\partial V_2} dV_2 + \cdots + \frac{\partial W}{\partial V_m} dV_m = 0 \tag{12.27}$$

* Quadrilateral mesh configurations are also widely employed in the FEM; however, they will be excluded from our discussion.

where $V_1, V_2, \ldots, V_m$ are the potentials at nodes $1, 2, \ldots, m$, respectively. Equation (12.27) can also be written as

$$dW = \left[\frac{\partial W}{\partial \mathbf{V}}\right]^T d\mathbf{V} = 0 \tag{12.28}$$

where

$$\frac{\partial W}{\partial \mathbf{V}} = \begin{bmatrix} \dfrac{\partial W}{\partial V_1} \\ \vdots \\ \dfrac{\partial W}{\partial V_m} \end{bmatrix} \quad \text{and} \quad d\mathbf{V} = \begin{bmatrix} dV_1 \\ \vdots \\ dV_m \end{bmatrix}$$

The elements of $d\mathbf{V}$ in (12.28) cannot be zero; therefore, the elements of $\dfrac{\partial W}{\partial \mathbf{V}}$ must be zero to minimize the energy functional. Thus,

$$\frac{\partial W}{\partial \mathbf{V}} = \begin{bmatrix} \dfrac{\partial W}{\partial V_1} \\ \vdots \\ \dfrac{\partial W}{\partial V_m} \end{bmatrix} = \begin{bmatrix} 0 \\ \vdots \\ 0 \end{bmatrix} = \mathbf{0} \tag{12.29}$$

The enlarged view of element e in Figure 12.13 is shown in Figure 12.14. The electric energy inside this element is

$$W^{(e)} = \frac{1}{2}\epsilon \int_{s^{(e)}} \left[\left(\frac{\partial V^{(e)}}{\partial x}\right)^2 + \left(\frac{\partial V^{(e)}}{\partial y}\right)^2\right] ds^{(e)} \tag{12.30}$$

where $V^{(e)}$ is the potential distribution inside the element e, and $s^{(e)}$ is the element area.

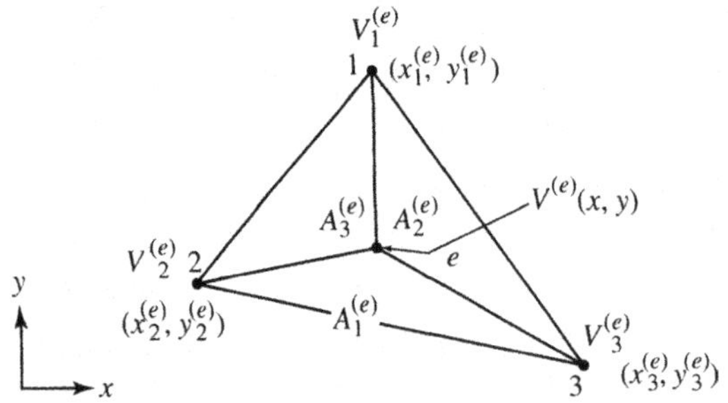

Figure 12.14 Coordinates and node potentials of a triangular mesh

Consequently, we can determine the total energy for the entire region from

$$W = W^{(1)} + W^{(2)} + \cdots + W^{(n)} = \sum_{e=1}^{n} W^{(e)}$$

or

$$W = \sum_{e=1}^{n} \frac{1}{2} \epsilon \int_{s^{(e)}} \mathbf{f}_e^T \mathbf{f}_e \, ds^{(e)} \tag{12.31}$$

where

$$\mathbf{f}_e = \begin{bmatrix} \dfrac{\partial V^{(e)}}{\partial x} \\ \dfrac{\partial V^{(e)}}{\partial y} \end{bmatrix}$$

Let us assume an approximate solution for the potential distribution inside element e as

$$V^{(e)}(x, y) = L_1^{(e)}(x, y)V_1^{(e)} + L_2^{(e)}(x, y)V_2^{(e)} + L_3^{(e)}(x, y)V_3^{(e)} \tag{12.32}$$

in terms of the shape functions $L_1^{(e)}(x, y)$, $L_2^{(e)}(x, y)$, $L_3^{(e)}(x, y)$ and node potentials $V_1^{(e)}$, $V_2^{(e)}$, $V_3^{(e)}$. The shape functions for a two-dimensional geometry are defined as

$$L_i^{(e)}(x, y) = \frac{A_i^{(e)}}{A^{(e)}} \qquad i = 1, 2, 3 \tag{12.33}$$

where $A^{(e)}$ is the area of element e, and A_i^e is the area of a section in element e, as shown in Figure 12.14. $A^{(e)}$ and $A_i^{(e)}$ can be calculated from

$$A^{(e)} = \frac{1}{2} \begin{vmatrix} 1 & x_i & y_i \\ 1 & x_{i+1} & y_{i+1} \\ 1 & x_{i+2} & y_{i+2} \end{vmatrix} \tag{12.34a}$$

and

$$A_i^{(e)} = \frac{1}{2} \begin{vmatrix} 1 & x & y \\ 1 & x_{i+1} & y_{i+1} \\ 1 & x_{i+2} & y_{i+2} \end{vmatrix} \qquad i = 1, 2, 3 \tag{12.34b}$$

We can evaluate $\dfrac{\partial V^{(e)}}{\partial x}$ and $\dfrac{\partial V^{(e)}}{\partial y}$ as

$$\frac{\partial V^{(e)}}{\partial x} = \frac{\partial L_1(x, y)}{\partial x} V_1 + \frac{\partial L_2(x, y)}{\partial x} V_2 + \frac{\partial L_3(x, y)}{\partial x} V_3 \tag{12.35a}$$

and

$$\frac{\partial V^{(e)}}{\partial y} = \frac{\partial L_1(x, y)}{\partial y} V_1 + \frac{\partial L_2(x, y)}{\partial y} V_2 + \frac{\partial L_3(x, y)}{\partial y} V_3 \tag{12.35b}$$

and reconstruct $\mathbf{f}_e$ as

$$\mathbf{f}_e = \mathbf{T}^{(e)}\mathbf{V}^{(e)} \tag{12.36}$$

where

$$\mathbf{T}^{(e)} = \begin{bmatrix} \dfrac{\partial L_1^{(e)}(x, y)}{\partial x} & \dfrac{\partial L_2^{(e)}(x, y)}{\partial x} & \dfrac{\partial L_3^{(e)}(x, y)}{\partial x} \\ \dfrac{\partial L_1^{(e)}(x, y)}{\partial y} & \dfrac{\partial L_2^{(e)}(x, y)}{\partial y} & \dfrac{\partial L_3^{(e)}(x, y)}{\partial y} \end{bmatrix}$$

and

$$\mathbf{V}^{(e)} = \begin{bmatrix} V_1^{(e)} \\ V_2^{(e)} \\ V_3^{(e)} \end{bmatrix}$$

By using (12.36), we can modify (12.31) as

$$W = \frac{1}{2}\epsilon \int_{s^{(e)}} \mathbf{V}^{(e)T}\mathbf{T}^{(e)T}\mathbf{T}^{(e)}\mathbf{V}^{(e)}\, ds^{(e)} \tag{12.37}$$

The partial derivatives of W with respect to the node potentials are

$$\frac{dW}{d\mathbf{V}} = \sum_{e=1}^{n} \epsilon \int_{s^{(e)}} \mathbf{T}^{(e)T}\mathbf{T}^{(e)}\mathbf{V}^{(e)}\, ds^{(e)} \tag{12.38}$$

Using (12.29) along with (12.38), we get

$$\sum_{e=1}^{n} \int_{s^{(e)}} \mathbf{T}^{(e)T}\mathbf{T}^{(e)}\mathbf{V}^{(e)}\, ds^{(e)} = \mathbf{0} \tag{12.39}$$

from which we can determine the node potentials.

The following example illustrates the preceding mathematical development.

EXAMPLE 12.4 Consider Example 12.1, and determine the potential distribution using the FEM.

Solution Let us number the triangular meshes and all the nodes, as shown in Figure 12.15. There are eight elements and nine nodes in the model. We will use two numbering systems. The numbering of the nodes as

Figure 12.15

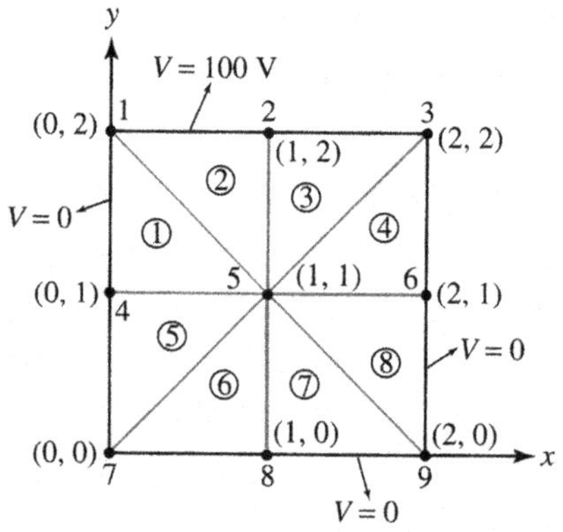

1 through 9 will be called the *global numbering system*. However, the nodes of each element will also be numbered 1 through 3; this is called the *local numbering system*. In this problem potentials at every node are given except the one at node 5 – this is the only unknown potential to be determined with the given mesh configuration. We will solve this unknown potential at node 5 using (12.39).

First of all we have to evaluate the shape functions for the triangular elements, using (12.33) and (12.34).

$$L_1(x, y) = \frac{(x_2y_3 - y_2x_3) + (y_2 - y_3)x + (x_3 - x_2)y}{2A^{(e)}} \tag{12.40}$$

$$L_2(x, y) = \frac{(x_3y_1 - y_3x_1) + (y_3 - y_1)x + (x_1 - x_3)y}{2A^{(e)}} \tag{12.41}$$

$$L_3(x, y) = \frac{(x_1y_1 - y_1x_2) + (y_1 - y_2)x + (x_2 - x_1)y}{2A^{(e)}} \tag{12.42}$$

where $A^{(e)} = (y_2 - y_3)x_1 + (y_3 - y_1)x_2 + (y_1 - y_2)x_3$.

In equations (12.40) through (12.42) all the coordinates are based on the local numbering system.

We can now construct $\mathbf{T}^{(e)}$ as follows:

$$\mathbf{T}^{(e)} = \frac{1}{2A^{(e)}} \begin{bmatrix} y_2^{(e)} - y_3^{(e)} & y_3^{(e)} - y_1^{(e)} & y_1^{(e)} - y_2^{(e)} \\ x_3^{(e)} - x_2^{(e)} & x_1^{(e)} - x_3^{(e)} & x_2^{(e)} - x_1^{(e)} \end{bmatrix}$$

Element 1:	Local Numbers (LN)	Global Numbers (GN)	*x*	*y*
	1	1	0	2
	2	4	0	1
	3	5	1	1

$$\mathbf{T}^{(1)} = \begin{bmatrix} 0 & -1 & 1 \\ 1 & -1 & 0 \end{bmatrix} \qquad \mathbf{T}^{(1)T} = \begin{bmatrix} 0 & 1 \\ -1 & -1 \\ 1 & 0 \end{bmatrix}$$

$$\mathbf{U}^{(1)} = \mathbf{T}^{(1)T}\mathbf{T}^{(1)} = \begin{bmatrix} 1 & -1 & 0 \\ -1 & 2 & -1 \\ 0 & -1 & 1 \end{bmatrix}$$

$$\int_{s^{(1)}} \mathbf{U}^{(1)}\mathbf{V}^{(1)} ds^{(1)} = A^{(1)}\mathbf{U}^{(1)}\mathbf{V}^{(1)} = \begin{bmatrix} 0.5 & -0.5 & 0 \\ -0.5 & 1 & -0.5 \\ 0 & -0.5 & 0.5 \end{bmatrix} \begin{bmatrix} V_1 \\ V_4 \\ V_5 \end{bmatrix} \tag{12.43}$$

Element 2:	LN	GN	x	y
	1	1	0	2
	2	2	1	2
	3	5	1	1

$$\mathbf{T}^{(2)} = \begin{bmatrix} 1 & -1 & 0 \\ 0 & -1 & 1 \end{bmatrix} \qquad \mathbf{T}^{(2)T} = \begin{bmatrix} 1 & 0 \\ -1 & -1 \\ 0 & 1 \end{bmatrix}$$

$$\mathbf{U}^{(2)} = \mathbf{T}^{(2)T}\mathbf{T}^{(2)} = \begin{bmatrix} 1 & -1 & 0 \\ -1 & 2 & -1 \\ 0 & -1 & 1 \end{bmatrix}$$

$$\int_{s^{(2)}} \mathbf{U}^{(2)}\mathbf{V}^{(2)} ds^{(2)} = A^{(2)}\mathbf{U}^{(2)}\mathbf{V}^{(2)} = \begin{bmatrix} 0.5 & -0.5 & 0 \\ -0.5 & 1 & -0.5 \\ 0 & -0.5 & 0.5 \end{bmatrix} \begin{bmatrix} V_1 \\ V_2 \\ V_5 \end{bmatrix} \tag{12.44}$$

Element 3:	LN	GN	x	y
	1	2	1	2
	2	3	2	2
	3	5	1	1

$$\mathbf{T}^{(3)} = \begin{bmatrix} 1 & -1 & 0 \\ -1 & 0 & 1 \end{bmatrix} \qquad \mathrm{T}^{(3)T} = \begin{bmatrix} 1 & -1 \\ -1 & 0 \\ 0 & 1 \end{bmatrix}$$

$$\mathbf{U}^{(3)} = \mathbf{T}^{(3)T}\mathbf{T}^{(3)} = \begin{bmatrix} 2 & -1 & -1 \\ -1 & 1 & 0 \\ 1 & 0 & 1 \end{bmatrix}$$

$$\int_{s^{(3)}} \mathbf{U}^{(3)}\mathbf{V}^{(3)} ds^{(3)} = A^{(3)}\mathbf{U}^{(3)}\mathbf{V}^{(3)} = \begin{bmatrix} 1 & -0.5 & -0.5 \\ -0.5 & 0.5 & 0 \\ -0.5 & 0 & 0.5 \end{bmatrix} \begin{bmatrix} V_2 \\ V_3 \\ V_5 \end{bmatrix} \tag{12.45}$$

Element 4:	LN	GN	x	y
	1	3	2	2
	2	5	1	1
	3	6	2	1

$$\mathbf{T}^{(4)} = \begin{bmatrix} 0 & -1 & 1 \\ 1 & 0 & -1 \end{bmatrix} \qquad \mathbf{T}^{(4)T} = \begin{bmatrix} 0 & 1 \\ -1 & 0 \\ 1 & -1 \end{bmatrix}$$

$$\mathbf{U}^{(4)} = \mathbf{T}^{(4)T}\mathbf{T}^{(4)} = \begin{bmatrix} 1 & 0 & -1 \\ 0 & 1 & -1 \\ -1 & -1 & 2 \end{bmatrix}$$

$$\int_{s^{(4)}} \mathbf{U}^{(4)}\mathbf{V}^{(4)}\, ds^{(4)} = A^{(4)}\mathbf{U}^{(4)}\mathbf{V}^{(4)} = \begin{bmatrix} 0.5 & 0 & -0.5 \\ 0 & 0.5 & -0.5 \\ -0.5 & -0.5 & 1 \end{bmatrix} \begin{bmatrix} V_3 \\ V_5 \\ V_6 \end{bmatrix} \tag{12.46}$$

Element 5:	LN	GN	x	y
	1	4	0	1
	2	5	1	1
	3	7	0	0

$$\mathbf{T}^{(5)} = \begin{bmatrix} 1 & -1 & 0 \\ -1 & 0 & 1 \end{bmatrix} \qquad \mathbf{T}^{(5)T} = \begin{bmatrix} 1 & -1 \\ -1 & 0 \\ 0 & 1 \end{bmatrix}$$

$$\mathbf{U}^{(5)} = \mathbf{T}^{(5)T}\mathbf{T}^{(5)} = \begin{bmatrix} 2 & -1 & -1 \\ -1 & 1 & 0 \\ -1 & 0 & 1 \end{bmatrix}$$

$$\int_{s^{(5)}} \mathbf{U}^{(5)}\mathbf{V}^{(5)} ds^{(5)} = A^{(5)}\mathbf{U}^{(5)}\mathbf{V}^{(5)} = \begin{bmatrix} 1 & -0.5 & -0.5 \\ -0.5 & 0.5 & 0 \\ -0.5 & 0 & 0.5 \end{bmatrix} \begin{bmatrix} V_4 \\ V_5 \\ V_7 \end{bmatrix} \tag{12.47}$$

Element 6:	LN	GN	x	y
	1	5	1	1
	2	7	0	0
	3	8	1	0

$$\mathbf{T}^{(6)} = \begin{bmatrix} 1 & -1 & 1 \\ 1 & 0 & -1 \end{bmatrix} \qquad \mathbf{T}^{(6)T} = \begin{bmatrix} 0 & 1 \\ -1 & 0 \\ 1 & -1 \end{bmatrix}$$

$$\mathbf{U}^{(6)} = \mathbf{T}^{(6)T}\mathbf{T}^{(6)} = \begin{bmatrix} 2 & -1 & -1 \\ -1 & 1 & 0 \\ -1 & 0 & 1 \end{bmatrix}$$

$$\int_{s^{(6)}} \mathbf{U}^{(6)}\mathbf{V}^{(6)}\,ds^{(6)} = A^{(6)}\mathbf{U}^{(6)}\mathbf{V}^{(6)} = \begin{bmatrix} 1 & -0.5 & -0.5 \\ -0.5 & 0.5 & 0 \\ -0.5 & 0 & 0.5 \end{bmatrix} \begin{bmatrix} V_4 \\ V_7 \\ V_8 \end{bmatrix} \tag{12.48}$$

Element 7:	**LN**	**GN**	*x*	*y*
	1	5	1	1
	2	8	1	0
	3	9	2	0

$$\mathbf{T}^{(7)} = \begin{bmatrix} 0 & -1 & 1 \\ 1 & -1 & 0 \end{bmatrix} \quad \mathbf{T}^{(7)T} = \begin{bmatrix} 0 & 1 \\ -1 & -1 \\ 1 & 0 \end{bmatrix}$$

$$\mathbf{U}^{(7)} = \mathbf{T}^{(7)T}\mathbf{T}^{(7)} = \begin{bmatrix} 1 & -1 & 0 \\ -1 & 2 & -1 \\ 0 & -1 & 1 \end{bmatrix}$$

$$\int_{s^{(7)}} \mathbf{U}^{(7)}\mathbf{V}^{(7)}ds^{(7)} = A^{(7)}\mathbf{U}^{(7)}\mathbf{V}^{(7)} = \begin{bmatrix} 0.5 & -0.5 & 0 \\ -0.5 & 1 & -0.5 \\ 0 & -0.5 & 0.5 \end{bmatrix} \begin{bmatrix} V_5 \\ V_8 \\ V_9 \end{bmatrix} \tag{12.49}$$

Element 8:	**LN**	**GN**	*x*	*y*
	1	5	1	1
	2	6	2	1
	3	9	2	0

$$\mathbf{T}^{(8)} = \begin{bmatrix} 1 & -1 & 0 \\ 0 & -1 & 1 \end{bmatrix} \quad \mathbf{T}^{(8)T} = \begin{bmatrix} 1 & 0 \\ -1 & -1 \\ 0 & 1 \end{bmatrix}$$

$$\mathbf{U}^{(8)} = \mathbf{T}^{(8)T}\mathbf{T}^{(8)} = \begin{bmatrix} 1 & -1 & 0 \\ -1 & 2 & -1 \\ 0 & -1 & 1 \end{bmatrix}$$

$$\int_{s^{(8)}} \mathbf{U}^{(8)}\mathbf{V}^{(8)}\,ds^{(8)} = A^{(8)}\mathbf{U}^{(8)}\mathbf{V}^{(8)} = \begin{bmatrix} 0.5 & -0.5 & 0 \\ -0.5 & 1 & -0.5 \\ 0 & -0.5 & 0.5 \end{bmatrix} \begin{bmatrix} V_5 \\ V_6 \\ V_9 \end{bmatrix} \tag{12.50}$$

In order to be able to form equation (12.39), we have to augment the matrices and vectors in (12.43)–(12.50) as follows:

Element 1:

$$
\begin{array}{c}
\text{Node} \\ \\ 1\\2\\3\\4\\5\\6\\7\\8\\9
\end{array}
\begin{array}{c}
\begin{array}{ccccccccc} 1 & 2 & 3 & 4 & 5 & 6 & 7 & 8 & 9 \end{array} \\ \\
\left[\begin{array}{ccccccccc}
0.5 & 0.0 & 0.0 & -0.5 & 0.0 & 0.0 & 0.0 & 0.0 & 0.0 \\
0.0 & 0.0 & 0.0 & 0.0 & 0.0 & 0.0 & 0.0 & 0.0 & 0.0 \\
0.0 & 0.0 & 0.0 & 0.0 & 0.0 & 0.0 & 0.0 & 0.0 & 0.0 \\
-0.5 & 0.0 & 0.0 & 1.0 & -0.5 & 0.0 & 0.0 & 0.0 & 0.0 \\
0.0 & 0.0 & 0.0 & -0.5 & 0.5 & 0.0 & 0.0 & 0.0 & 0.0 \\
 & & & & & & & & \\
 & 0.0 & & & & & 0.0 & & \\
 & & & & & & & & \\
 & & & & & & & &
\end{array}\right]
\end{array}
\begin{bmatrix} V_1 \\ V_2 \\ V_3 \\ V_4 \\ V_5 \\ V_6 \\ V_7 \\ V_8 \\ V_9 \end{bmatrix}
\quad (12.51)
$$

Element 2:

$$
\begin{array}{c}
\text{Node} \\ \\ 1\\2\\3\\4\\5\\6\\7\\8\\9
\end{array}
\begin{array}{c}
\begin{array}{ccccccccc} 1 & 2 & 3 & 4 & 5 & 6 & 7 & 8 & 9 \end{array} \\ \\
\left[\begin{array}{ccccccccc}
0.5 & -0.5 & 0.0 & 0.0 & 0.0 & 0.0 & 0.0 & 0.0 & 0.0 \\
-0.5 & 1.0 & 0.0 & 0.0 & -0.5 & 0.0 & 0.0 & 0.0 & 0.0 \\
0.0 & 0.0 & 0.0 & 0.0 & 0.0 & 0.0 & 0.0 & 0.0 & 0.0 \\
0.0 & 0.0 & 0.0 & 0.0 & 0.0 & 0.0 & 0.0 & 0.0 & 0.0 \\
0.0 & -0.5 & 0.0 & 0.0 & 0.5 & 0.0 & 0.0 & 0.0 & 0.0 \\
 & & & & & & & & \\
 & 0.0 & & & & & 0.0 & & \\
 & & & & & & & & \\
 & & & & & & & &
\end{array}\right]
\end{array}
\begin{bmatrix} V_1 \\ V_2 \\ V_3 \\ V_4 \\ V_5 \\ V_6 \\ V_7 \\ V_8 \\ V_9 \end{bmatrix}
\quad (12.52)
$$

Element 3:

$$
\begin{array}{c}
\text{Node} \\ \\ 1\\2\\3\\4\\5\\6\\7\\8\\9
\end{array}
\begin{array}{c}
\begin{array}{ccccccccc} 1 & 2 & 3 & 4 & 5 & 6 & 7 & 8 & 9 \end{array} \\ \\
\left[\begin{array}{ccccccccc}
0.0 & 0.0 & 0.0 & 0.0 & 0.0 & 0.0 & 0.0 & 0.0 & 0.0 \\
0.0 & 1.0 & -0.5 & 0.0 & 0.5 & 0.0 & 0.0 & 0.0 & 0.0 \\
0.0 & -0.5 & 0.5 & 0.0 & 0.0 & 0.0 & 0.0 & 0.0 & 0.0 \\
0.0 & 0.0 & 0.0 & 0.0 & 0.0 & 0.0 & 0.0 & 0.0 & 0.0 \\
0.0 & -0.5 & 0.0 & 0.0 & 0.5 & 0.0 & 0.0 & 0.0 & 0.0 \\
 & & & & & & & & \\
 & 0.0 & & & & & 0.0 & & \\
 & & & & & & & & \\
 & & & & & & & &
\end{array}\right]
\end{array}
\begin{bmatrix} V_1 \\ V_2 \\ V_3 \\ V_4 \\ V_5 \\ V_6 \\ V_7 \\ V_8 \\ V_9 \end{bmatrix}
\quad (12.53)
$$

Element 4:

$$
\begin{array}{c}
\text{Node} \\
1 \\ 2 \\ 3 \\ 4 \\ 5 \\ 6 \\ 7 \\ 8 \\ 9
\end{array}
\begin{array}{c}
\begin{array}{ccccccccc} 1 & 2 & 3 & 4 & 5 & 6 & 7 & 8 & 9 \end{array} \\
\left[\begin{array}{ccccccccc}
0.0 & 0.0 & 0.0 & 0.0 & 0.0 & 0.0 & 0.0 & 0.0 & 0.0 \\
0.0 & 0.0 & 0.0 & 0.0 & 0.0 & 0.0 & 0.0 & 0.0 & 0.0 \\
0.0 & 0.0 & 0.5 & 0.0 & 0.0 & -0.5 & 0.0 & 0.0 & 0.0 \\
0.0 & 0.0 & 0.0 & 0.0 & 0.0 & 0.0 & 0.0 & 0.0 & 0.0 \\
0.0 & 0.0 & 0.0 & 0.0 & 0.5 & -0.5 & 0.0 & 0.0 & 0.0 \\
0.0 & 0.0 & -0.5 & 0.0 & -0.5 & 1.0 & 0.0 & 0.0 & 0.0 \\
 & & & & & & & & \\
 & 0.0 & & & & & 0.0 & & \\
 & & & & & & & &
\end{array}\right]
\end{array}
\left[\begin{array}{c} V_1 \\ V_2 \\ V_3 \\ V_4 \\ V_5 \\ V_6 \\ V_7 \\ V_8 \\ V_9 \end{array}\right]
\qquad (12.54)
$$

Element 5:

$$
\begin{array}{c}
\text{Node} \\
1 \\ 2 \\ 3 \\ 4 \\ 5 \\ 6 \\ 7 \\ 8 \\ 9
\end{array}
\begin{array}{c}
\begin{array}{ccccccccc} 1 & 2 & 3 & 4 & 5 & 6 & 7 & 8 & 9 \end{array} \\
\left[\begin{array}{ccccccccc}
0.0 & 0.0 & 0.0 & 0.0 & 0.0 & 0.0 & 0.0 & 0.0 & 0.0 \\
0.0 & 0.0 & 0.0 & 0.0 & 0.0 & 0.0 & 0.0 & 0.0 & 0.0 \\
0.0 & 0.0 & 0.0 & 0.0 & 0.0 & 0.0 & 0.0 & 0.0 & 0.0 \\
0.0 & 0.0 & 0.0 & 1.0 & -0.5 & 0.0 & -0.5 & 0.0 & 0.0 \\
0.0 & 0.0 & 0.0 & -0.5 & 0.5 & 0.0 & 0.0 & 0.0 & 0.0 \\
0.0 & 0.0 & 0.0 & 0.0 & 0.0 & 0.0 & 0.0 & 0.0 & 0.0 \\
0.0 & 0.0 & 0.0 & -0.5 & 0.0 & 0.0 & 0.5 & 0.0 & 0.0 \\
0.0 & 0.0 & 0.0 & 0.0 & 0.0 & 0.0 & 0.0 & 0.0 & 0.0 \\
0.0 & 0.0 & 0.0 & 0.0 & 0.0 & 0.0 & 0.0 & 0.0 & 0.0
\end{array}\right]
\end{array}
\left[\begin{array}{c} V_1 \\ V_2 \\ V_3 \\ V_4 \\ V_5 \\ V_6 \\ V_7 \\ V_8 \\ V_9 \end{array}\right]
\qquad (12.55)
$$

Element 6:

$$
\begin{array}{c}
\text{Node} \\
1 \\ 2 \\ 3 \\ 4 \\ 5 \\ 6 \\ 7 \\ 8 \\ 9
\end{array}
\begin{array}{c}
\begin{array}{ccccccccc} 1 & 2 & 3 & 4 & 5 & 6 & 7 & 8 & 9 \end{array} \\
\left[\begin{array}{ccccccccc}
 & & & & & & & & \\
 & 0.0 & & & & & 0.0 & & \\
 & & & & & & & & \\
 & & & & & & & & \\
0.0 & 0.0 & 0.0 & 0.0 & 0.5 & 0.0 & 0.0 & -0.5 & 0.0 \\
0.0 & 0.0 & 0.0 & 0.0 & 0.0 & 0.0 & 0.0 & 0.0 & 0.0 \\
0.0 & 0.0 & 0.0 & 0.0 & 0.0 & 0.0 & 0.5 & -0.5 & 0.0 \\
0.0 & 0.0 & 0.0 & 0.0 & -0.5 & 0.0 & -0.5 & 1.0 & 0.0 \\
0.0 & 0.0 & 0.0 & 0.0 & 0.0 & 0.0 & 0.0 & 0.0 & 0.0
\end{array}\right]
\end{array}
\left[\begin{array}{c} V_1 \\ V_2 \\ V_3 \\ V_4 \\ V_5 \\ V_6 \\ V_7 \\ V_8 \\ V_9 \end{array}\right]
\qquad (12.56)
$$

Element 7:

$$
\begin{array}{c}
\text{Node} \\ 1 \\ 2 \\ 3 \\ 4 \\ 5 \\ 6 \\ 7 \\ 8 \\ 9
\end{array}
\begin{array}{c}
\begin{array}{ccccccccc} 1 & 2 & 3 & 4 & 5 & 6 & 7 & 8 & 9 \end{array} \\
\left[\begin{array}{ccccccccc}
 & 0.0 & & & & & 0.0 & & \\
 & & & & & & & & \\
 & & & & & & & & \\
 & & & & & & & & \\
0.0 & 0.0 & 0.0 & 0.0 & 0.5 & 0.0 & 0.0 & -0.5 & 0.0 \\
0.0 & 0.0 & 0.0 & 0.0 & 0.0 & 0.0 & 0.0 & 0.0 & 0.0 \\
0.0 & 0.0 & 0.0 & 0.0 & 0.0 & 0.0 & 0.0 & 0.0 & 0.0 \\
0.0 & 0.0 & 0.0 & 0.0 & -0.5 & 0.0 & 0.0 & 1.0 & -0.5 \\
0.0 & 0.0 & 0.0 & 0.0 & 0.0 & 0.0 & 0.0 & -0.5 & 0.5
\end{array}\right]
\end{array}
\begin{bmatrix} V_1 \\ V_2 \\ V_3 \\ V_4 \\ V_5 \\ V_6 \\ V_7 \\ V_8 \\ V_9 \end{bmatrix}
\tag{12.57}
$$

Element 8:

$$
\begin{array}{c}
\text{Node} \\ 1 \\ 2 \\ 3 \\ 4 \\ 5 \\ 6 \\ 7 \\ 8 \\ 9
\end{array}
\begin{array}{c}
\begin{array}{ccccccccc} 1 & 2 & 3 & 4 & 5 & 6 & 7 & 8 & 9 \end{array} \\
\left[\begin{array}{ccccccccc}
 & 0.0 & & & & & 0.0 & & \\
 & & & & & & & & \\
 & & & & & & & & \\
 & & & & & & & & \\
0.0 & 0.0 & 0.0 & 0.0 & 0.5 & -0.5 & 0.0 & 0.0 & 0.0 \\
0.0 & 0.0 & 0.0 & 0.0 & -0.5 & 1.0 & 0.0 & 0.0 & -0.5 \\
0.0 & 0.0 & 0.0 & 0.0 & 0.0 & 0.0 & 0.0 & 0.0 & 0.0 \\
0.0 & 0.0 & 0.0 & 0.0 & 0.0 & 0.0 & 0.0 & 0.0 & 0.0 \\
0.0 & 0.0 & 0.0 & 0.0 & 0.0 & -0.5 & 0.0 & 0.0 & 0.5
\end{array}\right]
\end{array}
\begin{bmatrix} V_1 \\ V_2 \\ V_3 \\ V_4 \\ V_5 \\ V_6 \\ V_7 \\ V_8 \\ V_9 \end{bmatrix}
\tag{12.58}
$$

By adding (12.51)–(12.58), we can obtain

$$
\begin{array}{c}
\text{Node} \\ 1 \\ 2 \\ 3 \\ 4 \\ 5 \\ 6 \\ 7 \\ 8 \\ 9
\end{array}
\begin{array}{c}
\begin{array}{ccccccccc} 1 & 2 & 3 & 4 & 5 & 6 & 7 & 8 & 9 \end{array} \\
\left[\begin{array}{ccccccccc}
1.0 & -0.5 & 0.0 & -0.5 & 0.0 & 0.0 & 0.0 & 0.0 & 0.0 \\
-0.5 & 2.0 & -0.5 & 0.0 & -1.0 & 0.0 & 0.0 & 0.0 & 0.0 \\
0.0 & -0.5 & 1.0 & 0.0 & 0.0 & -0.5 & 0.0 & 0.0 & 0.0 \\
-0.5 & 0.0 & 0.0 & 2.0 & -1.0 & 0.0 & -0.5 & 0.0 & 0.0 \\
0.0 & -1.0 & 0.0 & -1.0 & 4.0 & -1.0 & 0.0 & -1.0 & 0.0 \\
0.0 & 0.0 & -0.5 & 0.0 & -1.0 & 2.0 & 0.0 & 0.0 & -0.5 \\
0.0 & 0.0 & 0.0 & -0.5 & 0.0 & 0.0 & 1.0 & -0.5 & 0.0 \\
0.0 & 0.0 & 0.0 & 0.0 & -1.0 & 0.0 & -0.5 & 2.0 & -0.5 \\
0.0 & 0.0 & 0.0 & 0.0 & 0.0 & -5.0 & 0.0 & -0.5 & 1.0
\end{array}\right]
\end{array}
\begin{bmatrix} V_1 \\ V_2 \\ V_3 \\ V_4 \\ V_5 \\ V_6 \\ V_7 \\ V_8 \\ V_9 \end{bmatrix}
= 0
\tag{12.59}
$$

From (12.59) we obtain the unknown potential at node 5 as

$V_5 = 25$ V • • •

EXAMPLE 12.5 Write a finite-element computer program for the problem in Example 12.4.

Solution A computer program is written using the same mesh configuration shown in Figure 12.15. A listing of the computer program in FORTRAN language is given in Appendix B. With this program, the potential at node 5 is also obtained as 25 V. • • •

EXAMPLE 12.6 Determine the potential at 2.8 cm from the center of the coaxial cable shown in Figure 12.16 by using the FEM. Compare your numerical results with the analytical solution.

Figure 12.16

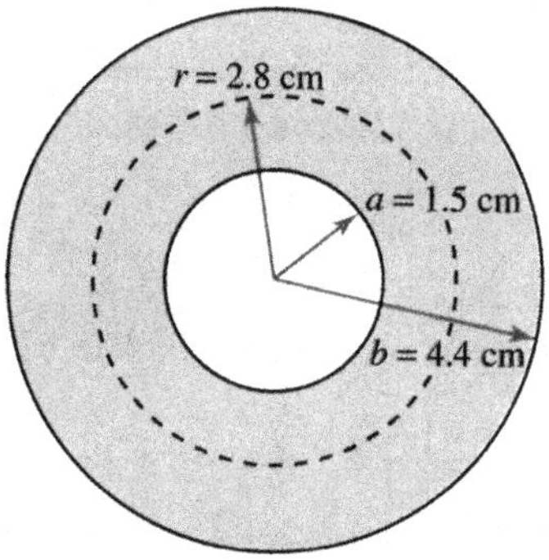

Solution There is no free charge between the inner and outer sheaths of the coaxial cable. Therefore, the solution of Laplace's equation gives us the required voltage calculation. Because the coaxial cable has an axisymmetric geometry, we can consider only one quarter of the cable, and generate the meshes as shown in Figure 12.17. In this model there are eight elements

Figure 12.17

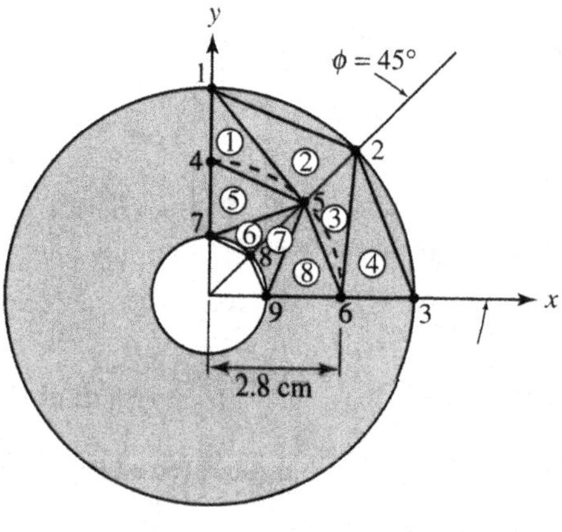

and nine nodes. Potentials V_4, V_5, and V_6 are the unknown potentials, and the potentials at the other nodes are specified as boundary conditions. With the help of the computer program FEM.CX, which is listed in Appendix B, we can calculate the unknown potentials as $V_4 = 42$ V, $V_5 = 41$ V, and $V_6 = 42$ V.

The analytical solution is determined as

$$V(2.8\,\text{cm}) = 100\frac{\ln\dfrac{4.4}{2.8}}{\ln\dfrac{4.4}{1.5}} = 42\ \text{V}$$

You can see that the numerical and analytical results are almost the same except at node 5. The discrepancy on potential node 5 is due to the discretization of the medium. • • •

12.4 Method of moments

In this section we will examine a technique referred to as the method of moments (MOM). which has a wide variety of applications for the analysis of electromagnetic fields. The method is conceptually rather simple; it basically employs the equations of unknown fields in integral form to determine the field distribution in a given medium. In our discussion, we will use the potential equation in integral form to evaluate the field distribution.

In electrostatics, the potential distribution at a point (x, y, z) due to a charge distribution at (x', y', z') is expressed as

$$V(x, y, z) = \frac{1}{4\pi\epsilon}\int_v \frac{\rho_v(x', y', z')\,dv'}{R} \tag{12.60}$$

Here $\rho_v(x', y', z')$ is essentially the source for the potential distribution, and R is the distance between points (x, y, z) and (x', y', z'). However, $\rho_v(x', y', z')$ is generally unknown; instead, the distribution of the potential is given in the source region. Consequently, in order to predict the potential distribution everywhere in space, we have to estimate the charge distribution ρ_v in the source region.

Let us assume a solution for $\rho_v(x', y', z')$ as

$$\begin{aligned}\rho_v(x', y', z') &= \alpha_1\rho_1(x', y', z') + \alpha_2\rho_2(x', y', z') + \cdots \\ &\quad + \alpha_n\rho_n(x', y', z') \\ &= \sum_{i=1}^{n} \alpha_i\rho_i(x', y', z')\end{aligned} \tag{12.61}$$

where $[\rho_i(x', y', z')]_{i=1}^n$ are preselected charge distributions at some discrete locations in the source region, and $[\alpha_i]_{i=1}^n$ are unknown coefficients yet to be determined.

Substitution of (12.61) into (12.60) yields

$$V_j = V(x_j, y_j, z_j) = \frac{1}{4\pi\epsilon}\int_{v'} \frac{\sum_{i=1}^{n} \alpha_i \rho_i(x', y', z')}{|R|} dv' \tag{12.62}$$

or

$$V_j = \sum_{i=1}^{n} \alpha_i \frac{1}{4\pi\epsilon}\int_{v_i'} \frac{\rho_i(x', y', z')}{|R_{ji}|} dv_i' \tag{12.63}$$

where $j = 1, 2, \ldots, n$. Hence, $V(x, y, z)$ is represented in terms of the linear combination of potentials

$$V_{ji} = \frac{1}{4\pi\epsilon}\int_{v_i'} \frac{\rho_i(x', y', z')}{|R_{ji}|} dv'_i \qquad i = 1, 2, \ldots, n \tag{12.64}$$

as

$$V_j = \sum_{i=1}^{n} \alpha_i V_{ji} \tag{12.65}$$

at locations $[\rho_i(x', y', z')]_{i=1}^{n}$. Since $V(x, y, z)$ is known within the source region, the unknown coefficients $\alpha_1, \alpha_2, \ldots, \alpha_n$ can be determined from

$$\begin{aligned}
V_1 &= \alpha_1 V_{11} + \alpha_2 V_{12} + \cdots + \alpha_n V_{1n} \\
V_2 &= \alpha_1 V_{21} + \alpha_2 V_{22} + \cdots + \alpha_n V_{2n} \\
&\vdots \\
V_j &= \alpha_1 V_{j1} + \alpha_2 V_{j2} + \cdots + \alpha_n V_{jn} \\
&\vdots \\
V_n &= \alpha_1 V_{n1} + \alpha_2 V_{n2} + \cdots + \alpha_n V_{nn}
\end{aligned}$$

or in matrix form,

$$\begin{bmatrix} V_1 \\ \vdots \\ V_j \\ \vdots \\ V_n \end{bmatrix} = \begin{bmatrix} V_{11} & & V_{1j} & & V_{1n} \\ \vdots & \ddots & \vdots & \ddots & \vdots \\ V_{j1} & & V_{jj} & & V_{jn} \\ \vdots & \ddots & \vdots & \ddots & \vdots \\ V_{n1} & & V_{nj} & & V_{nn} \end{bmatrix} \begin{bmatrix} \alpha_1 \\ \vdots \\ \alpha_j \\ \vdots \\ \alpha_n \end{bmatrix} \tag{12.66}$$

By using (12.61), we can now construct the charge distribution $\rho_v(x', y', z')$ in the source region. We can then use this distribution to predict the potential distribution at any point in space by employing (12.62).

EXAMPLE 12.7

A 20-cm-long thin cylindrical conductor of 1-mm radius is maintained at a potential of 1 V. Determine the charge distribution along the conductor using the method of moments.

Solution Figure 12.18 shows the conductor geometry. Because of the symmetry, we can reduce the problem to a two-dimensional case, and assume that the charge is concentrated along the axis of symmetry as a line charge with a density of ρ_ℓ, as illustrated in Figure 12.19.

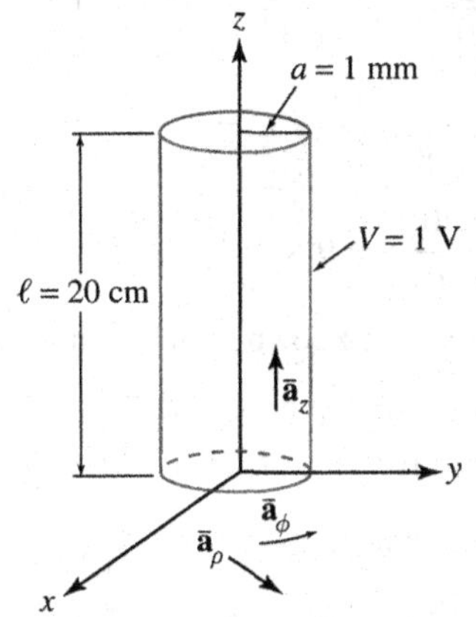

Figure 12.18 The conductor geometry for Example 12.7

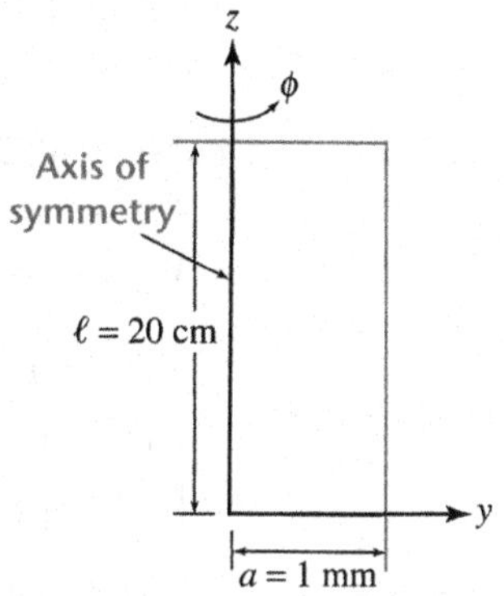

Figure 12.19 Two-dimensional model of the conductor (Example 12.7)

Let us divide the conductor into two elements, also known as cells, as indicated in Figure 12.20, and assume a line charge of unity in each element concentrated at the center of that element; i.e., $\rho_1 = 1$ and $\rho_2 = 1$. The distances R_{11}, R_{12}, R_{21}, and R_{22} can be calculated as

$$R_{11} = a = 0.001 \text{ m}$$

$$R_{12} = \sqrt{0.001^2 + 0.1^2} \text{ m}$$

$$R_{22} = a = 0.001 \text{ m}$$

$$R_{21} = \sqrt{0.001^2 + 0.1^2} \text{ m}$$

Figure 12.20 Conductor model with two elements

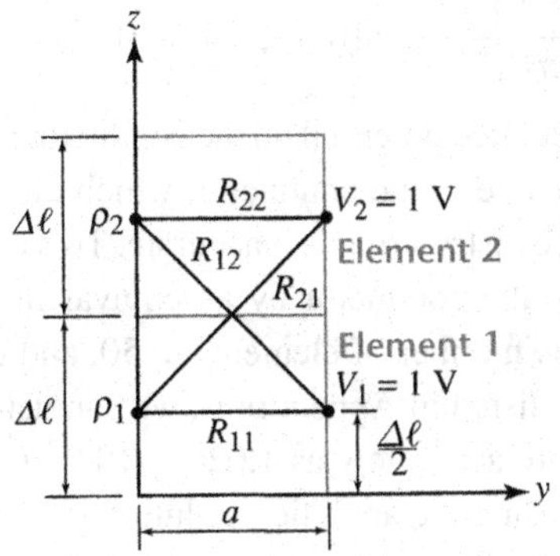

In our calculations, we also assume that the line charge in each element remains the same within the corresponding element. Now we can calculate the V_{ji}'s using (12.64) as

$$V_{11} = \frac{1}{4\pi\epsilon}\frac{1 \times 0.1}{0.001} = 9 \times 10^{11}$$

$$V_{12} = \frac{1}{4\pi\epsilon}\frac{1 \times 0.1}{\sqrt{0.001^2 + 0.1^2}} = 8.99 \times 10^9$$

$$V_{21} = \frac{1}{4\pi\epsilon}\frac{1 \times 0.1}{\sqrt{0.001^2 + 0.1^2}} = 8.99 \times 10^9$$

$$V_{22} = \frac{1}{4\pi\epsilon}\frac{1 \times 0.1}{0.001} = 9 \times 10^{11}$$

Substitution of all the known variables into (12.66) yields

$$\begin{bmatrix} 1 \\ 1 \end{bmatrix} = \begin{bmatrix} 9 \times 10^{11} & 8.99 \times 10^9 \\ 8.99 \times 10^9 & 9 \times 10^{11} \end{bmatrix} \begin{bmatrix} \alpha_1 \\ \alpha_2 \end{bmatrix}$$

from which we obtain

$$\alpha_1 = 1.1 \times 10^{-12}. \qquad \alpha_2 = 1.1 \times 10^{-12}$$

and the corresponding charge densities as

$$\rho_1 = 1 \times 1.1 \times 10^{-12}\ \text{C/m}. \qquad \rho_2 = 1 \times 1.1 \times 10^{-12}\ \text{C/m}$$

Let us calculate the potential at the midpoint and on the surface of the conductor, using the potential expression and the charge distribution obtained from the application of the method of moments. The potential at $(x, 0, 0)$ due to a finite line charge, as shown in Figure 12.21, is

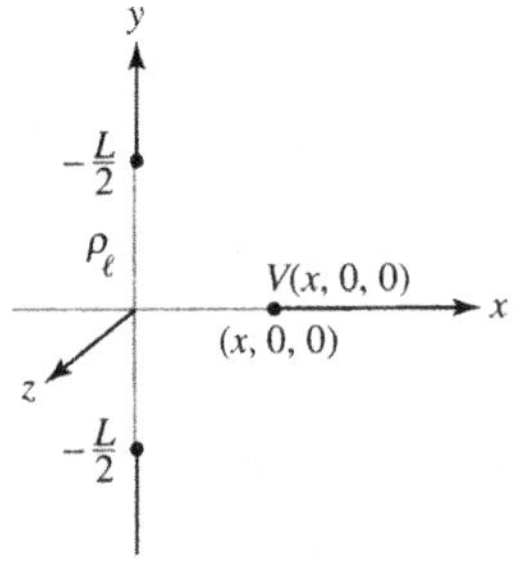

Figure 12.21 Potential at (x, 0, 0) due to finite line charge ρ_ℓ

$$V = \frac{\rho_1}{2\pi\epsilon_0}\left\{\ln\left[\frac{L}{2} + \sqrt{\left(\frac{L}{2}\right)^2 + x^2}\right] - \ln x\right\} \qquad (12.67)$$

Because the average $\rho_1 = 1.1 \times 10^{-12}$ C/m

$$V = \frac{1.1 \times 10^{-12}}{2\pi\epsilon_0}[\ln(10 + \sqrt{10^2 + 0.1^2}) - \ln 0.1] = 0.105\ \text{V}$$

The calculated potential on the conductor surface is only 10.5% of the applied voltage to the conductor, which clearly tells us that the calculated charge distribution is inaccurate. Two elements were not sufficient to simulate the conductor by an equivalent charge distribution. If we increase the number of elements to 50, and use the computer program MOM.CD, listed in Appendix B, we can determine the average charge density more accurately as 1.026×10^{-11} C/m. We can verify the accuracy by using the analytical solution in (12.67):

$$V = \frac{1.026 \times 10^{-11}}{2\pi\epsilon_0}[\ln(10 + \sqrt{10^2 + 0.1^2}) - \ln 0.1] = 0.98\ \text{V}$$

• • •

12.5 Summary

The design of an electromagnetic system requires the analysis of electric and/or magnetic fields in order to efficiently utilize the materials used in that system. Laplace's or Poisson's equations, along with some boundary conditions, are usually sufficient mathematical models for the analysis of static or quasistatic field problems. However, at high-frequency operation, the wave equation is a more accurate mathematical model for quantifying the electric and magnetic fields in a system.

Because of the irregularities present in most practical engineering problems, the solution of the mathematical equations poses a difficulty: a closed-form solution is usually not available. However, the use of computers facilitates the utilization of numerical techniques such as the *finite-difference method (FDM), finite-element method (FEM)*, and *method of moments (MOM)* to evaluate the field distribution in practical, yet complicated, geometries.

The finite-difference method basically divides the solution domain into some finite number of discrete points, and replaces the partial differential equations with a set of difference equations. Of course, the solution is not an exact solution, but an approximate one. The mesh size of the discretized solution domain is a measure of the accuracy of the solution – the smaller the mesh size, the better the accuracy. An iterative technique known as the *successive over-relaxation method* is a very useful way of solving the difference equations of the FDM. The proper acceleration factor expedites the solution of the equation significantly.

The finite-element method is another technique to solve electromagnetic field problems numerically. It is an optimization method that basically minimizes the total energy stored in the system subject to some constraints dictated by the boundary conditions. One of the most important advantages of the FEM is that it can treat complicated boundary conditions with almost no difficulties. Another important advantage is that it can handle the analysis of fields in multimaterials quite easily.

In problems with open boundaries, the method of moments appears to be the best choice to determine the electric and magnetic fields. This method uses the general integral equations, basically the retarded potential equations, so we do not really need to define the solution with finite boundaries. However, this technique requires the knowledge of the charge or current distribution on the existing boundaries of the problem, and often this information is not available. However, when the potential is given on the boundary, we can predict the distribution of the charge or the current on the boundary numerically by dividing the boundary into a number of elements, sometimes referred to as *boundary elements.* We can then determine the field distribution everywhere in the system.

12.6 Review questions

12.1 Why do we need the evaluation of electric and magnetic fields?
12.2 Which equations mathematically represent the static and quasistatic fields?
12.3 Can we use Laplace's equation to determine the electric fields in systems operating at 100 MHz?
12.4 Why do we need numerical techniques to solve the field problems?
12.5 What is the finite-difference method?
12.6 Why do we use the successive over-relaxation method?
12.7 What is the role of the acceleration factor in SOR?
12.8 Does the acceleration factor affect the solution of the field distribution?
12.9 What is the finite-element method?
12.10 Explain the procedures in the FEM.
12.11 What is the advantage of using a triangular mesh in the FEM?
12.12 What are the major advantages of the FEM?
12.13 Explain the method of moments.
12.14 In what situations do we mostly use the MOM?
12.15 What does the accuracy depend on in evaluating the fields by using the MOM?

12.7 Exercises

12.1 Write the coefficient matrix **A** to determine the potential distribution in the geometry given in Figure 12.22 using the finite-difference method.

Figure 12.22

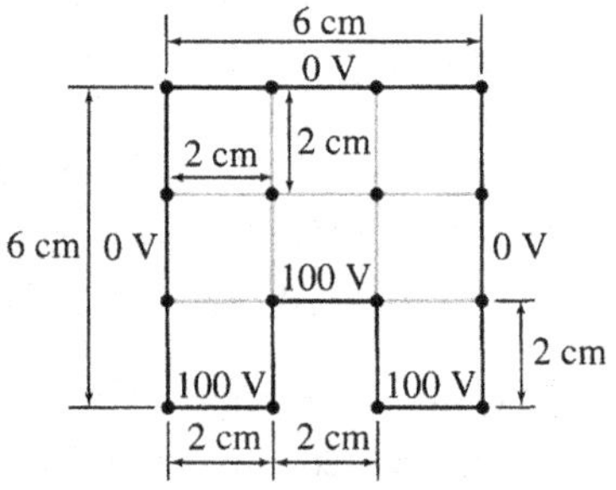

12.2 Perform two iterations in an attempt to solve the potential distribution in the geometry shown in Figure 12.22 using SOR in conjunction with the finite-difference method.

12.3 Figure 12.23 shows an enclosed transmission line with two infinitely long parallel conductors. Determine the potential distribution in this configuration by writing a simple FEM computer program.

12.4 Figure 12.24 illustrates a coaxial geometry with rectangular conductors. Determine the potential distribution between the conductors by developing an FEM computer program.

Figure 12.23

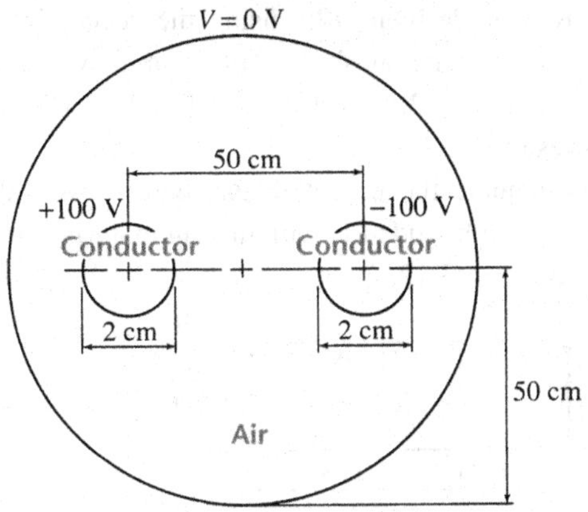

Figure 12.24

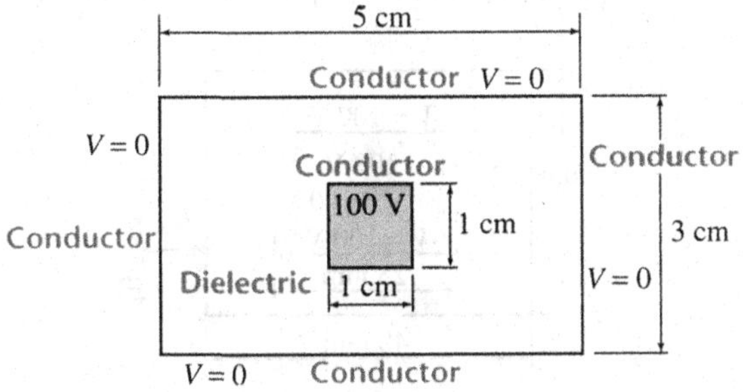

12.5 Write a computer program using MOM to determine the potential distribution everywhere for Example 12.7. Consider the number of elements as 2, 4, 10, and 20 in your program, and comment on their effect in the potential distribution.

12.8 Problems

12.1 The geometry in Figure P12.1 is given in the xy plane. Determine the potential distribution within the bounded medium using the FDM without successive over-relaxation. Consider a square mesh size of (a) 10 mm and (b) 5 mm. Compare your results.

Figure P12.1

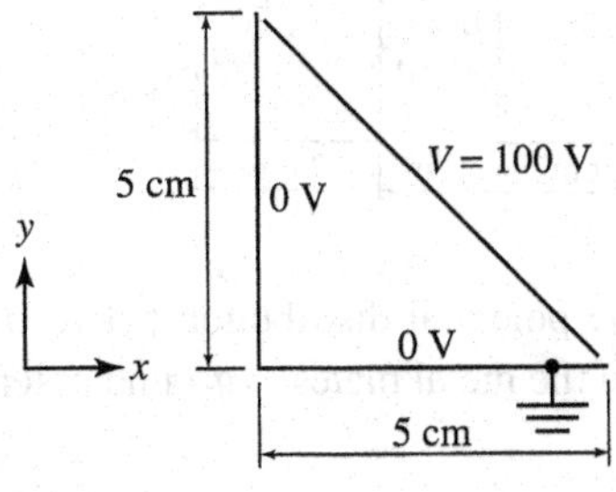

12.2 Repeat Problem 12.1 using the successive over-relaxation technique for a square mesh size of 10 mm with acceleration factors of $\alpha = 1.0$, $\alpha = 1.2$, $\alpha = 1.4$, $\alpha = 1.5$, $\alpha = 1.7$, and $\alpha = 1.9$. Discuss your results.

12.3 Calculate the potential distribution using the FDM on the dielectric surface bounded by conducting matarials at different potentials shown in Figure P12.3.

Figure P12.3

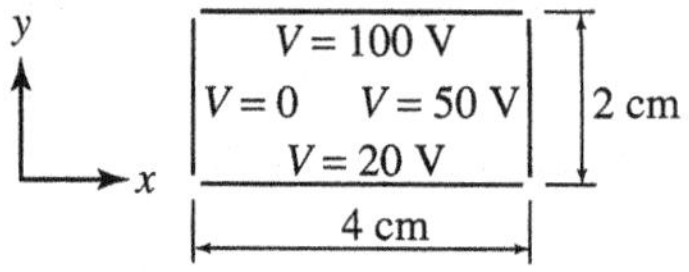

12.4 Determine the equipotential lines between the 100-V and 0-V potential conductors in Figure P12.4 using the FEM.

Figure P12.4

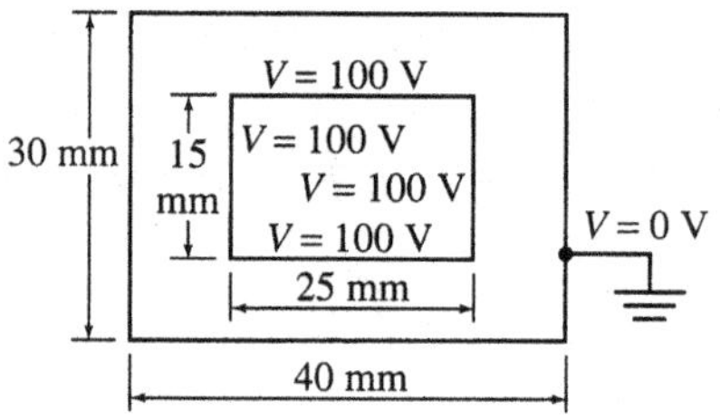

12.5 The stator and rotor teeth arrangement of an electromechanical device is given in Figure P12.5. Determine the distribution of scalar magnetic potential between the stator and rotor teeth and inside the slots of the given system, using the FEM. Assume that the magnetic regions are infinitely permeable.

Figure P12.5

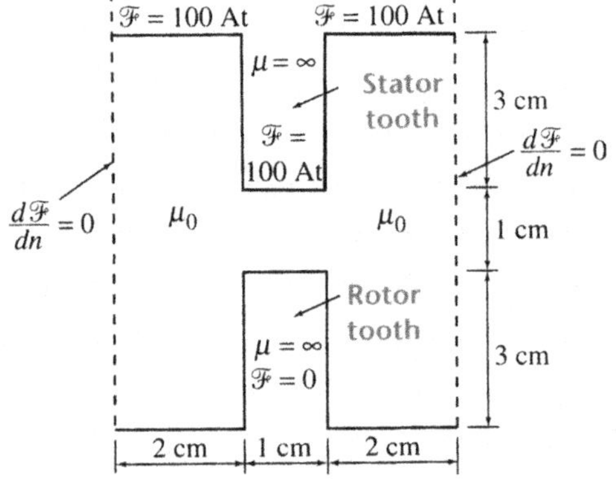

12.6 Calculate the potential distribution between the cylindrical conductor at 100 V and the metal plates at ground potential in Figure P12.6 using the FEM.

Figure P12.6

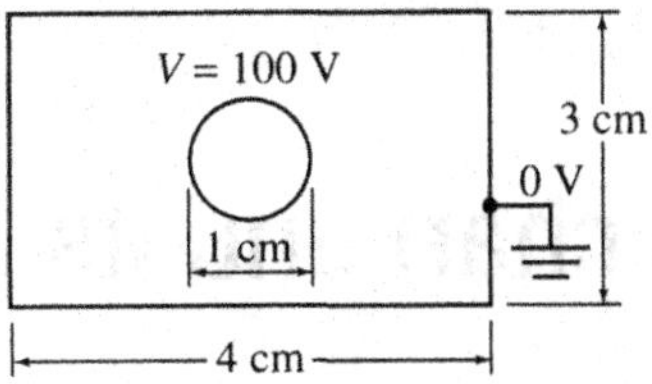

12.7 Determine the equipotential and electric field lines between the two round conductors given in Figure P12.7 using the method of moments.

Figure P12.7

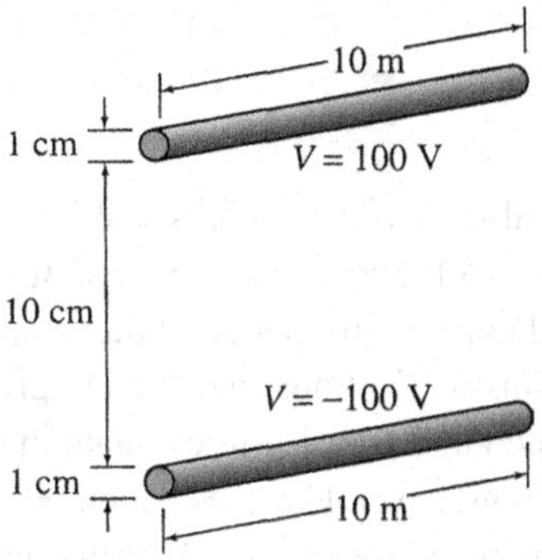

12.8 A cross-sectional view of a 100-mm-long microstrip is given in Figure P12.8. Determine the potential and electric field distributions between the conductors of this strip line using the method of moments.

Figure P12.8

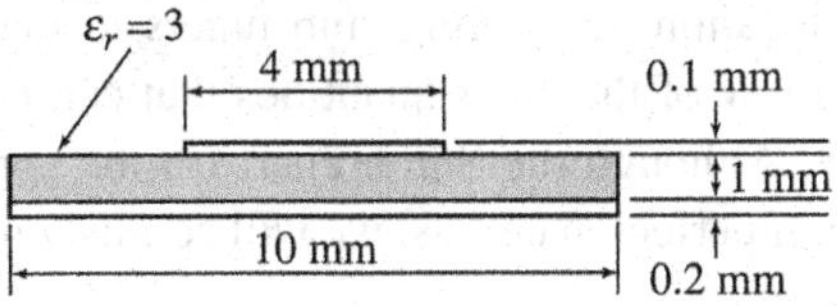

APPENDIX A

Smith chart and its applications

A.1 Introduction

The sinusoidal steady-state analysis of transmission lines requires calculations that involve complex numbers. Before the efficient usability of pocket calculators and digital computers, these calculations were quite time consuming and tedious. As a result, graphical analysis techniques were developed and adopted to evaluate the performance of transmission lines. Among several graphical methods, the Smith chart has gained the most popularity over the years. Although the computational speed is tremendously high with today's fast computers, the Smith chart still retains its popularity, mainly because it easily allows the user to have a quick physical interpretation of what is happening at any point along the transmission line. In addition to determining the input impedance at any location along the line, the voltage reflection coefficient, VSWR, and location for placing stub tuners to match transmission lines are but a few of the other quantities that can be obtained from the Smith chart. Although the Smith chart can be applied to transmission lines with imperfect materials, we will confine our study only to the lossless lines.

A.2 Smith chart

The Smith chart yields a relationship between the input impedance $\hat{Z}_{\text{in}}(z)$ at any point along a transmission line and the voltage reflection coefficient $\hat{\rho}(z)$ at that point. To obtain such a relationship we transform the input impedance from the impedance plane (RX plane) to the plane of the voltage reflection coefficient as we now explain.

The input impedance of a transmission line at any point, shown in Figure A.1, can be determined from (9.69) as

$$\hat{Z}_{\text{in}}(z) = \frac{\tilde{V}(z)}{\tilde{I}(z)} = \hat{Z}_c\left[\frac{1+\hat{\rho}(z)}{1-\hat{\rho}(z)}\right] \tag{A.1}$$

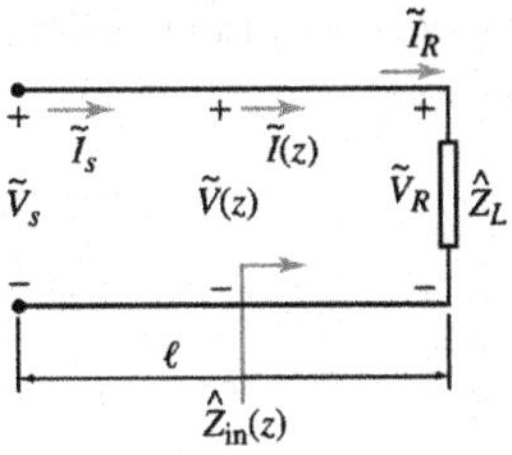

Figure A.1 A transmission line

in terms of the characteristic impedance $\hat{Z}_c$ of the line, and the voltage reflection coefficient $\hat{\rho}(z)$. Note that for a lossless line $\hat{Z}_c$ is purely resistive and is denoted by R_c. We can express $\hat{\rho}(z)$ as

$$\hat{\rho}(z) = \rho(z)\angle\phi = a + jb \tag{A.2}$$

The magnitude of the reflection coefficient traces a circle because $\rho^2(z) = a^2 + b^2$ is the equation of a circle centered at the origin with a radius of $\rho(z)$. Note that a and b are the projections of $\hat{\rho}(z)$ on the real and imaginary axes of the complex ρ plane shown in Figure A.2a. The voltage reflection coefficient varies between -1 and 1, hence the plane of $\hat{\rho}(z)$ is confined within the boundary of a unit circle, as illustrated in Figure A.2b. In other words, $0 < \rho(z) < 1$.

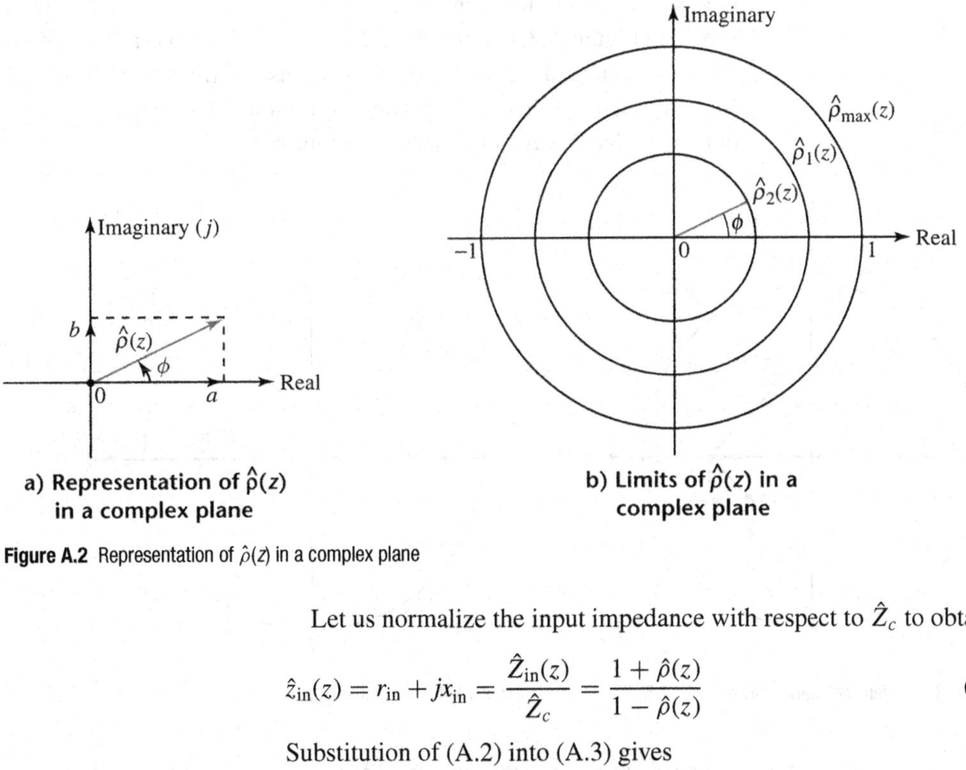

Figure A.2 Representation of $\hat{\rho}(z)$ in a complex plane

Let us normalize the input impedance with respect to $\hat{Z}_c$ to obtain

$$\hat{z}_{in}(z) = r_{in} + jx_{in} = \frac{\hat{Z}_{in}(z)}{\hat{Z}_c} = \frac{1 + \hat{\rho}(z)}{1 - \hat{\rho}(z)} \tag{A.3}$$

Substitution of (A.2) into (A.3) gives

$$r_{in} + jx_{in} = \frac{1 + a + jb}{1 - a - jb} \tag{A.4}$$

From the real and imaginary parts of (A.4) we can determine

$$r_{in} = \frac{1 - a^2 - b^2}{(1 - a)^2 + b^2} \tag{A.5}$$

and

$$x_{in} = \frac{2b}{(1 - a)^2 + b^2} \tag{A.6}$$

in terms of the real and imaginary parts of $\hat{\rho}(z)$. We can arrange (A.5) and (A.6) in such a way that we obtain two equations,

$$\left(a - \frac{r_{in}}{r_{in}+1}\right)^2 + b^2 = \frac{1}{(r_{in}+1)^2} \tag{A.7}$$

for constant r_{in}, and

$$(a-1)^2 + \left(b - \frac{1}{x_{in}}\right)^2 = \frac{1}{x_{in}^2} \tag{A.8}$$

for constant x_{in}. Each of the two equations above, (A.7) and (A.8), represents a circle in the $\hat{\rho}$ plane.

The circle given by equation (A.7) is traced for a constant r_{in} with a radius $1/(r_{in}+1)$, and a center located at $a = r_{in}/(r_{in}+1)$ and $b = 0$, as shown in Figure A.3. On the other hand, equation (A.8) is a circle of radius $1/x_{in}$ centered at $a = 1$ and $b = 1/x_{in}$, as constructed in Figure A.4 for constant x_{in}. When we superimpose Figure A.4 on Figure A.3, we obtain the *Smith chart* illustrated in Figure A.5.

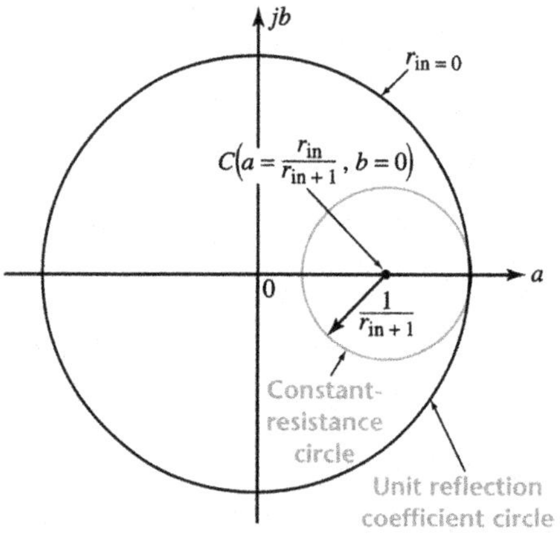

Figure A.3 Constant-resistance circle

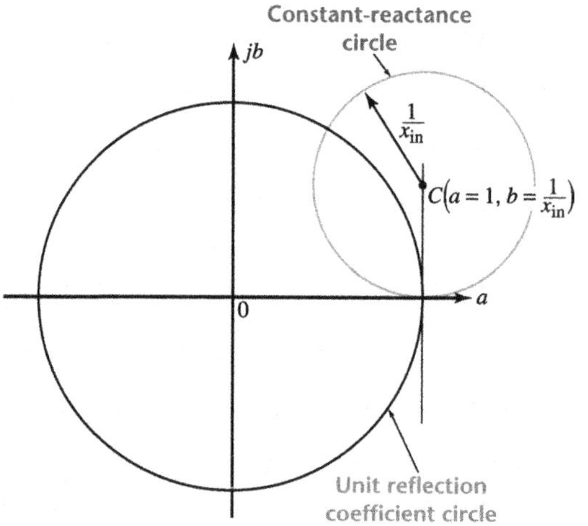

Figure A.4 Constant-reactance circle

The intersection of the two curves for r_{in} and x_{in} at point P on the Smith chart in Figure A.5 is the normalized input impedance at that point of the transmission line. Because x_{in} can be positive (inductive impedance) or negative (capacitive impedance), the curves for x_{in} are drawn for both positive and negative values of x_{in}. The Smith chart shown in Figure A.6 (page 624) is obtained by drawing circles in accordance with (A.7) and (A.8) for various values of r_{in} and x_{in}. In Figure A.5, the length of the radial line from the center of the $\hat{\rho}$ plane to the point P gives the magnitude of the reflection coefficient. The phase angle is measured with respect to the real axis. If D_1 is the radial distance from the center

Figure A.5 Resistance and reactance circles

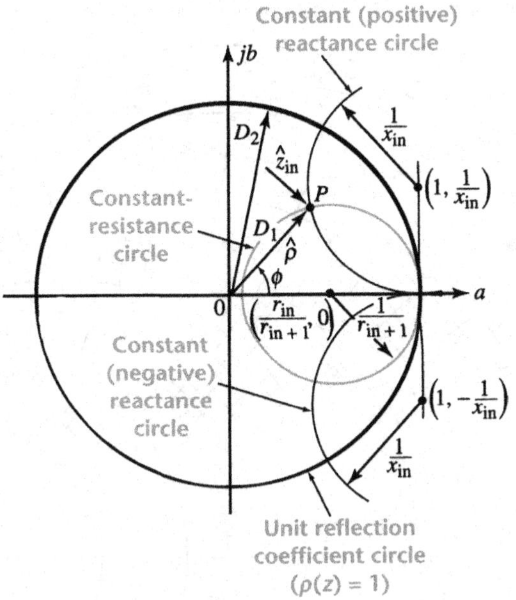

to the point P, and D_2 is the radial distance of the $\rho = 1$ circle, then

$$\rho(z) = \frac{D_1}{D_2}$$

The angular scale outside the periphery of the standard Smith chart given in Figure A.6 is for the angle ϕ of the reflection coefficient $\hat{\rho}(z) = \rho(\ell)e^{j[2\beta(z-\ell)-\theta]}$. Because angle ϕ in (A.2) is related to $2\beta(z - \ell)$, ϕ can be also expressed in terms of the length of the transmission line as a fraction, or as a multiple of the signal wavelength. In the Smith chart shown in Figure A.6, there are two outer scales besides the angular scale. The one labeled as Wavelengths Toward Generator is used when the observation point moves toward the generator along the transmission line. The other one, labeled as Wavelengths Toward Load, is used when the observation point on the transmission line moves toward the load. The following examples are intended to illustrate the use of the Smith chart for lossless transmission lines.

EXAMPLE A.1

Locate the following normalized impedances in the Smith chart: $\hat{z}_1 = 0.3 + j0.1$, $\hat{z}_2 = 0.2 - j0.3$, $\hat{z}_3 = j0.4$, $\hat{z}_4 = 0.6$, $\hat{z}_5 = 0 + j0$, $\hat{z}_6 = \infty + j\infty$.

Solution In order to locate the normalized impedance $\hat{z}_1 = 0.3 + j0.1$ on the Smith chart (Figure A.7) we proceed as follows. Locate $r_{in} = 0.3$ circle. Then locate the arc for $x_{in} = 0.1$ circle. The intersections of these circles marks the location of $\hat{z}_1 = 0.3 + j0.1$ impedance. The locations of the other normalized impedances are also marked in a similar fashion.

• • •

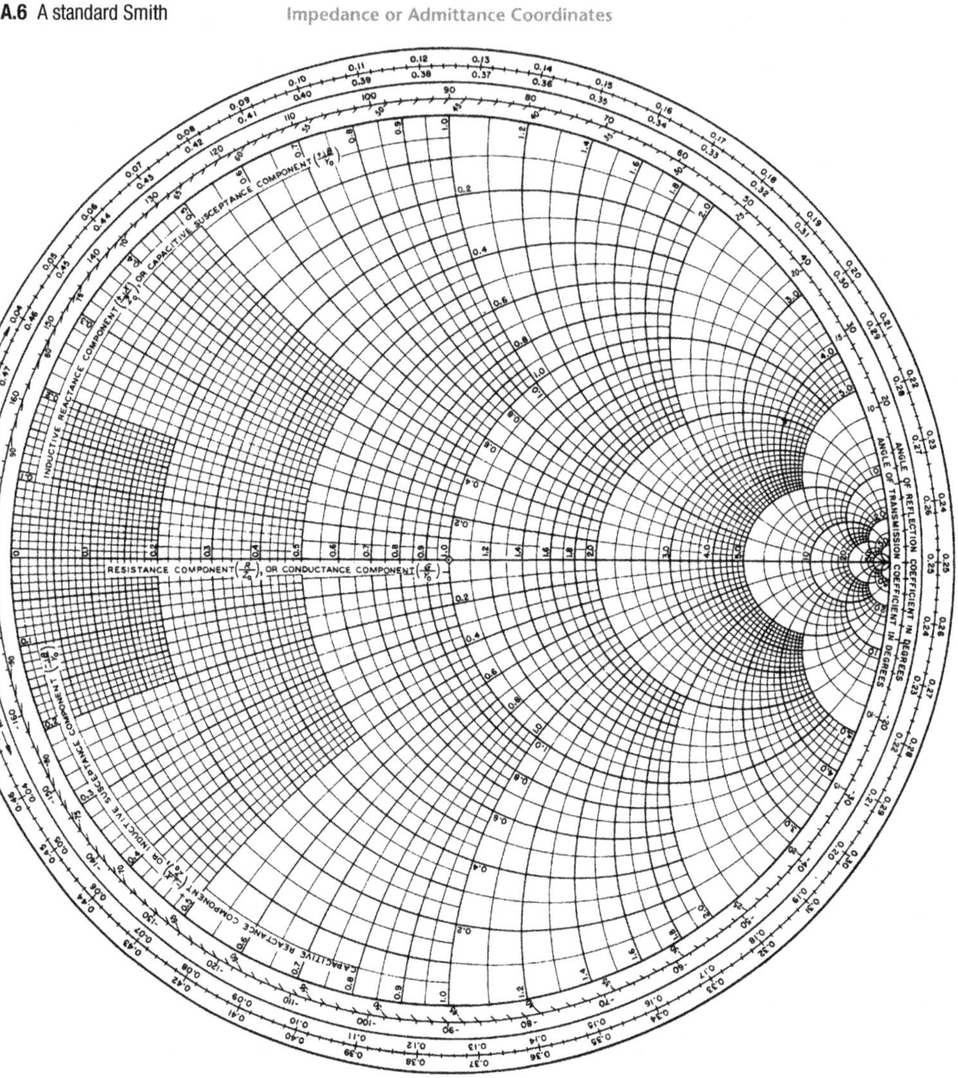

Figure A.6 A standard Smith chart

EXAMPLE A.2

The input impedance of a 50-Ω, 60-m-long transmission line is $\hat{Z}_{in} = 50 + j50\ \Omega$ at $z = 50$ m. Determine the voltage reflection coefficient at this point.

Solution The normalized input impedance is

$$\hat{z}_{in} = \frac{50 + j50}{50} = 1 + 1j$$

and can be located on the Smith chart as the intersection of the $r_{in} = 1$ and $x_{in} = 1$ circles, as shown in Figure A.8. By measuring the radial

Figure A.7 Smith chart for Example A.1

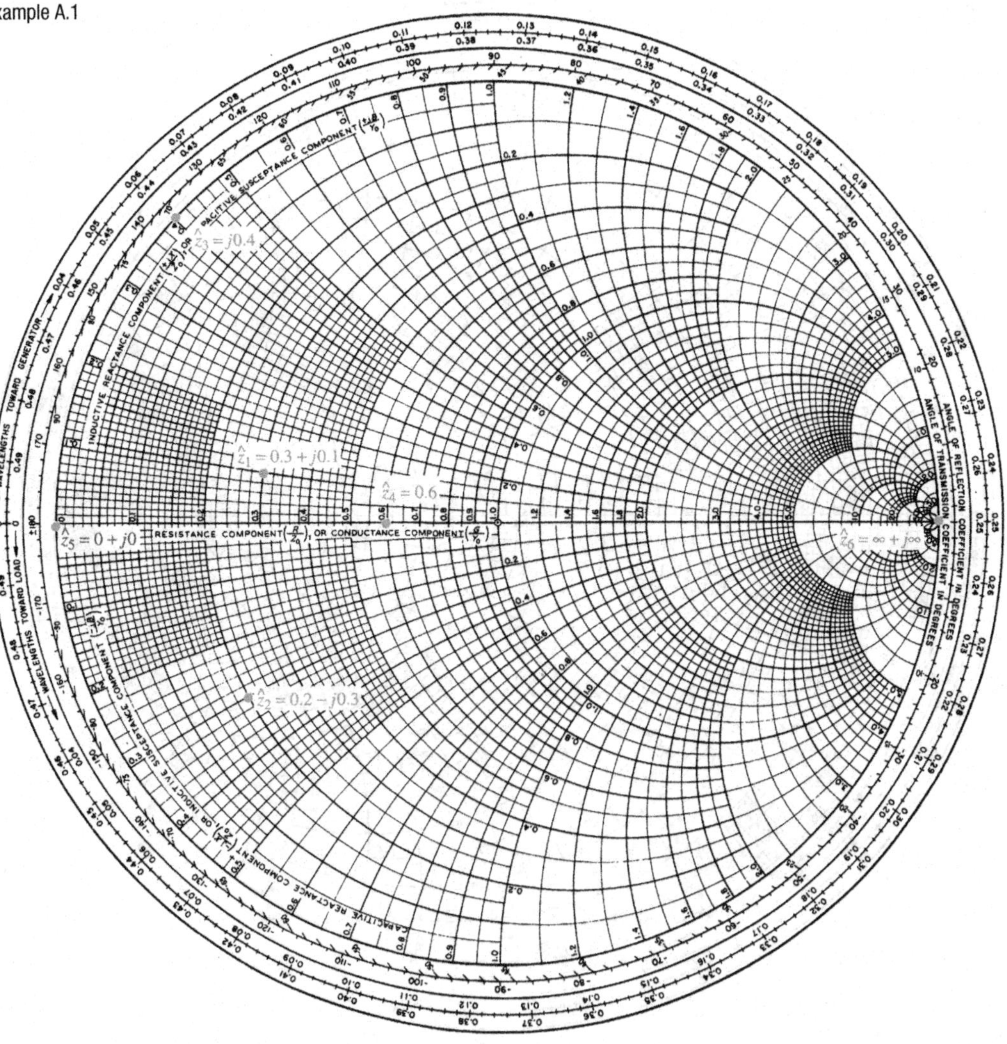

distances*, we find the reflection coefficient to be

$$\rho(50\,\text{m}) = \frac{D_1}{D_2} = \frac{38\,\text{mm}}{84\,\text{mm}} = 0.45$$

The angle ϕ of the reflection coefficient; i.e., the angle between the a axis (the real axis) and the radial line passing through $\hat{z}_{\text{in}} = 1 + j1$, is 63°. Thus the reflection coefficient is $\hat{\rho}(50\,\text{m}) = 0.45\underline{/63°}$.

Let us verify this result using the exact equation for the reflection coefficient. From (A.1) we can write

$$50 + 50j = 50\,\frac{1 + \hat{\rho}(50\,\text{m})}{1 - \hat{\rho}(50\,\text{m})}$$

* The radial distance measurements may differ from those used in the text due to the Smith chart's size.

Figure A.8 Smith chart for Example A.2

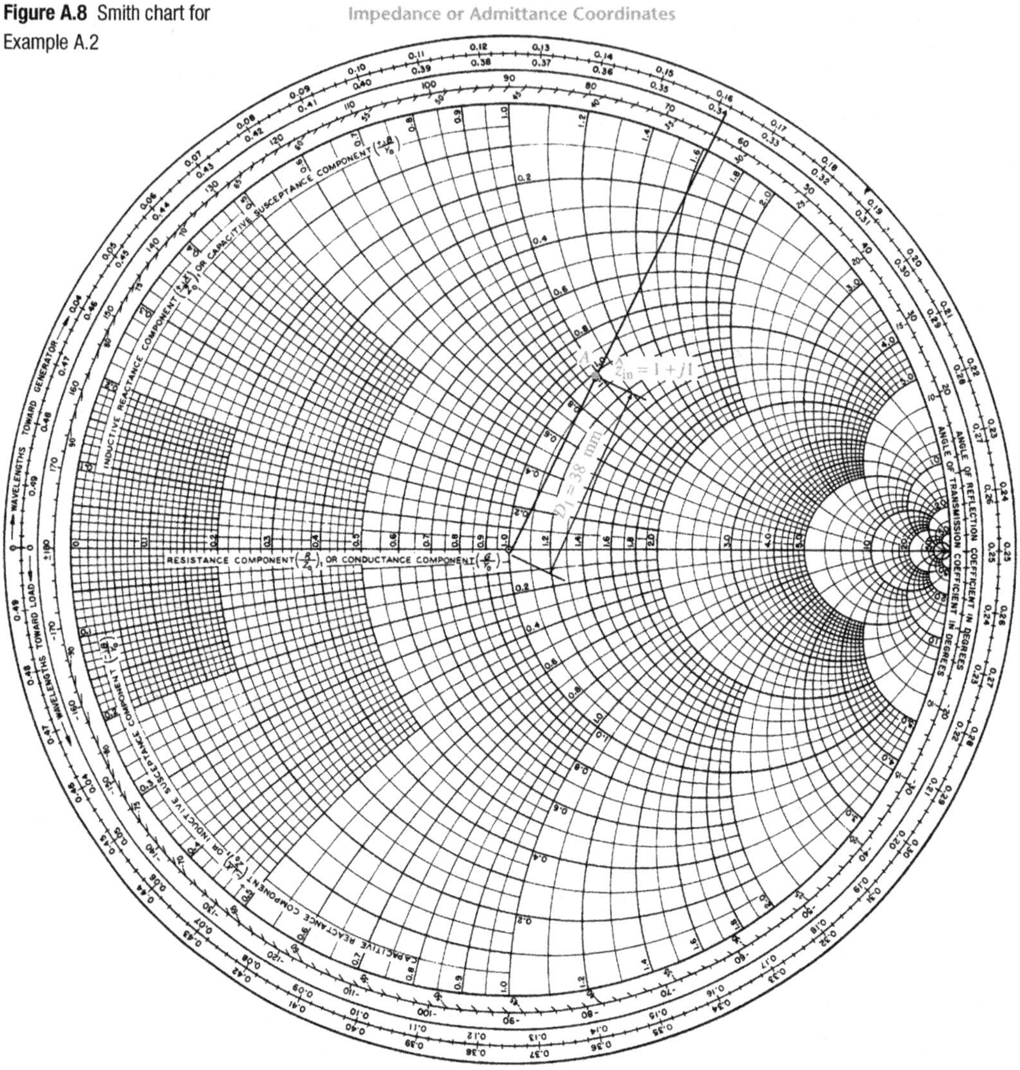

or

$$\hat{\rho}(50\,\text{m}) = \frac{j50}{100 + j50}$$
$$= \frac{j50}{2 + j}$$
$$= 0.447\underline{/63.43^\circ}$$

• • •

EXAMPLE A.3

The input impedances of a $\lambda/8$-long, 50-Ω transmission line are $\hat{Z}_1 = 25 + j100\ \Omega$, $\hat{Z}_2 = 10 - j50\ \Omega$, $\hat{Z}_3 = 100 + j0\ \Omega$, and $\hat{Z}_4 = 0 + j50\ \Omega$ when various load impedances are connected at the other end. In each case, determine the load impedance and the reflection coefficient at the input and load ends.

Solution The normalized input impedances are

$$\hat{z}_1 = \frac{25 + j100}{50} = 0.5 + j2, \qquad \hat{z}_2 = \frac{10 - j50}{50} = 0.2 + j0.1$$

$$\hat{z}_3 = \frac{100 + j0}{50} = 2 + j0, \qquad \hat{z}_4 = \frac{0 + j50}{50} = 0 + j1$$

and are plotted on the Smith chart as shown in Figures A.9a, b, c, and d.

As the magnitude of the reflection coefficient is the same at any point along a lossless transmission line, a circle can be drawn for each different case. The impedance at any point on the line lies on this circle. In this problem, because load impedances and the receiving-end reflection coefficients are required, we use the scale Wavelengths Toward Load and follow the corresponding reflection coefficient circle in the counterlockwise direction for $\lambda/8$. From Figure A.9a–d the normalized load impedances are $\hat{z}_{L1} = 0.11 + j0.35$, $\hat{z}_{L2} = 0.45 - j1.17$, $\hat{z}_{L3} = 0.8 + j0.6$, and $\hat{z}_{L4} = 0 + j0$. The actual load impedances can be obtained from $\hat{Z}_L = \hat{Z}_c \hat{z}_L$. Thus,

$$\hat{Z}_{L1} = 5.5 + j17.5\ \Omega, \qquad \hat{Z}_{L2} = 22.5 - j56.5\ \Omega$$

$$\hat{Z}_{L3} = 40 + j30\ \Omega, \qquad \hat{Z}_{L4} = 0 + j0\ \Omega$$

The voltage reflection coefficients at the sending end are

$$\hat{\rho}_{S1} = 0.83\underline{/51^\circ}, \quad \hat{\rho}_{S2} = 0.67\underline{/192^\circ}, \quad \hat{\rho}_{S3} = 0.34\underline{/0^\circ}, \quad \hat{\rho}_{S4} = 1\underline{/90^\circ}$$

Those at the receiving end are

$$\hat{\rho}_{R1} = 0.83\underline{/141^\circ}, \quad \hat{\rho}_{R2} = 0.67\underline{/-78^\circ}, \quad \hat{\rho}_{R3} = 0.34\underline{/90^\circ},$$

$$\hat{\rho}_{R4} = 1\underline{/180^\circ}$$

The theoretical calculations of the load impedances and the reflection coefficients are left to the reader as an exercise. • • •

EXAMPLE A.4

Determine the load impedance and the reflection coefficient at the receiving end of the transmission line given in Example A.2 if the transit time of the line is 540 ns, and the operating frequency is 10 MHz.

Solution From Chapter 9, equation (9.71), the reflection coefficient is

$$\hat{\rho}(z) = \hat{\rho}_R e^{-j2\beta(\ell - z)}$$

In Example A.2 we determined the reflection coefficient at $z = 50$ m as

$$\hat{\rho}(50 \text{ m}) = 0.45\underline{/63^\circ}$$

The receiving-end reflection coefficient is

$$\hat{\rho}_R = \hat{\rho}(z) e^{j2\beta(\ell - z)}$$

Figure A.9a Smith chart for Example A.3

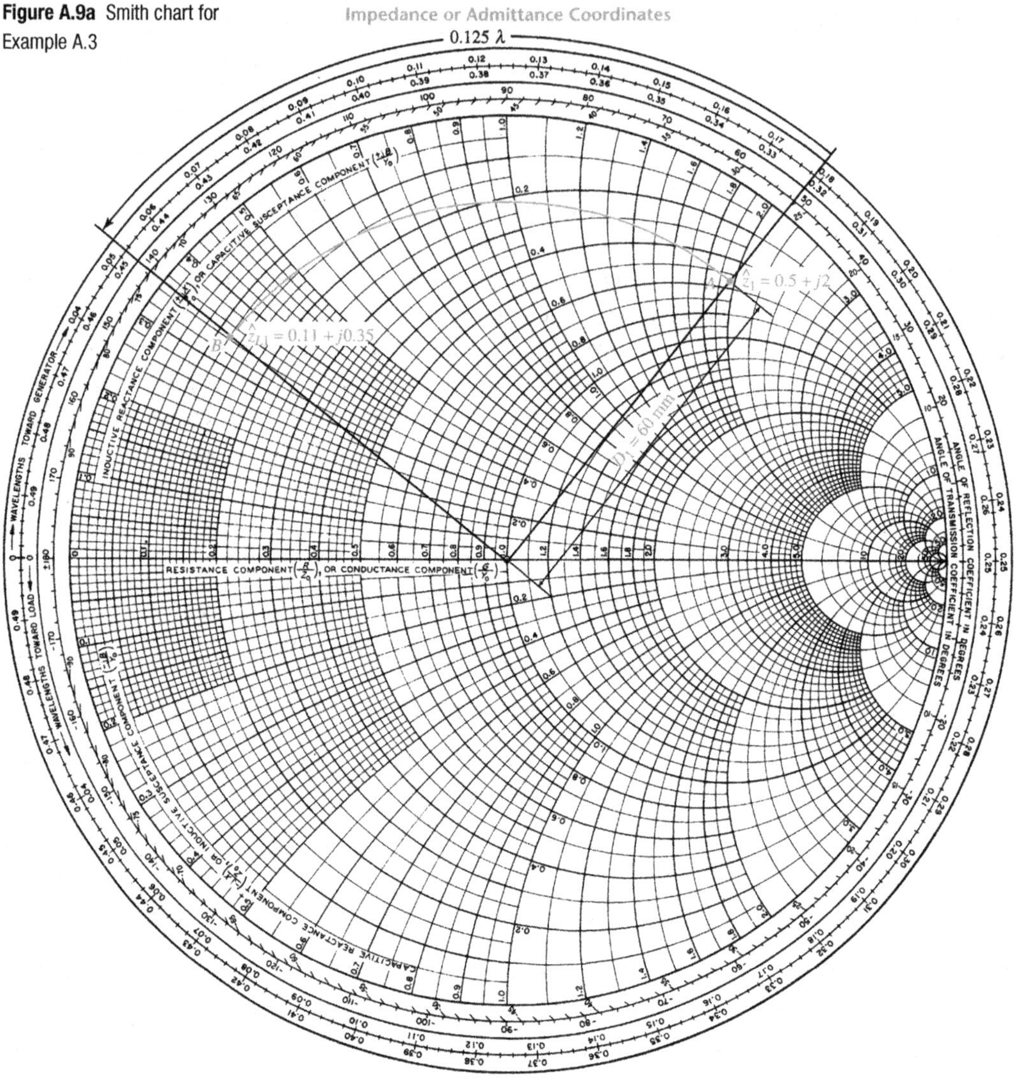

(a) $\hat{Z}_1 = 25 + j100\ \Omega$

The phase constant is

$$\beta = \frac{\omega}{u_p} = \frac{2\pi \times 10 \times 10^6}{60/540 \times 10^{-9}} = 565.9 \times 10^{-3}\ \text{rad/m}$$

The wavelength is

$$\lambda = \frac{2\pi}{\beta} = \frac{2\pi}{565.49 \times 10^{-3}} = 11.11\ \text{m}$$

and

$$\frac{\ell - z}{\lambda} = \frac{60 - 50}{11.11} \Rightarrow (\ell - z) = 0.9\lambda$$

Figure A.9b

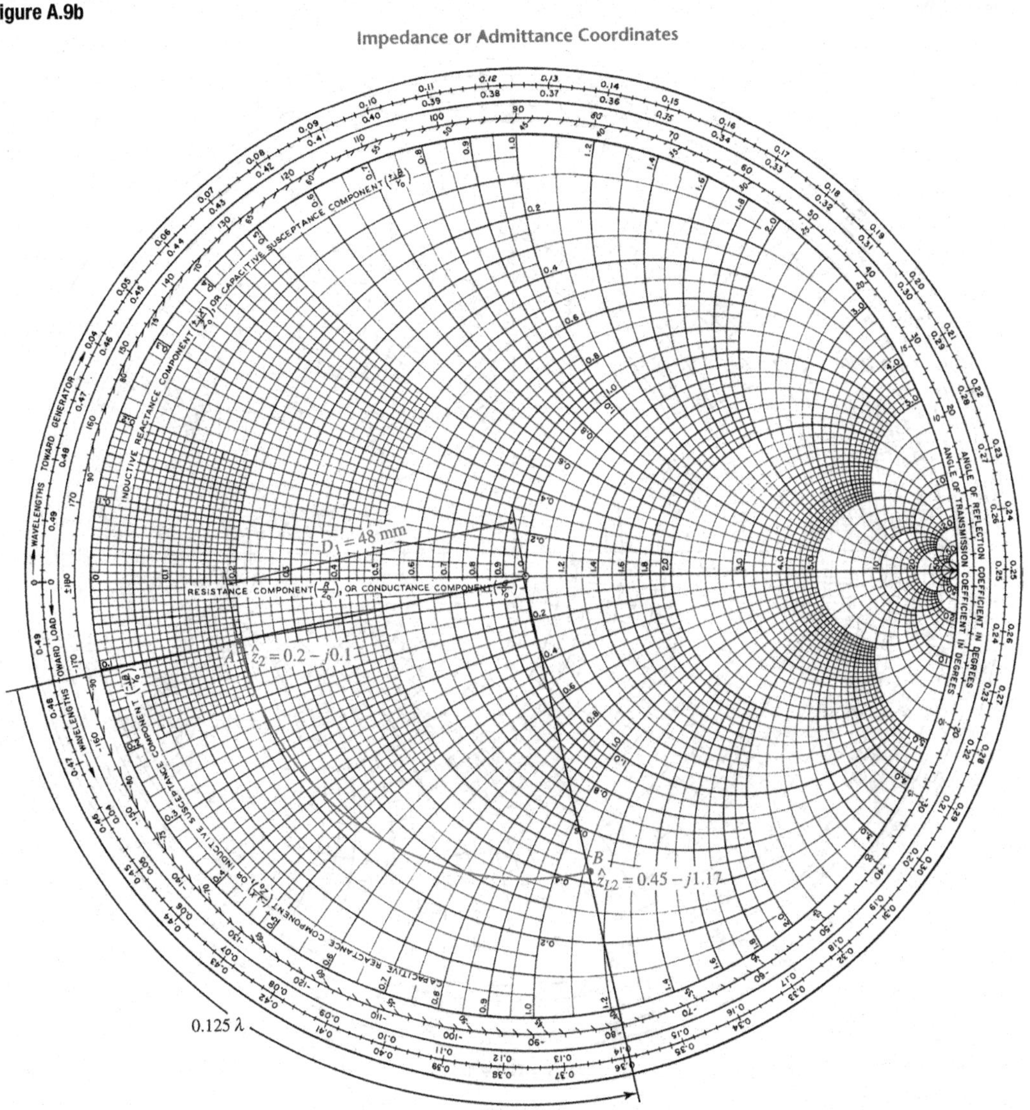

(b) $\hat{Z}_2 = 10 - j50\ \Omega$

As the magnitude of the reflection coefficient is the same at any point on the transmission line, we can draw a circle corresponding to $\rho = 0.45$. The impedance at any point z on the transmission line must lie on this circle.

Because the reflection coefficient is required at the receiving end, we will use the scale Wavelengths Toward Load and follow the $\rho = 0.45$ circle in the counterclockwise direction for 0.9λ. This can easily be done as follows.

a) Read the present value of Wavelengths Toward Load from the chart. In this case, it is 0.338λ.

b) Now add 0.9λ to this value to obtain 1.238λ.

Figure A.9c

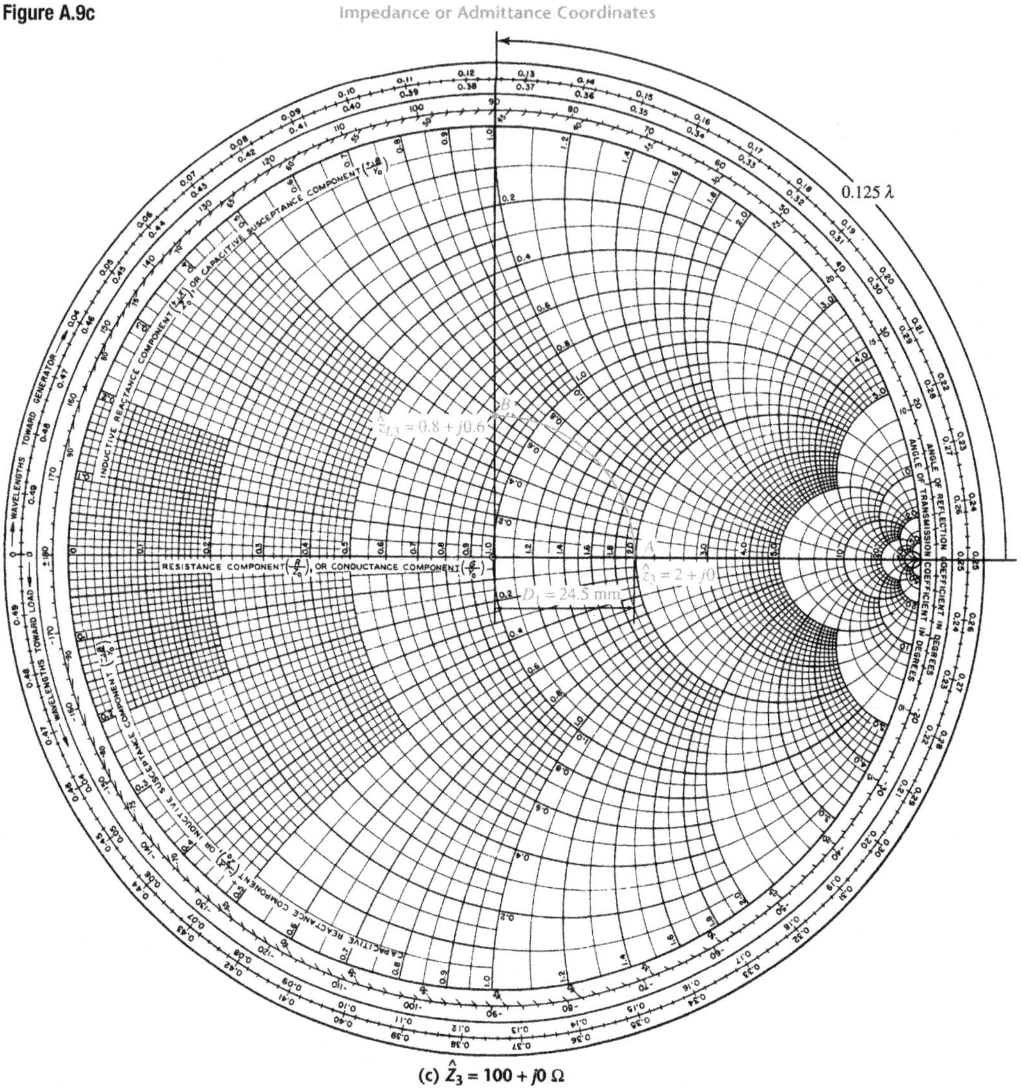

(c) $\hat{Z}_3 = 100 + j0\ \Omega$

c) Since the chart repeats every 0.5λ, subtract $n(0.5\lambda)$ from the total distance to obtain the distance less than 0.5λ. In this case, we will subtract 1λ corresponding to $n = 2$ and obtain 0.238λ.

d) Locate 0.238λ on the chart, move toward the load, and draw a radial line.

e) The intersection of the radial line with the $\rho = 0.45$ circle gives the normalized load impedance and the reflection coefficient at that point.

From the graph, the reflection coefficient at the receiving end is $\hat{\rho}_R = 0.45\underline{/-8^\circ}$. The normalized load impedance is determined to be $\hat{z}_L = 2.5 - j0.4$, as indicated in Figure A.10 (page 632). The actual value of the load impedance is $\hat{Z}_L = 50(2.5 - j0.4) = 125 - j20\ \Omega$. • • •

Figure A.9d

Impedance or Admittance Coordinates

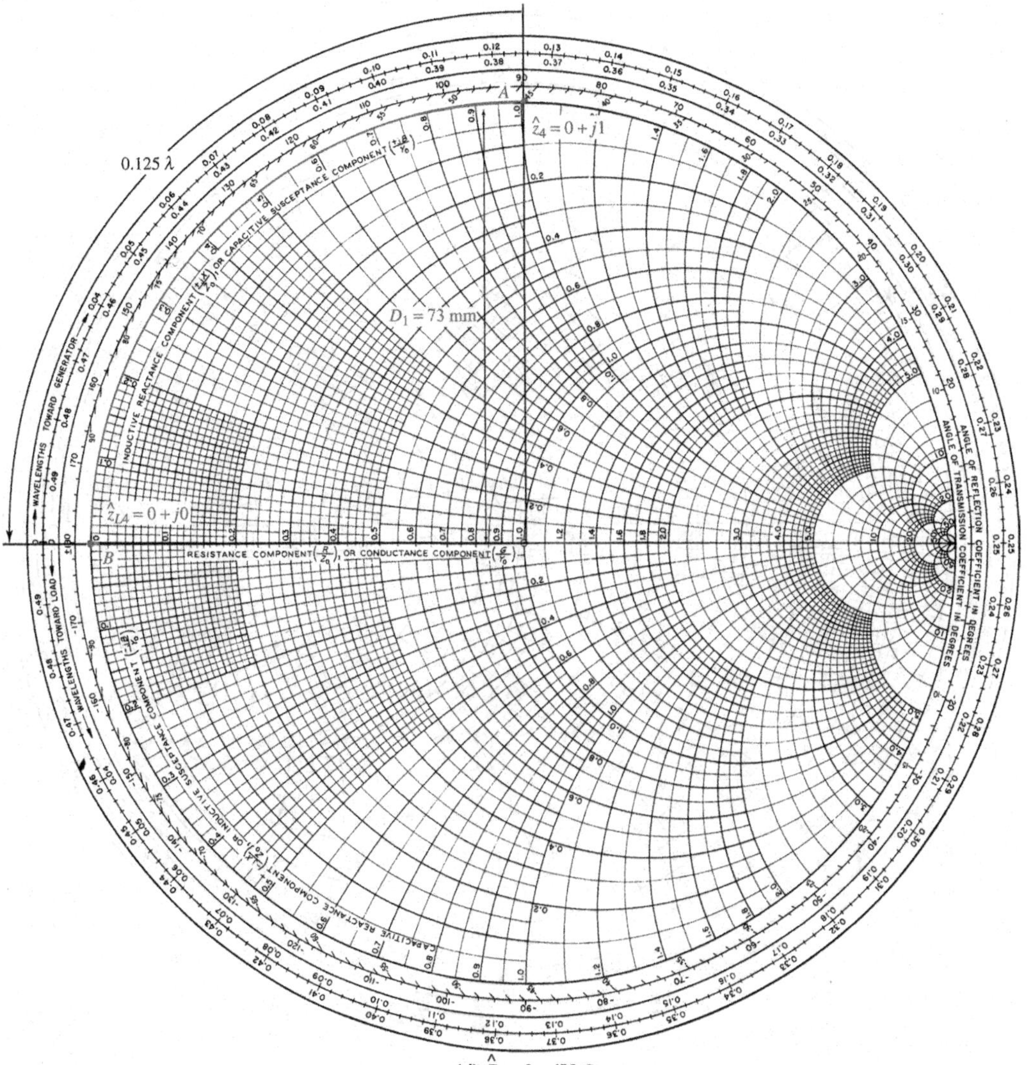

(d) $\hat{Z}_A = 0 + j50\ \Omega$

A.3 Determination of VSWR using the Smith chart

In Chapter 9 we derived an expression for the voltage standing wave ratio (VSWR) in a transmission line as

$$VSWR = \frac{1 + \rho_R}{1 - \rho_R} \tag{A.9}$$

When there is a voltage maximum at a certain location in a transmission line, then

$$1 + \hat{\rho}(z) = 1 + \rho(z) \tag{A.10}$$

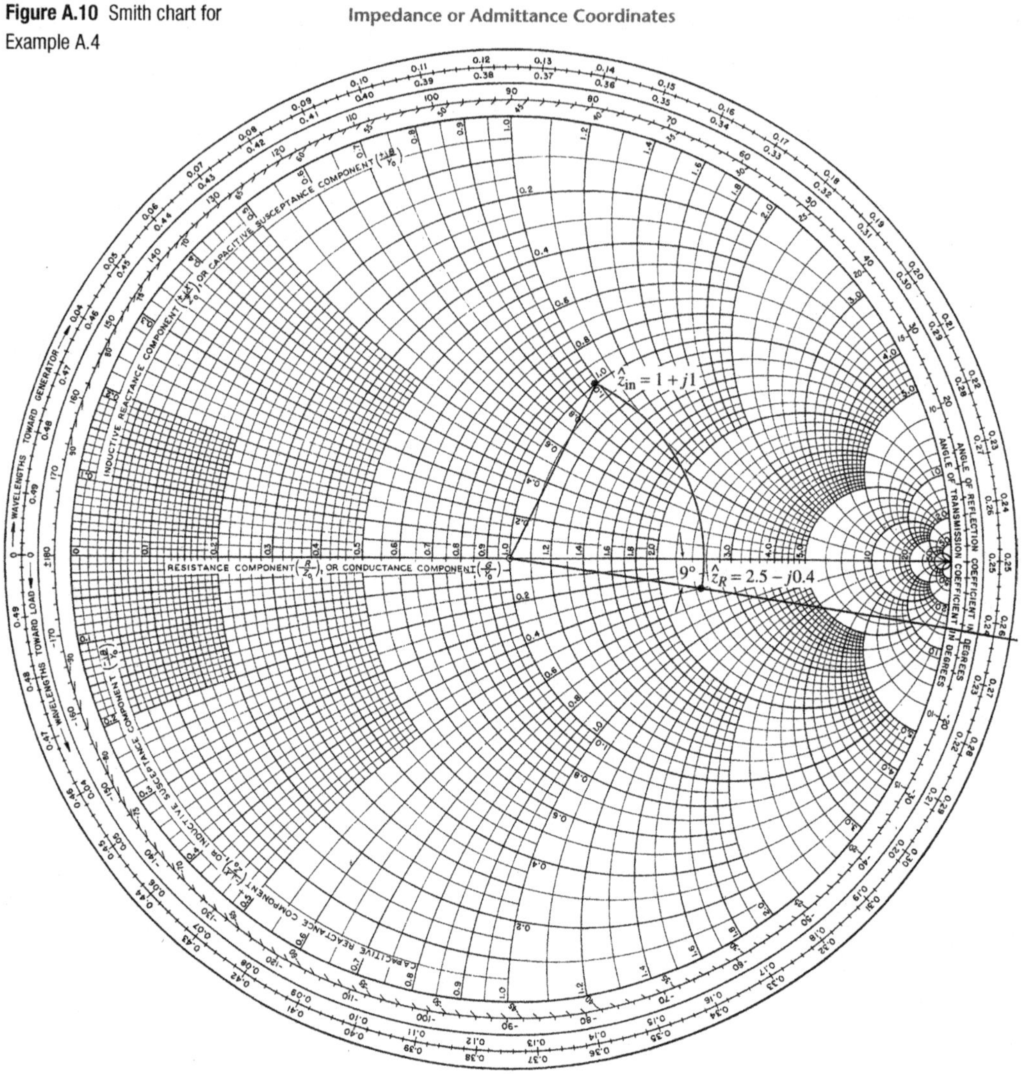

Figure A.10 Smith chart for Example A.4

In a lossless transmission line we know that the magnitude of the reflection coefficient is the same at every point along the transmission line; that is,

$$\rho(z) = \rho_R \tag{A.11}$$

where ρ_R is the magnitude of the reflection coefficient at the receiving end of the line. Also, at the location along the transmission line where there is a voltage maximum, the normalized input impedance is written as

$$\hat{z}_{\text{in}} = \frac{1+\rho(z)}{1-\rho(z)} = \frac{1+\rho_R}{1-\rho_R} = r_{\max} \tag{A.12}$$

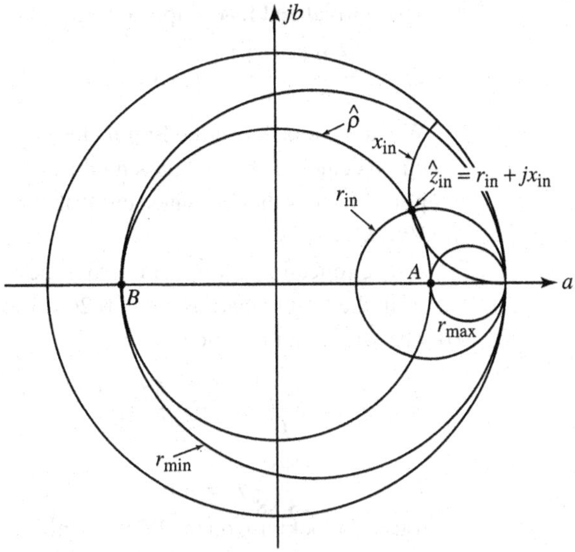

Figure A.11 Determination of VSWR from Smith chart

The normalized input impedance in (A.12) is a real quantity, and yields the maximum normalized input resistance. From (A.9) and (A.12), it is evident that

$$VSWR = r_{max} \tag{A.13}$$

After locating the normalized input impedance $\hat{z}_{in}$ shown in Figure A.11, we draw the voltage reflection coefficient circle that passes through $\hat{z}_{in}$. This circle intersects the a axis at points A and B, as shown in the figure. The resistance circle tangent to point A yields r_{max} and the VSWR is equal to r_{max}. The smaller resistance at point B is r_{min}, which is the reciprocal of r_{max} and corresponds to the voltage minimum. Thus, we can also state that $VSWR = 1/r_{min}$.

EXAMPLE A.5

A 50-m-long, 75-Ω transmission line with a 0.18-μs transit time is terminated in a load operating at 20 MHz. The voltage and current magnitudes are recorded as 50 V and 1 A respectively, at the load, and the current was observed to lead the voltage by an angle of 20°. (a) Determine the VSWR in the transmission line and the voltage reflection coefficient across the terminals of the load. (b) Calculate the input impedance and the corresponding voltage reflection coefficient at the sending end of the line. Assume that the transmission line is lossless and that the input impedances of the measuring devices do not cause any measurement errors.

Solution a) The load impedance at the receiving end of the transmission line is

$$\hat{Z}_L = \frac{50\underline{/0^\circ}}{1\underline{/20^\circ}} = 50\underline{/-20^\circ} = 46.98 - j17.10\ \Omega$$

The normalized load impedance is

$$\hat{z}_L = \frac{46.98 - j17.10}{75} = 0.63 - j0.23$$

When we plot $\hat{z}_L$ on the Smith chart in Figure A.12 (page 635), the circle centered at 0 and passing through $\hat{z}_L$ intersects the real axis at point A. Thus the resistance circle that passes through point A yields $VSWR = 1.72$.

The reflection coefficient across the load terminals is determined from the Smith chart as $\hat{\rho}_R = 0.26\underline{/-141^\circ}$.

b) The wavelength of the line is

$$\lambda = \frac{u_p}{f} = \frac{\ell/t_t}{f} = \frac{50/0.18 \times 10^{-6}}{20 \times 10^6} = 13.88 \text{ m}$$

Thus, $\ell = \frac{50}{13.88}\lambda = 3.6\lambda$. In the Smith chart given in Figure A.12, rotating clockwise on the Wavelengths Toward Generator scale, 3.6λ (or 0.1λ) yields a normalized input impedance of $\hat{z}_{in} = 0.62 + j0.19$.

The actual input impedance is

$$\hat{Z}_{in} = 75(0.62 + j0.19) = 46.5 + j14.25\ \Omega$$

The voltage reflection coefficient at the input is determined similarly to step (a) as $\hat{\rho}_{in} = 0.26\underline{/147^\circ}$. • • •

A.4 Admittance of an impedance using the Smith chart

The normalized input impedance at a point in a transmission line is expressed as

$$\hat{z}_{in}(z) = \frac{1 + \hat{\rho}(z)}{1 - \hat{\rho}(z)} \tag{A.14}$$

in terms of the reflection coefficient $\hat{\rho}(z) = \rho(z)\underline{/\phi}$. Because $\hat{y}_{in}(z) = 1/\hat{z}_{in}(z)$, we can also write

$$\hat{y}_{in}(z) = \frac{1 - \rho(z)\underline{/\phi}}{1 + \rho(z)\underline{/\phi}}$$

or

$$\hat{y}_{in}(z) = \frac{1 + \rho(z)\underline{/\phi + 180^\circ}}{1 - \rho(z)\underline{/\phi + 180^\circ}} = \frac{1 + \rho(z)\underline{/\theta}}{1 - \rho(z)\underline{/\theta}} \tag{A.15}$$

where $\theta = \phi + 180^\circ$.

From (A.14) and (A.15) it is clear that $\hat{y}_{in}(z)$ lies on the ρ circle at a point 180° away from $\hat{z}_{in}(z)$. Thus $\hat{y}_{in}(z)$ can be easily obtained from

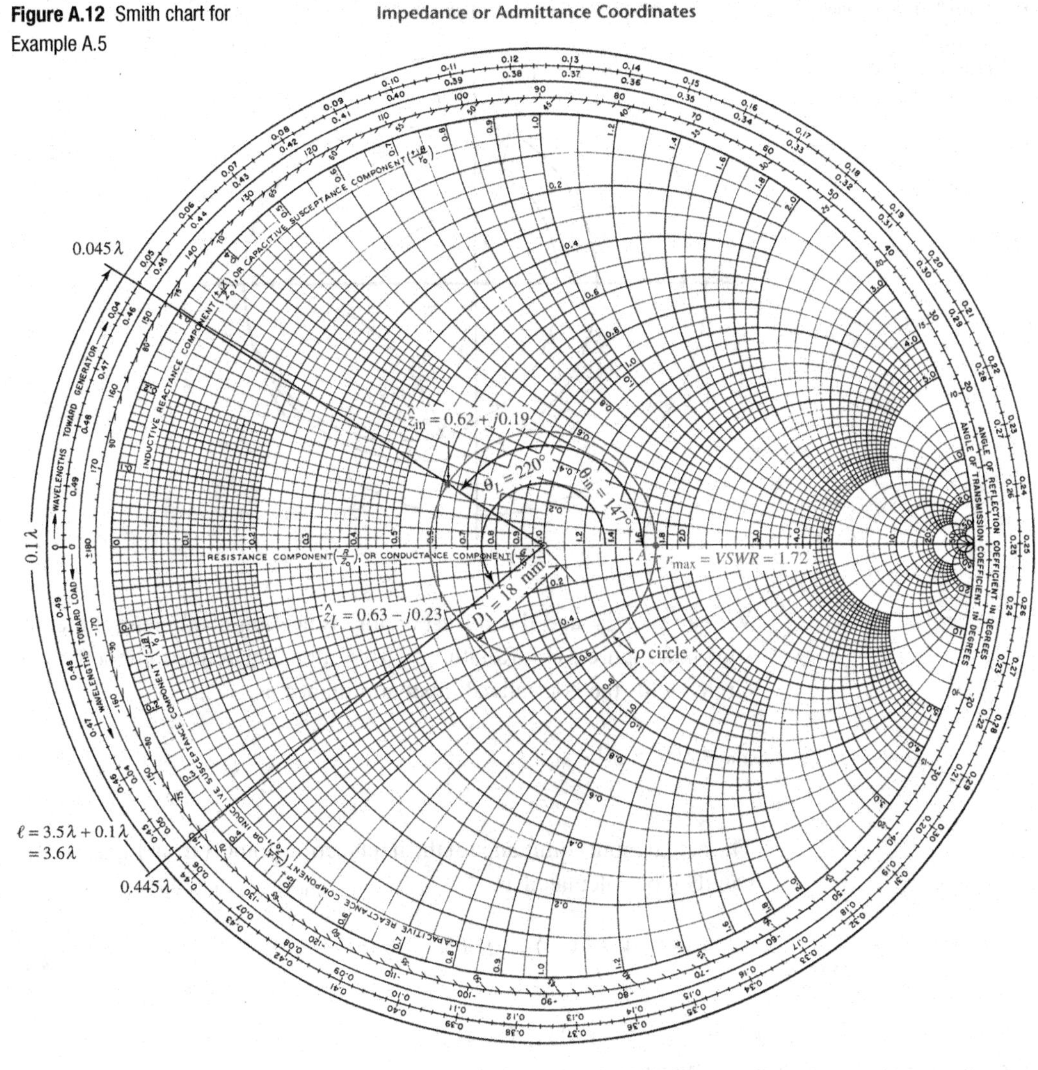

Figure A.12 Smith chart for Example A.5

the Smith chart by considering the resistance and reactance circles as, respectively, the normalized conductance g_{in} and susceptance b_{in} of the input admittance $\hat{y}_{\text{in}}(z)$, as shown in Figure A.13.

EXAMPLE A.6

The input impedance of a 50-Ω transmission line is given as $\hat{Z}_{\text{in}} = 100 + j50\ \Omega$ at the sending end of the line. Determine the corresponding admittance.

Solution The normalized input impedance is

$$\hat{z}_{\text{in}} = \frac{100 + j50}{50} = 2 + j1$$

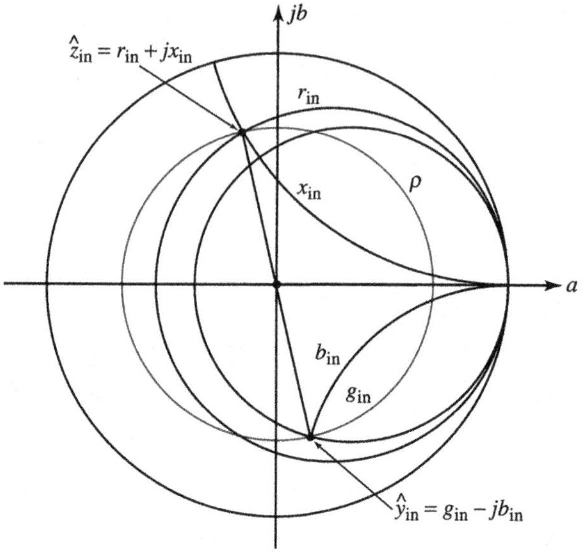

Figure A.13 Determination of the admittance of an impedance using the Smith chart

We can plot $\hat{z}_{in}$ and the corresponding reflection-coefficient circle, as shown in Figure A.14. Rotating $\hat{z}_{in}$ 180° along the reflection-coefficient circle yields the normalized input admittance as

$$\hat{y}_{in} = 0.4 - j0.2$$

The actual input admittance at the intersection point *A* from Figure A.14 can then be calculated as

$$\hat{Y}_{in} = \frac{1}{50}(0.4 - j0.2) = 0.008 - j0.004 \text{ S}$$

• • •

A.5 Impedance matching with shunt stub lines

In Chapter 9 we observed that the standing waves occur along a transmission line when it is connected to a load having an impedance other than the characteristic impedance of the line. We also know that the standing waves disappear when the load is perfectly matched to the transmission line.

In Chapter 9 we also discussed how to match a load to the transmission line using a short-circuited stub line. The length of this short-circuited stub line and its location along the transmission line are the two key parameters that help us achieve the matching. These two parameters can be conveniently obtained from the Smith chart. When the matching is accomplished, the input impedance at the connection terminals will be exactly equal to the characteristic impedance of the line.

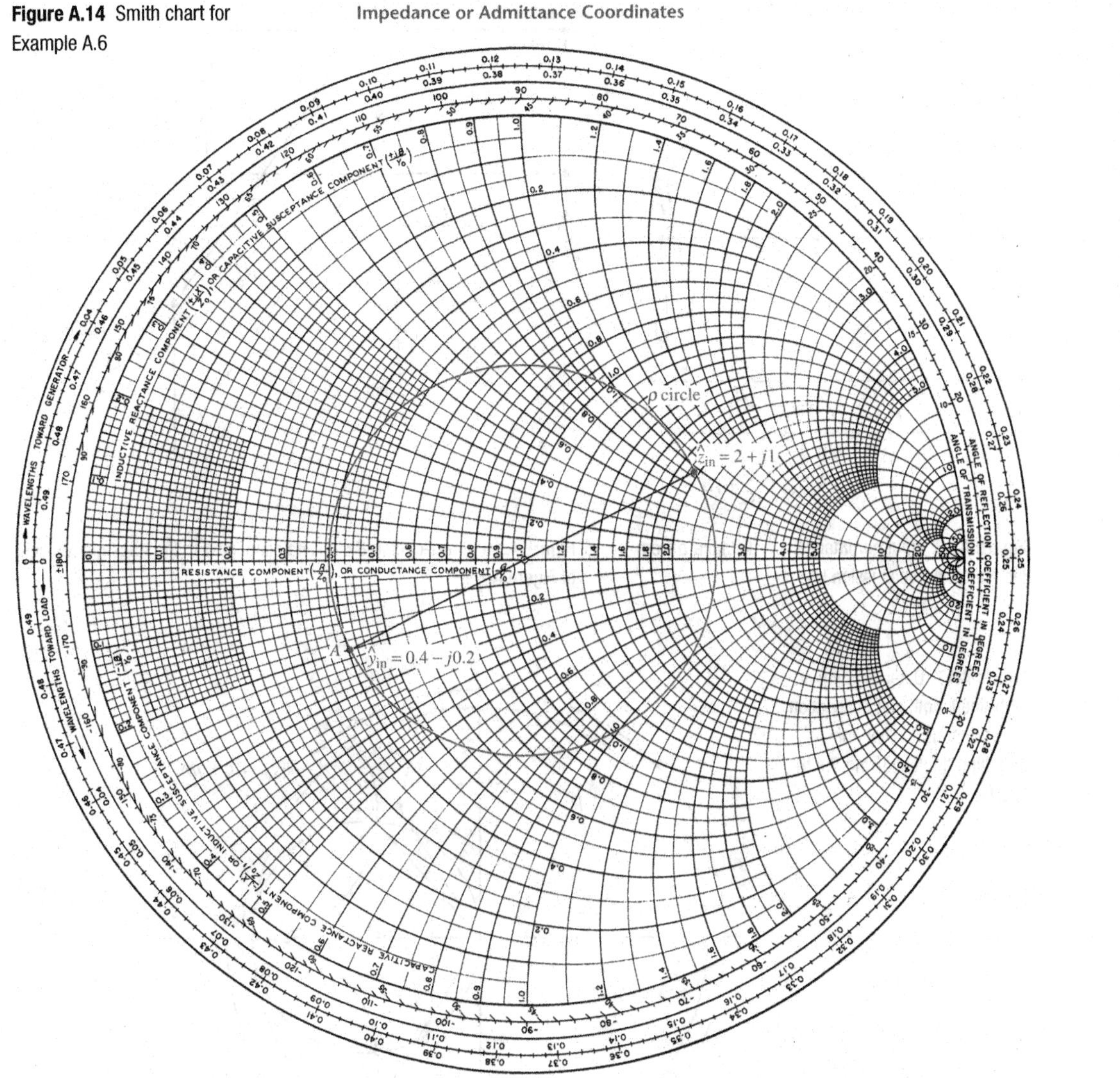

Figure A.14 Smith chart for Example A.6

Let us assume that the input admittance at a distance d from the load in a lossless transmission line is

$$\hat{Y}_{\text{line}} = \frac{1}{R_c} + jB$$

as indicated in Figure A.15. A stub line of input admittance $\hat{Y}_{\text{stub}} = -jB$ is to be connected, as shown in Figure A.16, at point D in order to match the load to the transmission line. Consequently, the input impedance at point D equals the characteristic impedance of the transmission line.

Let us now determine the length ℓ_s, of a short-circuited stub line and its location on the transmission line using the Smith chart.

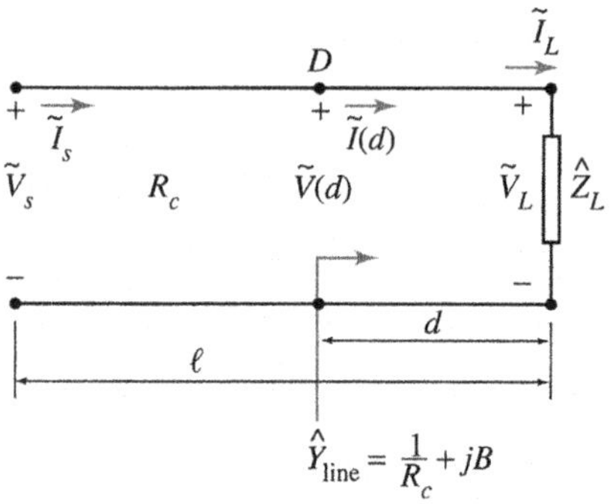

Figure A.15 Input admittance at a distance *d* from the load in a transmission line

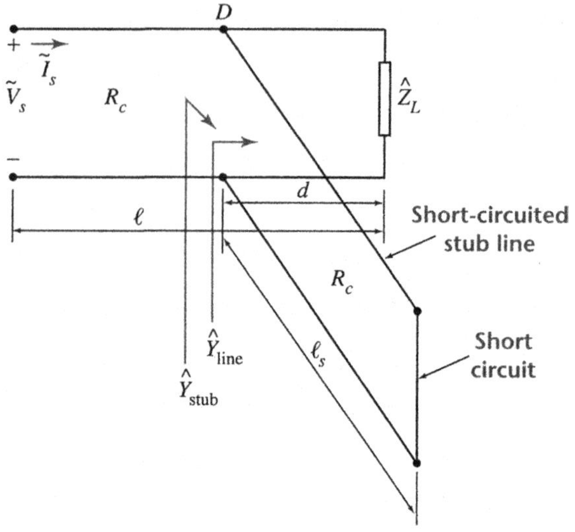

Figure A.16 Short-circuited stub line connected to the transmission line of Figure A.15

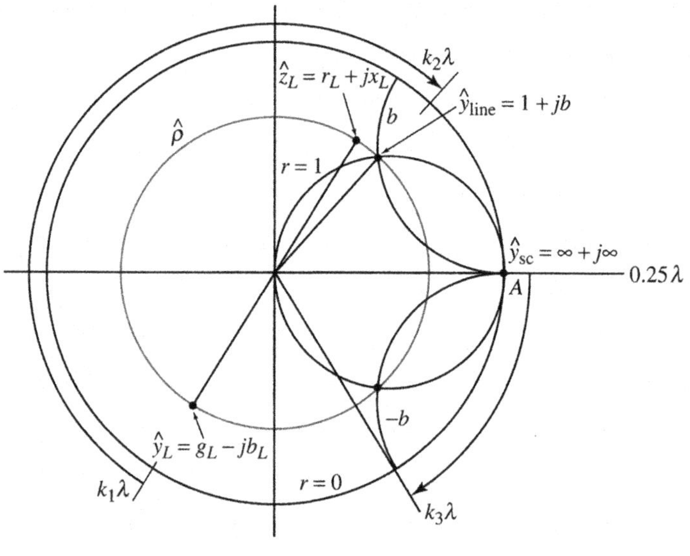

Figure A.17 Determination of stub-line length and its location on a transmission line for matching the load using Smith chart

a) Locate the normalized load impedance $\hat{z}_L = r_L + jx_L$ on the Smith chart and then rotate its position 180° about the origin of the chart to determine the admittance of the load (see Figure A.17).
b) Because the normalized input admittance at the stub line connection is

$$\hat{y}_{line} = R_c\left(\frac{1}{R_c} + jB\right) = 1 + jb$$

move clockwise along the line (Wavelengths Toward Generator) to find the $r = 1$ circle. The minimum distance from the load to the stub line connection is

$$d = k_2\lambda - k_1\lambda$$

c) Because the stub line is short-circuited at its termination, the normalized load admittance is

$$\hat{y}_{sc} = \infty + j\infty$$

as located at point A on the Smith chart. The normalized input admittance of the stub line is

$$\hat{y}_{stub} = -jb$$

Thus, we move along the stub line toward the generator (clockwise on the Smith chart) on the $r = 0$ circle to determine the length of the stub line as

$$\ell_s = k_3\lambda - 0.25\lambda$$

as shown in Figure A.17.

We can obtain many solutions to the problem; however, we usually prefer the shortest distances for actual implementations.

EXAMPLE A.7

A 100-Ω, 200-m-long, lossless transmission line operates at 10 MHz and is terminated in an impedance of $50 - j200$ Ω. The transit time of the line is 1 μs. Determine the length and the location of a short-circuited stub line.

Solution The normalized load impedance is

$$\hat{z}_L = \frac{50 - j200}{100} = 0.5 - j2$$

We can plot $\hat{z}_L$ and obtain the corresponding $\hat{y}_L$ as shown in Figure A.18. When we move from the load toward the generator to intersect the $r = 1$ circle, the normalized input admittance becomes

$$\hat{y}_{line} = 1 + j2.8$$

at the stub line connection point D. Thus, the distance from the load to the stub line connection on the transmission line is

$$d = 0.20\lambda - 0.07\lambda = 0.13\lambda$$

in terms of the wavelength. The velocity of propagation is $u_p = 200/1 \times 10^{-6} = 2 \times 10^8$ m/s along the transmission line. Consequently,

$$\lambda = \frac{u_p}{f} = \frac{2 \times 10^8}{10 \times 10^6} = 20 \text{ m}$$

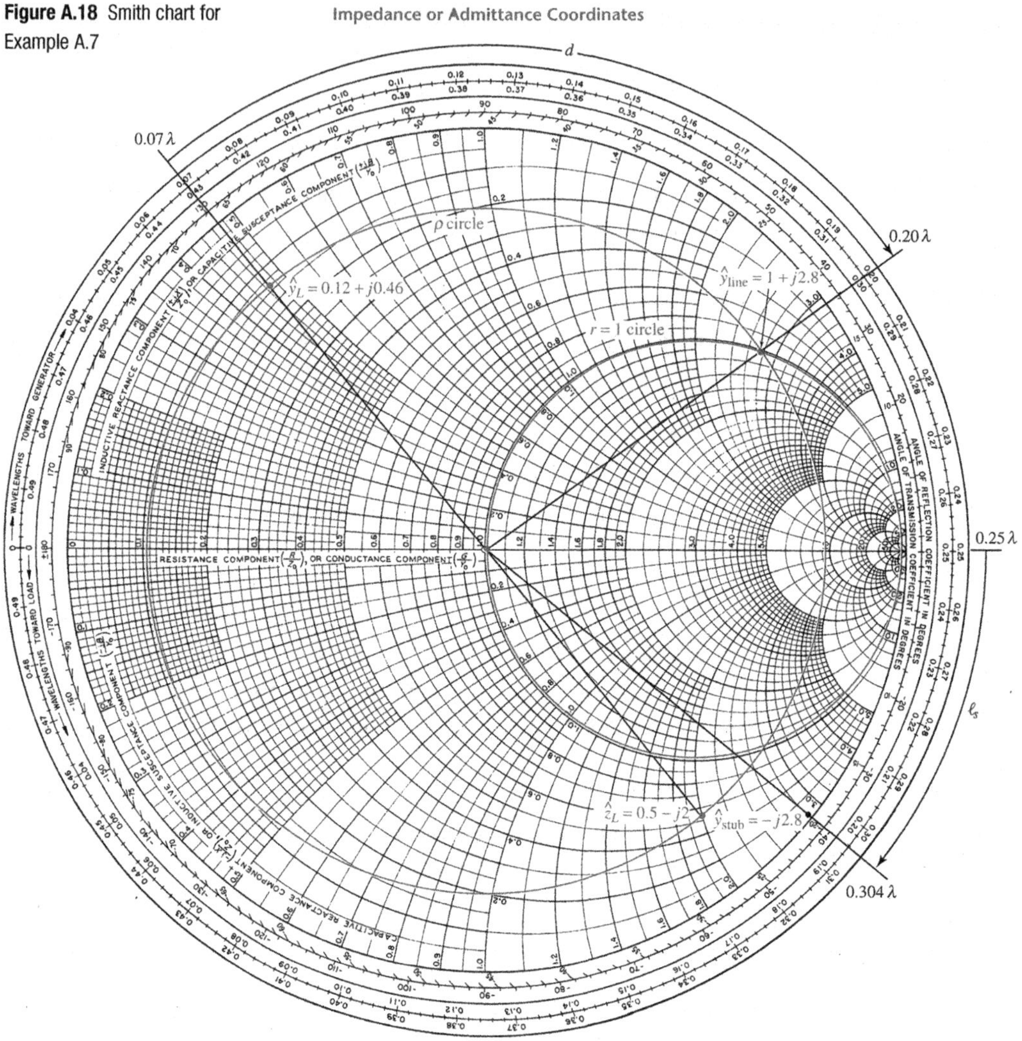

Figure A.18 Smith chart for Example A.7

The distance from the load where the stub line is to be connected is

$$d = 0.13 \times 20 = 2.6 \text{ m}$$

When we move from the short circuit load ($\hat{y}_{sc} = \infty + \infty_j$) along the stub line toward the generator to find $\hat{y}_{stub} = -j2.8$, we arrive at the connection point D to the transmission line. Thus, the length of the stub line is

$$\ell_s = 0.304\lambda - 0.25\lambda = 0.054\lambda$$

in terms of the wavelength, and

$$\ell_s = 0.054 \times 20 = 1.08 \text{ m}$$

• • •

A.6 Problems

A.1 Locate the following impedance in the Smith chart as normalized with respect to 50 Ω. $\hat{Z}_1 = 40 + j30\,\Omega$, $\hat{Z}_2 = 25 - j36\,\Omega$, $\hat{Z}_3 = j50\,\Omega$, $\hat{Z}_4 = 60\,\Omega$.

A.2 Determine the admittance of $\hat{Z} = 120 + j280\,\Omega$ with respect to $\hat{Z}_c = 100\,\Omega$ using the Smith chart.

A.3 The input impedances of a $\lambda/4$-long, 75-Ω transmission line are $\hat{Z}_1 = 50 + j25\,\Omega$ and $\hat{Z}_2 = 40 - j40\,\Omega$ when various load impedances are connected at the receiving end. In each case, determine the load impedance and the reflection coefficient at the sending and receiving ends.

A.4 The input impedance in a 50-Ω, 25-m-long transmission line is $\hat{Z}_{\text{in}} = 45 + j60\,\Omega$ at $z = 12$ m. Determine the load impedance and the reflection coefficient at the receiving end if the phase velocity is 2.6×10^8 m/s, and the operating frequency is 5 MHz.

A.5 A 30-m-long, 90-Ω, lossless transmission line with a phase velocity of 2.8×10^8 m/s is terminated in a load operating at 10 MHz. The load impedance is $60\underline{/15^\circ}\,\Omega$. (a) Determine the VSWR in the transmission line and the voltage reflection coefficient across the terminals of the load. (b) Calculate the input impedance and the corresponding voltage reflection coefficient at the sending end of the line.

A.6 A 2-m-long, lossless transmission line with $\hat{Z}_c = 75\,\Omega$ and $u_p = 2.6 \times 10^8$ m/s is terminated in a load of $\hat{Z}_L = 120 + j90\,\Omega$. If the operating voltage across the load, in time domain, is $v_R(t) = 150\cos(1.26 \times 10^8 t)$ V, calculate (a) the reflection coefficient $\hat{\rho}(z)$, and (b) the VSWR using the Smith chart.

A.7 A 50-m-long lossless transmission line with $L_\ell = 0.5\,\mu$H/m and $C_\ell = 50$ pF/m is subjected to a voltage of $v_S(t) = 280\cos(6.28 \times 10^7 t)$ V when the load at the receiving end is 250 Ω. Determine (a) the reflection coefficient at the receiving end, and (b) the current at the sending end using the Smith chart.

A.8 The voltage and current waveforms across the input terminals of a 75-Ω, lossless coaxial cable are $v_S(t) = 50\cos(10^7 t)$ V and $i_S(t) = 0.5\cos(10^7 t - 22^\circ)$ A, respectively. The load of this cable is a quarter-wavelength monopole antenna with an impedance of $\hat{Z}_{\text{ant}} = 36.5 + j21.25\,\Omega$. Determine the shortest possible length of this cable when $u_p = 2.5 \times 10^8$ m/s.

A.9 A half-wavelength, transmitting dipole antenna with an impedance of $\hat{Z}_{\text{ant}} = 73 + j42.5\,\Omega$ is fed via a 50-m-long, 50-Ω coaxial cable. The input voltage to the line is $\tilde{V}_s = 100\underline{/0^\circ}$ V (rms). The transit time of the cable is 0.1 μs and the operating frequency is 950 kHz. Determine the current at the input end of the cable.

A.10 A 50-Ω, lossless coaxial cable transmits a signal from a voltage supply having a resistance of $R_G = 25\,\Omega$ to a load with an impedance of

$\hat{Z}_L = 35 - j20\,\Omega$. Determine the shortest possible length of this cable in terms of λ to transfer the maximum average power from the supply to the load.

A.11 In order to match the coaxial cable to the load in Problem A.8, determine the length and the location of a short-circuited stub line.

A.12 Is the coaxial cable of Problem A.9 perfectly matched to the dipole antenna? If not, design a short-circuited stub line to match the coaxial cable to the load.

A.13 A 100-Ω, 15-m-long, lossless transmission line with a phase velocity of 2.8×10^8 m/s is connected to a load having an impedance of $Z_L = 250\,\Omega$. The open-circuit voltage of the supply that feeds the line is $v_G(t) = 35\cos(8 \times 10^7 t)$ V and its internal impedance is $\hat{Z}_G = 20 - j10\,\Omega$. Calculate the input impedance, voltage, current, and power at (a) the sending end, (b) the receiving end, (c) 5 m from the supply, (d) 5 m from the load, and (e) determine the voltage drop along the line.

A.14 A 50-Ω, 100-m-long, lossless transmission line is terminated into a load having an impedance of $40 - j100\,\Omega$. The transit time of the line is 0.5 μs. In order to match the transmission line to the load, determine the length and the location of a short-circuited stub line if the operating frequency is 20 MHz.

A.15 A 15-m-long air-filled coaxial cable operating at 125 MHz is terminated into a $150 + j225\,\Omega$. The diameters of the inner and outer conductors are 2.5 mm and 6 mm, respectively. In order to perfectly match the termination, what should be the length and location of a short-circuited stub line?

APPENDIX B

Computer programs for various problems

B.1 Computer program for Example 12.3

```
C  FINITE DIFFERENCE PROGRAM.
C  EXAMPLE 12.3
      DIMENSION V(50,50),MESH(50,50)
      OPEN(UNIT=6,FILE='F.O',STATUS='NEW')
C  SET BOUNDARY VALUES AND MESH ADDRESSES
C  SET ACCELERATION FACTOR AL.
      AL=1.
      DO 5 I=1,30
      DO 5 J=1,13
      V(I,J)=50.
    5 MESH(I,J)=2
      DO 6 J=1,13
      V(1,J)=0.
      MESH(1,J)=0
      V(30,J)=0.
    6 MESH(30,J)=0
      DO 7 I=1,30
      V(I,1)=0.
      MESH(I,1)=0
      V(I,13)=0.
    7 MESH(I,13)=0
      DO 8 I=4,27
      DO 8 J=3,4
      V(I,J)=0.
    8 MESH(I,J)=0
      DO 9 I=6,25
      DO 9 J=7,10
      V(I,J)=100.
    9 MESH(I,J)=1
```

```
C  ITERATION COUNT FOR S.O.R
      IT=0
   60 IT=IT+1
      DIFMAX=0.0
      DO 10 I=1,30
      DO 10 J=1,13
      M=MESH(I,J)
      IF(M.LT.2) GO TO 10
      VOLD=V(I,J)
      V(I,J)=VOLD+(AL/4.)*(V(I-1,J)+V(I,J-1)
      +V(I+1,J)+V(I,J+1)-4.*V(I,J))1)
      DIF=V(I,J)-VOLD
      DIF=ABS(DIF)
C  ERROR CHECK
      IF (DIF.GT.DIFMAX) DIFMAX=DIF
   10 CONTINUE
      WRITE(6,300) IT,DIFMAX
  300 FORMAT(1X,I3,F10.4)
      IF(IT.GT.50) GO TO 70
      IF(DIFMAX.GT.0.1) GO TO 60
      WRITE(6,200) ((V(I,J),J=1,13),I=1,30)
  200 FORMAT(1X,13F6.1)
   70 STOP
      END
```

It	Difax
1	33.3333
2	15.2344
3	8.5659
4	4.9125
5	3.5134
6	2.6380
7	2.0372
8	1.5849
9	1.2613
10	0.9874
11	0.7750
12	0.6146
13	0.4841
14	0.3798
15	0.2971
16	0.2320
17	0.1810
18	0.1418
19	0.1112
20	0.0871

POTENTIAL VALUES IN THE TRANSFORMER

0.0	0.0	0.0	0.0	0.0	0.0	0.0	0.0	0.0	0.0	0.0	0.0	0.0
0.0	0.4	1.1	2.7	5.7	9.1	12.0	13.5	13.4	11.7	8.6	4.5	0.0
0.0	0.5	1.3	3.8	11.0	18.8	25.2	28.4	28.3	24.8	18.1	9.4	0.0
0.0	0.1	0.0	0.0	15.7	29.9	41.4	46.6	46.5	41.1	29.4	15.1	0.0
0.0	0.0	0.0	0.0	21.8	43.6	63.8	70.1	70.1	63.6	43.4	21.6	0.0
0.0	0.0	0.0	0.0	27.9	58.9	100.0	100.0	100.0	100.0	58.8	27.8	0.0
0.0	0.0	0.0	0.0	31.1	63.9	100.0	100.0	100.0	100.0	63.9	31.0	0.0
0.0	0.0	0.0	0.0	32.4	65.7	100.0	100.0	100.0	100.0	65.6	32.4	0.0
0.0	0.0	0.0	0.0	33.0	66.3	100.0	100.0	100.0	100.0	66.3	33.0	0.0
0.0	0.0	0.0	0.0	33.2	66.5	100.0	100.0	100.0	100.0	66.5	33.2	0.0
0.0	0.0	0.0	0.0	33.3	66.6	100.0	100.0	100.0	100.0	66.6	33.3	0.0
0.0	0.0	0.0	0.0	33.3	66.6	100.0	100.0	100.0	100.0	66.6	33.3	0.0
0.0	0.0	0.0	0.0	33.3	66.7	100.0	100.0	100.0	100.0	66.7	33.3	0.0
0.0	0.0	0.0	0.0	33.3	66.7	100.0	100.0	100.0	100.0	66.7	33.3	0.0
0.0	0.0	0.0	0.0	33.3	66.7	100.0	100.0	100.0	100.0	66.7	33.3	0.0
0.0	0.0	0.0	0.0	33.3	66.7	100.0	100.0	100.0	100.0	66.7	33.3	0.0
0.0	0.0	0.0	0.0	33.3	66.7	100.0	100.0	100.0	100.0	66.7	33.3	0.0
0.0	0.0	0.0	0.0	33.3	66.7	100.0	100.0	100.0	100.0	66.7	33.3	0.0
0.0	0.0	0.0	0.0	33.3	66.6	100.0	100.0	100.0	100.0	66.6	33.3	0.0
0.0	0.0	0.0	0.0	33.3	66.6	100.0	100.0	100.0	100.0	66.6	33.3	0.0
0.0	0.0	0.0	0.0	33.2	66.5	100.0	100.0	100.0	100.0	66.5	33.2	0.0
0.0	0.0	0.0	0.0	33.0	66.3	100.0	100.0	100.0	100.0	66.3	33.0	0.0
0.0	0.0	0.0	0.0	32.5	65.7	100.0	100.0	100.0	100.0	65.7	32.4	0.0
0.0	0.0	0.0	0.0	31.1	63.9	100.0	100.0	100.0	100.0	63.9	31.0	0.0
0.0	0.0	0.0	0.0	28.0	58.9	100.0	100.0	100.0	100.0	58.8	27.9	0.0
0.0	0.0	0.0	0.0	21.9	43.7	63.9	70.3	70.2	63.7	43.4	21.6	0.0
0.0	0.1	0.0	0.0	15.8	30.0	41.5	46.8	46.7	41.2	29.5	15.1	0.0
0.0	0.5	1.4	3.8	11.1	18.9	25.3	28.5	28.4	24.9	18.1	9.4	0.0
0.0	0.4	1.1	2.7	5.7	9.2	12.0	13.5	13.4	11.7	8.6	4.5	0.0
0.0	0.0	0.0	0.0	0.0	0.0	0.0	0.0	0.0	0.0	0.0	0.0	0.0

B.2 Computer program for Example 12.5

```
C FINITE ELEMENT PROGRAM.
       DIMENSION X(20),Y(20),NE(20,20),TI(20,20),TIN
        (20,20),
      1U(20),XL(20,3),YL (20,3),B (20,2,3),IB
        (3,3),AREA (20),
      2P(20,3,3),BT(20,3,2),A(20,3,3), AA(20,20,
        20),S(20,20),
      3SA(20,20),TTI(20,20),SL(20,20),SAR(20,20),
        R(20),UN(20),
      4BB(20,20),SLINV(20,20)
       LM=20
```

```
C
C MN:  NUMBER OF NODES
C ME:  NUMBER OF ELEMENTS
C MS:  NUMBER OF NODES AT WHICH THE POTENTIALS ARE
         KNOWN
C MU:  NUMBER OF NODES AT WHICH THE POTENTIALS ARE
         UNKNOWN
C X:   X-COORDINATE
C Y:   Y-COORDINATE
C
      READ(5,110) MN
      READ(5,110) ME
      READ(5,110) MS
      READ(5,110) MU
  110 FORMAT(I3)
      IMU=2*MU
      DO 1 I=1,MN
      READ(5,100) X(I)
      READ(5,100) Y(I)
  100 FORMAT(F10.3)
    1 CONTINUE
C
C NE:  CONNECTION MATRIX WITH ELEMENTS WITH THEIR
         RELATED NODES.
C      IF THE NODE NUMBER IS RELATED TO AN ELEMENT,
         THEN NE=1,
C      OTHERWISE IT IS ZERO.
C
      READ(5,101) ((NE(I,J),J=1,MN),I=1,ME)
  101 FORMAT(16I1)
      READ(5,102) ((TI(I,J),J=1,MN),I=1,MU)
      READ(5,103) ((TIN(I,J),J=1,MS),I=1,MN)
  102 FORMAT(16F4.1)
  103 FORMAT(12F4.1)
C
C U: KNOWN POTENTIAL VALUES.
C
      DO 2 I=1,MS
      READ(5,100) U(I)
    2 CONTINUE
C
C IB: A WORK MATRIX
C
      READ(5,105) ((IB(I,J),J=1,3),I=1,3)
  105 FORMAT(3I2)
C
C  TRANSFORMATION FROM GLOBAL TO LOCAL COORDINATES.
```

```
C
      DO 5 I=1,ME
      K=0
      DO 5 J=1,MN
      IF(NE(I,J).EQ.0) GO TO 5
      K=K+1
      XL(I,K)=NE(I,J)*X(J)
      VL(I,K)=NE(I,J)*Y(J)
    5 CONTINUE
      WRITE(6,600) ((XL(I,J),J=1,3),I=1,ME)
  600 FORMAT(1X,'XL',3F5.1)
      WRITE(6,601) ((YL(I,J),J=1,3),I=1,ME)
  601 FORMAT(1X,'YL',3F5.1)
C
C  CALCULATION OF ELEMENT AREAS
C
      DO 22 K=1,ME
      AREA(K)=0.0
      DO 6 I=1,3
      B(K,1,I)=0.0
      B(K,2,I)=0.0
      DO 7 J=1,3
      B(K,1,I)=B(K,1,I)+YL(K,J)*IB(J,I)
      B(K,2,I)=B(K,2,I)-XL(K,J)*IB(J,I)
    7 CONTINUE
      AREA(K)=AREA(K)+B(K,1,I)*XL(K,I)
    6 CONTINUE
      IF(AREA(K).LT.0.0) AREA(K)= -1.*AREA(K)
   22 CONTINUE
      WRITE(6,605)(((B(K,J,I),I=1,3),J=1,2),
        K=1,ME)
  605 FORMAT(1X,'B',3F5.1)
      WRITE(6,606) (AREA(K),K=1,ME)
  606 FORMAT(1X,'TA',8F5.1)
      DO 8 K=1,ME
      DO 8 I=1,3
      B(K,1,I)=B(K,1,I)/AREA(K)
    8 B(K,2,I)=B(K,2,I)/AREA(K)
      DO 10 K=1,ME
      DO 10 I=1,2
      DO 10 J=1,3
   10 BT(K,J,I)=B(K,I,J)
      WRITE(6,230)(((BT (K,I,J),J=1,2),I=1,3),
      K=1,ME)
  230 FORMAT(1X,'BT',2F5.1)
      DO 9 K=1,ME
      DO 9 I=1,3
```

```
      DO 9 J=1,3
      P(K,I,J)=0.0
      DO 9 L=1,2
      P(K,I,J)=P(K,I,J)+BT(K,I,L)*B(K,L,J)
    9 CONTINUE
      WRITE(6,650) (((P(K,I,J),J=1,3), I= 1,3),
        K=1,ME)
  650 FORMAT (1X,'P',3F5.1)
      DO 11 K=1,ME
      DO 11 I=1,3
      DO 11 J=1,3
   11 A(K,I,J)=P(K,I,J)*AREA(K)/2.
      WRITE(6,701) (((A (K,I,J),J=1,3),I=1,3),
        K=1,ME)
  701 FORMAT(1X,'A',3F7.2)
C  AUGMENTING THE MATRIX.
      DO 12 L=1,ME
      I=0
      DO 12 J=1,MN
      IF(NE(L,J).EQ.0) GO TO 17
      I=I+1
   17 K=0
      DO 12 M=1,MN
      IF((NE(L,J)*NE(L,M)).EQ.0) GO TO 18
      K=K+1
      AA(L,J,M)=A(L,I,K)
      GO TO 12
   18 AA(L,J,M)=0.0
   12 CONTINUE
      WRITE(6,700) (((AA(K,I,J),J=1,MN),I=1,MN),
        K=1,ME)
  700 FORMAT(1X,'AA',9F7.2)
C  ADDING A'S.
      DO 15 I=1,MN
      DO 15 J=1,MN
      S(I,J)=0.0
      DO 15 L=1,ME
   15 S(I,J)=S(I,J)+AA(L,I,J)
      WRITE(6,500) ((S(I,J),J=1,MN),I=1,
        MN)
  500 FORMAT(1X,'S',9F7.2)
C  SOLVING THE EQUATIONS.
      WRITE(6,499) ((TI(I,J),J=1,MN),I=1,MU)
  499 FORMAT(1X,'IT',9F4.1)
      CALL MMULT(TI,S,SA,MU,MN,MN,LM)
      WRITE(6,501) ((SA(I,J),J=1,MN),I=1,MU)
  501 FORMAT(1X,'SA',9F7.2)
      CALL TRMAT(TI,TTI,MU,MN,LM)
      WRITE(6,498)((TTI(I,J),J=1,MU),I=1,MN)
```

```
  498 FORMAT(1X,'ITT',F4.1)
      CALL MMULT(SA,TTI,SL,MU,MN,MU,LM)
      WRITE(6,503) ((SL(I,J),J=1,MN),I=1,MU)
  503 FORMAT(1X,'SL',F7.2)
      CALL MMULT(SA,TIN,SAR,MU,MN,MS,LM)
      WRITE(6,497) ((TIN(I,J),J=1,MS),I=1,MN)
  497 FORMAT(1X,'INT',8F4.1)
      WRITE(6,504) ((SAR(I,J),J=1,MS),I=1,MU)
  504 FORMAT(1X,'SAR',8F7.2)
      CALL MVULT(SAR,U,R,MU,MS,LM)
      WRITE(6,505) (R(I),I=1,MU)
  505 FORMAT(1X,'R',F7.2)
      CALL CMINV(SL,BB,SLINV,MU,IMU,LM)
      CALL MVULT(SLINV,R,UN,MU,MU,LM)
      WRITE(6,400) (UN(I),I=1,MU)
  400 FORMAT(1X,'UN',F10.3)
      STOP
      END
C VARIOUS USEFUL SUBROUTINES
      SUBROUTINE MMULT(A,B,C,M,N,L,LM)
      DIMENSION A(LM,LM),B(LM,LM),C(LM,LM)
      DO 5 I=1,M
      DO 5 J=1,L
      C(I,J)=0.0
      DO 5 K=1,N
      C(I,J)=C(I,J)+A(I,K)*B(K,J)
    5 CONTINUE
      RETURN
      END
      SUBROUTINE MVULT(A,B,C,M,N,LM)
      DIMENSION A(LM,LM),B(LM),C(LM)
      DO 5 I=1,M
      C(I)=0.0
      DO 5 J=1,N
      C(I)=C(I)+ A(I,J)*B(J)
    5 CONTINUE
      RETURN
      END
      SUBROUTINE TRMAT(A,B,M,N,LM)
      DIMENSION A(LM,LM),B(LM,LM)
      DO 5 I=1,M
      DO 5 J=1,N
      B(J,I)=A(I,J)
    5 CONTINUE
      RETURN
      END
      SUBROUTINE CMINV(A,B,C,M,N,LM)
      DIMENSION A(LM,LM),B(LM,LM),C(LM,LM)
      DO 6 I=1,M
```

```
      DO 6 J=1,M
      IF(I. EQ. J) GO TO 11
      A(I,M+J)=0.0
      GO TO 6
 11   A(I,M+J)=1.0
  6   CONTINUE
      DO 5 J=1,M
      DO 5 I=1,M
      IF(I.EQ.J) GO TO 5
      IF(A(J,J).EQ.0.) GO TO 28
      GO TO 38
 28   DO 20 K=1,N
      B(J,K)=A(J+1,K)
 20   B(J+1,K)=A(J,K)
      DO 25 K=1,N
      A(J,K)=B(J,K)
 25   A(J+1,K)=B(J+1,K)
 38   PIVOT=A(I,J)/A(J,J)
      DO 15 K=1,N
      A(I,K)=A(I,K)-(PIVOT)*A(J,K)
 15   CONTINUE
  5   CONTINUE
      DO 10 I=1,M
      DO 10 J=1,N
      C(I,J)=A(I,J)/A(I,I)
 10   CONTINUE
      DO 7 I=1,M
      DO 7 J=1,M
  7   C(I,J)=C (I,M+J)
      RETURN
      END
```

Data File

```
009
008
008
001
0.
2.
1.
2.
2.
2.
0.
1.
1.
1.
2.
```

```
1.
0.
0.
1.
0.
2.
0.
100110000
110010000
011010000
001011000
000110100
000010110
000010011
000011001
0.      0.    0.   0.   1.   0.   0.   0.   0.
-1.     0.    0.   0.   0.   0.   0.   0.
0.      -1.   0.   0.   0.   0.   0.   0.
0.      0.    -1.  0.   0.   0.   0.   0.
0.      0.    0.   -1.  0.   0.   0.   0.
0.      0.    0.   0.   0.   0.   0.   0.
0.      0.    0.   0.   -1.  0.   0.   0.
0.      0.    0.   0.   0.   -1.  0.   0.
0.      0.    0.   0.   0.   0.   -1.  0.
0.      0.    0.   0.   0.   0.   0.   -1.
1.
1.
1.
0.
0.
0.
0.
0.
0.      -101
010     -1
-1010
```

B.3 Computer program for Example 12.6

```
      DIMENSION MR(35),TI(35,35),TIN(35,35),U(35)
      DIMENSION NE(35,35),X(35),Y(35),
     1XL(35,3),YL(35,3),B(35,2,3),IB(3,3),
       AREA(35),
     2P(35,3,3),BT(35,3,2),A(35,3,3),AA(35,35,35),
       S(35,35),
     3SA(35,35),TTI(35,35),SL(35,35),SAR(35,35),
       R(35),UN(35),
```

```
      4BB(35,35),SLINV(35,35)
       LM=35
C  FI: ANGULAR INCREMENT IN DEGREES.
C  RO: OUTER SHEATH RADIUS IN CM.
C  RI: INNER CONDUCTOR RADIUS IN CM.
C   G: RADIAL INCREMENT IN CM.
C   V: POTENTIAL VALUES.
       READ(5,200) FI
       READ(5,200) RO
       READ(5,200) RI
       READ(5,200) G
       READ(5,200) V1
       READ(5,200) V2
       READ(5,200) V3
       READ(5,200) V4
  200  FORMAT(F10.3)
C
C
      CALL MG(FI,RO,RI,G,LM,NE,X,Y,NJ,NKK,MN,ME)
      CALL BC(MR,NJ,NKK,MN,TI,TIN,U,LM,V1,V2,V3,
        V4,MU 1,MS)
C
      CALL FINEL(X,Y,NE,TI,TIN,U,MU,MS,ME,MN,LM,
      UN,XL,YL,B,IB,AREA,P,BT 1,A,AA,S,SA,TTI,SL,
        SAR,R,BB,SLINV)
      STOP
      END
C
      SUBROUTINE BC(MR,NJ,NKK,MN,TI,TIN,U,LM,V1,
        V2,V3 1,V4,MU,MS)
C   A SUBROUTINE THAT SETS THE BOUNDARIES.
      DIMENSION MR(LM),TI(LM,LM),TIN(LM,LM),U(LM)
      NX1=NJ
      NX2=NJ
      NY1=1
      NY2=NKK
C
      MS=0
      MU=0
C
      DO 21 I=1,NJ
      DO 21 J=1,NKK
      IJ=NJ* (J-1)
      MR (I+IJ)=1
      IF(I.LE.NX1.AND.J.LE.NY1) GO TO 10
      IF(I.GE.NX2.AND.J.LE.NY1) GO TO 11
      IF(I.LE.NX1.AND.J.GE.NY2) GO TO 12
```

```
      IF(I.GE.NX2.AND.J.GE.NY2) GO TO 13
      GO TO 21
   10 MS=MS+1
      MR(I+IJ)=2
      GO TO 21
   11 MS=MS+1
      MR(I+IJ)=3
      GO TO 21
   12 MS=MS+1
      MR(I+IJ)=4
      GO TO 21
   13 MS=MS+1
      MR(I+IJ)=5
   21 CONTINUE
C
      MS=0
      DO 30 I=1,MN
      IF(MR(I)-2) 30,31,31
   31 MS=MS+1
      TIN(I,MS)= -1.
      IF(MR(I)-3) 32,33,34
   32 U(MS)=V1
      GO TO 30
   33 U(MS)=V2
      GO TO 30
   34 IF(MR(I)-4) 30,35,36
   35 U(MS)=V3
      GO TO 30
   36 U(MS)=V4
   30 CONTINUE
C
      MU=0
      DO 40 I=1,MN
      IF(MR(I)-1) 40,41,40
   41 MU=MU+1
      TI(MU,I)=1.
   40 CONTINUE
C
      WRITE(6,103)
  103 FORMAT(1X,'U')
      WRITE(6,100) (U(I),I=1,MS)
  100 FORMAT(1X,F10.3)
      WRITE(6,110) (MR(I),I=1,MN)
  110 FORMAT(1X,'MR',I3)
      RETURN
      END
```

```
C
      SUBROUTINE MG(FI,RO,RI,G,LM,NE,X,Y,NJ,NKK,
        MN,ME)
C  A SUBROUTINE THAT GENERATES MESHES.
      DIMENSION NE(LM,LM),X(LM),Y(LM)
      NJ=90./FI
      PI=3.14159
      FI=FI*PI/180.
      NJ=NJ+1
      NK= (RO-RI)/G
      NJJ=NJ-1
      ME=2*NJJ*NK
      NKK=NK+1
      MN=NJ*NKK
      NRC=NJ/2
      NRCC=NRC*2
      GG=1.6
      IF(NRCC.EQ.NJ) GO TO 11
      NMC=NRC
      GO TO 16
   11 NMC=NRC-1
C
   16 DO 21 I=1,NJ
      DO 21 J=1,NKK
      IJ=NJ*(J-1)
      RR=RO-(GG*(J-1))
      IF(J.EQ.NKK) RR=RI
      THETA=(PI/2)-(I-1)*FI
      X(I+IJ)=RR*COS(THETA)
      Y(I+IJ)=RR*SIN(THETA)
   21 CONTINUE
C
      WRITE(6,300) NJ,NK,NJJ,ME,NKK,MN,NRC,NMC
  300 FORMAT(1X,8I3)
      DO 5 I=1,NRC
      NERO=1+(I-1)*4
      NFRO=1+(I-1)*2
      NSRO=1+NJ+(I-1)*2
      NTRO=NSRO+1
C
      NERE=2+(I-1)*4
      NFRE=NFRO
      NSRE=NFRE+1
      NTRE=NSRE+NJ
C
      DO 10 K=1,NK
      NE(NERO,NFRO)=1
      NE(NERO,NSRO)=1
```

```
      NE(NERO,NTRO)=1
      NERO=NERO+2*(NJ-1)
      NFRO=NFRO+NJ
      NSRO=NSRO+NJ
C
      KI=K/2
      KS=KI*2
      IF(KS.EQ.K) NTRO=NTRO+2*NJ
 10   CONTINUE
C
      DO 15 K=1,NK
      NE(NERE,NFRE)=1
      NE(NERE,NSRE)=1
      NE(NERE,NTRE)=1
      NERE=NERE+2*(NJ-1)
C
      KI=K/2
      KS=KI*2
      IF(KS.EQ.K) GO TO 20
      NFRE=NFRE+2*NJ
      NSRE=NSRE+2*NJ
      GO TO 15
 20   NTRE=NTRE+2*NJ
 15   CONTINUE
  5   CONTINUE
C
C
      DO 8 I=1,NMC
      NEMO=3+(I-1)*4
      NFMO=3+(I-1)*2
      NSMO=NFMO-1
      NTMO=NSMO+NJ
C
      NEME=4+(I-1)*4
      NFME=NFMO
      NSME=NFME+NJ
      NTME=NSME-1
C
      DO 18 K=1,NK
      NE(NEMO,NFMO)=1
      NE(NEMO,NSMO)=1
      NE(NEMO,NTMO)=1
      NEMO=NEMO+2*(NJ-1)
C
      KI=K/2
      KS=KI*2
      IF(KS.EQ.K) GO TO 28
      NFMO=NFMO+2*NJ
```

```
      NSMO=NSMO+2*NJ
      GO TO 18
   28 NTMO=NTMO+2*NJ
   18 CONTINUE
C
      DO 19 K=1,NK
      NE(NEME,NFME)=1
      NE(NEME,NSME)=1
      NE(NEME,NTME)=1
      NEME=NEME+2*(NJ-1)
      NFME=NFME+NJ
      NSME=NSME+NJ
C
      KI=K/2
      KS=KI*2
      IF(KS.EQ.K) NTME=NTME+2*NJ
   19 CONTINUE
    8 CONTINUE
      DO 9 I=1,MN
      WRITE(6,110) X(I),Y(I)
  110 FORMAT(1X,2F10.3)
    9 CONTINUE
      WRITE(6,100) ((NE(I,J),J=1,MN),I=1,ME)
  100 FORMAT(1X,'NE',20I2)
      RETURN
      END
C
C
      SUBROUTINE FINEL(X,Y,NE,TI,TIN,U,MU,MS,ME,
        MN,LM,UN,XL,YL,B,IB,AREA1,P,BT,A,AA,S,SA,
        TTI,SL,SAR,R,BB,SLINV)
C
C FINITE ELEMENT PROGRAM.
       DIMENSION X(LM),Y(LM),NE(LM,LM),TI(LM,LM),
        TIN(LM,LM),
      1U(LM),XL(LM,3),YL(LM,3),B(LM,2,3),
        IB(3,3),AREA(LM),
      2P(LM,3,3),BT(LM,3,2),A(LM,3,3),AA(LM,LM,
        LM),S(LM,LM),
      3SA(LM,LM),TTI(LM,LM),SL(LM,LM),SAR(LM,LM),
        R(LM),UN(LM),
      4BB(LM,LM),SLINV(LM,LM)
       IMU=2*MU
       IB(1,1)=0
       IB(1,2)= -1
       IB(1,3)=1
       IB(2,1)=1
       IB(2,2)=0
```

```
      IB(2,3)=-1
      IB(3,1)=-1
      IB(3,2)=1
      IB(3,3)=0
C TRANSFORMATION FROM GLOBAL TO LOCAL COORDINATES.
      DO 5 I=1,ME
      K=0
      DO 5 J=1,MN
      IF(NE(I,J).EQ.0) GO TO 5
      K=K+1
      XL(I,K)=NE(I,J)*X(J)
      YL(I,K)=NE(I,J)*Y(J)
    5 CONTINUE
      WRITE(6,600) ((XL(I,J),J=1,3),I=1,ME)
  600 FORMAT(1X,'XL',3F5.1)
      WRITE(6,601) ((YL(I,J),J=1,3),I=1,ME)
  601 FORMAT(1X,'YL',3F5.1)
      DO 22 K=1,ME
      AREA(K)=0.0
      DO 6 I=1,3
      B(K,1,I)=0.0
      B(K,2,I)=0.0
      DO 7 J=1,3
      B(K,1,I)=B(K,1,I)+YL(K,J)*IB(J,I)
      B(K,2,I)=B(K,2,I)-XL(K,J)*IB(J,I)
    7 CONTINUE
      AREA(K)=AREA(K)+B(K,1,I)*XL(K,I)
    6 CONTINUE
      IF(AREA(K).LT.0.0) AREA(K)=-1.*AREA(K)
   22 CONTINUE
      WRITE(6,606) (AREA(K),K=1,ME)
  606 FORMAT(1X,'TA',8F5.1)
      DO 8 K=1,ME
      DO 8 I=1,3
      B(K,1,I)=B(K,1,I)/AREA(K)
    8 B(K,2,I)=B(K,2,I)/AREA(K)
      DO 10 K=1,ME
      DO 10 I=1,2
      DO 10 J=1,3
   10 BT(K,J,I)=B(K,I,J)
      DO 9 K=1,ME
      DO 9 I=1,3
      DO 9 J=1,3
      P(K,I,J)=0.0
      DO 9 L=1,2
      P(K,I,J)=P(K,I,J)+BT(K,I,L)*B(K,L,J)
    9 CONTINUE
      DO 11 K=1,ME
```

```
      DO 11 I=1,3
      DO 11 J=1,3
   11 A(K,I,J)=P(K,I,J)*AREA(K)/2.
C AUGMENTING THE MATRIX.
      DO 12 L=1,ME
      I=0
      DO 12 J=1,MN
      IF(NE(L,J).EQ.0) GO TO 17
      I=I+1
   17 K=0
      DO 12 M=1,MN
      IF((NE(L,J)*NE(L,M)).EQ.0) GO TO 18
      K=K+1
      AA(L,J,M)=A(L,I,K)
      GO TO 12
   18 AA(L,J,M)=0.0
   12 CONTINUE
C ADDING A'S.
      DO 15 I=1,MN
      DO 15 J=1,MN
      S(I,J)=0.0
      DO 15 L=1,ME
   15 S(I,J)=S(I,J)+AA(L,I,J)
C SOLVING THE EQUATIONS.
      CALL MMULT(TI,S,SA,MU,MN,MN,LM)
      CALL TRMAT(TI,TTI,MU,MN,LM)
      CALL MMULT(SA,TTI,SL,MU,MN,MU,LM)
      CALL MMULT(SA,TIN,SAR,MU,MN,MS,LM)
      CALL MVULT(SAR,U,R,MU,MS,LM)
      CALL CMINV(SL,BB,SLINV,MU,IMU,LM)
      CALL MVULT(SLINV,R,UN,MU,MU,LM)
      WRITE(6,400) (UN(I),I=1,MU)
  400 FORMAT(1X,'UN',F10.3)
      RETURN
      END
      SUBROUTINE MMULT(A,B,C,M,N,L,LM)
      DIMENSION A(LM,LM),B(LM,LM),C(LM,LM)
      DO 5 I=1,M
      DO 5 J=1,L
      C(I,J)=0.0
      DO 5 K=1,N
      C(I,J)=C(I,J)+A(I,K)*B(K,J)
    5 CONTINUE
      RETURN
      END
      SUBROUTINE MVULT(A,B,C,M,N,LM)
      DIMENSION A(LM,LM),B(LM),C(LM)
      DO 5 I=1,M
```

```
      C(I)=0.0
      DO 5 J=1,N
      C(I)=C(I)+A(I,J)*B(J)
    5 CONTINUE
      RETURN
      END
      SUBROUTINE TRMAT(A,B,M,N,LM)
      DIMENSION A(LM,LM),B(LM,LM)
      DO 5 I=1,M
      DO 5 J=1,N
      B(J,I)=A(I,J)
    5 CONTINUE
      RETURN
      END
      SUBROUTINE CMINV(A,B,C,M,N,LM)
      DIMENSION A(LM,LM),B(LM,LM),C(LM,LM)
      DO 6 I=1,M
      DO 6 J=1,M
      IF(I.EQ.J) GO TO 11
      A(I,M+J)=0.0
      GO TO 6
   11 A(I,M+J)=1.0
    6 CONTINUE
      DO 5 J=1,M
      DO 5 I=1,M
      IF(I.EQ.J) GO TO 5
      IF(A(J,J).EQ.0.) GO TO 28
      GO TO 38
   28 DO 20 K=1,N
      B(J,K)=A(J+1,K)
   20 B(J+1,K)=A(J,K)
      DO 25 K=1,N
      A(J,K)=B(J,K)
   25 A(J+1,K)=B(J+1,K)
   38 PIVOT=A(I,J)/A(J,J)
      DO 15 K=1,N
      A(I,K)=A(I,K)-(PIVOT)*A(J,K)
   15 CONTINUE
    5 CONTINUE
      DO 10 I=1,M
      DO10 J=1,N
      C(I,J)=A(I,J)/A(I,I)
   10 CONTINUE
      DO 7 I=1,M
      DO 7 J=1,M
    7 C(I,J)=C(I,M+J)
      RETURN
      END
```

B.4 Computer program for Example 12.7

```
C  METHOD OF MOMENT(MOM.CD)
      DIMENSION V(20,40),VB(20),AL(20),R(20,20),
        RO(20),ROE(20),
     1VD(20),VINV(20,40),VV(20,40),RI(20,20),
      VM(20)
      LM=20
      ML=2*LM
      READ(5,100)N
  100 FORMAT(I2)
      NN=2*N
      READ(5,110) VC
      READ(5,110) ROL
      READ(5,110) RA
      READ(5,110) BOY
      READ(5,110) DA
      READ(5,110) AD
  110 FORMAT(F15.5)
      DL=BOY/N
      DO 1 I=1,N
      VB(I)=VC
    1 RO(I)=ROL
      DO 2 J=1,N
      DO 2 I=1,N
      F=1.*(I-J)
      F=ABS(F)
      RSQ=RA**2.+(F*DL)**2.
      R(J,I)=SQRT(RSQ)
      WRITE(6,300) R(J,I),RO(I),DL
      V(J,I)=9.E09*RO(I)*DL/R(J,I)
      WRITE(6,300) V(J,I)
  300 FORMAT(1X,3E15.3)
    2 CONTINUE
      WRITE(6,300) ((V(I,J),J=1,N),I=1,N)
      CALL CMINV(V,VV,VINV,N,NN,LM,ML)
      CALL MVULT(VINV,VB,AL,N,N,LM)
      ROS=0.
      DO 5 I=1,N
      ROS=ROS+AL(I)*RO(I)
    5 CONTINUE
      DO 6 I=1,N
    6 ROE(I)=AL(I)*RO(I)
      WRITE(6,300) ROS
      WRITE(6,300)(ROE(I),I=1,N)
      DO 18 I=1,N
      DO 18 J=1,N
   18 RI(I,J)=1./R(I,J)
```

```
      CALL MVULT(RI,ROE,VD,N,N,LM)
      DO 19 I=1,N
   19 VD(I)=9.E09*DL*VD(I)
      WRITE(6,200) ((R(I,J),J=1,N),I=1,N)
  200 FORMAT(1X,'R',5E15.3)
      WRITE(6,220) (VB(I),I=1,N)
  220 FORMAT(1X,'VB',5E15.3)
      WRITE(6,230) (AL(I),I=1,N)
  230 FORMAT(1X,'AL',5E15.3)
      WRITE(6,250) (VD(I),I=1,N)
  250 FORMAT(1X,'VD',5E15.3)
   32 RA=RA+DA
      DO 22 J=1,N
      DO 22 I=1,N
      F=1.*(I-J)
      F=ABS(F)
      RSQ=RA**2.+(F*DL)**2.
      R(J,I)=SQRT(RSQ)
      RI(J,I)=1./R(J,I)
      WRITE(6,300) R(J,I),RO(I),DL
   22 CONTINUE
      CALL MVULT(RI,ROE,VM,N,N,LM)
      DO 23 I=1,N
   23 VM(I)=9.E09*VM(I)*DL
      WRITE(6,290)(VM(I),I=1,N)
  290 FORMAT(1X,'VM',5E15.3)
      IF(RA.LE.AD) GO TO 32
      STOP
      END
      SUBROUTINE MVULT(A,B,C,M,N,LM)
      DIMENSION A(LM,LM),B(LM),C(LM)
      DO 5 I=1,M
      C(I)=0.0
      DO 5 J=1,N
      C(I)=C(I)+A(I,J)*B(J)
    5 CONTINUE
      RETURN
      END
      SUBROUTINE CMINV(A,B,C,M,N,LM,ML)
      DIMENSION A(LM,ML),B(LM,ML),C(LM,ML)
      DO 6 I=1,M
      DO 6 J=1,M
      IF(I.EQ.J) GO TO 11
      A(I,M+J)=0.0
      GO TO 6
   11 A(I,M+J)=1.0
    6 CONTINUE
      DO 5 J=1,M
```

```
      DO 5 I=1,M
      IF(I.EQ.J) GO TO 5
      IF(A(J,J).EQ.O.) GO TO 28
      GO TO 38
   28 DO 20 K=1,N
      B(J,K)=A(J+1,K)
   20 B(J+1,K)=A(J,K)
      DO 25 K=1,N
      A(J,K)=B(J,K)
   25 A(J+1,K)=B(J+1,K)
   38 PIVOT=A(I,J)/A(J,J)
      DO 15 K=1,N
      A(I,K)=A(I,K)-(PIVOT)*A(J,K)
   15 CONTINUE
    5 CONTINUE
      DO 10 I=1,M
      DO10 J=1,N
      C(I,J)=A(I,J)/A(I,I)
   10 CONTINUE
      DO 7 I=1,M
      DO 7 J=1,M
    7 C(I,J)=C(I,M+J)
      RETURN
      END
```

APPENDIX C

Useful mathematical tables

C.1 A brief list of series

$$(1+x)^n = 1 + nx + \frac{n(n-1)}{2!}x^2 + \frac{n(n-1)(n-2)}{3!}x^3 + \cdots \quad |x| < 1$$
$$(1-x)^n = 1 - nx + \frac{n(n-1)}{2!}x^2 - \frac{n(n-1)(n-2)}{3!}x^3 + \cdots \quad |x| < 1$$
$$(1-x)^{-n} = 1 + nx + \frac{n(n+1)}{2!}x^2 + \frac{n(n+1)(n+2)}{3!}x^3 + \cdots \quad |x| < 1$$
$$1 + \tfrac{1}{2} + \tfrac{1}{3} + \tfrac{1}{4} + \cdots = \infty$$
$$1 - \tfrac{1}{2} + \tfrac{1}{3} - \tfrac{1}{4} + \cdots = \ln(2)$$
$$1 - \frac{1}{3} + \frac{1}{5} - \frac{1}{7} + \cdots = \frac{\pi}{4}$$
$$1 + \frac{1}{2^2} + \frac{1}{3^2} + \frac{1}{4^2} + \cdots = \frac{\pi^2}{6}$$
$$1 - \frac{1}{2^2} + \frac{1}{3^2} - \frac{1}{4^2} + \cdots = \frac{\pi^2}{12}$$
$$1 + \frac{1}{3^2} + \frac{1}{5^2} + \frac{1}{7^2} + \cdots = \frac{\pi^2}{8}$$
$$\sin(x) = x - \frac{x^3}{3!} + \frac{x^5}{5!} - \frac{x^7}{7!} + \cdots$$
$$\cos(x) = 1 - \frac{x^2}{2!} + \frac{x^4}{4!} - \frac{x^6}{6!} + \cdots$$
$$\ln(1+x) = \sum_{n=1}^{\infty}(-1)^{n+1}\frac{x^n}{n} \qquad \text{for all } x$$

C.2 A list of trigonometric identities

$$e^{\theta} = \cosh(\theta) + \sinh(\theta) = 1 + \theta + \frac{\theta^2}{2!} + \frac{\theta^3}{3!} + \frac{\theta^4}{4!} + \cdots$$
$$e^{j\theta} = \cos(\theta) + j\sin(\theta) \qquad \text{where } j = \sqrt{-1}$$
$$\cosh(\theta) = \tfrac{1}{2}[e^{\theta} + e^{-\theta}]$$

$$\sinh(\theta) = \tfrac{1}{2}[e^{\theta} - e^{-\theta}]$$
$$\cos(\theta) = \tfrac{1}{2}[e^{j\theta} + e^{-j\theta}]$$
$$\sin(\theta) = \tfrac{1}{2j}[e^{j\theta} - e^{-j\theta}]$$
$$\sin(-\alpha) = -\sin(\alpha) \qquad \sin(\alpha) = \cos(\alpha - \pi/2)$$
$$\cos(-\alpha) = \cos(\alpha) \qquad \cos(\alpha) = -\sin(\alpha - \pi/2)$$
$$\cosh(j\alpha) = \cos(\alpha)$$
$$\sinh(j\alpha) = j\sin(\alpha)$$
$$\cos(j\beta) = \cosh(\beta)$$
$$\sin(j\beta) = j\sinh(\beta)$$
$$\sinh(\alpha + \beta) = \sinh(\alpha)\cosh(\beta) + \cosh(\alpha)\sinh(\beta)$$
$$\cosh(\alpha + \beta) = \cosh(\alpha)\cosh(\beta) + \sinh(\alpha)\sinh(\beta)$$
$$\sinh(\alpha + j\beta) = \sinh(\alpha)\cos(\beta) + j\cosh(\alpha)\sin(\beta)$$
$$\cosh(\alpha + j\beta) = \cosh(\alpha)\cos(\beta) + j\sinh(\alpha)\sin(\beta)$$
$$\sin(\alpha + j\beta) = \sin(\alpha)\cosh(\beta) + j\cos(\alpha)\sinh(\beta)$$
$$\sin(\alpha - j\beta) = \sin(\alpha)\cosh(\beta) - j\cos(\alpha)\sin(\beta)$$
$$\cos(\alpha + j\beta) = \cos(\alpha)\cosh(\beta) - j\sin(\alpha)\sinh(\beta)$$
$$\cos(\alpha - j\beta) = \cos(\alpha)\cosh(\beta) + j\sin(\alpha)\sinh(\beta)$$
$$\sin(\alpha + \beta) = \sin(\alpha)\cos(\beta) + \cos(\alpha)\sin(\beta)$$
$$\cos(\alpha + \beta) = \cos(\alpha)\cos(\beta) - \sin(\alpha)\sin(\beta)$$
$$\sin(2\alpha) = 2\sin(\alpha)\cos(\alpha)$$
$$\sin(3\alpha) = 3\sin(\alpha) - 4\sin^3(\alpha)$$
$$\begin{aligned}\cos(2\alpha) &= \cos^2(\alpha) - \sin^2(\alpha)\\ &= 2\cos^2(\alpha) - 1\\ &= 1 - 2\sin^2(\alpha)\end{aligned}$$
$$\cos(3\alpha) = 4\cos^3(\alpha) - 3\cos(\alpha)$$
$$\sin^2(\alpha) + \cos^2(\alpha) = 1$$
$$1 + \tan^2(\alpha) = \sec^2(\alpha) \qquad 1 + \cot^2(\alpha) = \csc^2(\alpha)$$
$$\sin^2(\alpha) = \tfrac{1}{2}(1 - \cos(2\alpha))$$
$$\cos^2(\alpha) = \tfrac{1}{2}(1 + \cos(2\alpha))$$
$$\sin^3(\alpha) = \tfrac{1}{4}(3\sin(\alpha) - \sin(3\alpha))$$
$$\cos^3(\alpha) = \tfrac{1}{4}(3\cos(\alpha) + \cos(3\alpha))$$
$$2\sin(\alpha)\cos(\beta) = \sin(\alpha + \beta) + \sin(\alpha - \beta)$$
$$2\cos(\alpha)\cos(\beta) = \cos(\alpha + \beta) + \cos(\alpha - \beta)$$
$$2\sin(\alpha)\sin(\beta) = \cos(\alpha - \beta) - \cos(\alpha + \beta)$$
$$\tan(\alpha + \beta) = \frac{\tan(\alpha) + \tan(\beta)}{1 - \tan(\alpha)\tan(\beta)}$$

C.3 A list of indefinite integrals

In the list of integrals that follows, C is simply a constant of integration.

Let $X = \sqrt{a^2 + x^2}$

$$\int x^{1/2}\,dx = \frac{2}{3}x^{3/2} + C$$

$$\int \frac{dx}{\sqrt{x}} = 2\sqrt{x} + C$$

$$\int X\,dx = \frac{1}{2}x\,X + \frac{a^2}{2}\ln|x + X| + C$$

$$\int xX\,dx = \frac{1}{3}X^3 + C$$

$$\int \frac{dx}{X} = \ln[x + X] + C$$

$$\int \frac{dx}{X^3} = \frac{1}{a^2}\frac{x}{X} + C$$

$$\int \frac{dx}{X^5} = \frac{1}{a^4}\left[\frac{x}{X} - \frac{1}{3}\frac{x^3}{X^3}\right] + C$$

$$\int \frac{x\,dx}{X} = X + C$$

$$\int \frac{x\,dx}{X^3} = -\frac{1}{X} + C$$

$$\int \frac{x\,dx}{X^5} = -\frac{1}{3X^3} + C$$

$$\int \frac{dx}{a^2 + x^2} = \frac{1}{a}\tan^{-1}(x/a) + C$$

$$\int \frac{dx}{(a^2 + x^2)^2} = \frac{x}{2a^2(a^2 + x^2)} + \frac{1}{2a^3}\tan^{-1}(x/a) + C$$

$$\int \frac{x\,dx}{a^2 + x^2} = \frac{1}{2}\ln|a^2 + x^2| + C$$

$$\int \frac{x\,dx}{(a^2 + x^2)^2} = -\frac{1}{2(a^2 + x^2)} + C$$

$$\int \frac{dx}{a^2 - x^2} = \frac{1}{2a}\ln|(a + x)/(a - x)| + C = \frac{1}{a}\tanh^{-1}(x/a) + C$$

$$\int \frac{x\,dx}{(a^2 - x^2)} = -\frac{1}{2}\ln|a^2 - x^2| + C$$

$$\int \sin(ax)\,dx = -\frac{1}{a}\cos(ax) + C$$

$$\int \cos(ax)\,dx = \frac{1}{a}\sin(ax) + C$$

$$\int \sin^2(ax)\,dx = \frac{x}{2} - \frac{\sin(2ax)}{4a} + C$$

$$\int \cos^2(ax)\,dx = \frac{x}{2} + \frac{\sin(2ax)}{4a} + C$$

$$\int \sin(ax)\cos(bx)\,dx = -\frac{\cos(a+b)x}{2(a+b)} - \frac{\cos(a-b)x}{2(a-b)}, \quad a \neq \pm b$$

$$\int \sin(ax)\sin(bx)\,dx = \frac{\sin(a-b)x}{2(a-b)} - \frac{\sin(a+b)x}{2(a+b)}, \quad a \neq \pm b$$

$$\int \cos(ax)\cos(bx)\,dx = \frac{\sin(a-b)x}{2(a-b)} + \frac{\sin(a+b)x}{2(a+b)}, \quad a \neq \pm b$$

$$\int \sin(ax)\cos(ax)\,dx = -\frac{\cos(2ax)}{4a} + C$$

$$\int \sin^n(ax)\cos(ax)\,dx = \frac{\sin^{n+1}(ax)}{(n+1)a} + C, \quad n \neq -1$$

$$\int \tan(ax)\,dx = -\frac{1}{a}\ln|\cos(ax)| + C$$

$$\int \cot(ax)\,dx = \frac{1}{a}\ln|\sin(ax)| + C$$

$$\int x\sin(ax)\,dx = \frac{1}{a^2}\sin(ax) - \frac{x}{a}\cos(ax) + C$$

$$\int x\cos(ax)\,dx = \frac{1}{a^2}\cos(ax) + \frac{x}{a}\sin(ax) + C$$

$$\int \tan^2(ax)\,dx = \frac{1}{a}\tan(ax) - x + C$$

$$\int \cot^2(ax)\,dx = -\frac{1}{a}\cot(ax) - x + C$$

$$\int e^{ax}\,dx = \frac{1}{a}e^{ax} + C$$

$$\int b^{ax}\,dx = \frac{1}{a\ln(b)}b^{ax} + C$$

$$\int x\,e^{ax}\,dx = \frac{e^{ax}}{a^2}(ax-1) + C$$

$$\int x^n e^{ax}\,dx = \frac{1}{a}x^n e^{ax} - \frac{n}{a}\int x^{n-1}e^{ax}\,dx$$

$$\int e^{ax}\sin(bx)\,dx = \frac{e^{ax}}{a^2+b^2}[a\sin(bx) - b\cos(bx)] + C$$

$$\int e^{ax}\cos(bx)\,dx = \frac{e^{ax}}{a^2+b^2}[a\cos(bx) + b\sin(bx)] + C$$

$$\int \ln(ax)\,dx = x\ln(ax) - x + C$$

$$\int x^n\ln(ax)\,dx = \frac{x^{n+1}}{n+1}\ln(ax) - \frac{x^{n+1}}{(n+1)^2} + C \qquad n \neq -1$$

$$\int \frac{1}{x}\ln(ax)\,dx = \frac{1}{2}[\ln(ax)]^2 + C$$

$$\int \sinh(ax)\,dx = \frac{1}{a}\cosh(ax) + C$$

$$\int \cosh(ax)\,dx = \frac{1}{a}\sinh(ax) + C$$

$$\int \tanh(ax)\,dx = \frac{1}{a}\ln[\cosh(ax)] + C$$

$$\int \coth(ax)\,dx = \frac{1}{a}\ln|\sinh(ax)| + C$$

$$\int \operatorname{sech}(ax)\,dx = \frac{1}{a}\sin^{-1}[\tanh(ax)] + C$$

$$\int \operatorname{csch}(ax)\,dx = \frac{1}{a}\ln|\tanh(ax/2)| + C$$

$$\int \sinh^2(ax)\,dx = \frac{\sinh(2ax)}{4a} - \frac{x}{2} + C$$

$$\int \cosh^2(ax)\,dx = \frac{\sinh(2ax)}{4a} + \frac{x}{2} + C$$

$$\int \tanh^2(ax) = x - \frac{1}{a}\tanh(ax) + C$$

$$\int \coth^2(ax)\,dx = x - \frac{1}{a}\coth(ax) + C$$

$$\int \operatorname{sech}^2(ax)\,dx = \frac{1}{a}\tanh(ax) + C$$

$$\int \operatorname{csch}^2(ax)\,dx = -\frac{1}{a}\coth(ax) + C$$

C.4 A partial list of definite integrals

$$\int_0^\infty e^{-ax}\,dx = \frac{1}{a} \qquad (a > 0)$$

$$\int_0^\infty xe^{-ax}\,dx = \frac{1}{a^2} \qquad (a > 0)$$

$$\int_0^\infty x^2e^{-ax}\,dx = \frac{2}{a^3} \qquad (a > 0)$$

$$\int_0^\infty x^ne^{-ax}\,dx = \frac{n!}{a^{n+1}} \qquad (a > 0, n > -1)$$

$$\int_0^\infty x^{1/2}e^{-ax}\,dx = \frac{1}{2a}\sqrt{\pi/a} \qquad (a > 0)$$

$$\int_0^\infty x^{-1/2}e^{-ax}\,dx = \sqrt{\pi/a} \qquad (a > 0)$$

$$\int_0^\infty e^{-ax}\sin(bx)\,dx = \frac{b}{a^2+b^2} \qquad (a > 0)$$

$$\int_0^\infty e^{-ax}\cos(bx)\,dx = \frac{a}{a^2+b^2} \qquad (a > 0)$$

$$\int_0^\infty xe^{-ax}\sin(bx)\,dx = \frac{2ab}{(a^2+b^2)^2} \qquad (a > 0)$$

$$\int_0^\infty xe^{-ax}\cos(bx)\,dx = \frac{a^2-b^2}{(a^2+b^2)^2} \qquad (a>0)$$

$$\int_0^{2\pi} \sin(ax)\,dx = 0 \qquad (a=1,2,3,\ldots)$$

$$\int_0^{2\pi} \cos(ax)\,dx = 0 \qquad (a=1,2,3,\ldots)$$

$$\int_0^{2\pi} \sin^2(ax)\,dx = \pi \qquad (a=1,2,3,\ldots)$$

$$\int_0^{2\pi} \cos^2(ax)\,dx = \pi \qquad (a=1,2,3,\ldots)$$

$$\int_0^{\pi} \cos(ax)\,dx = 0 \qquad (a=1,2,3,\ldots)$$

$$\int_0^{\pi} \sin(ax)\,dx = \frac{1}{a}[1-\cos(a\pi)] \qquad (a=1,2,3,\ldots)$$

$$\int_0^{\pi} \sin^2(ax)\,dx = \frac{\pi}{2} \qquad (a=1,2,3,\ldots)$$

$$\int_0^{\pi} \cos^2(ax)\,dx = \frac{\pi}{2} \qquad (a=1,2,3,\ldots)$$

$$\int_0^{\pi} \sin(ax)\sin(bx) = 0 \qquad a\neq b\ (a \text{ and } b \text{ are integers})$$

$$\int_0^{\pi} \cos(ax)\cos(bx) = 0 \qquad a\neq b\ (a \text{ and } b \text{ are integers})$$

$$\begin{aligned}\int_0^{\pi} \sin(ax)\cos(bx) &= 0 \qquad a=b\ (a \text{ and } b \text{ are integers})\\ &= 0 \qquad a\neq b \text{ but } (a+b) \text{ even}\\ &= \frac{2a}{a^2-b^2} \qquad a\neq b \text{ but } (a+b) \text{ odd}\end{aligned}$$

$$\int_0^{\pi/2} \sin(ax)\,dx = \frac{1}{a}[1-\cos(a\pi/2)]$$

$$\int_0^{\pi/2} \cos(ax)\,dx = \frac{1}{a}\sin(a\pi/2)$$

$$\int_0^{\pi/2} \sin^2(ax)\,dx = \frac{\pi}{4} \qquad (a=1,2,3,\ldots)$$

$$\int_0^{\pi/2} \cos^2(ax)\,dx = \frac{\pi}{4} \qquad (a=1,2,3,\ldots)$$

C.5 Frequency bands and their designations

Band Number	Frequency Range	Basic Description	Typical Service
2	30–300 Hz	ELF Extremely low frequency	Electric power
3	300–3000 Hz	VF Voice frequency	
4	3–30 kHz	VLF Very-low frequency	Navigation, sonar
5	30–300 kHz	LF Low frequency	Radio beacons, navigational aids
6	300–3000 kHz	MF Medium frequency	AM broadcasting, maritime radio, coast guard communication, direction finding
7	3–30 MHz	HF High frequency	Telephone, telegraph, facsimile, short-wave international broadcasting, amateur radio, citizen's band
8	30–300 MHz	VHF Very-high frequency	Television, police, FM broadcasting, air craft control, navigational aids, taxicab mobile radio
9	300–3000 MHz	UHF Ultra-high frequency	Television, satellite communication, surveillance radar, navigational aids
10	3–30 GHz	SHF Super-high frequency	Airborne radar, microwave links, common-carrier land mobile communication, satellite communication
11	30–300 GHz	EHF Extremely high frequency	Radar, experimental

C.6 Some exact and approximate expressions for TEM waves in lossy media

	Exact	For good dielectric $\frac{\sigma}{\omega\varepsilon} << 1$	For good Conductor $\frac{\sigma}{\omega\varepsilon} >> 1$
Attenuation constant (Np/m)	$\alpha = Re\left(j\omega\sqrt{\mu\varepsilon\left(1 - j\frac{\sigma}{\omega\varepsilon}\right)}\right)$	$\alpha = \frac{\sigma}{2}\sqrt{\frac{\mu}{\varepsilon}}$	$\alpha = \sqrt{\frac{\omega\mu\sigma}{2}}$
Phase constant (rad/m)	$\beta = Im\left(j\omega\sqrt{\mu\varepsilon\left(1 - j\frac{\sigma}{\omega\varepsilon}\right)}\right)$	$\beta = \omega\sqrt{\mu\varepsilon}$	$\beta = \sqrt{\frac{\omega\mu\sigma}{2}}$
Intrinsic impedance (Ω)	$\hat{\eta} = \sqrt{\frac{\mu}{\hat{\varepsilon}}} = \sqrt{\frac{j\omega\mu}{\sigma + j\omega\varepsilon}}$	$\eta = \sqrt{\frac{\mu}{\varepsilon}}$	$\hat{\eta} = (1 + j)\sqrt{\frac{\omega\mu}{2\sigma}}$
Wavelength (m)	$\lambda = \frac{2\pi}{\beta}$	$\lambda = \frac{2\pi}{\omega\sqrt{\mu\varepsilon}}$	$\lambda = 2\pi\sqrt{\frac{2}{\omega\mu\sigma}}$
Wave velocity	$u_p = \frac{\omega}{\beta}$	$u_p = \frac{1}{\sqrt{\mu\varepsilon}}$	$u_p = \sqrt{\frac{2\omega}{\mu\sigma}}$
Skin depth (m)	$\delta_c = \frac{1}{\alpha}$	$\delta_c = \frac{2}{\sigma}\sqrt{\frac{\varepsilon}{\mu}}$	$\delta_c = \sqrt{\frac{2}{\omega\mu\sigma}}$

C.7 Some physical constants

Constant	Symbol	Value
Velocity of light in vacuum	c	2.988×10^8 m / s
Electronic charge (magnitude)	$\lvert e \rvert$	1.602×10^{-19} C
Electronic mass	m	9.109×10^{-31} kg
Electronic charge to mass ratio	$\lvert e \rvert / m$	1.759×10^{11} C / kg
Permeability of free space	μ_0	$4\pi \times 10^{-7}$ H / m
Permittivity of free space	ε_0	$8.854 \times 10^{-12} \approx \frac{10^{-9}}{36\pi}$ F / m
Electron volt (energy)	$\lvert e \rvert V$	1.602×10^{-19} J
Boltzmann constant	k	1.381×10^{-23} J / K
Planck's constant	H	6.626×10^{-34} J. s

Index